T0342324

THE EVOLUTION OF THE HUMAN HEAD

The Evolution of the Human Head

Daniel E. Lieberman

THE BELKNAP PRESS OF
HARVARD UNIVERSITY PRESS

Cambridge, Massachusetts
London, England
2011

Library of Congress Cataloging-in-Publication Data

Lieberman, Daniel, 1964–
The evolution of the human head / Daniel E. Lieberman.
    p.   cm.
  Includes bibliographical references and index.
  ISBN 978-0-674-04636-8 (alk. paper)
    1. Head—Evolution.   2. Head—Growth.   3. Human evolution.   I. Title.
QM535.L525   2010
612.'91—dc22   2010020868

To my wife, Tonia, and my daughter, Eleanor
Of all the heads in the world, theirs are my favorites

Contents

Preface: Heads Up

This book is both a review and an argument that addresses three interrelated questions. First, why does the human head look the way it does? Second, how do heads in general, and the human head in particular, function and grow as an ensemble? And third, how is it that something as complex and vital as the head can be so variable and evolvable?

These are broad questions that integrate several fields, including craniofacial anatomy, craniofacial development, biomechanics, and paleoanthropology. Because these subjects are not always taught as an ensemble, I assume no prior knowledge in any of them, and I have tried to make each chapter accessible to upper-level undergraduates in fields such as biology and anthropology. I hope that the book will be of use to paleoanthropologists, craniofacial anatomists, developmental biologists, functional morphologists, and anyone else interested in human evolution, craniofacial development and function, and evolutionary biology.

Readers may wish to approach this book in different ways depending on their background and interests. Following an introductory chapter on some theoretical and practical problems relevant to studying and thinking about the evolution of the human head, the book is divided into three sections, each of which can be read on its own. Section 1 is on development: it includes chapters on the biology of bone and other skeletal tissues (Chapter 2); craniofacial embryology (Chapter 3); fetal and postnatal growth (Chapter 4); and morphological integration (Chapter 5). Section 2 is on the functional morphology of the head and includes chapters on the structure and physiology of the brain (Chapter 6); chewing (Chapter 7); swallowing, breathing, and vocalizing (Chapter 8); balance and locomotion (Chapter 9); and vision, smell, taste, and hearing (Chapter 10). Section 3 applies the information in first two sections to three major transitions in human evolution: the origin and diversification of early hominins, mostly the australopiths (Chapter 11); the origins of the genus *Homo* (Chapter 12); and the origins of *Homo sapiens* (Chapter 13). A final chapter attempts to tie together some basic ideas about tinkering, complexity, and natural selection and proposes some speculative hypotheses about the major selective forces in the evolution of the human head.

Even though the book has a modular organization (much like the head), it is also an integrated whole that attempts to make some arguments. By considering craniofacial development, function and evolution, we can gain insights into how and why the human head is the way it is, and how complexly integrated structures like heads work and evolve. I make three basic arguments. First, the way the head is integrated during development—from different structures sharing the same walls, from functional effects on regional growth, and from shared developmental pathways—permits the head to be highly functional, adaptable, and changeable over evolutionary time. Second, in spite of the head's complexity, just a few major developmental transformations underlie many key transformations in hominin craniofacial evolution. And, third, although some of these shifts may be chance occurrences, many reflect intense selection for novel behavioral shifts that helped make humans the way we are.

A few caveats. I am not an expert in all the subjects I have tried to cover, and the general topic of how human heads work and evolved is too broad for even a lengthy book. Scholars have written volumes on each of the subjects that this book treats in a single chapter or less. I hope readers will forgive the lack of comprehensive coverage of many anatomical features and topics. Some very complex subjects, such as how the brain works, or how we speak and swallow, have been relegated to just sections of chapters. Other topics, such as how teeth grow and function, have been tucked into various chapters. Chapters 11–13, which review the fossil record of human craniofacial evolution, have required particularly painful sacrifices for the sake of brevity.

Note on the Use of the Term *Human*

The term *human* can mean different things to different humans at different times. *Human* is often used to refer only to *Homo sapiens,* but sometimes the term refers to all hominins (animals more closely related to *Homo sapiens* than to extant apes), to all species in the genus *Homo,* or to all species of late *Homo* (such as *H. neanderthalensis* and *H. heidelbergensis*). Qualifying humans as "archaic" or "modern" makes sense if one considers *H. sapiens* to be a very diverse species that includes both "archaic" subspecies, such as the Neanderthals, and "anatomically modern" subspecies, such as ourselves. Although this taxonomic hypothesis has been disproved (see chapter 13), the terms have persisted.

For the sake of clarity, I use the term *human* in this book to mean just *H. sapiens*. But in cases where different interpretations of the word could lead to confusion between *H. sapiens* and other species of archaic *Homo,* such as *H. neanderthalensis* and *H. heidelbergensis,* I employ either the term *modern human* or the formal species name, *H. sapiens.*

1

A Tinkered Ape?

Darwinian Man, though well-behav'd
At best is only a monkey shav'd!

w. s. GILBERT, *Princess Ida,* 1884

The 1968 movie *Planet of the Apes* is nothing more than a Hollywood fantasy, but its premise reveals some deeply held views about what it means to be human. The movie imagines a postapocalyptic world in which humans have lost the ability to speak and think, and the other great apes have evolved to be like twentieth-century humans. The apes are bipedal, pair bonded, and religious; they speak English, they have dominion over the earth and all its creatures, and their society is organized into distinct classes with orangutans as administrators, gorillas as police, and chimpanzees as scientists. According to the movie's evolutionary logic, if humans were to devolve, then some other species of ape would then evolve to fill our place.

One shouldn't take science fiction movies too seriously, but the idea that apes could evolve to resemble humans is not entirely far-fetched. In fact, we have recognized this possibility since the time of Charles Darwin. What amuses me about the movie—after my willing suspension of disbelief—is that the imagined humanlike apes of the future walk and talk like us but otherwise retain apelike heads (given allowances for the human actors in ape suits who play them). In other words, *Planet of the Apes* imagines that apes could evolve to become humanlike without much selection acting on head shape.

It is true that many important unique human features lie below the neck, enabling us to walk and run bipedally, to fabricate and manipulate tools, to throw, to sweat profusely, and so on. But one would be hard pressed not to admit that human heads differ greatly from those of other mammals, including great apes, in many crucial ways. Our brains are more than five times larger than predicted by body size. Most mammals have long, narrow, tube-shaped heads, but ours are wide and short and round Humans alone have a forehead and no snout. The human neck (which is short and nearly vertical) attaches to the center rather than back of the cranium. We have a rounded tongue that descends far into the neck. We have an external nose that projects out from the face with downwardly oriented nostrils. The whites of our eyes are large and highly visible to others. We lack fur on most of the head. We have a chin below the lower jaw. We sweat profusely from the head, especially the scalp and face. We breathe through the mouth when exercising vigorously. We talk. And, of course, our cognition is more complex.

If one of the key questions of science is "Why are humans the way we are?" then an essential part of tackling that question is to ask, "Why is the human head the way it is?"

There are many ways to study the evolution of the human head, but a principal starting point is to compare human heads with those of the great apes, particularly the African great apes: common chimpanzees (*Pan troglodytes*), pygmy chimpanzees (*Pan paniscus,* also known as bonobos), and gorillas (*Gorilla gorilla*). Comparing the heads of humans and these species is especially informative not only because chimpanzees are our closest living relatives, but also because of evidence that the last common ancestor (LCA) of humans and chimpanzees must have resembled an African great ape in a number of anatomical respects. To explain this inference, it is useful to begin with previous ideas about the origins of the human lineage. A generation ago, it was widely believed that the African great apes were more closely related to each other than to humans. This view, depicted in Figure 1.1a, made sense given the many apparently derived (novel) similarities among these apes. Chimpanzees, bonobos, and gorillas share a unique, distinctive mode of terrestrial locomotion—knuckle walking —and they are anatomically similar in many ways. In fact, size accounts for many (but not all) of the differences between the African great apes. If one enlarged a chimp's skull, preserving the scaling relationships between size and shape, the result would be something that partly resembles a female gorilla skull (B. Shea, 1983, 1985; Berge and Penin, 2004). This scaling effect is illustrated in Figure 1.2 (adapted from Berge and Penin, 2004), which graphs overall shape (measured here using multivariate methods) of gorilla and chimpanzee skulls against size. The shapes of chimp and gorilla skulls scale similarly rela-

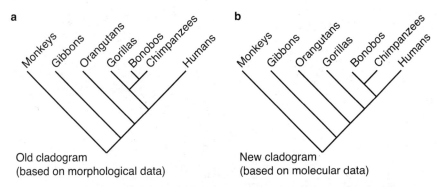

Figure 1.1. Cladograms (diagrams of evolutionary relationships based on shared derived characteristics) for the great apes. Left: old cladogram based on morphological data. Right: new and correct cladogram based on molecular data, showing that humans and chimpanzees are more closely related to each other than to gorillas.

tive to size, and, for a given size, the gorillas have slightly more developed shapes. Male gorillas' skulls, however, have large crests and other super-structures which cause them to diverge significantly from chimps and female gorillas.

In other words, the heads of gorillas are in many ways (but not completely) like those of big chimpanzees. Size contrasts also help explain some other key differences between the two species, such as diet. In general, bigger animals tend to have relatively lower metabolic rates, and bigger herbivorous mammals tend to eat foods with a higher percentage of fiber. Gorillas and chimps fit this trend: even though gorillas prefer fruit, their diet is generally much higher in fiber than the chimpanzee diet because they eat many leaves, shoots, and stems (McNeilage, 2001; Doran et al., 2002). So, when it was thought erroneously that gorillas and chimps shared a single LCA, it was commonly inferred that the LCA of humans and the African great apes was a sort of generalized ape, lacking most of the specializations evident in the other great apes. The paucity of the fossil record for ape and early hominin evolution provided no evidence to disprove this scenario.[1]

---

1. A *hominin* is the informal term for any organism more closely related to a human than to a chimpanzee. The old term, *hominid,* was correct according to the rules of taxonomic nomenclature by which humans were thought to be in a different family (the Hominidae) from chimps and gorillas (the Pongidae). Because humans and chimps are actually most closely related to each other, those species more closely related to humans than to chimps are reclassified into the tribe Hominini, which means that the correct informal term for them (by Linnaean rules) is *hominin,* not *hominid.*

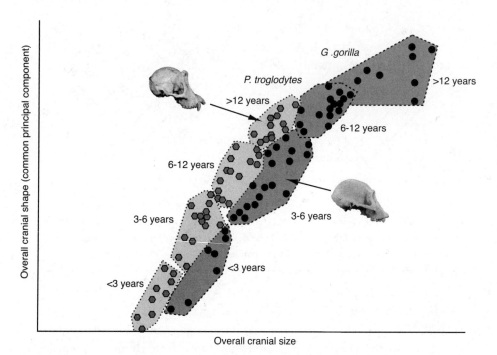

Figure 1.2. Geometric morphometric analysis (adapted from Berge and Penin, 2004) of chimpanzee and gorilla crania depicting how a multivariate measure of overall cranial shape scales with overall size. The measure of shape used here (a common principal component, discussed in Chapter 5) explains approximately 65 percent of the overall shape variance in the sample, which includes adults, subadults, and juveniles of each species (grouped by age in shaded areas). The slope of the relationship between size and shape in gorillas, not including the adult males, is nearly the same as in chimpanzees. However, for a given size and age, gorillas tend to have a higher (more adult) shape score than chimpanzees, hence the general resemblance in shape between an adult male chimp (left) and a juvenile male gorilla (right).

Molecular evidence that first emerged in the 1950s, but which became prevalent in the 1980s, has disproved this scenario (for reviews, see Pilbeam, 1996; Pilbeam and Young, 2004). Analyses of ape and human molecular data, illustrated in Figure 1.1, unambiguously indicate that humans and chimpanzees share a single LCA and that humans and chimps are more closely related to each other than either species is to gorillas (Shosani et al., 1996; Ruvolo, 1997; Chen and Li, 2001; Steiper and Young, 2006). Efforts to date this divergence require information on proportional differences between human and ape genetic sequences and on at least one estimated time of divergence from the fossil record to calibrate the molecular clock. For primates, this divergence has most commonly been the split between the orangutans and the African great apes. Because both sources of data have error, 5 and 10 million years ago (mya) are reasonable lower and upper bounds of the divergence between chimps and humans.

The ramifications of the revised ape and human tree are profound for thinking about human evolution. We can now appreciate that the many similarities between gorillas and chimps, once considered derived, are more likely to be ancestral. That is, unless knuckle walking and the many morphological similarities evolved independently in chimps and gorillas, then the LCA of the two species was probably chimplike or gorillalike in its anatomy (which amounts to nearly the same thing). Extensive convergence between chimps and gorillas is possible, but improbable. It therefore follows that the ancestor of humans and chimpanzees is likely to have been somewhat like chimps and gorillas in terms of morphology. Of course, this hypothesis does not mean that the genus *Pan* has not changed in many respects, as it too diverged from our LCA, and we cannot assume that the LCA behaved just like today's chimps or gorillas. In fact, without fossils, we cannot even test whether the LCA was chimp-sized or gorilla-sized. These shortcomings notwithstanding, if we put all the evidence together, it is reasonable to hypothesize that *Pan* and *Gorilla* diverged from a common ancestor more than eight million years ago. Gorillas likely underwent selection for larger body size, probably to enable them to fall back on low-quality, high-fiber diets in habitats with strong seasonal shortfalls in fruit availability. Meanwhile the chimp-human lineage continued to evolve separately, eventually splitting.

Hypotheses about the LCA, of course, are not without controversy, partly because rain forests are poor habitats for fossil preservation, leaving us a scant fossil record of the African great apes from this time period (the end of the Miocene). The only currently known fossil evidence of chimpanzee or gorilla evolution includes just a few 12 million-year-old teeth that might belong to a gorilla ancestor (Suwa et al., 2007), and a few chimpanzee teeth that are 500,000 years old (McBrearty and Jablonski, 2005). And, until the mid-1990s, the known hominin fossils more than four million years old could fit in a shoebox. Recently, however, we have been blessed by the discoveries of several very early fossils currently assigned to three hominin genera: *Sahelanthropus* (Brunet et al., 2002, 2005), *Ardipithecus* (T. White et al., 1994; Haile-Selassie, 2001; Haile-

Selassie et al., 2004; Semaw et al., 2005; White et al., 2009) and *Orrorin* (Senut et al., 2001). Significantly, these fossils share a few key, derived features typical of hominins: they were probably bipedal, had smaller canines (in males), and had slightly larger and more thickly enameled cheek teeth than chimps. In a number of other respects, however, these early species resemble chimpanzees in some aspects of their anatomy, such as other features of the teeth and the shape of the braincase (see Chapter 11). Thus they look like plausible intermediates between an apelike LCA and the australopiths.

That said, some researchers think that the earliest hominins evolved from an ancestor that was unlike any extant great ape, that most similarities between chimps and gorillas (and even orangs) are therefore independently evolved, and that hominins descended from a generalized ape that was neither a knuckle walker nor skilled in vertical suspension and climbing (White et al., 2009; Lovejoy et al., 2009a–d). The major basis for this speculation comes from a number of features described in *Ardipithecus ramidus,* which is dated to 4.4 mya. These hypothesized primitive features include modest projection of the lower face (prognathism), lack of knuckle walking specializations in the wrist, and limbs with monkeylike proportions that are hypothesized to lack adaptations for climbing and suspension. However, to what extent these features of *Ar. ramidus* are primitive rather than derived needs further careful assessment, particularly in a hominin that is 4.4 million years old and thus appeared at least two to three million years after the divergence of hominins from the LCA. In addition, *Sahelanthropus, Ardipithecus,* and *Australopithecus* resemble chimpanzees in numerous anatomical features of the skull as well as the postcranium (see Chapter 11). There are many unresolved questions about how *Ar. ramidus* and *Sahelanthropus* are related to other apes and hominins.

Putting the molecular and fossil evidence together, we can reasonably infer that the LCA of humans and chimpanzees probably resembled the extant African great apes in quite a few aspects of its anatomy. This is not to say, however, that our closest relatives, the chimpanzees, are living fossils that did not undergo their own evolutionary changes. It will be interesting to find more fossil evidence of chimpanzees and other African great apes, but, given all the available facts, one can predict that any morphological changes that occurred in the chimp lineage over the past six million years were much less extensive than those that occurred in the hominin lineage.

Regardless of the exact nature of the LCA, one of the most fundamental questions to ask about human evolution is how and why we came to be morphologically different from the other great apes. This question is becoming especially interesting as we compare the human and chimpanzee genomes. Even though the two genomes are almost 98 percent identical, chimps and humans differ by approximately 35 million base pairs. Most of these differences are in noncoding DNA, but approximately three million of the differences may affect protein-coding genes or other functional areas of the genome (Chimpanzee Sequencing and Analysis Consortium, 2005).[2] A major effort is therefore under way to determine which genes changed during our evolutionary history, when, and how these changes transformed us into modern humans. Discovering the genes and how they work will enrich our understanding of human evolution immeasurably; but genes alone will not answer all our questions, because modern humans didn't evolve directly from chimpanzees. Instead, we know from the fossil record that the hominin lineage includes numerous and varied species. To understand how natural selection helped make us human, we also need to understand the order, context, and consequences of the various evolutionary transformations that occurred.

Which brings us back to heads. As noted above, many important morphological differences between humans and the African great apes are above the neck. Distinctive human features, which range from a large brain to small front teeth and a protruding nose, emphasize the intensity of evolutionary change in the head, and many of them suggest strong levels of natural selection. Thanks to the fossil record, we now have a decent idea of the major stages over which these changes occurred. The first hominins probably differed from the LCA by reorienting the head on a vertical neck and by decreasing the size of the canines and incisors while increasing the surface area and thickness of the cheek teeth. Some of these hominins had descendants with even larger, more thickly enameled cheek teeth, along with massive faces and large chewing muscles. Later descendants evolved bigger brains, more balanced heads, smaller teeth, and vertical faces. Finally, with the origin of our species, *H. sapiens,* we evolved a rounder head; a smaller, retracted face; and a chin.

It is interesting to document what changes occurred and when, but it is also rewarding to figure out how and why these shifts occurred and what selection pressures drove them. To address such questions requires that we consider how

---

2. Noncoding DNA (DNA that doesn't code for proteins) was long presumed to be neutral, but we now know that much of it—we don't know how much—is functional, helping either to regulate gene transcription or to specify noncoding RNAs that are involved in various acts of gene regulation.

the head develops and functions in an appropriate evolutionary context. This is a challenge because heads are such closely integrated structures that shifts in the way one organ or region grows and functions can have widespread effects on the growth and function of other parts of the head. For example, increases in brain size alter how the braincase and the face grow. Larger brains also change the balance of the head, the range of sound frequencies an organism hears, aspects of chewing biomechanics, and other functional characteristics. Thus it is sometimes difficult to know whether specific shifts were adaptations (features selected because they improved an organism's ability to survive and reproduce), by-products of other shifts that were selected for different reasons, or simply the result of random evolutionary change. As an example, we can (and will) ask whether the evolution of a shorter face and the loss of a snout in the genus *Homo* was an adaptation for speech, for increasing turbulence in the nasal cavity, for balancing the head, for some combination of the above, or perhaps the result of neutral (unselected) evolutionary change. Testing hypotheses about such transformations requires us to know when and how the shifts occurred and their effects on how hominins performed tasks that influenced fitness. In turn, a better understanding of the interactions between genotype, development, morphology, and function can help us define species and infer their evolutionary relationships.

Why Heads Are So Complex

One of the challenges of studying how the human head evolved, and how it develops and functions, is its sheer complexity. This complexity has three different sources. The first is the range of critical functions the head performs. The head not only houses and protects the brain and other organs but also participates in respiration, thermoregulation, vocalization, locomotion, vision, chewing, swallowing, tasting, smelling, hearing, and balance. Almost every particle entering your body, either to nourish you or to provide information about the world, enters via your head, and almost every activity involves something going on in your head. Furthermore, the diverse components of the head that perform these functions not only share many of the same spaces and structural supports but must also grow and change together throughout the entire life cycle—from embryo to adult—without any serious loss or compromise of function. For example, as the various parts of the head grow, they change the shape of the face and hence the position of the teeth in the upper and lower jaws. But teeth need to fit together perfectly for occlusion. Thus, the lower and upper jaws have to grow in such a way as to accommodate shifts in the relative position of the lower and upper teeth to permit precise, proper, and forceful occlusion between all the lower and upper teeth. No engineer could possibly design such a dynamically complex and compact machine.

A second source of complexity in the head comes from the number and diversity of its components. A hallmark of all organisms is that they are modular —comprised of distinct, partially independent units. Modularity is evident at every level of an organism's structure, from the genotype (base pairs, codons, genes, chromosomes) to the phenotype (cells, tissues, organs, systems, regions) (see Wagner, 1996; Schlosser and Wagner, 2005; Gilbert, 2006). Although modules are generally less abundant and discrete at higher levels of organization (the genotype is more modular than the phenotype, and cells are more modular than organs), it should be clear that heads have an impressive number of modules, all packed into a comparatively small space. For example, at a very macro level, an adult human head has more than 20 bones, up to 32 teeth, a brain (itself comprising many parts), several sensory organs, dozens of muscles, and many glands, all supported by a host of nerves, veins, arteries, ligaments, and other structures. In addition, most of these modules comprise many component modules at finer levels of structure (e.g., regions, tissues, cells).

A third source of complexity in the head is its degree of integration. Integration, the counterpart of modularity, describes the way in which different components (modules) of a system are combined into a whole. Integration, a general property of all organisms, is manifest throughout the body both functionally (when the interaction among several morphological elements affects their combined performance) and developmentally (when morphological elements interact during their formation, or are directed by a common external source) (Cheverud, 1996). It is possible, though hard to prove, that heads are more intensely integrated than any other region of the body. Integration occurs in several ways. At a simple, structural level, integration is evident from shared walls and spaces. As examples, the roof of the oral cavity is also the floor of the nose; the top of the face is also the floor of the brain; and the pharynx is a major passageway for air, food, liquid, sound, and mucus. Thus changes to the form or function of any one component of the head cannot help but affect the form and function of others. Integration is inevitable because of the ways that heads develop and grow. Many of the genes that regulate the growth and development of the head influence different organs and tissues at different stages of ontogeny (a phenomenon termed *pleiotropy*). Further, as cells divide and migrate during embryogenesis, they interact copiously, inducing each other to differentiate, die, or proliferate, leading to myriad interconnections and correlations. Finally, integration occurs because many functions are not localized but instead involve several disparate parts of the head. For example, when chewing, one generates forces not only in the tooth crowns but also in the tooth roots, the periodontal ligament that attaches the roots to the jaws, the places where chewing muscles attach to the skull, the temporomandibular joint, and elsewhere. Consequently, behaviors such as chewing, balancing, running, and speaking interrelate seemingly disparate parts of the head into integrated functional complexes.

Integration is typically quantified statistically in terms of patterns of correlation and covariation (see chapter 5). In some studies, it seems as if everything correlates with everything else in the head. The size of the brain correlates to varying extents with the size and shape of the eyes, mouth, nose, and so on; enlarging the teeth causes many correlated changes throughout the skull. Another effect of integration is that mutations that disrupt growth in one part of the head often have indirect effects on growth in other parts of the head, and vice versa. For example, children with cartilage growth defects in the cranial base often develop concave faces with malocclusions and abnormally shaped braincases; likewise, children with prematurely fused bones in the cranial vault (craniosynostoses) can develop several malformations of the face and cranial base.

Considering the many modules of the head and how these modules function and are integrated is key to studying how the human head works and how and why it evolved. But the complexity that derives from multifunctionality, modularity, and integration has both good and bad consequences. One downside of such complexity is to hinder efforts to make simple inferences about how the head evolved. For one, widespread integration among modules makes it difficult to test whether a given derived feature was favored by natural selection, was a byproduct of selection on some other correlated feature, or evolved from random, unselected evolutionary changes. In addition, complex integration can frustrate efforts to test hypotheses about evolutionary relationships. Because heads have so many highly integrated modules, one cannot easily define independent features that develop in utero or after birth. Thus when the adults of two species share a similar feature such as a sagittal crest, the similarity could result from having the same last common ancestor, but the similarity could also result from independently evolving a suite of features that correlate with a convergently evolved feature, such as large teeth.[3]

Although the interplay between modularity and integration poses some challenges for evolutionary biologists, it has benefits, too. Without lots of modularity and integration, complex biological structures such as heads could not grow, function, and change. As a thought experiment, imagine the engineering challenges of "growing" a car from the size of a child's toy to full size. One would not only have to program the car's many modules to grow appropriately and in concert but also program each module to adjust itself to the changes of all the car's other modules. Any mismatch in the size, shape or position of the gears, wires, pistons, or axles could have disastrous effects. Engineers haven't come close to growing a car, let alone building something like a head, whose components are not only orders of magnitude more complex than those of a car in terms of structure and function but also grow severalfold in size and can adapt marvelously to varied circumstances.

In other words, heads (like many other complex biological structures) "work" not *in spite* of complexity, but *because* the many modules that comprise the head are complexly integrated in a special way. Every component of the head interacts with its neighbors throughout life, especially during growth and development: they stimulate each other to grow faster, slower, differently, and so on. These interactions happen in various ways. Sometimes tissues induce one another by secreting chemical signals that turn another tissue's genes on or off; sometimes they respond similarly or differently to the same circulating hormones; sometimes they physically push or pull on each other. Many of these interactions are direct, caused by proximity. It is no coincidence that almost every bony wall in the skull interacts with more than one tissue or functional space. For instance, the lateral wall of the nose is the medial wall of the eye, and the front of the braincase is also the top of the face. By sharing walls of bone, growth in one region can accommodate growth in a neighboring region and vice versa. But many interactions are both direct and indirect. Growth in one region can affect growth in other, distant regions through various intermediate interactions. For example, changing the angle of the cranial base alters the orientation of the upper face, which then shifts the orientation of the mid-

3. This is an example of functional integration, whereby features co-vary in some fashion because they are functionally related (in this case via large chewing muscles to generate high forces). Another form of integration is developmental, usually via pleiotropy. For example, the gene for a signaling factor called protein ectodysplasin (*Eda*) has been found to influence not only tooth size but also tooth cusp number and cusp shape, making these features nonindependent (Kangas et al., 2004).

dle and lower parts of the face, which realigns the tooth row, and so on. Hormones also permit distant regions to change in concert.

Complexity, Evolvability, and Tinkering

The Paradox of Complexity

How is something as complex and vital as the head so *evolvable* (i.e., with considerable capacity for evolutionary change)? From an engineer's (or a creationist's) perspective, great evolvability seems counterintuitive. Any complex feature that is indispensable for survival should be difficult to change via an agentless process such as natural selection. For one, it seems reasonable to expect that complex, highly intricate objects will function less well if they are modified, because even small alterations are likely to have potentially deleterious consequences. A second problem is that natural selection should act strongly to winnow out any modifications to complex structures like heads that reduce functionality. Heads perform so many vital tasks—such as breathing, smelling, swallowing, seeing, and hearing—that mutations which lead to even a slight impairment of any function will likely cause a decrease in fitness. Thus one might expect heads to be highly constrained and conservative.

Yet heads are far from conservative, especially when viewed over the course of human evolution. If we compare primates in general and the human lineage in particular, heads appear to be no less morphologically conservative than the rest of the body, and possibly even more evolvable. As examples, most of the differences among species of australopiths (e.g., *Au. afarensis, Au, africanus,* and *Au. boisei*) are cranial, not postcranial, and you are probably more similar to a Neanderthal below the neck than above. In addition, the record of human evolution provides extensive evidence for seemingly major modifications to the head that apparently have not compromised its most critical functions. For instance, modern humans have a unique throat in which the larynx descends relative to the soft palate, presumably to improve the ability to speak. Dropping the larynx, however, rearranged various components of the throat that function during swallowing: as a result, humans swallow somewhat differently and a bit less safely than other mammals (see Chapter 8). Given the importance of swallowing for survival, one would have imagined natural selection to have selected against hominins with even a slightly lower larynx in spite of any potential acoustic advantages. Any hominin less able to swallow safely would be more likely die young regardless of its ability to speak clearly.

In fact the expectation that natural selection cannot cope with modifications to complex systems has led some creationists to challenge the theory of evolution, especially when applied to humans (Behe, 1996).

There are two nonexclusive solutions to the paradox of complexity. The first is that heads may have evolved such exuberant variation not *in spite* of but *because* of the considerable intensity of natural selection on heads. Because heads perform so many critical functions, it is possible that variations in the head are more likely to have effects on fitness than variations in other regions of the body. Put differently, there are more targets of selection in the head than elsewhere because the head participates in more functions. According to this logic, variations in limb design will mostly be unselected (neutral) or negatively selected unless they improve an organism's ability to move about, but variations in the head are more likely to have a chance of being selected positively as long as they improve net performance in hearing, vision, chewing, thinking, balance, or other such functions without commensurate cost to other functions. If the intensity of positive selection is greater than the intensity of stabilizing selection, then heads will undergo more evolution. I am not sure this hypothesis is correct. There is little evidence to support the conjecture that the intensity of selection is greater on the head than on other parts of the body. Moreover, comparisons of genetic and craniofacial variation among human populations suggest that most changes in craniofacial form among recent humans occurred from random, unselected genetic change rather than from natural selection (Relethford, 1994; Roseman, 2004; Roseman and Weaver, 2004). In addition, even if selection were greater on the head, then one might also expect constraints on heads to be greater as well, because of the likelihood that any mutation would lessen the head's ability to perform one or more of its vital functions.

A second nonexclusive solution to the paradox of complexity is that heads may be more evolvable than other parts of the body precisely because of the nature of their complexity. This complexity, described above, may facilitate evolvability in several ways. First, not only do heads have many modules, but these modules are also intensely integrated in terms of development, structure, and function. Because evolvability theoretically occurs in rough proportion to the degree of modularity (e.g., the genotype is more evolvable than the phenotype), and heads have so many modules, then heads may be especially prone to evolutionary modifications. In particular, changes to the size, the shape, or the relative timing of development of each of the head's many modules offer a variety of opportunities for change. There are many examples in the head to consider, including teeth, which probably originated from scales (M. Smith and Johanson, 2003) and then diversified spectacularly through many variations in number, size, shape, and so on (see Jernvall, 1995). The same is also true for other highly modular parts of the head, including the brain and the face.

Given the interplay between modularity and integration, another, related contributor to the head's evolvability may be its high level of integration, both developmental and functional. Mechanisms of integration (reviewed in Chapter 5) allow modules to vary yet still grow and function effectively within the context of an organism. In fact, modules come with their own integrative mechanisms, in the forms of particular molecular and developmental pathways (Wilkins, 2002) and physical interactions between adjoining cells or tissues. One dramatic, well-known example of this phenomenon is the eyeless (ey) gene in fruit flies. Expression of the gene (a transcription factor) turns on a cascade of other genes that assemble an eye. Normally the transcription factor is expressed at the right time in the right places, but when ey is artificially expressed in an odd place like a leg, a properly formed eye nonetheless develops (Halder et al., 1995). No transformations we see in human evolution are as dramatic as an eye on a leg, but other novel shifts, such as bigger brains, thicker tooth crowns, and shorter snouts permit a viable, functional "organism" because the shifts often occur within the context of existing series of integrative processes at many levels of development. Integration often permits the organism to adjust to the novelty.

Tinkering

Put together, the combination of diverse functions, high modularity, and plentiful integration creates many opportunities for evolutionary change in the head. Much of this change occurs via *tinkering*. In a famous essay, François Jacob (1977) made a useful analogy between evolutionary change and tinkering (*bricolage* in French). Biologists often liken natural selection to the process of engineering, but this analogy is problematic. When engineers make something, they usually have a particular goal in mind, they design and build the object using a priori plans and principles, and they create novel components to fulfill specified functions. None of these processes are characteristic of evolution. A far better analogy for how evolution works is the way tinkers create and modify objects opportunistically, using whatever happens to be available and convenient but not necessarily designed or created for that function. Such tinkering is an apt description of natural selection, an agentless process that can only take advantage of heritable variations made available by random mutations. Mutations that happen to generate phenotypic variations that also improve an organism's reproductive success are more likely to be passed on to the next generation. Jacob's analogy of evolution by tinkering helps explain several key emergent properties of evolutionary change, including the tendency of organisms to function and to be highly integrated (see Lieberman and Hall, 2007). When new organisms make new use of preexisting or modified modules, these tinkered novelties often tend to work because they are made of modules that already function appropriately and come with existing mechanisms for adjusting to one another. In other words, tinkering takes advantage of modularity and leads to integration (and probably vice versa to some extent, too).

It should be evident that the head provides countless opportunities for tinkering because it offers many elements with which to tinker and many functions that potentially confer a selective advantage. Moreover, because the head is highly integrated both functionally and developmentally, such changes to the head are reasonably likely to work because heads can accommodate a great deal of variation. Perhaps it is not surprising that we see so much variations among heads, including the position and angle of the neck's attachment to the cranial base, the size of the brain, the size of the teeth, the length of the face, the size of the orbits, and so on.

This book explores the premise that the many derived features that make the human head distinctive occurred from tinkering, beginning with modifications to a common ancestor with the great apes. To reiterate, because evolvability is an emergent property of function, modularity, and integration, heads may be surprisingly evolvable for the simple reason that they consist of many functionally important modules integrated in a special way to accommodate one another.[4] Put differently, heads are especially evolvable *because of* the complex way they are integrated rather than *in spite of* their complexity. Further, the head's functional importance serves only to intensify the role of natural se-

---

4. The idea that the evolution of complexity is made possible by the emergent properties of modularity and integration is an old one (for reviews, see Wagner and Altenberg, 1996; Gerhart and Kirschner, 1997; Gilbert and Bolker, 2000). Recently, however, the idea has garnered considerable attention, thanks to the burgeoning field of evolutionary developmental biology ("evo-devo"). Evo-devo asks how evolution occurs from modifications to development. Although the field initially focused on macroevolutionary comparisons among distantly related model organisms such as fruit flies, zebra fish, and mice, there is a growing focus on microevolutionary comparisons among closely related species or individuals within populations (see Lieberman and Hall, 2007).

lection as an agent of tinkering. Therefore, it is useful to think about how the head functions and develops as an ensemble instead of focusing on particular parts and functions.

To address why the human head looks the way it does, key questions include the following: What are the major evolvable modules of the head, and how, where and when did they change? What mechanisms integrate the head's basic modules? How do the head's modules function? Were the particular transformations in head morphology we observe during human evolution the result of changes to modules, changes to processes of integration among modules, or some combination of the two? How did these changes affect function? What other effects did these shifts have on development and function? And to what extent and why might these changes have been favored by natural selection?

The human head is in many ways an ideal test case for thinking about these sorts of questions. Although ethical considerations prevent the use of humans and apes in experimental studies of embryonic development, the human head affords other advantages for studying the evolution of complexity. First, the human head is one of the best-studied complex biological structures. We have a wealth of data on the anatomy, development, and variation of almost every aspect of the head in modern humans, in our closest relatives the chimpanzees and other apes, and (for the skull and teeth) in the fossil hominin record. People sometimes joke that there are more researchers studying human evolution than there are fossil hominins, but this is untrue. The hominin fossil record is remarkable, including thousands of fossils from all over the world, many of which are well dated and from sites whose archaeological and environmental contexts have been carefully studied.[5] In addition, although we cannot do experiments on human craniofacial development, we have an exceptionally rich and well-documented record of "natural experiments" provided by genetic syndromes and clinical studies. A large proportion of genetic syndromes are manifested at least partially in the head.

Another advantage of using the human head to consider the evolution of complex structures is our ability to consider aspects of function, an essential ingredient of evolutionary change that is often not considered by studies of how evolution occurs through modifications to processes of development. Because of its clinical importance, we know much about how the human head

---

5. In truth, the hominin fossil record becomes rich only starting about 3.5 million years ago; the first few million years remain poorly sampled.

functions in terms of brain growth and physiology, mastication, swallowing, hearing, vision, smell, taste, speech, respiration, balance, and so on.

In short, the evolution of the human head merits special study from an integrated developmental and functional perspective for several reasons, not the least of which is that the human head has changed substantially since our lineage diverged from the chimpanzees. To understand how and why humans came to be the way we are, we need to understand how our heads came to be the way they are. But, more generally, the evolution of the human head also provides a useful opportunity for exploring how nature tinkers with development in ways that affect function and permit the evolution of complex structures.

As noted in the preface, this is a book about heads rather than just skulls. Even so, an especially useful starting point—the focus of the next chapter—is the skeletal tissues of the head, especially bone, which provide the structural framework for the head and play a special role in most aspects of tinkering.

2

The Skeletal Tissues of the Head

His bones are as strong as pieces of brass,
his bones are like bars of iron.

JOB 40:17–19

Imagine you are an engineer who wants to design something complex, like a robot. But there is a catch: your robot has to be able to grow from the size of a pea to a bowling ball without any loss of function. In addition, the robot has to be able to modify its own shape and capacities over many decades to accommodate a wide range of functional alterations, such as changes in the terrain in which it moves, the kind of fuel it uses, its manipulative abilities, and so on. Designing and building such a robot would be an impossible challenge for many reasons. One of the biggest problems is what substance to use for the robot's frame. The substance must be strong and stiff, but also highly modifiable so that the robot can alter its shape and size as it grows. The substance also needs to come in modular units that the robot can replace or modify as they wear out and as the robot's functions change.

No engineer is capable of making a complex growing, multifunctional robot, but the head is a superb example of how evolution has achieved such feats via tinkering, opportunistically using just a few components in novel ways. These ancient tissues—mostly bone, but also cartilage, tendon, ligament, enamel, dentine, and cementum—make up the essential framework of the skull, anchoring muscles and providing spaces and protection for all the diverse organs and tissues of the head, such as the brain, eyes, and muscles, that permit the head to function. In addition to being stiff, strong, and seemingly solid, bone and other skeletal tissues are also dynamic, able to change in shape, size, and structure in response to a wide variety of genetic and environmental stimuli. Bone is an intrinsically integrative tissue, well suited for tinkering.

Although bone is just one of many tissues that make up the head, it is useful to start our exploration of the how the human head grows and develops with a review of some key features of bone and other skeletal tissues (later chapters will review the structure and function of other important tissues, such as the brain). There are three reasons to focus initially on bone, cartilage, and teeth. First, most major ontogenetic and evolutionary changes to the head include modifications to the skeletal structures that house and anchor all the head's other tissues. Second, many properties of bone and cartilage permit the skull to modify itself during ontogeny, thus accommodating changes in various organs and functional spaces in the head. This modifiability also makes the skull highly evolvable. Third, bones, teeth, and their fossilized remnants provide most of the information we have about evolutionary changes to the head in extinct hominins. Accordingly, this chapter begins with a review of the structure and composition of bone, teeth, and other skeletal tissues. I then summarize the ways in which bones grow and derive their morphology, and I review the major genetic and non-genetic processes that influence bone growth. Next, I consider some principles of bone biomechanics and the dynamic interactions between bone and its mechanical environment. Finally, I conclude with a general model of how bone interacts with organs and functional spaces to accommodate change during ontogeny and over the course of evolution.

But first a caveat. Skeletal tissues are so varied across the animal kingdom that the following is a review of bone, cartilage, and dental tissue mostly as they pertain to the head in just humans and other mammals. For a thorough review of vertebrate skeletal tissues, see Hall (2005).

Skeletal Tissue Structure and Composition

Bone Tissue Synthesis and Composition

At a basic level, bone differs little between and within mammals. Yet because of the way it can modify its growth processes, bone tissue also varies widely in shape, histology, and function (see Figure 2.1). To explore how different growth processes can generate variations in skull shape and function, a first topic to review is how bone grows, beginning at the ultrastructural level.

Bone is a widely useful tissue largely because of its ultrastructure, a combination of mineral and organic components. It is synthesized by special cells known as osteoblasts in a two-step process. First, osteoblasts produce an organic matrix, osteoid, that consists mostly of collagen (see Figure 2.3). Collagen is composed of long chains of molecules wound together into a rope-like, triple-helix structure. These molecules are bound into larger fibrils by chemical bonds (cross linkages), which combine to create even larger fibers. Like rope, collagen fibers are flexible and elastic, providing resistance to tension but not to compression along their long axes. The arrangement of collagen fibers determines many of the tissue's structural and mechanical qualities. Collagen makes up about 60 percent of the tissue's dry weight.

After osteoblasts lay down a collagen matrix, they then stiffen the matrix with mineral, which makes up about 40 percent of bone's dry weight. Typically, spaces in the collagen matrix are initially filled with water and then replaced by tiny crystals of bone mineral (crystallites) that condense from a supersaturated solution of calcium phosphate secreted by the osteoblasts. These crystallites precipitate within spaces between the fibrils via a phase change, in much the same way that ice forms from water (see Weiner and Traub, 1992). Osteoblasts have specialized organelles, Golgi apparatuses, that produce and extrude little membrane-bound packages of supersaturated calcium phosphate. These packages can be left behind as the osteoblasts advance along the osteoid front so that biomineralization can occur at a distance from osteoblasts. Biomineralization occurs with the assistance of various noncollagenous proteins, also produced by osteoblasts, which help regulate the organization and mineralization of the collagen fibrils.

Once bone cells cease to synthesize mineral or collagen, some die and are resorbed, but others transform into different cell types that help regulate bone structure and function (see Figure 2.9). For example, osteoblasts on the once active bone-forming surface spread out and form a cellular network of bone-lining cells. These have several functions, including helping to regulate mineral exchange between blood and bone (Parfitt, 1987). In addition, many osteoblasts become entombed in bone tissue and persist as osteocytes. Osteocytes change shape, generating long, asymmetrical processes (canaliculi) that communicate with other osteocytes, osteoblasts, and bone-lining cells via gap-junctions (see Figure 2.9). Osteocyte function is poorly understood, but it may include the sensation of mechanical strain or bone fractures through changes in fluid pressure or flow in the canaliculi (Weinbaum et al., 1994; Verborgt et al., 2000).

Cartilage, Tendon, and Ligament

A relative of bone is cartilage, a stiff, less mineralized connective tissue that is synthesized by cells called chondrocytes. Cartilage has important differences from bone (Figure 2.1). Cartilage typically has a less organized organic matrix

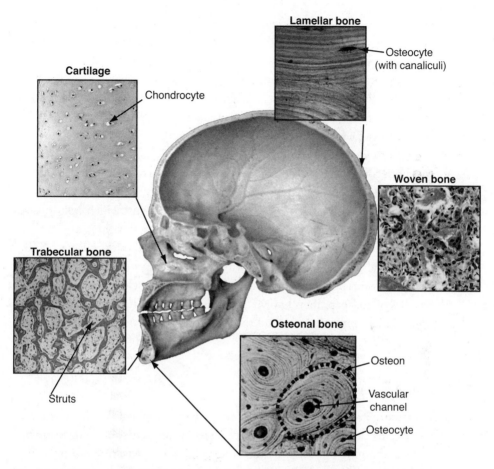

Figure 2.1. Different types of skeletal tissues. Cartilage is poorly organized, less mineralized than bone, and higher in water content. Bone comes in many different histological categories. With the exception of woven bone (which is rare in the adult skull, except following injury), most bone is organized in layers known as lamellae. The bone surrounding the cranial vault is often organized in parallel layers. Osteonal bone is formed of organized concentric layers around a central vascular channel; trabecular (spongy) bone consists of very small layers organized into interconnecting struts.

of collagen; it contains other fibers, such as elastin (an elastic protein), not present in bone; and it has many heavy, complex, and highly charged molecules (proteoglycans) that bind to water molecules. Thus by weight, cartilage has more water than bone (as much as 70 percent). Because water molecules don't compress well, this water makes cartilage highly resistant to compression and to a lesser extent to shearing and tension. Mammals have different types of cartilage (see Hall, 2005), the most common of which is hyaline cartilage, an opalescent tissue that covers articular surfaces in joints, absorbing compression and reducing friction, and which also participates in much bone growth. Other kinds of cartilage, such as yellow elastic and white fibrocartilage, vary in density and elasticity and are found elsewhere in the head, in structures such as the ear lobes, the larynx, and the epiglottis.

Two other connective tissues to note are tendons and ligaments. These tissues both consist primarily of longitudinally oriented collagen bundles but differ slightly in structure, composition, and function. Ligaments mostly link bones together to restrict movement at joints; tendons, which link muscles with bones, are more elastic. Both resist tension along their long axes and can act like springs. In addition, both tissues contain stretch-sensitive mechanoreceptors that sense and transmit information on changes in tissue length to the central nervous system. Ligaments, tendons and muscles attach to bone via bundles of collagen fibers known as Sharpey's fibers. These fibers project into bone, which mineralizes around them, and they hold both hard and soft tissues together.

Dental Tissues

Tooth form varies widely, but each tooth has a crown and at least one root comprising three hard tissues. The crown has a core of dentine with an enamel surface; the root is mostly dentine and pulp, with a covering of cementum. Each tissue is histologically distinct and is synthesized by its own type of cell (see Figure 2.2).

Enamel, which is almost 95 percent hydroxyapatite by dry weight, is deposited in a two-step process by ameloblasts. These narrow, cylindrical cells first synthesize an organic matrix (amelogenin and other matrix proteins) that then facilitates the nucleation of mineral, often in roughly 5 μm wide prisms, each made up of a tightly packed array of crystallites. The tooth crown grows as ameloblasts migrate away from the enamel-dentine junction, leaving a record of the cellular activity underlying its growth (see Figure 2.2). Circadian fluctuations in enamel secretion cause slight changes in the enamel prisms that are visible in thin section under a microscope as light and dark cross striations (Osborn, 1981). Every 6–12 cross striations, the tissue is marked by coarser, wider incremental lines (striae of Retzius) that represent periodic slowdowns in enamel secretion (Dean 1987; Bromage, 1991); in modern humans, the av-

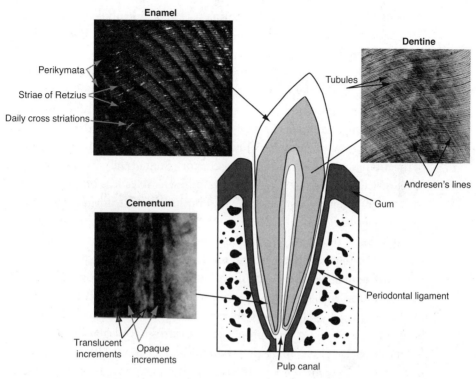

Figure 2.2. Tooth showing characteristic histology of the three major types of dental hard tissues: enamel, dentine, and cementum. The three tissues have different kinds of incremental structures. Image of enamel courtesy of C. Dean; image of dentine courtesy of T. Smith; cementum increments from Lieberman, 1993.

erage periodicity is 8–9 (the range is 6–12); in other primates, periodicities range from 2 to 12. Each striation records the temporary surface of the crown during development. By counting the striae in a thin section, or their visible outcroppings on the surface of unworn teeth (perikymata), one can determine with considerable precision the rate and process by which each crown was formed (Bromage and Dean, 1985; Dean, 2006; T. Smith et al., 2006).

Dentine more closely resembles bone, consisting of about 75 percent mineral and 20 percent collagen by dry weight. Whereas enamel is secreted outward from the inside of the crown, dentine-forming cells (odontoblasts) secrete the dentine toward the inside of the crown and root. Odontoblasts migrate toward the center of the pulp cavity, depositing tissue as they proceed in a two-phase process similar to the way osteoblasts synthesize bone. As they form tissue, the cells elongate, creating a cylindrical tubule that represents their path from the root surface (or from the enamel-dentine junction in the crown) to the tooth center (Figure 2.2). Once the tooth has erupted and the root is fully formed, odontoblasts continue to synthesize dentine slowly, eventually filling up the pulp cavity. Other cells from the surrounding gum, including nerves and blood vessels, migrate into the root. Dentine also leaves behind incremental markings that enable reconstruction of the pattern and rate of root growth. Andresen's lines, roughly perpendicular to the direction of the tubules, represent a temporary secretory front of the growing tooth root (Shellis, 1981; Dean, 1995). Between Andresen's lines it is often possible to see smaller incremental structures, von Ebner's lines, that form daily in many species (Yilmaz et al., 1977; Bromage, 1991; Dean, 1995; Klevezal, 1996).

Cementum, the Cinderella of dental tissues, is deposited by cementoblasts (very similar to osteoblasts) around tooth roots along the margin of the periodontal ligament (PDL). Cementum has none of the histological beauty of dentine and enamel, but it has the critical function of holding each tooth root in place (Figure 2.2). Histologically, cementum is similar to poorly organized bone, but it contains a high proportion of extrinsic collagen fibers bundles, Sharpey's fibers. These fibers are produced by cells in the PDL, and become mineralized ("cemented") into the edge of the tooth root by cementoblasts, thus anchoring the tooth to the PDL. In order to maintain a tooth in position for effective chewing, cementum grows gradually and continuously around the root throughout the tooth's life. Like the other dental tissues, cementum grows in increments that can help estimate a tooth's age and sometimes indicates the season of death (Lieberman, 1993, 1994).

How Bones Grow

How do bones grow to attain different sizes and shapes? And how do genes, cellular interactions, and mechanical stimuli influence the variety of sizes and shapes of bones found in the skull both throughout ontogeny and among species?

At a microscopic level, bone grows as either woven or lamellar (see Figure 2.1). Woven bone is deposited rapidly, often early in skeletal growth or in response to fracture. It consists of coarse, loosely packed collagen fibers of varying sizes that are poorly organized. Lamellar bone is more structured, largely because it is deposited more slowly as osteoblasts lay down and then mineralize collagen fibers in parallel arrays forming layers, called lamellae (Menton et al., 1984). Collagen fiber orientation is consistent within each lamella but varies between lamellae, forming a plywoodlike arrangement that creates strength in multiple planes (Giraud-Guille, 1988). Lamellae are separated by thin (1–2 μm), highly mineralized cement lines. These lines are structurally weaker than the surrounding bone but may increase the bone's overall elasticity and halt the spread of cracks through the tissue.

There are many kinds of lamellar bone that vary because of their rate of deposition and degree of vascularization (see de Ricqlès et al., 1991). Most common in the skull is circumlamellar bone, which osteoblasts deposit slowly in layers (3–10 μm thick) from the membranes surrounding each bone. More rapidly formed layers of bone, with more vascular channels, are called plexiform. In section, plexiform bone looks like a brick wall, because it has a three-dimensional interlacing network of vascular canals in multiple directions. Finally, lamellar bone also forms in tubelike structures, called osteons, which consist of up to 20 small, concentric lamellae around a central neurovascular channel (Figure 2.1). Osteons sometimes form new, primary bone around blood vessels, but they are most common as secondary osteons. These form through a process called Haversian remodeling, in which osteoclasts (bone-resorbing cells) first tunnel through old bone, releasing enzymes and acids that dissolve the tissue. Osteoblasts then follow behind the osteoclasts, filling in the tunnel with concentric lamellae of bone around a central vascular channel (see R. B. Martin et al., 1998). The result is a series of cylindrical structures (as shown in Figure 2.1) that spiral through bones. Secondary osteons are not as strong as primary lamellar bone, but they are stronger than damaged bone that has accumulated numerous microcracks (see below), and they may strengthen a bone by reorienting collagen in the principal directions of strain (Currey, 2002).

At a macroscopic level (visible to the naked eye), lamellae form two major types of bone, compact and trabecular (also known as spongy or cancellous). Compact bone, which is mostly solid but usually contains many vascular channels, makes up most of the skull. Trabecular bone, comprised of many tiny struts known as trabeculae, is mostly present within joints, but it is also present in several parts of the skull, such as in the middle of the cranial vault bones, around the tooth roots, and in the mandibular condyles. Despite their outward differences, both compact and trabecular bone form through the same major processes (intramembranous and endochondral growth), and once fully formed, both types of bone can change shape through a related set of processes known as remodeling. The combination of these processes permits bones to form almost any shape.

Intramembranous Growth

Almost all the skull (except the cranial base) grows through intramembranous ossification. The key property of intramembranous bones is that they form within the membranes that surround various organs, tissues, or spaces such as the brain or the oral cavity. Accordingly, the shapes of these organs or spaces have a strong influence on the bone's initial shape. As Figure 2.3 illustrates, intramembranous bones begin in the embryo at primary centers of ossification, places where osteoblasts are induced to differentiate from a precursor tissue called mesenchyme (a type of embryonic stem cell that can differentiate into many kinds of connective tissue). When so induced, mesenchyme surrounding developing bones condenses into a dense outer fibrous tissue known as periosteum.[1] In the periosteum, mesenchymal cells differentiate into osteoblasts that synthesize bone within the membrane, first producing and then mineralizing a matrix of collagen fibers (osteoid) as described above. Initially the osteoid is deposited in spicules (small struts) of bone interspersed with vascular channels. The spicules then accrue, eventually forming a sheet of bone within the periosteum (Figure 2.3).

Intramembranous bone growth is stimulated mainly by the tissues the bone encapsulates. Intramembranous bones can grow thicker from bone deposition along their surfaces, but they change shape primarily by depositing bone along their edges in sutures. Sutures are essentially articulations between bones, consisting of a fibrous, periosteumlike tissue, replete with mesenchymal cells (Alberius and Friede, 1992). Organs such as the brain and eyeball grow within a capsule of intramembranous bone, expanding like a balloon inside. The tension generated pushes the sutures of the capsule apart and stimulates osteoblasts to synthesize bone through various pathways (discussed below). Thus, even individuals with tiny brains (a condition known as microcephalus) or overly large brains (e.g., from hydrocephalus, a condition caused by too much cerebrospinal fluid) have cranial vault bones that fit snugly surround

---

1. Periosteum is often called endosteum when found on the inside of long bones, or pericranium and endocranium on the external and internal surfaces of the cranial vault.

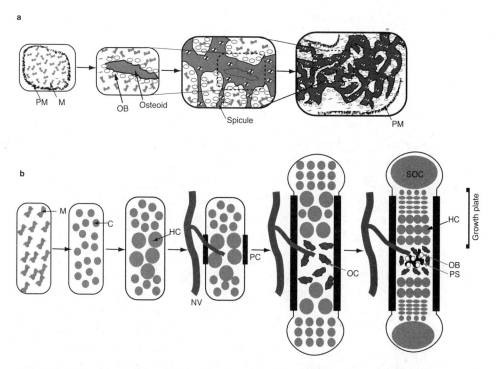

Figure 2.3. Schematic of different types of bone growth. (a) *Intramembranous growth* occurs in primary ossification centers in which a periosteum membrane (PM) surrounds a condensation of mesenchymal cells (M). Osteoblasts (OB) then differentiate from mesenchymal cells and synthesize spicules of unmineralized bone matrix (osteoid), which then become mineralized, eventually forming a layer of bone within the membrane. The shape of the membrane determines the initial shape of the bone. (b) *Endochondral growth* begins with a "preform" mass (anlage) of mesenchymal cells (M), which then differentiate into chondrocytes (C). These chondrocytes become hypertrophic (HC) near the center of the anlage, where they are invaded by a nutrient vessel (NV). Osteoclast cells (OC) resorb tissue in the center and are succeeded by osteoblasts, which deposit spicules of bone (primary spongiosa, PS); at the same time, osteoblasts build a periosteal collar (PC) around the outside of the bone shaft (also termed a diaphysis). Finally, chondrocytes line up into columns in the growth plate, where they undergo a sequence of differentiation, proliferation, and hypertrophy, displacing the secondary ossification center (SOC) away from the rest of the bone.

their brains whatever their size and shape. In addition, premature fusion of the cranial vault's sutures, known as craniosynostosis, causes potentially serious cranial malformations because the bones cannot grow properly around the expanding brain (see Chapters 4 and 5).

Endochondral Growth

Endochondral bone growth involves an intermediary tissue, cartilage (Figure 2.3b). Most of the postcranial skeleton (with the exception of the clavicle) grows endochondrally. In the skull, endochondral growth occurs just in the cranial base.

In endochondral growth each bone begins as a tiny model (*anlage*) of mesenchymal tissue that roughly approximates the shape of the future bone (see Figure 2.3). The anlage differentiates into cartilage cells that sequentially proliferate, mature, expand and synthesize cartilage matrix (for a review see Hall, 2005). The cartilaginous precursor, which is surrounded by a mesenchymal membrane, the perichondrium (similar to a periosteum), is replaced by bone in a complex series of steps illustrated in the bottom of Figure 2.3. First, cartilage cells in the center of the precursor expand and secrete a form of cartilage matrix that induces a nutrient blood vessel to invade. There, a primary ossification center forms at which chondroclasts first resorb cartilage, and then osteoblasts synthesize spicules of spongy bone (the primary spongiosa). At the same time, osteoblasts differentiate in the surrounding perichondrial membrane to form a periosteal collar of compact bone. In addition, secondary ossification centers form at the ends of the bone. Between the primary and secondary ossification centers are growth plates in which chondrocytes can rapidly synthesize cartilage from the growing end, while coordinated activity of osteoclasts and osteoblasts then replaces the cartilage with bone at the other side (as shown in Figure 2.3). In this way, a bone can elongate quite rapidly while maintaining fully functional joints.

Postcranial long bones typically have growth plates at both ends, but the endochondral bones of the cranial base elongate by interstitial growth within synchondroses. Synchondroses are similar to epiphyseal growth plates but grow from a central plate of proliferating cartilage cells, bordered on either side by bone-forming zones in which bone tissue replaces cartilage in a process similar to endochondral ossification (Rönning, 1995). Synchondroses and growth plates both fuse when cartilage production ceases (discussed below). Note that cranial bones, unlike postcranial bones, do not develop marrow cavities.

Remodeling

Intramembranous and endochondral ossification processes differ critically in that the initial shape of an endochondral bone is determined by the shape of its cartilage precursor, whereas the initial shape of an intramembranous bone is influenced largely by the shape of the organs or spaces it surrounds. However, once bone is formed, the distinction between the two processes becomes less important, and it is more useful to focus on the processes by which both types of bones change shape in response to cellular signals and forces exerted by other bones and tissues. These changes are generally termed *remodeling* (not to be confused with Haversian remodeling, which alters bone histology by replacing old bone tissue with new tissue in the form of secondary osteons). Figure 2.4 illustrates the main three categories of remodeling processes, all of which involve the coordinated combination of bone deposition and bone replacement: drift, displacement, and rotation (see also Enlow, 1963, 1977, 1990).

Drift

Drift occurs when osteoblasts add new bone tissue on one side of a bone in a depository field, while osteoclasts remove bone tissue on the other side in a resorptive field (Figure 2.4a). This process allows entire walls of bone to shift position without losing function. In a drifting bone, active depository and resorptive fields tend to be opposite each other (Enlow, 1990). If the rates of bone addition and resorption are the same, then the bone will "drift" in space without becoming thicker, in a manner that resembles moving a log pile by adding logs at one end and removing them at the other. Many bony walls in the skull move by drift. For example, the floor of the nose and the roof of the palate are the same wall of bone, but both regions grow downward through a drift process in which the roof of the palate is depository and the roof of the floor of the nose is resorptive. A titanium ball glued into the bone of the roof of the mouth in a growing animal eventually finds itself in the nose as the palate drifts downward around it (Enlow and Bang, 1956). If the rates or spatial distributions of depository and resorptive growth fields are not symmetrical, then the wall of bone can become thicker or thinner, or change shape.

Displacement

Displacement occurs when a bone grows or remodels on one end but does not resorb at the other end, thus pushing the bone as a whole away from other structures (Figure 2.4b). For example, the jaws mostly elongate from the deposition of bone along their posterior ends (in the mandible, this growth occurs at the condyle), displacing the teeth forward relative to the brain and other structures. Displacement can also move bones that are not themselves growing (an example of secondary displacement is the displacement of the toes

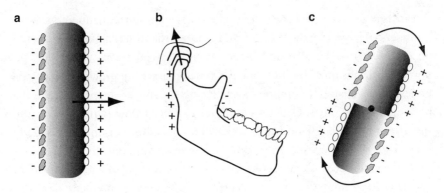

Figure 2.4. Types of bone remodeling. (a) Drift occurs as a result of the removal of bone on a resorption surface by osteoclasts, combined with the addition of bone by osteoblasts on an opposing depository surface. Together, these processes cause the bone to "move" in space in the direction of deposition. (b) Displacement occurs as a result of growth in one location (e.g., the mandibular condyle), which then moves (displaces) the rest of the bone away from other regions (e.g., the temporal bone). (c) Rotation occurs as a result of reversed depository and resorptive fields on either side of a central axis.

away from the pelvis as the femur elongates). Note also that the combined processes of displacement and drift cause many bones to lengthen and widen simultaneously, generating complex growth patterns. The ramus of the lower jaw, for example, has a depository growth field on the posterior margin and a resorptive field on the anterior margin, so that the ramus drifts posteriorly as the mandible displaces forward (as shown in Figure 2.4b).

Rotation

Many parts of the skull not only expand and change shape but also rotate relative to each other as a result of a combination of displacement and drift. Rotations occur when active depository and resorptive fields on opposite surfaces of a bone are positioned next to active resorptive and depository fields, respectively, creating an axis of rotation (Figure 2.4c). In essence, the bone drifts in opposite directions on either side of the axis. Rotations may also occur in a hingelike manner in synchondroses through more growth at one margin than at the other.

Growth and Scaling

At some level, all variations in bone shape derive from differences in the rate, timing and pattern of skeletal tissue synthesis and resorption in growth fields. But what processes control these activities? Do all bones start out more or less the same in size and shape and then vary because they subsequently grow differently? Or do different bones have initial, intrinsic differences in morphology that persist or become accentuated during growth? To address these questions it is useful to consider ways in which patterns of growth and development differ among species and how these differences influence morphology. This framework will facilitate an exploration of how patterns of bone growth and development are regulated by genes, by interactions with neighboring tissues, and by other aspects of a bone's environment. Identifying the morphological variations that result from different patterns of growth permits explanations of the *processes* that generate these different patterns.

Important variables in these processes include time, size, shape, and form. Time, of course, is a temporal dimension that can be measured in absolute or relative units. Size, shape, and form are descriptors of morphology. Technically, the *form* of something like a skull can be described in terms of its *size*— a measure of quantity—and its *shape*—a measure of spatial configuration that is independent of size. According to this framework, *size* and *shape* are different parameters, whose combination is *form*. A tennis ball and a basketball have the same shape, but differ in size; liter-sized bottles of wine and orange juice have the same size but different shapes.

Using these terms, we can formally define *growth* as the relationship between time and size, *scaling* as the relationship between size and shape, and *development* as the relationship between time and shape (see Zollikofer and Ponce de León, 2004). Each relationship is important for generating different patterns of morphological variation.

Time and Size (Growth Trajectories)

In general, organs and tissues grow bigger as we mature. But they can do so via trajectories that vary in onset, offset, and rate of growth. There are two major postnatal trajectories in the human head: neural and skeletal (Figure 2.5a). The brain and its associated capsules, the neurocranium and cranial base, grow especially fast during the first two postnatal years and then reach adult size within six to eight years. In contrast, most of the face grows more slowly, along with the rest of the skeleton, with spurts after birth and during adolescence, reaching adult size between ages 14 and 20. The dentition has its own peculiar trajectory, growing more or less continuously from birth until age 16–18. Growth trajectories differ among species. In chimpanzees, the neural trajectory is approximately complete by 3 years of age, and the skeletal trajectory is complete by approximately 12 years (Schultz, 1960; Leigh, 1992). In addi-

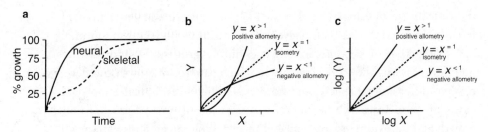

Figure 2.5. (a) Neural versus skeletal growth trajectories. (b) and (c) Unlogged and logged (linearized) allometric trajectories, showing differences between positive allometry, negative allometry, and isometry.

tion, different components of craniofacial regions sometimes have varying trajectories, thus affecting the interactions that can occur between growing structures. In the face, for example, the eyeballs (which, embryologically, are derivatives of the brain) have a rapid neural growth trajectory, but the bones around the eye interact with other components of the face (e.g., the nasal cavity) that have a more skeletal growth trajectory.

Size and Shape (Allometry)

Many structures change shape as they grow. Any scaling relationship between size and shape can be quantified as

$$Y = kX^b$$

where Y is a measure of shape, such as the ratio of two linear measurements (e.g., orbit height to orbit width), X is some measure of size (often body mass), k is a constant, and b is the coefficient of scaling between X and Y. When $b = 1$, then X scales linearly relative to Y and their relationship is said to be *isometric;* when b > 1, then X scales with *positive allometry* relative to $Y;$ and when $b < 1$, then X scales with *negative allometry* relative to Y. Figure 2.5b illustrates examples of each type of scaling relationship. Because curvilinear exponential lines are mathematically cumbersome, it is useful to linearize the above equation by taking the logarithm of both sides:

$$\log Y = \log k + b \cdot \log X$$

conveniently converting the coefficient of allometry, *b,* into the linearized equation's slope, and the constant, *k,* into the intercept (as shown in Figure

2.5c). Note the need to keep track of the dimensionality of Y and X. Body mass is related to volume and is thus a function of cubed linear dimensions. For example, in a group of spheres of different size, the isometric scaling relationship between their volumes (X) and their radii (Y) would be: $Y = X^{0.33}$. If we convert X to a linear dimension as its cube root, then the equation's coefficient of allometry is 1. Similarly, areas are a function of squared linear dimensions, so the isometric scaling relationship between an area, Y, and a length, X, is $Y = X^{0.50}$.

Allometric scaling is a powerful concept for exploring evolutionary change because one way organisms change shape is by varying in size (Gould, 1977). Allometry also allows one to correct for the effects of scaling when examining shape differences among organisms and to test hypotheses about constraints (limitations of variation) and functional relationships. As we compare shapes and sizes of different structures, it is useful to distinguish three kinds of allometry: *ontogenetic allometries* are changes in scaling during growth within a species; *interspecific allometries* compare scaling relationships among organisms at the same ontogenetic stage (usually adults) from different species; and *intraspecific allometries* compare scaling relationships among organisms from the same stage within a species. In general, ontogenetic scaling relationships have higher coefficients of allometry, and intraspecific scaling relationships have lower coefficients of allometry. Note also that interspecific allometries differ depending on the range of taxa included in an analysis. For example, there is a well-known negative allometry between body mass and brain mass in vertebrates in which the coefficient of allometry, b, varies depending on the range of species included (Martin and Harvey, 1985).

Time and Shape (Heterochrony)

Many morphological changes during ontogeny can be explained by scaling, and it follows that alterations to ontogenetic allometries are a powerful engine for generating evolutionary change. Shapes often vary as a result of modifications in size between ancestor and descendant species. Analyses of allometry alone, however, don't consider how changes in timing contribute to changes in size, and hence in shape.[2] A bone can grow to be bigger and therefore differently shaped in one species compared to another because it grows faster,

---

2. Because organisms get bigger as they mature, it is common to use size as a proxy for age when comparing ontogenetic size-shape relationships of two species. One problem with this approach is that patterns of growth may differ between the species, obscuring changes in rate or timing.

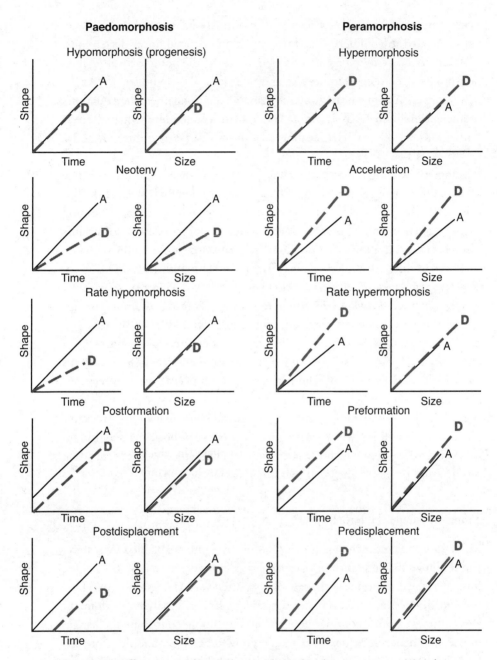

Figure 2.6. Different types of heterochronic relationships between ancestor (A) and descendant (D) species expressed in terms of shape versus size, and shape versus time (modified from Lieberman et al., 2007). As described in the text, the two major types of heterochrony are *paedomorphosis,* in which the descendant species is underdeveloped in

size or shape relative to the ancestor, and *peramorphosis,* in which the descendant species transcends its ancestor in terms of size or shape. For example, a neotenous descendant is the same size as the adult ancestor but is underdeveloped in terms of shape. This figure oversimplifies heterochronic patterns because multiple growth trajectories can arrive at the same endpoint in the relationship between age and shape. For example, two species may arrive at the same adult relationship between shape and age, but one species may do so via growth spurts, whereas the other may have a constant rate of growth. For simplicity, I have plotted all relationships between age and shape as linear.

starts growing earlier, or grows for longer. Thus, a useful complement to analyses of allometry is the study of changes in time versus shape between ancestor and descendant species, termed heterochrony (a term coined by the embryologist Ernst Haeckel in 1866).

Studies of heterochrony can be confusing to follow because different methods of defining the relationship between shape and time have led to alternative terms. As illustrated in Figure 2.6, there are two major categories of heterochronic change: *paedomorphosis* refers to an adult descendant that resembles a juvenile ancestor (a less developed descendant), and *peramorphosis* refers to an adult descendant that transcends an adult ancestor (an overdeveloped descendant). Paedomorphosis and peramorphosis can be subdivided further into several heterochronic patterns (Gould, 1977; Rice, 1997; Alba, 2002) that differ in terms of the rate and timing of size-shape relationships. For example, a paedomorphic descendant can have a reduced slope of allometry without any changes in timing, leading to *neoteny* (in which adults of the descendant species are as large as the ancestors but have a more juvenilized shape). Alternatively, a descendant species can have the same allometry as its ancestor but cease growth earlier (a phenomenon termed *progenesis* or *time hypomorphosis*), or grow more slowly (*rate hypomorphosis*). Finally, a descendant species may have the same allometry as its ancestor but be paedomorphic because it started at a less developed shape (*postformation*), or because it started developing later (*postdisplacement*). Figure 2.6 plots the peramorphic equivalents of neoteny (*acceleration*), hypomorphosis (*hypermorphosis*), rate hypomorphosis (*rate hypermorphosis*), postformation (*preformation*) and postdisplacement (*predisplacement*).

With such different concepts and unwieldy terms, it is tempting to avoid the subject of heterochrony altogether. That would be a mistake, however, as there is abundant evidence that heterochronic shifts are a major cause of evolutionary transformations. One classic case is the shape difference evident between the crania of common chimpanzees (*Pan troglodytes*) and bonobos (*Pan paniscus*). In terms of shape, adult bonobo crania closely resemble sub-adult chimpanzee crania; this observation has led to the suggestion that they are paedomorphic (e.g., Shea, 1983; 1992; Williams et al., 2003; Mitteroecker et al., 2004; but see Godfrey and Sutherland, 1996). My colleagues and I recently reanalyzed this problem in a sample of chimpanzee and bonobo crania, using dental staging criteria to estimate age and multivariate analyses to quantify size and shape (Lieberman et al., 2007a). According to our analysis, some of the shape differences between bonobos and chimpanzees can be attributed to paedomorphosis (specifically postformation, that is, paedomorphosis due to initial shape underdevelopment). But paedomorphosis can explain only a modest proportion of the variation between the two species. One reason is that scaling relationships among different components of a skull are often disassociated. That is, one region of a descendant organism may evolve relative to the ancestor via one pattern of heterochrony, but another region might have evolved via a different heterochronic pattern, or by a mechanism unrelated to heterochrony. Modern human skulls are an excellent example of this problem. Gould (1977) considered *H. sapiens* skulls to be neotenic versions of chimpanzees because, superficially, adult human skull shape seems to resemble that of large juvenile chimps. As Chapter 5 discusses, the *H. sapiens* braincase is an example of peramorphosis, whereas the face is an example of paedomorphosis (Shea, 1989).

Finally, it is useful to remember that allometric and heterochronic patterns of scaling cannot explain all aspects of shape change. Heads and skulls can also change via the repositioning of regions (heterotopy), by evolutionary novelty, or by the addition of bone and other tissues in patterns that are unrelated to the phenomenon of scaling.

Genetic and Cellular Regulation of Bone Development and Growth

So far, we have reviewed the basic processes by which bones grow at sutures, synchondroses, and growth fields, and how different patterns of growth and scaling can generate many (although not all) variations in form. Although these processes and patterns may seem unrelated, almost every variation in skull form is regulated by a shared set of mechanisms that stimulate or inhibit the differentiation of osteoblasts, chondrocytes, and osteoclasts in specific growth sites, and by a further set of mechanisms that influence the onset, offset, and rate of activity of these cells at each site. The next step is to discuss what mechanisms regulate these patterns and processes in the first place. To grow something complex growth like a skull, the various units need to form in the right place and time, and with the right size and shape. Each unit may have some sort of blueprint, but the growth and development of each component of the skull could not possibly be completely preprogrammed. Because of its dynamic complexity, the head must be able to accommodate its many bones, organs, and other tissues as they change in size, shape and relative position throughout ontogeny. These components need to interact dynamically during development to generate a functionally integrated whole. These interactions must also allow some degree of latitude for functionally competent variations to develop, thus permitting tinkering to occur over evolutionary timescales.

We are a long way from comprehending all the processes that influence the patterns of growth and development outlined above for each bone in the skull. Consider, for example, a relatively insignificant morphological feature such as the chin. This feature, a projection of bone at the base of the front of the mandible, grows from a combination of processes that regulate depository and regulatory growth fields: the lower front of the mandible grows forward from deposition, whereas the bone above it (around the roots of the incisors) is resorbed, leaving behind a projecting shelf of bone (Enlow, 1990; Kantomaa and Rönning, 1991). Although we know where these different growth fields are, and something about how osteoblasts and osteoclasts are regulated to be active or inactive, we are relatively ignorant about how the cells in each region of the mandible and elsewhere in the skeleton know their place and position, and when to turn on or off.

In spite of our ignorance about many details of what makes a human skull grow the way it does (and thus differ from, say, a chimpanzee skull), we have some understanding of the genetic and cellular processes that regulate skeletal growth. These processes act at three hierarchical levels that are useful to review before we examine human craniofacial growth and development in Chapters 3–5. The first step is *patterning:* the creation of units in the right place and time with information about their eventual identity as particular bones or other tissues. The second step is *morphogenesis,* in which particular cells are activated to differentiate and form particular cartilages or bones. The third step is *growth,* in which chondrocytes, osteoblasts, osteoclasts, and other such cells function to change the size of a given bone by growing and resorbing tissue at specific sites and rates. None of these steps occurs in isolation. Instead, they occur in an iterative and integrative fashion through many interactions among multiple cell types, both within and between regions, as well as from various stimuli from the environment, including mechanical loading. Given the especially important biomechanical functions of bone, we will discuss separately the effects of mechanical loading on skeletal growth.

Patterning

Studies of gene expression in chicks, mice, and other creatures combined with analyses of human skeletal growth disorders provide a general framework of the early stages of skeletal growth that determine the position and identity of precursor tissues. As noted above, all the bones and cartilages that make up the embryonic skull differentiate from mesenchyme, a tissue comprising precursor skeletal cells that can give rise to osteoblasts, chondrocytes, and other cells that synthesize connective tissues. Cranial mesenchyme has two different origins: some derives from cells that begin in the middle layer of the embryo, called mesoderm; the rest comes from a special group cells, the neural crest, which originates within the top layer of the embryo, the ectoderm, as it forms the neural tube. (For more on these cell types, see Chapter 3.)

Mesenchyme creates units in the head in different ways, depending in part on location and whether the cells originate from mesoderm or the neural crest. As detailed in Chapter 3, the posterior cranial base originates from segments of mesodermal mesenchyme. Some of these segments (which contribute to the occipital bone) come from somites, units of mesenchyme that comprise most of the vertebral column. Other segments derive from mesenchymal condensations that form on either side of the embryo's midline. At this point, it is useful to know that many condensations derive their identity from a class of genes known as homeobox (or *Hox*) genes, whose sequence of expression patterns many parts of the body along particular axes (e.g., from head to tail). Other condensations get their identity from interactions with nearby tissues.

The rest of the head does not arise from mesoderm but instead comes from mesenchymal cells from the neural crest. As I discuss in Chapter 3, these important cells are highly migratory (they move along prescribed pathways) and can give rise not only to mesenchymal precursors of bone and cartilage but also to a wide range of other tissues, including much of the face and the nervous system (Hall, 1999). The cells of the neural crest portions of the face also become segmented and patterned, but through different processes from the mesodermal somites (see Chapter 3).

Morphogenesis

Once mesenchymal segments that form the regions (or units) of the head exist in appropriate number and have identity and position, a series of additional genes and signaling factors work together to transform them into organs such as bones. These processes are complex and variable, but for the purposes of our review they may be simplified into several stages. First, as noted by Atchley and Hall (1991), five major factors determine the size of each condensation: (1) the number of mesenchymal stem cells initially present, which is largely a function of the number of cells that migrated there in the first place; (2) the time of initiation of the condensation; (3) the rate of cell division within the condensation; (4) the percentage of cells in the condensation that can divide; and (5) the rate of cell death in the condensation. The importance of condensation size is evident from a number of experimental observations (see Cohen, 2000b). For example, overly small condensations never develop into bones, and overly large bones develop from overly large condensations (Cottrill et al., 1987; Atchley and Hall, 1991).

The next crucial step is the regulation of cell fates within each condensation. In some cases, what happens to a cell depends on how it was patterned (e.g., by *Hox* genes). Cell fate also gets determined by spatially and temporally specific interactions between cells, a process known as induction. Induction occurs when one cell sends a signal to another cell that influences its fate. In the case of neural crest mesenchyme, as cells migrate from their point of origin, they often come into contact with various epithelial (lining) cells, which sometimes emit signaling molecules that can influence mesenchymal cell fate. For example, as streams of neural crest cells migrate from the back of the embryo's head toward the face, they receive signals from the cranial ectoderm cells that they pass along the way. These inductive interactions (reviewed in Chapter 3) help determine what part of the face a given set of cells will eventually form. For instance, cells that eventually generate the jaw receive a different set of inductive signals than cells that become part of the nose or neck. Countless other inductive interactions—many evident from transplant experiments— exist throughout the skull to determine the fate of different condensations. For example, transplanting epithelial cells from the incisor region of the jaw to the molar region causes the local mesenchymal cells to develop into an incisor rather than a molar (Peters and Balling, 1999; Jernvall and Thesleff, 2000).

Once a condensation is in place and its fate is determined, the next stage of morphogenesis is the differentiation and regulation of osteoblasts, chondrocytes, osteoclasts, and other such cells to generate the precursor tissues that actually form the skeletal element in question. As noted above, these processes differ somewhat in intramembranous and endochondral bones. In intramembranous bones, mesenchymal cells differentiate directly into osteoblasts, which then synthesize and mineralize bone tissue. Many genes regulate this process (see Karsenty and Wagner, 2002), but a particularly important gene that regulates osteoblast differentiation is *Runx-2* (also known as *Cbfa1*). Deletion of this gene in mice leads to boneless pups (Komori et al., 1997), and elevated levels of *Runx-2* expression stimulate increased rates of bone production (Ducy et al., 1999). In contrast, endochondral bones first form a cartilage precursor, which then transforms through many steps into a bone (see above). One key step involves the differentiation and proliferation of chondrocytes via the Sox9 transcription factor (see Olsen et al., 2000). Precursors then grow through the differentiation and hypertrophy of chondrocytes and through the secretion of cartilage matrix. Chondrocytes themselves secrete factors that stimulate the invasion of the precursor by blood vessels, which bring with them osteoblasts as well as osteoclasts (Gerber et al., 1999).

Growth and Development

After a skeletal condensation has formed and undergone the appropriate morphogenetic transformations to become a given unit, the future bone or cartilage grows and develops to its appropriate size and shape. This stage, which generates a large proportion of the variation in bone size and shape, occurs at a regional scale within bones to define growth fields and regulate their activity. In particular, mesenchymal cells in sutures, synchondroses, periosteum, and other growth sites differentiate into skeletal tissue-forming cells such as osteoblasts or chondrocytes, and these cells then regulate the activity of osteoclasts. In addition, other signals regulate the proliferation and activity of bone-forming cells to influence the rate and timing of deposition and resorption at particular locations. Regulation of local growth occurs through interactions between the genes that cause skeletogenic cells to synthesize, resorb, or otherwise modify skeletal tissue, and stimuli from other genes or cells. Such interactions between cells and their environment (which includes other cells) are generally categorized as epigenetic interactions, of which there are several types that regulate most bone growth.

One important form of epigenetic interaction, already noted above, is the induction of skeletal cells by neighboring cells. For example, the membrane surrounding the brain, the dura mater, induces activity by osteoblasts in the sutures of the cranial vault and in growth fields along the surfaces of cranial vault bones (Spector et al., 2002). Regions of the dura mater can also induce cartilage growth in the synchondroses of the cranial base (Yu et al., 1997). The different expression of signaling molecules by various portions of the dura mater may underlie some regional differences in cranial vault and cranial base activity and may themselves be regulated by signals generated by the brain itself. Indeed, in a famous experiment, reorienting an embryonic chick brain by 180° caused a similar reversal in the shape of the underlying cranial base (Schowing, 1968).

As growth proceeds, many more epigenetic interactions occur between bones, cartilages, and their surrounding tissues. Thus bone near an eye, a muscle, or a nerve, grows differently than bone near another organ or tissue. Muscle-bone interactions occur throughout the skeleton (for a fascinating example, see Matsuoka et al., 2005) and are particularly well studied in the mandible. Herring and Lakars (1982) for example, showed that the mandibular condyle in the mouse embryo fails to form in the absence of the lateral pterygoid muscle, whereas overgrowth of the muscle results in an unusually large condyle (Hall, 1978). Another example is the orbit. The developing eye is initially supported by a cartilage, the scleral cartilage, induced to form at the margin of the cornea (Rohrer and Stell, 1994). The orbit then forms from a half dozen intramembranous bones around the growing eyeball. Gruesome experiments that drain the fluid from growing eyeballs in young animals generate misshaped orbital skeletons whose shapes correspond to smaller eyeballs but whose differentiation and development are otherwise normal (Weiss and Amprino, 1940; Coulombre, 1965). These and other experiments illustrate the importance of secondary epigenetic interactions between bones and neighboring tissues. Such interactions preserve function during complex growth, even when the growth involves many different tissues that grow at different rates under different mechanisms of control (Moss, 1968). These interactions also permit morphological variations that range from minor differences between members of the same species to novel phenotypes.

Additional categories of epigenetic interactions influencing morphogenesis include systemic hormones, growth factors, and the effects of mechanical loading (discussed separately below). Many of these interactions occur throughout life, especially those involving hormones and growth factors, which often work in concert. Hormones, produced by glands such as the thyroid or the gonads, circulate throughout the body. Growth factors (a loosely defined subcategory of compounds known as cytokines) are usually larger molecules mostly produced by skeletal cells themselves, which have effects only within the cell or on nearby cells. In most cases, hormones and growth factors bind with specific receptors in the cell membrane or nucleus. Activated receptors then trigger a cellular response through mechanisms such as altering gene transcription, altering ion transport into or out of the cell, activating or inhibiting intracellular enzymes, stimulating protein synthesis, or inducing cell proliferation. The concentration and abundance of receptors, and the synergism or antagonism between different growth factors or receptors, affect cellular responses in various ways.

Many diverse hormones and growth factors, summarized in Table 2.1, influence skeletal cell activity. These variously up-regulate (stimulate differentiation, proliferation, or synthesis) or down-regulate (cause cell death, inhibit proliferation, or inhibit synthesis) chondrocytes and osteoblasts; an additional and large set of factors regulates osteoclast activity. As a key example, consider the IGFs (insulin-like growth factors), which regulate much bone growth. IGF-I is synthesized by osteoblasts and liver cells; IGF-II is produced only by osteoblasts. Both IGFs stimulate or speed up the rate of collagen matrix production in osteoblasts and chondrocytes. Experimental studies demonstrate

Table 2.1: Major hormones and growth factors affecting skeletal growth

Hormone	Major effects on skeletal cells
Estrogen	Up-regulates osteoblasts
	Up-regulates chondrocytes at moderate levels
	Down-regulates chondrocytes at high levels
	Down-regulates osteoclasts
Testosterone	Up-regulates chondrocytes and osteoblasts at moderate levels
Vitamin D	Up-regulates osteoblast and chondrocytes (promotes proliferation)
Thyroid hormone (TH)	Up-regulates osteoblasts and chondrocytes at normal levels
	Up-regulates osteoclasts at high levels
Parathyroid hormone (PTH)	Up-regulates osteoclasts
Growth hormone (GH)	Up-regulates osteoblast, chondrocyte, and osteoclast activity
IGF-I	Up-regulates osteoblasts/chondrocytes
Calcitonin	Down-regulates osteoclasts
	Up-regulates osteoblasts
Cortisol	Down-regulates osteoblasts and chondrocytes
	Up-regulates osteoclasts
Insulin	Up-regulates osteoblasts and chondrocytes

that IGF-I induces endochondral bone formation by increasing chondrocyte mitosis rates (e.g., in growth plates), and inducing osteoblasts to synthesize collagen and other proteins (Clemens and Chernausek, 2004). IGF-I is primarily stimulated by growth hormone (see below) but is also regulated by other hormones such as estrogen, insulin, cortisol, and parathyroid hormone. IGF-II mostly stimulates osteoblasts to synthesize collagen and other proteins in response to parathyroid hormone (Schmid, 1995).

A few systemic hormones summarized in Table 2.1 that regulate skeletal growth merit special mention because their effects are so major. One is growth hormone (GH), which the anterior pituitary gland synthesizes in response to hormones produced by the hypothalamus. GH (which is released in pulses every 3–4 hours) regulates carbohydrate and protein metabolism in the liver, but it also induces mitosis and protein synthesis in skeletal cells by triggering IGF-I production, helping make bones longer and thicker; GH sometimes stimulates osteoclasts to resorb bone (Ohlsson, et al., 1998). The effects of GH on bone growth occur throughout skeletal growth but are particularly evident during the adolescent growth spurt. Other hormones that influence bone growth include glucocorticoids (e.g., cortisol), insulin, estrogen, and testosterone, parathyroid hormone (PTH), and calcitonin. These hormones sometimes stimulate and sometimes inhibit growth, often by acting in pairs as antagonists. For example, the parathyroid gland synthesizes PTH to increase concentrations of calcium in blood plasma by stimulating osteoclasts to resorb bone (Canalis et al., 1989; T. L. McCarthy et al., 2000; Whitfield et al., 2002). In contrast, cells in the thyroid gland produce calcitonin, which reduces calcium levels in blood plasma by down-regulating osteoclasts (Findlay and Sexton, 2005).

The effects of the sex steroids on bone growth are especially interesting. At low levels, estrogen and testosterone up-regulate osteoblasts, inducing differentiation and bone synthesis. But at high doses, estrogen (which males convert from testosterone in growth plates via the enzyme aromatase) causes chondrocyte cell death. Thus surges of estrogen or testosterone during puberty cause fusion of bone growth plates and the cessation of growth in height (Compston, 2001). Males who lack either estrogen receptors or the aromatase enzyme have delayed epiphyseal closure and skeletal maturation, resulting in very tall stature (E. P. Smith et al., 1994; Morishima et al., 1995). Estrogen also helps regulate osteoclast activity so that declines in estrogen levels in women after menopause contribute to increased rates of osteoporosis (Compston, 2001; Lanyon et al., 2004).

Mechanical Loading

What is the relationship between the mechanical properties of bone and the ways that applied forces influence bone growth? Ideas about how bones respond to mechanical loading sometimes fall into one of two extreme views. One, which is clearly wrong, is that bone is a relatively inert, unresponsive tissue whose morphological properties result almost entirely from genetic programming. The other extreme view, also problematic, is that bone is a highly plastic tissue whose shape and histology are mostly determined by epigenetic responses to applied loads. As is often the case, the truth lies somewhere between the two views. Bones do have a strong genetic component to their growth and development, but a set of complex and constrained interactions between bone cells and their mechanical environment can influence bone morphology, particularly while the skeleton is still growing.

To understand how, when, and to what extent bone growth responds epigenetically to applied loads, it useful to consider several major steps in the process. First, applying a force generates small deformations that vary depending on the nature of the force and on the shape and structure of the bone. Second, bone cells sense these strains and transduce the information into a biochemical signal that regulates bone forming or resorbing cells. These responses then sometimes alter how a bone grows or develops. The effect depends on a variety of factors, including age and the nature of the strain signal.

Stresses and Strains

The most basic function of bone is to be stiff (to resist deformation from an applied load). In solid mechanics, the stress (σ) applied to a bone is defined as the amount of force (F) per unit area (A):

$$\sigma = F/A$$

and is usually measured in newtons/m^2 (or pascals). In turn, applied stress generates deformation, measured as strain (ε), a dimensionless unit defined as the change in length (ΔL) that occurs relative to the original length of the object (L):

$$\varepsilon = \Delta L/L$$

Positive strains are tensile, and negative strains are compressive. Note that forces always generate both tensile and compressive strains at right angles (orthogonal planes) to each other. If you compress an object along its long axis, it simultaneously becomes wider. Bones, however, are usually are bent, twisted, or sheared, and each of these types of strain causes distinctive patterns of tension and compression, as shown in Figure 2.7.

Two aspects of a bone's shape and structure determine the nature of the strains a given force generates. The first is the material properties of the bone itself, which is largely a function of variations in histology and degree of mineralization. As noted above, bone has both organic and mineral components that make it stiff and flexible. Elasticity, the relative stiffness or flexibility of the bone, is defined as σ/ε and is measured (confusingly) in the same units as stress (see Figure 2.7).[3] At most typical stresses, bone is linearly elastic and fairly stiff (bones rarely deform more than 0.3 percent). At a certain level of stress (σ_y), however, bone begins to deform so much that its structure breaks down permanently, eventually leading to mechanical failure at σ_f. Thus to avoid breaking, a bone must grow sufficiently strong to remain sufficiently stiff.

The second property that determines how much and what kind of strain is generated is a bone's overall shape. Because stress is a ratio of force to area, it follows that the larger the area over which a force is applied, the smaller the amount of stress and hence strain. For a bone tensed or compressed along its long axis (Figs 2.8a,b), increasing the cross-section of the bone increases the area over which the force is applied, thus lowering the strain generated. For a bone that is bent (Figure 2.8c), the most effective way to decrease strain is to add bone in the plane of bending (think of a roof beam that is taller than it is wide). In a bone that is twisted around a particular axis (Figure 2.8d), distributing bone around this axis most effectively decreases strains. Finally, in a bone that is sheared (Figure 2.8e), adding bone in the plane of shearing most effectively decreases strains.

Mechanotransduction

When a bone is subject to some combination and magnitude of strains, it sometimes senses how much and in what way it is strained and then recruits an appropriate cellular response. These mechanisms are poorly understood. One probable mechanism is that osteocytes function as strain receptors and transducers (Marotti, 1996). Recall from above, that osteocytes are osteoblasts

---

3. Both are usually measured in pascals (N/m^2). This unit confusion occurs because strain is dimensionless.

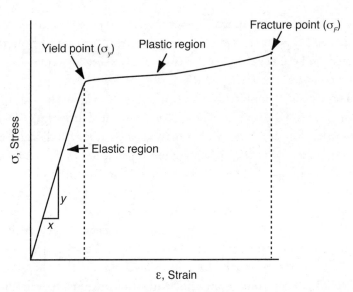

Figure 2.7. Mechanical properties of a tissue such as bone, graphed as the relationship between stress (σ, force per unit area) and strain (ε, change in length). Within the elastic region of deformation, there is a linear relationship between stress and strain (quantified as the modulus of elasticity, E) up to a yield point, σ_y. Above σ_y, deformation becomes plastic, with little additional stress leading to failure (σ_f).

that have become entombed in bone matrix during growth (see Figure 2.9). Once entombed, they develop dozens of long processes, canaliculi (also visible in Figure 2.9), which radiate out from the cell body in every direction and connect with other osteocytes to form a complex known as the connected cellular network (CCN). The CCN is like a nervous system throughout the bone, connecting cells within the bone to each other and to cells on the surface in the periosteum (Cowin and Moss, 2001; Pearson and Lieberman, 2004). Numerous studies suggest that the CCN detects strains, possibly in multiple ways. One hypothesis is that every time a bone is strained, osteocytes sense fluid flow within the canaliculi (Weinbaum et al., 1994; Cowin et al., 1995; Bakker et al., 2001, 2003). Further, osteocytes may detect small cracks (microcracks, as shown in Figure 2.9) in bone tissue, perhaps because such cracks induce cell death (Mori and Burr, 1993; R. B. Martin et al., 1998; R. B. Martin, 2000; Dodd et al., 1999). Bones may also detect strain via receptors in the periosteum by sensing small electrical charges that strains generate within bone tissue (Cowin and Moss, 2001), and by specialized filaments that project off the cell body (Whitfield, 2003).

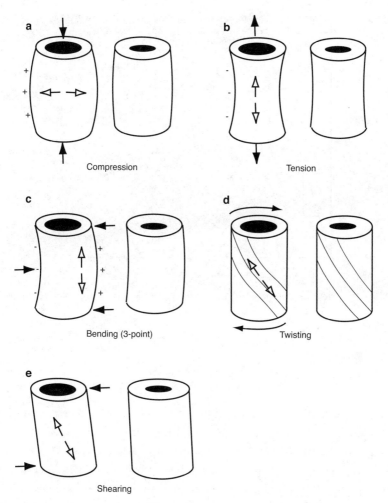

Figure 2.8. Different types of stress regimes, including (a) compression, (b) tension, (c) bending, (d) twisting, and (e) shearing. Solid arrows indicate directions of applied forces; hollow arrows indicate directions of tension (compression is always orthogonal to tension). To the right of each "stressed" object is an example of an object with greater resistance to that type of load.

One question that arises is, what aspects of strain does bone actually sense? Strain has many parameters, including magnitude, frequency, type (tension or compression), and temporal spacing, that apparently affect how a bone responds to loading. In terms of strain magnitude, meta-analyses of experimental studies show that peak strains on the periosteal surface of bones during

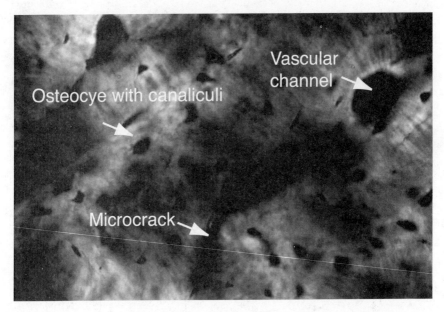

Figure 2.9. Microscope photo of a section of cortical bone, showing osteocytes with canaliculi, a vascular channel, and a microcrack (adapted from Pearson and Lieberman, 2004).

"vigorous" activity (e.g., running, flying, eating hard food) in vertebrates tends to be between 0.2 percent and 0.3 percent (Rubin and Lanyon, 1984, 1985; Rubin 1984; Biewener, 1991). This apparent threshold suggests that bones somehow grow to keep maximum deformation below a certain level. In addition, experimental studies on growing birds show that functionally equivalent sites (bone midshafts) maintain similar magnitudes and orientations of strain despite tenfold and threefold increases in mass and length, respectively (Biewener et al., 1986; Biewener and Bertram, 1993; but see Main and Biewener, 2004). In other words, bone responses to loading may function to maintain equilibrium. Other studies also suggest that bones may be tuned to maintain a constant total amount of strain per unit time (Carter and Beaupré, 2001), with different responses depending on the percentage of tension versus compression (Frost, 1986, 1987, 1990). The frequency and temporal spacing of loading events also matters. Because bones cells rapidly become saturated once they are loaded, intermittent bouts of moderate mechanical loading elicit a greater response than a single bout of more intense loading (Robling et al., 2002).

Variable responses to loading

Regardless of how bones sense strain, and what parameters are most important, applied mechanical loads lead to one of four potential outcomes. First, strains can stimulate osteoblasts in the periosteal and endosteal membranes to add new bone (hereafter termed *modeling*). Second, strains can stimulate resorption by recruiting osteoclasts to remove bone along a surface. Third, strains may stimulate bone turnover via Haversian remodeling. And, fourth, strains often elicit no response (quiescence), either because the signal was not sufficient or because a response was inhibited. There has been much debate about when, how much, and to what extent these four alternative responses may be activated. First, let's review we know about kinds of loading that do and do not elicit bone growth responses.

Modeling

Many studies have found that trabecular and cortical bone respond most actively and predictably to loading prior to skeletal maturity, particularly during the adolescent growth spurt. One well-known study analyzed cortical bone asymmetry in the upper arm of professional tennis players (a clever test, as the right and left arms of each player are genetically identical but the racket arm is loaded much more than the other arm). Athletes who started playing tennis between the ages of 8 and 14 years had approximately 50 percent thicker cortical bone in the playing than in the nonplaying arm. Those who started playing tennis after late adolescence (between ages 15 and 30) had much smaller increases in racket-arm growth and half as much asymmetry (Jones et al., 1977; Trinkaus et al., 1994; Ruff et al., 1994). Other comparative studies have replicated these results (see Carter and Beaupré, 2001; Currey, 2002; Pearson and Lieberman, 2004). In addition, several experiments on trabecular bone have shown that bony struts preferentially align themselves in the orientation of compression, a phenomenon sometimes known as Wolff's law, after the German anatomist who in 1892 recognized the dynamic (but hardly lawlike) relationship between bone structure and bone function (Biewener et al., 1996; Currey, 2002; Pontzer et al., 2006).[4] Studies have also tried to tease apart how variations in the magnitude, frequency, and temporal spacing of strain and in age, sex, and bone location affect responses to strain (see Bertram

---

4. This idea had already been developed by Roux (1881) and von Meyer (1867), but the term *Wolff's law* has stuck.

and Swartz, 1991; Martin et al., 1998; Carter and Beaupré, 2001; Currey, 2002). These experiments indicate that many factors influence how much, where, and when loading stimulates bone growth. Yet in all cases, adults tend to grow little or no extra bone in response to loading. The likely causes for this aging phenomenon are that adult mesenchymal cells have a diminished capacity to differentiate into osteoblasts (Nishida et al., 1999) and that adult osteoblasts and osteocytes are less sensitive to mechanical and other signals (Turner et al., 1995; Stanford et al., 2000; Donahue et al., 2001).

Often a bone adds tissue in response to loading in locations that strengthen the bone by increasing its resistance to bending and compression. (Chapters 6 and 7 explore the biomechanics of these responses in relation to protecting the brain and chewing.) The response is typically neither simple nor always optimal from a biomechanical perspective. Adding mass usually augments bone strength, but not necessarily in the most beneficial location. Consequently, a bone's cross-sectional shape is a poor predictor of how it was loaded (Lieberman et al., 2003, 2004c; Main and Biewener, 2004; Devlin and Lieberman, 2007). One reason is that increases in bone mass are constrained by location, especially as the skeleton ages. Although it is mechanically optimal to add mass on the outside of a bone to increase its resistance to bending or twisting, older individuals usually cannot add bone this way; instead they increase bone mass internally, mostly by slowing the rate of marrow cavity resorption (Ruff et al., 1994; Bass et al., 2002). This observation highlights the fact that bones have many functions, and resisting loading is only one of them.

Resorption and Turnover

Recall that when bones are loaded, they can develop small cracks (see Figure 2.9) and other defects that decrease their strength. The only mechanism bones have to repair themselves in response to damage or to prevent failure is Haversian remodeling, a coordinated resorption of bone by osteoclasts followed by the deposition of new bone. Haversian remodeling also may help to realign collagen along the axes of tension that a bone experiences (see Martin et al., 1998; Currey, 2002). As with growth, mechanical loading stimulates Haversian remodeling primarily in young animals (Kerley, 1965; Bouvier and Hylander, 1996; Lieberman et al., 2003). As animals age, Haversian remodeling rates generally increase regardless of loading conditions (Frost, 1990; Pearson and Lieberman, 2004). Rising remodeling rates may be a mechanism to compensate for mature individuals' reduced ability to build new bone, ensuring that the skeleton does not fill up with damaged, weakened bone.

In adults, mechanical loading from exercise affects the rate of bone loss from osteoclastic resorption by down-regulating osteoclast activity (Frost, 1999; Raisz, 1999; Bikle et al., 2003). Reducing applied loads relative to habitual levels can stimulate bone resorption in juveniles and adults. Over the long term, mammals who lose muscles and tendons in specific limbs (or who have lost the nerve supply to those muscles) lose up to 50 percent of the mass of the affected bones (Uhthoff and Jaworski, 1978; Uhthoff et al., 1985; Jaworski et al., 1980). In the skull, loss of dental function can decrease jaw height by up to 50 percent (Carlsson and Persson, 1967; Israel, 1973). Both rodent and human astronauts lose bone mass in the strain-free, nongravity environment of space, and patients who are confined to their beds for long periods face similar problems (Tilton et al., 1980; Jee et al., 1983; Robling et al., 2006). The pernicious consequence of bone loss from disuse is that it is hard if not impossible for adults to regrow lost bone.

A final point to remember about aging is that the effects of mechanical loading on bone growth change with age as mechanostransductive pathways in osteoblasts (but not osteoclasts) become senescent. Consequently, humans like other mammals tend to reach peak bone mass near the end of the skeletal growth period as a result of a combination of many genetic and environmental factors. Once the skeleton has finished growing, size and shape cannot change very much.

Nutrition

One other environmental influence on bone growth needs brief mention: nutrition. Synthesizing bone requires copious amounts of calcium and protein, so diets low in these and other nutrients (e.g., vitamin D) diminish the capacity to grow a healthy, normal skeleton. Adequate calcium combined with exercise has a synergistic effect on growth (Heaney and Weaver, 2005), but there is no evidence that diets with surplus calcium induce additional bone growth. As we age, maintaining an appropriate amount of calcium in the diet slows the rate of bone resorption, but surplus dietary calcium does not cause the skeleton to add bone mass once it is lost (see New, 2001).

Modularity in the Skull and the Functional Matrix Hypothesis

So far, I have discussed how bones and other skeletal tissues develop, grow and function. Chapters 3–5 apply these processes and properties to the formation of the embryonic head and its fetal and postnatal growth. But first we must grapple with one more key property of the skeleton in general and the skull in particular: its ability to form modules along with other tissues, organs, and functional spaces (e.g., sinuses). As Chapter 1 discusses, modules are a fundamental characteristic of organisms at many levels of development, from the genetic level (e.g., base pairs, genes) to the phenotypic level (e.g., tissues, organs, regions, systems). Modules permit evolvability because they are partially independent and autonomous, discrete, and transformable.

Bones seem like perfect examples of modules: they are discrete and independent, have some degree of developmental autonomy, and so on. For this reason, many anatomy textbooks treat each bone of the skull as a discrete unit. But many cranial bones are not truly independent modules during embryogenesis, and in terms of function they often participate with organs and functional spaces to form more complex modules comprising portions of bones along with their associated tissues and spaces.

The idea that bones are not basic modular units in the head was originally developed by the Dutch anatomist C. J. Van der Klaauw (1948–52), and then expanded and refined by Melvin Moss and colleagues. Instead of thinking of the head in developmental terms as a series of connected bones, muscles, organs, and other tissues, these scholars viewed the head as a series of *functional matrices,* defined as "genetically determined and functionally maintained" soft tissues and the spaces they occupy (Moss, 1968:69). Functional matrices in the head include spaces and organs that function in olfaction, hearing, balance, swallowing, vocalization, and respiration, and, of course, the central nervous system. Major functional matrices include the eyeballs, the nasal cavity, the oral cavity, tooth roots, and the brain. According to Moss, the skeletal tissue that encloses each functional matrix (a *functional cranial component*) derives its shape principally from the shape and functions of the soft tissue and spaces it encloses. In other words, even though bones are stiff, strong and seemingly immobile, their shapes and positions are actually determined by soft tissues such as the brain and the eyeballs and by spaces such as the pharynx. In this sense, bones integrate the head by accommodating the forms of other modular units. A corollary of the hypothesis is that to change a head, either during ontogeny or over evolutionary time, one primarily needs to alter the spaces and organs around which bones form, not vice versa.

The functional matrix hypothesis (FMH) has been influential but also controversial. Its best support comes from craniofacial disorders and experimental studies. As detailed in Chapters 3–6, the cranial vault bones grow to surround the brain, rather than the skull's determining the size or shape of the brain. Thus a microcephalic individual with a tiny brain has a correspondingly small cranial vault that fits precisely around the brain, and a hydrocephalic individual with too much cerebrospinal fluid has a bulging cranial vault that fits around the brain and its membranes. In addition, infants lacking eyes develop tiny orbits with abnormally small upper faces because the eyeball normally pushes out on the bones that form the orbital cavity walls (see Chapter 10). And children with swollen adenoid glands, who cannot breathe though the nose, can develop abnormally small nasal cavities with correspondingly deformed midfaces (see Chapter 8).

The FMH is an important concept and critical to understanding how heads grow and function. But in some respects it may be overstated. First, it is not true that bones have no intrinsic genetic regulation. This is especially the case for endochondral bones. When Murray and Huxley, in a classic early experiment, grafted a fragment of a chick limb bud into an older embryo, the graft developed into a recognizable femur even though it was unconnected to the rest of the skeleton (Murray, 1936). Intramembranous bones probably have much less intrinsic growth potential (see Hall, 2005), but it is an oversimplification to conclude that they have none at all.

Another problem with the FMH is that few functional matrices and their skeletal components are as independent as the FMH would suggest. Most parts of the skull participate in more than one functional matrix. As previously noted, the floor of the anterior cranial fossa is also the roof of the orbits, the inner walls of the orbits are the outer walls of the nose, the base of the nose is the roof of the mouth, and so on. Thus it should not be surprising that correlation analyses have identified significant interactions between bones that cross multiple functional matrices (e.g., Cheverud, 1982, 1989, 1995; Zelditch, 1988; Zelditch and Carmichael, 1989; Zelditch et al., 1990, 1995; Zelditch and Fink, 1995; Hallgrímsson et al., 2007a). Put differently, most bones participate in more than one functional matrix, making it hard to define truly discrete "functional cranial components." In addition, evidence for integration unexplained by the FMH is sometimes evident from patterns of invariance. For example, regardless of the size of the brain, eyes, oral cavity, and other functional matrices, the long axis of the orbits in all mammals is always nearly 90° relative to a line that defines the back of the face in the midline (Enlow, 1990; McCarthy and Lieberman, 2001: Chapter 5). How and why such correlations and constraints arise remains poorly understood, but they suggest that the processes by which the many components of the skull are assembled are sufficiently complex to make the task of defining truly independent units a quixotic endeavor.

The FMH, even if only partly correct, has profound implications for how the basic properties of bone and other skeletal tissues permit the head to grow and evolve by tinkering. Bones are stiff, strong, highly modifiable, and displaceable, and structurally they participate in multiple functional matrices. As such, they have an intrinsic capacity to adapt to many alternative configurations. Let's say a gene evolves to generate a bigger brain. If bone morphology were highly genetic, then such a mutation would have to coevolve with other mutations that made the cranial vault bones bigger in order to accommodate variations in orbit position, midface orientation, and so on. However, the primacy of functional matrices often enables the head to accommodate such changes with relative ease. Because of integrative processes, a bigger brain not only makes the braincase bigger but also pushes and pulls other the organs and tissues of the head into new positions. These functional matrices, in turn, push and pull their surrounding skeletal capsules into new positions so that everything can fit into a new arrangement.

It follows that to transform the ontogeny of a great ape's head into a modern human's may not be as difficult as one might imagine. Although many aspects of the head differ markedly between the two species, not every morphological shift needs to be programmed genetically. To be sure, one needs to shorten the face, reorient the foramen magnum, enlarge the vault, and so on, but the genetic bases for many such shifts lie in genes that act directly on the size and shape of organs such as the eyes, brain, and tongue. The development of the bones partly follows along with these shifts, permitting the whole head to develop as an integrated ensemble.

Coda

The first animals that evolved a simple head more than 450 million years ago set the stage for much subsequent innovation. Heads are adaptively useful because their unique organs and structures allow organisms to interact with the world in complex ways unavailable to headless creatures. It was fortunate, if not crucial, that early vertebrates used bone and its related hard tissues to form the skeleton of the head, permitting considerable evolvability. Bone is an ideal tissue for developmental and evolutionary tinkering because it is so multifunctional and adaptable. In addition to being stiff and strong, bone can change its shape in diverse and profound ways in response to many signals from both the genome and the environment. Because of these attributes, natural selection has been able to act not only on variations in the skeleton of the head itself but also on variations in the other components of the head whose shape and function are so agreeably accommodated by the skeleton.

As we continue our exploration of the evolution of the human head, our focus on bone and other skeletal tissues will wax and wane but will never fade from view, because these tissues, especially bone, play a vital role in everything the head does. Understanding how bone functions, grows, and adjusts to its environment provides much information about evolutionary changes in the head. In addition, bones and teeth, which are preserved after death so much better than the rest of the head, provide the major direct evidence of what heads were like in nonextant creatures. This bias sometimes leads to a slight degree of "osteocentrism," making all but the skull and teeth almost invisible to the eye of the morphologist or paleontologist. It makes sense to focus on the skeletal evidence, which is mostly what we have available to us. But it is useful to keep in mind that bone is an exquisitely tinkerable tissue whose many variations in size and shape derive from a complex interplay between inter-actions with neighboring tissues, other environmental stimuli, and genes.

3

Setting the Stage:
Embryonic Development of the Head

Those who have never seen a living Martian can scarcely imagine
the strange horror of its appearance. The peculiar V-shaped
mouth, . . . the absence of brow ridges, the absence of a chin.

H. G. WELLS, *The War of the Worlds,* 1898

It is hardly coincidental that space aliens tend to resemble giant embryos. Embryos are strangely familiar-looking life forms, yet unlike anything we see on a daily basis. Their immaturity and prematurity remind us that the life forms we observe sample only a tiny fraction of the many variations that potentially could exist. A vertebrate embryo is so generalized that it might become almost anything: you, me, a chicken, or some science-fiction fantasy with a gigantic brain.

This chapter reviews how the basic structure of the human head is set up during the embryonic period (the first 60 days of life), and how differences in embryonic development may help set the stage for some of the contrasts between modern human, hominin, and other nonhuman primate heads. As one might expect, the sequence of events that transform a single cell into the body's many organs and tissues, all in the right place and time, is an intricate and complex story. These transformations involve many cell types that change names, positions, and identities as they differentiate, move, and interact with each other in diverse ways. Embryonic development, moreover, involves the expression of thousands of genes that appear repeatedly in different contexts.

Although the complexity of craniofacial embryogenesis makes the subject hard to comprehend, the same complexity makes possible evolutionary transformations from creatures like fish to mice, chimps, and humans. As Chapter

1 proposes, certain aspects of complexity permit evolvability—the ability for natural selection to take advantage of existing variations to favor novel functional and potentially adaptive changes to the phenotype. Some useful evolutionary novelties are made possible by apparently simple changes to gene regulation during embryogenesis. These shifts can account for how new body parts such as limbs were modified from old parts such as fins (for general reviews, see Carroll, 2005; Kirschner and Gerhart, 2005). But what about the smaller, subtler shifts that are responsible for differences between a chimpanzee and a human, or between a Neanderthal and a modern human? Which of these "tinkerings" occur during embryogenesis, and which occur later in fetal and postnatal life? This is an especially difficult question to answer because we will never find embryonic hominin fossils, and for obvious ethical reasons we cannot do comparative, experimental embryology on humans and chimpanzees.

For many years, embryology mostly concerned biologists interested in differences between very disparate species. This focus on macroevolution partly stemmed from the popularity of what is known as von Baer's law. Proposed in 1828 by the German anatomist Karl Ernst von Baer, this "law" posits that structures present early in development are more widely distributed among animals than structures that are present late in development. Von Baer did not believe in evolution, and its chief application to evolutionary theory by Ernst Haeckel—the idea that ontogeny recapitulates phylogeny—has been refuted as a general principle in several ways (see Gould, 1977). Yet a persistent application of von Baer's law to evolutionary thinking is the notion that early developmental stages tend to be similar among different vertebrate species and that distinctive features of species tend to arise later rather than earlier in development. In other words, microevolutionary changes are thought to occur through alterations of terminal rather than early stages of development. Consequently, students of human evolution often devoted little attention to embryology.

Developmental studies have slowly chipped away at von Baer's idea, suggesting how much there is to learn about human evolution from the study of embryology. For example, many variations in limb morphology among closely related species, such as differences in the number of digits, have been shown to occur through early shifts in development when organs are generated (see Richardson, 1999). In addition, studies of organisms as diverse as frogs and sea urchins reveal that early developmental changes may have little observable effect on the anatomy of an embryo and become apparent only later in development (Raff, 1996). We also have hints that some of the differences we see among hominins arise from prenatal transformations, possibly during embryogenesis. The remains of fossil Neanderthal neonates possess distinctive characteristics that distinguish them from modern human neonates (Rak et al., 1994; 2002; Maureille, 2002; Ponce de León et al., 2008). In addition, comparisons of Neanderthal and modern human cranial ontogenies suggest different patterns of growth in the two species that are already distinct by birth (Ponce de León and Zollikofer, 2001). Divergent patterns of prenatal growth may also generate differences in facial shape among closely related species of apes as well as australopiths (Ackermann and Krovitz, 2002).

It is therefore worth learning some of the complexities of human craniofacial embryology. This chapter has two goals. The first is to summarize some key embryonic processes that set up the basic plan of the head. As discussed in Chapter 2, the developmental processes that generate complex morphologies such as the head can be divided into three main hierarchical levels: patterning, morphogenesis, and growth. To reiterate, *patterning* generates the head's basic units, each with its own position and identity; *morphogenesis* creates the head's tissues and organs; and these units then mature through subsequent *growth,* changing shape and size in response to a combination of intrinsic and extrinsic influences. We will examine all three levels of development in terms of the head's initial formation within the body and then in terms of the embryogenesis of the cranial base, the vault, and the face.

The second goal of this chapter is to provide a basis for hypotheses (mostly presented in Chapters 11–13) about which aspects of embryonic growth were subject to tinkering during human evolution. From the broad perspective of comparative biology, the variations in head anatomy evident among modern humans, great apes, and extinct species of hominins are subtle. These heads share the same basic plan with the same components (brain, organs of sense, bones, muscles and so on) arranged in essentially the same way, and with only slight variations in spatial relationships and relative size. Even so, there are many discernible and significant morphological differences between newborn human and extant great ape heads, all of which must have occurred via minor alterations to the same basic set of processes. Many key shifts in craniofacial form among these species probably have their roots in changes that occur during embryogenesis. Although we may never be able to perceive or document them directly for chimps and extinct hominins, we nonetheless benefit from considering them, especially to test hypotheses about their developmental bases using less direct sources of evidence, such as changes in patterns of covariation.

A few prefatory caveats. First, embryology is a venerable field with a challenging lexicon. Many of the stages and units that rapidly emerge, shift, and disappear have abstruse names (e.g., *prosencephalic organizing center, stomodeum,* and *sclerotome*) with few clues to their meaning for those not familiar with Greek and Latin. I have done my best to use such terms as sparingly as possible, and to define them as they appear and in the glossary. Second, what follows is inevitably only a cursory review of the major events of embryonic development in the head, with a focus on the origins of and interactions among tissues that play especially important roles. Readers interested in more complete and detailed accounts of the many events and mechanisms that occur during embryogenesis should consult some of the monographs and textbooks devoted to various aspects of this subject (e.g., Wolpert, 1991; Hanken and Hall, 1993; Dixon et al., 1997; W. J. Larson, 2001; Sperber, 2001; O'Rahilly and Muller, 2001; Meikle, 2002; Moore and Persaud, 2008). Finally, here are some tips to avoid confusing directionality in the embryo. In an adult human, direction always refers to standard anatomical position, in which the body is standing with the face and palms facing forward. Directionality in an embryo generally refers to an embryo lying on its ventral (stomach) surface with the dorsal (back) surface upward. *Cranial* describes structures closer to the head end, and *caudal* describes structures toward the tail end; in the head, structures toward the face and away from the dorsal surface are *rostral* ("beak"). Thus the embryonic rostro-dorsal axis becomes an antero-posterior axis in an adult; the cranio-caudal axis becomes the superior-inferior axis.

Early Stages of Embryonic Growth (from Plates to Tubes)

Many important structures in the human head begin to form around the third week after conception, but it is useful to summarize briefly a few earlier events that set up the embryo's basic body plan. During the first few days after fertilization, cells in the embryo divide repeatedly, developing from a single-celled conceptus to a spherical structure of about 100 cells (the blastocyst). Just prior to implantation in the uterus, the blastocyst includes an inner cell mass (the embryoblast) that will become the embryo, and a spherical outer layer (the trophoblast) that interacts with the uterus to form the placenta (see Figure 3.1a–d). After implantation—about eight days after conception—the embryoblast differentiates into two layers: the epiblast, which gives rise to the embryo; and the hypoblast, a temporary structure that contributes along with

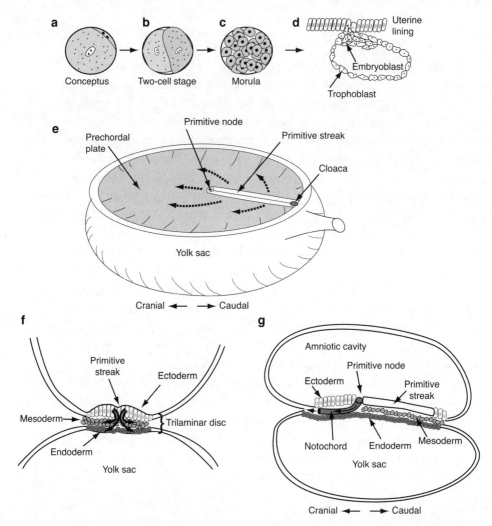

Figure 3.1. Early stages of embryogenesis through gastrulation (adapted from Gilbert, 1989; O'Rahilly and Muller, 2001; Sperber, 2001). Prior to implantation, the embryo, at first a single cell (a), divides many times (b), eventually forming a ball-shaped mass of cells, termed a morula (c). By the time of implantation into the uterus (d), the embryo differentiates into a *trophoblast,* which provides various support structures, and a mass of cells, the *embryoblast,* which will form the embryo itself. Gastrulation (e–g) occurs as cells from the upper layer of the embryoblast (initially called the epiblast) differentiate and dive beneath the primitive streak. In the dorsal view of the embryonic disc at approximately 16 days, (e), arrows indicate the forward and lateral directions of cell migration between the epiblast and hypoblast layers. Transverse (f) and sagittal (g) sections of the embryonic disc show how mesodermal and endodermal cells stream downward, forward and laterally from the primitive streak, forming the trilaminar disc with endoderm on the bottom, mesoderm in the middle, and ectoderm on the top. The sagittal section also shows how mesodermal cells in the midline condense into the notochord. The trilaminar disc is situated between the amniotic sac and the yolk sac.

derivatives of the trophoblast to the extraembryonic membranes that support the embryo (the amnion, chorion, yolk sac and allantois).

Initially, the epiblast and hypoblast grow slowly compared to the trophoblast. But during the second week of life, an especially critical series of events occurs, known collectively as gastrulation (after the Latin word *gastrula,* for stomach). Gastrulation creates three layers of cells—ectoderm, mesoderm and endoderm—which give rise to all the tissues of the body (Figure 3.1d–e). Gastrulation begins as cells from the epiblast form a narrow groove with swellings on each side, called the primitive streak. The primitive streak runs from the future tail (caudal) end of the embryo to a slightly elevated region near the center, the *primitive node,* which organizes cells that contribute to the trunk and head. As the diagrams in Figure 3.1d–e illustrate, the cells of the primitive streak become distinct from the upper layer of the embryo, now termed *ectoderm,* and migrate underneath and forward to form two new layers: a middle *mesoderm,* and a lower *endoderm.* The result is a pear-shaped disk with three layers that are suspended between a yolk sac and the amniotic sac (which surrounds and protects the embryo). Gastrulation also creates a central axis for the embryo. As migrating, differentiating cells plunge below the embryo's surface to form the mesoderm and endoderm, some of them coalesce into a rodlike mass of cells called the notochord (see Fig. 3.1d–e). The notochord eventually serves as the initial connective tissue of the axial skeleton and as an inductive center for the future embryonic head and the central nervous system.

The biologist Lewis Wolpert (1991:12) has quipped: "It is not birth, marriage, or death, but gastrulation, which is truly the most important time in your life." Wolpert's hyperbole may have some validity because, thanks to gastrulation, the embryo by the end of the third week has three layers of cells— the ectoderm, the mesoderm and the endoderm—from which the entire organism develops around a central notochordal axis. Cells in each layer differentiate into specific tissues and organs. Ectoderm becomes most of the organs and structures that interact with the outside world: the nervous system, many of the organs of sense (ear, nose, eye), and the skin and its appendages (hair, sweat glands, etc.). Ectodermal cells also form most of the bones and muscles of the head, tooth enamel, and several glands. Endoderm becomes the epithelial lining of the gastrointestinal, respiratory, and urinary systems, along with the exocrine secretory cells of the liver and pancreas. Mesoderm becomes most of the musculoskeletal system (except in the face, braincase, and part of the cranial base), the dermal and subdermal layers of the skin (except in the face and parts of the scalp), tooth roots and related tissues, and the vascular and urogenital systems. Mesoderm also contributes to all the organs that are lined with endoderm.

How do the three cell lines go from an ovoid trilaminar disk to generate an entire embryo? As outlined in Chapter 2, they do so by a combination of patterning, morphogenesis, and growth. All of these processes involve sequential, inductive interactions between adjacent tissues. First, the ectoderm. Beginning in the third week, the ectoderm differentiates into several portions, including a central neural plate. This plate of cells (the neuroectoderm) runs along the long axis of the embryo above the notochord. As its edges expand, cells in the plate grow upward (dorsally) and inward (medially) until they meet in the midline to form the neural tube (see Figure 3.2). This process, termed neurulation, begins in the center of the embryo with the tail and brain ends of the tube remaining open (as neuropores) much longer. One particularly important group of cells in the neural tube needs to be singled out for attention: the neural crest (see Figure 3.2). These neurectodermal cells are highly migratory and capable of differentiating into many tissues, including neurons and their supporting cells, pigment cells, and (in the head only) skeletal and connective tissue cells. In the head (Figure 3.2b), neural crest cells form along the lateral margin of the neural plate and migrate outward before it finishes folding into a tube; in the spinal region they migrate outward after the tube is formed (Figure 3.2c). The neural crest becomes a sort of fourth germ layer, a key innovation that evolved about 500 million years ago in the first vertebrates, which made possible the development of the head in general and the skull in particular (see Gans and Northcutt, 1983; Hall, 1998).

As the neural tube forms, three thickenings develop on the surface at its cranial end: the olfactory (nasal), otic (ear), and lens (eye) placodes. These epithelial cell clusters become structures of the nose, inner ears, and eyes. Epithelial cells in these placodes and elsewhere also act as signaling centers that are necessary to induce transformations in neighboring mesenchymal cells to stimulate craniofacial skeletogenesis (see Hall and Miyake, 2000).

Let's now turn briefly to the mesoderm and endoderm. We last left the mesoderm as a thin sheet of cells on either side of a central stiffened rod, the notochord, between the ectoderm and endoderm. After the neural tube and neural crest differentiate within the ectoderm, the mesoderm begins to thicken on either side of the midline notochord. At the same time, folds of endoderm from the cranial and caudal ends of the embryo move toward each other to form the front (foregut) and back (hindgut) portions of the gut tube (Figure 3.2). The midgut initially remains open to the yolk sac but eventually closes

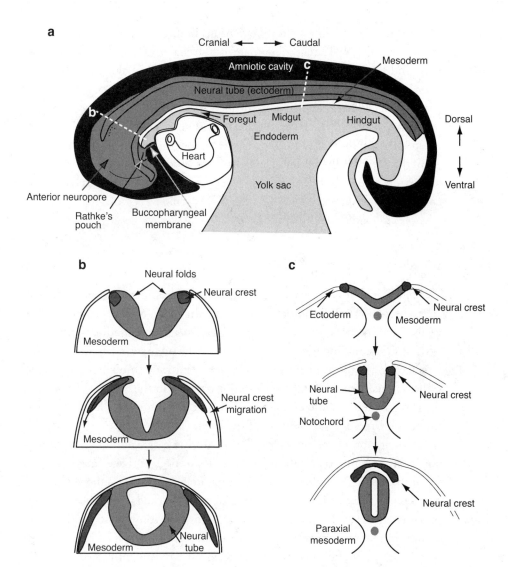

Figure 3.2. (a) Curled-up embryo, approximately 25 days old, showing the basic vertebrate body plan with an endodermal tube, a central layer of mesoderm, and a neural tube. Note the buccopharyngeal membrane at the site of the cephalic entrance to the foregut. Neurulation occurs as the neural plate folds into a tube differently in the spinal region (b) than in the brain (c). In both, special ectodermal cells on the dorsal margins of the neural tube migrate away from the tube, forming the neural crest, but in the spine, mesoderm on either side of the notochord differentiates into segments of paraxial mesoderm. (Adapted from Sperber, 2001; and Copp et al., 2003.)

off, forming a continuous gut tube. Various buds branch out from the endodermal tube that will later differentiate into several organs including the lungs,
liver, and pancreas. The thickened portion of mesoderm alongside the notochord is termed *paraxial mesoderm*. Laterally, the mesoderm forms a thin plate,
with two layers separated by a fluid-filled space (the coelomic cavity, which
forms the spaces in which the gut, heart, and lungs grow). The lower layer of
lateral-plate mesoderm is attached to the endoderm and forms the outer layers of the gut; the upper layer of lateral-plate mesoderm lines the ectoderm lateral to the neural ectoderm. These layers also fold inward around the gut,
creating a tube-shaped embryo. As the three germ layers grow, the embryo as
a whole curls up around its yolk sac.

Initial Segmentation of the Head

The first stages of embryogenesis set up the most rudimentary, albeit critical,
elements of the basic chordate body plan. In order to generate something
vastly more complex, the embryo next needs to generate modular units, each
with position and identity. Segmentation of the mesoderm begins in the occipital region of the head and extends caudally into and throughout the trunk.
Other cranial structures that are anterior (rostral) to the occipital region undergo segmentation differently from the trunk. For this reason, let's begin with
the simpler, mesodermal portion of the head.

Segmentation of the mesoderm begins during gastrulation, when numerous loosely organized clumps of embryonic connective tissue (mesenchyme)
begin to coalesce along either side of the notochord. These somewhat amor-
phous units become further organized in a cranio-caudal sequence with approximately 42 to 44 discrete segments, known as somites (see Figure 3.3).
Somite formation occurs through the reciprocal expression of several genes
that work as a kind of oscillating clock.[1] Somites mostly form the axial skeleton and its associated segmental muscles and dermis, but the first four somites
were co-opted long ago by early vertebrates to form part of the occipital bone
in the head. Thus, the seven cervical vertebrae derive from somites 5–11. After
that, humans typically have 12 thoracic, 5 lumbar, 5 sacral, and 4 tail (caudal)

---

1. The clock works through a kind of molecular oscillation in which gene expression in one set
of genes drives activity and expression in another set of genes, inducing the formation of physically
distinct somites; the clock is turned off by a wave of gene expression that travels in the opposite direction (Palmeirim et al., 1997; Forsberg et al., 1998; McGrew et al., 1998). Changes in the rate and
duration of these waves helps regulate somite number and size (Gomez et al., 2008).

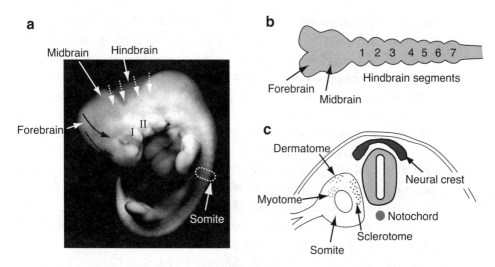

Figure 3.3. (a, b) Segmentation of the embryonic head showing the flexures of the neural tube into a forebrain (prosencephalon), midbrain (mesencephalon), and hindbrain (rhombencephalon), each sending streams of cells toward the face (as shown by the arrows). Note the two first branchial arches (I, II). (c) In the spine, somites of paraxial mesoderm further subdivide into cell types that form skeletal tissue (sclerotome), muscle (myotome), and the deep layers of the skin (dermatome).

somites, but there is much variation among individuals (and sometimes between the left and right sides) in the caudal boundaries between types of somites (see Pilbeam, 2004). There is also considerable variation among species. A typical chimpanzee, for example, has 13 thoracic, 3 lumbar, 6 sacral, and 3 tail (caudal) somites.

Variations in somite identity reflect how an extremely important category of genes, the homeobox genes, pattern various kinds of segments. Homeobox genes encode a family of proteins (transcription factors) that share a short sequence (≈180 base pairs), known as the homeodomain, which can bind to specific regulatory regions of DNA in other genes. Homeobox genes thus regulate how certain genes are transcribed.[2] There are many homeobox genes, including a group known as the *Hox* genes, which pattern, among other things,

---

2. Transcription is the process by which a sequence of DNA in the nucleus is "read" and then copied to a series of similar molecules called messenger RNA (mRNA). The mRNA then is transported to a protein-making organelle of the cell, the ribosome, where it is "translated" to create the protein coded by the DNA sequence.

the axial somites. There are actually four clusters of *Hox* genes (A–D), each with some combination of 13 variant genes (known as paralogues). The diversity of *Hox* genes probably arose from duplication events that occurred prior to the origin of vertebrates (Holland, 2000). An elegant property of *Hox* genes is that their spatial organization on the chromosomes reflects their pattern of expression along the cranial-caudal axis of the embryo. Studies of *Hox* expression in a wide variety of animals show that each somite expresses a distinct combination of *Hox* genes, reflecting position and determining identity (Bürglin, 1994; Burke et al., 1995). *Hox* genes, however, pattern more than just the axial somites, illustrating how evolution frequently uses modules in novel ways when given the chance. *Hox* genes also pattern the long axes of the limbs and some divisions of the neural tube. Moreover, the *Hox* cluster is just one of many classes of homeobox genes.

One last word on somite organization before we turn to the segmentation of the rest of the head. After the somites form, the paraxial mesoderm in each segment, including the four occipital somites that contribute to the head, differentiates into three cell types (Figure 3.3c). The mesoderm closest to the notochord forms precursor skeletal tissue (sclerotome) that gives rise to the cartilages of the axial skeleton and the cranial base around the foramen magnum. The rest of each somite's mesoderm differentiates into mobile precursor cells for the skin and other connective tissues (dermatome) and for muscle (myotome).

As emphasized above, only the first four occipital somites contribute to the head. The rest of the head's units come from a combination of unsegmented paraxial mesoderm, endoderm, and ectoderm. Ectoderm, which contributes the lion's share of the head's cells, has two major sources: the neural tube and the neural crest. How these various cell lines interact and form organized structures in the head is a complex story that involves many migrating streams of cells that derive their identity from a combination of patterning (e.g., via homeobox gene expression) and from inductive interactions with cells that they contact.

The story starts with the neural tube. Just as the paraxial mesoderm forms into segmented swellings, the neural tube also divides into segments (Figure 3.3a,b). Segmentation of the neural tube begins before it closes, but becomes evident in the third week, when the cephalic end of the future brain develops several swellings: the forebrain (prosencephalon), the midbrain (mesencephalon), and the hindbrain (rhombencephalon). These regions of the brain continue to subdivide (a process discussed in more detail in Chapter 6). How the neural segments are patterned is not yet fully understood, but it partially involves the *Hox* gene cluster, which is expressed in the developing hindbrain (see Trainor and Krumlauf, 2000).[3] Other homeobox genes, including the *Otx* and *Emx* clusters, pattern the more anterior portions of the brain, including the forebrain (Murakami et al., 2005).

Another critical source of patterning in the head that contributes to its basic integrated structure derives from cell migrations, especially the neural crest cells. Recall from above that these special neuroectodermal stem cells differentiate from the dorsal margin of the neural folds before it becomes a tube. During the third week, these stem cells begin to migrate from the lateral edges of the neural fold. Some of these cells migrate into the head as streams of cells that proliferate and differentiate into mesenchymal cells between the surface ectoderm and the deeper mesoderm (see Figure 3.3). The more rostral streams migrate over the forebrain, between the neural epithelium and the surface ectoderm, giving rise to segments known as the frontonasal prominences. These prominences contribute to the middle and upper parts of the face. The more caudal segments that originate adjacent to the hindbrain (the rhombencephalon) contribute to special segments known as the branchial arches that generate the lower face, the pharynx, and the front of the neck. Labeling experiments in mice suggest that the interactions between these various streams of mesenchymal cells and the tissues they encounter along their journeys provide critical opportunities for further inductions, which continue to differentiate each stream of migrating cells (Tam and Trainor, 1994; Trainor and Tam, 1995). We will outline these aspects of cranial embryogenesis below in their regional contexts.

The above summary highlights only a few of the steps by which the head is initially patterned. One conclusion to draw is that some aspects of the head's patterning derive from the independent segmentation of the mesoderm and ectoderm, which in turn patterns the neural crest cells. Further patterning arises from inductive interactions between streams of neural crest cells and other neural and mesodermal cells as they migrate and contact one another. Finally, homeobox genes pattern additional axes of the face. For example, *Hoxa* and the homeobox cluster *Dlx* are both involved in patterning portions of the face that derive from the branchial arches (for reviews, see Olsen et al., 2000; Depew et al., 2005). Stepping back from the details, the point to keep

---

3. *Hoxa-1, b-1,* and *d-1* are expressed at the boundary of hindbrain segments 3 and 4; *Hoxa-3, b-3,* and *d-3* are expressed at the boundary of hindbrain segments 4 and 5 (Trainor and Krumlauf, 2000).

in mind is that the embryonic head is segmented from its earliest stages by mechanisms that also cause the head to be highly integrated. In other words, the basic units of the head do not develop on their own and then interact once formed. Instead, they form through a reiterative set of processes that involve many sequential and well-integrated epigenetic interactions, thus permitting the basic regions of the head to simultaneously form and derive their identity.

At this point in our summary of embryogenesis, we have an embryo with pre-cursor regions that are ready to form the different regions of the head, but they have not yet begun to differentiate into specific tissues and organs. The next step, therefore, is to explore the morphogenesis of the head's major organs and skeletal components. To simplify, we will focus on the embryonic develop-ment of each of the skull's major regions in relation to the organs they house, such as the brain. Although there are different ways to divide up the skull and its adjacent structures, anatomists generally distinguish three major re-gions: the cranial base, the cranial vault, and the face. Each develops from different regions of the embryo, and although they become increasingly inte-grated during ontogeny, they always remain somewhat structurally indepen-dent (Cheverud, 1982, 1989; Hallgrímsson et al., 2007a).

The Embryonic Cranial Base (Chondrocranium)

We begin our examination of craniofacial morphogenesis with the cranial base (basicranium), whose cartilaginous precursors are collectively termed the chon-drocranium. Structurally, the cranial base is the foundation of the skull: it con-nects the skull to the axial skeleton, and it forms the central platform upon which the brain grows, and below which the face grows. The intermediate po-sition of the cranial base between the brain and the face accounts for its special role as a region that helps integrate the skull (Lieberman et al., 2000b, 2008; Hallgrímsson et al., 2007a). In addition, from an evolutionary perspective, the cranial base is the most primitive part of the skull (de Beer, 1937), and includes almost all the portions that ossify endochondrally (see Chapter 2). Although the chondrocranium is structurally complex and difficult to visualize, it is worth learning some basic details about how it forms. In an adult, the cranial base in-cludes parts of just five bones (the occipital, temporal, sphenoid, and ethmoid), but the chondrocranium has about a dozen pairs of precursor cartilages.

Morphogenesis

About 28 days post-conception, after the primordia of the brain, eyes and cranial nerves have formed, a shelf of mesenchyme coalesces beneath the brain. By 40 days post-conception, this mesenchymal plate begins to differentiate into many pairs of distinct condensations, which create the cartilages of the chondrocranium (see Figure 3.4). How these condensations are induced and how their morphogenesis is regulated are only partially known. In general terms, each mesenchymal condensation that forms below the brain becomes a distinct segment through a series of inductive interactions with other mesenchymal cells, with ectodermal tissue around the brain, and with mesodermal cells (Thorogood, 1988; Johnston and Bronsky, 2002). Many of these cell lines are already patterned by gene clusters such as *Hox, Dlx,* and *Msx* (Lufkin et al., 1992; Robinson and Mahon, 1994; Vieille-Grosjean et al., 1997, Schilling and Thorogood, 2000; Depew et al., 2002). In addition, the cells forming the mesenchymal plate below the brain have different embryonic origins and thus behave differently. Fate-mapping studies in chick and mouse embryos reveal that the notochord is an important boundary. Mesenchymal cells alongside the rostral tip of the notochord derive from paraxial mesoderm; in contrast, mesenchymal cells in front of (rostral to) the notochord originate not from the mesoderm but from the neural crest (Noden, 1991; Le Douarin et al., 1993; Couly et al., 1996; Kuratani, 2005). Thus cartilages that form in front of and behind this division are termed *prechordal* and *parachordal* (or *postchordal*), respectively. The location of the prechordal-parachordal division may differ slightly in mammals. In mice, this boundary, which matches the neural crest–mesoderm boundary, is between the hypophyseal and parachordal cartilages in the future spheno-occipital synchondrosis (McBratney-Owen et al., 2008).[4]

Each of the mesenchymal condensations that make up the chondrocranium begins to form distinct cartilages. To comprehend the eventual shape of the cranial base and its effects on overall cranial shape, it can be useful to know the origin and fate of each cartilage. In an effort to simplify, I divide these cartilages into three groups: parachordal, prechordal, and sensory (see Figure 3.4).

---

4. There are exceptions, however, to this boundary, including small paired cartilages related to the eye (the hypochiasmatic cartilages), which originate entirely from mesoderm but are situated prechordally among neural crest–derived cartilages (McBratney-Owen et al., 2008).

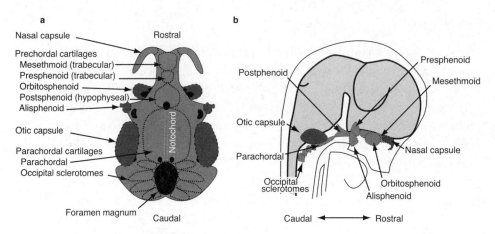

Figure 3.4. Development of the chondrocranium, showing the initial condensations and cartilages. (a) Superior view. (b) Lateral view.

Parachordal

A pair of parachordal cartilages, mesodermal in origin, originates on either side of the rostral end of the notochord (Barteczko and Jacob, 1999). It is possible that the notochord induces them to condense. In primitive vertebrates, the parachordal cartilages form almost the entire cranial base below the brain, known as the basal plate (de Beer, 1937). In higher vertebrates such as mammals, the parachordal cartilages are relatively smaller, and they fuse to become the basioccipital and the anterior third of the occipital condyles. The rest of the occipital forms behind (dorsal to) the parachordal cartilages from four initial pairs of condensations that derive from the first four somites (see also Figure 3.3b). These condensations, which form a ring around the foramen magnum, are termed the occipital sclerotomes (sclerotome is the paraxial mesoderm tissue that condenses next to the notochord to form vertebral bodies). A pair of exoccipital endochondral ossification centers then form in this region. The exoccipitals include the posterior two-thirds of the occipital condyles and the occipital plates on either side of the foramen magnum. Initially, a thin band of cartilage separates the left and right exoccipitals, but eventually they fuse to form a single plate. Later the exoccipitals fuse with intramembranous portions of the occipital (see below).

Prechordal

The prechordal group of cartilages begins as two pairs of condensations: an anterior pair of trabecular condensations that forms much of the nasal region below the front of the brain, and a posterior pair of hypophyseal condensations that forms just below the pituitary gland (also called the hypophysis). The morphogenesis of the hypophyseal region starts as an epithelial folding of the primitive mouth (the stomodeum; see below), which, in combination with the floor of the forebrain, differentiates into a major signaling center known as Rathke's pouch (Dattani and Robinson, 2000). This structure (indicated in Figure 3.2a) in turn most likely induces mesenchyme in the hypohyseal condensation to form a pair of cartilages that eventually become the posterior part of the sphenoid body (the postsphenoid). These later merge with a pair of presphenoid cartilages that bud off from the trabecular condensation. The longer anterior (rostral) portion of the trabecular condensation fuses into a vertical plate of cartilage, the mesethmoid. Much of the mesethmoid persists as the cartilaginous nasal septum, but its rostral end ossifies as the cribriform plate below the olfactory bulbs. The cribriform plate becomes a sievelike plate of bone around the olfactory nerves. The derivatives of the mesethmoid eventually fuse with the nasal cartilages to form the ethmoid bone (see below).

Two other pairs of prechordal cartilages also form lateral to the central hypophyseal cartilages. One pair, the alisphenoids, condenses beneath the temporal lobes of the brain and later gives rise to the greater wings of sphenoid. The other pair, the orbitosphenoids, forms in association with the eyes. In primitive vertebrates, the orbitosphenoids form an optic capsule around the orbits but do not fuse with the chondrocranium (de Beer, 1937); in higher vertebrates, the orbitosphenoids become the lesser (orbital) wings of sphenoid bone, in part providing a passageway for the ocular nerves between the brain and the eyeballs.

Sensory (Otic and Olfactory)

Other parts of the chondrocranium form around the olfactory (nasal) and otic (inner ear) sense organs. The inner ear begins as paired otic placodes—specialized areas of thickening in epithelial cells derived from ectoderm. The otic placodes form pits and then closed cysts that give rise to both vestibular (organs of balance) and cochlear (organ of hearing) portions. An otic capsule of cartilage then condenses from mesenchyme around the inner ear, eventually ossifying into the petrous portions of the temporal bone (the ear is discussed more fully in Chapter 10). The nasal capsule (ectethmoid) condenses to form a boxlike cartilage below the olfactory bulbs. The nasal capsule later fuses with the vertically oriented mesethmoid (see above), which separates the left and right sides of the nasal cavity in the sagittal plane below the cribriform plate. Much later, the lateral walls of the ectethmoid ossify to form the bones of the nasal cavity, but the median nasal septum remains mostly cartilage, except at its lower margin. where it becomes the plow-shaped vomer bone that sits on the floor of the nasal cavity.

Ossification of the Chondrocranium

Once formed, the chondrocranium grows into a butterfly-shaped platform below the brain (see Figures 3.4 and 4.1). Because the chondrocranium originates from many cartilages that form around the growing brain, it retains numerous holes (foramina) determined by the prior presence of nerves and blood vessels. The biggest hole is the foramen magnum (Latin for "great hole") in the middle of the occipital sclerotomes, through which the spinal cord passes. Other conduits persist around the cranial nerves and the many vascular channels that bring blood to and from the brain. Chapter 4 returns to these foramina.

Transformation of the chondrocranium into the basicranium occurs through at least 41 ossification centers that appear within the cartilages after about eight weeks in utero (Shapiro and Robinson, 1980; Kjaer, 1990). As discussed in Chapter 2, ossification of the cranial base occurs through the sequential replacement of cartilage by bone in several steps, including the formation of a periosteal collar, hypertrophy and then death of the chondrocytes, deposition of bone by osteoblasts, and the formation of secondary ossification centers that are not part of the original cartilage primordium. Three particularly important growth centers (synchondroses) persist in the cranial base: the spheno-occipital (between the occipital and sphenoid bones); the mid-sphenoidal (within the sphenoid bone between the former presphenoid and postsphenoid cartilages); and the spheno-ethmoid (between the ethmoid and sphenoid bones). Other cartilaginous growth sites fuse earlier (e.g., between the petrous portion of the temporal and the basioccipital bone). I will have much to say about these synchondroses in later chapters.

During the fetal period, ossification begins toward the caudal end of the chondrocranium and proceeds rostrally and laterally, eventually forming the four bones of the basicranium—the occipital, sphenoid, temporal, and ethmoid —all of which also include some intramembranous elements. The complex sequence in which the fetal basicranium ossifies from the chondrocranium helps one to understand the eventual shape of the cranial base (for detailed reviews, see Kjaer, 1990; Lieberman et al., 2000b; Scheuer and Black, 2000;

Sperber, 2001). Here I summarize some basic details by region, starting with the caudal region. Figures 3.4 and 4.1 illustrate many of these elements.

Occipital

The occipital forms in several portions. The occipital plate, which forms most of the nuchal region behind and lateral to the foramen magnum, derives from paired exoccipital ossification centers on either side of the foramen magnum, including the posterior two-thirds of each occipital condyle (Srivastava, 1992). The anteriormost pair of occipital sclerotomes ossifies on either side of the foramen magnum and fuses with the basioccipital, the derivative of the parachordal cartilage, beneath the brain stem.

Sphenoid

The sphenoid is the most complex bone in the skull because its central position requires it to adjoin many other bones and because it fuses from many diverse elements. In the center below the pituitary gland is the sphenoid body, which fuses from the basisphenoid (derivative of the hypophyseal cartilages) and the presphenoid (derivative of the trabecular cartilages). The posterior margin of the basisphenoid remains a growth plate—the spheno-occipital synchondrosis—and the mid-sphenoidal synchondrosis lies between the basisphenoid and presphenoid bodies. The anterior portion of the presphenoid cartilage retains its trabecular nature to form the mesethmoid part of the ethmoid bone. Lateral to the midline, ossification centers form in the alisphenoid and orbitosphenoid cartilages, eventually fusing them to the sphenoid body as the greater and lesser wings of the sphenoid, respectively (Kodama, 1976a–c; Sasaki and Kodama, 1976). The greater wings cradle the temporal lobes, and the lesser wings surround the optic nerves, overlying a passage through which several important nerves and vessels travel (the superior orbital fissure).

Ethmoid

The ethmoid forms the center of the anterior cranial floor—beneath the olfactory bulbs and the midline of the frontal lobe—and most of the nasal cavity. The surface of the ethmoid forms around the olfactory nerve fibers that project down from the olfactory bulbs, creating the sievelike cribriform plate. Each side of the ethmoid ossifies from three centers in the mesethmoid (anterior derivative of the trabecular cartilage) and the nasal capsule cartilages (Hoyte, 1991). An additional pair of cartilaginous ossification centers detaches from the ectethmoid to form a pair of scrolled bones, the inferior nasal conchae, inside the nasal cavity. One detail to note is the formation of the foramen cae-cum ("blind opening"). The dura mater membrane that covers the neural tube is in contact with ectoderm on the surface of the embryo at the dorsal, ante-rior edge of the ethmoid. This contact persists for some time, causing a tem-porary, tubelike projection of the membrane to the top of the face (at the frontonasal suture). Later, as the face grows around this projection, it gets sealed off, forming the foramen caecum at the end of the cranial base. In-complete closure of this dural projection can lead to brain hernias (Sperber, 2001).

Temporals

The temporals, which form much of the lateral aspect of the cranial base, de-velop from many components, most of which are intramembranous, includ-ing the squamous, tympanic, and zygomatic regions (see below). Only the petrous and styloid portions are endochondral, ossifying from approximately 21 ossification centers (Shapiro and Robinson, 1980; Sperber, 2001). As noted above, the petrous portion of the temporal is induced by the otocyst. At about 16 weeks in utero, osteogenesis of the petrous bone is induced around the semicircular canals and the cochlea (Bamiou et al., 2001). The center of the petrosal bone is a uniquely dense, avascular layer of woven bone that never re-models during life, thus enabling it to protect and insulate the delicate organs of balance and hearing (Doden and Halves, 1984).

The Embryonic Cranial Vault (Desmocranium)

The adult neurocranium consists of the parietals, with contributions from the occipital, frontal, temporal, and sphenoid bones. The main function of the cranial vault is to cover and protect the brain. Not surprisingly, regulation of the growth and development of the neurocranium and its embryonic precur-sor, the desmocranium, is dominated from the start by interactions with the brain, but these bones must also connect and integrate with the cranial base below and the face in front.

Morphogenesis

The intramembranous bones of the desmocranium begin to develop soon af-ter the skull base. It used to be thought that the bones of the vault differentiate within the ectomeninx membrane that surrounds the developing brain, but labeling studies in mice have shown that these bones actually begin as primordia at the edge of the cranial base and then migrate upward toward the apex of the head (Yoshida et al., 2008). As the primordia grow, they thicken and develop a trabecular structure. Another previously held idea based on studies of chicks was that the mesenchyme throughout the neurocranium is of neural crest origin (see Noden, 1986). However, studies in mice indicate a different pattern in mammals (Morriss-Kay, 2001; Jiang et al., 2002; Yoshida et al., 2008): the frontal bone and a portion of the temporal (the squamous portion on the side of the vault) derive entirely from the neural crest; the parietals derive from mesoderm, and the squamous occipital (sometimes called the interparietal) is of mixed neural crest and mesodermal origin (see Figure 3.5a). Just as the spheno-occipital synchondrosis marks the boundary between parts of the cranial base derived from mesoderm and neural crest cells, the coronal suture marks the boundary between parts of the vault derived from mesoderm and neural crest cells.

Bones in the cranial vault grow very differently from those of the cranial base. Regardless of their origin, mesenchymal cells around the cranial vault are induced to differentiate into two periosteal (osteogenic) membranes around the cranial vault primordia, an inner endocranium and an outer ectocranium, The neurocranial bones then grow via intramembranous ossification between these two membranes that lie above the ectomeninx and the brain (see Chapter 2). The growing brain, its membranous sheath, and the bones of the neurocranium grow as integrated units in large part due to sutures, which constitute the major sites of growth in the cranial vault. In fact, several lines of evidence indicate that the two major sutures responsible for a large proportion of cranial vault growth, the sagittal and coronal sutures, act as key signaling centers that coordinate cranial vault osteogenesis with growth of the brain. Tensile forces placed on the sutures by regions of the growing brain stimulate the production of signaling factors that regulate osteogenic cells within the sutures. As the brain expands, the dura mater (the outer, fibrous layer of the ectomeninx) produces factors (e.g., Fgf2) that bind to receptors in the sutures, stimulating bone formation via other transcription factors (e.g., Twist and Msx-2) (see Opperman, 2000; Morriss-Kay and Wilkie, 2005). Coordination of regional brain growth with suture activity is evident from congenital growth abnormalities in which sutures fuse prematurely or fail to form (craniosynostosis). In such cases, compensatory growth typically occurs in other sutures in

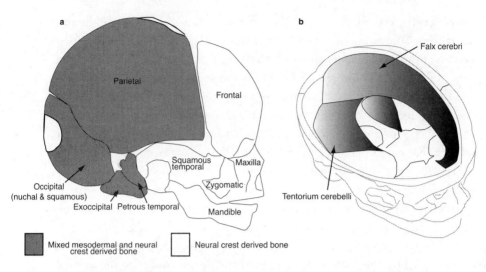

Figure 3.5. (a) Neural crest–derived versus mesodermal-derived portions of head in lateral view (adapted from Morriss-Kay, 2001), extrapolated from studies in mice. (b) The two major dural bands of the brain, showing their relationship to the neurocranium and basicranium.

predictable ways (Richtsmeier, 2002). For example individuals with a prematurely fused sagittal suture tend to develop long, narrow vaults (see Figure 4.8). In addition, anencephalic embryos that lack a brain develop a primordial cranial base but have almost no development of the vault (Figure 3.6a).

Another way in which physical forces integrate the growth of the entire brain and desmocranium is via the dura mater, which has two major foldlike projections (septa) of fibrous collagen bands: the tentorium cerebelli and the falx cerebri (Figure 3.5b). These sheets of dural membrane attach to the cranial base and to the inside of the cranial vault along sutures, separating regions of the brain (Friede, 1981; Sperber, 2001). The tentorium cerebelli forms a semihorizontal, trampoline-like structure that supports the upper part of the brain. It attaches to the petrous crests of the temporals and the lesser wings of the sphenoid (the anterior clinoid processes), and extends around the rim of the posterior cranial fossa, attaching to a bump on the inside of the occipital (at the boundary between the occipital's endochondral and intramembranous components). The falx cerebri originates along the midline of the tentorium and inserts into the dura along the midsagittal line of the cranium, extending all the way to the vertical plate of the ethmoid. Thus, as the brain grows, the

Figure 3.6. Examples of craniofacial anomalies: (a) anencephaly; (b) cleft palate; (c) holoprosencephaly (from Sperber, 2001; used with permission). See text for details.

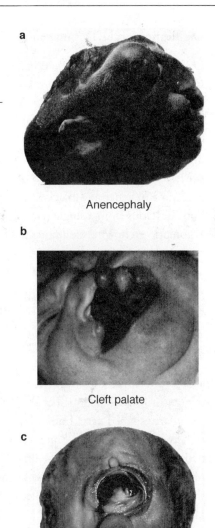

Anencephaly

Cleft palate

Holoprosencephaly

septa (known collectively as the dural bands) are thought to pull directly on the sutures of the growing neurocranium so that shape of the brain influences the shape of the neurocranium and vice versa (Moss and Young, 1960). Chapters 4–6 considers these bands more.

Ossification of the Desmocranium

Ossification of the bones that make up the cranial vault begins as woven bone is rapidly deposited in numerous centers that first appear after eight weeks in the uterus. Each parietal bone derives from two intramembranous ossification centers. Both sides of the frontal bone (which fuse after birth) form from a single ossification center near the pole of the each frontal lobe above the orbits. Ossification extends from this center to create the zygomatic and nasal processes, possibly as the result of secondary ossification centers in these regions (see Scheuer and Black, 2000). The squamous portion of the temporal bone develops from a single ossification center above the posterior root of the zygomatic arch. The squamous temporal fuses with the other intramembranous part of the temporal, the tympanic ring, that derives from four centers at the external margin of the auditory canal. These fuse shortly before birth with the petrous part of the temporal. The sphenoid is mostly endochondral (see above) but includes three intramembranous portions, each with a separate ossification center: the flat portions (squamae) of the greater wings (lateral to the temporal lobes) and the two pterygoid plates (medial and lateral). Finally, the intramembranous part of the occipital (the squamous occipital) arises from two ossification centers, one on each side of the occipital lobe above the tentorium cerebelli (Pal et al., 1984; Srivastava, 1992). In addition, small, irregularly shaped extra bones, known as Wormian bones, occasionally form in between the standard bones of the vault.[5] These appear to have some genetic basis (Hauser and DeStephano, 1989) but may also form in response to mechanical stimuli. Hydrocephalic individuals often develop Wormian bones that grow to surround outpocketings of cerebrospinal fluid (Forrester et al., 1966).

The Embryonic Face (Splanchnocranium)

The face is probably the most variable, evolvable part of the skull, in part because it is also the most complex and the least constrained. The face incorporates approximately 20 bones, which derive from many more growth centers. These bones grow intramembranously below the brain and cranial base, around the eyes, tongue, pharynx, teeth, and other structures. In order to accommo-

---

5. Wormian bones typically have vermicular or worm-shaped edges, but they are actually named after the Danish physician Olaus Worm (1588–1654).

date so many structures into a functional whole, the face needs to be highly integrated. Because the face forms initially around the anterior end of the oral cavity and pharynx, which develop from the foregut, the embryonic face is called the splanchnocranium (*splanchno* means "gut" in Greek). In fetal and postnatal life, the same region's name is Latinized to *viscerocranium* (see Dart, 1922).

Early Morphogenesis

The embryonic face originates through a series of cell movements and inter-actions among various segmented units. Before considering this sequence in detail, it may help to elucidate three organizing principles. First, most of the face's bones derive from two general sets of migrating cells. The frontonasal group migrates from the top of the head to form the upper half of the face, and the branchial arches migrate from the back of the head around both sides of the head like a set of collars to contribute to the lower face, the neck, and the face's various muscles. Second, almost all the facial bones coalesce around particular spaces and organs, especially the end of the foregut, which becomes the pharynx. Third, the appropriate differentiation of the face's many com-ponents depends on a combination of prepatterning and induction by signal-ing centers.

To trace the origin of the two groups of neural crest cells that form the face and their relationship to the pharynx, we need to go back to gastrulation. Af-ter gastrulation, the cranial end of the trilaminar germ disk includes a thick-ened mass of endodermal cells that projects beyond the layer of mesoderm (see Figure 3.2). This mass of cells forms the top a two-layered membrane (the oropharyngeal membrane) with the endodermal cells that surround the most cranial part of the foregut. The oropharyngeal membrane thus lies above a de-pression at the end of the pharynx called the stomodeum (*stomo* is Greek for mouth). The stomodeum is actually an empty space at the future location of the mouth.

The early face initially grows from neural crest cells that migrate toward and around the stomodeum in several different streams (as shown in Figure 3.3). Differentiation of each stream of cells occurs from interactions between neu-ral crest and epithelial cells at one of two signaling centers (Figure 3.7). The first is the prosencephalic organizing center, which derives from mesoderm in the primitive streak. After gastrulation, these cells migrate to a spot below the primitive forebrain (the prosencephalon), slightly beyond (rostral to) the tip

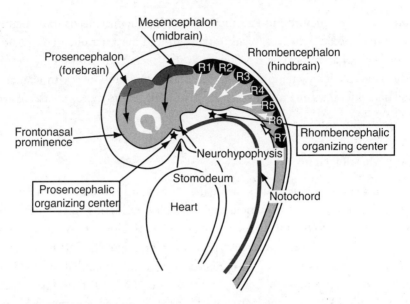

Figure 3.7. Schematic view of three-week-old embryo, showing the prosencephalic and rhombencephalic organizing centers in relation to the approximate pathways of the neural crest migrations from locations adjacent to rhombomeres (R1–R7) that contribute to the face.

of the notochord. The prosencephalic signaling center produces signaling factors (Shh and Fgf8) that induce neural crest cells surrounding the forebrain to differentiate into mesenchymal cells that make up the upper third of the face above the stomodeum (Schneider et al., 2001). These cells become the frontonasal prominence around the eyes and the nose (Figures 3.7 and 3.8); they also form parts of the ear.

The other streams of cells make up the middle and lower portions of the face around the sides and bottom of the foregut. These cells have a different origin, and most are influenced by the rhombencephalic organizing center. The hindbrain (rhombencephalon) differentiates and divides into seven segments, rhombomeres, whose identity is specified by *Hox* genes (Noden, 1983; Lumsden and Krumlauf, 1996). Neural crest cells tend to migrate; those initially adjacent to each rhombomere are no exception, as diagrammed in Figure 3.7. These cells stream ventrally around the side of the embryo toward the stomodeum and future face. As they migrate, they interact with other cells (notably, endoderm that surrounds the aortic arches, ectoderm on the surface of the neural tube, and paraxial mesoderm) to form five pairs of distinct bi-

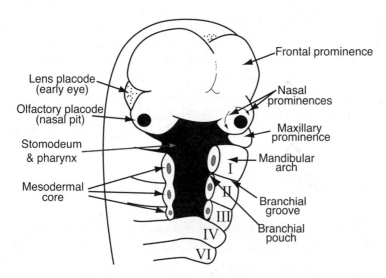

Figure 3.8. Schematic view of four-week-old embryo, showing basic anatomy of the facial prominences and branchial arches (identified with roman numerals). The frontonasal prominences interact with the lens and olfactory placodes to form the facial and nasal prominences. The maxillary prominences grow around the pharynx toward the ventral midline between the nasal prominences and the first branchial arch. (Adapted from Sperber, 2001: figure 4.1.)

lateral swellings, known as the branchial arches (Figure 3.8) (Trainor and Krumlauf, 2001). Each pair of arches grows like a collar around the foregut, eventually merging at the ventral midline of the embryo. The branchial arches (numbered I, II, III, IV, and VI because the fifth arch never forms in higher vertebrates) give rise to the lower face, most of the pharynx, and the middle and external portions of the ears (for review, see Hall and Hörstadius, 1988). Each arch is a semidistinct segment, like a little expeditionary force, containing all the equipment necessary to grow a bit of face or neck. At the heart of the first arch is a central cartilage core around which skeletal and other connective tissues grow (but not through any kind of endochondral process). The mesodermal component of each arch contributes muscle-forming groups of cells (myomeres); other mesodermal cells form vascular channels within each arch (these branch off from the aortic arch). In addition, each arch has sensory and motor fibers that derive from a specific cranial nerve (nerves V, VII, IX, and X). Outgrowths of the gut generate pouches that initially separate each arch on the inner (endodermal) surface; externally, the arches are separated by grooves (shown in Figure 3.8). The endodermal lining of each arch gives rise to the lining of pharynx, along with various other structures that differentiate from the branchial pouches (e.g., the tympanic membrane and several glands, such as the thymus and parathyroid). The ectodermal surface of each arch contributes epithelial cells that become organs such as dental enamel, taste buds, and salivary glands. The arches initially remain separate but later fuse and combine. Table 3.1 summarizes the components and fate of each branchial arch.

Formation of the Face

Let's look more carefully at how the face forms from a series of units, termed prominences. One major prominence, the frontonasal prominence, is initially organized by the prosencephalic center and forms the upper part of the face around the eyes and nose and above the oral cavity (as shown in Figure 3.8). From its earliest stages of development, the frontonasal prominence has critical interactions with the lens and olfactory placodes, thickened regions of ectoderm that originate in the neural folds. Placodes are sites of induction. At about four weeks in utero, the lens placodes differentiate as outpocketings of the ectoderm above the outer, lateral margins of the forebrain. Once the eyes start forming (a process reviewed in Chapters 6 and 10), the frontonasal prominence grows downward, mostly in between them. Another pair of neu-

Table 3.1: Major derivatives of the branchial arches

Arch	Ectodermal groove	Endodermal pouch	Skeletal	Muscles	Cranial Nerves
I	External auditory meatus, External ear cartilages	Auditory tube, middle ear tympanum	Meckel's cartilage, malleus, incus	Masticatory, Tensor tympani, mylohyoid, anterior digastric	V
II	Disappears	Tonsilar fossa	Reichert's cartilage	Facial, Stapedius, Stylohyoid, Posterior digastric	VII
IV	Disappears	Superior parathyroid	Thyroid and laryngeal cartilages	Pharyngeal constrictors, Levator palatini	X
VI	Disappears		Cricoid and aretynoid cartilages	Larynx muscles	X

rally derived placodes, the olfactory placodes, is located between the eyes
and above the stomodeum. These placodes induce the formation of the olfactory epithelium and also induce the surrounding region of the frontonasal
prominence to differentiate into two swellings, the median and lateral nasal
prominences. As Figure 3.8 shows, these two divisions of the frontonasal
prominence grow in a horseshoe shape around depressions on each side of the
future nose (the nasal pits). Eventually they fuse around the nasal openings.
Thus the frontal prominence and its divisions create the entire upper face
around the eyes and nose. In addition, the frontonasal prominence contributes
to the incisor-bearing region of the front of the palate, the future premaxilla.

Facial morphogenesis below and to the sides of the stomodeum occurs differently. As the frontonasal prominence grows down around the eyes and nose,
two other pairs of prominences—maxillary and mandibular—also sweep
around the sides of the face toward the ventral midline of the embryo (Figure
3.8). It was once believed that these prominences were both derivatives of the
first branchial arch, but studies reveal that they have different cellular origins,
and that homeobox genes (in this case, *Dlx5* and *Dlx6*) give them distinct segmental identities (Depew et al., 2002). The mandibular prominences arise
from the first branchial arch, but the maxillary prominences largely derive
from cells from the postoptic region that lie above (cranial to) the first arch
(Lee et al., 2004). Regardless of their origin, both prominences stream around
the lateral margins of the stomodeum: the maxillary prominences create much
of the midface, fusing below the nose to form much of the palate, upper jaw,
and upper lip; the lower, mandibular prominences are the primordia for the
lower jaw.

Fusion of the maxillary prominence with the medial and lateral nasal
processes is a tricky business. The left and right maxillary processes grow toward the front of the embryo above the stomodeum, eventually fusing in the
midline. As they do so, they fuse to the median and lateral nasal processes
above (Diewert, 1983). This first fusion creates the primary palate at the front
of the mouth, thereby beginning the formation of separate oral and nasal cavities. The formation of the secondary palate above the rest of the oral cavity
occurs differently, via the outgrowth of the palatal shelves, which grow inward
(medially) from the maxillary prominences. These shelves initially point downward, but as they grow they elevate (fold upward) and eventually meet in the
midline when they are approximately horizontal. Fusion of the palate occurs
through disintegration of epithelial cells on the surfaces of the prominences
when they contact each other, causing the underlying mesenchymal cells to

merge. Failure of the facial prominences or the palatal shelves to merge, due to insufficient growth or death of the surface epithelial cells following contact, causes various kinds of congenital clefting of the upper lip and palate (Figure 3.6b). Most frequently (about one in 800 births), the medial nasal and maxillary processes do not successfully fuse on one side of the face, causing a unilateral cleft lip (see Johnston and Bronsky, 2002; Marazita, 2002).

As we consider how the head becomes integrated, it is useful to keep in mind the relationship between the facial prominences and the major organs and spaces in the face. From the earliest phases of facial growth, the facial prominences not only grow and fuse with each other but also grow *around* various functional spaces and organs. In fact, the major sensory organs of the face (all epithelial in origin) play essential inductive roles that regulate facial morphogenesis. The most important of these functional capsules are the divisions of the brain, the eyes, the nose, the ears, the oral cavity, and the teeth. As an example, in the upper face, the frontonasal process grows around the base of the forebrain and around the eyes and the nose. The lens and olfactory placodes initially differentiate from the neural tube as fields of ectodermal cells in the anterior neural plate. Expression of the *Cyclops* gene causes this region to split in two, forming two cerebral hemispheres. Thus each side of the forebrain develops its own primordial eye in the middle of the frontonasal prominence, connected via a nerve stalk to the lateral margin of the forebrain (Hatta et al., 1994). The paired olfactory bulbs differentiate from the forebrain in a somewhat similar manner. Consequently, the forebrain, olfactory bulbs, and eyes not only participate in the induction of several precursor condensations of the cranial base (described above) but also induce around them the intramembranous bones of the upper face. For example, the paired placodes most likely induce mesenchymal condensations in the cartilaginous cranial base below the brain and behind the eyes (as described above) as well as the differentiation of the median and lateral nasal prominences.

In other words, from the earliest stages, the upper face grows through the coordinated activity of a group of units—the anterior cranial base, the forebrain, the eyes, the nose, and the frontonasal prominences—that remain morphologically distinct to some extent but influence each other's growth and development through epigenetic interactions. Several craniofacial growth disorders provide dramatic evidence for this integration. One example is holoprosencephaly (Figure 3.6c), which is caused by defects in the prosencephalic organizing center that regulates division of the brain into left and right lobes (see Muenke and Beachy, 2000). In extreme cases, the embryo has only one lens placode and one olfactory placode, hence a single eye (cyclopia) and a single-nostril nose (cebocephaly). Milder forms cause incomplete separation of the lens placodes, causing the face to have closely set eyes (hypotelorism) or median clefting of the lip and palate (see Johnston and Bronsky, 2002). In addition, there are no survivable congenital defects in which the lens or olfactory placodes fail to develop (Gorlin et al., 2001). From a developmental standpoint, such defects would be unviable because differentiation of the lens and olfactory placodes are so essential for inducing later morphogenetic events.

The lower and middle portions of the face also develop around several organs and functional spaces, most notably the nasal and oral cavities and the teeth. Whereas the internal nose mostly derives endochondrally from cartilages of the cranial base (especially the mesethmoid and ectethmoid), the external portions of the nose all derive from intramembranous bone and cartilages that differentiate from the frontonasal prominence. The cavity formed by the nasal pits (created by the median and lateral nasal processes) is initially separated from the stomodeum by the palate (the merged maxillary prominences), and by an oronasal membrane that later disintegrates. Epithelial tissues play a role in all these morphogenetic activities. For example, the nasal septum, which develops from the vertical plate of the mesethmoid, not only separates the two olfactory bulbs atop the forming cribriform plate, but also contains specialized epithelial cells that become the vomeronasal organ—a vestigial chemosensory organ (T. D. Smith et al., 2001; see also Chapter 10).

Jaws and Ears

The jaws have a special pattern of morphogenesis. The upper and lower jaws develop from the maxillary and mandibular prominences, respectively. Each of these prominences contains several elements, including a central cartilaginous rod around which intramembranous bone grows. I focus here on the development of the mandible, which is better studied and provides to some extent a useful model for understanding the morphogenesis of the maxilla.

The mesenchymal cells that contribute to the mandible's central cartilaginous rod (Meckel's cartilage) and its associated condensations derive from a mixture of neural crest cells and non-neural crest cells from the paraxial mesoderm (Couly et al., 1996; Kongtes and Lumsden, 1996; Chai et al., 2000).[6]

---

6. Meckel's partial equivalent in the maxilla is sometimes thought to be the dorsal (palatopterygoquadrate) cartilage, but this cartilage in amniotes may derive from the same precursor as Meckel's cartilage (Cerny et al, 2004).

The majority of cells in the mandibular arch come from neural crest cells that migrated from the first two rhombomeres (R1 and R2), with only the posteriormost elements deriving from R4. Thus some degree of patterning within the mandible occurs long before the neural crest cells migrate to the future region of the face. These neural crest cells constitute the outside of the arches, and paraxial mesoderm fills the core (Mina, 2001). As Figure 3.9 illustrates, at least seven developmentally and functionally distinct condensations of mesenchyme then coalesce around Meckel's cartilage: the symphysis, the condylar process, the coronoid process, the angular process, the ramus, the molar alveolar process, and the incisor alveolar process.

Atchley and Hall (1991) proposed that overall mandibular shape derives from the combined effects of the different patterns of morphogenesis within each condensation. These patterns are partly a function of the size of each condensation, combined with specific developmental programs that arise from spatially dependent, inductive interactions that determine the developmental fate of each unit's mesenchymal cells. For example, cells in Meckel's cartilage differentiate solely into chondroblasts to produce cartilage. In contrast, cells in the ramal region differentiate almost exclusively into osteoblasts that produce the ramus; and cells in the alveolar processes differentiate into either osteoblasts that produce the alveolar bone that surrounds the tooth roots or the odontoblasts and cementoblasts that produce the tooth roots (see the section below on tooth morphogenesis). The pattern of expression of transcription factors and growth factors that regulates the identity and morphogenesis of these specific regions is currently the subject of much research (see Mina, 2001). The nature and timing of the inductions are significant because they probably are the source of many small-scale shifts (tinkering) in mandibular form of the sort that arise between species. For example, MacDonald and Hall (2001) showed that differences in the timing of the epithelium-induced initiation of osteogenesis in the mandible of three mouse strains caused changes in relative amounts of growth and hence in the shape of their mandibles.

The first branchial arch, which gives rise to the mandible, also contributes to the middle and outer ear, and to the pharynx in conjunction with the other branchial arches and grooves (listed in Table 3.1). The morphogenesis of the pharynx, ear, and neck is too complex to review here in depth, but a few details are worth noting. First, the spaces and tissues of the middle and external ear derive from both the first and second arches. Between arches I and II is an internal pouch and an external groove (Figure 3.8). The central auditory tube

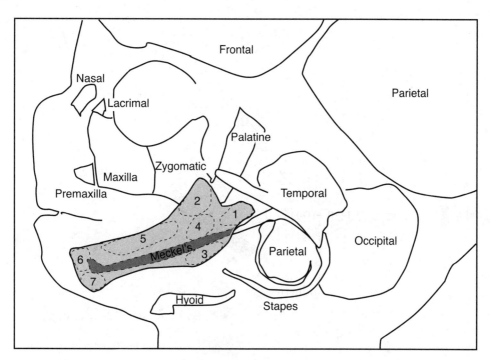

Figure 3.9. 12-week-old embryonic face in lateral view, showing major ossification centers of the face and highlighting those of the mandible, in which seven ossification centers form around Meckel's cartilage: (1) condylar; (2) coronoid; (3) angular; (4) ramus; (5) molar alveolar; (6) incisor alveolar; (7) symphyseal.

of the middle ear derives from the first branchial pouch, and the cartilages of the external ear derive from swellings ("auricular hillocks") that arise in the ectoderm of the arches surrounding the first branchial groove (see Figure 3.10). Within the middle ear, the heads of two tiny middle ear bones, the malleus and incus, derive from the central cartilages of arch I (the malleus is a remnant of Meckel's, and the incus is probably a remnant of the dorsal cartilage). The rest of malleus and incus (inferior processes) along with the third middle ear bone, the stapes, derive from the central cartilage of arch II (Reichert's cartilage). All other key bits of the neck and lower face derive from various portions of the branchial arches as outlined in Table 3.1. Note that the muscles, nerves and arteries of the face and pharynx are segmented via the branchial arches. For example, the muscles of mastication derive from arch I, and the muscles of facial expression derive from arch II.

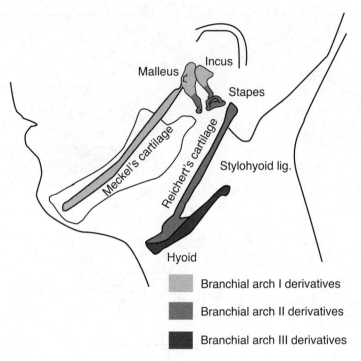

Figure 3.10. Embryonic origins of structures (described in the text) in the neck and middle ear.

Tooth Morphogenesis

Tooth morphogenesis occurs in conjunction with the development of the mandibular and maxillary arches. Like much of the head, tooth formation involves lots of reciprocal signaling between mesenchymal and epithelial cell lines. However, teeth have a different evolutionary origin from the rest of the skull. Instead of developing from bones, teeth first evolved as scales overlying the jaws of primitive fishes. These scales were then co-opted to form denticles, sharp placodes in the skin that are sensitive to osmolarity, electric charge, and temperature. This unique history is reflected in the distinctive composition, function, and embryology of teeth, which originate in part from ectodermal cells, with contributions from neural crest and mesodermal cells. The several major stages by which teeth form are summarized in Figure 3.11.

Placode Stage

Tooth morphogenesis is initiated as interactions among different cell lines in each division of the first branchial arch. Recall that branchial arches have a

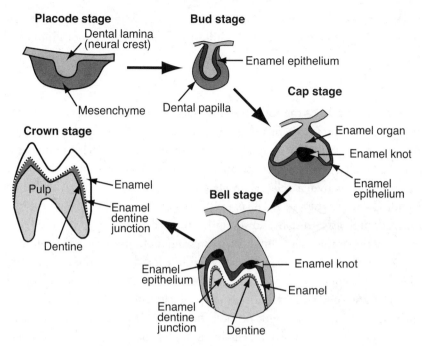

Figure 3.11. Stages of tooth morphogenesis described in the text.

surrounding layer of epithelial cells that are ectodermal in origin. This epithelial layer, the dental lamina, grows downward and folds inward into the front and back of each mandibular arch. Note that the epithelial cells that fold into the anterior portion of the upper jaw appear to derive from the frontonasal prominence rather than from the maxillary prominence (Depew et al., 1999). The epithelial cells then grow and coalesce into two horseshoe-shaped, parallel arches in each jaw.

Bud Stage

Teeth begin to develop as buds from interactions between cells in the dental lamina and underlying cells in the mesenchyme. Initially some cells in the dental lamina express growth factors (e.g., Fgf8) that activate the expression of the transcription factor Pax-9 in the adjacent mesenchyme; other regions of the dental lamina express other signals (e.g., BMP-2, BMP-4), which inhibit the expression of Pax-9 elsewhere in the mesenchyme (Neubuser et al., 1997). Regions where Pax-9 is expressed induce the formation of a tooth bud, an invagination of epithelial cells from the lamina into the underlying mesenchyme (see Figure 3.11); regions where BMPs (bone mor-phogenetic proteins) are secreted inhibit the formation of a tooth bud (Ker-anen et al., 1999).

Cap Stage

The secretory cells of the tooth begin to take shape during the cap stage. Al-though patterning in the epithelium determines the position of each tooth bud, patterning in the mesenchyme determines the identity of each tooth (Lums-den, 1988; Sharpe, 1995; Peters and Balling, 1999). After a tooth bud begins to form, mesenchymal cells accumulate into a little clump called the dental papilla (Latin for "nipple"). The papilla later gives rise to the dentine-forming cells, odontoblasts. Initially, however, the mesenchymal cells in the tooth bud take over the primary signaling role, expressing molecules (e.g., BMP-4, Msx-1, and Pax-9) that regulate rates of cell division and death in the overlying ep-ithelial cells, collectively called the enamel organ (Mandler and Neubuser, 2001). Transplant studies indicate that prepatterning of cells in the mes-enchyme influences what type of tooth the epithelial cells will create. For ex-ample, transplanting molar mesenchyme under incisor epithelium causes a molar rather than an incisor to develop (Kollar and Baird, 1970).

Bell Stage

During the bell stage of development, the boundary between the epithelial cells of the enamel organ and the mesenchymal cells of the dental papilla takes shape. The enamel organ has several layers: the outer layer expresses signaling factors, which cause the inner epithelial layer to buckle in three dimensions through variations in regional rates of cell division.[7] Buckling and folding of the epithelial cap determines the shape of the enamel-dentine junction (EDJ), which underlies the future enamel crown. Genetic control of the initial shape of the inner enamel epithelium occurs via signaling centers in the epithelium called *enamel knots* (Jernvall et al., 1994). Enamel knots act at the future tip of each dentine horn. Knots work to a large extent from coordinated expres-sion of FGFs (fibroblast growth factors) that stimulate cell proliferation and division around the knot, and BMP-4, which inhibits cell growth around the knot (for review, see Jernvall and Jung, 2000).

---

7. Think of a knitted fabric: regions where the density of stitches is higher will buckle relative to regions with fewer stitches per unit area.

Crown Stage

During the crown stage, the cells of the inner enamel epithelium stop dividing, and cells in the dental papilla adjacent to the inner enamel epithelium differentiate into odontoblasts and begin to synthesize dentine toward the center of the tooth. Soon after, cells in the inner enamel epithelium differentiate into ameloblast cells that begin to secrete enamel in the opposite direction, away from the dental papilla. See Chapter 2 for a brief description of how dentine and enamel grow.

Throughout tooth morphogenesis, the jaw also grows around the tooth. Tooth growth is not independent of mandibular or maxillary arch growth but instead involves some epigenetic interactions between the tooth root and condensations of the alveolar processes of both the upper and lower jaws. Presence of a tooth root, which expresses key signaling molecules (e.g., the transcription factor Msx-1), has an effect on stimulating ossification of its surrounding alveolar condensations. Mice in which Msx-1 is knocked out develop no teeth and alveolar bone but have otherwise normal mandibles (Satokata and Maas, 1994).

Ossification of the Splanchnocranium

The many bones of the face begin to ossify intramembranously, starting at about six weeks in utero. The signaling mechanisms by which ossification centers are induced within various mesenchymal condensations are not entirely known (or particularly interesting), but there has been substantial work on the pattern of ossification. Some key elements are illustrated in Figure 3.9.

In the jaws, as noted above, Meckel's cartilage, which lies at the center of each mandibular arch, does not ossify but instead serves as a sort of template around which woven bone from the various condensations begins to form. Blood vessels from the newly formed bone then penetrate the cartilage, bringing cells that break down the cartilage; the remnant space becomes the center of the marrow cavity of the jaw. The posterior end of Meckel's however, ossifies into part of the malleus and incus bones of the middle ear (see above). Other parts of Meckel's also contribute to additional structures in the jaw and neck.

The other parts of the face begin to ossify at around eight weeks in utero. Each side of the maxilla fuses from four distinct ossification centers (the maxillary body, zygomatic, orbitonasal, and nasopalatine) (Kjaer, 1989). Anterior to the maxilla is the premaxilla, initially a separate bone (some dispute this), which forms the base of the anterior part of the nasal cavity. Eventually the premaxilla and maxilla fuse, forming a continuous, horseshoe-shaped shelf of bone around the tooth roots, the alveolar crest (Behrents and Harris, 1991). Although the posterior midline part of the palate remains unossified as the soft palate, a pair of palatine bones ossifies at the back corners of the maxilla (see Figure 3.9). The zygomatic ossification center creates the zygomatic arch to connect the rest of the temporal bone to the face, and distinct intramembranous ossification centers posterior to the maxilla also contribute to bone formation in the greater wings of the sphenoid and in the medial and lateral pterygoid plates of the sphenoid (Gasser, 1967).

Early on, the supraorbital region spatially connects the upper face around the eyes with the anterior cranial fossa, as described above. In addition, the nasal region connects the middle face with the skull base around the ethmoid. Distinct intramembranous facial bones, each of which derives from separate ossification centers, surround the nasal cavity laterally (the nasal bones, the lacrimal bones, and the perpendicular plates of the palatine bones), and inferiorly (the premaxilla, the maxilla, and the horizontal plates of the palatine bones). The ethmoid bone makes up the rest of the nasal cavity, contributing intracartilaginous bone to the midline floor of the cranial fossa above, and part of the wall of the orbital cavity on the side. The ethmoid also forms two curling projections on each lateral side of the cavity, the superior and middle nasal conchae, and a thin nasal septum, which divides the nasal cavity in half. The inferior nasal conchae at the base of the nose are separate endochondral bones.

The initial relationships of the face to the cranial base and cranial vault in the embryo persist throughout ontogeny (as later chapters will detail). By the end of the embryonic stage, the sphenoid and ethmoid bones connect the midface with the cranial base; the upper face articulates with the cranial base through the ethmoid bone in the anterior cranial fossa and with the cranial vault through the frontal bone. Although the lower face grows downward from the middle face, it articulates with the basicranium via the mandible at the temporomandibular joint because of the common embryonic origin of the mandible and the outer ear via the first branchial arch. These spatial relationships and interactions continue throughout life, helping integrate the growth of multiple regions into a functional whole.

Embryological Bases for Tinkering in the Skull

To conclude, let us step back from all the details we have just covered and ask where and how shifts in cranial form are likely to have arisen during embryogenesis over the course of human evolution. Chapters 11–13 return to this topic in the context of particular hypotheses about specific changes in cranial shape in human evolution. For the time being, it suffices to emphasize that ape, modern human, and hominin heads do not differ in terms of their basic craniofacial plan. These heads all have the same components (bones, organs, sutures, growth sites) arranged in the same basic way, with only variations in the shape and relative size of components and their spatial relationships. In addition, as Chapters 4 and 5 show, many variations in form involve changes in the size of the brain, the angle of the cranial base, the size of the face, and the position of the face relative to the brain.

The hierarchy of developmental processes that take place during the embryonic period gives us some clues about where and how such shifts occur so that changes in the genotype cause changes in the phenotype. Because the development of the mandible is the best understood, I use it as an exemplar of these potential developmental shifts, but the principles are general.

First come the "big gun" genes that determine basic fields and patterns. After the formation of the basic germ layers (endoderm, mesoderm, ectoderm, and neural crest), expression of homeobox genes such as *Hox, Pax,* and *Dlx* act to determine the position and identity of the basic parts of the body. We know, for example, that *Dlx 5/6* specifies the identity of the mandible versus the maxilla, and we strongly suspect that other homeobox genes determine identity of regions within the mandible (such as patterning in the dental lamina). These patterning genes are probably quite functionally conservative in mammalian evolution and offer little opportunity for tinkering through mutations in the genes themselves. One possibility, however, is that alterations in their expression might affect morphogenesis subtly by influencing the size of particular regions or the kinds of receptors they contain, thus modifying how they will later respond to various growth-inducing stimuli such as hormones and growth factors.

In the second level of morphogenesis come two types of shifts. First, epigenetic interactions between tissues influence regional cell fates. Often, these occur via epithelial-mesenchymal interactions. For example, epithelial cells in the dental lamina of the mandible interact with mesenchyme in the mesoderm to determine the position of teeth by inducing the tooth bud to differentiate. These epigenetic interactions are frequently reciprocal, increasing both the variability and specificity that results from induction. Again, in the case of the mandible, prepatterning of the mesoderm by homeobox genes determines what kind of tooth bud is formed.

The second type of shift that occurs within regions is through changes in *cis*-regulatory elements of genes, by which ensembles of transcription factors affect developmental pathways (see Carroll et al., 2001). There are many examples of how changes in *cis*-regulatory elements modify local morphogenesis. In the mandible, for example, many transcription factors that affect the growth of condensation centers are regionally specific (reviewed in Hall, 2003). For example, mice with knockouts of the *goosecoid* transcription factor have a major reduction in the size of the condensations that form the coronoid and angular processes of the mandible, but not the rest of the jaw (Rivera-Pérez et al., 1995). In contrast, knockout studies show that the growth factor TGFβ-2 influences the size of the condylar, angular, and coronoid processes (J. F. Martin et al., 1995), and Msx-1 is necessary for the growth of the teeth and the alveolar processes (Satokata and Maas, 1994). One can therefore easily imagine that shifts in mandibular shape (or craniofacial shape in general) could occur simply through alterations of the expression or reception of these transcription factors within particular condensations. As noted in Chapter 2, changes in the relative size of particular condensations can occur through regulation of the *rate* of growth in condensations (faster or slower) or the *timing* of growth. Such changes should lead to predictable heterochronic shifts in growth that may be apparent postnatally but whose basis could be determined in embryogenesis.

Finally, interactions between tissues must have a range of effects on the growth of particular organs, or even regions of organs. Skeletal elements do not grow in isolation from each other or from other tissues but interact spatially with muscles, sensory organs, and other tissues—all of which can have epigenetic interactions with each other through induction and the actions of physical forces. Muscles, for example, appear to be necessary for normal bone growth in the embryo. In a classic instance, the absence of the lateral pterygoid muscle (which inserts on the mandibular condyle) causes the absence of a condyle, and hyperactivity of the lateral pterygoid causes a hypertrophic condyle, because mechanical stimulation of the condylar growth center is necessary for normal growth (Hall, 1978). Many muscles, the tongue, the teeth, and other tissues, such as nerves, all likely exert various effects on local growth of the mandible. Although these sorts of epigenetic interactions presumably act at all stages of development, one expects a relationship between the stage of an interaction and the magnitude of its effect. If there is any truth to von Baer's law, then epigenetic influences during embryogenesis probably have more dramatic effects than those that occur later in ontogeny.

In conclusion, if we examine the grand sequence of developmental events of bones such as a mandible, a parietal, or a sphenoid—from the formation

of germ layers to the factors that influence the shape of skeletal regions—
embryonic condensations appear to be the most evolvable modular unit of
morphogenetic change (Hall, 2003). The basic arrangements of condensations
in closely related species, such as chimps and modern humans, are the same.
But the growth of these condensations can be altered to change their relative
size and shape at several hierarchical levels of development. First, changes in
homeobox gene expression can alter the initial sizes and relative locations of
condensations. Second, changes in transcriptional regulation or induction of
cells in condensations can influence their intrinsic growth processes and how
they respond to epigenetic stimuli. And third, physical interactions between
skeletal condensations and neighboring tissues can influence rate and timing
of growth.

We will return to this model of embryogenetic change later to hypothesize
about the developmental bases for particular shifts in human evolution. But
first we need to examine postnatal growth and function in the skull in order
to understand what units changed in the first place.

4

Modular Growth of the Fetal
and Postnatal Head

In the head of the young orang, we find the childlike
and gracious features of man.

E. GEOFFROY SAINTE-HILAIRE, 1836

Chapter 3 reviews how the basic building blocks of the head were set up during the embryonic period. Our next step is to consider the head's growth and
development during the fetal and postnatal periods in order to ask two questions. First, how do the many components of the head manage to function
and accommodate one another as they change in shape while growing in size
by several orders of magnitude? Second, how do human heads grow differently
from those of apes and other nonhuman primates?

In a way, the first question is easier to address. Heads cope with complexity during growth through the interplay between modularity and integration.
The embryonic stage generates the basic modules of the head such as the brain,
the eyes, and the jaws. These modules—each comprising a further subset of
modules—are partially autonomous both genetically and phenotypically.
However, they interact substantially not only during embryogenesis but also
later in ontogeny via many mechanisms. The skeletal framework of the head
plays a particularly important role in these dynamic interactions by setting up
bony walls that serve as capsules around organs, and platforms from which
various structures grow. As Chapter 2 emphasizes, bones are stiff and strong,
but they are also interactive and movable. So as the eyes, brain, nasal cavity
and other functional units of the head grow and function, they each influence
the morphology of neighboring walls of bone, many of which they share in

common. The combination of these and other processes helps generate a highly integrated head.

The different ways in which humans and other apes, such as chimpanzees, carry out this complex interplay of modular growth and morphological integration is not a trivial topic. By birth, a chimp and a human already have distinctively shaped heads, and as they mature, their heads become increasingly different. A chimp develops a long snout, marked brow ridges, and an elongated braincase, and its foramen magnum migrates toward the back of the skull. In contrast, a human develops an enormous, globular braincase, much of which expands behind a centrally placed foramen magnum. The modern human face also remains relatively short, vertical, and tucked beneath the frontal lobes. Given evidence that the morphology of the last common ancestor (LCA) of humans and chimps probably resembled the African great apes in some respects (see Chapter 1), it is reasonable to infer that the LCA and the African great apes share similar (though not identical) patterns of growth and development. This inference prompts the question of what major developmental differences evolved in the hominin lineage to give humans snoutless, nearly spherical heads and other distinctive features, such as an external nose and a chin. Many of these differences have interesting functional consequences, which later chapters will consider. But first, it is valuable to ask how these differences are achieved by changes in growth and development. (Remember, the terms *growth* and *development* are used here in a restricted sense to denote changes in size and shape, respectively.)

There are several ways to consider what makes an adult human's head different from those of other apes and monkeys. The first is to compare variation in shape and size between juveniles and adults of different species. If the adult forms differ, then their growth and development must have differed, often in predictable ways. This comparative approach is especially useful for addressing questions about taxonomy and the evolutionary relationships among species. But comparisons of adult forms alone are insufficient to test hypotheses about how and why the species differ. One also needs information about ontogeny to address how particular transformations occurred. The most common approach is to quantify and describe the pattern of ontogenetic differences among forms, ideally ancestor versus descendant species. Such comparisons permit one to test hypotheses of scaling (allometry) and changes in the rate and timing of growth (heterochrony) (reviewed in Chapter 2). Ontogenetic patterns also contain information about integration in terms of the strength and patterns of correlation and covariation among various components. One way to use such data is to tease apart independent aspects of growth and development, and thus help distinguish adaptations (features that were targets of natural selection) from shifts that were not selected directly, but which changed from selection acting on other aspects of the organism (often called spandrels, after an analogy made by Gould and Lewontin, 1979).

Analyses of ontogenetic patterns are important in the study of evolutionary transformations, but they do not necessarily provide information on developmental mechanisms. Consider two species that grow long, projecting snouts. One species might grow a long snout by adding bone along the back of the snout, displacing the lower face forward; the other species might grow a similarly long snout by adding bone from within a suture in the middle of the snout, or from adding bone at the front of the snout. In such cases, analyses of pattern might suggest similarities, but analysis of mechanism would reveal differences. That is, different processes of growth and development can produce similar or identical patterns of morphology.

If we want to know how human and great ape heads came to differ, then we need information on both the patterns and processes by which heads grow. Obviously, we did not transform directly from the last common ancestor into modern humans, but it is nonetheless useful to ask how one would alter the ontogeny of an apelike LCA to make it grow and develop like a human. Below I ask when, how, and why each of these transformations occurred.

This chapter is the first of a pair of chapters on fetal and postnatal growth. This chapter focuses on the modules of the head, reviewing key processes by which major anatomical components grow within and around the skeletal framework of the skull. Because the head has so many units, each with its own skeletal capsule (e.g., the brain and the neurocranium, the eyeballs and the orbits, the oral cavity and the jaws), I adopt a *regional* approach, considering separately the basicranium, neurocranium, and face. For each region, I first review some basic anatomy in a developmental context, briefly compare the basic pattern of growth in modern humans and nonhuman primates (especially chimpanzees), and then examine the mechanisms of growth that contribute to differences in craniofacial growth and development in humans and other primates. Interested readers may wish to consult more detailed accounts of how the skull grows in humans (e.g. Enlow, 1990; Ranly, 1984; Moyers, 1988; Scheuer and Black, 2000), and nonhuman primates (Sirianni and Swindler, 1985; Schneiderman, 1992). Keep in mind, however, that regional descriptions of craniofacial growth have limitations, the biggest of which is that they do not address how the head's many units are integrated into a functional ensemble. Thus, Chapter 5 considers how the head as a whole is integrated, and how processes of integration differ between humans and nonhuman primates.

Time and Ontogeny

Before we compare growth and development in humans and nonhuman primates, it is useful to consider how to deal with time and ontogeny. The two concepts are different but related. Ontogeny is the series of stages over which an organism develops and grows—an unfolding of events in time. It is therefore quantified either in absolute units of time, such as days and years following a given event like conception or birth, or in relative units, such as stages (see Hall and Miyake, 1995). Both approaches have their uses. For many questions, absolute age is the preferable way to measure ontogenetic time. However, when comparing postnatal ages among species, one has to assume that the time of birth is an appropriately comparable moment for species in terms of ontogeny. When comparing humans and apes, this assumption may be problematic, given that humans are born less mature than apes. In addition, we lack accurate age data for many samples, including wild chimpanzees. For this reason, ontogenetic time is often measured in relative stages rather than absolute ages, thereby enabling comparisons of samples at similar, equivalent periods in development. In particular, postnatal ontogeny is typically divided into six loosely defined stages of development: the neonatal stage, infancy, childhood, the juvenile stage, adolescence, and adulthood (for discussion, see Bogin, 2001). Each stage has different characteristics that are reasonably comparable among species, and many of these general stages of growth are approximately recorded by the eruption of the permanent teeth (summarized in Table 4.1). In all primates, the first permanent molars (M1) erupt approximately at the end of the neural growth period (about 3 years in chimpanzees

Table 4.1: Ages (very approximate) of dental stages in humans and chimpanzees

Stage	Criteria	Humans	Chimpanzees
I	Prior to eruption of M1	0–6 years	0–3 years
II	After eruption of M1, before eruption of M2	6–12 years	3–6 years
III	After eruption of M2, before eruption of M3	12–18 years	6–12 years
IV	After eruption of M3	>18 years	>12 years

and 6 years in humans). The third permanent molars (M3) erupt approximately at the end of the skeletal growth trajectory (about 12 years in chimpanzees and 16–18 years in humans). The second molars (M2) erupt more or less halfway between these two events. However, there is enormous variation in these ages and their developmental correlations (see Dean, 2006).

Note that different organ systems grow at different rates and in different trajectories, as illustrated in Figure 2.5a. In the head, the face and pharynx typically grow along with the rest of the body in a skeletal growth trajectory, with rapid growth through infancy, slower growth during the childhood and juvenile periods, and more rapid growth during adolescence until the eruption of the M3s (following an S-shaped curve). The trajectory of neural growth, along with that of the cranial vault and basicranium, is different, with the most rapid growth in the first few years; it then tails off, with brain growth 95 percent complete in terms of mass and volume by the time of M1 eruption. Sensory organs that are outgrowths of the brain, such as the eyes and olfactory bulbs, also typically grow even more rapidly than the brain.

Differences in growth trajectories affect integration. Two general principles are useful to remember. First, different units can have unequal influences on each other, and these effects depend in part on their growth trajectory. For example, the size and shape of the growing brain directly influence craniofacial form only until the brain's growth is complete. After that, changes in craniofacial form mostly occur from facial growth and development, but in ways that may be constrained by neurocranial form. The second principle is that interactions among components with different trajectories often lead to contrasting scaling relationships. For example, brain volume scales to body mass in a different way than face volume scales to body mass (see Vinicius, 2005).

Regional Growth of the Cranial Base

Let's start our regional exploration of patterns and processes of craniofacial growth with the cranial base. If any part of the head physically acts as an "integrator," then it must be the cranial base, which is sandwiched in the center of the head between the brain and the face. Both the brain and face interact directly with the cranial base, and indirectly with each other via the cranial base. These interactions, however, are unequal in terms of strength and timing. The brain, the largest organ in the head, grows on top of the cranial base in a neural growth trajectory, completing most of its enlargement by the time of the eruption of the first permanent molars. In contrast, the face and pharynx include many organs and spaces that grow below and in front of the cranial base in a slower, more extended skeletal growth trajectory. The face finishes growing about the time that the third permanent molars erupt. In addition, most of the crucial nerves and blood vessels that connect the brain with the rest of the body traverse the cranial base. These multiple roles, in combination with the difficulty of visualizing its structure externally, make the cranial base especially challenging to study.

Anatomy

When we left the embryonic cranial base (the chondrocranium) at the end of Chapter 3, its many cartilages had condensed into parts of the four major bones that form the basicranium: the occipital, temporal, sphenoid, and ethmoid. Many texts describe basicranial anatomy by reviewing each bone separately. An alternative, more integrative approach is to review how these bones contribute to the three major concavities that cradle the base of the brain (Figure 4.1a): the posterior, middle, and anterior cranial fossae (PCF, MCF, and ACF).

Posterior Cranial Fossa (PCF)

This fossa, the deepest and largest, houses the hindbrain (the cerebellum and brain stem), and includes three main components: (1) the derivatives of the exoccipital somites that form around the foramen magnum; (2) the spheno-occipital clivus (slope) that runs below the brain stem in the midline from the foramen magnum to the hypophyseal fossa; and (3) the petrous pyramids. Note that the clivus includes parts of the occipital (the basioccipital) and the sphenoid (the basisphenoid), and that the petrous pyramids are part of the temporal bone.

Middle Cranial Fossa (MCF)

Viewed from above (see Figure 4.1a), the MCF is shaped like a butterfly, with wings that support the temporal lobes of the brain and a central sphenoid body, consisting mainly of the hypophyseal fossa that cradles the pituitary gland. Recall from Chapter 3 that the anterior part of the sphenoid body derives from the presphenoid cartilage, and the posterior part derives from the postsphenoid cartilage (also known as the basisphenoid). The hypophyseal fossa is often called the called the Turkish saddle (sella turcica) because its raised anterior and posterior margins, the tuberculum sellae and dorsum sel-

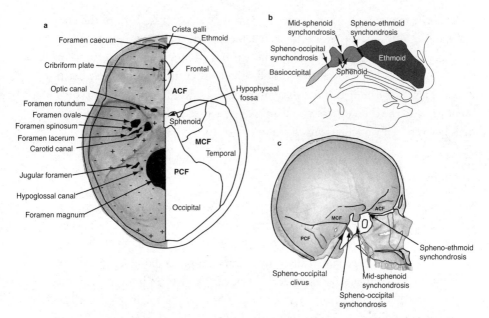

Figure 4.1. The cranial base. (a) Superior view showing endocranial fossa, major foramina and skeletal elements, and the distribution of depository (+) and resorptive (–) fields in the shaded left half and the bones in the unshaded right half. (b) Midline view of the fetal cranium showing how these chondrocranial cartilages ossify into five bones (the temporal is not shown), with three synchondroses in the midline separating the four parts: basioccipital, the postsphenoid, the anterior sphenoid, and the ethmoid. (c) Adult midline configuration. The mid-sphenoid and spheno-ethmoid synchondroses fuse much earlier than the spheno-occipital synchondrosis.

lae, respectively, give it a saddle shape. The bulbous presphenoid body below the fossa is mostly a sinus (a hollow cavity). Two pairs of wings project laterally from the sphenoid body. The greater wings (derived from the alisphenoid cartilages) support much of the temporal lobes; the margins of the lesser wings (derived from the orbitophenoid cartilages) form the boundary between the MCF and the ACF. Several important structures lie on the inferior surface of the MCF, including the glenoid fossa (often called the mandibular fossa), where the mandible articulates with the cranium (part of the temporal bone).

Anterior Cranial Fossa (ACF)

The ACF, which underlies the frontal lobes and olfactory bulbs, includes parts of the frontal, sphenoid, and ethmoid bones (Figure 4.1). The posterior part of the ACF is the lesser wing of the sphenoid, and the anterior part in the midline is the cribriform plate of the ethmoid. Variable amounts of the frontal bone (a mostly intramembranous, neurocranial part of the skull) form the rest of the floor of the anterior cranial base. Unlike nonhuman primates, humans have no intervening portion of the frontal bone between the cribriform plate and the lesser wings of the sphenoid (Aiello and Dean, 1990; McCarthy, 2001). An ossified portion of the ethmoid plate, the crista galli ("cock's comb"), rises in the midline of the cribriform plate, posterior to the foramen caecum (the "blind hole"). As Chapter 3 described, the foramen caecum is the remnant of the end of the neural tube and thus marks the anteriormost point on the midline cranial base.

Cranial Base Angle (CBA)

In primitive vertebrates, the cranial base is a somewhat flat, oval-shaped plate (albeit highly perforated) at the base of the skull (de Beer, 1937), but in higher vertebrates, especially humans, the PCF, MCF, and ACF lie at angles relative to each other, with the synchondroses acting like hinges between each fossa in the midline (Figure 4.1c). There is no single way to quantify these angles (for a review, see Lieberman and McCarthy, 1999). The most common measure, CBA1, quantifies the angle between two midsagittal lines (Figure 4.2): the first is between the anterior margin of the foramen magnum (basion) to the center of the hypophyseal fossa (sella); the second is from sella to the foramen caecum. CBA1 thus measures the angle between the prechordal and parachordal portions of the cranial base.[1] As I discuss below, a key difference between humans and nonhuman primates is that in humans CBA1 rapidly flexes after birth in a neural growth trajectory, whereas in nonhuman primates this angle slowly extends in a skeletal growth trajectory. Adult humans thus have a much more flexed cranial base than other mammals, including apes.

Foramina

The cranial base has almost a dozen holes (labeled in Figure 4.1a) through which nerves and blood vessels traverse between the brain and either the face or the rest of the body. These foramina all lie near the midline center of the cranial base in an oval-shaped region close to the brain stem, protecting these vital connections from injury. This region grows rapidly, reaching adult size soon after birth.

---

1. The notochord ends just below the hypophyseal fossa; hence portions of the cranial base anterior or posterior to this spot are prechordal or parachordal (sometimes called postchordal), respectively.

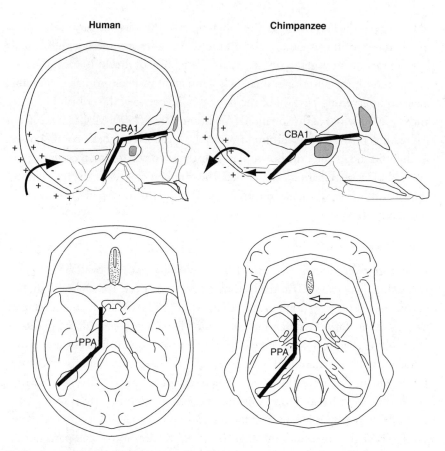

Figure 4.2. Cranial base angles and occipital growth patterns in a modern human (left) and a chimpanzee (right). The angle of the cranial base (CBA1) is much more flexed in humans than chimpanzees, and the angle of the petrous pyramids (PPA) relative to the midsagittal plane is more coronal (perpendicular to sagittal) in humans than in chimpanzees. In addition, the nuchal plane rotates horizontally in humans, with little posterior drift of the foramen magnum; in chimpanzees, the nuchal plane rotates vertically, and there is substantial posterior drift of the foramen magnum (bold arrows indicate rotations). Note also the intervening portion of the frontal bone in the midline (hollow arrow) in the chimpanzee but not the human cranial base.

Patterns of Growth and Development in the Basicranium

In general, the basicranium grows and changes shape in several major ways: the three fossae elongate anteroposteriorly, widen mediolaterally, and grow deeper (inferiorly). In addition, the cranial base angle flexes or extends in the midsagittal plane, and the combination of mediolateral and anteroposterior growth rotates the petrous temporals around a vertical axis. Figure 4.3 com-

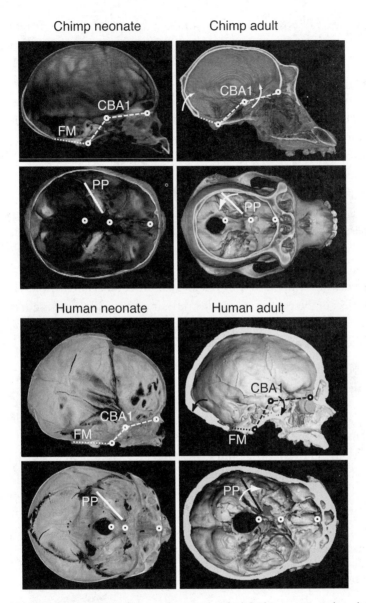

Figure 4.3. Midsagittal and dorsal views of neonatal and adult chimpanzees and modern humans, highlighting differences in cranial base growth. Circles show position of basion, sella, and foramen caecum; also shown are the cranial base angle (CBA1), the foramen magnum (FM), and the petrous portion of the temporal (PP). In chimpanzees, the cranial base extends, the petrous rotates toward the sagittal plane, and the nuchal plane and foramen magnum rotate vertically; but in modern humans the cranial base flexes, the petrous rotates coronally, and the nuchal plane rotates slightly horizontally.

pares the overall postnatal patterns of cranial base shape changes in humans and chimpanzees. In chimps, the cranial base as a whole becomes flatter and more elongated (mediolaterally narrow and anteroposteriorly long). These changes occur primarily because the cranial base extends and the posterior (parachordal) portion of the cranial base becomes as long as, if not longer than, the prechordal cranial base. Additionally, in nonhuman primates, the foramen magnum migrates posteriorly in the cranium, and the petrous portions of the temporal bones rotate from a more coronal (mediolateral) to a more sagittal (anteroposterior) orientation.

The pattern of basicranial growth in humans is considerably different, leading to a very differently shaped cranial base. Most critically, the human cranial base flexes rather than extends after birth, and the anterior (prechordal) portion of the cranial base becomes relatively longer rather than shorter (Figure 4.4). In addition, the human foramen magnum remains near the center of the cranium, and the petrous temporals swing out into a more coronal orientation. The combined effect of these differences is that the human cranial base does not elongate and narrow like an ape's but remains relatively short and wide, with considerable differences in the relative angle of the three middle cranial fossae.

Processes of Growth and Development in the Basicranium

How do the mechanisms of basicranial growth and development differ between humans and nonhuman primates? Further, how does the cranial base grow in different ways while simultaneously performing several key roles? As noted above, the basicranium must (1) articulate the head with the neck securely, and with some degree of movement; (2) provide a commodious platform for the brain and all its vital connections; (3) provide a platform from which the face grows; (4) articulate the jaw with the cranium; and (5) accommodate the pharynx, which runs through the face and neck.

To evaluate how the basicranium grows differently in humans and other primates while permitting these many challenging functions, I consider first the basic mechanisms of cranial base growth and their variations. Later, mostly in Chapter 5, I explore how these processes help integrate the head into a functional whole.

Anteroposterior Growth

Three processes elongate the cranial base. First, most midline growth occurs in the three major synchondroses (see Figures 3.4 and 4.1). The posteriormost is the spheno-occipital synchondrosis (SOS), which separates the basioccipi-

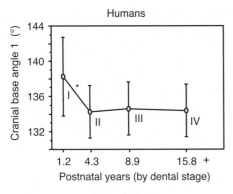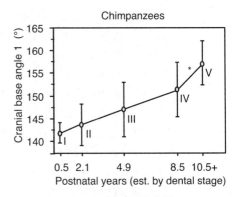

Figure 4.4. Changes in angle of the cranial base (CBA1) over time in modern humans (left) and chimpanzees (right). The modern human cranial base rapidly flexes after birth in a neural growth trajectory, but the chimpanzee cranial base slowly extends in a skeletal growth trajectory. The development of each species is divided into five dental stages (I, prior to the eruption of the deciduous canines; II, after the eruption of the deciduous canines and before the eruption of the first permanent molars (M1); III, after M1 eruption and before second molar (M2) eruption; IV, after M2 eruption and before third molar (M3) eruption; V, after M3 eruption). Asterisks indicate significant differences between stages. Modified from Lieberman and McCarthy (1999).

tal and sphenoid within the clivus below the brain stem. In the basicranium's center is the mid-sphenoid synchondrosis (MSS), which separates the pre-chordal and postchordal (parachordal) portions of the sphenoid body. In front is the spheno-ethmoid synchondrosis (SES), which separates the sphenoid and the ethmoid. Within the synchondroses, endochondral ossification (described in Chapter 2) elongates the basioccipital, the sphenoid body, and the ethmoid in the midline of the cranial base. Synchondroseal growth also causes antero-posterior displacement of structures relative to each other.

Not all cranial base elongation occurs at the synchondroses. The dominant mode of basicranial elongation lateral to the midline occurs from deposition within sutures that have some coronal orientation (e.g., the frontosphenoid, the sphenotemporal, and the occipitomastoid). In addition, the posterior cranial base (the basioccipital) also elongates slightly from drift of the foramen magnum (Figure 4.2) as bone is added along the anterior margin of the foramen, and resorbed along the posterior margin.

Mediolateral Growth

The cranial base widens as it elongates in two ways. First, each cranial fossa widens at its lateral margins from drift in which the external and internal sur-

faces of the flat tables of bone (called squamae) are depository and resorptive, respectively (see Figure 4.5). Second, the three fossae also widen from
intramembranous bone growth in sutures that have some component of anteroposterior orientation (e.g., the fronto-ethmoid in the ACF, the sphenotemporal in the MCF, and the occipito-mastoid in the PCF). The sphenoid
body, which lies in the center of everything, grows relatively little after birth
(Kodama 1976a,b,c). A good overall measure of cranial base width is the distance between the two external acoustic meatuses (biporionic width).

Superoinferior Growth

The three endocranial fossae also become deeper, mostly from drift, because
most of the endocranial floor is a resorptive growth field, and the inferior surface of the cranial base is a depository growth field, as shown in Figures 4.1
and 4.5 (Duterloo and Enlow, 1970; Enlow, 1990). However, the boundaries
between the fossae—the petrous temporal, which separates the PCF and
MCF; and the anterior clinoid processes of the sphenoid's lesser wing, which
separate the MCF and ACF—remain depository surfaces and do not drift inferiorly. In addition, most inferior drift in the cranial base occurs in the PCF.
Drift is probably minimal in the ACF because the floor of the anterior cranial
base is the roof of the orbits. An exception within the ACF is the cribriform
plate, which sinks slightly in humans (Moss, 1963) and more extensively in
many nonhuman primates, forming a "deep olfactory pit" at the front of the
cranial base (Cameron, 1930; McCarthy, 2001). Inferior drift of the MCF occurs primarily in the greater wings of the sphenoid beneath the temporal lobes.

Angulation in the Sagittal Plane

One of the most important categories of cranial growth includes changes in
the cranial base angle (CBA), the orientation of the anterior (prechordal) cranial base relative to the posterior (parachordal) cranial base in the midsagittal
plane (see Fig. 4.2). Technically, flexion occurs when the CBA becomes more
acute, and extension occurs when the CBA becomes more obtuse. Measures
of CBA, however, describe the combined effects of changes in all three synchondroses, so they should never be considered a simple, single character.

   Measures of the orientation of the anterior and posterior portions of the basicranium such as CBA1 (defined above) change dynamically during fetal and
postnatal growth. The pattern of change is different in humans than in other
primates. The cranial base in humans and nonhuman primates first flexes rapidly during the embryonic period and then extends (retroflexes) during much

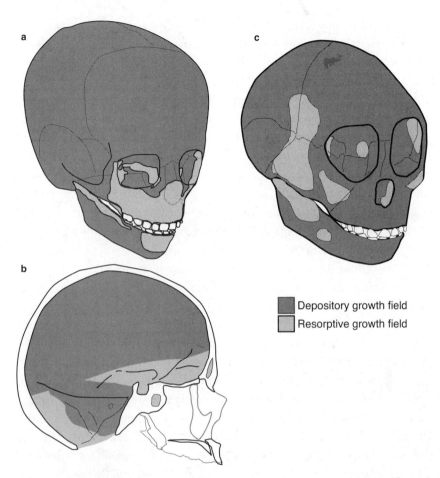

Figure 4.5. Map of the distribution of depository and resorptive growth fields in modern human (a, b) and nonhuman primate (c) skulls (not to scale). A major difference between the two species is that the human midface and incisor region is resorptive. Note also the reversal line in the upper versus lower parts of the endocranium, which are resorptive. Figure adapted from Dutterloo and Enlow, 1970.

of fetal growth by approximately 9–15° (Jeffery and Spoor, 2002, 2004a; Jeffery, 2003; Morimoto et al., 2008). As Figure 4.4 shows, the postnatal human cranial base then flexes again rapidly in the first two years by approximately 10° (thus in a rapid neural growth trajectory), after which it remains stable; in contrast, the nonhuman primate cranial base slowly extends postnatally (thus in a skeletal growth trajectory). In *Pan troglodytes,* the cranial base extends approximately 15° between birth and age 10 (Lieberman and McCarthy, 1999).

In spite of its importance, much about cranial base angulation is incompletely understood, including the mechanisms of growth and the stimuli that influence flexion and extension. It is difficult to be definitive, but some degree of cranial base angulation probably occurs interstitially within synchondroses through a hingelike action characterized by differential cartilage growth. Flexion occurs from more rapid chondrogenesis in the upper than the lower aspect of the synchondrosis, and extension results from the opposite pattern. Flexion and extension probably also occur from rotations caused by drift (see Chapter 2, Figure 2.4). Growth studies in macaques show that depository and resorptive growth fields differ in the bones on either side of a synchondrosis, causing rotations around an axis through the synchondrosis (Michejda, 1971, 1972a,b; Michejda and Lamey, 1971; Giles et al., 1981).

In Chapter 5, I consider how extensively cranial base angulation may be affected by intrinsically regulated growth in different synchondroses, as opposed to epigenetic responses to the growth of the brain above and the face below. Regardless of the relative importance of these stimuli, the synchondroses participate differently in cranial base growth prenatally and postnatally, with important differences between humans and nonhuman primates (Hofer, 1960; Hofer and Spatz, 1963; Sirianni and Newell-Morris, 1980; Diewert, 1985; Aganastopolou et al., 1988; van den Eynde et al., 1992). The SOS, which remains unfused until the end of facial growth, is probably the most active synchondrosis, flexing in humans and extending in nonhuman primates (Björk, 1955; Scott, 1958; Melsen, 1969; Sirianni and Van Ness, 1978). The SOS also contributes to relatively less elongation in the human cranial base, keeping the basisphenoid relatively short (Dean and Wood, 1984). The MSS fuses prenatally in humans (Ford, 1958) but remains open for longer in nonhuman primates and may be a site of postnatal extension (Scott, 1958; Hofer and Spatz, 1963; Michejda, 1971, 1972a, 1972b; but see Lager, 1958; Melsen 1971; Giles et al., 1981). Finally, the SES remains patent in humans during the neural growth period, probably as a site of cranial base elongation, but fuses near birth in most nonhuman primates (Scott, 1958; Michejda and Lamey, 1971).

Other ontogenetic changes affect the cranial base angle but are not actually processes of angulation. These include posterior drift of the foramen magnum (see above), inferior drift of the cribriform plate relative to the anterior cranial base (Moss, 1963), and remodeling of the sella turcica, which causes its center (sella) to move slightly posteriorly (Baume, 1957; Shapiro, 1960; Latham, 1972).

Petrous Rotation

The width of the cranial base has an important influence on craniofacial shape and vice versa. One way the cranial base widens is through the rotation of the petrous portions of the temporal bones (often called the petrous pyramids) along the boundary between the MCF and PCF. Petrous rotation occurs almost entirely on the endocranial surface. In humans, the petrous pyramids begin in a somewhat sagittal orientation but swing out during prenatal ontogeny into a more coronal orientation, approximately 135–140° relative to the midline (Jeffery and Spoor, 2002). Postnatally, the petrous pyramids rotate coronally by another 10–15° to an average of 125° (Spoor, 1997) (see Figs 4.2 and 4.3). In contrast, in *P. troglodytes* the petrous temporals rotate sagittally during ontogeny from an orientation of about 125° to 135° relative to the midsagittal (see Figure 4.3). A likely proximate mechanism for coronal rotation in humans is deposition within the obliquely angled occipitomastoid suture or the spheno-petrosal synchondrosis. What drives the rotation is less clear, but a good hypothesis is that the rotation is a mechanism to accommodate increases in brain volume within the PCF when there is limited or no elongation of the SOS (Dean and Wood, 1984; Dean, 1988). Thus in nonhuman primates, in which the basioccipital lengthens substantially during postnatal ontogeny and endocranial volume (ECV) remains relatively small, the petrous pyramids become more sagittally oriented; but in bipedal hominins, whose basioccipital remains shorter in order to position the foramen magnum more anteriorly (see Chapters 9 and 11), the petrous pyramids may rotate coronally to accommodate increases in ECV by widening the PCF. Spoor (1997) found that, across a broad sample of primates, the petrous pyramids are more coronally oriented in primates with larger brains relative to cranial base length. However, Jeffery and Spoor (2002) found that petrous pyramid orientation is independent of hindbrain volume in fetal humans, at the stage when brain growth is most rapid. More research is needed to determine what drives petrous rotation.

Position and Orientation of the Foramen Magnum

A last significant aspect of cranial base growth and development is how the foramen magnum fits within the rest of the basicranium. The foramen magnum differs markedly in position and orientation between humans and apes. In chimpanzees, the foramen magnum lies near the end of the skull base and is oriented somewhat backward; in humans, it is positioned closer to the center of the skull base and oriented downward and slightly forward (Figure 4.2). These contrasts, regardless of their functional consequences for posture and locomotion (discussed in Chapter 9), are partly attributable to several different patterns of neurocranial and facial growth. More of the PCF expands behind the foramen magnum in humans because of a relatively larger brain combined with a different pattern of nuchal plane rotation. In addition, more of the skull is positioned in front of the foramen magnum in chimpanzees because of more facial projection and elongation. However, the foramen magnum is also positioned slightly differently *within* the cranial base in humans compared with chimps and other nonhuman primates. Often this difference is illustrated by comparing the position of the anteriormost point on the foramen magnum, basion, relative to a line drawn between the left and right carotid foramina (through which the internal carotid arteries enter the cranium) or the top of the left and right external auditory meatuses in each temporal bone. There is much variation, but basion usually falls approximately along the bicarotid or biporionic lines in humans and is typically about a centimeter behind either line in chimpanzees (Dean and Wood, 1982a; Ahern, 2005). In addition, the angle between the inferior surface of the basioccipital and the plane of the foramen magnum is close to horizontal (about 160°) in humans but is more flexed (about 130°) in chimpanzees.

Several factors account for these differences, helping to position the foramen magnum more anteriorly and to orient it more downward in humans. First, more growth within the SOS and more posterior drift of the foramen magnum make the basioccipital relatively longer in nonhuman primates, thus displacing the foramen magnum posteriorly relative to other bones such as the temporals (Michejda, 1971; Giles et al., 1981). In addition, nonhuman primates such as chimpanzees have a more extended (obtuse) CBA, which positions the foramen magnum posteriorly relative to the anterior cranial fossa and face and also rotates it acutely relative to the rest of the cranial base (Lieberman and McCarthy, 1999).

Regional Growth of the Cranial Vault

The cranial vault, technically the neurocranium, is often considered in conjunction with the basicranium, because together they encapsulate the brain. However, as Chapter 3 describes, the neurocranium has a different embryological origin than the basicranium, and it ossifies intramembranously rather than endochondrally. In addition, neurocranial development and growth interacts more intensely with the brain because the vault bones form between the pericranial and endocranial membranes outside the dura mater membrane surrounding the brain (see Chapter 3). Brain expansion generates tension in the dura mater, which then stimulates bone deposition within various sutures of the vault.

How the neurocranium grows is of interest because of the many differences in cranial vault shape between modern humans and other primates, including apes and other hominins. Many key differences relate to brain size. The average neonatal brain volume is 330 cm³ in humans but 130–150 cm³ in chimpanzees (Zuckerman, 1928; Schultz, 1969; DeSilva and Lesnik, 2006). Postnatally, the human brain also grows faster and for a longer period than those of nonhuman primates. In modern humans, brain volume and cranial capacity are 96 percent complete between ages 6 and 7, between 1150 cm³ and 1450 cm³ (Tobias, 1971; Lavelle et al., 1977).[2] In contrast, the chimpanzee brain reaches adult volume (350–400 cm³) at approximately 3 years, and macaques reach adult brain volume (80–110 cm³) at about 1.6 years (Sirianni and Swindler, 1985).

Brain size has several effects on neurocranial and overall cranial shape. In hominin evolution, brain expansion has mostly been made possible by a wider and more flexed cranial base and a rounder, more spherical cranial vault. But brain size alone is not sufficient to account for all the unique derived features of the modern human braincase. In Neanderthals, the brain is as large as or even larger than those of modern humans, but the cranial vault is more elongate and less globular. These differences occur because many aspects of cranial vault shape arise from interactions among the bones of the cranial vault with the brain, the basicranium, and parts of the face.

Anatomy

Neurocranial anatomy involves only five bones (Figure 4.6). Three of them—the parietals, most of the frontal, and most of the occipital—make up the majority of the cranial vault. The temporal and sphenoid bones also contribute flat (squamous) intramembranous elements along the side. Each bone corresponds roughly to the similarly named lobes of the brain it partially

---

2. Not all this increase in capacity is brain tissue, which occupies roughly 94 percent of the cranial volume at birth but only 80 percent when it is fully formed because of the steady increase in size and volume of the membranes, vessels, nerves, and cerebrospinal fluid that supply and protect the brain.

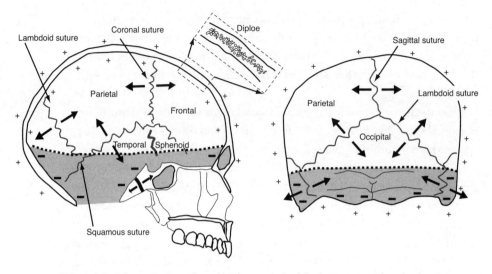

Figure 4.6. Schematic views (lateral and posterior) of the human cranial vault, showing major bones, sutures, depository (+) and resorptive (−) fields, and vectors of growth. Modified from Lieberman et al., 2000a.

covers. The frontal bone covers most of the frontal lobes, the parietal bones roughly cover the parietal lobes and some of the frontal lobes, the temporal and sphenoid squamae are lateral to the temporal lobes, and the occipital bone surrounds the occipital lobes along with the cerebellum. As noted above, these bones (except the parietals) also contribute to the three endocranial fossae.

Given their intramembranous origin between the endocranial and pericranial membranes, the neurocranial bones all have inner and outer tables of lamellar bone, often with an intermediate layer of trabecular bone, the diploe (Greek for "double" or "folded"). Most neurocranial growth occurs in a few sutures, illustrated in Figure 4.6. The sagittal suture separates the two parietal bones, the coronal suture separates the frontal and parietal bones, the lambdoid suture separates the occipital and parietal bones, and the squamous sutures separate the temporal and parietal bones. Other, smaller sutures also exist (e.g., occipitomastoid, sphenoparietal, sphenofrontal, sphenotemporal, parietomastoid); one suture, the metopic, which separates the left and right sides of the frontal, fuses by age 3. Sometimes little extra (Wormian) bones form within the lambdoid suture (see Chapter 6).

Patterns of Growth and Development

The neurocranium grows and changes shape in four major ways: it elongates anteroposteriorly, it widens mediolaterally, it grows taller (superiorly), and it becomes thicker. Together, the first three of these vectors of growth generate variation in the overall sphericity of the neurocranium. In addition, although neurocranial growth occurs from brain expansion, the vault bones (except the parietals) also participate in the growth of the face and the cranial base. Such connections integrate the neurocranium with the face and basicranium in ways that affect the cranial base angle, the position of the foramen magnum, and upper facial shape (notably the brow ridges).

Figure 4.7 compares the overall postnatal patterns of neurocranial shape changes in humans and common chimpanzees. Both species are born with relatively spherical vaults, although the human's is already a little more globular. Leaving aside size contrasts, the most obvious difference between the two species is that as the chimp neurocranium grows, it becomes relatively long and narrow, whereas the human neurocranium remains relatively wide, tall, and spherical. Note also that, in humans, the PCF expands considerably behind the foramen magnum and the nuchal plane rotates horizontally, whereas in chimps the nuchal plane rotates vertically. Consequently, the foramen magnum remains close to the center of the skull base in humans but migrates toward the posterior margin in chimps (see Figure 4.2). Finally, the front of the neurocranium in humans remains positioned above the orbits; but in African apes, the face projects forward substantially in front of the neurocranium, bringing with it the brow ridges above the orbits. As the chimp face moves forward relative to the braincase, it leaves behind a narrowed region, a postorbital constriction, between the upper face and the ACF. The combined effect of these differences is that the chimp neurocranium grows (like the cranial base) to be relatively longer, narrower, and flatter than in humans, whereas the much larger human neurocranium (about three times the volume) remains more spherical than in chimps.

Processes of Growth and Development

Three mechanisms account for most of the variation in neurocranial growth: (1) growth within sutures; (2) rotations and displacements that occur from drift; and (3) thickening of the vault bones. Aspects of neurocranial growth related to the brow ridges is considered below in the context of facial growth.

Chimp neonate   **Chimp adult**

Human neonate   **Human adult**

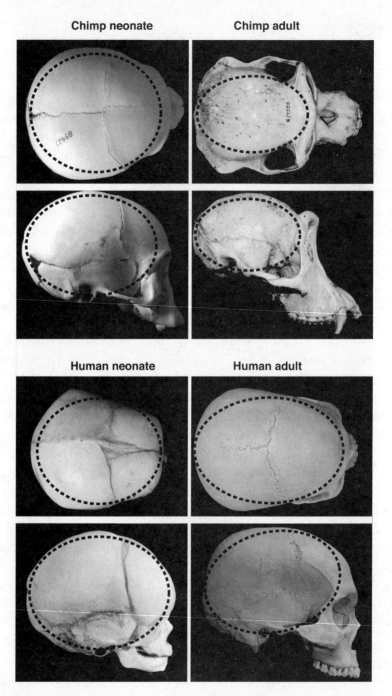

Figure 4.7. Dorsal and lateral views of neonatal and adult chimpanzees and modern humans, highlighting differences in vault growth. In chimpanzees, the vault becomes much more elongated and smaller relative to face size, but in humans the vault remains more spherical and larger relative to face size.

Sutural Growth

Recall from Chapter 3 that the cranial vault bones begin as rapidly deposited woven bone in ossification centers between two membranes. Bone secreted by the outer pericranium membrane becomes the outer table of the vault; bone secreted by the inner endocranium membrane becomes the inner table. Later a layer of spongy bone, the diploe, develops between these two layers (see below).

Once the tables of bone begin to form, almost all changes in vault size and shape occur either in sutures or in their precursors, flexible membranes called fontanelles. Fontanelles permit rapid stretching and deformation as the brain expands more rapidly than bone can grow. Sutures and fontanelles are both derivatives of mesenchymal cells that interact with the brain via the meningeal membranes, especially the dura mater. To what extent intrasutural growth is caused by intrinsic as opposed to extrinsic stimuli is not entirely known, but most evidence suggests that the dominant stimulus for sutural growth is intracranial pressure (ICP) generated by increases in brain size or fluids within the cranial cavity (Moss, 1960; Cohen, 2000a). Higher ICPs stretch the dura mater, which then expresses signaling factors (e.g., Fgf2) that either activate or inhibit osteoblasts in the sutures (see Opperman, 2000; Wilkie and Morriss-Kay, 2001). These pathways may differ slightly in sutures with disparate embryonic origins (see Chapter 3).

Rates of bone deposition in sutures and fontanelles are mostly stimulated by the effects of ICP on the dura mater and perhaps by the sutures and fontanelles directly, but the direction of neurocranial growth is constrained by suture and fontanelle orientation. Anteroposterior growth occurs in sutures with coronal orientations, especially the coronal suture, the occipitomastoid suture, the sphenotemporal suture, and the lambdoid suture. Mediolateral growth occurs directly in the sagittal and metopic sutures. The sagittal suture fuses after brain growth is complete; the metopic suture fuses within a year of birth in most primates, including humans (Schultz, 1940; Krogman, 1969; Shapiro and Robinson, 1980). Mediolateral growth also occurs in other sutures with sagittal components to their orientation, such as the occipitomastoid and sphenotemporal sutures. Superoinferior growth occurs in sutures with some transverse component, especially the squamous and lambdoid sutures.

Fontanelles are also major growth sites during fetal and early postnatal growth. The paired sphenoid and mastoid fontanelles permit superoinferior growth; and the anterior and posterior fontanelles participate in anteroposterior and mediolateral growth. Osteogenic activity in sutures usually ceases long before they completely fuse through remodeling. The pattern and timing of suture fusion is highly variable and differs among species, with chimpanzees and humans being more similar to each other than to gorillas (Cray et al., 2008). The endocranial sutures fuse between 1 and 15 years in macaques and between 12 and 55 years in humans (Falk et al., 1989).

The complex and varied epigenetic interactions between brain and sutural growth are illustrated by syndromes that involve premature fusion of cranial vault sutures, termed craniosynostoses (Figure 4.8). Craniosynostoses have different causes (e.g., mutations in signaling factors or signaling receptors), and varied effects (Richtsmeier, 2002). However, some predictable trends are evident (see Cohen, 2000a). Premature fusion of the sagittal suture, for example, restricts mediolateral growth of the vault and often leads to compensatory elongation of the skull (scaphocephaly) from growth in the coronal and lambdoid sutures. Premature fusion of the coronal or lambdoid sutures restricts anteroposterior growth of the vault, leading to compensatory lateral growth (usually nonsymmetric) of the vault and face (plagiocephaly). Artificial head binding provides additional evidence for the interactions between brain and neurocranial growth. In some cultures, adults restrict anteroposterior growth of an infant's vault by strapping pads or boards to the front and back of the head, or they restrict mediolateral growth by wrapping boards on the side of the head (see Fig. 4.8d). These practices typically cause some degree of compensatory growth in sutures whose activity is unrestricted (Antón et al., 1992; Cheverud et al., 1992).

If brain size and shape have such a strong influence on cranial vault growth, what determines brain shape? This is a difficult question to answer, as it is not technically or ethically possible to grow a brain in vitro or to perturb experimentally other parts of the head that influence brain shape. Theoretically, there are four major factors: (1) the intrinsic shape of the brain itself (e.g., the relative size of the various lobes); (2) the shape of the cranial base on which the brain sits (e.g., the shape of the endocranial fossae and the angle of the cranial base); (3) the shape of the face (e.g., facial width, projection); and (4) the dural bands. Chapters 5 and 6 consider all these factors more explicitly, but the dural bands merit consideration here because of their intimate connection with several key sutures. The dura mater has several folds (septa), the dural bands, which are thick bands of connective tissue that subdivide the cranial cavity and connect several fontanelles and sutures in the vault with structures in the cranial base (Figure 3.5b). Two bands are especially important. First, the tentorium cerebelli forms a semihorizontal shelf that stretches like the roof of a tent from the lesser wings of the sphenoid and the petrous pyramid mar-

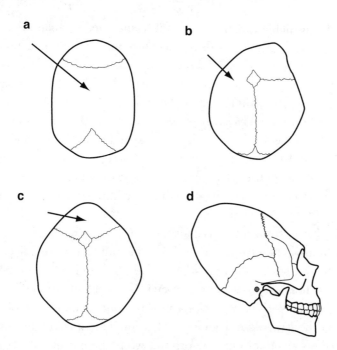

Figure 4.8. Examples of craniosynostoses and head binding. (a) Scaphocephaly (premature fusion of the sagittal suture); (b) plagiocephaly (premature fusion of the coronal suture); (c) trigonocephaly (premature fusion of the metopic suture); (d) example of head binding in which vertical and lateral growth was restricted, elongating the vault posteriorly. (Collections of Peabody Museum, Harvard University.)

gins all the way to the back of the occipital. Second, the falx cerebri connects the midline of the tentorium cerebelli above the PCF to the sagittal and metopic sutures and to the ethmoid's crista galli. The dural bands are thought to partially constrain and hence to influence the shape of the expanding brain and the cranial vault (Friede, 1981). During fetal growth in humans, the tentorium cerebelli rotates inferiorly as much as 60° as the neocortex expands, rotating the whole PCF with it (Moss et al., 1956). It has been hypothesized that a brain would grow into a perfect hemisphere without the dural bands (Moss, 1975).

Drift

As noted previously, bone deposition or resorption occurs in the two periosteal layers that cover the intramembranous bones of the vault, an inner endocranial membrane and an outer ectocranial membrane. The pattern of depository and resorptive growth fields is specific within the vault, possibly because of differential patterning of the mesenchymal precursors of the periosteal membranes above and below the tentorium cerebelli. In the upper half of the vault, above the tentorium cerebelli, both the internal and external surfaces of the vault bones are depository growth fields (Duterloo and Enlow, 1970). Thus the upper half of the vault does not expand through drift and can only get thicker (Ohtsuki, 1977). In contrast, the lower portion of the vault, below the tentorium cerebelli, grows mostly through drift because the inner (endocranial) surface is a resorptive growth field and the outer (ectocranial) surface is depository (Enlow, 1990).

Different patterns of drift account for one key difference in vault growth between human and nonhuman primates: the rotation of the nuchal plane. The external surface of the nuchal plane is the attachment site of most neck muscles; the internal surface of the plane is below the insertion of the tentorium cerebelli on the internal occipital protruberance and thus part of the PCF that houses the cerebellum (Figure 4.1). The internal surface of the human occipital below the tentorium is resorptive, while the outside is a depository growth field (see Figure 4.5), so that it drifts outward, rotating horizontally (Enlow, 1968, 1990; Srivastava, 1992). In nonhuman primates, these growth fields are reversed so that most of the outside surface of the nuchal plane is resorptive and the inside is depository, causing the occipital to rotate vertically (Duterloo and Enlow, 1970). This critical difference also affects the position of the posterior margin of the foramen magnum, which is closer to the center of the skull in humans than in nonhuman primates (Figures 4.2 and 4.3). This growth field reversal may have played a role in the origins of bipedalism in the first hominins (see Chapter 11).

Thickening

Vault bones usually have three major layers. In the upper vault, where both the inner and outer surfaces are depository, the endocranial and ectocranial aspects of the bone consist of well-organized lamellar bone, known respectively as the inner and outer tables. As the vault becomes vascularized, osteoclasts invade the central region between the tables, creating the diploe, as shown in Figure 4.6. The diploe also contains a rich network of veins (see Chapter 6). The bones of the vault gradually thicken in a skeletal rather than neural growth trajectory, reaching adult thickness during late adolescence (A. F. Roche, 1953;

Brown et al., 1979). Thickening occurs in the upper vault from bone deposition in both the inner and outer tables, and in the lower vault from differentially more outer table deposition than inner table resorption (Moss and Young, 1960). As Chapter 6 discusses, cranial vault thickness varies considerably in different parts of the vault, among different individuals, and between species (Lieberman, 1996).

Superstructures

A few final aspects of neurocranial growth to consider are superstructures that serve as muscle attachment sites. Several muscles insert on the external surface of the vault, notably the temporalis muscle on the side and a series of paired neck (nuchal) muscles on the lower part of the occiput (the nuchal plane) and the mastoid process. (The temporalis is discussed in more detail in Chapter 7 and nuchal muscles in Chapter 9.) In many mammals, including humans and chimps, the edge of the origins of these muscles can leave a distinct, raised line. For example, the upper edge of the temporalis creates the temporal line, a raised semicircle that starts behind the lateral margin of the orbit, curves up toward the sagittal suture, and then curves away toward the mastoid region behind the earhole. In orangutans, gorillas, and some fossil hominins (see Chapter 11), the temporalis is larger than the surface area available on the vault, and toward the end of the growth period the muscles generate raised crests for extra attachment sites. Presumably these crests grow because the deep layers of the muscle insert into the periosteal membrane (Sakka, 1984). Thus as the muscles expand, they may pull the membrane with them, perhaps inducing bone formation in the crest.[3] Nuchal crests develop in a similar manner for the superficial nuchal muscles (mostly the trapezius) in some hominins (Dean, 1985a). Sometimes the nuchal crest merges with the crest formed by the posterior temporalis, forming a compound temporonuchal crest.

The mastoid process is a slightly different superstructure. This cone-shaped region of bone projects inferiorly from the back of the petrous part of the temporal, behind the external auditory meatus. Its roughened surface area is the attachment site for three neck muscles: the splenius capitis, the longissimus capitis, and the sternocleidomastoid. The inside of the mastoid becomes filled

---

3. It is possible that growth in the crest itself plays some role; this hypothesis is supported by the observation that the fibrous insertion of the temporalis muscle does not always extend all the way to the top of the crest in gorillas (Sakka, 1984).

with air cells. The mastoid is a highly variable structure, much larger in humans than great apes, and usually much larger in males than females following the adolescent growth spurt.

Regional Growth in the Face

The face is highly derived in humans, and it is also (not coincidentally) the most complex part of the head. The many components of the face have to accommodate growth and function in an impressive set of organs and spaces, including the eyes, nose, mouth, pharynx, muscles of mastication, and teeth. As traditionally defined, the facial skeleton (the viscerocranium) includes more than a dozen intramembranous bones that grow downward and forward from the cranial base around these organs and spaces. Two parts of the basicranium also participate extensively in facial growth: the ethmoid, which contributes to the nasal cavity; and the temporal portion of the temporomandibular joint (TMJ), which has a major role in mandibular growth.

General Structure of the Face

Because the face has so many components, I begin with an overview of its general design and some basic principles that help to elucidate how its components grow and fit together. Recall from Chapter 2 that much of the head's structure, especially in the face, arises from the way it grows in concert with numerous organs and spaces, sometimes called *functional matrices,* each surrounded by skeletal capsules, called *functional cranial components* (Moss, 1960, 1968, 1997a–d; Moss and Salentjin, 1969). So far, this chapter has focused on the neurocranial and basicranial components of the skeletal capsule that surrounds the largest functional matrix of the head, the brain. The face, in contrast, is a jumble of functional matrices with at least half a dozen capsules that surround a diverse set of organs and spaces, including the eyes, nose, oral cavity, pharynx, and the teeth. Thus rather than review the face one bone at a time, it makes sense to review facial anatomy one matrix at a time. But as we proceed, it is critical to keep in mind that *few, if any, of the face's skeletal capsules are independent.* The floor of the anterior cranial fossa is the roof of the orbits; the lateral walls of the nasal cavity are the part of the medial wall of the orbits; the roof of the oral cavity is the floor of the nasal cavity; and so on. For this reason, it is useful to divide the face into three general anatomical regions based loosely on the functional units that the facial skeleton encapsulates: the upper face is the region around the eyes and below the frontal lobes of the brain, the middle face mostly surrounds the nasal cavity, and the lower face surrounds the oral cavity. Again, these regions and their boundaries are not discrete.

A second point to consider is the critical role of epigenetic interactions in helping to integrate the face as it grows and develops. Because the major capsular divisions of the face share the same intramembranous walls (for instance, the floor of the nasal cavity is the roof of the oral cavity), facial growth inevitably requires many interactions among neighboring units to accommodate their simultaneous growth. Integration occurs throughout ontogeny through a variety of processes, including the effects of mechanical forces generated by organ growth and activities such as chewing and respiration (discussed in Chapters 7 and 8). Facial growth has to accommodate not only different components of the face but also the cranial base and neurocranium. The face grows downward and forward from the cranial base and thus is constrained by many aspects of middle and anterior cranial fossa shape and position (a subject Chapter 5 discusses). In addition, the mandible grows forward from and articulates with the cranial base, requiring many aspects of skull growth to be coordinated and integrated to prevent malocclusion between the lower and upper jaws.

Finally, the face grows at a different pace from other parts of the head. Unlike the cranial vault and base, the face has a skeletal growth trajectory that continues until the end of the adolescent growth spurt. In humans, the basicranium and neurocranium reach adult size between 6 and 7 postnatal years, at least 10 years before the face. Thus, integration between the face and the rest of the skull is probably unequal. Although facial growth certainly influences neurocranial and basicranial growth, the reverse influence is probably greater because the face grows downward and forward from the cranial base and around the eyes. In other words, the brain, the eyes, and the cranial base provide a sort of template from which much of the facial skeleton grows.

Patterns of Facial Growth and Development

The mammalian face grows downward from the anterior cranial base and fossa and forward from the MCF. Most mammals thus have faces that are approximately as wide as the neurocranium and basicranium and grow predominantly forward (anteriorly or rostrally) and downward (inferiorly or ventrally) (see Hallgrímsson et al., 2007a). Figure 4.9 summarizes some basic differences

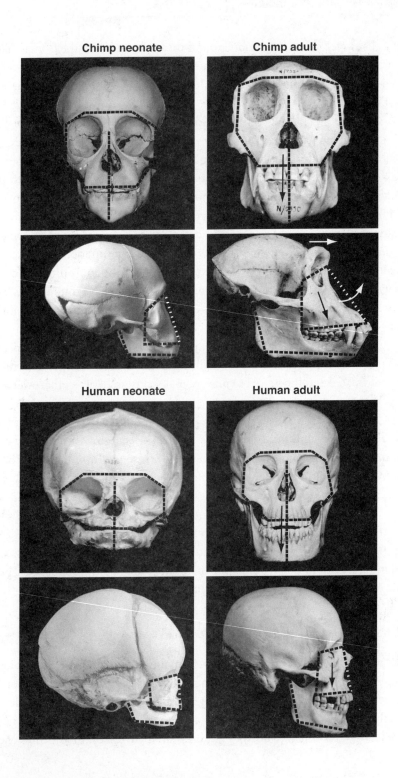

Chimp neonate

Chimp adult

Human neonate

Human adult

Figure 4.9. Anterior and lateral views of neonatal and adult chimpanzee and modern human skulls, summarizing some key differences in patterns of facial growth. The boundaries of the upper face, the height of the face (nasion to gnathion [the base of the mandibular symphysis]), and the angle of the mandible are highlighted. In both species, the face grows taller and wider, but in chimpanzees, the face rotates dorsally (becomes more prognathic) and grows anteriorly relative to the vault (becomes more projecting). In modern humans, the face remains orthognathic and nonprojecting, primarily growing superoinferiorly.

between humans and chimps in the postnatal growth of the face. The most dramatic differences, seen in lateral view, relate to the fact that the modern human face is smaller relative to overall cranial size because it grows less inferiorly and anteriorly from the neurocranium and basicranium. These differences are already evident in neonates and become exaggerated in postnatal growth as the modern human upper face remains largely tucked beneath the ACF, and the inferior margin of the maxilla does not grow much below an imaginary plane at the base of the PCF (depicted in Figure 4.9). In contrast, chimp facial growth is characterized by a combination of facial *projection* and *prognathism*. Facial projection describes the anterior position of the front of the face relative to the front of the cranial base and neurocranium; prognathism describes the protrusion of the lower face relative to the upper face. In chimps, prognathism occurs from the combination of a relatively long lower face, especially the premaxillary region, combined with an extended cranial base (note these processes are reduced in bonobos, leading to a less prognathic face than in common chimpanzees). The chimp's face is also relatively taller, so that the lower margin of the maxilla grows well below the lower margin of the PCF.

Processes of Facial Growth and Development

Here I consider the anatomy and growth processes of the human and nonhuman primate face in the context of major organs and spaces, plus their surrounding skeletal capsules. I proceed from top to bottom, beginning with the upper face around the ACF and orbits, the middle face around the nasal cavity, and the lower face around the oral cavity. I conclude with a discussion of the temporomandibular joint. The anatomy, growth and development of many of the organs, spaces, and structures in the face are treated later (e.g., teeth in Chapters 3 and 7, the muscles of mastication in Chapter 7, the pharynx in Chapter 8, and the eyeballs in Chapter 10).

Anterior Cranial Fossa and Supraorbitals

The top of the face, which is below and in front of the frontal lobes and encapsulates the orbits, is substantially different in humans compared to other apes and hominins. Most obviously, modern humans usually lack any substantial brow ridge (or supraorbital torus, the roughly horizontal bar of bone above the orbits and in front of the frontal lobe). The absence of a brow ridge is caused primarily by the relatively small size of the face and by a lack of upper facial projection (more on this Chapter 5). Modern humans, in fact, are unique in having orbits positioned almost entirely beneath the frontal lobes. Minimal upper facial projection and a relatively spherical neurocranium orient the modern human forehead almost vertically above the top of the orbits.

Different architectural relationships of the orbits relative to the frontal lobes are accommodated by variations in the frontal bone's components, each of which derives from separate ossification centers in the frontonasal prominence (see Chapter 3). The neurocranial part of the frontal that encapsulates the frontal lobe has two tables of bone. The inner table ossifies from the endocranial membrane and thus attaches directly to the frontal lobe. This layer of bone lines the front of the ACF, and extends continuously to form most of the floor of the ACF on either side of the ethmoid (also the roof of each orbital cavity). The outer table of the frontal bone is more complex. With the exception of the frontal squama, the frontal's outer table is really part of the face. These facial components (Figure 4.10) include the brow ridge (supraorbital torus) above the orbits; a pair of processes lateral to each orbit (pars lateralis) that are part of the lateral margins of the upper face; and a short central process (pars nasalis) that projects toward the nose. The human brow ridge is distinctive because it can be divided into three highly variable portions: (1) the supraciliary arches above the medial and central part of each orbit; (2) the supraorbital trigones at the lateral end of the orbits; and (3) the glabella, a central bulge.

A crucial point about the upper face is that as the brain, orbits, and the rest of the upper face grow and develop with the frontal bone in their midst, each part of the frontal bone grows in different ways, reflecting different functional matrices. The inner table of the frontal around the brain (the ACF floor and the frontal squama) grows mostly in response to expansion of the frontal lobe. This growth involves bone deposition within sutures (e.g., metopic, sphenofrontal), and drift from bone resorption on the endocranial surface and bone deposition on the ectocranial surface (Duterloo and Enlow, 1970). The rest of the frontal bone grows somewhat independently of the inner table, in conjunction with the face and the orbits. Some of this growth is initially stim-

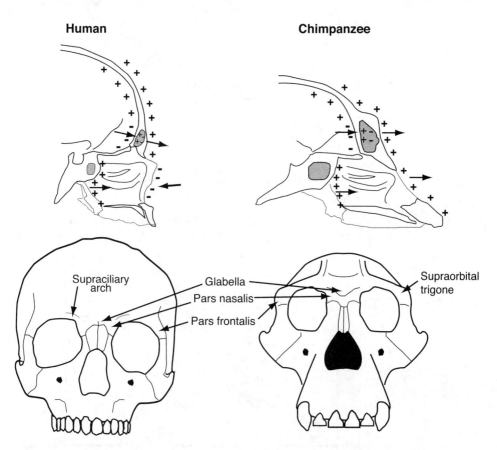

Figure 4.10. Comparison of basic anatomy of the supraorbital torus and facial growth patterns in modern humans (left) and chimpanzees (right). In both species, the upper face around the orbits drifts forward, but considerably more facial projection occurs in chimpanzees. The midface is displaced forward in both species by deposition along its posterior margin. A major difference is that the anterior surface of the midface is mostly a resorptive field (–) in humans but a depository (+) field in chimpanzees, helping contribute to a longer snout (rostrum) in the latter.

ulated by expansion of the eyeballs while they are still growing (see below), but the rest occurs from the growth of the face as a whole, forward and downward relative to the ACF (Figure 4.10).

Much of the space between the two tables of frontal bone becomes the frontal sinuses. As the outer table of the frontal grows forward from the inner table, osteoclasts from the ethmoidal sinus (see below) migrate into the space between the two tables, resorbing bone bilaterally to hollow out the frontal si-

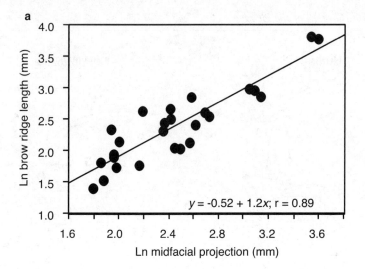

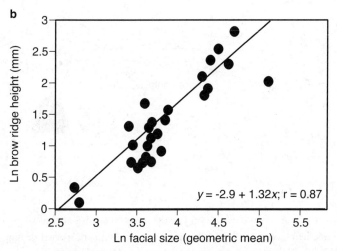

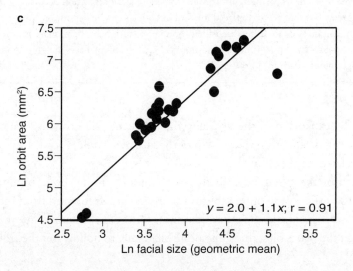

Figure 4.11. Scaling relationships: (a) between brow ridge length and midfacial projection (nasion–foramen caecum); (b) between brow ridge height and facial size (a geometric mean of facial height, length, and width); (c) between orbit area and a geometric mean of facial size. Orbital area and facial size scale with strong positive allometry (the slope of isometry between area and volume is 0.67). Data are from adult males and females of 15 different species of Old World monkeys and apes (for details, see Lieberman, 2000).

nuses (W. A. Brown et al., 1984). Thus sinus expansion does not cause brow ridge growth but is a byproduct of facial projection. Sinuses are often asymmetrical and grow in a skeletal growth trajectory along with the rest of the face (Tillier, 1977). As Figure 4.11 shows, across primates, brow ridge length scales with slight positive allometry with upper facial projection (defined as the distance from the foramen caecum to nasion, the midpoint of the frontonasal suture), and brow ridge thickness (superoinferior height at midorbit) also scales with positive allometry to overall facial size (Lieberman, 2000).

Orbits

The eyeballs develop as outgrowths from the embryonic forebrain below the ACF and in front of the MCF. Chapter 10 considers ocular development and function in detail. At this point, a few facts are useful. First, the eyeballs play a key role in many aspects of early orbital growth and development. The epithelial layer of the embryonic eye (the optic sclera) initiates development of the lesser wing of the sphenoid, which then grows around the optic nerve (Sperber, 2001). Further, as the embryonic face forms, the divisions of the frontonasal and maxillary processes migrate over and around each lens placode to create the upper and middle face around the eyes (see Chapter 3). Thus, eyes develop below the frontal lobes, in front of the temporal lobes, and in the midst of many morphogenetic precursors of the cranial base and facial skeleton. In all, six bones—the frontal, zygomatic, sphenoid, maxilla, ethmoid, and lacrimal—contribute walls of bone that arise from distinct primary ossification centers around each eyeball, eventually fusing to create the orbital skeleton (see Figure 3.9). An advantage of this intricate design is to orient sutures in almost every direction in the orbit, permitting growth in many directions. Not surprisingly, bone shapes of the bones in the orbital walls tend to be highly variable, especially the lacrimal bone, which acts as a sort of space filler between the frontal, ethmoid, maxillary, and sphenoid contributions to the medial orbital walls (Enlow, 1990). The superior orbital fissure contributes to the orbital floor, forming a gap between the MCF and the back of the middle face (the pterygopalatine fossa) through which nerves and vessels traverse between the brain and face (see below).

The orbital skeleton expands through drift and displacement around the eyeball. Drift primarily expands the orbit outward (laterally) because the internal surface of the lateral orbital wall is resorptive and the external surface is depository (Enlow, 1966). Most orbital expansion, however, is through intrasutural growth that causes displacement. For the first two postnatal years, this growth is partly stimulated by tension exerted by the eyeball itself. Congenital and experimentally induced deformations of the eyeball produce predictable deformations of the orbit. For example, hypertension (from too much eyeball growth) and hypotension (from too little eyeball growth) lead to overly large or small orbits, respectively (Coulombre et al. 1962; Mustarde, 1971). Experiments in which the eyeballs of young rabbits were removed or inflated show similar, predictable effects (Sarnat, 1982). However, eyeball growth alone does not stimulate all early orbital growth, and by birth the orbits are already much larger than the eyeballs. The eyeball of a modern human neonate is approximately 3.2 cm^3, less than half (45 percent) the volume of the orbit; in adults, eyeball volume averages 8.2 cm^3, approximately one-third of the orbit's volume (24 cm^3) (Schultz, 1940). In adult chimpanzees, mean eyeball size is just a little smaller than in humans, about 6 cm^3, but the eyeballs fit within an orbit that is approximately the same size as a human's (Schultz, 1940). In both species, extraocular muscles and fat deposits fill most of the rest of the orbit (see Chapter 10).

Viewed together, the orbital skeleton initially grows around the eyeball but becomes increasingly independent of the eyeballs, whose growth ceases much earlier in ontogeny than the rest of the face. Most postnatal expansion of the orbit occurs as the lower and outer rims of the orbit expand inferiorly and laterally with the maxilla, and the upper part of the frontal expands laterally with the anterior cranial fossa (Denis et al., 1992, 1993). These patterns account for several scaling relationships, which Chapter 10 considers further. Notably, across primates, eyeball size scales with negative allometry relative to both brain and body size (Kirk, 2004), but orbit area scales with strong positive allometry to facial size. Thus, primates with relatively big faces, including hominins such as the Neanderthals, have relatively big orbits.

Nasal Cavity

The nasal cavity is a byzantine box of bones—the ethmoid, sphenoid, nasal conchae, maxilla, palatine, and vomer—around a tubelike space used primarily for airflow. Let's look at the cavity's components before considering its overall growth. The roof and medial wall of the nasal cavity are basicranial in origin but grow in the midst of the face in a skeletal growth trajectory. As reviewed in Chapter 3, the rostral portions of the paired trabecular cartilages contribute to two parts of the ethmoid below the olfactory bulbs: the ectethmoid and mesethmoid. The ectethmoid becomes the central nasal capsule, a roughly cubic structure filled with dozens of small sinus cavities (so complex that they are termed the ethmoid labyrinth). The top of the ectethmoid, below the olfactory bulbs, becomes the cribriform plate, a sievelike rectangle with numerous perforations for the olfactory nerves. The sloping posterior portion of the nasal cavity's roof behind the cribriform plate is the underbelly (inferior surface) of the sphenoid body (Figure 4.12). The mesethmoid begins as a perpendicular plate in the midsagittal plane and then differentiates into three structures in the midsagittal plane that divide the nose bilaterally: the perpendicular plate of the ethmoid at the top, the cartilaginous nasal septum in the middle, and the plow-shaped vomer at the bottom. A pad of fatty tissue separates the nasal septum and vomer, preventing their fusion.

The nasal cavity's lateral walls are irregularly shaped because of three paired projections, the nasal conchae (see Figure 4.12). The upper two conchae are delicate medial projections of the ethmoidal labyrinth (thus endochondral in origin). The lower conchae are distinct intramembranous bones projecting from the nasal cavity's lateral walls (made up of the maxilla and the palatine). Each concha is covered with a highly vascularized, mucous-rich epithelium, creating three narrow channels (meatuses) through which air flows (see Chapter 8). The floor of the nasal cavity (also the roof of the palate) fuses from the conjunction of the maxilla, the palatine, and the premaxilla in the midline. The precursors of these bones meet in the midline below the nasal cavity through medial migration of the maxillary prominences and downward migration of the frontonasal prominence (see Chapter 3). Incomplete fusion causes various forms of cleft palates.

Although the ethmoid forms the top and middle wall of the nasal cavity, the maxilla forms most of the bone on the sides and bottom. These portions of the maxilla fuse from several initially distinct pairs of ossification centers, all of which participate in different functional matrices (Figure 4.12). On each side, the central, basal portion develops around the infraorbital nerve below the orbits; the nasal portion forms medially along the lateral wall of the nasal cavity; the orbital portion forms superiorly as the majority of the floor of the orbital cavity; and the alveolar process of the maxilla forms below the maxil-

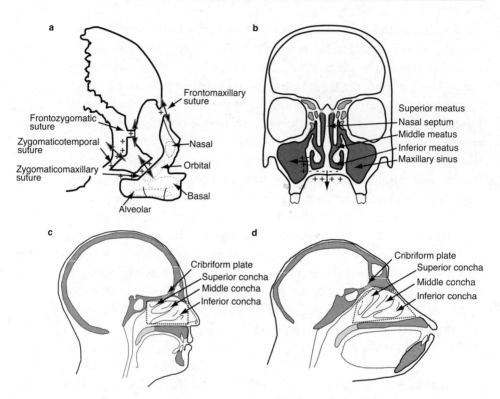

Figure 4.12. Schematic view of growth of the human midface and the nasal cavity. (a) The maxilla, palatine, and ethmoid bones (the ethmomaxillary complex) grow downward and forward from displacement along the posterior and superior aspects; also depicted are the different portions of the maxilla. Major sutures of growth include the zygomaticomaxillary, frontonasal, frontozygomatic, and zygomaticotemporal. (b) The lateral and inferior margins of the nasal cavity also drift laterally and inferiorly, respectively. Differences in facial growth (cranial base angulation, facial projection) cause the superior and inferior margins of the nasal cavity to be more parallel in humans (c) than in chimpanzees (d).

lary body around the tooth roots. The infraorbital portion of the maxilla lateral to the nasal cavity is filled with a sinus (see below).

If one compares nasal cavity anatomy in adult humans and chimps, some interesting differences are evident (see Figure 4.12c, d). Most obviously, the human nasal cavity is roughly rectilinear, with the roof (the cribriform plate) and floor (the palate) of similar length and nearly parallel. In a chimp, the roof and the floor are aligned more obliquely, and the anterior portion of the relatively longer palate projects well in front of the bridge of the nose. These contrasts reflect variations in the basic pattern of midfacial growth common to all primates. The nasal cavity roof elongates along with the rest of the cranial base (see above), but the nonhuman primate midface grows primarily through a combination of displacement (via intrasutural growth) and drift. Most displacement occurs through bone deposition along the back of the maxilla relative to the sphenoid (a fat pad separates them). Adding bone to the back of the maxilla displaces the entire midface (sometimes called the ethmomaxillary complex) forward relative to the MCF. Some displacement also occurs from bone growth within those sutures of the palate and the lateral and medial walls of the nasal cavity that have some coronal or transverse orientation (e.g., palatomaxillary, incisive, maxillolacrimal). The other mechanism of midfacial elongation is drift, but this differs in humans and nonhuman primates (Figure 4.5; see also below). In the nonhuman primates that have been studied, the infraorbital portions of the maxilla are depository growth fields, thus permitting drift to contribute to midfacial elongation; in humans, this surface is a resorptive field limiting midfacial elongation and prognathism (Duterloo and Enlow, 1970).

The midface also becomes taller through several mechanisms. One is drift of the floor of the nasal cavity, a resorptive field, whose opposite surface is the roof of the palate, a depository field (see Figure 4.12). The bottom of the nasal cavity therefore drifts downward, expanding into space formerly occupied by the oral cavity. If one glues a titanium ball to the roof of the oral cavity, it eventually ends up in the nose. Growth within sutures that have some horizontal orientation (pterygopalatine, frontomaxillary, zygomaticomaxillary, and zygomaticotemporal also shown in Figure 4.12) also helps make the midface taller (Weinmann and Sicher, 1941). But there is debate over the extent to which such growth actively pushes the face downward or responds to other forces that pull the face downward. The debate centers on the median nasal wall, which has three components: the perpendicular plate of the ethmoid, the nasal septum, and the vomer. The nasal septum, which is cartilage, may grow endochondrally, like a synchondrosis. To the extent that this is true, then cartilage growth in the septum may drive midfacial growth, pushing the nasal cavity downward and forward and thereby generating tension in other sutures (J. H. Scott, 1958, 1967). Some evidence supports this hypothesis. Removal of the nasal septum inhibits midfacial growth in rabbits (Sarnat and Wexler, 1967), and nasal septum defects inhibit midfacial growth in humans and nonhumans (Grymer and Bosch, 1997; Adamopoulos et al., 1994; Verwoerd et al., 1979, 1980). However, the nasal septum is unlikely to drive all nasal cavity growth, because growth in the rest of face (especially the orbits and oral cavity) also carries along the midface, pulling it laterally, inferiorly and anteriorly (Moss and Salentjin, 1969; Enlow, 1990).

The midface becomes wider as it gets taller and longer. The nasal cavity floor widens from growth in the midpalatal suture. Bone surfaces that line the lateral and inferior walls of the ethmoid are resorptive, and the opposite surfaces are depository (Enlow, 1968, 1990). Similarly, portions of the maxilla and palatine bones that form the lateral walls of the nasal cavity relocate laterally through drift because their inner nasal surfaces are resorptive and their outer surfaces are depository.

We now have enough information to diagnose what growth processes make the human nasal cavity so rectilinear. As shown in Figures 4.9 and 4.12, the roof of the nasal cavity is relatively longer in humans than in nonhuman primates such as chimps because it also forms the midline of the anterior cranial base. Humans have a relatively longer anterior cranial base to accommodate a relatively larger brain. At the same time, while the roof of the human nasal cavity is relatively longer, the floor is relatively shorter. This is because the floor of the nose, which is also the roof of the palate, grows to be less prognathic in humans than in nonhumans, for two reasons. First, as previously noted, the surface of the midface is a resorptive field in humans but a depository growth field in nonhuman primates (Figure 4.13). Second, the human cranial base is more flexed, thereby tilting less of the lower face outward in front of the upper face. Thus, in chimps the midface grows forward through drift and rotates outward. But in humans, the midface remains vertical and drifts posteriorly even as it is displaced forward by growth along the back of the face (see below).

Another source of variation in the nose is the relationship between the maxilla, premaxilla, and vomer (Figure 4.13). In humans, the nasal floor is sometimes described as "flat" or "smooth" because the premaxillary and maxillary portions of the nasal floor are continuous. In apes this configuration is variable, but most chimps have a "stepped" nasal floor in which the maxillary portion is lower than the premaxilla (McCollum et al., 1993). Part of this difference is explained by the thickness of the palatal shelves. But the vomer may also play an interesting role. In apes with stepped nasal floors, the vomer does not extend onto the premaxilla either at all or very much, but in humans (and some fossil hominins) with smooth nasal floors, the vomer extends onto the premaxilla (McCollum et al., 1993) (Figure 4.13). This difference is significant given the intrinsic growth potential of the nasal septum, which pushes down the lower portion of the face somewhat perpendicular to the anterior

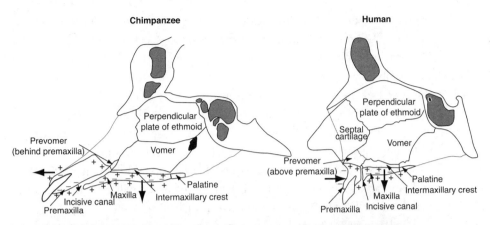

Figure 4.13. Differences in the configuration and growth of the palate and nasal cavity in chimpanzees (left) and humans (right). In chimpanzees, the premaxilla is larger, more angled, and higher, creating a stepped nasal sill in front of the incisive canal. In humans, a portion of the vomer inserts on the premaxilla, which is smaller, more ventrally rotated, and in line with the maxilla. In both species, the palate grows inferiorly from drift.

cranial base (see above). Consequently, insertion of the vomer onto the premaxilla makes these downward forces act not only on the maxilla but also the premaxilla, helping rotate the front of the snout downward (thus reducing prognathism) and flattening the nasal floor (McCollum, 1999, 2000).

Finally, a brief comment on the external nasal cavity of humans, a unique derived feature of the genus *Homo.* This part of the nose mostly functions as a vestibular space through which air must make an approximately 90° turn when entering or exiting the nose (see Chapter 8). The external nose consists mainly of five pairs of cartilages (there are more smaller ones) that make up the lateral and dorsal margins. The midline cartilage is a continuation of the nasal septum. A pair of nasal bones (easily broken) forms the bridge of the nose.

Paranasal Sinuses and Zygomatic Arches

The midface has four pairs of paranasal sinuses housed in the ethmoid, frontal, sphenoid, and maxillary bones. These spaces grow as osteoclasts migrate outward from the middle and superior meatuses, removing bone as they go (a process termed *pneumatization*). The mucosal lining of the nasal cavity then follows the osteoclasts, covering the sinus walls. The sinuses begin to expand during the fetal period, beginning with the ethmoid sinus, which expands laterally from the middle meatus, filling the ethmoid with holes (air cells). The frontal sinus grows from osteoclasts that migrate upward from the superior meatus. The sphenoid sinus grows from osteoclasts that migrate posteriorly from the ethmoid into the sphenoid body. Finally, the maxillary sinus grows from cells that migrate inferiorly and laterally from the middle meatus into the center of the maxilla. This sinus grows rapidly during fetal and postnatal ontogeny, keeping pace with rest of the facial growth (Ennis, 1937; McGowan et al., 1993). The function of sinuses is debated (see Chapter 8). In addition to lightening the face and acting as resonance chambers, sinuses may help to reduce the volume of bone turnover necessary during drift (Blaney, 1990; Rae and Koppe, 2004). For example, expansion of the maxillary sinus helps the maxilla to drift laterally and inferiorly (Lund, 1988).

The last part of the midface to note is the pair of zygomatic arches, which includes portions of the zygomatic, temporal, and maxillary bones. Initially the anterior (zygomatic) and posterior (temporal) parts of the arch are separate, but they grow toward each other to form a sutural junction late in the fetal period (Bach-Peterson and Kjaer, 1993). As the brain expands and the face grows forward relative to the neurocranium, the arch elongates primarily in the temporozygomatic suture. The zygomatic arch also thickens and drifts laterally because its medial surface is resorptive and its lateral surface is depository (Enlow, 1990).

Oral Cavity (Mandibular and Maxillary Arches)

The lower face derives almost entirely from divisions of the first branchial arch around the oral cavity and pharynx. Bones that form the upper part of this region—the palate and maxillary arch—include the premaxilla in front, the maxilla in the middle and to the sides, and the horizontal portion of the palatine in the back.[4] The posterior portion of the palatine cartilage never ossifies but becomes the soft palate. Traces of the incisive suture between the maxilla and premaxilla disappear slowly during postnatal ontogeny, except for one hole, the incisive foramen (which transmits the nasopalatine nerve and the sphenopalatine vein). The maxillary arch itself forms a "crest" of alveolar bone around the tooth buds and, later, the developing tooth crowns and roots (see Chapter 3). Because the palate is the floor of the nasal cavity and the maxillary arch grows down from the margins of the palate, the growth of the upper

---

4. The premaxilla derives from the frontonasal process. Given conflicting evidence for a distinct premaxillary ossification center, some anatomists do not consider the premaxilla a separate bone. This latter view is probably wrong.

jaw is stimulated to a large extent by the growth of the nasal cavity and the rest of the middle face (the ethmomaxillary complex). At the beginning of the fetal period, the maxillary arch is shaped like a small triangle below the nose (Shapiro and Robinson, 1980); then, as the palate and midface grow anteriorly and laterally and the teeth begin to develop, they pull the arch with them into a parabolic shape and cause the alveolar process to increase in height.

The mandible grows in a very different manner from the palate and maxillary arch: instead of being connected to the face above it, the mandible articulates with the cranial base and must grow autonomously to match the size and shape of the upper jaw. Recall from Chapter 3 that the mandible ossifies intramembranously lateral to Meckel's cartilage, initially around the alveolar nerve and the primordial tooth germs, to form the alveolar process (Figures 3.9, 4.14). Later, centers of ossification spread to form the main part of the mandible, the future corpus and ramus. In addition, secondary accessory cartilages form in several corners of the mandible: the condyles at the site of the future TMJ (discussed separately below), the coronoid processes, and the angular processes. Cartilage nodules also form in the symphysis region (Goret-Nicaise and Dhem, 1982), which remains a cartilaginous plate until 4–12 postnatal months (Scheuer and Black, 2000). These cartilages are replaced by intramembranous bone, with the notable exception of the condylar processes. Although the mandible becomes a single bone, it grows initially as subunits around the central corpus (Figure 3.9), each with its own functional matrix (see Atchley and Hall, 1991). The alveolar process grows around the teeth; the symphyseal region unites the two halves of the mandible (in most mammals, the two sides of the mandible remain distinct, independently mobile bones); the angular process grows within the insertions of the masseter and medial pterygoid muscles at the base of the ramus; the coronoid process grows within the insertion of the temporalis muscle at the top of the ramus; and the condylar process connects the mandible to the cranial base and grows within the insertion of the lateral condyle (see below). Contractions of the masticatory muscles, both prenatally and postnatally, are crucial for stimulating proper growth of some of these elements (see Hall, 1978; Herring and Lakars, 1982; Vinkka-Puhakka and Thesleff, 1993; S. K. Lee et al., 2001).

The way the growth of the mandible, palate, and maxillary arch is integrated merits consideration because the functional demands of proper occlusion require the upper and lower jaws to fit together perfectly even though they grow independently (more on this topic in Chapter 7). Notably, the maxillary arch grows downward from the middle face and forward from the MCF,

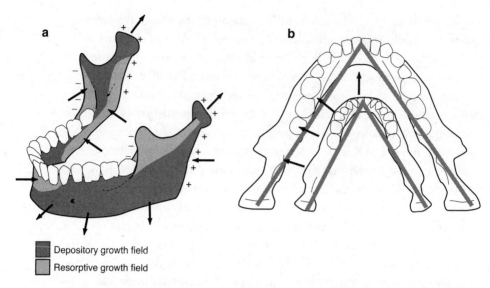

Depository growth field
Resorptive growth field

Figure 4.14. (a) Schematic view of mandibular growth (modified from Enlow, 1990) showing depository (+) and resorptive (–) fields. (b) These cause the mandible to expand like a growing V from displacement along the posterior end and drift along the two corpora. The ramus elongates from deposition along its posterior margin, combined with resorption along its anterior margin. In addition, humans uniquely develop a chin from resorption of bone along the anterior alveolar surface.

but the mandible grows forward entirely from the TMJ in the PCF. This difference means that human and nonhuman primate mandibles must compensate in somewhat different ways for the different patterns of maxillary arch growth. As shown in Figure 4.9, the palate and maxillary arch in chimpanzees grow to be much longer and more prognathic relative to the middle face, particularly in the premaxillary region, which protrudes convexly below the nasal cavity. Chimpanzees thus have long maxillary arches with a space (diastema) between the canines and the incisors. The human palate and maxillary arch, in contrast, is shorter and less prognathic, more retracted relative to the cranial base, and with much less space for the teeth. Thus humans must grow shorter mandibles, with a relatively shorter ramus that is more perpendicular with respect to the corpus, and they lack a diastema.

How the mandible and maxilla match each other's growth is still only partially understood. In the embryo, the mandible is initially longer than the maxilla, but by the fetal period, the mandible generally lags behind the maxilla (*retrognathism*). Thereafter, the mandible usually remains retrognathic until occlusion begins within the first postnatal year (see Sperber, 2001). Because the mandible generally appears to keep up with the maxilla, not vice versa, it makes sense to consider the latter's growth first. During fetal and postnatal ontogeny in all primates, the palate and the maxillary arch widen, lengthen, and become taller in concert with the midface, through a combination of drift and displacement. Vertical growth occurs mostly from bone deposition along the maxilla's upper margin (e.g., the zygomaticomaxillary suture), which displaces the midface and lower face inferiorly (Weinmann and Sicher, 1941; Baume, 1951; Craven, 1956; Sirianni and Swindler, 1979). The maxillary arch and palate also grow downward from the nasal region through drift, as the floor of the nasal cavity and the roof of the oral cavity are resorptive and depository growth fields, respectively.

Forward growth of the lower face, which occurs though several mechanisms, poses a challenge for proper integration of the maxillary and mandibular arches. The dominant mechanism of anterior growth of the lower face is bone deposition along its posterior margins, especially at the maxillary tuberosities. These important growth sites are situated at the back of the maxillary arch above the alveolar processes and in front of the pterygoid processes of the sphenoid (Figure 4.12). Bone deposition at the maxillary tuberosities (and in fact along the entire posterior margin of the maxilla) lengthens the lower and middle portions of the face, displacing the entire region (the ethmomaxillary complex) forward relative to the middle cranial fossa (Enlow, 1990). In other words, the lower and middle face grows forward from the back. A second source of anterior growth of the lower face occurs through bone deposition in coronally oriented sutures, which elongate the palate. Fusion of the incisive suture between the premaxilla and maxilla is highly variable but generally occurs before the adolescent growth spurt (Scheuer and Black, 2000).

A final mechanism of anterior growth of the maxillary arch is drift (see Figure 4.5). In nonhuman primates, drift of the lower and middle portions of the face occurs in which the anterior surfaces of the maxilla and premaxilla are depository, and the internal surface (inside the maxillary sinus) is resorptive (Duterloo and Enlow, 1970). In contrast, humans are unique in having a resorptive growth field on the front of the maxilla, which restricts lower facial elongation by counteracting displacements that occur at the maxillary tuberosities and elsewhere. In other words, humans lack a snout in part because of bone resorption along the front of the face.

Mandibular growth is similar to upper-jaw growth in some respects and different in others (Figure 4.14). One difference is in how they elongate. The human maxilla lengthens by a combination of deposition on the maxillary tuberosities and intrasutural growth. In contrast, the mandible elongates to a considerable extent from bone deposition at its posterior end, displacing the whole mandible forward relative to the glenoid fossae of the temporal bone. The posterior margin of the ramus and the mandibular condyle (see below for details) are the two major sites for mandibular elongation. The ramus, which roughly spans the anteroposterior length of the pharynx behind the palate, thus gets longer from the back, but bone resorption on its anterior edge stabilizes its length (R. J. Smith, 1984; Enlow, 1990). Resorption of the anterior ramus border thus creates room for the molars to erupt (put differently, most bone in the corpus was previously part of the ramus). Seen from above, the mandible also lengthens and widens like an expanding V from a pattern of drift in which most of its lateral (outer) surface is a depository growth field, and most of the medial (inner) surface is a resorptive field (Duterloo and Enlow, 1970; Enlow, 1990). In addition, the ramus becomes taller (helping it keep up with inferior growth of the midface that pushes the maxillary arch downward) from growth within the condyle. The inside surface of the condylar process is depository and the outside surface is resorptive, so bone that was once part of the condyle's head becomes part of the neck (Enlow and Harris, 1964). The corpus also gets taller from bone deposition on its inferior surface.

Differences between human and nonhuman primate mandibular growth patterns during fetal and postnatal ontogeny thus make sense in terms of the shape and position of the maxillary arch (or dental arcade). As noted above, compared to nonhuman primates, humans grow relatively shorter maxillary arches that are positioned below a more retracted face and which are rotated more horizontally (ventrally) below the rest of the face and ACF. Consequently, humans need to grow a relatively shorter mandible with a relatively shorter mandibular ramus and corpus (see Humphrey et al., 1999; Schmittbuhl et al., 2007). In addition, because the maxillary arch is rotated more horizontally relative to the cranial base in humans than in chimpanzees, the human mandible has to accommodate this rotation by orienting the corpus more perpendicularly to the ramus (see Figure 4.9). That said, some mandibular rotation during growth is required in all primates because the maxilla descends relative to the MCF, reducing the angle between them (Nielsen et al., 1989; Schneiderman, 1992). Thus, as the mandibular ramus gets taller, longer, and wider, it also rotates clockwise (when viewed from the right) because its anterior edge is depository at the top and resorptive at the bottom, while its posterior edge is resorptive at the top and depository at the bottom (Enlow, 1990).

The coordination of mandibular and maxillary growth is poorly understood. Because there is an obvious selective disadvantage to improperly coordinated growth of the two jaws, one expects some degree of integration to occur from genetically regulated growth processes via mechanisms such as pleiotropy (Olson and Miller, 1958; Cheverud, 1982). However, malocclusions from mismatched mandibular and maxillary arch growth are common, especially in postindustrial populations. Available data on heritability in the lower face suggest that genetic factors explain less than 50 percent of the variation (Harris and Smith, 1980; Corruccini et al., 1990; Corruccini, 1999). As discussed in Chapter 7, nongenetic stimuli from chewing play an important role in generating growth to integrate the mandibular and maxillary arches properly (reviewed in Herring, 1993; Lieberman et al., 2004a). Hypothetically, fine-tuning of tooth position occurs mostly in the alveolar process, whereas overall coordination of mandibular size relative to maxillary size occurs mostly in the mandibular condyles, which experience significant loading during chewing (Hylander and Bays, 1978). Animals fed hard food have faster rates of condylar growth than control animals fed soft food (Shaw and Molyneux, 1994; Tuominen et al., 1994). In turn, mandibular condyle growth elongates the mandible as described above, displacing it away from MCF.

One last component of lower facial growth deserves mention: the chin or mental eminence. This unique derived feature of modern humans appears prenatally from a novel configuration of growth fields in the symphysis (Figures 4.5 and 4.14). The anterior margin of the symphysis is entirely a depository growth field in nonhuman primates and hominins, but in modern humans, the surface of the alveolar process below the anterior teeth is a resorptive field (Duterloo and Enlow, 1970; Bromage, 1990). Bone is thus resorbed on either side of the symphysis, hollowing out two mental fossae and leaving an upside-down T-shaped protuberance below, projecting out from the symphysis (Buschang et al., 1992; Schwartz and Tattersall, 2000).

The Temporomandibular Joint

The temporomandibular joint (TMJ), the mandible's articulation with the cranial base, deserves special consideration given the unusual and complex nature of its development, anatomy, and function. Some of these characteristics are a product of the TMJ's evolutionary and developmental origin. At the very beginning of the fetal period, the future jaw opens and closes at a primitive, temporary joint *within* Meckel's cartilage between the future malleus and incus. This configuration, which disappears after several weeks, is essentially homologous to the reptilian jaw joint. The mammalian TMJ, however, does not develop from these vestigial structures in Meckel's cartilage but instead arises from two separate mesenchymal cell masses: a temporal mass in the otic capsule of the cranial base and a condylar mass in the secondary condylar cartilage of the mandible (see above). The temporal and condylar cell masses grow toward each other, while the intervening mesenchymal tissue differentiates into several layers of fibrous tissue in which the joint cavities and the articular disk form (Bach-Peterson et al., 1994). Muscle activity is necessary for the formation of the joint capsule within the fibrous zone between the temporal and condylar cartilages (Van der Linden et al., 1987, Takahashi et al., 1995). In contrast to other joints, which consist of hyaline cartilage (see Chapter 2), the articular surfaces of the mandibular and temporal portions of the TMJ consist of fibrous cartilage secreted by chondrocytes that differentiate in the periosteumlike sheath surrounding the bone. In addition, mesenchymal cells surrounding the joint differentiate to form the surrounding fibrous joint capsule and ligamentous structures.

The newborn TMJ is unstable, largely because its basicranial portion is narrow and flat, with no articular fossa or eminence to restrict movements of the condyle. Following the onset of mastication and the eruption of the dentition, the temporal surface of the TMJ becomes increasingly S-shaped, with a U-shaped articular eminence at the anterior end of the glenoid fossa and a short postglenoid tubercle that projects inferiorly at the posterior end of the fossa (see Figure 7.6). The temporal surface of the TMJ grows through a combination of endochondral and intramembranous growth. As Chapter 7 discusses, growth of the eminence appears to be stimulated in part by mechanical loading of the joint. The same is true for the mandibular condyle, which grows as a secondary ossification center, not only elongating the condylar neck but also widening and lengthening the condyle (Dibbets, 1990).

Chapter 7 considers TMJ function and the significance of variations in TMJ shape. Here, it is worth noting a few major differences in TMJ anatomy between humans and great apes, including the concavity of the temporal surface and the position of the joint itself relative to other structures in the skull. In great apes, the TMJ is relatively long, wide, and flat, and it juts out laterally from the lateral wall of the middle cranial fossa; in humans, the TMJ is mediolaterally compressed and has more topographic relief, with a deeper glenoid fossa, a more vertical tympanic, and more pronounced articular eminence (Sherwood et al., 2002; Lockwood et al., 2002).

Coda

Well, there you have it: a region-by-region description of the key patterns and processes of craniofacial growth in humans, with occasional comparisons to other primates, especially chimps. The number, diversity, and intricacy of the many organs, bones, spaces, sutures, synchondroses, and growth fields reviewed above are testimony to natural selection's powers of assembling and modifying complex things. In some respects, the head is like a Gordian knot of tissues, bones, organs, and spaces that are all tied and jumbled together. Although I have taken pains to discuss how various walls of bone are shared between organs and spaces, you are probably left with little sense of how the many modular units of the head all manage to fit together and function during growth on more than a regional level. In other words, I have not discussed how the head is integrated, and how patterns of integration differ in humans and other great apes. The task of the next chapter is to use the information presented in this chapter to consider how the head grows into a fully functional ensemble and how these integrative mechanisms relate to some of the differences between humans and our closest relatives.

5

Integration of the Head during Fetal and Postnatal Growth

Instead evolution played Mr. Potato Head, putting different combinations of features on ancient hominids then letting them vanish until a later species evolved them.

SHARON BEGLEY, "The New Science of Human Evolution," 2007

We frequently simplify complex things in order to understand them. This practice, however, can sometimes lead to unwarranted oversimplifications. One way we oversimplify is to make rudimentary analogies, such as "The brain works like a computer," and then assume that aspects of computer design can actually explain aspects of brain design. Another is to assume one-dimensional cause-and-effect relationships from observations, such as "Gene X causes a big brain." Although aspects of the brain may be computerlike, and gene X may contribute to encephalization, such oversimplifications can be problematic because they inappropriately and illogically attempt to convert intrinsically messy, contingent processes into neat, predictive ones (see Mayr, 1988).

To oversimplify is to err, but this doesn't mean that complex phenomena cannot be usefully partitioned into simpler components. One useful way to simplify complex biological things like heads is to divide them into modules. Organisms are naturally and necessarily modular at many hierarchical levels, from the genome (base pairs, codons, genes, chromosomes), to the phenotype (proteins, cells, tissues, systems, organisms). Further, modules have several key properties that promote evolvability: they are partially autonomous, both genetically and physically; they undergo distinct transformations; and they connect and interact with other modules at different hierarchical levels (see Raff, 1996). And so it is with heads, whose many modules were described and discussed in Chapters 3 and 4 in terms of their embryological origins and the processes by which they grow and develop during fetal and postnatal ontogeny. Although both chapters emphasized how these modules derive from a combination of genetic and epigenetic interactions, one can usefully describe the head as a complex amalgamation of modules such as the brain, eyes, mouth, and pharynx, each with a skeletal capsule, and a set of nerves, blood vessels, muscles and ligaments to connect them. A next logical step might be to posit that evolutionary change can occur by altering one module or another, as in a game of Mr. Potato Head (a popular children's toy in which a plastic potato can be decorated with various noses, eyes, mustaches, and other features to make a nearly endless variety of faces). According to this analogy, evolutionary processes can generate and favor variant species of hominins, like experiments with differing sizes of teeth, faces, and brains.

The problem with the Mr. Potato Head analogy is that although the structures of the head may be modular, they are not simple, interchangeable units that can be swapped or exchanged according to whim. They need to be grown, and they need to function together as an ensemble, accommodating each other through shared bony walls, using the same spaces, experiencing the same environmental stimuli, and so on. So, although the head is extremely modular, it is also extremely integrated. Modularity and integration are counterparts, and both are equally fundamental to evolvability. As Chapter 1 argues, modular organization of the head and the ways in which modules interact impose many constraints, but one reason the head may be so evolvable is that intense, widespread integration permits the head's many modules to change in ways that can accommodate each other and still function effectively.

Integration occurs most fundamentally at the organismal level in two ways: *functional integration,* which describes interactions among components that affect an organism's performance; and *developmental integration,* which describes interactions among components during growth and development (Cheverud, 1996). Developmental integration itself occurs through both genetic and epigenetic mechanisms. *Genetic integration* occurs via pleiotropy, when one gene has effects on multiple tissues, or via linkage, when genes lie near each other on a chromosome, causing selection on one gene to carry with it the effects of other genes (Cheverud, 1982, 1996). *Epigenetic integration* occurs mostly through physical and growth-factor-mediated interactions during ontogeny and through shared responses to similar environmental stimuli (West-Eberhard, 2003). The combination of genetic and epigenetic integration permits modules to fit and work together (to be functionally integrated) as they grow. Further, because genes vary in populations, genetic integration is a phenomenon evident at the population level and leads to *evolutionary integration,* the coordinated evolution of morphological traits (Cheverud, 1996).

Widespread integration permits an individual's head to grow and function but also has implications for change at the populational level, and thus for evolution. When measured within a population or species, morphological integration usually manifests itself as widespread correlation and covariation among diverse anatomical units, or sometimes as constrained scaling relationships. As examples, in humans and other primates, the widths of the neurocranium, basicranium, and face are always highly correlated to each other, and the orientation of the back of the face is always nearly perpendicular to the floor of the anterior cranial fossa and the orientation of the orbits (see below). Both covariation and constraint have profound evolutionary consequences. A key one is that a few major changes to the development of particular components can, through integration, have multiple, distributed effects throughout the head. For example, expansion of the brain influences the growth and development of the neurocranium and also affects the basicranium and various parts of the face. It follows that to transform the head of a great-ape embryo or fetus so that it grows into an adult human, it is not necessary to change how every aspect of the head grows. A little tinkering may go a long way.

Such effects may seem counterintuitive, but they are an expected emergent property of self-assembled objects that develop ontogenetically (Wagner, 1996). Complex, engineered things like cars do not have this property. Imagine transforming a small racing car like a Porsche into a minivan. Because cars are modular and assembled, one would have to change just about everything to accomplish this transformation: the chassis, the engine, the windows, the seats, and so on. Probably very little of the Porsche could be used in the minivan. However, to transform a fetal ape's head so that it grows into an adult human's rather than an adult chimpanzee's theoretically requires much less change because of integrative mechanisms that operate during ontogeny. In fact, the overarching hypothesis I consider is that a sizable proportion of the obvious shape differences between the species could probably be achieved through a relatively small number of major changes early in ontogeny, such as growing a bigger brain, a shorter and wider cranial base, and a smaller face (Lieberman et al., 2002, 2004b). As our tinkered ape skull grows and develops, interactions among its units will then transform and distribute the effects of these key ontogenetic shifts to many parts of the skull.

The goal of this chapter, therefore, is to consider how the many parts of the head fit, grow, and function together in an integrated way. I do so with some trepidation because integration is a rather abstract topic to study and discuss. On the one hand, there are good analytical tools for detecting and measuring integration. It is most commonly analyzed statistically by quantifying patterns of correlation, covariation, and scaling. On the other hand, integration is more than just a set of quantifiable patterns: it is a set of dynamic processes. These patterns, and the processes that generate them, raise two questions. First, what major mechanisms integrate the head's various modules during ontogeny? Second, how did these mechanisms change during human evolution? Another way of asking the second question is, what changes in integration occurred to transform ontogenies between ancestral and descendant species?

To address these questions, this chapter first reviews how integration can be quantified in structures such as the head. We then examine overall patterns of integration in a human head and in an ape's head (focusing on the common chimpanzee) and explore several major processes by which the major components of the head are integrated differently during ontogeny in the two species. As noted above, integration can occur via both genetic and epigenetic means. Both are important, but we will consider primarily epigenetic mechanisms and patterns of morphological change for two reasons. The first is that although we know pleiotropy and linkage to be sources of integration, we do not yet know enough about the genes and molecular mechanisms that regulate most aspects of craniofacial morphogenesis to explore their integrative roles in much detail.

My dog's head is an example of this problem. She is a hybrid: half border collie and maybe half chow. In most respects, she is a charming blend of parental phenotypes, but she'll never win a beauty contest because her lower jaw is too long relative to her upper jaw (she has a class III malocclusion). The Cornell dog experiments of the 1920s and 1930s generated similar examples of poor integration by crossing very disparate dog breeds such as salukis (a tall, long-faced breed) and Pekingese (a tiny, short-faced breed). A proportion of the second generation (F2) of these crosses also had malocclusions indicating partially separate genetic influences on upper versus lower jaw length (Stockard, 1941). Such studies, in combination with observations of the phenotypic effects of particular genetic mutations in model organisms like mice, give us glimpses into how combinations of genes contribute to various aspects of morphological integration. But we don't yet know which genes and pathways are

responsible for malocclusions, let alone for most other aspects of morphological integration relevant to human evolution.

The second reason to focus on mechanisms and patterns of morphological
integration is their importance from an evolutionary perspective. As Chapter
1 proposed, there is good reason to believe that heads work effectively and are
highly evolvable in part because of numerous interactions among modules
during development. My dog's malocclusion appears to have no negative effect on her ability to chew, possibly because her skull has compensated during
growth in many ways for whatever minor genetic deficiencies or incompati-
bilities she inherited. In addition, there turns out to be a strong correspondence between phenotypic and genotypic measures of integration (Cheverud,
1988). Thus although we cannot yet identify which genes are responsible for
most aspects of genetic integration, analyses of morphological integration provide useful clues about where to look.

Quantifying Integration in the Head

Before considering how patterns of craniofacial integration might have changed
in human evolution, and what processes may be responsible for these shifts, it
is useful to review some alternative, complementary ways to quantify integration (for a more complete review, see Chernoff and Magwene, 1999). Note
that some methods focus on patterns, and others focus on processes.

Scaling

Heads change size (volume and mass) as they grow, and their constituent modules vary in size and shape as they grow in a coordinated fashion. Thus a basic way to test many hypotheses of integration is to compare different modules'
allometries (size-related changes in shape) or heterochronies (changes in the
rate, timing, and appearance of size and shape) (see Chapter 2). There are several ways to do this. First, it is reasonable to hypothesize that if different modules of an organism are influenced by the same developmental stimuli or
programs that relate shape, size, and time, then these modules are likely to
share the same allometry. For example, if brain expansion drives endocranial
fossa growth, then the volume of each endocranial fossa should scale similarly
to brain volume during ontogeny. The shapes of particular modules can be
quantified simply with a few measurements or ratios, or more complexly us-

ing multivariate data-reduction methods such as principal components analysis (PCA, described below). The latter permit comparisons of shared covariation with respect to both size (allometry) and time (heterochrony). Comparisons of allometries among species can also be used to look at different hierarchical levels of integration (Zelditch and Fink, 1996). Within a head, some components will scale differently and with varying degrees of strength at different spatial scales. Nasal dimensions, for example, can have different scaling relationships relative to body mass as a whole, to general regions (e.g., neurocranial or facial volume), or to the size of local neighboring tissues (e.g., maxillary length or frontal lobe volume). As patterns of integration change or stay constant between species, so might these allometries.

Patterns of Correlation and Covariation

Integration among different components of the head can occur through many processes, such as the effects of scaling, hormonal influences, pleiotropy, and regional interactions. Thus another way to look at craniofacial integration is to compare patterns of correlation and covariation of various parts of the head (remember that correlations are size-standardized measures of the strength of covariation between two variables). This approach, pioneered by Olson and Miller (1958), was later combined with a population genetics framework in a series of landmark papers on primate skulls by James Cheverud (1982, 1988, 1995, 1996). Two of Cheverud's findings are especially important. First, Cheverud found that patterns of correlation and covariation in the basicranium, neurocranium, and face are partially independent, suggesting that they act as partially independent units. Second, Cheverud found that patterns of genetic and phenotypic variation correlate strongly with each other. Given that most traits do not have high heritabilities (genetic factors typically account for less than 50 percent of the variation in most skeletal traits), Cheverud concluded that genetic (mostly pleiotropic) and environmental stimuli have broadly comparable effects on growth and development of the phenotype (but see Roff, 1995). In other words, environmental effects may produce correlated responses throughout the cranium via the same epigenetic mechanisms as genetic effects.

Many studies have followed from Cheverud's insights, including several on primate crania (e.g., Marroig and Cheverud, 2001; Strait, 2001; González-José et al., 2004; Ackermann, 2005; Bastir and Rosas, 2005; Hallgrímsson et al.,

2007a). They all corroborate the basic finding that the basicranium, neuro-cranium, and face are semi-independent units that interact with each other to generate different overall patterns of craniofacial integration in different species. But there are some complications. Studies by Hallgrímsson and col-leagues (2004, 2006, 2007b) have found that disruptions to normal patterns of cranial growth increase phenotypic variability, sometimes with the effect of increasing rather than decreasing overall levels of integration. This seemingly counterintuitive effect occurs because many developmental processes (e.g., skeletal condensation size, muscle-bone interactions, brain growth) each con-tribute different patterns of craniofacial covariation. The covariation we mea-sure is actually the sum of these patterns. Hallgrímsson et al. (2007b) liken this summation to a palimpsest—a manuscript that has been partially erased and then written over so that old and new words are mixed together. Thus, if a key process such as brain growth is a major source of covariation, then mod-ifications to that process may increase the skull's overall degree of integration because its co-varying effects will account for a greater percentage of the total covariation present within the skull. As examples, mice with disrupted pat-terns of cranial base or brain growth often have more rather than less inte-grated skulls (Willmore et al., 2006; Hallgrímsson et al., 2007a, 2007b).

Why do varied patterns of covariation and degrees of integration matter? The answer is that some developmental processes appear to be more impor-tant than others in generating or constraining patterns of integration in the skull. Analyses of covariation in species such as humans and mice find that brain size and the widths of adjoining regions tend to be particularly impor-tant contributors to covariation (Lieberman et al., 2000a; Polanski and Fran-ciscus, 2006; Hallgrímsson et al., 2007a). Individuals with larger brains tend to have wider cranial bases, presumably because of the effects of brain size on cranial base growth. Moreover, there is a consistent pattern of covariation between the widths of adjoining regions, so that an individual with a wider neurocranium also has a wider cranial base and a wider face (and vice versa). This pattern makes developmental sense because the face grows forward and downward from the cranial base, and the brain grows above the cranial base, suggesting that the widths of these regions should constrain or interact with each other. Finally, individuals with wider faces also tend to have shorter faces, but the overall degree of integration in the face tends to be less than in the neurocranium and basicranium, especially in humans. These patterns of co-variation are important because the dominant changes in human evolution in-volve variations in brain size, cranial base angulation, and face size.

Ontogenetic Patterns of Integration

Another important method of studying craniofacial integration is to examine ontogenetic patterns of covariation—the degree to which two or more variables vary together over the course of ontogeny (Zelditch, 1988). This approach can be applied to several kinds of data, including caliper measurements, but is increasingly used in conjunction with three-dimensional landmark data in which shape is a multidimensional quantity (e.g., Bookstein et al., 2003; Strand-Vidarsdóttir et al., 2002; Ponce de León and Zollikofer, 2001; Ackermann and Krovitz, 2002; Cobb and O'Higgins, 2004; Lieberman et al., 2002, 2004b, 2007a; Mitteroecker et al., 2004, 2005; Bastir and Rosas, 2004b, 2005). Such studies on cross-sectional samples of human and ape skulls typically find that most ontogenetic variation is explained by just a few independent, multivariate measures of shape (e.g., principal components, PCs, which quantify a set of correlated shape changes) that co-vary during ontogeny. Multivariate measures, such as PCs of cranial data sets, rarely quantify discrete traits that can be interpreted individually (one needs to look at many PCs to avoid missing important aspects of variation), but specific PCs often capture particular suites of morphological change, such as facial elongation or neurocranial globularity. Further, many studies have found that just a handful of ontogenetic shifts often account for a large proportion of the transformations observed between closely related species. However, interpretations of these transformations differ, as shown in Figure 5.1. Several studies find that modern humans, Neanderthals, and apes have nearly parallel but different postnatal ontogenetic trajectories for multivariate measures of shape when graphed against time (Ponce de León and Zollikofer, 2001; Ackermann and Krovitz, 2002; Zollikofer and Ponce de León, 2004; see also Bastir et al., 2007). If so, then African apes, hominins, and modern humans undergo similar patterns of postnatal integration; they differ in adult shape primarily because of prenatal differences. However, other studies find that the ontogenetic trajectories of apes and humans are not so coincident (Cobb and O'Higgins, 2004; Mitteroecker et al., 2004, 2005; see below). Such results, especially those that consider integration among different regions of the skull, suggest that contrasts between humans and chimpanzees are products of different prenatal starting points as well as different postnatal trajectories.

Both viewpoints have some validity. In some respects (depending on what one is analyzing), African great ape and human ontogenies resemble each other because of fundamental similarities in how these species' skulls grow from the

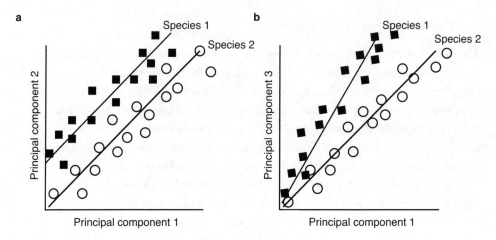

Figure 5.1. Comparison of multivariate allometry for two arbitrary species (represented by squares and circles). Note that PC1 and PC2 scale similarly, but PC1 and PC3 do not.

same set of sutures, synchondroses, and growth fields. However, the postnatal ontogenies of humans and great apes differ in substantial ways that derive not only from prenatal differences but also from derived mechanisms of growth and development of the sort considered in Chapter 4, such as more brain growth (both relative and absolute), reversals of depositional versus resorptive fields, and so on. There are several ways to quantify or visualize such differences. One method is to compare matrices of covariation for landmarks or other kinds of measurements in different species (see Marroig and Cheverud, 2001; Ackermann, 2005). Another method is to compare the slopes of ontogenetic trajectories of aspects of shape (e.g., angles, distances, ratios, or PCs) as shown in Figure 5.1. However, ontogenetic samples of two species may have parallel ontogenetic trajectories for two aspects of shape (e.g., brain versus orbit shape, or PC1 versus PC2), but the slopes of the trajectories may diverge significantly in other dimensions of shape change (e.g., brain versus nasal shape, or PC1 versus PC3) (Cobb and O'Higgins, 2004; Mitteroecker et al., 2004; Zollikofer and Ponce de León, 2004). We will return to this issue below.

The Functional Matrix Hypothesis

Multivariate analyses of shape change during ontogeny and studies of correlation and covariation are powerful tools for identifying patterns of integration, but they do not address the mechanisms by which these patterns occur.

Thus a complementary way to look at integration in the skull is to focus on how units interact. One such framework is the functional matrix hypothesis (FMH), originally developed by van der Klaauw (1948–52) and subsequently formulated by Melvin Moss (Moss and Young, 1960; Moss, 1968, 1997a–d; Moss and Salentijn, 1969). As Chapter 2 describes, the FMH is essentially a hypothesis about local interactions. It views the head as many functional matrices, defined as organs and spaces (such as the eyeball, brain, and muscle insertions), whose size and shape are very genetically regulated. In turn, these matrices have skeletal capsules (such as the orbit, neurocranium, and coronoid process) whose size and shape are primarily determined by the size and shape of the matrices they enclose. The FMH probably applies most strongly to the brain, whose size and shape conform closely to those of the endocranial fossae and the neurocranium (Richtsmeier et al., 2006). For example, the volume of the temporal lobe scales isometrically with the volume of the MCF (McCarthy, 2004). However, the FMH does not apply equally well to all parts of the skull (see Chapter 2). One problem is that disparate functional matrices can have different influences on particular aspects of morphology. For instance, the eyeball makes up only a small portion of the orbit (it scales with negative allometry) because the orbit is part of several functional matrices, of which the eyeball is only one (see Chapter 10). The orbit's roof is also the floor of the ACF, the orbit's posterior wall is partly the anterior wall of the MCF, the medial wall of the orbit is part of the nasal cavity's lateral wall, and so on.

Moss's FMH is nonetheless a useful conceptual framework to study integration if one expands the concept to distinguish *primary interactions* between a given functional matrix and its skeletal capsule and *secondary interactions* between a given functional matrix and other capsules it influences indirectly, via shared components of the skull (Moss, 1968, 1997d; McCarthy, 2004). If the FMH is considered in this broad sense, then the contents of capsules should have a strong primary influence on neighboring skeletal units and somewhat weaker secondary influences on more distant skeletal units. Such logic leads to complex predictions. For example, if the form of the frontal lobe directly influences the form of the floor of the ACF, which is also the roof of the orbit, and the form of the eyeball directly influences the medial wall of the orbital cavity, then the frontal lobe should scale more strongly (and more isometrically) with the orbit than with the nasal cavity. Analyses of covariation and scaling in primates for the most part uphold these predictions, although more strongly in the neurocranium and basicranium than in the face (McCarthy, 2004).

Part-Counterpart Analysis

A related way to consider how regions of the skull fit together is the part-counterpart principle formulated by Donald Enlow. Based largely on empirical observations, Enlow (1990) argued that the skull consists of a series of parts whose growth and development must conform spatially with structural counterparts (and vice versa). For example, the mandibular and maxillary arches are counterparts in the sense that growth in one must be matched by growth in the other. In turn, the maxillary arch is a counterpart to the roof of the nasal cavity, which is a counterpart to the floor of the ACF. As Figure 5.2 illustrates, Enlow argued that the skull has two major series of parts and counterparts, divided by a key conceptual boundary between the back of the face and the front of the MCF, the posterior maxillary (PM) plane (discussed below in detail) (Enlow and Azuma, 1975). The anterior set consists of the frontal lobes, the ACF, the nasal cavity, the palate, the maxillary and mandibular arches, and the mandibular corpus. The posterior compartment contains the temporal lobes, the MCF, and the pharynx behind the palate. Enlow suggested that the parts and counterparts are highly integrated within each compartment but relatively independent between compartments. Bastir and Rosas (2006) showed that the skull also has a midsagittal and a lateral set of compartments, each with their own parts and counterparts.

In a comprehensive analysis of primates, McCarthy (2004) found a strong degree of correlation between parts and counterparts within Enlow's anterior and posterior compartments in terms of widths rather than lengths. In particular, the widths of the anterior compartment tend to scale isometrically with adjacent counterparts, and with negative allometry with more distant parts of the same compartments. However, measurements of facial length tend to be independent of the rest of the skull.

Synthesis

There is no single way to measure integration in complex morphological units such as the head because integration is a complex phenomenon and because the methods one uses depend on the questions one asks. If one has a particular hypothesis about how patterns of covariation change between ancestral and descendant species, then comparisons of allometries and other interesting covariations are especially useful. Alternatively, if one wants to identify or compare patterns of covariation across ontogeny without a specific a priori hy-

Figure 5.2. Schematic view of Enlow's part-
counterpart model, showing two columns of major
units posterior and anterior to the posterior
maxillary (PM) plane. The lengths of the units
in the anterior and posterior columns roughly
correspond with each other. See text for details.

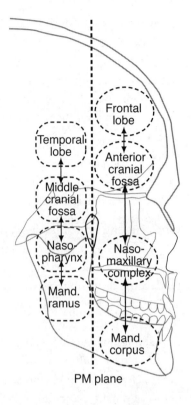

PM plane

pothesis, then multivariate analyses of covariation (e.g., principal components
analysis) are more useful tools. Since Olson and Miller 1958, there have been
dozens of studies of integration in human and primate skulls.

Note also that analyses of integration *patterns* address different questions
than do analyses of integration *processes.* Given the basic questions addressed
here—what major mechanisms integrate the head's various components during ontogeny, and how these mechanisms changed during human evolution—
the latter are especially important to consider. I do so in the context of both
Moss's FMH and Enlow's part-counterpart hypothesis. The two hypotheses
are complementary because they model different types of interactions among
partially discrete regions of the head. In addition, both Moss's and Enlow's
analyses indicate a general pattern similar to that of the analyses of covariation
reviewed above. That is, the head consists of many units that are initially partly
independent but which become increasingly integrated into a whole by a few
dominant interactions. It follows that to study craniofacial transformations
that occurred during human evolution, one needs to know which aspects of
the skull influence each other to what degree during ontogeny and by what

mechanisms. Our prime suspects are brain volume, the dimensions of the cranial base (notably, angle, width, and length), and key dimensions of the face (notably, width, height, and depth). As in Chapter 4, we will first consider similarities and differences in overall patterns of integration during great ape (mostly chimpanzee) and human ontogeny. We'll then return to consider how certain regional mechanisms reviewed in Chapter 4 may be responsible for a large proportion of the transformations that occurred during human cranial evolution.

Patterns of Craniofacial Integration in Chimpanzees and Humans

One could fill several books with analyses of craniofacial integration. A useful, succinct way to start is to summarize key similarities and differences in the patterns of overall integration in humans and our closest relative, the chimpanzee, by comparing the two species' ontogenies using multivariate methods. To do this, I use cross-sectional samples of common chimpanzees and humans to ask two general questions. First, how are patterns of ontogenetic integration similar and different in chimps and humans? Second, how would one have to alter a chimp's ontogeny to make its skull develop into an adult human's? However, before comparing patterns of craniofacial growth and integration humans and chimps, it is useful to begin with a short preamble on ways to measure size and shape.

Quantifying Size and Shape

For centuries, size and shape were most commonly quantified using tools such as calipers and protractors to measure distances and angles between landmarks (homologous points in skulls, such as the intersection of two sutures). One can quantify size in a number of ways with such measurements. Often one uses a particular measurement of general size such as the skull's maximum length, or one calculates the mean of a set of distances to reflect some notion of size. A better measure of an object's size is the geometric mean, the nth root of the product of n measures (see Jungers et al., 1995). However, the shapes of complex objects like skulls are not as easily captured using angles and linear distances. One common method is to decompose the skull into various ratios and angles of interest. For example, facial shape might be measured as the ratio of facial width and height or as the angle of the line between the top and bottom of the face relative to some other line. In addition, multivariate methods such as principal components analysis (PCA) allow one to consider these variables together. PCA is a data-reduction method that transforms a set of correlated variables into a smaller number of uncorrelated (statistically independent) variables called principal components (PCs). By definition, the first PC accounts for as much variability as possible in the data set, and each succeeding PC accounts for as much of the remaining variability as possible. Each PC is orthogonal to other PCs, and is essentially an independent aspect of variation within the sample that is statistically uncorrelated with other PCs.[1]

Linear distances and angles are tried-and-true methods of measuring things, but they have limitations for quantifying the size and shape of objects as complex as skulls. One drawback is that intricate shapes cannot be captured effectively by just a few measurements. If you took several dozen linear measurements and angles on a skull, you would not be able to reconstruct what the actually skull looks like. Fortunately, the last few years have seen a revolution in a method known as geometric morphometrics (GM), which uses new technologies to measure the coordinates of landmarks instead of measuring distances between landmarks. Rather than measure 190 distances for a set of 20 landmarks, one needs to record only the coordinates of the 20 landmarks (often using a CT scan or a machine that measures three-dimensional (3-D) coordinates in a calibrated space).[2] The use of 3-D coordinates saves time and allows one to calculate distances and angles between any pair of landmarks and reconstruct their spatial relationships. The scale of objects described by such data is naturally described by a quantity known as the centroid size (the square root of the sum of squared distances of a set of landmarks from the object's geometric center), which is uncorrelated with the shape of the landmark configuration. GM methods therefore permit one to measure and compare aspects of shape, size, and form (shape plus centroid size) among complex objects such as skulls more powerfully than is possible with interlandmark distances (see Dryden and Mardia, 1998; Zelditch et al., 2004).

GM analyses work by taking a set of coordinates, scaling them to a common centroid size, and then computing aspects of shape using a technique termed Procrustes analysis (after the mythical Greek bandit who either amputated or stretched the legs of his "guests" to make them fit his iron bed). In

---

1. One limitation of PCA is the arbitrary decomposition of data into orthogonal components. Biologically meaningful aspects of variation thus might be distributed among several PCs, and multiple biological factors can contribute to an individual PC. As with all methods that simplify data, PCs require careful interpretation.

2. For a set of n landmarks, the total number of interlandmark distances is $n(n - 1)/2$.

a Procrustes analysis, one scales the landmark sets to the same centroid size, translates them to a common origin, and then rotates them to minimize the sum of their squared landmark distances (Rohlf and Slice, 1990; Bookstein, 1991; Zelditch et al., 2004). The resulting Procrustes "superimposition" thus allows one to quantify shape independently of size as the deviation of a given object's landmarks from a reference configuration, typically the average landmark configuration of the entire sample. Each analysis thus defines its own "shape space." Aspects of shape that co-vary in a sample can then be quantified using PCA, as noted above.[3] The set of PCs may help the researcher visualize the key aspects of shape variability within a sample (O'Higgins, 2000b; Eble, 2002), but individual PCs should be interpreted with care, as they depend on the full covariance structure of the sample, and biologically interesting aspects of variation, such as allometry, may be distributed over more than one PC (see below and Figure 5.4). For analysis of form rather than shape variability, inclusion of the log of centroid size in the PCA results in PCs that collectively describe form variability. One might subsequently examine how shape or form varies with body mass, length, or other variables in relation to the question at hand.

As noted above, a powerful way to examine patterns of shape or form change in the skull during ontogeny is to extract the principal components of landmark variation from an ontogenetic sample. The sample might include one or more individuals measured successively at different ages (a longitudinal sample) or different individuals of different ages (a cross-sectional sample). In either case, the first few PCs may give a useful impression of those aspects of shape change that co-vary during ontogeny. That is, if one graphs PC1 versus PC2, as in Figure 5.3b, the individuals may vary in position along the axes (from left to right and top to bottom) in approximate ascending order of age. This is because PCA extracts linear components of variability that are ordered according to total explained variance, and ontogentic shape variability is usually large relative to other sources of variability. In addition, although the PCs of shape are derived from

---

3. In a 3-D landmark data set, the PCA rotates the landmark configurations using their covariance matrix in multidimensional space to identify independent (uncorrelated) linear combinations of geometric landmark shifts (warps) (for discussion, see O'Higgins, 2000a, 2000b; Zelditch et al., 2004). The transformation of one shape (a matrix of coordinates) into another shape (another matrix of coordinates) can be described completely, using linear algebra, by its eigenvalues and eigenvectors. An eigenvector is a vector whose direction is unchanged by that transformation. The factor by which the magnitude is scaled is called the eigenvalue of that vector.

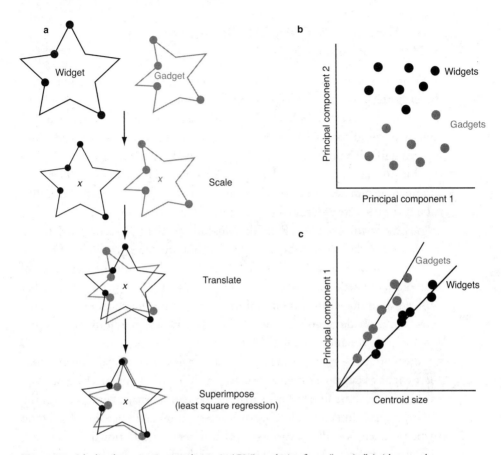

Figure 5.3. Idealized geometric morphometic (GM) analysis of two "species" (widgets and gadgets) with homologous landmarks (circles). (a) In a Procrustes superimposition, the "species" are scaled to the same centroid size, translated to the same origin (*x*), and then rotated to minimize the least-square distances between homologous landmarks. Principal components analysis of the landmark coordinates for samples of both species can reveal similarities and differences in the distribution of patterns of covariation (b) and in scaling (c) for different principal components of variation.

scaled landmark configurations, PC1 scores (and scores on the first few PCs) often correlate with measures of organismal scale, because during ontogeny animals often undergo allometric changes in shape as they get bigger. Therefore, plotting a given PC of shape change versus centroid size, as in Figure 5.3c, allows visualization of some aspect of multivariate allometry (see Chapter 2).

A GM Comparison of Craniofacial Ontogeny in Humans and Chimpanzees

Let's now look at a modest analysis of some similarities and differences between patterns of craniofacial ontogeny in chimps and humans, highlighting how these heads are integrated. These analyses are based on thirty-two 3-D landmark coordinates taken from cross-sectional samples of human and chimpanzee skulls. The human sample has 16 individuals, ranging from ages 0 to 18 (the Bosma collection; for details, see Shapiro and Richtsmeier, 1997). The chimpanzee sample includes 25 individuals ranging from a neonate to adults; all but the neonate are from a population of Liberian chimpanzee skulls (*Pan troglodytes verus*) from the Peabody Museum at Harvard University. Landmarks on the skulls were obtained using a computer to record the 3-D coordinates from CT scans of each skull. Many of the landmarks are sites of craniofacial growth (e.g., sutures, synchondroses, major growth fields), so that differences in shape also contain information about differences in how crania grow. For more details on the samples, landmarks, and methods used to analyze the data, see Lieberman et al (2007a).

After correcting for size using a Procrustes analysis (see above), one can summarize many key shape differences between chimps and humans by graphing, as in Figure 5.4, the first few principal components of shape computed for the pooled sample. In this case, PC1 explains a substantial 84 percent of the total shape variance, and PC2 explains another 4 percent (additional PCs explain diminishingly less variation, each with little information). Note that these two PCs describe differences between species as well as aspects of covariation that are *common* to both species (because the PCs were computed from the pooled sample); the wireframes at the extremes of each axis summarize aspects of shape quantified by each PC. Unsurprisingly, the first PC records aspects of difference between species, in that there is very little overlap in shape space between them. PC1 also represents aspects of shape change that co-vary most strongly with both size and age, particularly in humans. In both chimps and humans, younger individuals have lower PC1 values (they lie toward the left side of the graph), which describe more globular skulls with relatively smaller, more retracted faces; older individuals have higher PC1 values (toward the right), with more elongated braincases and longer, more projecting, and more prognathic faces. As Figure 5.4 shows, when corrected for size, adult human crania are closest to the chimpanzee neonate in aspects of shape described by PC1 (the degree of facial projection, length of the face, globularity of the

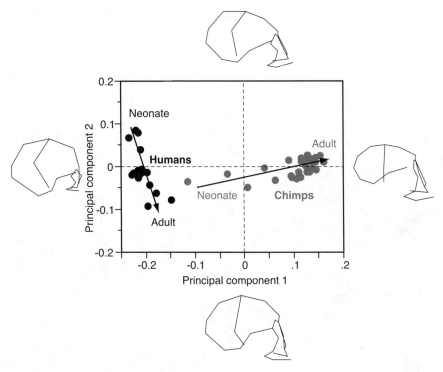

Figure 5.4. Principal components analysis of ontogenetic samples of chimpanzee and human crania, showing PC1 (which accounts for 84 percent of sample variance) and PC2 (4 percent of sample variance); the wireframe diagrams around the borders graph the extremes of shape variation for each PC. The analysis reveals profound differences in cranial anatomy with no overlap in PC1, which primarily measures globularity of the neurocranium and projection of the face. Note also the completely different trajectories of development for PC2, which primarily measures the angle of the cranial base and facial prognathism: adults have lower values in humans but higher values in chimpanzees.

braincase) and PC2 (cranial base angle, facial prognathism). Note that other information not shown here (12 percent of total sample variance) is contained in other PCs.

Scaling between PC1 scores and centroid size is summarized in Figure 5.5, which also plots PC1 against postnatal age. The human ages are known, but the chimp ages were estimated using data on the relationship between age and the degree of completion of tooth roots and crowns in zoo animals (Dean and Wood, 1981; Kuykendall, 1996). These graphs highlight how the multivariate allometry between size and those aspects of shape quantified by PC1 is dif-

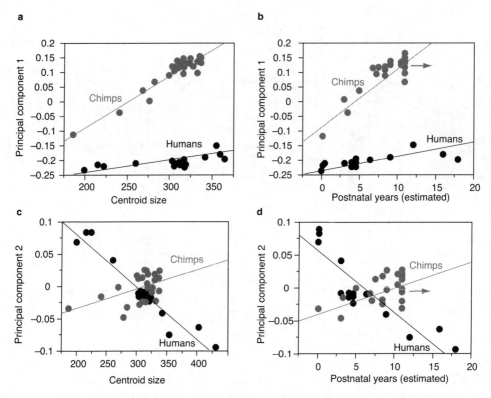

Figure 5.5. Scaling of PC1 and PC2 against centroid size (a, c) and age (b, d) for ontogenetic samples of chimpanzee and human crania. Humans have a lower slope and intercept for PC1 (which explains 84 percent of sample variance) than chimpanzees when plotted against both size and age (see Figure 2.6 for details). Note also that PC2 scales inversely with size and age in humans but not in chimpanzees.

ferent in humans and chimps. Chimps not only have a different growth trajectory in terms of slope and intercept but also undergo a greater degree of shape change relative to size (as reflected by a much steeper slope), and do so more rapidly than humans.

In the above analysis, PC1 explains 84 percent of the variance, leaving another 16 percent to be considered. With the exception of PC2, the additional PCs reflect variations between individuals that bear no obvious relationship to ontogeny. PC2 does record important changes that occur during ontogeny, although it correlates more poorly with size than PC1 in chimpanzees ($r^2 = 0.24$), but not humans ($r^2 = 0.93$). As shown in Figure 5.5c and d, younger humans have higher PC2 values, which indicate less flexed cranial bases and more prognathic faces; older humans have lower PC2 values, indicating more flexed cranial bases and more vertical faces. Significantly, this pattern is the reverse of what occurs in chimpanzee ontogeny, in which younger individuals have more flexed cranial bases and ventrally rotated faces (low PC2 scores), and older individuals have more open cranial bases and dorsally rotated faces (high PC2 scores). Thus the ontogenetic trajectory for PC1 and PC2 differs fundamentally in chimps and humans (Figures 5.5c,d).

The above GM analysis gives a snapshot of some general similarities and differences in patterns of ontogeny between chimps and humans and corroborates the findings of other published studies (notably Penin et al., 2002). As both species progress through postnatal ontogeny, the braincase becomes more elongated, and the face becomes relatively larger and more projecting. Thus, the two species share some patterns of integrated growth (quantified here by PC1), but with several important differences. First chimp and human crania already differ in shape by the time of birth. Second, chimp crania undergo considerably more postnatal shape change than humans. Third, some patterns of shape change differ during ontogeny in chimps and humans—one being the cranial base, which extends in chimpanzees but flexes in humans.

Can we use these landmark data to identify more precisely differences and similarities in craniofacial growth between the two species? A useful way of looking at the problem is to compute a transformation grid after the style first imagined by D'Arcy Thompson (1917), a pioneer of mathematical biology. To do this, the landmarks (in this case, after a Procrustes size correction) are plotted on a regular 2-D grid and "warped" from one configuration (the reference) to the other (the target) using a mathematical function called a thin-plate spline (TPS). Deformations of the grid squares reflect the deformation of the reference landmarks into the target, giving a sense of vectors (orientations and magnitudes) of shape change among the landmarks. Because most of the landmarks are key growth sites for the skull, it follows that differences in the transformation grid contain information about differences in growth processes. Figure 5.6 compares three transformations. Parts a and b compare the transformations from a neonatal chimp and human into an "average" adult chimp and human, respectively. Again, the chimpanzee undergoes much more shape change than the human, and the patterns of change are very different. As the chimp grows, its face becomes longer, taller, more prognathic, and more projecting; the cranial base both elongates and extends; the PCF and foramen magnum both grow posteriorly; and the nuchal plane rotates vertically. Overall, the chimp neurocranium becomes more lemon-shaped (relatively longer

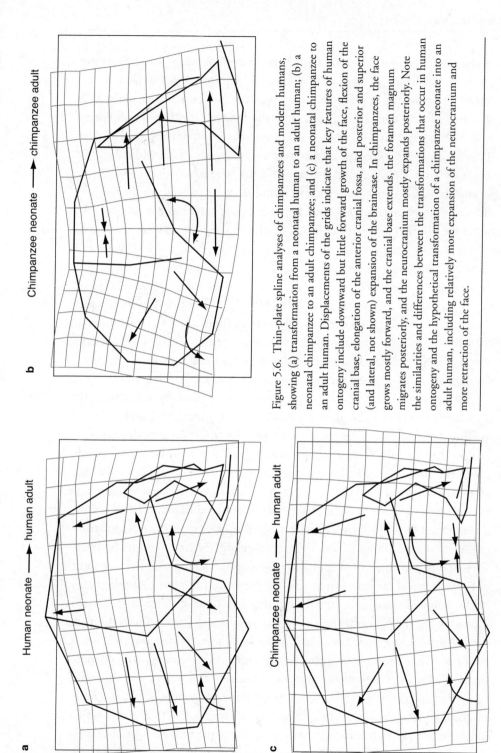

Figure 5.6. Thin-plate spline analyses of chimpanzees and modern humans, showing (a) transformation from a neonatal human to an adult human; (b) a neonatal chimpanzee to an adult chimpanzee; and (c) a neonatal chimpanzee to an adult human. Displacements of the grids indicate that key features of human ontogeny include downward but little forward growth of the face, flexion of the cranial base, elongation of the anterior cranial fossa, and posterior and superior (and lateral, not shown) expansion of the braincase. In chimpanzees, the face grows mostly forward, and the cranial base extends, the foramen magnum migrates posteriorly, and the neurocranium mostly expands posteriorly. Note the similarities and differences between the transformations that occur in human ontogeny and the hypothetical transformation of a chimpanzee neonate into an adult human, including relatively more expansion of the neurocranium and more retraction of the face.

a Human neonate ⟶ human adult

b Chimpanzee neonate ⟶ chimpanzee adult

c Chimpanzee neonate ⟶ human adult

from front to back and relatively smaller from top to bottom), and the upper part of the vault is "scrunched-up" (compressed) in the grid because it is relatively smaller in the adult than in the neonate.

In contrast, the human face grows mostly downward, not forward, with no upper facial projection; the cranial base both elongates and flexes; and the neurocranium expands like a balloon in all directions. I do not include a picture of the transformation grids for the crania projected into other planes, but the big difference you would see is that the human cranium undergoes relatively more mediolateral growth, especially in the MCF. Finally, Figure 5.6c examines the transformation from a neonatal chimpanzee into an adult human. This transformation shows less change than the human ontogenetic transformation in Figure 5.6a, except that the neurocranium expands relatively more and the face relatively less (especially anteoposteriorly). In addition, there is relatively less elongation of the oropharynx behind the palate because of less facial prognathism.

It has often been suggested that it takes relatively less shape change to transform a neonatal chimp or human skull into an adult human than into an adult chimpanzee. In addition, and as noted by Penin et al. (2002), on PC1 alone, adult humans resemble young chimps (again, see Figures 5.4 and 5.5). Leaving aside differences evident from other PCs (not always a good idea, as noted above), one might be tempted to conclude that human craniofacial form evolved through a heterochronic transformation by shifting the rate and timing of developmental patterns present in chimpanzees. Many authors, especially Gould (1977), have specifically hypothesized that the modern human head is an example of neoteny, in which adult modern humans retain the shape of the juvenile stage of an apelike ancestor but grow to a similar adult size. (For a review of heterochrony and neoteny, see Chapter 2.)

As with many complex phenomena, there is some truth to the idea that heterochrony was involved in human evolution, but a simple diagnosis of neoteny is wrong for several reasons. First, the adult modern human skull vaguely resembles a juvenile chimp's because of the combination of a relatively big brain and a relatively small face (an aspect of integration I discuss below in detail). But modern humans achieve this combination through two contrasting processes. The brain and the surrounding braincase grow through a combination of rate and time hypermorphosis (increased rate and duration of growth of the brain and its surrounding capsule). In contrast, the modern human face has a more prolonged growth trajectory than that of the chimp (see Figure 4.1), but growth proceeds at a vastly slower rate. Superficially, this is an example of time *hyper*morphosis combined with rate *hypo*morphosis, not neoteny (see Figure 2.6). Moreover, as we saw in Chapter 4, several processes of skull growth are fundamentally different in humans and chimps, precluding the use of just heterochrony as a framework for interpreting the superficial similarities we observe (Shea, 1989). Key derived differences in modern humans include flexion rather than extension of the cranial base, horizontal rather than vertical rotation of the nuchal plane, and resorption rather than growth along the front of the face.

Major Mechanisms of Craniofacial Integration

To summarize, the analyses and the covariation and GM studies discussed above suggest that four general suites of ontogenetic change are most important in the transition from a hypothetical African great ape to a human pattern of craniofacial ontogeny:

1. The brain becomes relatively larger and the neurocranium becomes more spherical in modern humans. This transformation occurs via relatively more superoinferior, posterior, and mediolateral growth of the brain combined with more anteroposterior and mediolateral growth of the endocranial fossa.
2. The modern human cranial base becomes more flexed.
3. The foramen magnum in modern humans stays closer to the center of the skull base, and the nuchal plane elongates posteriorly while rotating to a more horizontal position.
4. The modern human face remains relatively small and nonprognathic during ontogeny, growing mostly inferiorly and mediolaterally and remaining more retracted.

Leaving aside the issue of what developmental mechanisms make a face smaller, a cranial base shorter and wider, and a brain bigger, the next step is to consider how many ontogenetic shifts are responsible for the morphological differences between the heads of these two species. By now, my basic argument should be familiar: the intrinsic structure of the head and its mechanisms of growth generate a limited suite of interactions among units in the basicranium, neurocranium, and face. These integrative interactions cause a predictable, limited set of variable responses, thereby channeling craniofacial variation in predictable, limited ways.

One problem with elucidating these integrative interactions is that the mechanisms themselves are neither linear nor independent. One cannot describe them as simple *pathways* in which A influences B, which then influences C, and so on. Instead, they are complex, developmental *networks,* characterized by combinations of many pathways, each with multiple inputs and outputs, and with various interactions with other pathways (Wilkins, 2002). Put differently, these interactions take place at multiple levels of the developmental hierarchy. Some involve shared developmental-genetic pathways, others involve inductive interactions among modules during morphogenesis, still others involve mechanical interactions between neighboring tissues (e.g., muscle-bone interactions), and others involve shared physical displacements during growth. As an example, brain size influences cranial base flexion, which influences facial projection. But facial projection and cranial base flexion are also influenced in other ways by the size of the face, which also has additional effects on brow ridge shape, and so on. In other words, the logic of the way heads assemble themselves makes it impossible to describe a simple network of interactions that can explain all directions of variation in craniofacial shape. Instead, we must consider craniofacial shape as the complex combinatorial product of multiple, interdependent interactions.

There are many interactions—all with combinatorial effects—that account for the patterns of craniofacial integration documented above. Three general interactions, however, appear to have the greatest influence on differences between chimpanzee and human craniofacial growth (and thus of likely importance in the various transformations that occurred during hominin evolution): cranial base angulation, facial projection, and interactions between brain size and face size with the width and length of the cranial base.

Cranial Base Angulation

The angle of the cranial base is one of the most important structural factors that influences integration in the primate skull, and changes in how much and when the cranial base flexes or extends have played key roles in human evolution. Recall from Chapter 4 that the cranial base angle (CBA) quantifies the angle in the midsagittal plane between the prechordal and parachordal (postchordal) parts of the cranial base, and actually combines three angles at three synchondroses: spheno-occipital (SOS), mid-sphenoidal (MSS), and spheno-ethmoid (SES). In this sense, CBA is not a simple, single character but instead is a convenient way to quantify complex differences in the relative orientations of the three endocranial fossae in the sagittal plane. There are many ways to measure CBA, but here I primarily use CBA1, the angle between basion, sella, and the foramen caecum. Two questions concern us the most. First, what factors generate variations in CBA? Second, what are the structural consequences of variations in CBA?

Causes of Cranial Base Angulation

Determining what generates variations in CBA is no simple matter, because basicranial flexion and extension are dynamic processes influenced by several factors. The dominant factor is brain size, best expressed by the "spatial packing" hypothesis. This longstanding hypothesis is that flexion in the synchondroses of the midline cranial base increases the volume of the brain that can be accommodated above the basicranial platform (Bolk, 1926; Dabelow, 1929; Biegert, 1957, 1963; Hofer, 1969; Gould, 1977). Several lines of evidence support this hypothesis. First, there is comparative evidence that primates with larger endocranial volume relative to the length of the cranial base tend to have more flexed CBAs (Ross and Ravosa, 1993; Spoor, 1997; Lieberman et al., 2000b; McCarthy, 2001; Ross et al., 2004). Figure 5.7 illustrates this statistically significant relationship. Much of this flexion, moreover, appears to be related to expansion of the neocortex rather than the brain stem (Strait, 1999). Experimental studies that compared mutant mouse strains with variations in brain size relative to cranial base length also indicate that overall brain volume contributes to cranial base flexion. Hallgrímsson et al. (2007a) and Lieberman et al. (2008a) found that mutant mice with larger brains or relatively shorter cranial bases have significantly more flexed cranial bases than wild-type controls. In addition, a few studies have experimentally inhibited elongation of the SOS in animals (mostly rats), causing a more flexed cranial base, presumably because the cranial base could not extend as the face was growing (DuBrul and Laskin, 1961; Moss, 1976; Bütow, 1990; Reidenberg and Laitman, 1991). In most of these experiments, a more flexed cranial base was associated with a shorter posterior cranial base and a more rounded neurocranium (see, however, Young, 1959). Finally, "natural experiments," such as head binding, that artificially constrain growth of the cranial vault also tend to cause slightly but significantly more flexed CBAs (Antón, 1989; Cheverud et al., 1992; Kohn et al., 1993). In such cases, the cranial base probably flexes more to accommodate the growth of a brain whose expansion in other directions (e.g., anteroposterior or mediolateral) is constrained. Simi-

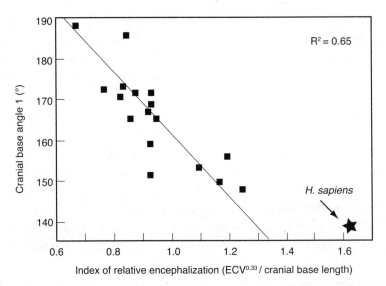

Figure 5.7. Plot of CBA1 versus the index of relative encephalization (IRE), the cube root of endocranial volume divided by the length of the cranial base (basion-sella + sella–foramen caecum), for a sample of adult primates. Primates with bigger brains relative to the length of the cranial base tend to have more flexed cranial bases, but modern humans fall significantly off the primate regression, with much less flexion than predicted by IRE. (Adapted from McCarthy, 2001.)

larly, microcephalic individuals with tiny brains tend to have less flexed cranial bases (Babineau and Kronman, 1969), and the cranial base in anencephalic individuals fails to flex prenatally (Trenouth, 1989).

Brain size, however, is not the only factor related to cranial base angulation. Figure 5.7 shows that only about 65 percent of the variation in CBA1 among adult primates can be accounted for by brain size relative to cranial base length (Ross and Ravosa, 1993; Lieberman et al., 2000b). Also, by some measures, modern humans do not have cranial bases that are as flexed as predicted by the relationship between brain size and the CBA among primates (Ross and Henneberg, 1995; but see McCarthy, 2001). Finally, ontogenetic studies show that the cranial base extends rather than flexes prenatally at times when brain growth is most rapid (Jeffery and Spoor, 2002, 2004a; Jeffery, 2003). The other factor most likely to influence CBA is the size of the face, including the pharynx. Just as the expanding brain on the upper surface of the cranial base is accommodated by more flexion, the growing face below the cranial base is accommodated by extension. This accommodation may occur to some extent via epigenetic processes as the face tenses the inferior surface of the anterior cranial base, inducing it to extend. Regardless of the mechanism, the effects of face size on the angle of the cranial base have been demonstrated in mice: in which relatively longer faces are associated with more extended cranial bases (Lieberman et al., 2008a). These interactions are less well studied in humans and other primates, but they probably matter. In modern humans, the cranial base flexes during the first two years, when the brain grows most rapidly relative to rate that the face grows; in contrast, the cranial base in chimpanzees is comparatively stable during this time, when both brain volume and facial prognathism increase rapidly in concert (Lieberman and McCarthy, 1999). In nonhuman primates, the cranial base then extends after the brain stops growing, and the face keeps projecting forward from the neurocranium; but in modern humans, CBA stays stable in part because the face remains short and tucked underneath the brain (see Figure 4.4b).

Just as cartilaginous growth plates drive long bone growth elsewhere in the skeleton, the synchondroses affect how the cranial base elongates, and they may also play key roles in regulating how much it flexes or extends (J. H. Scott, 1958). One likely mechanism is that cranial base flexion occurs through differential growth at the top and bottom of the synchondrosis, creating a hingelike mechanism (Sirianni and Van Ness, 1978). In addition, flexion may occur passively in the synchondroses from the brain pushing down on the posterior cranial base or the face pushing up on the anterior cranial base. This sort of spatial packing is thus likely to be influenced by the extent of elongation in different portions of the cranial base relative to increases in the volume of the brain above or the face below. If so, then it is possible that humans flex the cranial base postnatally because the SOS ceases to elongate at a same time when the brain expands rapidly. In other words, reduced basioccipital elongation is like binding the cranial base, forcing the cranial base to flex to accommodate the expanding brain. This hypothesis merits further study. Regardless, genes that influence cranial base elongation or angulation might be important mechanisms of genetic integration, but so far none have been identified.

Effects of Cranial Base Angulation

Regardless of what percentage of variation in CBA occurs as a result of brain growth, facial growth, and intrinsic cranial base growth, the other major issue to consider here is the integrative effects of cranial base angulation on overall craniofacial shape. These are several, but perhaps the most important is the influence of variations in CBA on facial orientation. Two fundamental constraints illustrated in Figure 5.8 underlie this pattern of integration. First, remember that the anterior portion of the cranial base is also part of the floor of the ACF and the roof of the orbits. Hence the orientation of the top of the face relative to the posterior cranial base is essentially the same as the orientation of the prechordal cranial base relative to the parachordal cranial base. The second constraint is that the posterior margin of the face, the junction between the MCF and the maxilla (hence the whole ethmomaxillary complex), is always nearly perpendicular to the orientation of the orbits (Enlow and Azuma, 1975). Specifically, if one models the orbits as a cone, then the orbit's central axis (the Neutral Horizontal Axis, NHA) is always within a few degrees of 90° relative to the posterior maxillary (PM) plane, the midsagittal projection of a line from the maxillary tuberosities to the anterior poles of the MCF (Enlow and Azuma, 1975; Enlow, 1990). The PM plane thus describes the posterior margin of the face where it grows forward from the MCF. In anthropoids, the mean angle between the PM plane and the NHA is 90.0° ± 0.38 s.d. (range 85.5°–92.5°), as is the case for prosimians and all other mammals so far studied (McCarthy and Lieberman, 2001).

Why the PM-NHA angle is always close to 90° is unknown, but the constraint's consequences for facial shape are profound because the NHA is always within a few degrees (<5° in anthropoids) of the floor of the ACF, itself the roof of the face. This means that the whole face rotates with the anterior cranial base as a block. If one flexes the cranial base, then the entire face rotates downward (ventrally) relative to the posterior cranial base (McCarthy and Lieberman, 2001, Bastir and Rosas, 2004a). As Figure 5.8 illustrates, ventral rotation of the face (the technical term is *klinorhynchy*), tucks more of the face below the ACF and the frontal lobes, and it shortens the nasopharynx and oropharynx behind the palate. Conversely, if one extends the cranial base, then the whole face must rotate upward (dorsally). Dorsal rotation (also known as *airorhynchy*) protrudes more of the face in front of the ACF and elongates the oropharynx and nasopharynx (see Figure 5.8c). In other words, the cranial base angle influences spatial separation between the upper face and the anterior cranial fossa (facial projection), and the angle of the lower face relative to the upper face (facial prognathism).

A second constraint (also seen in Fig. 5.8) relates the cranial base angle to facial orientation. In primates, the cribriform plate is usually close to perpendicular to the midsagittal profile of the face—a line from the bottom of the face, prosthion, to the front of the ACF (Enlow and Azuma, 1975). This an-

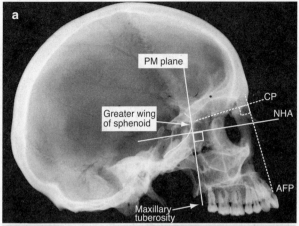

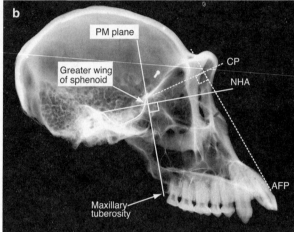

Figure 5.8. Major constraints on cranial shape and the facial block. The posterior maxillary (PM) plane is nearly always perpendicular to the neutral horizontal axis of the orbits (NHA) in modern humans (a), chimpanzees (b), and other species. Similarly, the plane of the cribriform plate (CP) tends to be nearly perpendicular to the anterior facial plane (AFP) in primates. Because the top of the face and the NHA are nearly coincident with the floor of the anterior cranial base (ACB), changes in the angle of the cranial base rotate the PM plane, the NHA, and the face as a whole relative to the posterior cranial fossa (c), affecting several aspects of cranial shape, including facial projection and the length of the nasopharynx.

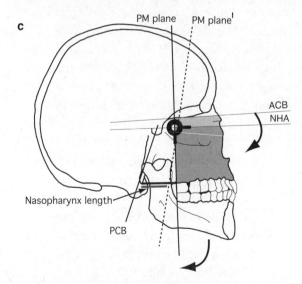

gle averages 90.8° ± 2.7 s.d. in monkeys, but with a greater range of variation (82°–98°) than the PM-NHA angle (McCarthy, 2004). This constraint may be adaptive because it orients the olfactory bulbs perpendicular to the nasal cavity, and is important for craniofacial integration because primates with more flexed cranial bases tend to have less prognathic faces.

Variation in CBA may also influence overall neurocranial shape. Primates with more flexed cranial bases have rounder brains, as measured by the angle of the brain stem to the long axis of the neocortex (Biegert, 1963; Hofer, 1969; Lieberman et al., 2000b; Ross et al., 2004). However, this relationship is partially confounded by circular logic because primates with larger brains have relatively larger neocortices (see Chapter 6), and the relative size of the cortical portion of the brain correlates strongly with CBA (Strait, 1999). In modern humans, brain volume is independent of CBA (Lieberman et al., 2000a).

One last aspect of skull shape potentially affected by the cranial base angle is the position and orientation of the foramen magnum. Many anatomists have noted that flexing the cranial base can rotate the foramen magnum closer to the center of the skull, orienting the foramen magnum more inferiorly than posteriorly (e.g., Bolk, 1910; Weidenreich, 1941, Schultz, 1942; Ashton and Zuckerman, 1956). This is true, but the position and angle of the foramen magnum are also influenced by other aspects of cranial shape such as nuchal plane orientation, posterior cranial base length, and the orientation of the orbits (see Chapter 9). Consequently, there is no simple, predictive relationship between CBA and either the position or the orientation of the foramen magnum (Strait and Ross, 1999; Lieberman et al., 2000b; Zollikofer et al., 2005).

In short, the angle of the cranial base has major effects on how different parts of the skull accommodate each other. A bigger brain and/or a shorter face is associated with a more flexed cranial base, a rounder brain case, and a more downwardly (ventrally) rotated face that is both less projecting and less prognathic. In addition, different combinations of brain size, face size, and other factors should lead to much variation in CBA. For example, big brains can sit on cranial bases that are more or less flexed depending on face size.

Facial Size and Projection

A second major aspect of craniofacial shape variation is facial projection, defined broadly as the position of the front of the face relative to the anterior cranial fossa and the anterior cranial base. A number of factors influence facial projection, including the cranial base angle, facial length, brain size, and the length of the anterior cranial base. In turn, facial projection influences the relative position of the face and neurocranium, and the shape of the entire supraorbital region.

Recall from Chapter 3, that the most anterior point on the cranial base is the foramen caecum, the remnant end of the neural tube. During fetal and early postnatal life, the primate brain grows in a neural trajectory, relatively more rapidly than the face, which remains tucked underneath the ACF. During this period, the rim of the orbits (the outer table of the frontal bone) remains spatially close to the foramen caecum. But after brain growth slows and then ceases, the face continues to grow forward in a skeletal trajectory. Some of this growth occurs because of bone deposition elsewhere in the face (especially the maxillary tuberosities) that displaces the whole face forward relative to the MCF. Other upper facial growth occurs through the process of drift, because the anterior surface of the primate frontal is a depository field and the inside is a resorptive field (Duterloo and Enlow, 1970). Consequently, the outer table of the frontal drifts forward independently of the inner table, which is affixed to the frontal lobe. As the upper face grows forward relative to the ACF, it pulls forward the supraorbital portions of the frontal bone above the orbits. The anatomist Franz Weidenreich was the first to observe that this separation (which he termed neuro-orbital disjunction) relates to supraorbital growth. Primates such as humans, whose orbits remain tucked below the ACF, retain anteroposteriorly short brow ridges, whereas primates such as chimpanzees, whose orbits are pulled forward relative to the ACF, grow relatively longer brow ridges (Weidenreich, 1941; Ravosa, 1988). As Figure 4.11 illustrates, across primates there is a highly correlated relationship and positive allometry between anteroposterior brow ridge length and projection of the upper face relative to the foramen caecum (Lieberman, 2000).

Figure 5.9 illustrates how four interrelated aspects of cranial growth influence variations in facial projection: cranial base angulation, brain growth, elongation of the anterior cranial base, and elongation of the face (Lieberman, 1998, 2000). As discussed above, the first of these, the cranial base angle, is especially important because a more extended cranial base rotates more of the face in front of the ACF, increasing upper facial projection (Figure 5.9c). The second factor, brain size, also matters because a relatively larger brain, including a larger frontal lobe, requires a relatively longer anterior cranial floor (frontal lobe size and ACF size scale isometrically) (McCarthy, 2004). Thus more of the face fits below the ACF in primates with relatively larger brains.

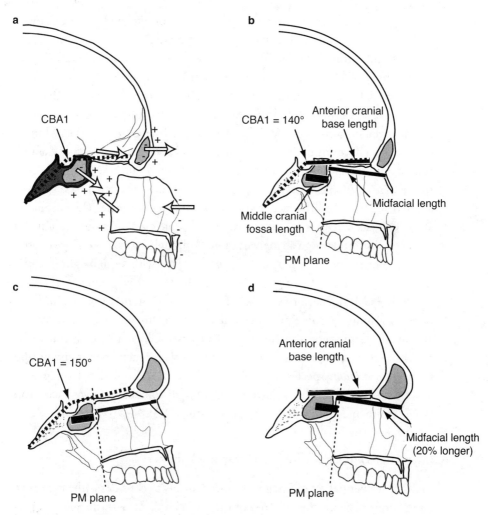

Figure 5.9. Midsagittal view of the cranium, showing how the middle and upper parts of the face grow forward from the cranial base and relative to the anterior cranial base (a). Projection of the face is largely influenced by the angle of the cranial base (CBA), the length of the anterior cranial base, and the length of the face (b). Greater projection of the face can arise from several configurations, including a more extended cranial base (c), or a longer face relative to the length of the anterior cranial base (d). (Modified from Lieberman, 2000).

Finally, the relative anteroposterior lengths of the anterior cranial base and the middle and upper face also influence facial projection. If the face is long relative to the anterior cranial base, then the front of the face must be positioned further forward relative to the end of the cranial base (Figure 5.9d). Relatively longer faces project more.

The four factors that influence facial projection are evident in the many variations in supraorbital shape among primates. Compared with chimpanzees and gorillas, modern humans have less projecting faces and hence shorter brow ridges, because we have relatively shorter faces, larger brains, longer cranial bases, and more flexed cranial bases.[4] Together, these differences place the orbits almost entirely beneath the ACF in modern humans, whereas they are well in front of the ACF in African apes (Lieberman, 2000). Modern humans also have less facial projection and hence shorter brow ridges than archaic hominins such as Neanderthals. Even though brain size is large in both species, *H. sapiens* have less facial projection because the cranial base is flexed about 15° more, and the face is about 15–20 percent shorter (Lieberman et al., 2000b; see Chapter 13). However, facial projection does not account for all aspects of brow ridge shape variation. Although brow ridge *length* correlates with facial projection, brow ridge *height* scales with facial height (see Figure 4.11). Primates, including hominins, with relatively taller faces have relatively taller, thicker brow ridges. There are also subtle variations in brow ridge continuity. In some primates, such as chimpanzees, the lateral, central, and medial components of the supraorbital (see Chapter 4) create a single, horizontal bar across the top of the face. In others, such as orangutans and some archaic *Homo* species (e.g., Neanderthals) the central glabellar region is lower relative to the orbital portions, creating distinct arches over each orbit (F. H. Smith and Raynard, 1980; Lieberman, 1995).

Interactions among Widths, Lengths, and Volumes

Another influential set of integrative mechanisms to consider is the collective interactions between the volumes of the brain and face with the area dimensions (widths and lengths) of the basicranial platform on which they grow. An analogy might be a useful way to introduce this idea. Imagine you are making bread. After kneading a ball of dough, you divide the ball in half, letting each rise in a different shallow pan. One pan is rectangular (long and narrow), and

---

4. A study on dogs provides an interesting glimpse of the sort of mechanism that could be responsible for rapid decreases in facial size. Fondon and Garner (2004) found that the length of repeated elements in the *Runx-2* gene, a transcription factor that regulates osteoblasts, correlates well with snout length in different breeds (actually, there are two different three-base sequences in the tandem repeat, and only one of them correlates well with snout length). It is unlikely that the same mechanism was involved in facial reduction in humans or other primates, but the finding illustrates how simple genetic changes can lead to rapid morphological transformations.

the other is square (short and wide). Both loaves will rise to the same volume, but in shapes constrained by the pans (rectangular or square). Further, if one pan's area is smaller, then its loaf will rise higher, creating a more dome-shaped, spherical loaf. A similar interaction occurs in the skull, but with some crucial differences. In our bread analogy, only one loaf rises in each pan, but the skull has two "loaves," the brain above and the face below. Moreover, in the skull, the basicranium (the brain's "pan") is *tilted* (diagonally when viewed from the side) so that the brain grows mostly upward (dorsally) and backward (posteriorly), while the face grows mostly downward (ventrally) and forward (anteriorly or rostrally). Finally, the basicranium has three endocranial fossae (pans), each at a different orientation, only two of which support the face (the MCF and ACF).

This double-layered configuration, illustrated in Figure 5.10, leads to several interactions and constraints. First, the brain and neurocranium grow above the endocranial fossae and cannot grow to be much wider than the basicranium, but they can expand dorsally as well as posteriorly (because the basicranial floor is tilted with the posterior end down). Second, the face grows downward from the ACF and forward from the MCF, and thus its posterior and superior margins cannot initially be much wider or narrower than these fossae. As the capsules of the face grow away from the cranial base, however, their dimensions (but not their orientations) become increasingly independent of the basicranium. Third, although both the length and width of the MCF and ACF interact with the neurocranium above and the face below, the widths of the fossae, especially the MCF, will co-vary most strongly with widths of the face and neurocranium. This is because the ethmomaxillary complex grows forward along its posterior margins by displacement against the MCF. Therefore the width of the midface correlates reasonably well with the width of the MCF, but its length and height are more independent of MCF or ACF dimensions. Finally, as brain volume increases during ontogeny, the sphericity of the neurocranium is influenced by the volume of the brain relative to the area of its platform. As Chapter 4 describes, prenatal brain growth probably stimulates the petrous portions of the temporal to rotate coronally, perhaps widening the cranial base when cranial base flexion or elongation is constrained (Dean and Wood, 1984). In addition, brain growth stimulates widening of the endocranial fossae through lateral drift. Thus, a shorter, wider cranial base should generate a more spherical (globular) brain case.

The tilted brain pan model of interactions helps explain some aspects of craniofacial variation. Notably, almost all mammals have long and narrow

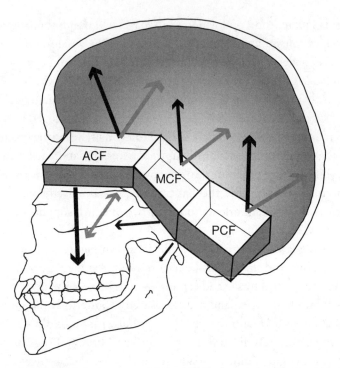

Figure 5.10. Tilted double brain pan model (from Lieberman et al., 2008a), which shows how the face grows downward and forward from the anterior and middle cranial fossae, and the brain grows upward, backward, and laterally from all three endocrinal fossae. See text for details on how these relationships constrain craniofacial shape.

skulls, because mammals also tend to have narrow, flat cranial bases and relatively small brains with little lateral and upward (dorsal) growth relative to the cranial base. Likewise, most mammals have long faces that grow forward relative to a narrow, extended cranial base with little inferior (ventral) or lateral growth. Quantitative support for the effects of these interactions comes from studies, which find that cranial widths account for a dominant proportion of covariation in the skulls of humans (Lieberman et al., 2000a) and apes (Polanski and Franciscus, 2006). These studies also find that faces are generally more independent from the basicranium than the neurocranium, presumably because the many capsular parts of the face grow in a more complex manner relative to the basicranium than to the neurocranium (and there are no facial equivalents of the dural bands). Benedikt Hallgrímsson and I also tested the model experimentally by comparing skull shapes in strains of mice from the same genetic background that differed solely because of independent, discrete mutations that affected either the face, cranial base, or brain (Hallgrímsson et al., 2007a; Lieberman et al., 2008a). As predicted, mice with relatively wider but shorter cranial bases have more spherical brain cases, along with relatively wider, shorter faces; in addition, mice with bigger brains relative to the area of the cranial base have more spherical brain cases.

Testing interactions between brain volume and cranial base area using comparative data is a challenge because it is difficult to find taxa with independent variations in basicranial width, basicranial length, and brain size that are not also confounded by evolutionary relationships (more closely related species tend to be more similar). In addition, most comparative analyses of the relationship between brain size and basicranial shape in primates have focused on midsagittal measurements. After correcting for phylogenetic effects, Ross and colleagues (2004) found no statistically significant correlation across primates between midsagittal roundedness of the neurocranium and brain volume relative to cranial base length or brain volume relative to the angle of the cranial base, but their study did not include measurements of cranial base widths. However, a few ontogenetic studies provide some corroborative evidence. Both chimpanzees and humans are born with rather spherical braincases (see Figure 5.3). Lieberman et al. (2002, 2004b) found that globularity of the cranial vault in chimpanzees initially increases before the eruption of the first permanent molars, as the brain grows relative to cranial base length, but that globularity subsequently decreases after brain growth ceases, and the cranial base extends and becomes relatively longer.

Coda

I began Chapter 4 by asking how human and nonhuman primate skulls grow, and how changes in craniofacial growth processes generate the major differences in head shape between humans and great apes such as chimps. For the most part, humans and other apes grow their skulls using the same basic mechanisms. After the bones of the head are set up during the embryonic stage, they grow in sutures, ossification centers, synchondroses, and fields of bone deposition. Some of this bone growth is stimulated by the organs and functional spaces that the skull's many bones encapsulate or attach to, including the brain, the eyeballs, the pharynx, and various muscles. Craniofacial growth is also stimulated by mechanical loading from actions such as chewing and breathing, and by growth factors, hormones, and other epigenetic mechanisms. And some growth must be regulated intrinsically through control of endochondral growth within the synchondroses.

In spite of the head's complexity, apes, humans and other creatures manage to grow heads that function well yet also vary substantially because of the many processes that integrate them. I suspect that it is the very complexity of heads, and the mechanisms that integrate them, that permit heads to be so varied and evolvable. By sharing common walls, common processes of growth, and common functions, different bony compartments of the head can usually adjust to each other dynamically during ontogeny, accommodating various changes in size, shape and position. In addition, the head has a basic plan, with a few key mechanisms that constrain how the whole thing fits together. These integrative mechanisms largely involve the cranial base, which lies between the face and braincase. Variations in the relative size, position, and shape of the brain and face are thus accommodated by variations in the angle of the cranial base, by variations in the widths of the endocranial fossae, and by the relative lengths of the parts of the cranial base. Not surprisingly, the same dynamic mechanisms also have key roles in evolution. For example, changes in cranial base flexion, which help accommodate a larger brain and a shorter face during ontogeny, were also important during human evolution as hominins became more encephalized and shorter faced. Chapters 11–13 will explore the evidence for these and other shifts in the fossil record.

It also follows that a few key differences in growth have had widespread and profound effects on overall head shape during human evolution, generating many of the transformations we observe between apes and humans. There is good reason to believe that some of these differences are embryonic in origin. In addition, several major differences in fetal and postnatal growth contribute to a wide array of differences in head shape. These include a larger brain and relatively longer anterior cranial base, a shorter face, and a more flexed cranial base. To this list we should also add smaller front teeth (especially the canines), a foramen magnum that does not drift posteriorly in the cranial base, and a nuchal plane that rotates horizontally rather than vertically. As noted above, none of these changes are independent: together, they produce a wide range of effects on craniofacial shape because of integration.

A few key questions arise from the conclusion that many, if not most, of the differences among human and ape skulls derive from just a handful of developmental shifts. A first question is, what are the genetic bases for these shifts? In spite of this question's importance, we are a long way from answering it, given our rudimentary understanding of the relationship between genotype and phenotype and of the roles that most genes play in craniofacial development. A second question is, why did these shifts occur? To address this evolutionary question, our next step is to explore in depth the many different functions of the skull, including how we chew, swallow, breathe, hear, speak, locomote, balance, and see. In addition, we need to examine these functions in the context of the fossil record of human evolution.

6

The Brain and the Skull

I would not be just a nuffin'
My head all full of stuffin'
My heart all full of pain.
I would dance and be merry,
Life would be a ding-a-derry
If I only had a brain!

E. Y. HARBURG, "If I Only Had a Brain," 1939

A massive brain is a quintessential human trait. A typical modern human brain has a volume of about 1,350 cm³, roughly four times the size of a male chimpanzee's. And although there is neither a strong nor a simple correlation between brain size and intelligence, few doubt that brain expansion was a major factor in the evolution of humans' remarkable cognitive capacities.[1] Even so, major increases in brain size occurred relatively late in our evolutionary history. Darwin, in his typically prescient way, suspected this. In *The Descent of Man* (1871) he inferred that brain expansion followed the transition to bipedalism, perhaps because upright walking permitted the origin of early tool making and favored anatomical changes in the head and neck that also led to speech. Other early and influential experts on human evolution believed otherwise. Most notably, the anatomist Grafton Elliot Smith (1912) argued that the reorganization and growth of the brain led the way in human evolution. It may have been Smith's views, widely shared by other scientists of the time, that inspired the infamous Piltdown forgery. This skull, "discovered" between 1908 and 1913 by Charles Dawson, consisted of some recent human cranial

---

1. Brain size and measures of cognitive performance have a correlation of approximately 0.3 (McDaniel, 2005).

vault fragments and a partial orangutan jaw whose teeth had been filed down —all cleverly stained and broken to look ancient (see F. Spencer, 1990; Miles, 2003). The forgery, an obvious fake to a modern eye with the benefit of hindsight, deceived many fine anatomists until 1953, when the fraud was exposed (Weiner, 1955), because it was exactly what some scholars were looking for in an early human ancestor: an ancient Englishman with the colossal brain like a modern human's combined with a primitive, apelike face.

Of course, what is most interesting about the human brain is not its size but how we use it to think, speak, remember, perceive, move, and regulate the body. In fact, one of the most interesting questions in science is how the human brain performs these tasks differently, and in some cases better, than other species' brains. Answering this question is obviously beyond the scope of this chapter, whose focus is more mundane. Given this book's general questions of how the head grows and functions as an ensemble and how the head's functions changed during human evolution, this chapter addresses three basic questions about human brain evolution. First, how big is the human brain, and what factors make it bigger than those of related species? Second, what are the physiological challenges and costs of brain size in terms of blood supply, metabolism, and thermoregulation? And third, what are the biomechanical challenges of protecting the brain within the skull? (Chapters 4 and 5 discuss the related question of how changes in brain size affect craniofacial growth and shape, and Chapters 12 and 13 consider how evolutionary changes in brain size relate to changes in hominin cranial shape.)

Although our primary interest is in the head as an integrated structure, it may be useful to begin with a brief review of the brain's basic anatomy and a few essential details about brain function. Readers interested in the functional anatomy and biology of the brain should consult more comprehensive reviews (e.g., Carter, 1999; Kandel et al., 2000; Joseph and Cardozo, 2004).

Anatomical and Functional Modules of the Brain

A single thought or action can activate more than a million cells, sometimes throughout the brain. These infinitely variable patterns of activation are hardly random but instead lead to a very special type of emergent complexity from the interplay between modularity and integration. In fact, the brain is arguably the most modular part of the body, from the level of individual cells to that of entire regions, such as the cerebrum or the frontal lobe. These structural modules, however, are so highly integrated (a given nerve cell can interact with thousands of other nerve cells) that anatomical descriptions of the brain do a very poor job of describing how it works. Thus another category of modules to consider in the brain are functional modules, or systems, in which multiple regions work together to perform specific functions, such as sensation, motor control, and memory.

Cells of the Brain

The central nervous system (CNS), including the brain, begins as a tube. Recall from Chapter 3 that during early embryonic development, ectodermal cells in the midline initially form a plate and then roll up into the neural tube (see Figure 3.2) along the embryo's long axis. The tail end of the neural tube becomes the spinal cord, and the cranial end becomes the brain. Both the spinal cord and brain continue to grow as a tube, but various folds and bulges later obscure their tubular origins. Cells along the tube's inner surface differentiate into neurons and various supporting (glial) cells that help maintain the brain. Although there are as many glial cells as neurons, neurons are the core of the CNS. Each neuron has several features (illustrated in Figure 6.1): a cell body with a nucleus, many processes (dendrites) that communicate information from other neurons to the cell body, and a single long process (the axon) that connects to other neurons. For many neurons, the cell body remains near the center of the neural tube, but the axon (sometimes a meter in length) migrates outward, along with glial cells, to make room for additional cells and to create a complex, interconnected network.

Neurons exchange electrochemical stimuli with each other and with other cells, mostly muscles and glands. Information travels from a neuron's nucleus to other cells via directional changes in the electric charge of the axon's membrane. These waves, known as action potentials, travel down an axon at lightning speed by a process that alters the balance of charged ions at a given location in the axon. Once an action potential reaches the axon's end, the signal is transmitted at a synapse to the dendrite of another nerve cell or to the membrane of a muscle or gland cell (see Figure 6.1). Synapses are voltage-triggered junctions: when the action potential hits an axon terminal, the membrane releases small neurotransmitter molecules that diffuse rapidly across the tiny space between the terminal and receptors in the next cell's membrane. These neurotransmitters bind to receptors, inducing a response (e.g., another action potential) in the next cell (see Kandel et al., 2000). Thus, at a basic level, neural information is coded and transmitted as variable frequencies of action

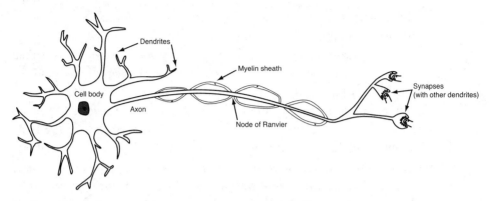

Figure 6.1. Simplified diagram of a neuron, showing basic elements.

potentials. The complexity of signaling is partly a function of the number of possible connections in the system. A human CNS is estimated to have approximately 100 billion neurons. Because neurons typically have thousands of dendrites each (the maximum is 400,000), a brain has as many as 1×10^{15} connections.

An important property of the CNS is speed. The rate at which a normal axon can conduct an action potential is a function of the square root of its diameter (Hodgkin, 1954), so that a 10-fold increase in conduction velocity requires a 100-fold increase in axon diameter (and a 1,000-fold increase in volume). Vertebrates have overcome this spatial constraint by wrapping axons in myelin, a fatty insulating substance. Myelin comes in sheaths a few millimeters in length, with intervening spaces (nodes of Ranvier), as shown in Figure 6.1. This nodal structure speeds up the rate at which an axon conducts an action potential by causing the electrochemical wave to jump almost instantly from node to node. Myelination allows the vertebrate CNS to be simultaneously fast and complex, with billions of neurons that would otherwise require impossibly thick axons. Regions full of myelinated axons are called white matter because of their fatty, whitish appearance; regions filled with neuronal cell bodies are called gray matter. Myelination lags behind neuronal growth: in humans, the process continues into early adulthood (S. C. Zhang et al., 1999).

Anatomical Modules of the Brain

The simplest way to group neurons into modules is by major anatomical region. From a phylogenetic perspective, the two most elementary modules are the brain stem and the forebrain. The former dominates the brain in lower vertebrates: it processes sensory information (with the exception of smell), maintains homeostasis, and controls movement through nerves in the spinal cord and the head. Higher vertebrates have evolved an enlarged forebrain, especially the cerebrum, which integrates sensory information and makes possible emotions, learning, cognition, and other complex behaviors.

These phylogenetic events are reflected in the arrangement of the basic components of the CNS during embryonic development, when multiple stages of segmentation generate a complex set of distinct yet integrated units. Very early signaling interactions with neighboring cells divide the neural tube into three parts (Chizhikov and Millen, 2004). The first, cranial division forms the brain (the encephalon), and the middle and posterior divisions become the spinal cord. By three weeks, the brain has further subdivided into three major segments, each surrounding the remaining central portion of the neural tube (see Figure 3.3). The anteriormost segment is the forebrain (prosencephalon); the next two segments, the midbrain (mesencephalon) and the hindbrain (rhombencephalon), form the brain stem. Divisions between the segments are evident as flexures. By the end of the fifth embryonic week, the forebrain and hindbrain further divide, creating the final major subdivisions of the CNS (Figure 6.2). The forebrain gives rise to the largest part of the brain, the cerebrum (the telencephalon), and a smaller set of paired structures immediately below, the diencephalon. The hindbrain subdivides into the pons and cerebellum (metencephalon), and the medulla oblongata (myelencephalon). The remaining caudal portion of the CNS is the spinal cord.

Differences in the relative size, structure, and function of the brain's major parts among living vertebrates reflect key transformations in vertebrate evolutionary history. In primitive vertebrates such as fish, the brain stem is the largest part of the brain, and the forebrain is diminutive, consisting mostly of the olfactory bulbs. These are connected to a pair of tiny bulges, the cerebrum, which relays information from the olfactory bulbs to the hindbrain (Northcutt, 1981). As higher vertebrates evolved from fishlike ancestors, the forebrain enlarged and subdivided substantially, creating regions that process information from the other sensory organs (see Johnson et al., 1994). Forebrain expansion occurred through a massive increase in the number of neurons and synapses. In fish and amphibians, most of the neuronal cell bodies (visible as gray matter) lie near the center of the cerebrum. But in larger-brained, more complex vertebrates, many neurons were pushed toward the outer surface of the cerebrum, forming an expanded outer layer called the cere-

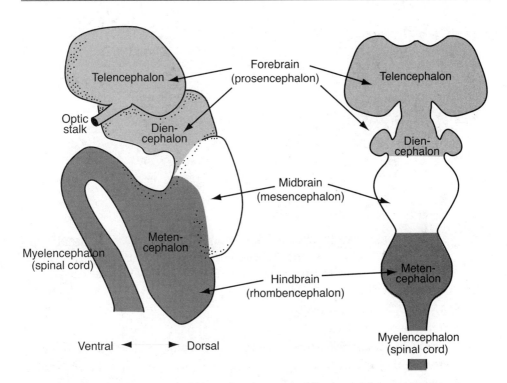

Figure 6.2. Lateral (left) and dorsal (right) views of a 5-week-old embryonic human brain. The brain still retains its initial tubelike shape, but the forebrain (prosencephalon) and the hindbrain (rhombencephalon) have begun to subdivide further into units.

bral cortex. One region of the anterior cerebral cortex, the neocortex, is expanded in mammals, and is especially larger in humans.

With this history in mind, lets review the basic anatomical structures of the brain stem and the forebrain and their position within the skull (summarized in Figure 6.3) before reviewing how these regions are integrated functionally into systems.

The Caudal Brain Stem

The most caudal (in a biped, inferior) portion of the brain stem is the medulla oblongata, an oblong-shaped continuation of the spinal cord that extends from the upper spine through the foramen magnum. The medulla oblongata is partly a highway of axons going to and from the brain; just above the medulla oblongata is the pons ("bridge"), which sits on top of the basioccipital bone. Together, the medulla and pons contain many vital clusters of cell

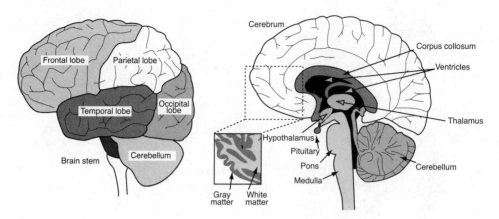

Figure 6.3. Basic structures of an adult human brain in left lateral view. Left: external surface. Right: midsagittal section. The outer surface consists of gray matter (cell bodies), whereas many inner structures of the cerebrum are dominated by white matter (axons covered in myelin).

nuclei (masses of gray matter) that regulate functions such as breathing, heartbeat, and blood pressure, as well as reflexes involved in swallowing, coughing, and sneezing. The pons also functions as a relay station for sensory nerve impulses on their way to the cerebellum.

Cerebellum

The cerebellum ("little brain") part of the hindbrain is a fist-shaped pair of lobes that extend behind the brain stem, occupying most of the floor of the posterior cranial fossa. The cerebellum has as many neurons as the rest of the brain combined. These neurons receive sensory input from the rest of the body and coordinate this information with outgoing motor commands from the cerebrum. The cerebellum thus helps you move as you intend to and to maintain posture and balance. Damage to the cerebellum disrupts most aspects of coordination. In mammals, the cerebellum also helps the forebrain to process sensory information involved in some cognitive tasks (Middleton and Strick, 2000). As the neocortex has expanded in human evolution, the size of the cerebellum has decreased relative to total brain size (A. H. Weaver, 2005).

Midbrain

The midbrain is a small component of the brain stem anterior to the pons. Its key roles include connecting the brain stem with portions of the forebrain and helping to coordinate body movements. Cell degeneration in a region of the midbrain called the substantia nigra ("black substance") is associated with Parkinson's syndrome (Corti et al., 2005). The midbrain also retains its ancient visual role by relaying signals that control eye movements.

Diencephalon

The diencephalon ("twin brain"), which sits atop the brain stem, is the smaller part of the forebrain. The eyes are technically outgrowths of the embryonic diencephalon (see Chapter 3), but this region of the brain is mostly a cluster of paired regulatory regions, including the thalamus, the epithalamus, and the hypothalamus. The thalamus ("inner chamber") acts as the conduit to the cerebrum for all sensory information, with the exception of smell. The thalamus also plays major roles in awareness and learning. The epithalamus, located on top of the thalamus, has several small structures, including the pea-sized pineal gland that secretes the hormone melatonin. Melatonin promotes sleep and helps regulate circadian rhythms. Another epithalamic structure is the habenula, a cluster of cell bodies that are involved in processing sensory information on pain, stress, smells, nutrition, and wakefulness and in the brain's positive and negative feedback systems. Finally, as its name implies, the hypothalamus lies below the thalamus. The hypothalamus has many functions: it controls the release of various hormones by the pituitary gland, thereby serving as the primary connection between the CNS and the endocrine system; it homeostatically regulates many autonomic functions of the body such as thirst, digestion, body temperature, heart rate, and blood pressure via a combination of hormonal and neuronal signals; and it interacts with the limbic system (see below) to regulate emotions such as aggression, pleasure, and sexual arousal.

Cerebrum

The largest part of the human brain is the cerebrum (or telencephalon), which performs most complex cognitive functions (Figure 6.3). Its structure is accordingly complex, with an outer layer of gray matter called the cerebral cortex, an inner region of white matter (mostly tracts of myelinated axons), and several clusters of gray matter deep within the white matter. The human cerebral cortex is about 2–5 mm thick. During the later stages of human embryonic development, the gray matter in the cerebral cortex grows much faster than the underlying white matter, causing it to buckle and fold over on itself, creating many bumps and ridges (gyri) separated by grooves (sulci). The surface area of the unfolded human cerebral cortex is about 2,400 cm^2 (about the same as a double bed). In general, the cerebral cortex receives and processes sensory information from organs such as the skin, eyes, and ears and is involved in many complex cognitive functions, including memory, perceptual awareness, thinking, language, and consciousness (see below).

The cerebral cortex itself is divided into two hemispheres, each with four lobes named for their covering bones (Figure 6.3). A few generalizations can be made about each lobe. The frontal lobes, which occupy the anterior cranial fossa, include areas that regulate voluntary movement, concentration, memory and recognition, motor aspects of speech, and emotions generated by the limbic system. The parietal lobes, which are separated from the frontal lobes by the central fissure, are particularly important for aspects of fine motor control, including speech and gesturing, and for processing visual and other somatic stimuli. The temporal lobes, which lie in the middle cranial fossa, are especially involved in various aspects of perception and memory, including the recognition of words, sounds, and visual images. The occipital lobes at the back of the brain process visual information.

The white matter within each cerebral hemisphere is mostly a mass of bundles of myelinated axons traveling to synapse with widely distributed neurons. Many axons synapse with other neurons in the cerebral cortex of the same hemisphere; others transmit impulses to the corresponding region of the opposite hemisphere; still others link neurons in the cerebral cortex to non-cortical neurons in the same hemisphere or to other parts of the brain, particularly the thalamus. There are so many connections between the two hemispheres that they are linked by a thick bundle of mostly myelinated fibers called the corpus callosum ("hard body"). Deep within the white matter portion of each cerebral hemisphere are three masses of gray matter, collectively termed the basal ganglia. These comprise several clusters of cell bodies, including the caudate nucleus, putamen, and globus pallidus, that lie deep in the cerebrum. These structures relay information from different parts of the cortex and play key roles in motor control, learning, and cognition.

Functional Modules of the Brain

Many of the anatomical structures described above can be partitioned into more specialized regions. The cortex can be mapped into multiple discrete areas that correlate with certain functions. As an example, the anteriormost part of the frontal lobe, the anterior prefrontal cortex, manages many cognitive processes, such as planning, and helps retrieve memories. Other well-known regions of the cortex include Broca's and Wernicke's areas (which are involved in language), the primary motor cortex (which plans and executes voluntary movements), and the primary visual cortex (which processes visual stimuli). Subdividing the brain into discrete regions is useful for understanding many aspects of brain function and evolution, but such maps need to be considered in the context of the brain's functional modules, which integrate different regions to generate perceptions, movements, emotions, and other cognitive functions.

At a very general level, one can recognize four major integrative systems: the sensory system monitors and interprets environmental stimuli (sights, sounds, smells, etc.); the motor system initiates and coordinates muscles to generate movement; the autonomic system regulates and controls the visceral organs; and the limbic system coordinates autonomic functions with consciousness and plays key roles in learning and in generating and regulating many emotions. Each of these systems involves diverse regions of the brain, many of which participate in multiple systems. The limbic system, for instance, is an integrated circle of cortical and subcortical structures along the inner margins of the cerebral hemispheres and around the top of the brain stem. In contrast, the motor system is centered in the brain at the posterior margin of the frontal lobe, the premotor cortex. This region, which has connections to other parts of the cortex, sends axons down to structures in the midbrain, where it connects with the cerebellum and with nerves in the spinal cord and thence to the rest of the body.

Reviewing the brain's functional systems is not central to this book's focus on the relationship of the brain to the skull in particular and to the head in general. But it may be useful to highlight a few basic principles. First, as noted above, many regions of the brain have specific functions, and some, such as the prefrontal cortex, have become relatively larger in humans than in apes. Expansion of these regions suggests natural selection for particular functions (e.g., planning and memory). Second, some neural functions "map" to brain structures. For example, cell bodies in the primary motor cortex send axons to the spinal cord that connect to muscles and control movement in conjunction with premotor areas of the cortex, as well as the basal ganglia and cerebellum. The cell bodies in the primary motor cortex are actually organized into a maplike representation of the body (known as a homunculus), with different areas corresponding to the feet, legs, arms and hands, nose, lips, and so on. The relative size of each area reflects the proportionate association of the cortex with parts of the body. Third, no region of the brain operates in isolation; instead it has many other connections within the brain, sometimes on the same side and sometimes on the opposing hemisphere. Many functions are lateralized to one hemisphere (e.g., language on the left, spatial manipulation and music on the right). Finally, a diverse array of pathways integrates the brain with the rest of the body. Many of these connections are neural (via the cranial nerves to the head, and the spinal cord to the rest of the body), but some incorporate other systems. For example, the hypothalamus has receptors that sense hormones such as steroids and insulin, and which can trigger further hormonal responses via connections to the pituitary gland.

Brain Size

How Big Is the Human Brain?

A first step in considering how increases in brain size during human evolution affected the function, shape, and growth of the head is to examine the size, both absolute and relative, of the brain and to consider what developmental processes may have driven its development in these directions.

Absolute brain size is usually measured either directly (via dissection) as a mass or a volume, or indirectly from the volume of the endocranial cavity (ECV). In such cases, brain mass is estimated from ECV by regression equations that relate volume to mass (a cubic centimeter of brain tissue weighs approximately 1.036 grams) and correct for the presence of other tissues and fluids in the braincase. For primates, this regression (from R. D. Martin, 1990) is:

$$\text{Brain mass} = 1.147 \cdot \text{ECV}^{0.976}$$

Table 6.1 summarizes data on brain size and body mass for a variety of adult primates (males and females averaged), including some fossil hominin taxa. The most obvious fact from Table 6.1 is the much greater absolute size of the human brain compared to that of other primates. Although chimpanzees and other apes have brains about three times larger than those of most monkeys, modern human brains are three to four times bigger than those of chimps. In addition, ECVs in hominins prior to about two million years ago are close to or only slightly larger than those of apes. As Chapters 11 and 12 discuss, early *Homo* has a slightly larger brain than apes, but really substantial increases in brain volume did not occur until about a million years ago, after which brains reached modern or near-modern sizes (Rightmire, 2004). Brain size in recent modern humans is highly variable, ranging from 1000 to 1700 cm³, but today's mean of about 1400 cm³ is absolutely smaller than in modern humans prior to 12,000 years ago and in other late Pleistocene species such as the Neanderthals. (Chapter 13 discusses this shift further.)

Table 6.1: Mean brain sizes in extant primates and fossil hominins

Species	Endocranial volume (cm³)	Brain mass (g)[a]	Body mass (g)	Encephalization quotient[b]
Apes				
Pan troglodytes	393	390.6	41,000	2.07
Gorilla gorilla	465	460.3	128,000	1.02
Pongo pygmaeus	418	414.8	64,000	1.56
Hylobates lar	100	102.7	5,442	2.52
H. syndactylus	123.7	126.4	10,775	1.85
Monkeys				
Macaca mulatta	83	85.6	4,600	2.39
Papio anubis	177	179.3	16,650	1.88
Colobus badius	61.6	64.0	8,617	1.11
Cebus apella	76.2	78.8	2,437	3.56
Saimiri sciureus	23.6	25.1	914	2.39
Callicebus moloch	18.3	19.6	1,078	1.64
Prosimians				
Tarsius spp	3	3.4	112	1.57
Indri indri	33.4	35.2	6,250	0.78
Lemur mongoz	23	24.5	1,669	1.47
Galago demidovii	2.6	2.9	63	2.12
Hominins				
Sahelanthropus tchadensis	365	363.4	40,000	1.96
Ardipithceus ramidus	300	300	3–50,000	1.4–2.0
Australopithecus afarensis	458	453.5	39,000	2.49
Au. africanus	461	456.4	34,000	2.78
Au. robustus	530	522.9	36,000	3.05
Au. boisei	472	467.0	41,000	2.47
Homo habilis	610	599.9	39,000	3.30
H. erectus	970	943.3	61,000	3.69
H. heidelbergensis	1260	1217.7	71,000	4.24
H. neanderthalensis	1488	1432.3	72,000	4.94
H. sapiens	1409	1358.0	64,000	5.12

Extant primate data from Martin (1990); fossil body mass estimates from Robson and Wood (2008); fossil endocranial volume estimates from Chapters 11–13.
[a]Brain mass estimated as $1.147 \cdot ECV^{0.976}$
[b]Encephalization quotient calculated as $BrM / 0.059 \cdot BoM^{0.76}$

Any analysis of brain size, however, needs to consider how it scales with body size, because bigger animals tend to have larger brains (a sperm whale weighs about 19,500 kg and has a brain weighing approximately 8kg, more than five times the mass of a human brain). The first researcher to analyze this problem comprehensively was Jerison (1973), who defined the *encephalization quotient* (EQ) as the ratio of the observed brain mass for a species relative to the brain mass predicted for its body size, based on an a priori assumption that, in mammals, brain mass scales to body mass to the power of 2/3. Jerison calculated EQ to be approximately 3 in chimpanzees and 8 in humans. Calculations of EQ, however, depend largely on the scaling relationship used to predict expected brain mass. Jerison's (1973) equation to predict brain mass (BrM, in grams) from body mass (BoM, in grams) was

$$\log_{10} \text{BrM} = 0.67 \, \log_{10} \text{BoM} + 2.08$$

From this equation, one can calculate that

$$EQ = \text{BrM} / (0.12 \cdot \text{BoM}^{0.67})$$

However, other studies (R. D. Martin, 1981, 1990) have empirically demonstrated that brain mass across mammals actually scales to body mass close to the power of 3/4:

$$\log_{10} \text{BrM} = 0.76 \, \log_{10} \text{BoM} + 1.77$$

yielding different EQ estimates, as shown in Table 6.1, which is based on the revised equation

$$EQ = \text{BrM} / (0.059 \cdot \text{BoM}^{0.76})$$

Keep in mind that other equations for EQ exist, because the scaling relationship between brain weight and body mass depends on the taxa included in the analysis. As noted by several analyses (e.g., Pilbeam and Gould, 1974; R. D. Martin, 1983, 1990; R. D. Martin and Harvey, 1985), the scaling exponent of regressions between brain and body mass for most broad comparative samples of mammal is approximately 0.75, but the intercepts of these lines differ. Among primates, monkeys and apes (haplorhines) have a similar slope but higher intercept than prosimians (lower primates), so that at a given body mass a typical monkey will have a brain twice as large as that of a prosimian (Martin, 1983). A more confounding issue is that the scaling exponent of the relationship between brain and body mass typically declines at lower taxonomic levels of analysis (Holloway and Post, 1982). Although the scaling exponent within anthropoids and primates is about 0.75, it is about 0.59 at the level of the subfamily, 0.45 at the level of the genus, and 0.25 at the level of the species (Martin and Harvey, 1985). So how one calculates EQ depends on the particular scaling relationship one is interested in. Finally, calculations of relative brain size suffer from various sources of error, especially estimates of body mass. Body mass for many primate species is known primarily from zoo specimens (which tend to be overweight) and therefore may be higher than average body masses of animals in natural habitats (Smith and Jungers, 1997). In addition, it is difficult to estimate body mass accurately in many fossil hominins (see McHenry, 1992; Hartwig-Scherer, 1993).

Regardless of the challenges of defining relative brain size, we may draw several reasonably reliable conclusions about EQ, which are illustrated in Figure 6.4. First, relative brain size in early hominins is only a little higher than in apes. Second, really major increases in hominin EQ did not coincide with the origins of the genus *Homo* but evolved later (see also Wood and Collard, 1999a; Rightmire, 2004). Third, modern human EQs are not much larger than those of closely related archaic *Homo* species, such as Neanderthals (Ruff et al., 1997; see Chapter 13). What EQ actually means in terms of intelligence, however, is ambiguous, and it would be unwise to read too much into subtle differences within and between species. Capuchin monkeys, for example, have a higher EQ than apes (Gibson et al., 2001), even though few primatologists would conclude that these New World monkeys are more intelligent than chimpanzees or gorillas.

What Makes Human Brains Bigger?

What makes the modern human brain absolutely and relatively large? Are human brains bigger because they grow faster, longer, or both? Much effort has been devoted to answering this question. An early hypothesis, proposed by Count (1947) and later elaborated by Finlay and Darlington (1995), was that humans have larger brains than apes or other mammals because we grow them for longer. This hypothesis turns out to be only partially true (see Figure 6.5). A chimpanzee brain reaches 95 percent of adult size by about 3–4 postnatal years, whereas a human reaches 95 percent of adult size approximately 6–7

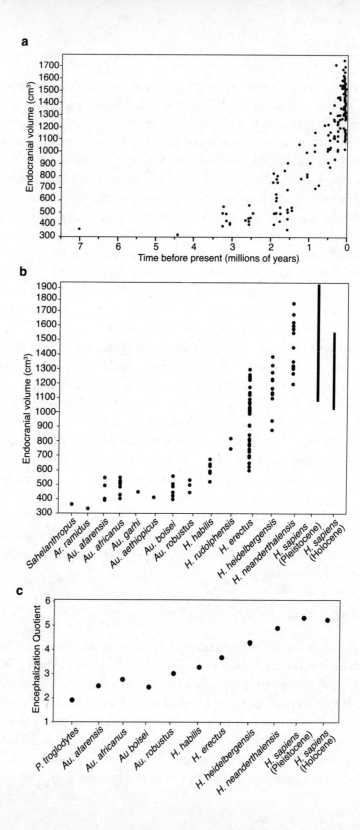

a

Endocranial volume (cm³) vs. Time before present (millions of years)

b

Endocranial volume (cm³)

Sahelanthropus
Ar. ramidus
Au. afarensis
Au. africanus
Au. garhi
Au. aethiopicus
Au. boisei
Au. robustus
H. habilis
H. rudolphensis
H. erectus
H. heidelbergensis
H. neanderthalensis
H. sapiens (Pleistocene)
H. sapiens (Holocene)

c

Encephalization Quotient

P. troglodytes
Au. afarensis
Au. africanus
Au boisei
Au. robustus
H. habilis
H. erectus
H. heidelbergensis
H. neanderthalensis
H. sapiens (Pleistocene)
H. sapiens (Holocene)

Figure 6.4. Endocranial volumes of fossil hominin crania plotted (a) against time and (b) by species. (c) Encephalization quotients for major species. Data primarily from Ruff et al., 1997, and Holloway et al., 2004. See text for details of how EQ was calculated.

years after birth. However, the extended duration of human brain growth is complicated by an even longer period of body growth, which is slower than that of chimpanzees and other primates (Bogin, 2001; Leigh, 2001). Whereas a chimpanzee attains adult body size by about 10–12 years, a human does not reach adult body size until 14–18 years (Walker et al., 2006a, 2006b). In other words, the combination of a more prolonged period of brain growth combined with a slower rate of body growth results in a higher brain-size to body-size ratio in adult humans than in chimpanzees. Thus rate must also be an important factor. In terms of volume per unit time, humans grow their brains more rapidly than chimpanzees, particularly in the first postnatal year, when the human brain continues to enlarge at a fetal rate of growth (Vrba, 1998; Rice, 2002; Leigh, 2004). Rapid brain growth combined with slow body growth leads to a more positive ontogenetic allometry in humans (Vinicius, 2005).

Although the human brain reaches nearly adult size by 6–7 years, it continues to change in structure. Some parts of the cortex (e.g., in the frontal) continue to myelinate and form connections until early adolescence and then start to thin as some of those connections are pruned (see Giedd, 2008).

Information on the distinctive rate and pattern of human brain growth does not address directly the more fundamental question of what, structurally, in human brains is actually getting bigger as they grow. The answer to this question will eventually come from detailed histological comparisons of human and ape brains. For the time being, we can make some educated, general guesses by examining scaling studies of the four major variables that might contribute to increases in brain volume: the number of neurons; the size of the neurons; the number of glial cells; and the number and size of the various connections between neurons (called the neuropil), comprising dendrites, axons, and glial cell processes.

One factor that makes a human brain bigger is the absolute number of neurons, which scales to gray matter volume (GMV) to the 2/3 power across mammals (Jerison, 1973; Passingham, 1973). A human brain has about 11.5×10^9 cortical neurons—probably about twice as many as a chimpanzee, which is estimated to have approximately 6.5×10^9 neurons (Haug, 1987). However,

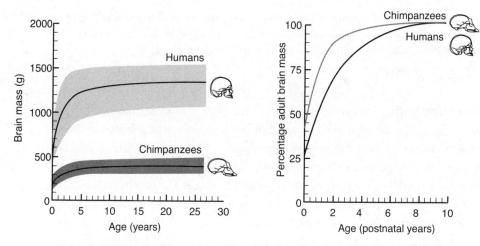

Figure 6.5. Ontogeny of absolute brain mass (left), and brain mass as percentage of adult brain mass (right) in modern humans and chimpanzees (adapted from Leigh, 2004, and Coqueugniot et al., 2004). Shading in (a) indicates 95 percent range of variation. Note that human brains grow both faster and over a longer period than those of chimpanzees.

cortical neuron density scales negatively to GMV (to the −1/3 power) and is thus predicted to be lower in humans than in chimps (Prothero, 1997). Synapse density is invariant with respect to GMV (Abeles, 1991) and is thus about the same for the two species. Putting these facts together, one can predict that the neuropil—the tangled web of dendrites and axons that form the brain's many connections—constitutes a greater relative proportion of GMV in the absolutely larger human brain. In this respect, absolute size may be more important than relative size when comparing similar species, such as apes and humans, because larger brains have absolutely more neurons and absolutely more connections (Gibson et al., 2001). That said, there is no evidence that human brains have relatively more connections per unit volume than those of chimpanzees: the number of synapses per neuron appears to be constant relative to brain size (Changizi, 2001). Large brains, however, may have more functionally specialized regions (Changizi and Shimojo, 2005). More on this issue below.

White matter, which mostly consists of myelinated axons that connect different regions of the cortex, yields a similar picture. Several studies confirm that white-matter volume (WMV) in the brain scales to GMV to the power of 4/3 (Frahm et al., 1982; Hofman, 1989; Changizi, 2001). At first glance, this positive allometry suggests that larger brains, such as those of humans, have more interconnections per neuron. But it turns out that the increase in WMV is close to what one would predict to maintain an invariant interconnectedness between neurons (which scale relative to GMV to the power of 2/3). Even so, there is some evidence that the proportion of white matter in the human cortex is slightly larger than would be predicted by scaling (Schenker et al., 2005). White-matter tracts may also play a role in shaping the brain during growth. Bundles of long axons are established early in development, but they must lengthen as the brain grows. Their intrinsic tensile properties may help pull the neocortex inward to help create the characteristic sulci and gyri of the brain's surface (Goldman and Galkin, 1978; Goldman-Rakic, 1980; Van Essen, 1997).

Putting together these data, it appears that human brains, by virtue of being absolutely larger than chimpanzee brains, have absolutely but not relatively more neurons and interconnections between neurons; further, human brains have relatively more white matter (mostly axons). More detailed histological studies between humans and chimpanzees are needed to test these differences more definitively and on a regional basis, as there may be differences in structure and relative size within certain regions (see below). It will also be interesting to determine which genes are responsible for overall and regional size increases. Given the constraints on the rate at which neurons undergo cell division during brain development, it is reasonable to hypothesize that this expansion was made possible by genes that enable an extra number of cell divisions early in neural growth (Rakic and Kornack, 2001; McKinney, 2002). Most genes that are expressed in the human brain do not bear any evidence of recent positive selection (Shi et al., 2006), but a few candidate genes have been identified that have undergone strong selection in recent evolutionary history and appear to regulate prenatal brain growth (for reviews, see Gilbert et al., 2005; Hill and Walsh, 2005). Several of these genes, such as *ASPM* and *MCPH*, have mutant forms that cause microcephaly, a congenital disease that leads to severely reduced brain size, although not necessarily with cognitive deficits (Kouprina et al., 2004; Mekel-Bobrov et al., 2005; Evans et al., 2005).

Fine detailed histological studies also reveal other differences between human and nonhuman brains, especially in the neocortex. The neocortex consists of columns of neurons, often in clusters of 80–100 cells. These columns turn out to be bigger, more complexly arranged, and with more neuropil in humans than in chimpanzees in regions such as the primary visual cortex (Preuss et al., 1999; Sherwood et al., 2003) and the planum temporale of the temporal lobe (Buxhoeveden and Casanova, 2002). In addition, there are probably many important differences between humans and apes in deeper parts of the cortex. One example may be von Economo neurons (VENs), a special kind of long, spindly neuron unique to humans and African apes (Nimchinsky et al., 1999; Allman et al., 2005). These neurons, which are thought to speed transmission rates between specific regions, appear to be larger and to form bigger clusters in humans than in chimps in the frontoinsular cortex (within the frontal lobes) and in the anterior cingulate cortex—both regions involved in social cognitive functions, such as feelings of trust, empathy, and guilt (Sherwood et al., 2006; Allman et al., 2001).

Which Parts of the Human Brain Grew Bigger?

Because different components of the brain have different functions as well as different effects on cranial shape, it is useful to ask which, if any, parts of the brain are bigger in modern humans than in nonhuman primates, such as chimpanzees. Put differently, is a human brain merely scaled up in size, with all components proportionally larger than in an ape's, or are certain regions of the human brain relatively bigger than others? This question has been addressed by several studies that measure regions of the brain using magnetic resonance imaging (MRI). One influential analysis by Finlay and Darlington (1995) used a large data set of 131 mammalian species (from Stephan et al., 1981). Using regression techniques, they found that the sizes of most components of the brain other than the olfactory bulbs scaled in the same way relative to overall brain size across all mammals. This finding suggests that the major differences between the brains of humans, chimps, bats, and squirrels are a consequence of evolutionary changes to brain size as a whole. Further, because the brain grows from repeated divisions of an initial pool of neural progenitor cells, these results suggest that proportional increases in particular components of the brain are a function of the initial size of the pool and the time available for cell divisions to occur. According to Finlay and Darlington (1995), structures generated later during brain development are proportionally bigger (see below), leading them to conclude that natural selection did not act on particular parts of the brain in a mosaic fashion but instead acted on the size of the brain as a whole.

One problem with these results is that they do not explain many species-specific differences in cognition and behavior. Subsequent studies, moreover, have reanalyzed the same data set and obtained different results. Barton and

Harvey (2000) and de Winter and Oxnard (2001) both found that there are scaling constraints in the sizes of particular components of the brain, but that these scaling relationships differ both between parts of the brain and in different groups of species. For example, increases in brain size within the bat and primate clades generated different scaling patterns, reflecting different types of selection on the brain. In addition, a few analyses have focused on just humans and apes. Although these analyses suffer from small sample sizes and the challenge of measuring lobe size accurately, they yield some important results, some of which Figure 6.6 illustrates. Most parts of the human neocortex, such as the frontal, are scaled-up versions of an ape's (Semendeferi et al., 2002; Bush and Allman, 2004), but there are some exceptions. The human occipital lobe is relatively smaller, in large part because of massive reductions (121 percent) in the size of the primary visual striate cortex (Stephan et al., 1981; Holloway et al., 2003). It is difficult to measure exactly how much smaller the human occipital is, but it follows that the parietal lobes are also relatively larger in humans (Holloway et al., 2003; Bruner et al., 2003). In addition, the human cerebellum is proportionally about 20 percent smaller than in apes; and the temporal lobe is as much as 25 percent larger (Semendeferi, 2001; Rilling and Seligman, 2002), primarily because of the presence of more white matter, suggesting more neuronal connections within the lobe itself (Schenker et al., 2005). On a finer scale, other differences exist within lobes. As noted above, the primary visual striate cortex of the occipital lobe is relatively smaller in humans, and the prefrontal cortex of the frontal lobe is about 6 percent larger (Semendeferi et al., 2001), possibly because of increases in white matter (Schoenemann et al., 2005; see also Schenker et al., 2005).

In short, natural selection has acted on brain size as a whole within the primate clade but also in a mosaic fashion on particular parts of the brain. Although the human neocortex is both absolutely and relatively larger than those of other primates, the volume of regions such as the temporal lobe has also undergone relative increases. This finding has several implications. On a proximate level, regional expansions influence skull shape. For example, a larger temporal lobe in modern humans may cause the middle cranial fossa to be larger (both wider and longer), and also may play a role in causing the modern human cranial base to flex more (see Chapter 5). This hypothesis is hard to test, but it is interesting to consider, because a longer anterior cranial base and a more flexed cranial base are both implicated in generating the characteristic shape of the modern human skull, with a retracted face and a rounded braincase (Lieberman et al., 2002, 2004b; see Chapter 13). In addition, sub-

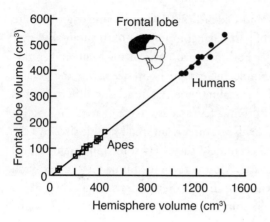

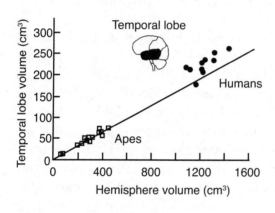

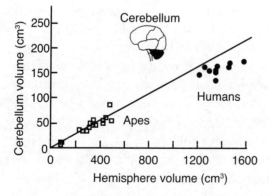

Figure 6.6. Scaling of (a) frontal lobe, (b) temporal lobe, and (c) cerebellum volume relative to hemisphere volume in apes and humans. (Adapted from Semendeferi et al., 2001:264–65).

tle increases in the relative size of particular regions offer clues about selection pressures that may have driven brain expansion. The temporal lobe, for example, is associated with complex functions relevant to the organization of sensory input, including language and various kinds of memory (see Carter, 1999). Similarly, the prefrontal cortex is implicated in many abstract cognitive functions such as language, reasoning, problem solving, social interactions, and planning (Wood and Grafman, 2003). Perhaps selection for various "higher" cognitive functions, including language, helped drive these increases.[2]

Taking Care of the Brain

Maintaining a big brain is a considerable physiological challenge because the brain is an expensive, thirsty, and delicate tissue. Like a power-hungry supercomputer, it consumes lots of energy, must be kept at just the right temperature, and needs to be vigilantly protected from physical damage. Moreover, these functions have to be carried out effectively and without interruption throughout ontogeny despite considerable changes in size and shape. Not surprisingly, the physical requirements of the brain have influenced many aspects of the head's anatomy and growth, and some of these demands are magnified in our relatively large brains.

Feeding the Brain

Various experiments have attempted to estimate how much energy the brain consumes. One early method was to measure the rate of oxygen uptake of tissue samples extracted from anesthetized animals and kept "alive" in a solution of blood (Martin and Fuhrman, 1955). Newer, more sophisticated in vivo methods directly measure rates of glucose and oxygen uptake in regions of the brain (see Poulsen et al., 1997). These studies indicate that an adult brain is metabolically very costly, consuming about 10 kcal/kg/hr. For comparison, the small intestines are about as costly, muscle is much less costly when at rest (0.5 kcal/kg/hr), and the heart is more expensive (18 kcal/kg/hr) because it never rests (Elia, 1992). Extrapolations from these data indicate that an adult human brain uses approximately 280–420 calories per day, about 20–30 percent of the body's metabolic energy when at rest (Clark and Sokoloff, 1999). These high costs, moreover, are constant regardless of activity (including intense thinking).

Brain is an expensive tissue primarily because of the continuous and high costs of propagating action potentials within axons and transmitting these sig-

---

2. Among its many roles, the temporal lobe has been shown to be active during intensely spiritual and religious thoughts; stimulation of the temporal lobe during surgery can induce spiritual emotions even in self-described atheists (Persinger, 2001). If so, then human spirituality may be partly a by-product of selection for other cognitive functions.

nals between neurons in synapses (Raichle and Gusnard, 2002). Every time a neuron fires, membrane-transport mechanisms consume energy to pump ions and neurotransmitters in and out of the cell; in addition, the brain constantly has to pump fuel and waste between the brain and bloodstream and synthesize the many proteins that help regulate brain function. These considerable energy demands are compounded by brain tissue's inability to store energy and the way it uses blood. Many tissues, such as muscle and liver, store energy in the form of glycogen (a carbohydrate that is easily broken down into glucose). The brain, however, has little glycogen storage capacity and thus requires the circulatory system to deliver an unceasing supply of both oxygen and glucose to provide energy (mostly as ATP, derived by the oxidative breakdown of glucose). Insufficient delivery of oxygen and glucose to the brain causes dizziness, convulsions, and loss of consciousness; brain cells deprived of oxygen and glucose for more than a few minutes cease to function and die. So brains constantly need vast quantities of blood: about 750–1000 mL/min at rest (Kandel et al., 2000), and much more during vigorous physical exercise (in order to keep cool) (Thomas et al., 1989). A typical adult human brain accounts for only about 2 percent of total body weight but uses about 12–15 percent of the total blood supply (approximately 4.7 L). In contrast, a chimp's brain makes up less than 1 percent of its body mass and is supplied by a somewhat smaller fraction of its blood supply, probably about 7–9 percent.

Supplying a bloodthirsty brain with so much blood presents a plumbing problem, especially in humans. Most of the blood travels into the brain through the two internal carotid arteries, which run up the sides of the neck and enter the base of the skull through the carotid canals in the petrous portion of the temporal. Each internal carotid artery typically transports about 250 mL of blood per minute (Chu et al., 2000). A smaller volume of blood (<100 mL/min) also enters the head through the vertebral arteries, which run through the transverse processes of the cervical vertebrae. At the point where these arteries enter the brain, just above the foramen magnum, they form a circular network of connected vessels, the cerebral arterial circle (also called the circle of Willis), illustrated in Figure 6.7. This network ensures that oxygenated blood can reach every part of the brain through multiple possible routes, increasing the chance of avoiding a permanently damaging stroke in the event of a partial interruption to the head's blood supply. Numerous arteries arise from the circle of Willis, branching repeatedly to bring blood to every part of the brain. As Figure 6.7 shows, the circle of Willis is slightly more elaborated in humans than in monkeys and other smaller-brained mammals (Kapoor et al., 2003).

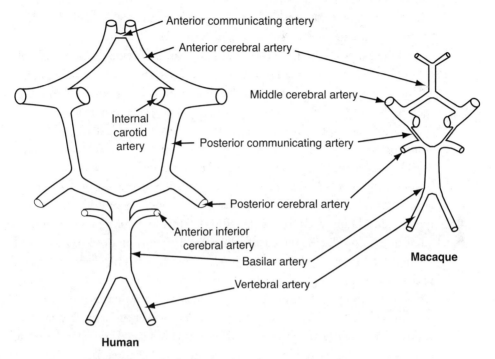

Figure 6.7. Circle of Willis (cerebral arterial circle) in humans and macaques (*Macaca mulatta*). The basic elements are similar, but the human circle has more branches (Modified from Kapoor et al., 2003).

An additional challenge in feeding the brain is the blood-brain barrier (BBB). Blood interacts with most tissues by diffusing through capillaries to supply cells directly with oxygen, glucose, hormones, and other circulating substances. The brain, however, minimizes contact between brain and blood cells, apparently to reduce the chances of harmful blood-borne substances reaching the brain. Capillaries deep in the brain have tight junctions that prevent blood from diffusing passively into the brain's extracellular fluid. Oxygen, glucose, and small, lipid-soluble molecules such as hormones, but not red blood cells, are capable of either squeezing or being pumped through the BBB. All other chemical interactions between the brain and the rest of the body have to occur via cerebrospinal fluid (CSF). This clear, colorless liquid bathes the entire CNS, including the brain. CSF is mostly produced in specialized, highly vascularized regions, the choroid plexus, which line a series of interconnected cavities in the brain known as the ventricles (see Figures 6.3 and 6.8). The ventricles, remnants of the center of the embryonic neural tube, are a kind of man-ufacturing, storage, and circulatory system for CSF.

CSF insulates the brain, nourishes brain cells, and removes wastes (Segal, 2000), but it comes at a high metabolic price. As in the BBB, the lining of the CSF in the choroid plexus permits the passage of just plasma, gases, and var-ious lipid-soluble substances (among them alcohol and heroin) from the blood to the CSF. Thus specialized active-transport mechanisms—all of which con-sume much ATP—are necessary to pump nonlipid soluble substances, such as glucose, into the CSF and pump wastes back out into the bloodstream. CSF is also costly to synthesize. New CSF is produced at the rate it is depleted, about 0.40 mL/minute, maintaining a constant volume of 80–150 mL in a typical adult brain (Standring et al., 2004). At this rate of replenishment, an adult human produces about half a liter per day, turning over 50 percent of the total CSF every 5–6 hours. Overproduction of CSF or blockage of the ven-tricles causes serious problems, notably hydrocephalus (water on the brain), which can produce intense intracranial pressure. Because CSF is contained within the brain and its surrounding meningeal covering, hydrocephalus in children often stimulates an expanded, abnormally shaped cranial vault as the bones of the skull try to form around the fluid (see Chapter 5).

What goes in must come out. Just as the brain is plentifully supplied with oxygen and glucose-rich fluids to feed its thirsty cells, it also has a large and intricate system of veins and spaces, the cranial sinuses, that act as drains (also illustrated in Figure 6.8). These conduits collect deoxygenated blood, used CSF (which perfuses into the circulatory system via the arachnoid granula-tions), and other waste products and channel them into a network of veins— the vertebral venous plexus—and the internal jugular vein. Note that the cranial sinuses and the vertebral venous plexus are a valveless network, which means that blood can flow in any direction through the system. Thus some blood can travel straight from the brain to the abdomen without first passing through the heart and lungs. Although the lack of valves may seem like an illogical de-sign, it protects the brain from temporary shifts in venous blood pressure caused by activities such as rapidly changing position or pressurizing the ab-domen and thorax (e.g., from straining to lift a heavy object). Trauma to the sinuses is often fatal.[3]

---

3. An illustrative case of the vital role of sinuses in the brain is provided by the amazing story of Phineas Gage. In 1848 Gage, a healthy 25-year-old construction foreman, was injured when an ac-cidental explosion propelled a massive, pointed iron bar through his skull. Amazingly, Gage survived even though the bar shot through his left cheek and his eyeball, then passed into his frontal lobe and

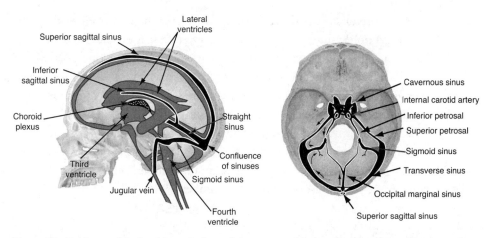

Figure 6.8. Drainage of the head (ventricular and sinus systems) in lateral and superior views.

A few details of the cranial sinuses (see Figure 6.8) merit some mention because they play important roles in circulation and thermoregulation. First, almost all the sinuses converge either directly or indirectly at one of two key locations: the cavernous sinuses that lie on either side of the sphenoid body and the "confluence" of sinuses (also called the torcular Herophili, which translates as "Herophilus's wine press") at the back of the occipital. The confluence receives blood from the top and front of the brain (via the superior sagittal sinus), and from the center of the brain (via the inferior sagittal sinus and the great cerebral vein). The cavernous sinuses mostly receive blood from the scalp, the face, and the eyes (more on this later). Importantly, the confluence and cavernous sinuses are linked to additional sinuses (the superior petrosals; see Figure 6.8). Therefore blood can flow to the internal jugular veins through many different possible routes from either the confluence or the cavernous sinuses. Typically, blood descends from the cavernous sinuses via the

out of the head (the bar was recovered and subsequently reunited with Gage's skull after he died in 1860). Although one study suggested that the bar passed through both hemispheres (Damasio et al., 1994), a reanalysis of Gage's skull showed that the bar passed only through his left frontal lobe, missing the superior sagittal sinus in the midline (Ratiu et al., 2004). Corroborating evidence comes from contemporary accounts of his injury, which document that Gage was transported upright in a cart back to his boarding house after the accident (Harlow, 1848). Had Gage's superior sagittal sinus been pierced, he probably would have died from massive bleeding or an air embolism caused by negative air pressure in the rigid venous sinuses.

inferior petrosal sinuses that run down the cranial base in front of the foramen magnum. In humans and apes, the largest drainage routes from the confluence are the two transverse sinuses that run obliquely along cranial wall toward the ears and then into the sigmoid sinuses, which wind their way down to enter the internal jugular. However, in about 2–4 percent of humans, the confluence is also drained by the occipital sinus, which runs down the back of the skull in the midsagittal plane. The occipital sinus then diverges into a pair of occipital marginal sinuses that encircle the foramen magnum, where they join the jugular. This latter pattern is common in some hominins, notably *Au. afarensis* and the robust australopiths (Kimbel, 1984; Falk, 1986).

Keeping the Brain Cool

Like the rest of the body, the brain needs to be maintained at or near 37°C (98.6°F), but the brain is less tolerant of temperature fluctuations than most organs. One potential problem is cold: brain function is impaired below 35°C, and temperatures below 30°C can be fatal to brain cells. A more common problem, however, is too much heat (hyperthermia): the brain can only briefly tolerate temperatures above 41°C (106°F–108°F) before suffering permanent damage or death. Because muscle contractions produce copious quantities of heat (core body temperatures during exercise can reach 42°C), heat stroke most often occurs from exercising in warm conditions (Hales et al., 1986). For example, running generates up to 10 times as much heat as walking. Yet in this respect, humans are remarkable. A sprinting cheetah generates so much heat that it has to stop running after about four minutes to avoid hyperthermia (Taylor and Rowntree, 1973), but fit humans can run for hours in astonishingly hot conditions.

Not surprisingly, the brain has many adaptive mechanisms to defend itself from excessive heat and cold. The first levels of defense are special thermoreceptors in the hypothalamus that sense the brain's temperature and activate appropriate warming or cooling mechanisms. When overly cold, these receptors trigger heat-generating (thermogenic) responses in the body, such as shivering, and they stimulate constriction of blood vessels near the body's surface, avoiding heat loss by sending blood preferentially through more internal routes. But, as noted above, excess heat is a more common concern for most mammal brains. Many adaptations to heat are behavioral: in hot temperatures, mammals (except mad dogs and Englishmen) generally seek shade, suppress their appetites, and decrease activity levels. Chimpanzees are no exception to this tendency. When temperatures in the shady forests in which they live ex-ceed 40°C, they appear to become uncomfortable and restrict their behavior (Richard Wrangham, personal communication).

Like most mammals, chimpanzees and other nonhuman primates cool their brains primarily by panting. When a mammal pants, it takes shallow, rapid breaths (in dogs, this rate is about 300–400 breaths per minute, about 10 times faster than the normal rate of respiration) (Schmidt-Nielsen, 1990). As exhaled hot air passes over the wet surfaces of the tongue and upper pharynx, water evaporates. This phase shift causes cooling because vaporization requires con-siderable heat—580 calories per gram of water—which is conducted from the moist membranes that cover the tongue and pharynx. This evaporation cools the superficial venous blood that circulates in copious quantities just below the surface of the tissues. Cooler blood then lowers body core temperature; some of it also lowers brain temperatures, thanks to some clever plumbing present in many mammals, including primates. Before cooled blood from the nasal cavity passes into the internal jugular vein, it travels through the cav-ernous sinuses on each side of the sphenoid body. The carotid artery, which brings hot, oxygenated blood up from the heart, also passes through the same sinus. This arrangement sets up a countercurrent heat-exchange system be-tween hot arterial blood and cooler venous blood (Cabanac and Caputa, 1979). In some mammals (not primates), the efficiency of this cooling system is increased dramatically by a carotid rete, a spidery network of arteries that vastly increases the surface area available for heat exchange (Baker and Chap-man, 1977; Mitchell et al., 2002).

Panting is obviously effective and widespread among mammals, but it has several disadvantages. The biggest problem with panting is that it occurs al-most entirely in the dead space of the upper pharynx (the mouth and throat), without any gas exchange occurring in the lungs. Rapid, shallow breathing is a critical adaptation for panting because it matches the natural resonant frequency of elastic tissues in the respiratory tract, minimizing the muscular effort needed (which would generate heat and thus be counterproductive) (Schmidt-Nielsen, 1990). Such breathing, however, prevents an animal from acquiring oxygen and clearing carbon dioxide from the lungs, potentially lead-ing to alkalosis. Thus panting animals need to alternate rapid, shallow breath-ing with periods of normal, slow respiration.

Humans don't pant (at least not to cool down) but cool the brain primarily by taking advantage of our phenomenal ability to sweat. Sweating, like pant-ing, cools through evaporation. Whereas panting acts only on the tongue and upper pharynx, sweating occurs on the skin wherever sweat (eccrine) glands excrete a dilute solution of water and salt that is exposed to air convection.[4] Thus sweating can turn the entire surface of the body into a cooling system. The rate and hence effectiveness of sweating are largely a function of air temperature, humidity, ratio of skin surface area to body volume, and rates of air convection. A typical human has as many as 5 million sweat glands (Kuno, 1956; Jablonski, 2006), and a good number of these are on the face and scalp. Unlike panting, sweating can occur independently of respiration. Quadrupeds cannot simultaneously pant and gallop, and most mammals are challenged by running long distances in hot weather because they cannot dump heat fast enough through panting. Dogs can gallop for many kilometers in cold but not hot temperatures (see Dill et al., 1933). Sweating, however, has some disadvantages. Sweat evaporation is hindered by long and dense fur (McArthur and Monteith, 1980). In addition, sweating requires prodigious quantities of water. Athletes may lose as much as 1–2 liters of fluid per hour through sweating during vigorous activity at high temperatures (Torii, 1995).

Humans are not the only mammals who sweat, but we have drastically elaborated on the sweating capabilities present in other primates. Whereas chimpanzees sweat mostly on the face, hands, feet, and scrotum (Whitford, 1976; Hiley, 1976), humans sweat prodigiously all over the body, including the forehead, neck and scalp. In fact, the human head is the sweatiest part of the body (Kuno, 1956; Cabanac and Brinnel, 1988). Because the superficial veins that drain the scalp and face flow directly into the cavernous sinuses, sweating through the head is an important local source of cooling for the brain, by means of the same countercurrent heat-exchange mechanism described above (Cabanac and Caputa, 1979). It has been suggested, moreover, that the cavernous sinus in humans is elaborated relative to other mammals (Cabanac, 1993). In addition, Cabanac and Brinnel (1985) have proposed that blood cooled by sweating in the scalp can flow backward into the diploic space and then into the cranial vault via a network of tiny veins. If so, cooled venous blood from the scalp would cool CSF and arterial blood in the brain. How much reverse flow occurs is hotly debated, as the tiny emissary veins probably contribute little blood to the diploic space, and evidence for reverse flow comes

---

4. Eccrine sweat glands differ from apocrine glands, which secrete lipids and proteins (thus causing some body odors). Nonprimate mammals have mostly apocrine glands, with just a few eccrine glands on the sole of the foot and the palm of the hand. Eccrine glands on the body are a derived feature of Old World monkeys, but their density and activity is vastly elaborated in humans.

only from studies of cadavers and from descriptions of the retelike bed of veins in the dura mater (Zenker and Kubik, 1996). I return to this issue in Chapters 9, 11, and 12 when I consider cranial cooling mechanisms and the evolution of bipedal walking and running.

One place where humans have kept our fur is on the top of the head. Hair on the head, despite limiting air convection, has an important role as a natural hat, reflecting solar radiation away from the brain. In a typical quadruped, about 20 percent of body surface is exposed to solar radiation throughout the day, but a biped in the midday sun exposes only about 7 percent of its body surface to radiation, mostly at the top of the head (Wheeler, 1984, 1985, 1991). Fur constrains air convection, but it can act as a potent reflector (Walsberg, 1988), so it is probably not coincidental that human fur has been retained in the one place where it does the most good. Indeed, one might conjecture that there is probably some thermoregulatory trade-off between the advantages of having hair on the head for reflection and the advantages of being bald for sweating.[5]

Protecting the Brain from Injuries

A critical function of the skull is to protect the brain from physical injury. This is not a trivial problem. The brain is very soft, about the consistency of a firm pudding, and also rather heavy, about 1.5 kg (not counting blood and CSF). Damage to the brain is a major cause of accidental death. Even so, heads can absorb extraordinary forces without experiencing injury. Professional soccer players, for example, regularly head a 0.43 kg ball traveling at speeds up to 60 km/hr (Kirkendall and Garrett, 2001; Naunheim et al., 2003). The skull must protect the brain by absorbing the force of impacts to keep the vault from fracturing and by cushioning the brain within the vault to prevent damage caused by accelerations of brain tissue.

The skull functions in part as a helmet. As Chapter 2 explains, when a force is applied to a bone, it generates stress (force per unit area), which in turn generates strain (deformation) that can lead to a fracture above some threshold. The force required to fracture the cranial vault therefore depends on a number of external factors (e.g., the force applied and the area over which the force is distributed) as well as internal factors (e.g., the shape and material proper-

---

5. Taking this logic to its extreme, Cabanac and Brinnel (1988) proposed that that male pattern baldness arose as an adaptation to enhance brain cooling in compensation for the growth of beards.

ties of the region affected). Although there are many kinds of skull fractures, grisly experiments on cadavers have found that a force as small as 73 N (about the force of walking into a wall) can fracture a cranial vault (Gurdjian et al., 1964). An adult human when falling can produce a force of 873 N when the head collides with the floor, more than enough to crack a skull. Fortunately, most falls and other impacts don't fracture the cranial vault, for two reasons. First (as anyone who has dropped a skull knows), the skull is pretty good at bouncing, so that a considerable proportion of the impact energy is transferred into movement of the head and neck. In addition, the cranial vault has a number of structural features that help it to absorb forces and resist fractures.

The vault's first line of physical defense is the scalp, which is potentially capable of absorbing up to 35 percent of the force of an impact (Gurdjian et al., 1964). Most of the skull's strength, however, comes from the vault itself, which is shaped something like a half shell or dome. From a biomechanical perspective, this shape is an interesting design. The most common natural domes are the shells of marine creatures (mollusks, turtles, sea urchins) that rarely experience very hard impacts (Vogel, 2003). Shells and domes are probably less common in terrestrial creatures because they are strong when subject to forces that are applied over large areas, but generally weak with respect to applied forces that are locally concentrated. When an external and concentrated force, such as a blow to the head, acts on a domed structure, it generates both bending and shearing stresses. For the most part, we can ignore the shearing stresses because the large, continuous surface area of the dome provides strong resistance to shearing. But bending stresses, illustrated in Figure 6.9, are a problem not only at the site of impact but over the entire dome, whose rounded contour avoids local stress concentrations. Engineers divide these domewide stresses into two components: meridional, which are compressive stresses that run up and down, like the lines of longitude on a globe; and parallel, which are tensional stresses that run circumferentially, like the lines of latitude. The magnitude of the compressive, meridional stresses, usually the biggest challenge, are described by the following formula:

$$\sigma_m = F/2\pi r_o (r_o - r_i) \sin^2\theta$$

where σ_m is the meridional stress at a given point, r_o and r_i are the dome's outside and inside radii, and ϕ is the angle between the meridian and the horizontal (0° at the top and 90° at the base) (Vogel, 2003). Since $\sin^2 (90°) = 1$, the stress at the equator is simply the loading force divided by the thickness

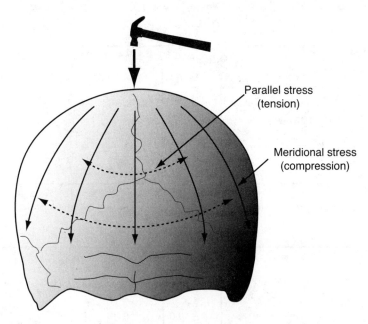

Figure 6.9. Comparison of meridional and parallel stresses in a dome-shaped skull.

of the wall; but the stresses rise toward the top of the dome as θ approaches 0° (to a theoretically infinite value at the top).

The above equation has two important implications for skull design. First, the top of the skull (or any likely site of impact) needs to be relatively thicker in order to be as strong as the base. Second, as vaults get bigger (and thus have larger radii), stresses in the walls will decrease in proportion to the square of the vault's radius. Thus, bigger domes are relatively stronger, as long as external forces scale with the linear dimensions of the dome; in fact, they should scale to the power of 1/2. This scaling relationship explains why domes are such wonderful architectural structures: for buildings, in which the forces applied are usually those of just the building itself; domes are one of the few structures that get stronger as they get bigger. But, sadly, cranial vaults are exposed to proportionally greater forces than buildings. The external forces to which heads are subject typically scale relative to mass or volume (a linear dimension cubed). Thus the walls need to become thicker relative to size (scale with positive allometry) to keep the same strength.

The analysis shown in Figure 6.10 indicates that cranial vault thickness generally scales with positive allometry relative to brain volume in primates, but

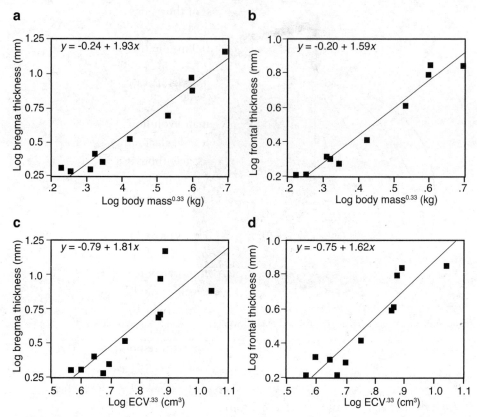

Figure 6.10. Allometries of cranial vault thickness relative to the cube root of body mass (top) and endocranial volume (ECV) (bottom) for two locations on the vault: bregma (the intersection of the coronal and sagittal sutures), and the midpoint of the frontal. Comparative primate data taken from Gauld, 1996. Note that the expected slope for isometry is 1.0 because the cube root of body mass and ECV were used.

close to the power of 1.75: thus it is greater than expected based on internal forces and greater than expected based on external forces. Vault thickness also scales with positive allometry relative to body mass. Figure 6.10 shows that vault thickness at locations near the top of the skull, such as bregma (the intersection of the sagittal and coronal sutures), scales with a higher coefficient of allometry than thickness at points lower on the vault, such as the frontal or temporal bones. In fact, in most primates, the skull is thickest at the two places where impacts are most likely to occur: the top of the skull and the back of the occipital. In modern humans, mean thickness of the vault at bregma and the occipital is 7.6 ± 1.6 mm and 14.0 ± 2.7 mm, respectively, but 7.0 ± 1.2 mm in the middle of the parietal and only 3.7 ± 0.9 mm in the middle of the temporal squama (Gauld, 1996). Interestingly, the vault bones of archaic *Homo* species, such as *H. erectus* and Neanderthals, fall well above the allometry between brain size and vault thickness (Gauld, 1996), whereas cranial vault thickness has declined significantly in human evolution. The earliest modern humans have vaults as thick as those of Neanderthals, and vaults have become 15–30 percent thinner mostly over the last few thousand years since the origins of agriculture (Lieberman, 1996). Why the vault has thinned is unclear. Boaz and Ciochon (2004) suggested that thick vaults in early *Homo* were driven by sexual selection in response to high levels of violence between males. Another possibility is that their skulls are thick because they are systemically robust, with thick bones throughout the skeleton. In one perplexing experiment (Lieberman, 1996), I found that juvenile mini-pigs forced to run on a treadmill everyday an hour a day for three months not only had thicker leg bones (which I was expecting) but also thicker vaults (which I was not expecting). I still don't know why the exercised swine had thicker skulls, but one possibility is that the exercise triggered a systemic growth response.

Regardless of how and why archaic *Homo* and early modern humans grew such thick vaults, their robust skulls would have been useful for withstanding the local effects of blows to the head. Local stresses (often called orbicular stresses) occur from inward bending of the wall at the point of impact, with compressive strains on the outside and tensile strains on the inside. Thus, at a local level, the vault behaves like a thin, curved beam in which resistance to bending is a function of the modulus of elasticity of the bone and the distribution of bone mass relative to the neutral axis (NA) of bending (the NA is the axis around which the bone bends). Thin beams have interesting mechanical properties because all the material is close to the NA. As elaborated in Chapter 7, the cross-sectional resistance to bending around the neutral axis of a given unit of bone is calculated as the second moment of area, *I*:

$$I = \Sigma_i^{1-}{}_N x_i^2 \cdot r_i^2$$

where is *x* is a given unit of bone, and *r* is its perpendicular distance to the NA. From this equation, one can see that thin beams inevitably have low values of *I*, because *r* is always small. This has two implications. On the one hand, thin beams are not very resistant to bending; but on the other hand, a given force generates relatively low strains in a thin beam for the same reason, namely that all the mass of the bone is close to the NA.

A further issue to consider is the sandwichlike structure of the cranial vault, which has both inner and outer tables of cortical bone on either side of a layer of trabecular bone, the diploe (see Chapter 4). As noted above, the cranial vault is typically between 5–10 mm thick, of which 2–4 mm is diploe (Lieberman, 1996; Lynnerup et al., 2005). The diploe probably has several functions. One is to reduce mass. Currey (2002) has shown that this arrangement optimizes strength relative to mass, with savings of about 15–20 percent. The lightweight trabecular bone in the middle has little mass but lies close to the NA and so experiences only tiny bending forces. In addition, the diploe provides a site for blood-cell production; it may help strengthen the cranial vault by making the bone more elastic; and it is possible that it plays a role in cooling the head.

Another point to consider is that the vault is not a single bone but instead consists of several curved plates separated by sutures. Sutures are growth sites, but they also play a mechanical role. A number of studies have shown that sutures transmit mostly compressive forces, but they also help cope with strains by dissipating tensile and shearing forces (Herring and Teng, 2000). Sutures also help a baby squeeze through the birth canal, probably the single most traumatic event a typical brain experiences (see below). However, as we age the sutures fuse, transforming the skull into a single, rigid element (see Chapter 4).

For the most part, this discussion of vault biomechanics has concentrated on high forces from rare but potentially disastrous impacts. Most of the time, however, the vault is loaded much less dramatically from benign functions such as chewing, walking, or running. A number of studies demonstrate that the braincase is overdesigned for these quotidian types of loading. In many of these studies, tiny strain gauges were glued during surgery to the surface of the skull in vivo in vaults of all sizes and shapes, including those of dogs (Iwasaki, 1989), primates (Behrents et al., 1978), armadillos (Lieberman, 1996), hyraxes (Lieberman et al., 2004a), pigs (Herring and Teng, 2000; Sun et al., 2004), and even one masochistic Englishman (Currey, 2002). In all these species, including the human, the strains measured during various activities such as chewing, running, and being hit on the head were minuscule (<100 units of microstrain), indicating that the braincase is minimally loaded during most of an individual's life. In fact, the strains measured are so low that elsewhere in the skeleton they would typically stimulate bone-resorbing cells (osteoclasts) to remove bone. However, as noted in Chapter 4, the braincase appears to be adapted to prevent such resorption under conditions of low strain. Most skull bones grow in part through the process of drift, in which osteoblasts deposit bone on the outer surface and osteoclasts remove bone on the inner surface, thereby maintaining a constant thickness of bone. However, the upper two-thirds of the braincase somehow inhibits the differentiation or recruitment of osteoclasts. Thus the volume of the braincase mostly expands by deposition within the sutures, and the bones of the vault can only get thicker, never thinner. This constraint is a good thing. If the braincase were liable to grow thinner, then it might resorb when subject to typically low strains, eventually becoming too thin to withstand rare but potentially fatal impacts.

Protecting the Brain from Concussions

Most injuries to the brain do not come from fractures of the skull but instead derive from accelerations and decelerations of brain tissue within the skull following an impact. When the skull is not fixed in position (as is usually the case), a considerable proportion of the impact energy is transferred into movement of the head and neck, often causing very high accelerations. The good news about these bounces is that they help prevent fractures. The bad news is that accelerations of the head, depending on their magnitude, duration and orientation, can cause serious, sometimes fatal injuries to brain tissues. Accelerations damage brain cells in a similar manner to the way that they can damage an egg yolk suspended in the white of the egg. Just as violent shaking can rupture the yolk without fracturing the eggshell, rapid jolts of the head generate shearing forces that can rupture tissues in the brain, especially arteries and the interface between white and gray matter (Leestma, 1988). The most clinically serious accelerations and decelerations are caused by angular rather than linear displacements of the brain. As noted by Holbourn (1943), deformations in the brain resemble those evident when shaking a bowl of porridge. If you shake the porridge back and forth, it may slosh about (not a problem for the brain, which is in a covered container), but rotating the bowl rapidly produces much more relative motion within the porridge (shear). In addition, because the acceleration of brain tissue caused by a blow is often highest on the opposite side of the head, brain injuries often occur on the side opposite to the site of trauma (these are known as contrecoup injuries). For this reason, blows to the front of the head sometimes damage the brain stem below the flexure of the brain in the MCF; these injuries can cause coma because this region controls many autonomic functions that maintain consciousness.

Although the human head has several structural adaptations to reduce the amount of energy transferred from impacts to the head into deformations within the brain, the large inertial mass of the human brain poses special challenges. One way the skull protects a big brain from shearing forces is by suspending it loosely within the vault in a slightly pressurized liquid bath. Three membranous layers, collectively termed the meninges (illustrated in Figure 6.11), make up this suspension system. The outermost layer, the dura mater ("hard mother"), is a tough, fibrous sheet of connective tissue, typically about half a millimeter thick, itself comprising two layers. The outer layer, against the bone, is the endocranium membrane. The inner, more fibrous layer not only surrounds the brain but also comprises the dural folds (see below) that separate parts of the brain. The venous sinuses lie between the two layers of the dura mater. The innermost layer of the meninges is the pia ("tender") mater. This layer is a thin and highly vascularized membrane that covers the brain tightly, like a glove. The pia mater dips into all the folds between gyri, carrying with it blood vessels. In between the pia and dura mater is the arachnoid ("spiderlike") mater, a delicate, transparent membrane separated from the deeper pia mater by a subarachnoid space that is filled with CSF. Importantly, the arachnoid mater is impermeable to CSF (Weller, 2005).

The three-layered design of the meninges is a fine example of "engineering" by natural selection. Because the outermost dural layer adheres to the braincase and the innermost layer adheres to the brain, the CSF between the pia mater and the arachnoid layer acts like a shock-absorbing, slightly pressurized water balloon that encapsulates the brain. Fluids are excellent shock absorbers because they are poorly compressible but highly displaceable (hence the popularity of waterbeds and hydraulic shock absorbers). Thus jolts to the head push the brain against its own form-fitting bag of fluid, damping accelerations and dissipating vibrations (see Bergsneider, 2001).

As noted above, the brain is more susceptible to damage from rotational than linear movements because the former are more likely to stretch arteries and other structures in the brain, especially those that traverse the meningeal layers (remember the porridge analogy above). The relationship between the force of impact, F, for a given linear displacement (x) and the resulting angular acceleration, α, is a function of its moment of inertia, I:

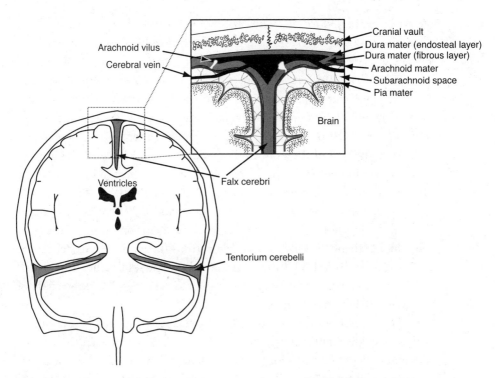

Figure 6.11. Meningeal layers of the brain in a coronal section through the plane of the foramen magnum. The dura mater and falx cerebri are part of the same tissue; in addition, posterior to the foramen magnum, the falx cerebri and tentorium cerebelli connect to each other.

$$F_x = I\alpha$$

For a sphere (which is a crude approximation of the human head),

$$I = 2/5 \; mr^2$$

where m is the mass and r is the radius from the center of the sphere. Thus, the force generated by a given rotational acceleration increases as a function of the brain's mass times the square of its distance from the center of rotation. A large human brain needs a good suspension system to restrain these movements.

The dura mater provides this suspension system. If you open a human skull and pry off the bones, the dura mater does not stick to the cranial vault. However, it is firmly tied to the cranial base and to a few key places in the vault where the dura folds back on itself to create thick rubbery sheets (septa) that divide the brain into three major regions (see Figure 6.11). One of these dural folds is the tentorium cerebelli, a tentlike sheet that covers the posterior cranial fossa. The tentorium attaches to crests on the floor of the braincase that separate the middle and posterior cranial fossae (the petrous crests of the temporal bone and the posterior clinoid processes of sphenoid). It then sweeps back along the wall of the vault on both sides and attaches around the confluence of sinuses on the inside of the occipital at the internal occipital crest. The tentorium probably acts as a sort of trampoline that keeps the cerebrum from pushing down on the mid- and hindbrain above the foramen magnum. The tentorium is not flat, but instead is tent-shaped in coronal view (the midline is higher than the walls). This angle permits the tentorium to deflect the mass of the cerebrum laterally toward the walls of the skull rather than inferiorly toward important structures below, such as the cerebellum, the circle of Willis, and the foramen magnum.

Another important dural fold is the falx cerebri (so named because of its sickle shape) that incompletely divides the two cerebral hemispheres in the midsagittal plane. The back of the falx attaches to the internal occipital protuberance, and the front is affixed to a protuberance known as the crista galli (cock's comb) just anterior to the cribriform plate. In addition, the base of the falx inserts along the midline of the tentorium, pulling it upward. The falx not only encloses the superior and inferior sagittal venous sinuses but also helps to prevent the two cerebral hemispheres from compressing each other during lateral accelerations of the head. Two other smaller dural septa also exist. The falx cerebelli divides two hemispheres of cerebellum much like the falx cerebri divides the cerebral hemispheres. In addition, a small flap of membrane, the diaphragma sellae, forms a roof over the hypophyseal fossa, protecting the pituitary and hypothalamus from being squashed by the overlying brain.

As noted above, the dura mater and the dural bands have proportionately more work to do in larger brains. This role probably explains why the dura mater averages about 0.2 mm thick at birth but gradually thickens to approximately 0.5 mm by age 18 (Ziablov, 1982). The dura is so thick in adult humans that it obscures many of the impressions of the sulci and gyri that one can see on the inner surface of the cranial vault in smaller-brained creatures, such as chimpanzees. There is some reason to suspect that the falx cerebri and tentorium cerebelli became thicker as the brain became larger in hominin evolution, at least as judged by the size of their insertions on the cranial floor (Aiello and Dean, 1990).

The dural bands help protect the brain, but they may also have other roles that have become increasingly important with the brain's expansion during human evolution. One possible role is to help the skull and brain grow together as an integrated, spherical unit. If you remove a brain from a skull, it slowly flattens out. According to a theory by Moss and Young (1960), the fibrous dural bands act like ropes in a circus tent, tethering the brain to key locations on the cranial base (reviewed above). Thus as the brain expands, the bands are tensed, pulling together the anterior and the posterior cranial base. Although hard to test, this theory is supported circumstantially by the distinctive shape ("clubbing") of the crista galli, which appears to be pulled up by the falx cerebri in response to tension exerted by the expanding brain (Moss, 1963). The tensile elastic properties of the white-matter tracts (see above) may also generate forces that help maintain a rounded brain (Van Essen, 1997).

Coda: Brain Size and Skull Shape in Human Evolution

Growing a large brain in the skull is an impressive challenge, given the four-fold increase in volume that occurs postnatally in humans (from about 330 cm^3 to 1350 cm^3) in concert with the considerable growth taking place in the face and cranial base. All this integrated growth, which happens at different rates, leads to substantial changes in shape that must occur without any loss of function. The same problems of integration are also true over evolutionary time, as adult hominin brains have expanded from less than 400 cm^3 to modern size and even larger (in the case of some Neanderthals). These increases, summarized in Table 6.1, have led to a variety of changes in cranial form, addressed in Chapter 5 and examined further in Chapters 11–13. To summarize, larger brains generally contribute to a rounder, more spherical cranial vault, a wider, longer, and (sometimes) more flexed cranial base, a wider face, and a more centrally located foramen magnum.

To what extent does a big brain pose special metabolic or structural challenges? One major cost is energy. An adult modern human needs to consume about 280–420 kcal of energy a day just for the brain's upkeep; in contrast, an adult chimpanzee brain needs only 100–120 kcal/day, and a typical early *H. erectus* brain (about 800 cm^3) would have needed about 200 kcal/day. Aiello and Wheeler (1995) have suggested that humans offset the increased cost of a large, expensive brain by decreasing the size of another very expensive tissue, the gut. Whereas average adult humans have a brain that is 851 g heavier than expected for a primate of our body mass, we have guts that are 781 g lighter than expected relative to body mass. Because gut size affects the rate of nutrient absorption, Aiello and Wheeler suggest that this trade-off also required an increase in dietary quality, such as more meat and fat. Larger brains combined with smaller guts may also have been made possible by eating other foods high in accessible energy, such as tubers, or by increasing the bioavailability of nutrients through cooking, although there is much debate about when cooking first evolved (see Chapter 7).

Another challenge pose by a bigger brain is thermoregulation. Although the brain itself doesn't generate much heat on its own, the larger the brain, the greater the volume of tissue that has to be kept cool. So larger brains need to be supplied with more blood to prevent heat stroke—a particularly critical problem during vigorous activities, such as walking or running in the midday sun. Humans have an elaborated suite of mechanisms to help cool the brain, including sweating in the upper face and scalp. Whether brain expansion itself favored or benefited from the evolution of these cooling mechanisms is unresolved. One hypothesis is that cranial "radiators" evolved prior to brain expansion to keep the hominin brain cool in hot, arid habitats when walking (Falk, 1990; Wheeler, 1985, 1991) and, even more crucially, when running (Bramble and Lieberman, 2004). These mechanisms first appear in early *Homo*, prior to substantial increases in brain size. If the radiator hypothesis is correct, then the thermoregulatory mechanisms that evolved to solve the demands posed by locomotion may have helped release a constraint on brain size, allowing hominins to keep their larger brains cool.

Another substantial cost of brain size is incurred during birth, as the big-brained human baby has to squeeze through the mother's pelvis, which has an unusually short, wide shape because of many modifications for bipedalism. In modern human females, the entrance into the birth canal at the top of the pelvis (the pelvic inlet) averages 113 mm in the sagittal plane and 122 mm in the transverse plane, but the average neonatal head is 124 long by 99 mm wide (Leutenegger, 1974). Thus the human neonate, whose brain volume is about twice that of a chimpanzee neonate's, must enter the birth canal sideways and then make a perilous 90° turn to pass through the pelvic outlet, which is longer sagittally than mediolaterally. The trip is a tight squeeze, and the baby typically faces dorsally, toward the mother's sacrum, while exiting the birth canal. In contrast, the chimpanzee pelvic inlet is 150 by 98 mm, much larger than the neonatal head, which is 56 by 72 mm (Schultz, 1969). A chimpanzee neonate enters and leaves the pelvis without changing orientation, facing ventrally toward the mother. Human birth thus takes much longer and, until recently, was a perilous process with a high rate of maternal mortality (see Rosenberg and Trevathan, 1996).

A final consideration is whether bigger brains pose greater mechanical demands on the skull itself. The answer is, in some ways yes, and in other ways no. One mechanical factor is strength. Bigger brains make the vault more dome-shaped, and domes become proportionally stronger and lighter as they get bigger. However, there is a strong positive allometry between vault thickness and body size across primates. Furthermore, the allometry is much greater than the expected scaling coefficient of 0.5, particularly in regions, such as the top and back of the vault, that experience the highest stresses, probably because external forces applied to the skull scale above body mass. Another mechanical factor is stabilization. A larger brain increases the mass of the head and hence the inertial forces that need to be counteracted to keep the head stable. As Chapter 9 discusses, most mammals stabilize the head using movements of the neck, which projects somewhat horizontally from the back of a cantilevered head, but bipedal hominins have short, vertical necks that attach near the center of the skull base. Hominins thus have had to evolve novel mechanisms for head stabilization. In this respect, increased brain size has been helpful because bigger-brained hominins have positioned much of the increased mass of the brain posterior to the foramen magnum, thereby improving the balance of the head. Indeed, humans probably have the most balanced head of any mammal, thanks to the combination of a big brain and a small, retracted face.

In conclusion, brain expansion in human evolution has come at a high price. Big brains cost more to feed, they are a challenge to keep at the right temperature, and they require special protection from injury during birth and the slings and arrows of outrageous fortune that follow. Big brains have also led to many changes in the overall shape and function of the head. Yet, somehow, in spite of these many challenges, they appear to have been worth the cost.

7

You Are How You Eat: Chewing and the Head

"In my youth," said his father, "I took to the law,
And argued each case with my wife;
And the muscular strength, which it gave to my jaw,
Has lasted the rest of my life."

LEWIS CARROLL, *Alice's Adventures in Wonderland*, 1885

What did you eat in the last 24 hours? It was probably cooked, chopped, blended, or otherwise highly processed. Although you may have lingered over a meal or two for social reasons, it is unlikely that you actually spent much time and effort getting the food into your mouth and breaking it down by chewing. For an ape, this brevity and ease of eating is highly unusual. A chimpanzee typically spends about half the day feeding, which means it also spends nearly half the day chewing (Wrangham, 1977). The many hours it spends putting food in its mouth and then chewing are partly a function of diet. Ape food is far less nutritious, tougher, and harder than what humans typically eat. Compared to hunter-gatherers, apes must eat larger quantities of food and chew more times with greater force to break down each mouthful prior to swallowing. The contrast is even more extreme when one compares ape diets with the highly processed cuisine of recent humans. These changes were certainly among the most profound transformations in human evolution.

Getting food into the mouth and chewing it prior to swallowing are vitally important aspects of cranial function, with strong selective consequences. This chapter reviews these functions and the ways they relate to head shape, growth, and variation (Chapter 8 considers swallowing). These relationships are interesting in their own right, but they are also important to understand for making inferences and testing hypotheses about how diet has changed in human evolution and how such changes may have influenced our oddly shaped heads. Of particular interest is why humans have a shallow, vertical, and snoutless face with small teeth. This configuration contrasts markedly with that of apes and other hominins, whose faces are generally longer relative to body size, less retracted, and, in the case of the australopiths, equipped with much bigger and more thickly enameled teeth. How much and why did these features evolve because of diet? And how can we use them to reconstruct key shifts in hominin diet, such as the transition to eating meat, to cooking, and to processing our foods in other intensive ways?

To explore these and other questions, we begin with a brief review of how teeth function mechanically to break down food. We'll then review the anatomy of the jaws, the temporomandibular joint, and the key muscles of mastication before describing how animals chew by moving the mandible. We next consider the role of variations in craniofacial shape for generating and resisting chewing forces. The chapter concludes with some thoughts on food processing and cooking and their profound effects on our food's mechanical and nutritional properties, how we chew, and how our heads grow.

The Basics of Chewing: Incision and Mastication

Humans use two major types of chews: incising with the front teeth and masticating with the postcanine teeth. Incision breaks food into pieces small enough to fit in the mouth. Sometimes the front teeth are also used to grip or strip items. Unlike mastication, incision often occurs on both sides of the jaw at the same time (it is bilateral) and occurs with the mandible drawn forward (protruded). In addition, during incision, muscles on both sides of the skull elevate (adduct) the mandible's front teeth upward against the stationary upper teeth, with little to no lateral movement. In contrast, mastication primarily breaks food items down mechanically into smaller pieces (comminution). Mastication is typically repetitive and rhythmic, with the tongue darting about between bites to move food around the oral cavity, from the front teeth to the premolar and molar region and finally to the back of the oral cavity to be swallowed. During mastication, the mandible is pulled both forward and backward (retracted). In addition, mastication in mammals is almost always unilateral, occurring on just one side of the mouth—the "working side" (the other side is the called "balancing" side), and involving a combination of vertical, anteroposterior, and lateral movements of the lower jaw.

Unilateral mastication, a key adaptation of mammals, has two major advantages that compensate for its complexity (Crompton, 1981). First, compared to bilateral chewing, unilateral mastication generates bite force more efficiently by recruiting both the working- and the balancing-side muscles to generate force at just one site on one side. Second, chewing food on one side of the jaw permits finely controlled movements as the lower teeth are pulled both upward (superiorly) and inward (medially) against the upper teeth with the food in between. The majority of food breakdown occurs just around the time of centric occlusion, when there is maximum contact between food and the tooth cusps (intercuspation) as the lower teeth reach the upper limit of their trajectory. There is some uncertainty as to how much food breakdown continues after maximum intercuspation, as the teeth disengage while the jaw starts to open. Hiiemae and Kay (1972; see also Kay and Hiiemae, 1974) proposed that mammals such as humans, with significant lateral jaw movements during chewing, have two phases of occlusion: during phase I, the lower teeth move upward and inward into centric occlusion; during phase II, the lower molars move slightly downward and medially out of centric occlusion (see Figure 7.1). Although it was originally thought that some grinding occurred during phase II, experimental data from macaques and baboons suggest that occlusal forces during phase II are too low to break food down significantly (Hylander et al., 1987; Wall et al., 2006).

Mastication and incision have important implications for head growth and function. First, both activities generate high bite force in a single location, the bite point, by recruiting muscles on both the working and the balancing sides of the skull. Bite forces are tricky to measure, particularly in nonhumans, but they can be substantial. Typical peak bite forces in adult humans can be as high as 200–450 newtons (N) (Hylander, 1977; Jenkins, 1978; Sasaki et al., 1989). And humans are probably wimpy biters compared to other apes. According to an estimate by Lucas et al. (1994), an orangutan may generate more than 2,000 N to break open hard seeds! The stresses from these forces can generate substantial strains in the face. Moreover, chewing is repetitive and frequent, subjecting the skull to high magnitudes of cumulative strain. It takes me about 50–70 chews before I can swallow a morsel of beef jerky two centimeters square—a food similar to what may have been consumed in the Paleolithic. It should therefore not be surprising that many aspects of skull design reflect adaptations to generate and resist these high, repetitive forces.

Another critical aspect of chewing is precision. Each chew generates high, dynamic forces at minute but specific locations. A human typically applies

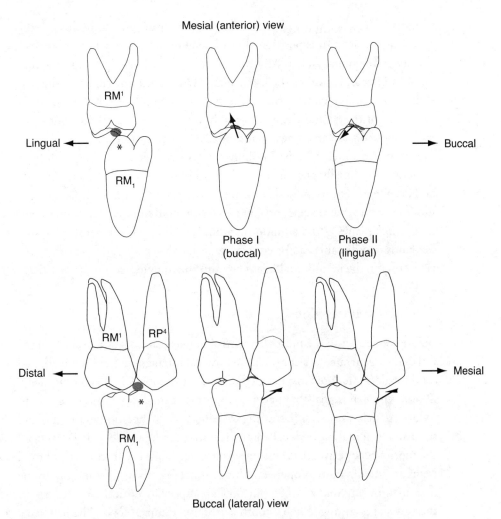

Figure 7.1. Tooth-food-tooth movements during occlusion (a right-side chew) from both mesial (anterior) and buccal (lateral) views. The protoconid of the RM_1 is marked with an asterisk. During phase I the protoconid moves mostly vertically and with high adductor force into the basin created by the RM^1 and RP^4; during phase II, the lower tooth moves more medially and inferiorly with less adductor force.

about 200–400 N of force at a premolar or first molar, across an area of about 100 mm^2 (Sasaki et al., 1989; Ellis et al., 1996). While generating this force, the cusps of the lower teeth must move precisely into the appropriate basins provided by the upper teeth. As an example, let us consider the path of a single cusp on a molar tooth: the most anterior (mesial) cusp on the medial (lingual) side of a lower first molar (Figure 7.1). Just prior to occlusion, this cusp (the metaconid) is positioned just lateral to the V-shaped space between the upper first molar, M^1, and fourth premolar, P^4 (superscripts and subscripts denote upper and lower teeth, respectively). During the power stroke, the cusp slides into a fossa about the size of the cusp, created by the space between the two cusps of the P^4 and the two mesial cusps of the M^1. This movement occurs under considerable upward and medially directed force. Errors of movement on the order of a few millimeters can render the chew ineffective. Different but equally precise movements occur with chews at other locations, and despite variations in food consistency, particle size, and other factors. Jaw movements are thus diverse and complex, with rapid simultaneous movements in all three planes at the temporomandibular joints. This dynamic precision demands considerable motor control, with much sensory feedback from the jaw muscles, the periodontal ligament, and the tongue (see Thexton, 1992).

Tooth-Food Interactions

Before we delve into the muscular control of mastication and incision and the forces they generate, let's consider the physical interactions between teeth and food (readers interested in a detailed review of this topic should consult Lucas, 2004). The major goals of chewing are to break food down mechanically to reduce particle sizes, thus increasing the efficiency of digestion in the gut, and to form a soft ball of food, called a bolus, that can be swallowed. Thus, from the tooth's perspective, the most important mechanical characteristics of food items are their fracture properties. These properties vary widely and cannot be described by any one variable, but they are especially critical with respect to the two stages during which a tooth acts upon a unit of food. The first stage is the initial deformation of the food in response to the applied load. This characteristic is best described by the modulus of elasticity, E, illustrated in Figure 7.2 (see also Chapter 2). For a given material, E is the ratio of stress to strain in compression or tension within the elastic range (the range in which the material will return to its initial shape after the load is released). Because the relationship between stress and strain is linear within this range, E is mathematically equivalent to the slope of the stress versus strain "curve." In colloquial terms, E is the material's "stiffness." The second stage of food breakdown occurs beyond a certain threshold of stress, σ_y, when the substance deforms plastically, meaning that it does not return to its original shape after removal of the load. If force continues to increase beyond σ_y, then eventually the ob-

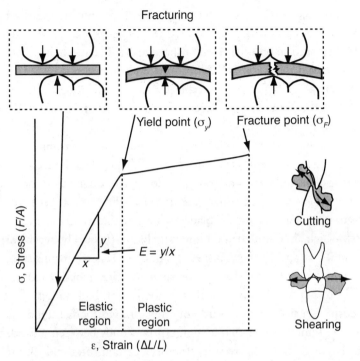

Figure 7.2. Many aspects of the mechanical properties of food are described by the stress-strain curve. Prior to the yield point (σ_y), applied stress (force per unit area, *F/A*) is linearly related to strain (change in length) in the elastic region. Stiff foods crack at the yield point, and typically little extra energy is required to fracture the object at σ_f. As the cartoons illustrate, stiff, force-limited foods are fractured primarily by crushing movements, but tough, displacement-limited foods are fractured primarily by cutting (slicing) or shearing movements.

ject will crack (fail) at σ_f. Note, however, that once a food particle begins to crack, then the crack needs to be propagated through the entire object. Thus another key variable to consider is toughness, *R*, defined as the work needed to propagate a crack through the object (calculated as the area under the force-displacement curve divided by crack area).

Most foods fail when the teeth apply sufficiently high forces, but it is useful to distinguish between foods that are force-limited and those that are displacement-limited (Lucas, 2004). Force-limited foods, such as a raw apple or cracker, are stiff and sometimes require high magnitudes of force to fracture; the mechanical properties of these foods are generally described by $ER^{0.5}$. Foods with a high value of $ER^{0.5}$ are typically considered "hard." In contrast, more compliant foods, such as meat or spinach, are not necessarily hard or stiff, but lots of energy may be required to propagate a crack through them because they need to be either torn or sliced apart. The mechanical properties of these displacement-limited foods are generally described by $(R/E)^{0.5}$. Foods with a high value of $(R/E)^{0.5}$ are typically considered "tough." Toughness is often due to the presence of collagen (in meat) or cellulose (which makes up the cell walls of many plants).

Teeth fracture food effectively because enamel is harder, stronger, and stiffer than anything we normally eat. Tooth size and shape affect the area over which a force is applied and how food is broken down. Although tooth form is enormously variable among mammals (see Jernvall, 1995), most teeth are combinations of three major elements: blades, which are sharp and narrow; cusps (or wedges), which are rounded and blunt; and basins (fossae) between cusps and blades (Hiiemae and Crompton, 1985). Human cheek teeth consist mostly of blunt cusps, which function well to break apart thick, blocky food items that are hard but not all that tough—that is, foods with high values of $ER^{0.5}$. A blunt cusp essentially cracks brittle or stress-limited foods trapped within a fossa in the same manner that a cook crushes foods between a pestle and a mortar. Foods crack when stress levels cross a certain threshold. But to break down tough, compliant foods with high values of $(R/E)^{0.5}$, teeth need sharp features to slice or continually reinitiate fractures as the food is spread across the tooth by loading. Because pointy features are ineffective at spreading cracks away from the initial break point, such teeth typically have bladelike elements (crests) in which the point is extended along a plane like a pair of scissors (see Lucas, 2004).

These principles help explain some basic differences among teeth. All primates have two kinds of cheek teeth, premolars and molars. Both premolars and molars are fairly low-cusped ("bunodont"), even more so in humans than in the African apes (Figure 7.3). Molars come in two successive sets (deciduous and permanent), whereas premolars form only in the permanent dentition. The two premolars in each quadrant of the mouth (denoted P3 and P4 because they are equivalent to the third and fourth premolars in mammals with four such teeth) are usually bicuspid, with not much of a fossa between the buccal and lingual cusps. Molars have a more complex shape, typically with four cusps in the maxillary molars (denoted by the suffix -cone) and five cusps in the mandibular molars (denoted by the suffix -conid). These cusps are labeled in Figure 7.4 for a typical upper and lower molar, but much variation exists, particularly in larger teeth. Minor cusps often grow on ridges between

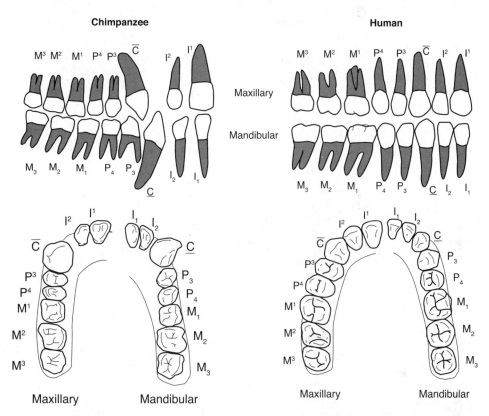

Figure 7.3. Permanent dentition of a chimpanzee (left) and a human (right) in buccal (top) and occlusal views (bottom). Modified from Aiello and Dean, 1990; Osborn, 1981.

major cusps (denoted by the suffix *-conule* or *-conulid*) or on a collar of enamel (the cingulum) that often surrounds the base of the crown (in which case they are denoted by the suffix *-style* or *-stylid*). The two deciduous molars are generally similar in occlusal shape, but thinner-enameled, smaller, and a little more bulbous. In humans, all premolars have a single root, except P^3, which has two. Human maxillary molars have three roots, and mandibular molars have two roots.

To break down tough foods, one needs teeth that can displace or cut. Mammals such as carnivores, grazers, and many folivores have various kinds of bladelike, shearing crests between the cusps in their postcanine teeth. Apes and humans, like most frugivorous primates, concentrate blades in the anterior dentition. In fact, anthropoid (monkey and ape) incisors are rather specialized

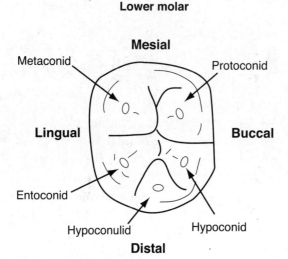

Lower molar

Figure 7.4. Occlusal view of the arrangement of cusps in a typical lower and upper human molar.

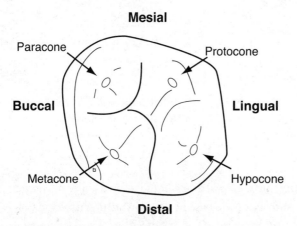

Upper molar

(Figure7.3). Whereas many mammals have small, peglike incisors for gripping foods, anthropoid incisors tend to be broad, spatulate, and bladelike, with long cutting edges. Apes and monkeys sometimes capitalize on these shapes for fracturing foods but more often use them for slicing, ripping, or peeling (Ungar, 1994, 1996). Consequently, the occlusal surfaces of the incisors can wear down substantially (Hylander, 1977; Ungar et al., 1997).

Human canines have more conical (diamond-shaped) crowns than incisors but otherwise generally resemble and function like incisors (they are termed *incisiform*) in the length and the shape of the occlusal surface. However, canines in nonhuman anthropoids are considerably more pointy and longer than the incisors, projecting well above (or below) the rest of the tooth row (as shown for chimpanzees in Figure 7.3). Primate canines have several functions. As weapons, they concentrate force in a small location for effective puncturing or gripping. They also function as blades, because the distal margin of the lower canine slides upward against the mesial margin of the upper canine. In addition, the distal margin of the upper canine in some primates, such as chimpanzees, forms a cutting edge (a shearing plane) against the mesial margin of the P_3, known as a C-P_3 honing complex. This kind of honing is absent in humans and reduced in the oldest known hominins (Haile-Selassie et al., 2004; Brunet et al., 2005; see Chapter 11).

Teeth also vary in the surface area and thickness of the enamel crown. To some extent, variations in crown surface area are a function of scaling: bigger animals have bigger teeth. But teeth don't scale to body size in an intuitively obvious way. If tooth surface area scaled with body size to maintain identical shapes at different sizes (geometric similarity), then one would expect tooth crown areas to scale with body mass to the power of 0.67 (because areas increase to the power of two, and body masses increase to the power of three, yielding a ratio of 2/3). However, postcanine tooth surface area among mammals scales with body mass to the power of 0.61, slightly but significantly below geometric similarity (Fortelius, 1985). One explanation for this negative allometry is mechanical. Lucas (2004) has argued from the perspective of fracture mechanics that tooth surface area should scale to body mass to the power of 0.5. The argument is based on the principle that larger food items of the same composition fracture at forces that scale to the power of 0.5 relative to the mass or volume of the food item (Lucas, 2004). The physics behind this principle is complex, but it essentially follows from the fact that once a crack is initiated in an object, large or small, little additional energy is needed to crack it all the way through unless the item is tough. Consequently, to chew larger food items, muscles in bigger animals need to increase bite force production only to the same power of 0.5 relative to body mass. Moreover, because the area of tooth-food contact should not increase relative to bite force, then tooth surface area should also scale to the power of 0.5 relative to body mass. Because the force that a muscle produces is a function of its cross-sectional area, then muscle cross-sectional areas should scale relative to body mass to power of 0.67. Nevertheless, as noted above, postcanine tooth surface area among mammals scales to the power of 0.61, significantly above 0.50 but also significantly below 0.67, suggesting that tooth size represents a compromise between geometric similarity and mechanical factors.

Postcanine tooth crown size, moreover, varies among species for reasons other than scaling. For example, the total surface area of the three permanent molars in any given quadrant (average of males and females) is about 3.5 cm^2 in chimpanzees, about 9 cm^2 in *Australopithecus boisei,* and approximately 4.5 cm^2 in early *Homo erectus,* but only 2.9 cm^2 in modern humans, despite overall similarity in body size (chimps are about 25 percent smaller) (Demes and Creel, 1988; Wood, 1991). Why do chimpanzees (or, for that matter, early hominins) have relatively larger tooth crowns than humans? One possible explanation is that tooth surface area influences the rate of food breakdown. Larger cusped teeth create more sites for food breakage, accommodate a greater range of food particle shapes, and perform better at cracking or grinding small items, such as seeds (Lucas, 2004). Among New World monkeys, frugivores typically have smaller teeth than leaf eaters or seed eaters (Kay, 1975, 1984). Another explanation, especially relevant to humans, is that smaller crowns save space. Modern humans have relatively short jaws, with barely enough room for the teeth. As faces became shorter, selection may have favored smaller teeth. A final contributing factor (explored at the end of this chapter) may be that permanent-tooth crown size is influenced very slightly by mechanical loading in the jaw.

One more functional interaction between teeth and food to consider is the wear that occurs from tooth-food-tooth contact. Tooth-tooth contact occurs mostly in the anterior dentition and is presumably one of the reasons that incisors and canines tend to have high crowns. In the cheek teeth, tooth-food contact wears down enamel, primarily from abrasion by silica particles in or on food, which grind teeth down like sandpaper. This wear can sometimes be beneficial. Many herbivores that need bladelike teeth to shear fibrous plant matter have thin enamel crowns in which exposed dentine wears faster than the enamel, creating sharp cutting edges; in contrast, mammals such as primates, which rely more on crushing, tend to have more low-cusped (bunodont) postcanine teeth that are covered with thicker enamel (Kay, 1975; Shellis and Hiiemae, 1986; Andrews and Martin, 1991). In these teeth, thicker enamel serves to protect the cusps from being worn out but may also be an adaptation to resist the higher stresses generated by larger bite forces (Martin, 1985; Lucas et al., 2008a,b). Detailed studies by Schwartz (2000) show that cusps more involved in cracking and crushing have relatively thicker enamel than cusps more involved in shearing and cutting.

Muscles and Joints

To explore how movement and forces are generated during incision and mastication—and how these forces and movements relate to head design—the next step is to review some basic anatomy of the masticatory muscles and the temporomandibular joint (TMJ). These structures permit several types of routine jaw movement: elevation (adduction) and depression (abduction), anteroposterior translation (protrusion and retraction of the mandible), and mediolateral translation (side-to-side movements of the mandible).

Five muscles, illustrated in Figure 7.5, are the mandible's primary movers. In humans and other apes, the largest is the temporalis, a fan-shaped muscle that originates from the side of the cranial vault, runs through the space between the zygomatic arches and the vault (the temporal fossa), and inserts on the mandible's coronoid process. The temporalis has diversely oriented fibers that differ by region. The anterior portion, whose fibers are mostly vertical, originates in front of the TMJ behind the brow ridge and along the anterior margin of the temporal fossa. The posterior portion has mostly horizontal fibers that originate behind the TMJ from the back of the vault. The middle portion has oblique fibers. In some apes and hominins, portions of the temporalis are so large relative to the vault's surface area that they originate from protruding crests that provide extra surface area for muscle attachment (Ashton and Zuckerman, 1956). These crests form along the sagittal suture, above the superior nuchal line at the back of the vault, or both. All three portions of the temporalis elevate the mandible, but the posterior and middle portions also retract the mandible, and the anteriormost portion can protrude it slightly.

The masseter originates on the zygomatic arch and inserts on the lateral surface of the ramus. The masseter has both deep and superficial compartments: the superficial portion originates from the anterior, inferior border of the zygomatic and inserts onto the angle of the ramus; the deep portion has a more transverse orientation, arising from the posterior, internal surface of the zygomatic arch and inserting along much of the lateral surface of the ramus. Both the deep and superficial masseters primarily elevate the mandible. Their different orientations, however, give them additional roles: the deep masseter on the working side can also act to pull the mandible laterally (see below), and the superficial masseters can help protrude the jaw slightly (Hylander and Johnson 1994).

Mirroring the masseter on the inside of the mandible is the medial pterygoid, most of which runs downward and backward from the medial aspect of

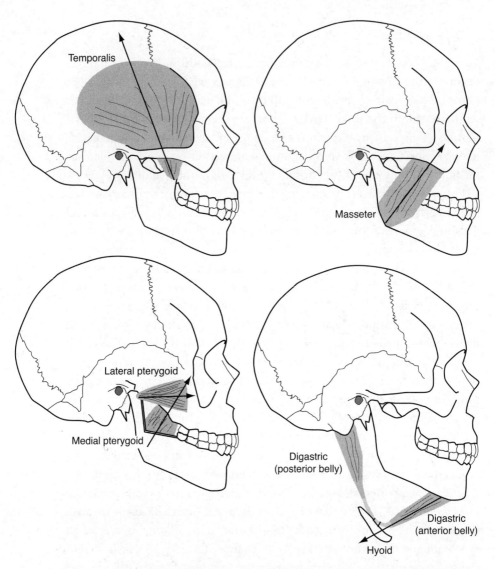

Figure 7.5. Major muscles of mastication, with arrows showing approximate orientation of the vector of force. The medial pterygoid is shown through a "window" cut into the ramus of the mandible.

the lateral pterygoid plate to the medial surface of the ramus; a small head of the medial pterygoid also arises from the back of the maxilla, just above the maxillary tuberosity. The medial pterygoid mostly elevates the mandible, but also pulls it medially on the working side. Together, the medial pterygoid and masseter muscles form a V-shaped sling around the ramus.

The lateral pterygoid arises from two heads, both of which insert onto the mandibular condyle. The much larger inferior head has its origin on the undersurface of the greater wing of the sphenoid and runs posterolaterally at a slight upward angle into a small pit on the anterior surface of the condylar neck. The superior head arises from the lateral surface of the temporal fossa and runs posterolaterally at a roughly horizontal angle to insert on the condylar neck, the articular disc, and the capsule surrounding the TMJ (described below). Both heads of the lateral pterygoid pull (protrude) the condyle forward, and their combined contractions help stabilize the condyle within the TMJ.

The digastric, so named for its "two bellies" joined by a tendon, is a major depressor (abductor) of the mandible (Figure 7.5). The anterior belly arises on the internal surface of the symphysis, from which it runs posteriorly, laterally, and obliquely; the posterior belly arises in the digastric fossa just medial to the mastoid process. An intermediate tendon connecting the two bellies runs through a loop of connective tissue affixed to the hyoid bone, thereby forming a pulley through which the tendon slides. The digastric muscle mainly depresses and retracts the mandible, but it can also raise the hyoid when the mandible is fixed. Most of the hyoid muscles of the neck below the tongue, such as the geniohyoid, also depress the mandible when the hyoid is fixed. The anatomy of these hyoid muscles is discussed in Chapter 8 in the context of swallowing.

The TMJ (Figure 7.6) is a paired joint with many components. The roof of the TMJ is a portion of the temporal bone whose central region is a transversely oriented, shallow groove, the glenoid fossa (sometimes called the mandibular or condylar fossa). The posterior and medial margins of the fossa are bordered by two small processes—the postglenoid and entoglenoid processes, respectively—that act like doorstops to prevent the condyle from moving too far posteriorly or medially. The anterior margin of the glenoid fossa is a sloping surface, the articular eminence, which merges continuously with a flat, horizontal surface, the preglenoid plane. In the "resting" position (when the teeth are just below their occlusal position), the mandibular condyles lie below the glenoid fossa. However, the condyles and the glenoid fossa do not articulate directly with each other. Instead, the condyle moves below an intervening mobile, fibrous pad called the articular disc. The articular disc has two layers: the upper layer is loosely attached to the temporal bone, and the lower layer is more securely attached to the condylar neck. In addition, the posterior portion of the disc is loose, vascular, and fibrous, while the anterior portion is stiffer and more tendonlike. When the mandible opens (abducts), the condyles are pulled forward and downward below the articular eminence.

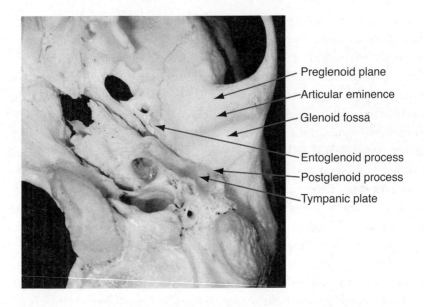

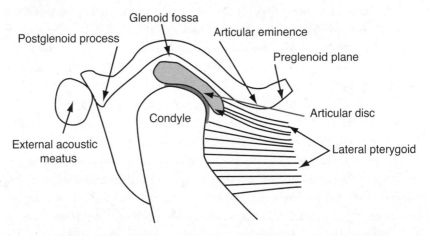

Figure 7.6. Anatomy of the temporomandibular joint (TMJ). Top: inferior view of the temporal bone, showing the major components of the glenoid fossa. Bottom: schematic lateral cross-section through the TMJ. Note that the condyle does not articulate directly with the glenoid fossa but instead with a movable articular disc.

As they slide forward, each condyle brings along its disc, whose fibrous posterior portion is drawn into the back of the joint capsule (Figure 7.6). When the jaw closes (adducts), the condyles are pulled backward and upward, stuffing the fibrous portion back into the space behind the fossa. Improper timing or control of the disc's movements can cause odd clicking sounds (crepitations) or pain.

The TMJ is highly mobile joint enclosed by a synovial joint capsule that needs to be protected from dislocation. Not surprisingly, many ligaments bind the condyle tightly to the cranium, mostly to limit anterior and lateral movements. These include a capsular ligament around the whole joint, a ligament that runs from the condylar neck to the outer edge of the glenoid fossa (temporomandibular ligament); and several extracapsular ligaments between the cranial base and the mandibular ramus (stylomandibular and sphenomandibular ligaments).

The Kinematics and Motor Control of Chewing

How do the masticatory muscles move the jaw during chewing? Given the variable and dynamic nature of incision and mastication, it helps to examine separately the closing and opening phases of each chewing cycle. These phases differ in several key respects during incision and mastication.

Opening

During the opening phase, the mandible depresses (abducts) from gravity and from contractions of the digastric and other muscles that attach to the hyoid. Depression causes the condyles to rotate around a mediolateral axis. As the condyles rotate, contractions of the lateral pterygoids also pull (translate) the condyles forward and downward below the articular eminence. The combination of translation and rotation at the condyle leads to an instantaneous center of rotation (the point where there is just rotation and no translation) that is inferior and slightly anterior to the condyle (near the insertion of the sphenomandibular ligament). During the opening phase of bilateral incision, both condyles translate synchronously; but during unilateral mastication, the timing of the left and right lateral pterygoid is staggered slightly, shifting the mandible slightly laterally toward the working side. You can feel these anterior and lateral movements by placing your fingers next to your TMJ as you open your jaw.

Closing

The closing phase of mastication has two stages: fast and slow close. Fast close in humans is typically initiated by the balancing-side medial pterygoids and

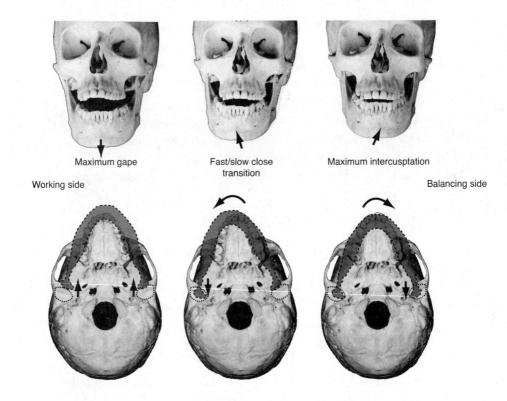

Maximum gape   Fast/slow close transition   Maximum intercusptation

Working side   Balancing side

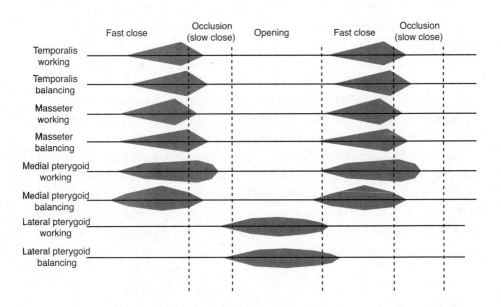

Figure 7.7. Movements of the mandible during mastication. Top: anterior view. Bottom: inferior view. At maximum gape (following the opening stroke), both condyles are translated forward as the jaw is depressed (abducted). During the first part of closing (fast close), the working-side condyle is retracted more posteriorly than the balancing-side condyle while the jaw elevates (adducts). During occlusion (slow close) the working-side condyle acts as a pivot point while the balancing-side condyle is retracted. This generates medial movement of the lower teeth as the jaw is adducted under high force. The bottom figure illustrates idealized activation patterns for the major working- and balancing-side jaw muscles. (Schematic summary based on EMG data from Miller, 1991.)

the working-side deep masseter, which not only begin to elevate the mandible but also shift it slightly toward the working side. As the jaw closes, the condyles begin to translate posteriorly, with more translation initially on the working side. At the end of fast close, the postcanine teeth of the working-side mandible lie just lateral to those of the maxilla. Slow close occurs as compartments of the medial pterygoid, masseter, and temporalis muscles contract in a different sequence, causing the mandible to continue to elevate with increasing force and the working-side teeth to move medially toward the midline of the skull. Both condyles rotate during closure, and medial movement takes place as the balancing-side condyle is pulled (translated) backward into the glenoid fossa (in other words, medial movement is largely a rotation of the mandible around a vertical axis that runs through the working-side condyle). The critical occlusion portion of slow close is called the power stroke, when the major jaw adductors (medial pterygoid, masseter, and temporalis) exert peak contractile force (Hylander et al., 1987). At this point the lower teeth on the working side are not only pressed up against the upper teeth by all the adductors but also dragged medially by differentially greater activity of certain muscles, especially the balancing-side deep masseter, and the working-side medial pterygoid and temporalis (see Weijs, 1994; Hylander 2006). The balancing-side teeth don't occlude because the mandibular arch is a little narrower than the maxillary arch.

One of the most important aspects of mastication from the perspective of skull design is that jaw elevator muscles contract on both the balancing and the working sides to produce bite force on just one side of the skull. This combination generates a complex, nonsymmetrical pattern of movement and loading, illustrated in Figure 7.7. In humans and other anthropoids, occlusal movements also have a substantial transverse component. These precise, forceful movements occur from differential timing and force production by the balancing- and working-side muscles. Consequently, during each chewing cycle the muscles shift roles, particularly those with some degree of horizontal orientation. For example, muscles that pull the jaw medially, such as the balancing-side deep masseter, must do more work, and their effort tends to peak after that of their paired counterparts (Weijs, 1994). Muscle contractions reach their peak during the power stroke, just before the teeth are brought into centric occlusion (Hylander et al., 1987). Thereafter, muscle activity and bone strain (see below) quickly decline during the so-called phase II, as the lower cheek teeth begin to move downward (Hylander et al., 1987; Wall et al., 2006).

The closing stroke differs during incision and mastication. In primates, the upper and lower incisors do not normally occlude during mastication because the mandibular arch is shorter than the maxillary arch. Therefore, the condyles must remain forwardly positioned on the articular eminence throughout the incisive closing stroke to protrude the jaw, allowing the upper and lower front teeth to make contact. Protrusion is maintained throughout incision by the lateral pterygoids (Miller, 1991). In addition, the balancing- and working-side adductor muscles contract more or less synchronously and with equal force during the incisive power stroke, limiting any mediolateral movement of the condyles or the mandible as a whole (Hylander, 1984).

The function of anteroposterior translation of the condyles has been something of a puzzle. Given all the pain that translation can cause when improperly coordinated, why not have a simple, hingelike joint? A number of benefits have been noted. For one, translation allows mastication without the incisors getting in the way, and incision without the cheek teeth occluding. In addition, at maximum gape, translation draws the back of the mandible forward, helping to prevent impingement of the pharynx (R. J. Smith, 1985), and reduces stretching of some of the jaw adductors (Hylander, 1978, 2006). Another explanation is that translation helps the teeth to function optimally during occlusion (Crompton et al., 2006). In most herbivorous mammals, including primates, the occlusal plane lies below the TMJ, as indicated by the distance R in Figure 7.8. Dropping the tooth row down helps to make space for larger masseter and medial pterygoid muscles but introduces an anterior component of tooth movement during occlusion in the absence of translation. Anterior movement is a problem because the shapes of most mammalian teeth don't accommodate anteroposterior movements (rodents' teeth are an exception). Translation solves the problem: posterior translation slides the condyle backward in the TMJ during jaw closing, eliminating any anterior trajectory of the mandibular teeth relative to the maxillary teeth during occlusion. In-

Figure 7.8. Geometry of translation. (a) In a simple jaw with a hingelike joint, the occlusal plane is in line with the TMJ, and the teeth occlude with a primarily vertical (orthal) direction. (b) Adding a ramus (R) introduces an anterior trajectory of the lower teeth relative to the upper teeth. (c) This trajectory is corrected by translating the joint posteriorly during adduction. (Adapted from Crompton et al., 2006.)

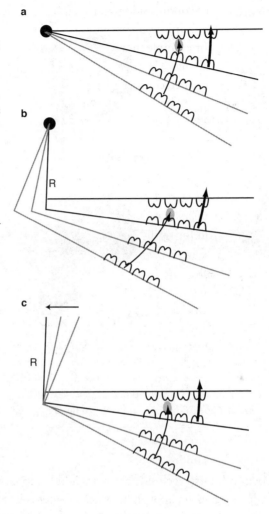

deed, the distance of the occlusal plane below the TMJ in different mammals predicts well how much translation occurs (Crompton et al., 2006).

Force Generation

We can now delve into how muscles generate forces during mastication and incision and how variations in skull shape affect the ability to generate these forces. (Later we will consider how the skull resists forces.) A good way to approach the topic is to start with a model of force generation by the major muscles in lateral and frontal views (Figures 7.9 and 7.10). Keep a few principles

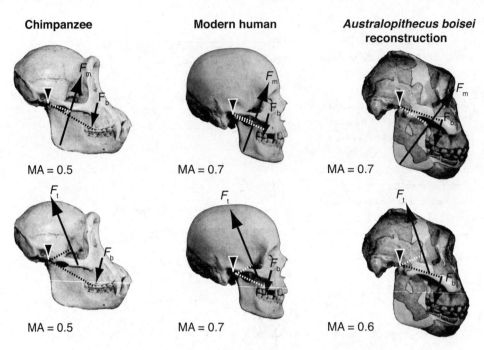

Figure 7.9. Approximate vectors of vertical adductor force and moment arms for the masseter (top) and temporalis (bottom) in a chimpanzee male (left), a modern human male (middle), and a reconstruction of an *Au. boisei* skull (right). The human and the robust australopith have a higher estimated mechanical advantage (MA) for a bite on the second molar than the chimpanzee, even though the chimpanzee can produce more total bite force.

in mind. First, each side of the mandible can be modeled as a simple beam in which forces are applied between a center of rotation (the TMJ), the symphysis, and (on the working side) the bite point. The force (m) a given muscle produces around a joint results in a torque, T_m:

$$T_m = F_m \cdot \text{lever arm}$$

where F is the contractile force of the muscle (generally proportional to its physiological cross-sectional area), and the lever arm is the *perpendicular* distance between the muscle's line of action and the center of rotation (COR). The bite also generates another torque, T_b:

$$T_b = F_b \cdot \text{load arm}$$

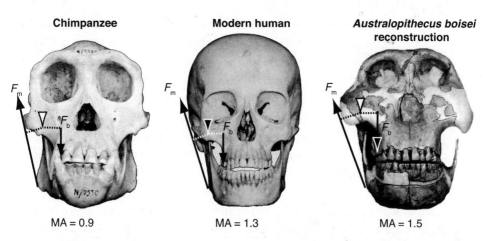

Figure 7.10. Approximate vectors of transverse adductor force and moment arms for the masseter in a chimpanzee male (left), a modern human male (middle), and a reconstruction of an *Au. boisei* skull (right). The human and especially the robust australopith have much greater MA for generating transverse force than the chimpanzee.

where F_b is the force of the bite, and the load arm is the *perpendicular* distance between the bite force vector and the COR. Because for every action there is an equal and opposite reaction, we can estimate bite force, F_b, as the muscle resultant F_r (the magnitude of the vector sum of all of the separate muscle forces) times the ratio of the muscle resultant lever arm divided by the load arm.

A muscle can produce both rotational and translational forces; their relative percentage is a function of muscle orientation. For example, the superficial masseter runs from the anterior part of the zygomatic arch to the angle of mandible. Thus the muscle force pulls the angle of the mandible both anteriorly and superiorly. This creates rotational forces that elevate the jaw but also anterior forces that can translate (pull) the mandible forward. A final point to remember is that a muscle's mechanical advantage (MA), the ratio of lever arm to load arm length, measures solely the *efficiency* of torque generation. A higher MA results in greater bite force per contraction, but at the expense of angular velocity. Natural selection does not always act to maximize MA.[1]

Let's now consider chewing forces in the sagittal plane by comparing the MAs of the major jaw adductors in lateral view for a chimpanzee, a modern

---

1. Consider a crocodile. Its first concern is to capture its prey in its mouth, requiring rapid jaw adduction. Crocodile jaws therefore have a low MA, thanks to a long snout with long load arms and short lever arms; they compensate for the low MA by having massive muscles.

human, and a robust australopith, as shown in Figure 7.9. All three skulls are scaled to the same size, and I have used approximately average muscle orientations to estimate the average lever arm for each muscle. Despite the imprecision of these estimates, several points should be evident. First, the greater degree of prognathism in the chimpanzee elongates the load arm for the muscles, thereby reducing MAs. Second, increasing the height of the TMJ above the occlusal plane (especially evident in the australopith), adds to the relative length of the masseter (and medial pterygoid) lever arms because of the anterior orientation of the muscle's line of action (R. J. Smith, 1978). The masseter's MA is also augmented considerably in the robust australopith by its anteriorly positioned zygomatic arch, which give the muscle a more oblique orientation. Modern humans have a high MA for both the masseter and temporalis because we have a retracted face that shortens the relative length of the load arm.

Frontal views of the masseter, shown in Figure 7.10, provide information about the efficiency of transverse force generation. Note that the much wider, flaring zygomatic of the robust australopith almost doubles the relative length of the masseter's MA in the coronal plane compared to that of the chimp. Modern humans also have a high MA for transverse forces because we have a relatively wider palate, which positions the tooth row closer to the TMJ, shortening the relative length of the load arm in the transverse plane.

An additional factor that affects bite forces is the location of the bite point relative to the muscle force resultant, essentially the combined orientation of all the muscle forces. (Because muscle force can be modeled as a vector, a quantity with direction and magnitude, the resultant of several muscles is calculated by adding their combined vectors.) Greaves (1978, 2000) noted that if equal force is generated with the working- and the balancing-side muscles, as Figure 7.11 illustrates, then the resultant of the adductor muscles probably lies near the midline of the dental arcade, just behind the third molar. This upward force must be balanced by downward forces at three places: the two condyles and the bite point. If the muscle force resultant lies outside this "triangle of support," then the system is unstable, causing the working-side condyle to pull out of the TMJ (termed *distraction*). An analogy is a three-legged stool that tips over if you sit on its edge and lean over, thus placing your center of gravity outside the triangle of support created by the three legs. As Figure 7.11 shows, modern humans with short, wide jaws have a potential difficulty with bites on the second and third molars, in which the resultant lies outside the theoretically predicted triangle. To prevent this unstable condition,

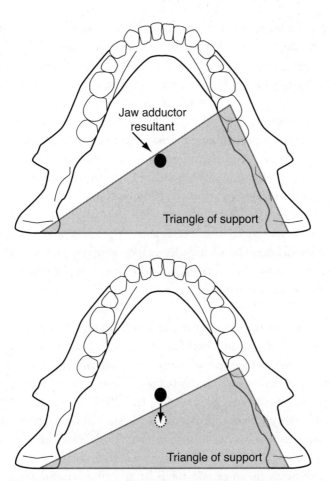

Figure 7.11. Triangle of support (gray area) for mastication in a modern human jaw for a bite on the second and third molars. To prevent the vector of the resultant from going outside the triangle of support on the M3 chew (bottom example), less force is recruited from the balancing-side muscles, moving the resultant posteriorly. (Adapted from Spencer, 1998.)

the balancing-side adductor force is decreased, thereby moving the resultant toward the working side at the expense of a reduced bite force (Spencer, 1998). Other structural factors evident in nonhuman primates, such as narrowing of the tooth rows relative to the TMJ, also help prevent distraction (Spencer, 1999).

Analyzing the MA of muscles provides only limited information (essentially efficiency estimates) about force production. As noted above, modern humans have higher masseter and temporalis MAs than great apes largely because of reduced prognathism, yet it is almost certain (even in the absence of good in vivo data for most species) that we produce absolutely less total occlusal force than other apes (Demes and Creel, 1988; Lucas, 2004). There are several potential explanations for this difference. The first is the proportion of type I (slow-twitch) versus type II (fast-twitch) muscle fibers. Type II fibers can generate several times the force of type I fibers, but they fatigue more rapidly. Whereas nonhuman primates (and most other mammals) tend to have a high percentage of large type II fibers, modern humans have a derived, inactivated version of the *MYH16* gene, which causes a substantial reduction in the size of the type II fibers (Rowlerson et al., 1983; Stedman et al., 2004) and hence smaller, less forceful muscles. The contractile force of a muscle is roughly proportional to its average cross-sectional area. We lack accurate cross-sectional area data for most masticatory muscles in apes, with one partial exception: the temporalis, whose maximum cross-sectional area is roughly approximated by the temporal fossa, the space between the zygomatic arch and the braincase (Weijs and Hillen, 1984). Because the temporalis is pennated (the fibers are at an angle and insert onto a central tendon), the actual physiological cross-sectional area (PCSA, the sum of the areas of all the fibers in a muscle) is greater than the temporal fossa area. Although body mass is smaller in chimpanzees than in modern humans, the mean temporal fossa area in chimpanzees is almost twice that of humans (Demes and Creel, 1988). Zygomatic arch length, a less reliable morphological proxy, also suggests that the cross-sectional area of the masseter is relatively smaller in humans than in the other great apes.

Bite force is not the only variable of interest from the perspective of mastication. Teeth break down certain foods by applying pressure (force per unit area) to food items, and this pressure in turn induces stress. Chimpanzees may bite with greater force than modern humans, but their molar-crown areas are also larger (by about 20–30 percent). As Demes and Creel (1988) showed in a clever analysis, illustrated in Figure 7.12, there is a highly correlated and isometric relationship between estimated bite force and molar-crown area among all apes, including humans and fossil hominins. In other words, the occlusal pressures produced during mastication do not necessarily differ much between these species, despite marked differences in muscle size and MA.[2] Perhaps this finding seems counterintuitive, but it shouldn't. As diets have changed during hominin evolution, most recently becoming softer and more processed, total

---

2. Forces are not generated over whole tooth crowns, but typically at cusps and in basins. So a proper analysis is needed to test how bite forces scale against cusp and basin areas.

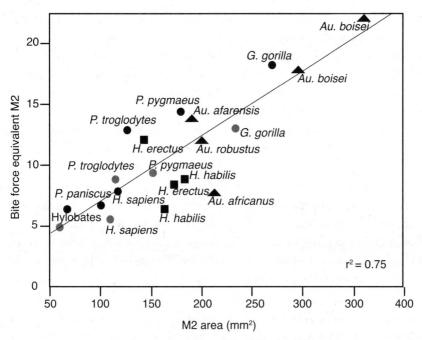

Figure 7.12. Bite force equivalents (not actual estimates of force in newtons) for adductor force from the masseter and temporalis for M2 chews, plotted against the crown area of the second molar (adapted from Demes and Creel, 1988) in different species. Bite forces increase linearly with tooth surface area, keeping occlusal pressures similar across species. Most of the data points above the line are males, and those below are females.

bite force production appears to have changed in concert with tooth size. This finding has different implications for chewing displacement-limited and stress-limited foods (see above). Generating high bite forces is usually not a critical factor for initiating cracks in compliant foods that are displacement-limited. In contrast, maximizing bite forces may have been important for fracturing hard or brittle foods that initially contact the teeth only at the cusp tips, thus generating high pressures by concentrating bite forces in small areas. Chapters 11–13 will return to the implications of correlated changes in bite force production and tooth size for reconstructing hominin.

A final comment about force generation. Leaving aside the issue of occlusal pressure and tooth size, let's assume we wanted to design a skull to maximize bite force. We would probably do several things. First, we might increase the proportion of type II (fast-twitch) muscle fibers, increase muscle pennation (augment the number of muscle fibers by arranging them at an angle relative to a central tendon), or add more muscle mass. The first two possibilities would have no direct effect on skull shape, but the last might lead to larger origin and insertion areas, expanding the angle of the ramus, the zygomatic, and the coronoid process, possibly resulting in cranial crests if the braincase were small. Second, we might maximize muscle MAs by reducing load arms and lengthening lever arms. This kind of engineering would involve shortening or retracting the face to position the TMJ as close as possible to the teeth and positioning the zygomatic arches as far forward and laterally as possible. In such a skull, one would also position the coronoid processes (where the temporalis inserts) as far forward as possible and elongate the mandibular ramus.

As a species, humans don't have particularly large masticatory muscles with high percentages of fast-twitch fibers. But we have, ironically, a particularly efficient design in terms of MA, even though we eat the least mechanically demanding diet of any primate and produce modest bite forces. In contrast, most primates, even gorillas and orangutans that generate high bite forces also have long snouts that decrease MAs. Maximizing the efficiency of bite force production is evidently not always selected for. Instead, when high bite forces are advantageous, selection sometimes favors big, strong muscles, probably because the biomechanical inefficiencies of long snouts may be outweighed by other benefits, such as distributing strains away from the braincase (see below). As I discuss later, the masticatory efficiency of the short human face may be an evolutionary "spandrel," a byproduct of other changes related to brain size. That said, there might be some evidence of selection for efficiency within modern humans. Higher MAs for jaw adductors are evident among some human populations, such as Inuits, that appear to generate higher bite forces (e.g., Spencer and Ungar, 2000).

Resistance to Chewing Forces in the Mandible

Given our typically soft and highly processed diets, it is hard to imagine chewing as a significant load-bearing activity. Yet humans can generate quite substantial bite forces (150–400 N) at a tooth's surface. And prior to modern culinary technology, the cumulative product of the magnitude and number of these chews must have exceeded current norms. High, repetitive bite forces matter because they generate strains that can be sufficient to damage bone microstructure (see Chapter 2). A monkey biting on a hard piece of food such as a nut can generate as much as 2,000 $\mu\varepsilon$ in the mandibular corpus (Ross,

2001), similar to the strain a full-sized horse typically generates in its tibia when trotting (Biewener and Taylor, 1986).[3]

How the human skull resists chewing forces is a significant but challenging issue to understand for at least four reasons. First, the loading patterns are complex. Chewing applies forces to the skull not only at the bite point but also at the TMJs and the origins and insertions of the many muscles involved. Second, the forces are dynamic. Peak forces typically occur at maximum intercuspation, but they are high throughout the power stroke, they change depending on bite location, and they differ between the working and the balancing sides. Third, the design and composition of the skull are complex, making it hard to model the ways bone, ligament, and other tissues resist applied loads. For example, when force is applied on a tooth in the upper jaw, that force passes through the crown, the tooth root, the periodontal ligament, the alveolar bone, and a series of irregularly shaped bones, each separated by sutures. Few simple biomechanical models of the skull are well suited to handling this complexity.

A final problem is that it is not possible to infer how a bone is loaded solely from external bone dimensions. Applied forces generate stress, which then generates strain (see Chapter 2). Because bones resist strain primarily through mass in the plane of deformation, variations in cross-sectional shapes are often assumed to reflect the loading patterns to which a bone is subjected. Although this is sometimes the case, patterns of loading and cross-sectional shape are not always well correlated, largely because bone shape reflects many influences, of which resistance to strain is only one (see Lieberman et al., 2004a). In addition, a bone's external dimensions do not always reflect the distribution of mass within the bone because of the presence of spaces, cancellous bone, and other tissues (Daegling and Grine, 1991). Thus, variations in external shape affect how a bone resists loading, but they do not necessarily predict how the bone was loaded in the first place.

Despite these caveats and complications, we can nonetheless investigate how the human skull withstands chewing forces by relating models of how forces are generated and resisted with experimental data on the strains these forces generate in nonhuman primates and other mammals. Experimental stud-

3. Strain (deformation) is typically measured in dimensionless units of microstrain, $\mu\varepsilon$.1,000$\mu\varepsilon$ equals a 0.01 percent change in length. Maximum in vivo strains recorded anywhere in the skeleton during vigorous activities rarely exceed 3,000 $\mu\varepsilon$ (Rubin and Lanyon, 1984).

ies mostly use strain gauges, tiny resistors that are glued to a bone's surface during surgery (see Biewener, 1992). When the animal eats, the electrical current in the resistor changes in proportion to the magnitude of strain at the bone's surface. Strain gauges measure the magnitude of strain as tension (ε_1, always positive) or compression (ε_2, always negative) and their orientation (tension and compression are always oriented perpendicular to each other). New, computationally intensive engineering approaches, such as finite-element modeling, also provide useful information (see below).

Let's start with the mandible, which is less complex than the cranium, and which has been studied comprehensively in nonhuman primates by W. Hylander and colleagues, whose many papers on the topic form the basis for the following summary. These experiments confirm models of the mandible as a U-shaped beam, illustrated in Figure 7.13. During a chew, the mandibular corpus and symphysis are subject to a combination of bending and twisting, with differences between the symphysis and the corpus on the working and balancing sides. On the working side, the corpus is primarily twisted because the major adductor that inserts on the ramus, the masseter, has a muscle force resultant with a substantial lateral component. As the masseter elevates the jaw, it also everts (pulls to the side) the inferior margin of the mandible, twisting the corpus around its long axis.[4] The same laterally directed forces also cause some lateral bending in the transverse plane (called "wishboning" because it resembles the strains made when pulling apart a cooked bird's sternum while making a wish). Finally, the corpus is also bent slightly in the parasagittal plane from the downward forces exerted at the condyles and the bite point. Behind the bite point, this bending compresses the inferior margin of the corpus; anterior to the bite point, the inferior margin of the working side corpus is tensed because of vertical forces transferred across the symphysis (Crompton, 1995; Lieberman et al., 2004b).

The balancing-side mandibular corpus experiences the same suite of strains, but with some differences that primarily reflect the absence of a bite point and differential muscle activity (resulting in less overall force, particularly in the transverse plane). The balancing-side corpus is mostly bent in the parasagittal plane from downward reaction forces at the condyle and those transferred across the symphysis. Because it is exposed to relatively less transverse force, the balancing-side corpus typically experiences less twisting than the working side; it is also wishboned slightly.

---

4. The medial pterygoid counteracts this twisting but does not prevent it entirely.

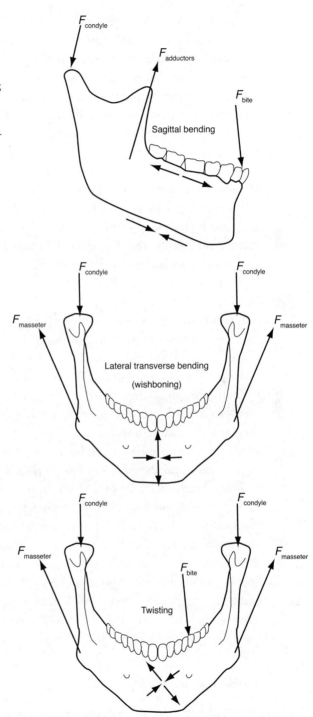

Figure 7.13. Schematic view of strains generated in the mandible during chewing. Top: sagittal bending (shown for the balancing-side mandible). Middle: wishboning (lateral transverse bending) strains. Bottom: twisting strains.

The symphysis is subject to a similar set of strains but is dominated in non-human primates by wishboning during the power stroke (Hylander, 1984). Wishboning is caused by laterally directed (transverse) forces from both the working and the balancing sides of the mandible that compress the anterior surface and tense the posterior surface of the symphysis. Twisting of both the mandibular corpora also causes the symphysis to bend a little in the transverse plane, tensing the inferior margin and compressing the superior margin. Finally, the symphysis is also sheared (deformed in a single plane, like warping a square into a parallelogram) from opposing upward forces on the balancing side and downward forces from the working side. It is worth noting that strains in the symphysis are affected by its fusion, a shared derived characteristic of apes and monkeys. Most mammals have unfused symphyses in which ligaments bind together the two halves of the mandible, permitting them to move somewhat independently in order to align the teeth properly for occlusion. Hylander and colleagues (Ravosa and Hylander, 1994; Hylander et al., 2000; Ravosa, 1999) pointed out that fusion of the two sides of the mandible is an adaptation to *strengthen* the symphysis (i.e., to avoid failure) in higher primates that transfer a greater percentage of adductor force from the balancing to the working side of the mandible than do most prosimians, such as galagos. Lieberman and Crompton (2000) noted that fusion is also an adaptation to *stiffen* the symphysis, increasing the efficiency with which forces in the transverse plane transfer across the symphysis. In this respect, anthropoids are convergent with mammals such as horses, pigs, camels, hyraxes, and wombats—all of which have a substantial transverse component of tooth movement during occlusion and also have fused symphyses (Crompton et al., 2008).

So far, I have described only strains during the power stroke during mastication. During incision, both the balancing- and the working-side corpora bend in the sagittal plane and twist around their long axes from eversion of the ventral margins, but they undergo much less transverse bending because medial movement is minimal (Hylander et al., 1998). In addition, during the opening stroke, both mandibles are bent slightly in the sagittal and transverse planes from adductor and medial transverse bending forces, respectively. These strains, however, are considerably lower than strains that occur during the power stroke (Hylander and Johnson, 1994).

Mandibular Shape and Strain

How well is the mandible designed to withstand the bending, twisting, and shearing strains it experiences during mastication? Here, a short digression on mechanics may be useful. Recall from Chapter 2 that bones function to a large extent like structural beams or tubes, resisting strain primarily by distributing mass in the plane of deformation.[5] Because bending compresses and tenses opposite surfaces of a beam around a neutral axis (NA), one can strengthen a beam by adding mass (making it thicker or denser) around this axis (see Figure 7.14). A handy example is the force needed to bend a book: the thicker or denser the book, the harder it is to bend; it is even harder to bend the book in the plane of the spine. Assuming bone density to be fairly constant, resistance to bending in a given plane can be usefully quantified as the second moment of area (I_{xx}) around the neutral axis, calculated by dividing a cross-section into many squares of dimension x^2, each located a distance y from the neutral axis:

$$I_{xx} = \Sigma_i^n x_i^2 \cdot y_i^2$$

This formula has several key implications, the most important of which is that a given area of bone is more resistant to bending the further away it is from the axis of bending, because I_{xx} is a function of each unit area (x^2) times its *squared* distance to the NA (y^2). Large but hollow objects can therefore be very strong. However, the external dimensions of a bone don't entirely reflect its overall resistance to bending because of variations in cortical bone thickness, density, presence of trabecular bone, and other aspects of its internal geometry (Daegling, 2002). In addition, bone is stronger in compression than in tension, making it particularly advantageous to add bone differentially on the tensed side.

Applying these principles, it follows that adding bone mass to the mandibular corpus in the sagittal plane resists bending in the sagittal plane. Such reinforcement could occur by making the corpus taller (see Figure 7.14). Wishboning would be resisted by making the corpus mediolaterally wider, particularly on its tensed medial surface. Similarly, wishboning of the symphysis would be resisted by making it deeper (anteroposteriorly thicker), particularly on the tensed internal surface.

Resistance to torsion, which is twisting around a central axis, is a little more complex but is best quantified in a section as the polar moment of inertia (J) around the axis of twisting (Vogel, 2003). J is calculated as the sum of I_{xx} around any two perpendicular axes. Thus, to make a tube more resistant to twisting, one can make it thicker or increase its diameter (again, the latter has

---

5. To a lesser extent, bones also augment their strength in other ways, such as increasing mineral density and aligning collagen fibers in the bone's axis of tension.

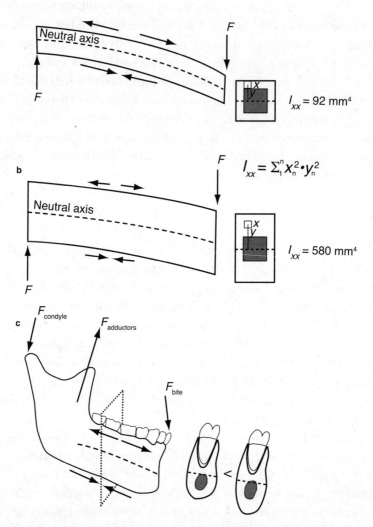

Figure 7.14. Resistance to bending in the mandible. (a) Schematic view of two-point bending, showing neutral axis (NA) and cross-section of the beam with the second moment of area (I_{xx}). (b) A taller beam distributes more mass around the NA, substantially increasing I_{xx}. (c) These principles applied to a schematic human mandible with two alternative cross-sections, in which the section on the right has more resistance to bending.

greater effect). It follows that twisting of the mandibular corpus and symphysis is resisted by adding mass anywhere to increase its thickness, width, and height. One complication, however, is that the mandible is not really tubular in shape because the upper portion is interrupted by the tooth in its socket, along with the periodontal ligament (see Figure 7.14). Does the corpus therefore function more like an O-shaped "closed" section or a U-shaped "open" section? Experiments by Daegling et al (1992), which examined the effects of torsion in jaws with and without these structures, indicate that the tooth, periodontal ligament, and alveolar bone each influence patterns of strain resistance differently but generally cause the mandibular corpus to behave as a closed section.

A number of studies have also investigated whether the structural principles by which mandibular geometry functions to withstand chewing strains can help explain observed variations within and between species. Do primates that chew harder food have more built-up corpora and symphyses? On a crude level of analysis, the answer appears to be yes. Overall scaling relationships between dimensions of the mandibular corpus and symphysis relative to jaw length make sense in terms of resistance to loading (Hylander, 1985, 1988). For example, within apes, larger-bodied species, such as gorillas and orangutans, have taller and wider corpora and symphyses (these variables scale with slight positive allometry) relative to jaw length than smaller species, such as gibbons (Ravosa, 2000). Several factors explain these shape differences, shown in Figure 7.15. First, larger animals tend to eat lower-quality food that is harder or tougher, requiring more masticatory force and chewing cycles. Larger animals also recruit relatively more force from the balancing-side muscles (Hylander et al., 1998). Finally, because larger species tend to have relatively narrower jaws, the inside surface of the symphysis will experience relatively higher wishboning strains that are counteracted primarily by increasing symphysis width, giving larger-bodied apes more curved symphyses (Hylander, 1985). That said, fine-scale analyses of jaw morphology appear to be unable to separate species by dietary differences. Daegling (2002) compared cross-sectional geometrical properties of the mandibular corpus in three species of monkeys that eat different diets. Despite expectations to the contrary, monkeys that chew harder or tougher diets do not have jaw shapes that confer greater resistance to bending or twisting, presumably because of other factors such as tooth size, facial shape, and individual variation.

Variation within species is another confounding factor. Differences in mandibular shape are sometimes considerable within species such as humans

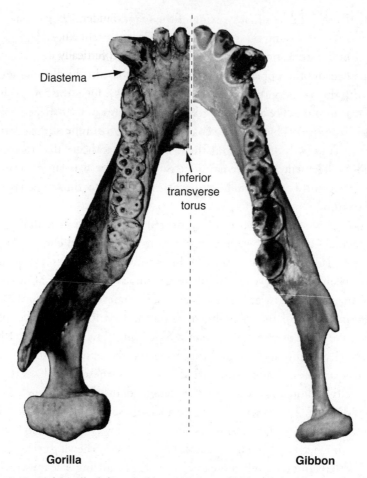

Figure 7.15. Male gorilla (left) and gibbon (right) mandibles, scaled to the same length. The gorilla mandible is more robust, with a relatively thicker corpus, a deeper inferior transverse torus, and a space (diastema) between the incisor and canine.

with varied diets, partly because of bone growth responses to different loading histories. Prior to skeletal maturity, bones can add mass in response to mechanical loading; in addition, loss of function can lead to local resorption. Thus chewing forces can have noticeable effects on jaw shape. The most responsive region of the jaw is the alveolar process that encapsulates the periodontal ligament (PDL) and the tooth roots. Alveolar bone is not only bent, twisted, and sheared along with the rest of the mandible by the forces discussed above, but it is also subject to intense local loading in the region of the bite point (Korioth and Versluis, 1997; Daegling and Hylander, 1997). Force from the tooth travels via collagen fibers around the tooth root, through the periodontal ligament, and into the alveolar socket. The vertically directed component of the bite force pushes the tooth downward in the alveolar socket, and the transverse component rotates the tooth within the socket, tensing and compressing opposite sides (Berkovitz et al., 1995). In turn, these strains (particularly the tensile ones) activate remodeling within the alveolar process around each tooth, helping to position teeth for proper occlusion (Davidovitch, 1991; Melsen, 2001). In the absence of teeth, the alveolar process gradually resorbs (Carlsson and Persson, 1967; Sugimura et al., 1984). There is abundant evidence that modern humans with larger bite forces or larger jaw muscle cross-sectional areas tend to have significantly taller and wider jaws, partly from increased alveolar process size (e.g., Ingervall and Helkimo, 1978; Heath, 1982; Weijs and Hillen, 1986; Kiliaridis et al., 1989; Kiliaridis, 1995; Raadsheer, et al., 1999; English et al., 2002).

Many factors can lead to significant correlations between jaw size and bite forces (Hannam and Wood, 1989), but experimental studies on other mammals confirm that the strains generated by mastication are influential. Rats that grow up chewing soft food have smaller jaw adductor muscles, generate lower mandibular strains, and have significantly shorter mandibles than rats that eat harder foods (Kiliaridis et al., 1986; Engström et al., 1986; Yamada and Kimmel, 1991). Studies on larger animals, such as primates, pigs, and hyraxes, show the same effects (Corruccini and Beecher, 1982, 1984; Beecher et al., 1983; Ciochon et al., 1997; Lieberman et al., 2004b). Figure 7.16 illustrates a typical example: a comparison of mean mandibular shape (the first two principal components, which explain 50 percent of the variation) in two groups of sibling pigs fed hard versus soft food for just three months (the diets were otherwise identical). After correcting for size, the pigs raised on hard food had relatively longer rami (by 9 percent), taller corpora (7 percent), and wider corpora (11 percent).

Three other regions in the mandible need to be discussed in terms of their biomechanical role in resisting chewing forces: the TMJ, the ramus, and the chin. For some time, it was argued that the TMJ was unloaded during mastication, in part because the tissues of the TMJ, especially the paper-thin temporal portion, seem too feeble to resist loading. This hypothesis was disproved experimentally by Hylander (1979), who showed that the condyles and articular disc can and do resist high magnitudes of stress. Indeed, the condyle's anatomy and development reflect this load-bearing function. Superficially, the

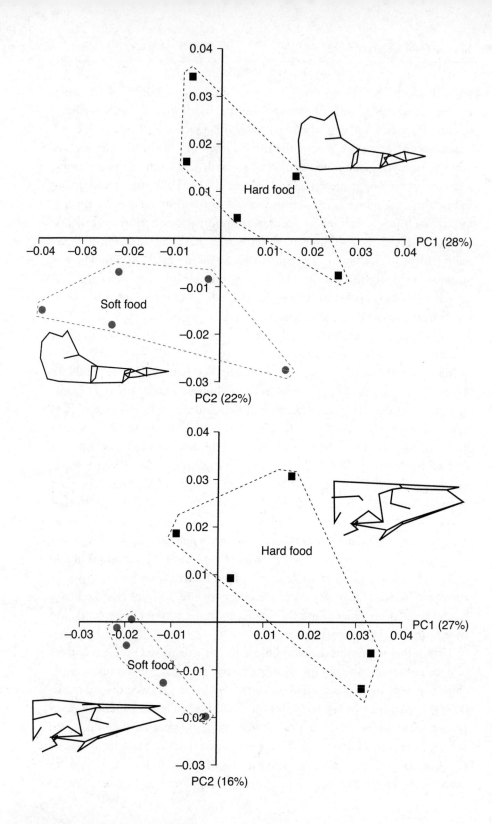

Figure 7.16. The two first principal components from a geometric morphometric analysis for the mandible (top) and the cranium (bottom) from 10 pigs fed nutritionally identical hard- and soft-food diets. Wireframes indicate average shape of the mandibles of the hard- and the soft-food groups (after size correction). The pigs fed a hard-food diet have relatively wider rami, taller and wider corpora, and taller, wider rostrums.

condyle resembles a typical postcranial epiphysis, in which the cartilaginous joint is separated from the diaphysis by a secondary growth plate that grows *interstitially* in response to systemic hormones. The condyle, however, has no growth plate. Instead, cartilage in the condyle grows *appositionally* from the surface of the condyle.[6] Thus, as the face grows downward and forward, the condyles must grow upward and backward, with differential growth on the superoposterior margin (Dibbets, 1990). The cartilage surface of the condyle not only functions to resist compression but is also stimulated to grow by compressive forces, largely mediated via the insulin growth factor IGF-I (Fuentes et al., 2002). Animals, such as rats and monkeys, that are fed softer food have condyles whose thickness and density are lower at the superoposterior margin than those of controls fed harder food (Bouvier and Hylander, 1982; Hinton, 1993; Kiliaridis et al., 1999). Not surprisingly, modern human populations that chew harder, less processed diets have slightly larger mandibular condyles and TMJs relative to overall cranial size (Hinton, 1983).

The mandibular ramus also responds to mechanical loading. As noted above, one function of the ramus is to position the corpus below the maxillary arch, anterior and inferior to the TMJ. In addition, the ramus—only a few millimeters thick in some places—provides insertion areas for the major jaw adductors: the masseter, temporalis, and medial pterygoid. As one might expect, these muscles have an important epigenetic influence on ramus shape. In one extreme experiment, Hohl (1983) surgically transposed attachments of the temporal and masseter to the posterior margin of the ramus in six juvenile macaques. Following surgery, their rami became shorter and less deep, and they lost the coronoid processes; the modified animals also developed smaller mandibles and faces, possibly because the reconfigured jaws were not capable of producing normal bite forces. Likewise, the size of the angle of the mandible

6. One consequence of this difference is manifested in achondroplasia, a disease in which disruption of the growth hormone pathway causes differentiated cartilage cells to fail to undergo mitosis. Achondroplastic dwarves have short limbs but very long mandibles.

(the gonial region) largely reflects the size of the masseter (Gionhaku and Lowe, 1989).

Perhaps it is fitting to end this discussion of the mandible with its own end: the chin. Unique to modern humans, the chin grows from resorption of the upper, alveolar portion of the symphysis, leaving behind a projecting inferior portion. Many biomechanical hypotheses have been proposed to explain the chin as an adaptation to resist strain (e.g., DuBrul and Sicher, 1954; Dobson and Trinkaus, 2002). None are very satisfactory, with the possible exception of Daegling's (1993) suggestion that the chin resists symphyseal bending in the transverse plane, caused by twisting of the two corpora. For a given bite force, twisting strains in the symphysis ought to be higher in humans than in nonhuman primates because of a wider mandibular arch, which increases torque. However, adding bone below rather than in front of the symphysis would counteract these strains more effectively. Moreover, computer modeling studies find that the presence or absence of a chin has little effect on the magnitude and pattern of strains that chewing generates (Ichim et al., 2006). In hindsight, such results should not be surprising: if the chin counteracts high strains, then one would expect its presence in other hominins, such as *H. erectus,* with similarly wide mandibles and who probably generated much higher bite forces. An alternative hypothesis, in need of further scrutiny, is that resorption of the upper alveolar margin helps to orient the lower incisors more vertically in modern humans whose faces are particularly short and retracted (Enlow, 1990) (see Chapter 13). Alternatively or additionally, the chin may be an example of sexual selection. Testing this last hypothesis is especially difficult, but the reader is encouraged to think of appropriate experiments.

Chewing Strains in the Human Cranium

How chewing strains are resisted in the human cranium is less well understood than resistance in the mandible. Our ignorance derives from several factors, the biggest of which is that the cranium must accommodate many more functions than the mandible, such as housing and protecting the brain and other organs of sense. In addition, the cranium is more complex morphologically than the mandible because it has so many bones, sutures, and odd-shaped cavities filled with various tissues, fluids, and air. A simple beam model that can reasonably describe the mandibular corpus obviously does not apply to the cranium as a whole. Another problem is the difficulty with generalizing experimental data on facial biomechanics in nonhuman primates and other mammals to humans because of the unique shape of the human face. Most nonhuman primates have a long, narrow face in which the tooth row lies in a snout (rostrum) well in front of the rest of the face. In contrast, the human face is a wide, tall, flat, and short, with no snout. Although these differences raise the possibility that strain patterns differ between humans and other mammals, testing for these differences is challenging without experimental in vivo strain data from humans (currently an impossibility). We can nonetheless make some tentative inferences about how strains are generated and resisted in the human cranium by first examining what we know about how the cranium is loaded and withstands strains in nonhuman primates and other mammals with varied facial shapes. We can then examine how the unique aspects of human facial shape may alter these patterns and look at the available evidence to test any predicted differences.

Models and Data

Figure 7.17 illustrates an oversimplified model of how regions of the nonhuman primate cranium are loaded. In lateral view, all the inferiorly directed forces generated by the muscles are applied at only the back of the face, whereas superiorly directed force is applied closer to the front of the face, especially for anterior bites. This arrangement resembles a tube-shaped beam. If the model is correct, then during anterior bites, in which muscles forces are generated evenly on the left and right sides of the skull, the snout should be sheared or bent upward, tensing the palate and compressing the anterior surface of the face (Endo, 1966; Hylander, 1977; Rak, 1983; Preuschoft et al., 1986; Ross, 2001).[7] Moreover, strains should be highest toward the posterior end of the rostrum, where the moment arm for bending is highest. Note, however, that the forces applied are not symmetrical during unilateral mastication. In such cases, the molar region is pushed upward only on the working side, and the back of the face is pulled downward on both sides by the masseter and other muscles, typically with more force on the working side (see above). As Figure 7.17 illustrates, this combination of forces should cause the face to act simultaneously like a sheared beam in the sagittal plane and like a twisted tube in the coronal plane (Greaves, 1985).

---

7. Shearing and bending are related but different. Bending a rectangular beam by applying forces at both ends (two-point bending) makes the beam U-shaped, compressing and tensing opposite surfaces. In very short beams, such forces mostly cause shearing by transforming the rectangular beam into a parallelogram rather than by bending it into a U shape.

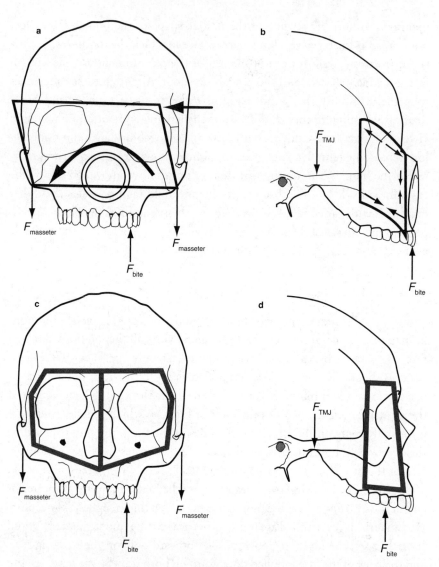

Figure 7.17. (a) Schematic models of twisting and shearing strains in the face from unilateral mastication. (b) Bending and shearing forces from incision or mastication. (c) and (d) Structural models for main columns or struts that provide resistance to these strains.

There are also nonbeam models of strain resistance in the face. Although the maxillary arch is U-shaped like the mandible, the body of the maxilla is mostly a hollow space (the sinus) with four pairs of projecting plates of bone: the palate, the nasal processes, the zygomatic processes, and the palatine processes. These processes, in turn, connect with processes of other bones to form framelike elements or struts around the nose and orbits (Figure 7.17). Another strut may be the nasal septum, which divides the nasal cavity in half (see Chapter 4). Several models (Endo, 1966; see also Rak, 1983; Russell and Thomason, 1992) predict that chewing therefore loads the face through these columnlike elements. For anterior bites, forces that load the front of the face possibly travel upward via columns of bone such as the orbital and nasal margins. If this model is correct, such columnar elements would be compressed axially (i.e., with no bending), causing bending in other parts of the face, such as the brow ridge. During unilateral mastication, the columns are predicted to behave nonsymmetrically, with vertical compression on the working side and tension on the balancing side.

Experimental strain data in nonhuman primates and other mammal faces reveal a complex picture that is partially consistent with the above models. The only clear conclusion one can reach is that there is no overall strain pattern in the cranium. Each region behaves differently. The snout, which mostly consists of the maxilla, is structurally the most tubular and beamlike part of the face. So far, the snout has not been studied comprehensively using strain gauges in primates, but the rostrum has been studied in vivo in pigs (Herring et al., 2001; Rafferty et al., 2003) and hyraxes (Lieberman et al., 2004a). These two species are an interesting contrast because they differ in their patterns of mastication and the shape of the snout. Pigs sometimes masticate bilaterally (with occlusal contact on both sides), and they have long snouts well in front of the orbits. In contrast, hyraxes—cat-sized creatures that are the closest living relative of elephants—masticate only unilaterally and have short snouts in which the postcanine teeth are mostly below the orbits. As one might expect, the pig snout is mostly sheared in the sagittal plane, with little evidence for twisting, but the hyrax snout is mostly twisted. These data tentatively suggest that the nonhuman primate snout (which is long but also subject to unilateral bite forces) is probably both twisted and sheared (see below).

The orbital plane of the face has been studied intensively by Ross and colleagues, who applied strain gauges to many challenging places in nonhuman primates, such as the internal walls and external margins of the orbits (Ross and Hylander, 1996, 2000; Ross, 2001). Most of these experiments indicate that the monkey midface behaves like a short beam that is sheared in the sagittal plane during anterior biting. During unilateral mastication, the orbital region is subject to a combination of shearing and twisting similar to the pattern described above for the hyrax snout. Ross (2001) found that mastication compresses the pillarlike region between the orbits, transferring stress upward. Ross also found that strains in the central region of the midface tend to be rather low compared to those closer to the occlusal plane. In contrast, strains in the beamlike zygomatic arch are much higher, thanks to bending caused by the masseter muscles that pull each arch downward (Herring and Mucci, 1991; Herring et al., 1996; Hylander, 1986; Ross, 2001; Lieberman et al., 2004a).

Strains in the upper part of the face have been measured in vivo in several mammals, such as pigs and hyraxes (Herring and Teng, 2000; Herring et al., 2001; Lieberman et al., 2004a) and many nonhuman primates (Hylander et al., 1991; Ross and Hylander, 1996; Ravosa et al., 2000a, 2000b; Ross, 2001). In all these creatures, upper facial strains tend to be low, even during chews that generate high strains in other regions, such as the mandible and zygomatic arch. In prosimians and nonprimates, such as pigs and hyraxes, unilateral mastication twists the upper face, as the tube model above predicts. However, the circumorbital region in some nonhuman anthropoids (owl monkeys, macaques, and baboons) shows little, albeit variable, evidence for twisting, despite predictions to the contrary, given their heavy recruitment of balancing-side adductor force (Hylander et al., 1991; Ross and Hylander, 1996). We do not understand the basis for these differences, but they may derive from a combination of variations in facial form such as snout length, retraction of the tooth row, recruitment of balancing-side muscles, and the extent to which the lateral and posterior margins of the orbits are composed of bone (see Lieberman et al., 2004a).

Despite variations in the orientations of facial strains, one pattern is clear: chewing generates a steep gradient of strain magnitudes, illustrated in Figure 7.18 (Hylander et al., 1991; Hylander and Johnson, 1994; Ross and Hylander, 1996; Ross, 2001). Strains are likely to be high in the alveolar process, major sites of muscle insertion, and possibly the palate (although the latter have never been measured in vivo); strains then decrease in approximate proportion to distance from the tooth row. Mean strain magnitudes from masticating hard food are typically moderate in magnitude in the midface, approximately 400–1,500 $\mu\varepsilon$ of shear. Compared to the midface, strain magnitudes from the same chews are at least 50 percent lower in the upper face, including the lateral margins of the orbits and the brow ridge (Hylander et al., 1991, 1992). Similarly, the wall of bone behind the orbits, the postorbital septum, also experiences low strains (Ross and Hylander, 1996; Ross, 2001). In other words, little stress transfers to the upper face during chewing. These data suggest that the rostrum, maxillary arch, and lower face do most of the work

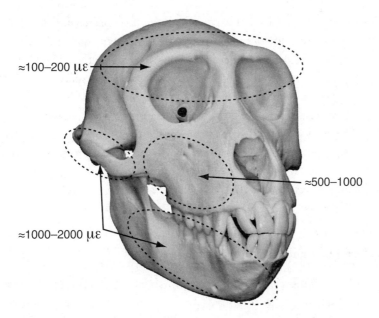

≈100–200 με

≈500–1000

≈1000–2000 με

Figure 7.18. Idealized view of the overall strain gradient in a primate face, based on studies noted in the text. The strain gradient misses local concentrations of strain that may occur in particular regions.

to counteract bite force strains and that features of the upper face, such as the brow ridge, are not beamlike adaptations or responses to resist high strains (contra suggestions by Endo, 1966, and Greaves, 1985).

One factor that may contribute to the facial strain gradient is sutures. Sutures are necessary for growth and fuse in some parts of the skull after growth is complete. In the human vault, for example, the sagittal and coronal sutures tend to fuse after 30 postnatal years (Meindl and Lovejoy, 1985). However, many facial sutures persist throughout life, with their collagen fibers likely functioning to absorb strains from attenuating tensile forces (Herring and Mucci, 1991). Strain gauges placed across sutures measure much higher strains during mastication, sometimes as much as 10-fold higher, than gauges placed on nearby bones (Herring and Teng, 2000; Lieberman et al., 2004a). In addition, strains transmitted from one bone to another typically decline across a suture by about 50 percent (Jaslow and Biewener, 1995). Some sutures also have complex, interdigitating serrated edges that function to counteract shearing between bones and to align collagen fibers in tension during bending (Moss and Young, 1960; Jaslow, 1990; Herring and Mucci, 1991).

A final aspect of cranial strain to consider is the effect of chewing on the vault. Various studies have analyzed this effect in nonhuman mammals (Behrents et al., 1978; Iwasaki, 1989; Lieberman, 1996; Sugimura et al., 1984; Herring et al., 2001; Lieberman et al., 2004a), as well as in one human (Currey, 2002). In all these species, including the human, the strains generated in the cranial vault by chewing are low, presumably because most stresses dissipate in the face before reaching the rest of the cranium. In addition, the dome-shaped structure of the vault effectively distributes any forces over large areas, resulting in low stresses (see Chapter 6). Even contractions of the temporalis muscle, which originates on the vault, have little effect on strain magnitudes at any location (Behrents et al., 1978).

Strain Resistance and the Human Face

Before examining more closely how the human face differs from those of nonhuman primates and other mammals, let's consider how faces in general resist chewing strains If the above data are credible, then we can reasonably infer that the human face as a whole is subject to a combination of several strain regimes: bending or shearing in the sagittal plane; twisting in the coronal plane; and perhaps some degree of axial compression along bony struts such as the nasal margins. In addition, many discrete regions of the face probably experience particular local bending strains. Such bending has been measured in the zygomatic arches, and it almost certainly occurs in the alveolar processes and the palate (Thomason and Russell, 1986).

The basic principle for resisting strain in the face is the same as elsewhere: to add mass in the plane of deformation (technically, to increase the second moments of area around axes of bending and polar moments of inertia around axes of twisting). Thus for bending and shearing forces in the rostrum, high strain magnitudes would be resisted by making the alveolar processes larger (especially taller), thickening the palate, and increasing the vertical height of the rostrum. Making the rostrum convex in the sagittal plane would reduce stress concentrations and increase second-moment areas (Demes, 1987). Any twisting forces in the rostrum would be countered by making the rostrum as a whole thicker, wider, and taller and by thickening the palate. In the orbital plane of the face, which is posterior to the rostrum, both shearing and bending forces would be counteracted by adding mass in the coronal plane, especially in the lower part of the face, where strains are highest. Such a design would lead to wide and tall infraorbital regions with flaring zygomatic arches, not unlike what we see in the face of a robust australopith (see Chapter 11). Finally, if bony pillars such as the nasal or orbital margins act as vertical struts to transfer compressive (or sometimes tensile) forces, then these struts will be reinforced by increases in cross-sectional area. Note, however, that the modifications to particular regions of the skull (e.g., making the rostrum thicker) can potentially alter how stresses and strains are transferred from one part of the skull to another (for an example, see Strait et al., 2007).

Experimental data on mechanical loading and facial growth in vivo provide, not unsurprisingly, only partial support for these predictions. Animals with long snouts that chew harder food grow larger faces than controls raised on soft food, mostly in the region around the palate and rostrum (e.g., Corruccini and Beecher, 1984; Ciochon et al., 1997; Lieberman et al., 2004a). Figure 7.16 compares the thickness and width of the rostrum in pigs raised on hard versus soft food. Because we know from experimental data (Herring et al., 2001; Rafferty et al., 2003) that the pig rostrum is mostly deformed from shearing and bending in the sagittal plane, a simple model predicts that the animals fed hard food will have a taller rostrum in the sagittal plane (particularly toward the posterior end of the rostrum). Indeed, the pigs raised on hard food have 7 percent taller rostrums, but they also have rostrums 11 percent wider than those of pigs raised on soft food (both differences are statistically significant); in other respects, they differ little in shape. Thus the pig face is either twisted more than we otherwise thought, or its growth in response to loading does not correspond entirely to our simple model's predictions. Analyses of other species also remind us that facial shape reflects a complex design that fulfills many functions, of which strain resistance is only one. Demes (1982), for example, showed that the locations of peak resistance to bending in the primate face coincide poorly with the locations of highest predicted actual bending.

And, finally, as discussed above, the human face differs from those of other mammals, including nonhuman primates, in several key respects: the absence of a rostrum, retraction of the face under the orbits, and a relatively wide and tall face that lies almost entirely in the coronal plane. Because we lack a rostrum, it is reasonable to predict that chewing forces subject the human face primarily to shearing during anterior bites and to some combination of twisting and shearing during unilateral chews on the posterior dentition. Both types of strain regimes would be countered by adding mass in the coronal plane, particularly in the lower face. As with other primates, the human face probably also has a strain gradient with the highest strains in the maxilla and much lower strains in the upper face. Unfortunately, we have little direct evidence to test these hypotheses because one cannot apply strain gauges to human faces, and there are no really appropriate experimental models. Limited experimental evidence comes from the nonhuman primates discussed above, and also from the hyrax, whose teeth are retracted underneath the orbits as in modern humans (Figure 7.19). When hyraxes masticate, they generate mostly twisting strains that follow a typical strain gradient, diminishing considerably in the upper face relative to the lower face (Lieberman et al., 2004a). Hyraxes fed harder food also have relatively more growth in the lower, posterior portion of the face. In addition, in vitro studies by Demes (1982, 1984) using photoelastic-coated models of the human cranium, suggest that the wide, tall, and flat human face probably withstands effectively any bending and twisting forces generated by chewing.

Finite-Element Models

Further advances in understanding how human and nonhuman primate faces function biomechanically during chewing will require improved models and new analytical methods. One especially promising method is finite element (FE) analysis, an engineering tool that takes advantage of computational power to model complex structures such as the cranium by breaking them down into many thousands of small elements, each with its own material properties (for reviews, see Ross, 2005; Richmond et al., 2005; Rayfield, 2007). The model can then be loaded virtually to calculate and graph the distribution of stresses as they travel from one element to another. FE models of skulls hold much promise, especially as they become more adept at dealing with variations in the mechanical properties of different cranial tissues (factors that traditional biomechanical models traditionally ignore) such as bone, cartilage, the periodontal ligament, sutures, and articular discs (e.g., Dechow and Hylander, 2000). FE models will also become more useful as we gather experimental data on how skulls are loaded dynamically during occlusion and acquire more in vivo strain data from model organisms to validate and refine the results of computer-generated models.

FE models have begun to generate insights, particularly when applied to discrete regions of the skull such as the mandible (Korioth and Versluis, 1997; Vollmer et al., 2000; Hirabayashi et al., 2002; Ichim et al., 2006). More ambitious models of entire crania have also been attempted for nonhuman pri-

Figure 7.19. Differences in the position of the face in a human, a hyrax, and a baboon (from Lieberman et al., 2004a). The human has no rostrum, and the postcanine teeth lie underneath the orbits and anterior cranial fossa. The hyrax has a rostrum, but most of the postcanine teeth are positioned behind the orbits. The baboon has a long rostrum overlying the premolars and anterior dentition.

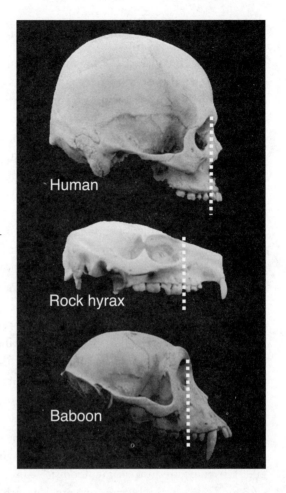

mates (Richmond et al., 2005; Strait et al., 2005, 2007), modern humans (e.g., Tanne et al., 1988; Jinushi et al., 1997; Rees, 2001; Gross et al., 2001), and fossil hominins (Strait et al., 2009). Although the results of these studies are preliminary, they generally suggest that the stresses generated by occlusion in the cranium are highest in the maxillary arch near the bite point and in the zygomatic arch and zygomatic root. Further, stresses dissipate from these regions via a steep but complex strain gradient through the infraorbital region and along the lateral margins of the orbit. In addition, and as predicted, the nonhuman primate face appears to function somewhat like a vertical plate in the coronal plane, particularly during chews on the posterior dentition (Strait et al., 2007).

Food Processing, Chewing, and Craniofacial Integration

Chapters 11–13 consider the major transformations in skull form that occurred over the course of human evolution and speculate on how these changes relate to shifts in diet. However, to conclude this chapter, Let's return to the topic with which we began: processing food through cooking and other means, and its effects on how we chew and how our skulls grow. The topic merits consideration here for two reasons. First, cooking and food processing has had profound effects on human diet as well as on human teeth, jaws, and other parts of the skull. Second, the effects of cooking and food processing on chewing offer an excellent example of the importance of integration between growth, form, and function among the skull's many components.

Because mammals chew primarily to break food items apart in order to swallow and digest them more easily, almost every aspect of the masticatory system needs to be considered when we examine how teeth break down food. Further, food processing methods, including cooking, have dramatic effects on the mechanical properties of food. Food processing can be divided into two major categories: mechanical processing (cutting, pounding, slicing, grinding), and cooking (the application of heat). Other species engage in limited food processing, but no species other than humans cooks its food, and all human cultures, with no exceptions, rely heavily on cooking and other forms of food processing, such as pounding, cutting, soaking and grinding. Indeed, Richard Wrangham and colleagues have suggested that cooking is actually a biological necessity for modern humans (Wrangham et al., 1999; Wrangham, 2009). Even postindustrial "raw-foodists"—usually well-to-do health fanatics with food processors who consume nutrient-rich domesticated foods—find it difficult to thrive without cooking (Wrangham and Conklin-Brittain, 2003). Mechanical processing of food may be even more of a biological necessity than cooking.

A reliance on mechanical food processing almost certainly predates cooking. Chimpanzees occasionally engage in a variety of food processing techniques, mostly pounding of nuts (Goodall, 1986; Boesch and Boesch, 1990). Early hominins almost certainly did the same, but the archaeological record suggests that the invention of simple flaked-stone tool technology, the Oldowan, dated to approximately 2.6 mya (de Heinzelin et al., 1999; Semaw et al., 2003), was probably a turning point in the evolution of human diets. Oldowan tools were used not only to cut meat and plant foods but also to pound open long bones to remove marrow and probably to tenderize tough foods, such as meat and tubers, by mechanical pounding (for details, see Schick and Toth, 1993; Klein, 2009). Slicing and mechanical tenderizing would have enabled hominins to cut foods into small pieces prior to mastication, thereby reducing the number of chews per unit of food. In addition, mechanical tenderizing breaks down cellulose in plants and collagen fibers in meat, making foods softer and less tough. In fact, it is reasonable to speculate that early hominins would have been unable to consume substantial portions of meat without the use of such technology. Because meat (especially from nondomesticated animals) is highly compliant and tough, it needs to be either sliced or ripped during mastication. Human and ape teeth lack the shearing crests necessary to break down raw unprocessed meat efficiently (imagine chewing a hunk of raw game). Chimps occasionally hunt and eat colobus monkeys, but they have a hard time chewing the meat despite being able to produce much higher bite forces than modern humans. A chimp can spend as many as 11 hours consuming a few kilograms of monkey flesh, yielding a meager return of approximately 380–400 kcal/hr (Goodall, 1986; Wrangham and Conklin-Brittain, 2003).

At some point, hominins added cooking to their arsenal of food processing techniques. The oldest archaeological evidence, in the form of hearths and burned bones that were apparently cooked, is from the Middle Paleolithic, about 250,000 years ago (James 1989; Brace 1995; Rowlett 2000; Goldberg et al., 2001). There is possible evidence for fire from 700,000 years ago (Goren-Inbar et al., 2004). Burned bones from older sites, such as Swartkrans (ca. 1.0–1.5 mya), are sometimes cited as evidence for fire (Brain and Sillen, 1988), but these may have resulted from brush fires. Yet absence of evidence isn't always evidence of absence, and there is some reason to hypothesize that cooking may be even older (Wrangham, 2009). Regardless, until pottery was invented in the Neolithic, the most common form of cooking was probably roasting in hearths and pits (Wandsnider, 1997).

Cooking has at least four major effects on food. The first is to improve nutrient accessibility. Heating partially breaks down and thus "predigests" many proteins and carbohydrates so that enzymes can more efficiently break them down in the gastrointestinal tract (Park and Russell, 2000; Bishnoi and Khetarpaul, 1994; Bhatty et al., 2000). Even a banana is about 50 percent more digestible after cooking (Englyst and Cummings, 1986). Second, cooking inactivates toxins that are naturally present in a wide variety of foods, particularly the nondomesticated things our Paleolithic ancestors ate; it also eliminates a wide range of harmful parasites and microbes that may be present in raw meat and seafood. Third, cooking extends food storage time, even for highly perishable items such as fish. And, finally, it makes food easier to chew.

The various methods of applying heat to meat and vegetables alter tissue-fracture properties by bursting cells, softening cellulose, and denaturing collagen. Figure 7.20 shows two examples: potato and beef. As one expects, cooking lowers the potato's modulus of elasticity (the slope of the stress-strain curve in the elastic region) by a large factor; boiling and roasting also lower the toughness of the potato's flesh and skin two- to fourfold (Lucas, 2004). Cooked vegetables are thus more tender and less stiff. Cooking affects meat somewhat differently. Raw meat is generally compliant and tough because its high collagen content makes the tissue rather elastic, thereby dissipating the force generated by biting, and preventing fracture. As Figure 7.20 shows, cooked meat is easier to chew because heat breaks down the collagen, making the tissue stiffer and allowing fractures to propagate farther than they would in raw meat (Purslow, 1985).

The changes caused by food processing and cooking have three biomechanical consequences for chewing. First, because cooked food is easier to fracture, it requires less force per chew. Most of us know this intuitively (compare the force you produce to eat a raw versus cooked carrot, or a rare versus cooked piece of steak). Second, eating cooked food requires fewer chews per unit of food. And, third, because cooked food yields nutrients more readily, less food is required to supply the same amount of nutrients. All things considered, cooking dramatically decreases the total amount of force generated and resisted by chewing. A simple experiment from my lab demonstrates this nicely. We asked a group of nine Harvard students to eat cubes of raw and roasted parsnips (5.5 grams each) while measuring the force generated by their masseter and temporalis muscles, using electromyogram (EMG) electrodes previously calibrated by having the students bite against a tiny force plate. On average, the students generated about 40 percent more force when chewing the raw parsnips. In addition, they required, on average, 51 percent more chews before they were ready to swallow the raw root. Finally, after a given number of chews, the cooked parsnips were broken down into significantly smaller particles. The combined effect of cooking was to require fewer chewing cycles, each of lower magnitude. Over many meals (not to mention a lifetime), the biomechanical effects of chewing even crudely roasted food certainly add up to an enormous difference in total loading. It is impossible to calculate precisely the effects of both mechanical processing and cooking, but a Paleolithic hunter-gatherer probably had to chew at least twice as much as a typical mod-

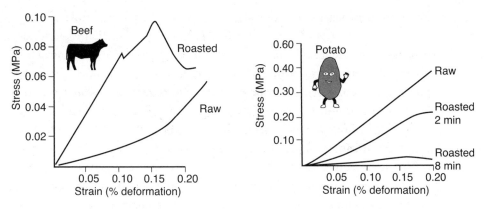

Figure 7.20. Effects of roasting on the material properties of meat and potato. Cooking makes the meat stiffer but the potato less stiff. (Adapted from Lucas, 2004.)

ern human, and the average force per chew might have been 30–50 percent higher. Put differently, one of the oddest things about modern human life is how little time and effort we spend chewing.

Mechanical processing and cooking almost certainly relate to decreases in tooth size and face size that occurred in the genus *Homo*. As Chapters 12 and 13 discusses in greater detail, the crown area of the posterior teeth in *Homo erectus* is about 25 percent smaller than in australopiths such as *Au. afarensis* or *Au. africanus* (McHenry 1988; Wood 1991, 1992). Postcanine teeth are another 12–15 percent smaller in *H. sapiens* than in *H. erectus* (Brace et al., 1991; McHenry and Coffing, 2000). Some researchers attribute shrinking teeth in the genus *Homo* to cooking. Peter Lucas (2004) has shown that to generate a similar fracture, tooth size should scale with food toughness to the power of 1/3. Applying his formula, we can calculate that a 50 percent reduction in food toughness would allow hominins to have 20 percent smaller crowns, and a fourfold decrease in toughness (the difference between a raw and a roasted potato) would permit nearly a 40 percent reduction in crown size. Boiling foods decreases their toughness even more (about sixfold), permitting approximately a 50 percent reduction in crown size. Obviously, actual tooth-size reductions have not been so extreme. One reason is that tooth size is probably strongly constrained by other aspects of facial size and shape; moreover, selection to decrease tooth size was probably not strong. Effective occlusion requires precise positioning of the teeth, so overly small teeth would lead to problematic gaps between teeth. In addition, cooking probably did not become prevalent until well after the major reductions in molar and premolar crown area occurred. I therefore suspect that dental size reductions occurred in two stages. The first stage was made possible by mechanical processing of food, such as cutting, slicing, and pounding, which would have appreciably reduced toughness, stiffness, and particle size (more research is needed to quantify how much processing would have affected the mechanical properties of Paleolithic foods). Later, perhaps in the Middle Paleolithic and Middle Stone Age (see Chapter 13), widespread adoption of cooking permitted further reductions in tooth size.

The effects of mechanical processing and cooking also go well beyond reductions in tooth size. One can have larger teeth than necessary (possibly useful on occasions when one has to eat raw food). However, as noted above, to break down raw, unprocessed foods, the masticatory muscles must generate higher occlusal forces, which the jaw and face must resist. Because mechanical processing and cooking substantially decrease the necessary number of chews and amount of force per chew, and because bones respond to mechanical loading during growth, these processes should have a substantial and integrated effect on how the skull grows. The effects of chewing processed food cannot be tested experimentally in humans, but there is much circumstantial evidence that jaws and faces do not grow to the same size that they used to *precisely* because of our softer, more processed diets.

One landmark study by Carlson and colleagues (Carlson, 1976; Carlson and Van Gerven, 1977) compared facial shape in Nubian populations that transitioned from foraging to farming (Figure 7.21). Although overall cranial size increased by about 6–9 percent over many millennia in this population, many facial dimensions became relatively smaller, some by as much as 21 percent, with the biggest decreases occurring in the dimensions of the mandible and the length of the masseter's origin on the zygomatic arch. Comparable decreases in maxillary arch size have also occurred over the past century among Australian aborigines and other populations who have transitioned to industrial lifestyles (see Lukacs, 1989; Corruccini, 1984, 1990), and similar but less extreme decreases in lower and midfacial size have been documented in populations prior to and after the Industrial Revolution (e.g., Varrela, 1992). These data suggest that humans raised on a softer diet experience less facial growth, primarily in the region around the dental arches, the part of the face most affected by masticatory loading (see above).

While there have been many studies (summarized earlier) on the effects of soft versus hard food on skull growth, the only attempt to test experimen-

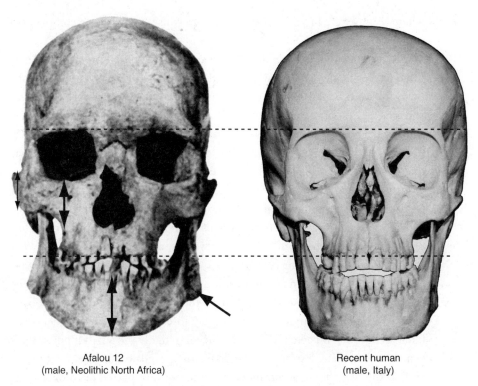

Afalou 12
(male, Neolithic North Africa)

Recent human
(male, Italy)

Figure 7.21. Comparison of a male Neolithic skull from North Africa, Afalou 12 (from Briggs, 1955), with a more recent human male (Peabody Museum, Harvard University). Both skulls are scaled to the same facial height (top of orbits to prosthion). The Neolithic skull has a relatively taller symphysis; a relatively taller infraorbital (malar) region; relatively taller, wider, and more flaring zygomatic arches; and a wider, more flaring gonial region of the mandible.

tally the effects of eating cooked food on craniofacial growth was our study of hyraxes, discussed above (Lieberman et al., 2004a). We studied hyraxes because, like modern humans, they have teeth that are retracted under their braincase, and because they have a rather humanlike pattern of chewing (Janis, 1983). We raised two groups of growing animals on a nutritionally identical diet of apples, carrots, sweet potato, kale, and rabbit chow, but the food was lightly microwaved for one group and dessicated for the other. Not surprisingly, the magnitude of strains measured in vivo was typically about twice as low in the group fed microwaved food. In other words, this group chewed with less force. In addition, the group fed microwaved food had about 10 percent less growth in many dimensions of the lower portions of the face where strains were highest. These results are consistent with the comparative evidence from humans.

One implication of these data is that cooking and food processing sometimes lead to inadequate facial growth, thus contributing to certain orthodontic problems. Dental crowding and various kinds of malocclusions (overbites and underbites) are far more prevalent in postindustrialized, urban populations than in rural, more agrarian populations (e.g., Corruccini et al., 1985; Olasoji and Odusanya, 2000), leading some orthodontists to speculate that the cause of these problems is inadequate coordination of facial growth in regions influenced by mechanical loading (Corruccini, 1984). The hypothesis is supported by experimental evidence on a variety of nonhuman animals. Monkeys and pigs raised for long periods on very soft diets typically develop serious malocclusions from overly narrow jaws, arched palates, and crowded teeth that are in the wrong orientation (Corruccini and Beecher, 1982, 1984; Ciochon et al., 1997). In addition, soft diets likely contribute to the high incidence of impacted wisdom teeth in recent humans. Most Americans have their wisdom teeth surgically removed because they are either impacted or likely to become impacted. Evidently, chewing highly processed modern diets generates insufficient strain to stimulate enough growth in the mandibular and maxillary arches to accommodate all the teeth. Impacted teeth were a significant health risk prior to modern orthodontic techniques and antibiotics, but they appear to have been quite rare until a few hundred years ago. In a meta-analysis of 62 studies, the incidence of wisdom-tooth impaction was 24.2 percent in twentieth-century humans but only 2.3 percent in preindustrial populations (Mucci, 1982; Corruccini, 1999). Perhaps we should feed our children harder and tougher food or encourage them to chew more gum (sugar-free, of course) to save on orthodontics bills. Another possible contributor to wisdom tooth (M3) impaction is the secular trend. Humans today are growing faster and maturing earlier than we used to, which could lead to a higher likelihood of impaction if skeletal growth has accelerated more than dental growth (see Bogin, 2001; Ellison, 2003).

Finally, it is interesting to speculate whether there are developmental mechanisms that integrate growth of the jaw with that of tooth crowns and roots. As previously discussed, tooth size and jaw size need to be coordinated. Mandibular and maxillary arches that are too small for the teeth cause dental crowding. Too much space leads to different problems because effective occlusion requires the precise alignment of lower relative to upper teeth. During a chew, the protoconid of a lower molar fits into the basin created between the mesial and distal cusps of two upper teeth. Too much space between teeth will thus hamper occlusion. Also, too much space creates nooks and crannies where food can lodge and then decay, leading to the build-up of plaque, which can cause infections and cavities. Finally, bigger crowns probably benefit from bigger roots.

Because teeth and jaws need to fit together so well, I suspect that several epigenetic mechanisms help integrate tooth growth, jaw growth, and tooth wear. Several lines of evidence support this hypothesis. First, tooth crown size is widely assumed to be highly genetic, but narrow-sense heritability (h^2) estimates for permanent tooth crown areas are only about 0.60 (Harris and Johnston, 1991). This means that environmental factors explain at least 40 percent of the variation in crown area. Second, permanent teeth have been getting smaller for thousands of years—averaging about a 1 percent decrease every 1,000 years (Brace et al., 1991)—but the trend is ongoing even today, especially in countries undergoing rapid technological changes, such as China (Harris et al., 2001). Third, children with cleft palate syndrome tend to have significant asymmetries in the size of their permanent but not their deciduous tooth crowns (Harris 1993). This difference makes sense from an epigenetic point of view because deciduous tooth crowns largely form before children start chewing, but the permanent tooth crowns form while they are chewing with their deciduous teeth. Perhaps this difference explains why there have been much greater decreases in permanent than in deciduous tooth crown size in human populations since the Ice Age (Hillson and Trinkaus, 2002).

How might tooth and jaw growth be integrated? One hypothesis, yet to be tested thoroughly, is that masticatory forces have a regional (paracrine) effect on permanent tooth crown and root growth. Strain induces osteoblasts in the alveolus and fibroblasts in the periodontal ligament to up-regulate growth factors, such as IGF-I and TGF-β that can stimulate osteoblasts in the region affected by strain to differentiate, proliferate, and secrete bone tissue (Yang and Karsenty, 2002). However there is also evidence that these factors up-regulate the cells that form crowns (ameloblasts) and roots (odontoblasts) in similar ways (Joseph et al., 1994; Unda et al., 2001). In other words, given the lack of any membrane to separate developing teeth from the hormonal milieu of the alveolus, it is hypothetically possible that chewing hard food causes not only jaws but also developing tooth roots and their crowns to become bigger. Thus chewing highly processed, cooked foods may have contributed to some of the declines observed in permanent but not deciduous tooth size during recent human evolution.

Coda

One of the most impressive things about chewing is how much of the head participates in the process. Because incision and mastication involve so many diverse aspects of head shape and function, it is not surprising that many variations in hominin craniofacial form relate either directly or indirectly to shifts in diet. Chapters 11–13 will return to this topic, but a next step is to consider how food moves from the head into the esophagus and then on to the stomach. Oral transport and swallowing involve a very different set of anatomical and physiological challenges. Many of these challenges are peculiar to humans because of the unique configuration of the human pharynx and its multiple functions as the passageway for liquid and food as well as an airway for respiration and vocalization.

8

Pharynx, Larynx, Tongue, and Lung

His tongue was too large for his mouth, which ever made him speak full in the mouth, and made him drink very uncomely, as if eating his drink, which came out into his cup on each side of his mouth.

ANTHONY WELDON, *The Court and Character of King James I,* 1650

I once had the pleasure of taking a seven-year-old traveling companion to the Museum of the Sewers of Paris (Le Musée des Égouts de Paris). It was a little malodorous, but it gave me a new appreciation of the importance of pipes. Pipes are as vital to cities as they are to bodies, and like most people, I usually take them for granted. In cities, pipes deliver water and power safely to thousands of buildings and then remove the waste. So it is with heads, in which fluids, solids, and gases must flow between the mouth and nose and either the gut or lungs through a series of pipelike spaces. Unless we have an obstruction, a sore throat, or an overly large tongue (like King James), we give these pipes little attention as we swallow, cough, breathe, sing, speak, smell, and gargle.

This chapter considers the anatomy and growth of the pharynx, nose, and oral cavity and their functional roles in swallowing, breathing, and vocalizing. The complex and still mysterious evolutionary history of the human pharynx and its related spaces has left us with several major peculiarities. In most mammals, much of the pharynx extends backward from the oral cavity at an obtuse angle in most postures, but this portion of the pharynx is nearly perpendicular to the oral cavity in humans. Second, the human pharynx has a short, rounded tongue that positions many key structures, such as the hyoid and larynx, low relative to the oral cavity, creating a large shared space for food, water, and air. This configuration requires us to swallow a little differently from other mammals and makes us susceptible to choking, but it contributes to our unusual capabilities for producing articulate speech. In addition, humans are unique among primates in lacking a snout and having an external nose with nostrils that face downward rather than forward. These features play important roles in swallowing, respiration, and speech, but they also relate to posture, thermoregulation, and other functions. However, because the sequence by which these capabilities evolved is not yet clear, it is hard to determine what selective forces helped to modify the human pharynx.

I begin by reviewing the general anatomy and growth of the pharynx. I next consider separately its primary functions, starting with the two most critical: swallowing and breathing. I then examine the acoustic nature of vocalizations and the specialized functional anatomy that makes articulate speech possible. I conclude with a discussion of how these functions are integrated and how respiratory adaptations related to bipedalism and facial shortening may have set the stage for features of the pharynx that are useful for speech.

Anatomy of the Pharynx

As you read this, you are passing air between your lungs and either your mouth or your nose; perhaps you are also sipping a beverage or munching on a snack. These substances are traveling along the pharynx, a 12–14-cm-long passageway that extends from your neck into your head, made up of various spaces, muscles, and other tissues.

Spaces

Trying to divide the pharynx into discrete spaces is a tricky business because the boundaries keep moving. Although the human pharynx is a dynamic space, anatomists traditionally recognize three partially distinct spaces, illustrated in Figure 8.1.

Nasopharynx

The uppermost space, above the soft palate and behind the two openings to the nasal cavity, is the nasopharynx (the openings, shown in Figure 8.4, are known as the choanae ["funnels"] or posterior nares). The roof of the nasopharynx is the sphenoid body, but the posterior and lateral walls of the nasopharynx are like curtains suspended from the cranial base. These walls consist of several layers, including an inner mucosal lining covering a tubular muscle, the superior pharyngeal constrictor. As noted in Chapter 3, the soft

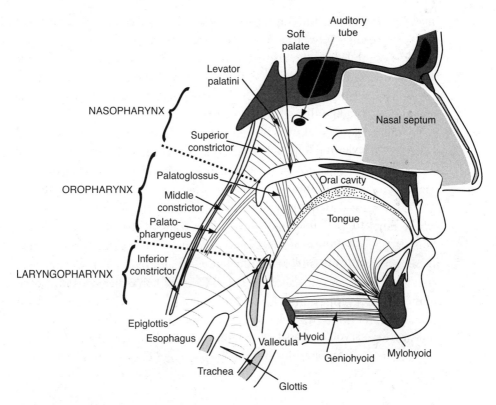

Figure 8.1. Midsagittal section showing major spaces and structures of the pharynx. The divisions between the nasopharynx and the oropharynx (the soft palate) and between the oropharynx and laryngopharynx (the epiglottis) move, making these boundaries dynamic and imprecise. Also shown are the orientations of some of the major muscles that elevate the soft palate (levator palatini), stiffen the fauces (palatoglossus), and pull the hyoid upward and forward (geniohyoid).

tissues surrounding the upper part of the pharynx develop, along with the inner ear, from the first two branchial arches. The pouch separating these arches persists as the auditory (Eustachian) tube, which runs from a pair of holes in the lateral walls of the nasopharynx to the middle ear. The auditory tube equalizes air pressure between the atmosphere and the middle ear, helping the tympanic membrane to vibrate freely (see Chapter 10).

Oropharynx

The middle portion of the pharynx is the oropharynx, much of which you can see by shining a flashlight into the throat of a person saying "Aaaaah." The human oropharynx is defined as the space behind the oral cavity above the tip of the epiglottis and below the soft palate (see Figures. 8.1 and 8.6). The soft palate is a mobile flap of tissue that extends behind the hard palate; its visible end is a diminutive, mucus-covered flap of muscle, the uvula ("little grape"). Elevating the soft palate seals off the nasopharynx from the oropharynx. The epiglottis is a paddle-shaped flap of cartilage at the root of the tongue. Normally the epiglottis sticks up, but during a swallow it flips down like the lid on a box, closing off the space between the vocal cords within the larynx. The posterior and lateral walls of the oropharynx, also covered by a mucosal lining, are composed of various sheets of muscle, including the middle pharyngeal constrictor and the palatopharyngeus muscle (an important sphincter whose function is highlighted below). The front of the oropharynx is the boundary between the oropharynx and the oral cavity. The transition between these two spaces is created primarily by the left and right palatoglossus muscles, which form an arch called the fauces ("throat") on either side of the back of the oral cavity. These muscles descend inferiorly and laterally from the hard palate and insert into the sides of the posterior tongue. When they contract, they not only elevate the back of the tongue but also narrow the fauces, helping to seal off the oral cavity and the oropharynx (Thexton and Crompton, 1998).

Laryngopharynx

The bottom portion of the pharynx is the laryngopharynx (sometimes called the hypopharynx) (Figure 8.1). The top of this space is the tip of the epiglottis, and the bottom is a narrowing below the cricoid cartilage where the pharynx becomes continuous with the esophagus. The laryngopharynx is also continuous with another tube, the larynx, at the top of the trachea. The inlet (or aditus) of the larynx is just behind the flexible wall of the epiglottis. On the other side of the epiglottis, between the epiglottis and the back of the tongue, is a little fold that creates a pocket, the vallecula ("little valley"). Other pouch-shaped spaces, the piriform ("pear-shaped") recesses, lie on either side of the aditus and epiglottis. The vallecula and piriform recesses are important spaces through which food moves during swallowing.

Larynx

The larynx is a cube-shaped box, 30–40 mm in each dimension, at the top of the trachea. It is housed in a series of cartilages suspended from the hyoid bone. The anatomy of the larynx and its cartilages is reviewed below in the context of vocalization; at this point, it is useful to point out the vocal folds (also called vocal cords) that stretch across the larynx. In most mammals, the vocal folds function primarily as a valve to seal off the trachea, thus keeping food out of the airways during swallowing and keeping air in or out of the lungs. The vocal folds also generate the energy source for vocalizations as the lungs push puffs of air through the glottis. The pitch of the vocalization is determined by the rate of these puffs, and the loudness is determined by the volume of air in each puff (more on this below).

Hyoid

The only bone intimately connected to the pharynx is the hyoid. This U-shaped floating bone lies at the base of the tongue, typically 2–3 centimeters below the lower margin of the mandible. The anterior portion of the hyoid, the body, has a semicircular shape that is convex anteriorly; from the body project two pairs of horns (greater and lesser) that partially encircle the top of the laryngopharynx (Figures 8.2 and 8.3). The hyoid is suspended from the skull by a series of suprahyoid muscles and ligaments, and from the hyoid is suspended the thyroid cartilage, the largest cartilage of the larynx. Because the hyoid and larynx are so tightly bound together, they are often called the *hyolaryngeal complex.* In an upright human breathing normally, the hyoid usually lies beneath the tongue, in front of the third cervical vertebra, and the larynx lies in front of the fourth to sixth cervical vertebrae (Haralabakis et al., 1993).[1] The position of the hyoid changes dynamically during speech, swallowing, and other functions, especially those that involve tongue and mandibular movements.

The Tongue and Other Muscles

I now review some major features of the tongue and the other muscles of the throat that generate the movements of swallowing, speaking, and chewing. The tongue fills much of the oral cavity and oropharynx. In humans, the tongue is round (anteroposteriorly short and superoinferiorly tall), but in most mammals, it is long, flat, and rectangular (Figure 8.3). The tongue has both intrinsic and extrinsic muscles (see Table 8.1). The intrinsic muscles, which run longitudinally, transversely and vertically, change the tongue's shape by

---

1. The hyolaryngeal complex tends to move cranially when the body is horizontally oriented (e.g., when someone is being imaged in an MRI or CT scanner) or after death. Hence measurements of hyoid and larynx position in cadavers and nonvertical subjects are subject to considerable error (depending on one's point of view).

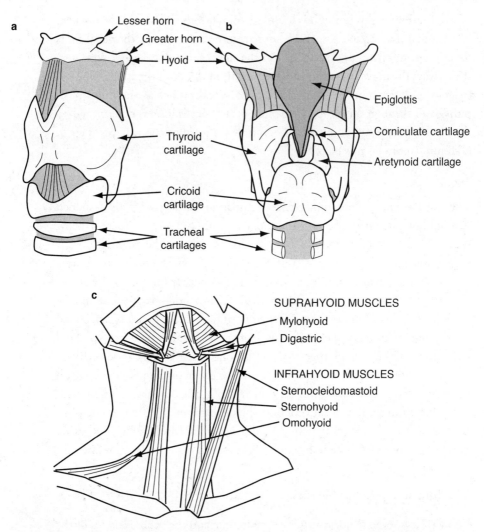

Figure 8.2. Anatomy of the hyoid and larynx: (a) from the front (in a slightly oblique view); (b) from the back, with the posterior part of the larynx sectioned; (c) with the major infrahyoid and suprahyoid muscles.

narrowing, shortening, curling, or rounding the muscle as a whole. In this respect, the tongue acts like a hydrostat, a fluid-filled body whose volume remains constant despite changes in shape. The tongue becomes longer and taller when it narrows, wider and taller when it shortens, and flatter and narrower when it lengthens. Extrinsic muscles also help change the tongue's shape,

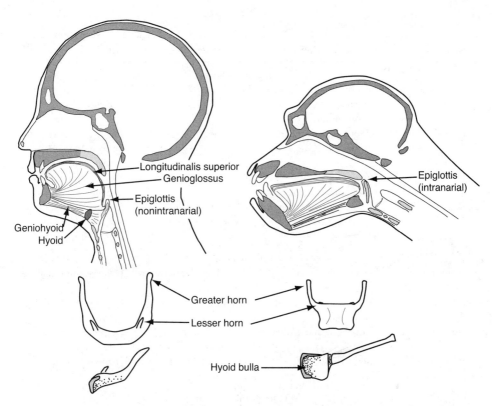

Figure 8.3. Comparison of tongue configuration of modern human (left) and chimpanzee (right) (based on Swindler and Wood, 1973). The human tongue is round and short; it positions the hyoid and larynx low in the pharynx, leading to an epiglottis that does not contact the soft palate (nonintranarial). In contrast, the chimpanzee tongue is long and high, setting up an intranarial epiglottis. Shown below are hyoids from each species in superior and midsagittal views. The chimpanzee hyoid has an expanded, hollowed body above the air sac (absent in humans).

and they move the entire tongue by pulling it downward, upward, backward, and forward. Some of these muscles also help to round the tongue or to create a troughlike depression in the center during movements that depress or elevate the tongue.

Another set of muscles, which originate on the cranial base, move the soft palate. The soft palate's main job is to flip up and stiffen, sealing off the nasopharynx from the oropharynx (failure to do so when simultaneously laughing and drinking sends liquid into the nose). The major elevator of the soft palate is the levator veli palatini, which arises on the inferior surface of the

Table 8.1:   Muscles of the neck

Muscle	Superior attachment	Inferior attachment	Action
Suprahyoids			
Mylohyoid	Mylohyoid line (along medial margin) of mandible	Hyoid body	Raises floor of mouth; elevates hyoid
Geniohyoid	Inferior posterior margin of symphysis	Hyoid body	Elevates and protracts hyoid; depresses mandible
Stylohyoid	Styloid process of temporal	Hyoid body	Elevates and retracts hyoid
Digastric	Inferior posterior margin of symphysis	Medial side of mastoid process	Elevates and stabilizes hyoid; depresses mandible
Infrahyoids			
Sternohyoid	Hyoid body	Medial clavicle, sternum	Depresses and stabilizes hyoid
Sternothyroid	Oblique line of thyroid cartilage	Sternum	Depresses and stabilizes hyoid and larynx
Thyrohyoid	Hyoid body and greater horn	Oblique line of thyroid cartilage	Depresses hyoid and elevates larynx
Omohyoid	Inferior hyoid body	Superior border of scapula	Depresses hyoid
Muscles of the pharynx			
Superior constrictor	Medial pterygoid plate	Median raphe of pharynx	Constricts upper pharynx
Middle constrictor	Stylohyoid ligament, greater and lesser hyoid horns	Median raphe of pharynx	Constricts oropharynx
Inferior constrictor	Oblique line of thyroid cartilage, side of cricoid cartilage	Median raphe of pharynx	Constricts laryngopharynx
Palatopharyngeus	Hard palate, aponeurosis of the soft palate	Posterior thyroid cartilage, lateral walls of pharynx	Closes channel (isthmus) between nasopharynx and oropharynx
Stylopharyngeus	Styloid process	Posterior and superior borders of thyroid cartilage	Elevates larynx
Extrinsic muscles of the tongue			
Genioglossus	Superior posterior margin of symphysis	Dorsum of tongue	Protracts tongue
Hyoglossus	Body and greater horn of hyoid	Side and inferior aspect of tongue	Depresses and retracts tongue

Table 8.1: Continued

Muscle	Superior attachment	Inferior attachment	Action
Styloglossus	Styloid process and stylohyoid ligament	Side and inferior aspect of tongue	Retracts tongue, creates midline trough
Palatoglossus		Side of tongue	Elevates posterior part of tongue
Muscles of the soft palate			
Levator palatini	Petrous temporal and cartilage of auditory tube	Aponeurosis of the soft palate	Elevates soft palate
Tensor veli palatini	Scaphoid fossa, sphenoid spine, and cartilage of auditory tube	Aponeurosis of the soft palate	Tenses soft palate

temporal bone and inserts onto a thick sheet of fascial tissue (the palatine aponeurosis) in the soft palate (Figure 8.4). Another muscle, the tensor veli palatini, arises on the medial pterygoid plate and the sphenoid spine but then hooks around a bony projection from the pterygoid (the hamulus) before inserting on the soft palate. Thanks to the hamulus, the tensor levi palatini pulls laterally against the soft palate, flattening the palate as it elevates (Figure 8.4). The palatoglossus muscles, which form the fauces, also insert in the aponeurosis of the soft palate (Figure 8.1). When they contract, they narrow the side walls of the back of the oral cavity (the fauces) and elevate the back of the tongue. These actions seal off the oral cavity from the oropharynx (you can feel this when you suck liquid through a straw). A final paired muscle to note is the palatopharyngeus, which extends laterally and posteriorly from the each side of palate, arching obliquely around the oropharynx (Figure 8.1). When these muscles contract, they act like a sphincter to narrow the oropharynx and to divide the oropharynx and nasopharynx (see Thexton and Crompton, 1998).

The hyoid, which lies at the base of the tongue, is at the center of action for most movements in the pharynx, floating like a pulley amid the roughly dozen muscles that can pull it up, down, backward, and forward (some of these muscles are illustrated in Figures 8.1 and 8.2). Six pairs of suprahyoid muscles pull it upward (see Table 8.1). These include muscles that run anteriorly to the inside of the chin, the mandible, and the tongue pull the hyoid both up and forward: the geniohyoid, the mylohyoid, the hyoglossus, and anterior belly of the digastric. Two other suprahyoid muscles pull it upward and backward: the stylohyoid and the posterior belly of the digastric. Another set of infrahyoid mus-

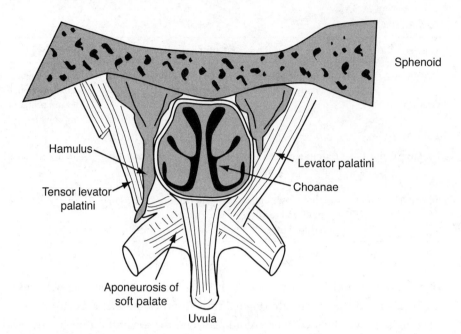

Figure 8.4. Schematic view of the soft palate and the entrance to the viewed from the back of the pharynx. The different lines of action of the levator palatini and tensor levator palatini are shown; the latter curves around the hamulus of the pterygoid before inserting on the aponeurosis of the soft palate.

cles runs below the hyoid and helps pull it downward. One of these, the sterno-hyoid, inserts on the sternum and the medial clavicle; another, the omohyoid, loops through a fascial sling in the clavicle and inserts onto the upper border of the scapula.[2] A final infrahyoid muscle, the thyrohyoid, connects the hyoid to the thyroid cartilage of the larynx below. Thus movements of the hyoid depend on the combined action and relative force generated by many diverse muscles. Often the infrahyoid muscles contract in conjunction with the suprahyoids to stabilize the hyoid's vertical position, enabling some of the suprahyoid muscles to pull it either forward or backward or to depress the mandible itself.

The last but not the least of the muscles of the pharynx are the pharyngeal constrictors (two of which are illustrated in Figure 8.1). These muscles were mentioned above, because the superior, middle, and inferior constrictors form the posterior and lateral walls of the nasopharynx, oropharynx, and laryngo-

2. The seemingly bizarre path of muscles that connect the hyoid to, of all places, the shoulder re-flects their evolutionary origin as part of the gill anatomy in fishes.

pharynx, respectively. The three constrictors overlap like a stack of hats, with the superior innermost and the inferior outermost; they partially merge in a fascial seam (raphe) along the posterior wall of the pharynx. The constrictors usually contract involuntarily in order (from the top) to narrow the pharynx, squeezing food downward like a boa constrictor swallowing a rat. In addition, they share a highly vascularized and slippery mucosal lining that helps food slide down and helps to modify the temperature and humidity of air (more on the mucosa below).

Other Tissues

The pharynx and oral cavity have other tissues tucked into their walls and spaces. Among these are the lymph glands, filled with immune cells, which we perceive mainly when they become infected and enlarged. The largest pair of these lymph glands is the tonsils. The adenoid tonsils lie in the roof and posterior wall of the nasopharynx. When swollen, they can obstruct airflow to the nasopharynx, requiring one to breathe through the mouth. Another pair of tonsils, the palatines, lie in the lateral wall of the oropharynx; these tonsils reach maximum size at puberty and typically become resorbed as one ages.

Two pairs of salivary glands (submandibular and lingual) lie under the tongue, and another pair (parotid) lies in the walls of cheeks. These glands produce saliva, a mixture of water, salt, mucus, and an enzyme, amylase. The water and mucus help moisten and lubricate food so it can stick together in a ball (a bolus) and slide down the pharynx; amylase helps break down carbohydrates. Because the salivary glands are controlled by the autonomic nervous system, a wide range of stimuli, including smells, images, and even thoughts, can trigger them to squirt saliva.

Finally, the pharynx is replete with sensory nerve cells. Many of these cells, which line the tongue and the pharyngeal walls, are sensitive to pressure and texture, and they trigger reflexes in the brain stem involved in chewing, swallowing, and gagging. Others sense chemicals and temperature. A large percentage of the tongue's sensory cells are housed in projections known as papillae, which come in a variety of forms. Chapter 10 considers the taste buds in the context of taste and olfaction.

Oral Transport and Swallowing

Let's now follow how foods and liquids pass through the oral cavity, into the pharynx, and then to the esophagus, and consider how the unique configuration of the human pharynx affects these movements. Oral transport and swallowing (technically "deglutition") are mostly involuntary and require a complex succession of reflexes. As we chew, the tongue darts about collecting and moving food from one tooth to another, storing it at the back of the oral cavity, and then pushing it into the oropharynx. During swallowing, more than a dozen muscles contract sequentially in a precise manner to squeeze food or liquid through the pharynx while simultaneously blocking off the larynx, nasopharynx, and oropharynx. Mistakes can be fatal if they lead to asphyxiation or to aspiration (breathing liquids or solids into the lung).

Oral Transport

The stages of oral transport and swallowing work as a cycle; each stage is regulated by neuromuscular "gates" that determine if and when food passes on to the next stage (Hiiemae, 2000). The process starts usually with the ingestion of food in the front of the mouth, often by incision. There, the tongue has its first opportunity to evaluate the food, which can be rejected by spitting it out. Consequently, the tip of the tongue is richly innervated with chemical and sensory receptors that sense if food seems odd-tasting, dirty, too hot, too cold, or otherwise noxious (Trulsson and Johansson, 2002; see also Chapter 10). If food passes this first gate, then stage I transport is initiated. The tongue retracts, pulling food into the postcanine region of the oral cavity. During mastication, the tongue constantly changes shape to mix food with saliva to form a bolus and to push the ball from side to side under the appropriate teeth. Tongue movements typically occur as the jaw opens and closes. Cheek muscles, the buccinators, work in concert with the tongue, usually by keeping food from spurting into the buccal cavity between the teeth and the cheek. In addition, riblike grooves (rugae) cover the roof of the oral cavity, providing resistance against which the tongue can push food. The rugae point slightly backward, keeping food from slipping forward toward the lips.

As chewing proceeds, the middle portion of the tongue functions as a second sensory gate. This part of the tongue has fewer taste buds than the tip but is loaded with sensory nerves that perceive particle size and texture. Humans, like many mammals, appear to be able to detect food particles as small as 1–2 mm in size (Lilliford, 1991; Prinz and Lucas, 1997). The brain stem uses such data, plus related information from sensory receptors in the periodontal ligament, the roof of the palate, and the TMJ, to coordinate movements during mastication and to trigger stage II transport. This sensory gate is biologically important because digestion works more efficiently if particles are small, but the cheek teeth will wear each other down if the particles are too small to prevent direct contact (Lucas, 2004). During stage II transport, the tongue contacts the hard palate in front of the food and then retracts (largely via contraction of the hyoglossus muscle), moving the zone of contact backward in a wave. This action pushes the food to the back of the oral cavity and through the relaxed fauces into the oropharynx (Palmer et al., 1997).

Chewed solid foods accumulate as a bolus in several spaces in the oropharynx. In most mammals, the bolus accumulates mostly in the pouch-shaped vallecula, just in front of the epiglottis, and to a lesser extent in the piriform recesses on either side of the epiglottis. In primates such as macaques and humans, however, the vallecula is small, permitting little accumulation (Thexton and Crompton, 1998; Hiiemae, 2000). Note also that for solid foods, stage I and II transport can occur more or less simultaneously as food is ingested, passed to the cheek teeth, chewed, and added to the bolus forming in the oropharynx. Liquids, however, tend to pass straight through for immediate swallowing.

Swallowing

The final sensory gate in the oropharynx triggers a swallow. Sensory receptors in the oropharynx and vallecula transmit information to the brain stem about various aspects of the bolus, including particle size, moistness, and bolus size (Lilliford, 1991; Thexton and Crompton, 1998). The brain stem integrates this information, triggering a swallow when conditions are deemed right.[3] Thus a large, "wet" bolus such as a raw oyster can be swallowed whole, but drier foods, such as nuts, need to be chewed to smaller sizes.

Deglutition itself happens in several rapid but precisely coordinated stages. After an initial trigger, which can be voluntary (as in swallowing a pill), a series of involuntary events then occur in quick succession. It is useful to divide these actions into four stages, illustrated in Figure 8.5a–d:

1. The tongue sweeps food back toward the fauces, which are stiffened by contractions of the palatoglossus and styloglossus muscles. The hyoid also begins to elevate as the suprahyoid muscles contract.

---

3. Consequently, anencephalic individuals who still have a brain stem can exhibit swallowing reflexes (Peleg and Goldman, 1978).

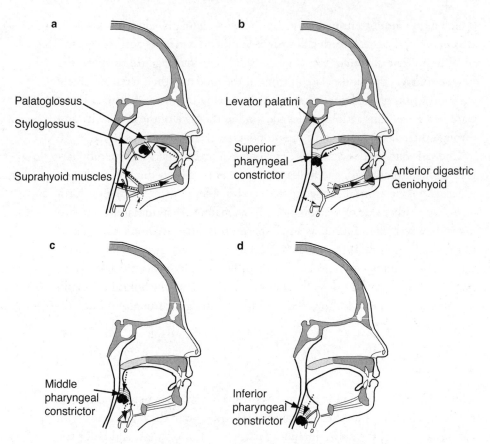

Figure 8.5. The four major stages of swallowing (deglutition), showing movements of the bolus and major actions. See text for details of what happens in each stage.

2. The oropharynx is simultaneously sealed off from the nasopharynx and oral cavity. The nasopharynx is closed off as the levator veli palatini and the tensor veli palatini elevate and stiffen the soft palate. The oral cavity is closed off as the pair of palatoglossus muscles contracts to narrow the fauces and retract and elevate the back of the tongue. At the same time, suprahyoid muscles, especially the geniohyoid and anterior digastric, elevate and pull forward the hyoid and larynx. The key action of hyolaryngeal protraction is to open the entry to the esophagus, the upper esophageal sphincter, at the base of the laryngopharynx, which is normally closed. Contractions of the superior pharyngeal constrictor push the bolus inferiorly.

3. Two actions seal off the larynx from the food pathway. First, the vocal folds are drawn together, closing the trachea's entrance. Second, the epiglottis rapidly flips down, about 150° in 1/15 second. Sealing the larynx prevents food from being pushed into the trachea, preventing breathing or vocalizing during swallowing. Contractions of the middle pharyngeal constrictor push the bolus inferiorly through the oropharynx.

4. Peristaltic contractions by the inferior pharyngeal constrictor push the food inferiorly through the upper esophageal sphincter and into the esophagus.

How does swallowing differ in humans and nonhuman primates? Although the basic anatomical components of the oropharynx are largely the same in humans and other mammals, there are a few critical differences in the arrangement and proportions of the structures, illustrated in Figures 8.3 and 8.6. Most important, in nonhumans, the pharynx contains a sort of tube within a tube, in which air flows directly from the trachea to the nasopharynx and vice versa. This inner tube exists because the nonhuman epiglottis overlies the soft palate when both flaps are in resting position (epiglottis up, soft palate down). Because the entry to the larynx, via the epiglottis, lies within the nasopharynx (known also as the nares), this inner tube arrangement is termed an *intranarial* larynx or epiglottis. The outer tube of the nonhuman oropharynx comprises the walls of the oropharynx, the piriform recesses, and the pharyngeal constrictors. The advantage of this nested tube configuration is that the airway can remain open while liquids are swallowed, with little risk of fluid moving from the outer to inner the tube and thus entering the larynx. X-ray studies show that when infant macaques swallow liquids or soft solids, the bolus passes laterally into the esophagus around an immobile, upright epiglottis; infant pigs pass milk laterally around the sides of the epiglottis, also without flipping it down (Crompton et al., 1997). However, even when these animals swallow liquids without flipping the epiglottis, they stop breathing for an instant because they close the vocal folds. Moreover, with the exception of the opossum, every adult mammal so far studied, including pigs, macaques, and humans, flips down the epiglottis as solid food passes into the esophagus over the epiglottis, requiring a brief cessation of airflow (Biewener et al., 1985; Larson and Herring, 1996; German et al., 1998).

One of the most critical anatomical differences in the pharynx between adult humans and other mammals is the lack an intranarial epiglottis (Negus, 1949). As noted above, the hyoid lies at the base of the tongue, and all the lar-

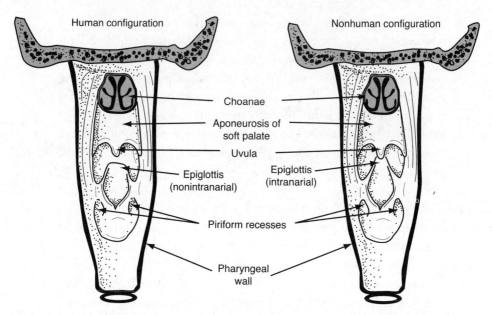

Figure 8.6. General anatomy of the pharynx when viewed from the back in a modern human (left) and a generalized nonhuman primate (right). Because the human epiglottis is descended relative to the soft palate, humans lack the "tube within a tube" configuration of the nonhuman primate that separates the airway from the foodway.

ynx's structures, including the epiglottis, are suspended from the hyoid. Because the modern human tongue is shorter, taller, and rounder than those of other mammals, the base of the human tongue, and hence the hyoid, is pushed down lower in the pharynx. By three months after birth (see below), the entire hyolaryngeal complex, including the epiglottis, is lower relative to the soft palate in modern humans than in any other mammal. As Figure 8.3 shows, when extended, the tip of the adult human epiglottis lies nearer the base of the tongue than its roof. We thus lack the tube within a tube so characteristic of other mammals. Consequently, air cannot pass directly from the nasopharynx into the larynx but instead must pass through the same pharyngeal space through which food travels after the tongue pushes the bolus through the fauces and into the oropharynx.

This novel configuration of the modern human pharynx has several ramifications. The first is that swallowing apparently requires more precision in humans than in other mammals. Humans, like other adult mammals must flip down the epiglottis and pull the entire hyolaryngeal complex upward and forward at just the right moment to open the upper esophageal sphincter. However, because the laryngopharynx in humans is so open and communicates directly with the opening of the trachea in the larynx, human deglutition requires relatively more vertical movement and precise timing of the larynx. Failure to coordinate properly epiglottal closure and hyolaryngeal elevation and protrusion creates the risk of swallowing disorders (dysphagia) or asphyxiation (Miller, 1982; Bosma, 1992; Cook et al., 1989; Kahrilas, 1997). Neither problem is trivial. About 500,000 Americans suffer from dysphagia, and choking is the fourth largest cause of unintentional-injury deaths in the United States (1.2 deaths per 100,000 people in 2007), about one-tenth the incidence of motor vehicle accident deaths (National Safety Council, 2009).

Other differences are also notable in the configuration of the human pharynx, especially in the tongue and its relationship to the shape of the pharynx. The human oropharynx and laryngopharynx are usually perpendicular to the oral cavity, whereas in most other mammals the angle is typically quite obtuse, sometimes nearly parallel (again, see Figure 8.3). This angle, of course, varies with posture. Apes and other nonhuman mammals can swallow when they adopt upright positions (such as sitting) that approximate a 90° angle; and humans can swallow food in a variety of nonstandard positions, including resting on all fours like a quadruped (Palmer, 1998) or even upside down (Dejaeger et al., 1994). So the human configuration is unlikely to be a simple consequence of reorienting the neck to become bipedal. But the low, nonintranarial larynx in humans does make sense in terms of some basic anatomical and functional constraints of the tongue. It is worth repeating that the hyoid is affixed to the root of the tongue. Thus, a long, flat tongue root sets up a high intranarial epiglottis, and a short, rounded tongue root sets up a low, nonintranarial epiglottis. In addition, to perform its many feats of oral transport and swallowing, the tongue of any mammal has to fill the oral cavity. The tongue must not only contact the entire roof of the oral cavity to sweep food backward during stage I and II transport, but it also needs to retract to push food into the oropharynx and form a seal with the fauces between the oral cavity and the oropharynx (Hiiemae and Palmer, 1999).

It therefore follows that the position of the larynx is constrained by the shape of the tongue. In nonhuman primates and other mammals, the oral cavity is long, and the tongue is long, flat, and rectangular in its resting position. Consequently, the hyolaryngeal complex is high, and the epiglottis is also high and relatively close to the soft palate (Figure 8.3). Humans, in contrast, have a shorter, taller oral cavity and a rounder tongue whose base is much lower relative to the palate. A rounder, deeper tongue sets up a nonintranarial larynx but still permits humans to push a bolus backward toward the low vallecula and piriform recesses. In other words, without a rounded tongue, it would be impossible to have a pharynx with a low, nonintranarial larynx; similarly, it would be impossible to have a high intranarial larynx with a rounded tongue that dips down into the pharynx. These constraints lead to the conclusion that the unique configuration of the human pharynx is a consequence of having a short oral cavity and an upright neck combined with a rounded tongue. It is reasonable to infer that when the oral cavity and the space behind it became shorter during human evolution, the tongue stayed the same size but changed shape by being pushed down into the pharynx (see below).

Finally, there is some reason to suspect that the descended human larynx requires the epiglottis to flip down via a novel mechanism not present in other mammals. What causes the epiglottis to flip in most mammals is unknown. It is often thought that solid food simply pushes the epiglottis down as the food is squeezed downward. Pressure exerted by the bolus may play some role in flipping the epiglottis, but this is not an entirely satisfying explanation, given evidence that the epiglottis remains closed for some time after food has passed over it (Larson and Herring, 1996). Thexton and Crompton (1998) proposed that epiglottis closure in nonhuman mammals may occur in part because the epiglottis is pushed downward—literally squeezed—by contractions of the palatopharyngeus muscles that arc around both sides of the raised epiglottis. In fact, in nonhuman mammals, the palatopharyngeus is really a sphincter-like, muscular extension of the soft palate through which the upright epiglottis pokes (Wood Jones, 1940). But, as Figure 8.1 illustrates, this sort of squeezing mechanism is unavailable to humans, whose epiglottis lies well below the palatopharyngeal sphincter. If that is the case, then what causes the human epiglottis to flip? One hypothesis is that two tiny muscles (the aryepiglottis and thyroepiglottis) perform this function (Ekberg and Sigurjonsson, 1982). This hypothesis is unlikely because individuals with damage to the nerve that innervates the larynx can still flip the epiglottis. Another possibility is that the base of the tongue in humans pushes the epiglottis down (Aiello and Dean, 1990). A final possibility, advocated by A. W. Crompton, is that the epiglottis snaps down passively from elastic forces generated elsewhere, perhaps by the thyroid cartilage when it is squeezed and elevated by other muscles (see Vandaele et al., 1995).

Another derived feature of humans, also present in apes, is the lack of a hyo-epiglottis muscle. In apes and humans the epiglottis is connected to the hyoid by a ligament (Kelemen, 1969), but in many mammals, such as dogs and horses, this ligament is a muscle. During heavy respiration, this muscle pulls the epiglottis forward against the back of the tongue, enlarging the opening to the trachea (Amis et al., 1996; Holcombe et al., 2002).

Pharyngeal Ontogeny and the Role of Pattern Generators

The problems posed by a nonintranarial larynx raise the question of why modern humans have such an odd, seemingly maladaptive configuration. A low human larynx combined with a short, rounded tongue is a likely adaptation for speech (see below), but an additional possibility is that laryngeal descent in humans is related to adaptations for bipedalism, such as an upright vertebral column and a shorter, more flexed cranial base. Before considering these ideas, we first need to review how the human pharynx grows and how the demands of swallowing constrain its growth. This is because human neonates (illustrated in Figure 8.7) do not resemble adults but instead have the typical nonhuman primate configuration, with an intranarial larynx (Negus, 1949; Sasaki et al., 1977; Bosma, 1986). This resemblance has led some researchers to infer that the high infant epiglottis enables newborns to breathe while suckling (Laitman et al., 1977), presumably because milk passes laterally around the epiglottis into the esophagus while air travels in the tube within a tube. Other researchers, however, have shown that airflow is interrupted when human infants swallow (Ardran et al., 1958; Wilson et al., 1981). That said, the infant arrangement does appear to facilitate almost continuous suckling with brief, intermittent breathing. But, as human infants grow, the hyolaryngeal complex, along with the root of the tongue and the epiglottis, "descends" relative to the soft palate, causing the loss of an intranarial larynx by about three postnatal months. As we grow (Figure 8.7), the vertical dimensions of the nasopharynx more than double in a skeletal growth trajectory, ceasing after the end of the pubertal growth spurt (Fitch and Giedd, 1999; Lieberman et al., 2001; Vorperian et al., 2005).

There are several possible explanations for why humans are born like other mammals with an intranarial larynx. One is that an intranarial epiglottis is an adaptation to help all mammals, including humans, make the sudden, dramatic transition from fetal to infant life. A fetus exchanges all gases via its mother, but the moment of birth requires the infant to begin breathing on its own. An intranarial epiglottis probably helps neonates make this vital switch successfully by ensuring an unobstructed airway from the nose to the trachea.

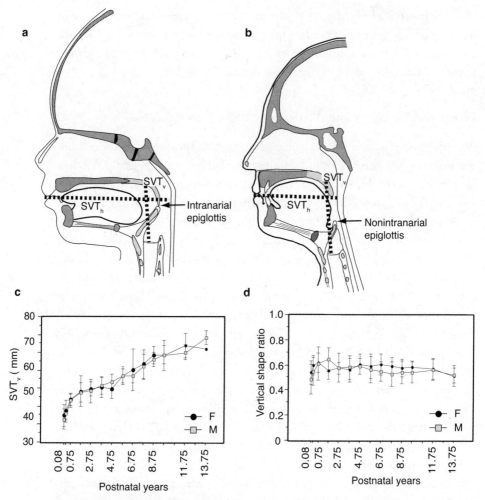

Figure 8.7. Comparison of (a) neonatal and (b) adult pharynx (both scaled to the same size), showing the different positions of the epiglottis relative to the soft palate and differences in the approximate lengths of the vertical and horizontal portions of the supralaryngeal vocal tract (SVT$_v$, and SVT$_h$). Although the vertical dimensions of the pharynx elongate substantially in postnatal humans, as shown in (c), growth occurs in such a way as to maintain similar vertical proportions in the SVT$_v$, as shown in (d), which graphs the ratio of the distance from the base of the mandible (gonion) to the palate divided by the total length of the SVT$_v$ (from Lieberman et al., 2001).

In addition, infants have only a limited degree of neuromuscular control, so an intranarial larynx decreases the risk of aspirating milk or saliva (Bamford et al., 1992).

Another factor is space. There is a very little space between the back of the oral cavity and the posterior cranial base and cervical vertebrae. Human neonates can accommodate a high larynx, but after birth their mandibular and maxillary arches grow rapidly at the same time that the cranial base begins to flex (Lieberman and McCarthy, 1999) and the cervical vertebrae begin to curve anteriorly (Scheuer and Black, 2000). The combination of all three processes decreases the space available for the hyolaryngeal complex, which perhaps descends to decrease the likelihood of obstructions. Indeed, improperly timed hyolaryngeal descent could be a factor in sudden infant death syndrome (SIDS), the largest cause of death in infants less than one year old in the United States. About 90 percent of SIDS deaths occur by six months, with most cases occurring between two and four months (Cai et al., 2005), approximately the time when the intranarial contact between the soft palate and epiglottis is lost.

Regardless of why the human hyoid and larynx descend during ontogeny (further discussed below), swallowing appears to constrain pharyngeal growth. Swallowing involves such a complex, rapid set of coordinated motions, demanding precise timing by more than a dozen muscles, that it cannot be achieved completely through voluntary neuromuscular control. Instead, swallowing is a sequence of reflexes, controlled by neuronal circuits between the brain stem and the muscles and sensory nerves of the pharynx. These kinds of circuits, termed central pattern generators (CPGs), are also requisite in walking, breathing, and other intricate rhythmic activities. Although CPGs are necessary to perform such functions effectively, they are difficult to reprogram (Marder and Bucher, 2001; Destexhe and Marder, 2004). CPGs thus may impose constraints on pharyngeal growth. We (Lieberman et al., 2001) tested this hypothesis using a longitudinal study of human growth from Denver conducted between 1931 and 1966, in which a large sample of children were radiographed regularly from birth until adulthood (McCammon, 1970). Although the pharynx grew along with the skull in these children in a skeletal growth trajectory, the vertical proportions of the pharynx were remarkably constant throughout the entire period of growth (Figure 8.7). In particular, the relative distances (computed as ratios) between the origin and insertion points of the major muscles in the pharynx did not vary between neonates and adults, despite other changes in shape that took place over the same period (Lieberman et al., 2001). In other words, the rate and pattern of hyolaryngeal descent apparently occurs in such a way as to maintain functional equivalence among the structures involved in swallowing, perhaps permitting the same CPG to operate with minimal adjustment during these changes.

This hypothesis needs to be tested more directly using muscle activity data from growing humans, but it is supported by other data from nonhuman mammals, including chimpanzees. Adult chimps have a high, intranarial larynx and thus have very different pharyngeal proportions from humans. Nonetheless, MRI and CT studies show that they too, grow in such a way as to maintain constant, albeit different, vertical proportions in the pharynx (Nishimura et al., 2003; Nishimura, 2005). In particular, as the chimpanzee pharynx grows, most vertical growth occurs between the larynx and the hyoid, because the chimpanzee hyoid remains high at the base of the tongue near the back of the mouth. Thus chimpanzee pharyngeal growth differs from that of humans by retaining a high hyoid and dropping the larynx farther relative to the hyoid. A similar pattern also appears to characterize pharyngeal growth in macaques (Flügel and Rohen, 1991).

A final point. Adult humans are not the only mammals to have a relatively low larynx. Large male red deer and fallow deer also have a low larynx, which they can pull down remarkably far in their necks when they roar (Fitch and Reby, 2001). However, the descended larynx in these animals results from a different anatomical configuration from that of humans. Whereas the human larynx is low because the base of the tongue, the hyoid, and the larynx are all lower relative to the palate and mandible, in deer the tongue and hyoid remain at the base of the mandible, but the rest of the larynx has been pulled down relative to the hyoid. Apparently big cats (lions and tigers) can also pull down the hyoid along with the base of the tongue when they roar. The low larynx of these animals lengthens their vocal tracts, making it seem as if their roars come from animals with bigger bodies (Reby et al., 2005).

Respiration

If you are reading this book while sitting, then you are probably breathing about twelve to sixteen times per minute, exchanging an average volume of 500 mL of air per breath. If you are reading this while exercising on a treadmill or a stationary bicycle, then your demand for oxygen is much greater, causing you to breathe more rapidly and deeply. As you inspire and expire, the pharynx, nose and oral cavity act as tubes through which air moves passively, driven by changes in pressure from the thoracic cavity. These spaces have additional respiratory functions. Most important, they humidify and (usually) warm the air going in. In order for the highly elastic and delicate walls of the lung to function properly, inspired air must be within a few degrees of core body temperature and nearly saturated with water vapor (Proctor, 1977; Cole, 1982). The pharynx, nose, and mouth also modify the air we breathe out. Depending on environmental conditions, we "dump" heat in an effort to cool down, or we recapture body heat and moisture from expired air in an effort to stay warm or conserve water. Finally, the airways of the head and neck act as an air filter, trapping and removing unwanted particles before they enter the lungs.

In many respects, the human pharynx, nose, and mouth perform these functions much as those of other mammals do. Tubes are tubes. Yet humans have a number of modifications not only for speech but also for respiration during vigorous, aerobic activities; these contribute to our impressive endurance capabilities. Chief among them is our uniquely shaped nose, with a prominent external vestibule combined with the lack of a snout. In addition, although we usually prefer to breathe through our noses like other mammals, we obligatorily switch to breathing through the nose and mouth or just the mouth when we exercise vigorously for a long time.

Moving Air through Tubes

To explore the function of the pharynx during respiration, let's begin with some principles about how air moves through tubes. When air flows in a tube, a difference of pressure exists between the two ends; its magnitude depends on the rate and pattern of flow. At low rates, airflow tends to be laminar: that is, air flows parallel to the walls of the tube (Figure 8.8). For our purposes, there are three characteristics of laminar flow to keep in mind. First, pressure is directly proportional to the flow rate:

$$P = kV$$

where P is pressure, V is the volume flow rate (the volume of fluid that passes through a given volume of space per unit time), and k is some constant. Second, for a straight, circular tube, the resistance to pressure, R, is given by the Hagen-Poiseuille equation:

$$R = 8nl \, / \, \pi r^4$$

where r is the tube's radius, n is the viscosity of the gas or fluid (its resistance to movement across a surface), and l is the tube's length. Because the viscosity of gases is very low, airflow resistance increases mostly in proportion to tube length and decreases to the power of 1/4 in proportion to the tube's radius. Thus doubling a tube's diameter decreases airflow resistance 16-fold! A third important characteristic of laminar flow is the gradient of velocities within the tube (Figure 8.8). In truly laminar flow, the velocity of air or liquid at the wall of the tube is zero, whereas air or liquid in the center of the tube travels at twice the velocity of the total average flow rate (Vogel, 2003). In other words, laminar flow creates a static "boundary zone" along the wall of the tube where flow rates asymptotically approach zero.

In contrast to laminar flow, turbulent flow is inherently unsteady. It is defined as the condition in which air or liquid flows in circulating vortices with no particular orientation relative to the walls of the tube or the average direction of flow (see Figure 8.8). Turbulent flow has several key properties that affect respiration and the function of the pharynx. First, for a given rate of flow, pressures are generally much higher in turbulent than in laminar conditions because they are proportional to the square of the volume flow rate ($P = kV^2$). Therefore turbulent flow tends to have much lower overall flow-rate velocities unless the air is under very high pressure. Second, the degree to which flow is laminar or turbulent is largely a function of the system's Reynolds number, Re, which describes the ratio of inertial to viscous forces as

$$Re = 2rvd/n$$

where r is the radius of the tube, v is the average velocity of the fluid, d is the density, and n is the viscosity. Re has implications for a wide range of phenomena, including how much fuel a jet airplane uses and how cream mixes with coffee. In considering the pharynx, the density and viscosity of air tend to be invariate and very small. Thus, Re for a given velocity of air in the head is mostly determined by the tube's radius. Textbooks often state that flow in a smooth-walled tube is laminar when Re is below 2,000; in fact, the transition to turbulent flow tends to occur at much lower Reynolds numbers when the tube is not perfectly straight, circular, and smooth-walled. In addition, turbulence rises in proportion to Re, and the shape of the tube has enormous effects (for a comprehensive review, see Vogel, 2003).

How do these principles influence airflow through the pharynx, nose, and mouth? First and foremost, there is a fundamental trade-off between turbu-

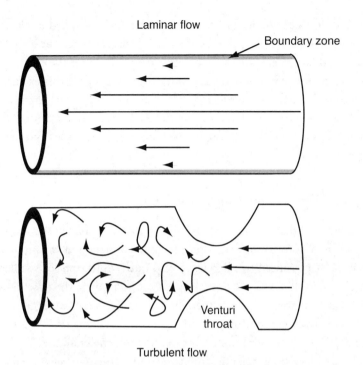

Figure 8.8. Comparison of laminar and turbulent flow in a pipe. Top: in conditions of laminar flow, the velocity of the gas (or liquid) is near 0 along the edge of the pipe but is twice the average flow rate in the center. Bottom: turbulent flow, here depicted as the result of a constriction (a Venturi throat), is characterized by less ordered circulation, with lower velocity relative to pressure.

lent and laminar flow. As a general design principle, it would seem preferable to pass air from the atmosphere to the lungs as much as possible via laminar flow, which minimizes the pressure needed to generate a given rate of airflow. Lower pressure means that the lungs have to work less hard and expend less energy both to inspire and to expire a given volume at a given rate. You feel this relationship as you breathe in terms of resistance, R, the opposition to flow. During laminar flow, airway resistance is quite low. However, when you exercise vigorously, flow rates increase, and the air becomes more turbulent as the Reynolds number increases. Turbulent flow patterns generate much higher resistance (very approximately proportional to r^5, but for turbulent flow there is no neat equation like the Hagen-Poiseuille), which means that the lungs must exert more pressure to produce the same flow rate.

A major cost of laminar flow for the pharynx, however, is the velocity gradient. As noted above, the velocity of airflow approaches zero near the wall of a tube. Because a major function of the airway is to alter the humidity and temperature of air on its way in and out of the lungs, and this exchange occurs on the surface of mucous membranes that line the nose and trachea, some degree of turbulence is required to prevent the formation of a boundary layer of immobile air, which would prevent any contact between air and the lining of the pharynx.

Using Your Nose: Humidifying and Warming Air

The chances are that you are now breathing through your nose—unless you are exercising vigorously or have some unwanted obstruction of your nasal passageway perhaps caused by allergies, a respiratory infection, or swollen adenoids. Humans, like most mammals, prefer to respire via the nose because it functions much more effectively than the oral cavity to exchange moisture and heat. This efficacy comes from several aspects of nasal anatomy, illustrated in Figure 8.9. The skeletal framework of the nose on the cranium, the piriform aperture, divides the human nose into two compartments, the internal and external nasal cavities. The internal nasal cavity, where most of the action occurs, is divided approximately in the midsagittal plane by the nasal septum. The nasal septum, made up of the perpendicular plate of the ethmoid, the vomer, and a cartilage plate (see Chapter 4), thus forms the medial wall of each chamber. The lateral walls are complex. The largest contributor to each lateral wall is the medial surface of the maxilla, which has a hole (the maxillary hiatus) that opens into the maxillary sinus (discussed below). This opening is partly closed off by various small pieces of bone (from the palatine, lacrimal, and ethmoid bones), and by three fragile, scroll-shaped projections: the inferior, middle, and superior nasal conchae (or turbinates). The inferior nasal concha is an independent bone; the middle and superior conchae both derive from the ethmoid.

Most of the internal nasal cavity, including the conchae, the medial nasal walls, the floor and the roof (with the exception of the olfactory plate) is swathed in a highly vascularized, mucus-secreting membrane, the respiratory epithelium (also called the nasal mucosa). This epithelial lining also runs into the nasal sinuses and covers much of the pharynx and trachea. Within the respiratory epithelium are various types of cells, including the euphemistically named goblet cells that produce mucus (sometimes at prodigious rates) and a

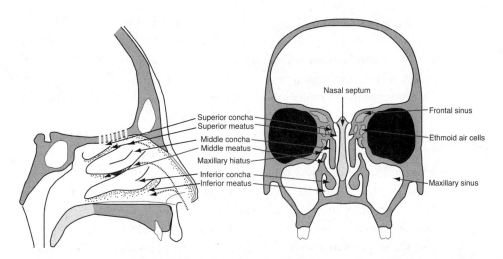

Figure 8.9. Schematic anatomy of the external and internal nasal cavities: (a) midsagittal section; (b) coronal section. The three nasal conchae (turbinates) set up three passageways (meatuses) in the internal nose.

superficial layer of hair cells (cilia). Among its various functions, the mucus acts like flypaper to trap dust and microbes, which immune cells then attack. The cilia constantly move mucus back into the nasopharynx at a rate of approximately 6 mm/minute; it then gets dumped into the oropharynx for swallowing (Sleigh et al., 1988). Regrettably, mucus occasionally spills forward into the external nasal cavity.

The primary functions of the internal nose occur via interactions between air and the mucus-covered surface of the respiratory epithelium. The respiratory epithelium is thickest over the three conchae where they form three separate channels: the inferior, middle, and superior meatuses (Figure 8.9). These projections more than double the surface area of the internal nose relative to its volume. As we breathe, we force air through the meatuses, mostly the inferior and middle ones, along the nasal floor, and along the nasal septum (Proctor, 1977; Scherer et al., 1989; Churchill et al., 2004). During the quarter second that it usually takes for air to pass between the nasal cavity and the pharynx, contact between air and the mucosa permits the exchange of water and heat. Humans occupy habitats with temperatures ranging from 49°C to −40°C, and 1 percent to 95 percent humidity. Regardless of these conditions, as air passes by the mucosa, water and heat are exchanged in a manner similar to the way coils function in air conditioners or electric heaters, bringing the air to approximately 75–80 percent humidity and 37°C (Cole, 1982). Sensors in the nose help the mucosa respond to demand. If inspired air is too dry, hot, or cold, then blood rushes into the nasal capillaries, expanding the tissue and increasing the capacity for exchange.[4] Because expired air tends to be warmer and more saturated than the mucosa, the nasal epithelium typically recaptures about one-third of this heat and moisture as air travels back to the outside world during expiration in temperate conditions (Cole, 1982).

The exchange mechanisms by which the nose regulates air humidity and temperature require direct contact between air and the surface of the epithelium. Three structural factors maximize these functions. First, exchange capacity is proportional to the ratio of the surface area of the respiratory epithelium relative to the volume of inspired or expired air. Because lung volume scales isometrically with body mass (Schmidt-Nielsen, 1984; Blaney, 1986), comparisons of respiratory epithelium surface area relative to body mass provide much the same information. This scaling relationship was analyzed by T. Owerkowicz (2004), who quantified the surface area of the nasal and tracheal respiratory epithelium in dozens of mammal and bird species. Owerkowicz found that the surface area of the nasal respiratory epithelium scales to body mass to approximately the power of 0.75 (Figure 8.10a). In general, animals adapted to particularly cold or arid habitats (e.g., camels and reindeer) have relatively high nasal surface areas; in contrast, animals with low metabolic rates (e.g., wombats), those adapted to conserve water (e.g., hyraxes), and those that live in moist, tropical habitats (e.g., primates) have relatively low nasal surface areas. As Figure 8.10a shows, nonhuman primates have much smaller nasal surface areas relative to body mass than most other mammals, but humans fall even farther below the mammalian line—with approximately 1/10 the expected value for a mammal our size! Humans also fall considerably below the mammalian regression line for tracheal surface area versus body mass (Figure 8.10b). The small surface area of the respiratory epithelium in humans—largely due to the lack of a snout plus a short neck—partly reflects our tropical primate ancestry. However, given evidence that the genus *Homo* evolved in arid rather than humid conditions, the small surface area of the epithelial tract in humans compared with that of nonhuman primates is likely a consequence of selection for a shorter face. A hypothesis, so

4. You may have noted that the two sides of the nose take turns (about 60–90 minutes each) to be dominant, with more airflow in one than in the other. This "nasal cycle" alternation, which helps protect the mucosa from desiccation, is caused by periodic fluctuations in blood flow rates. Other mammals also have a nasal cycle.

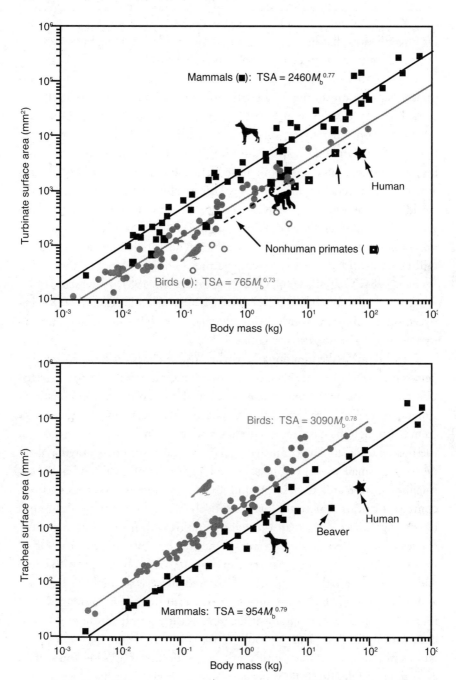

Figure 8.10. Graphs of the surface area of the turbinates (top) and trachea (bottom) relative to body mass (M_b) in birds and mammals (Owerkowicz, 2004). Both axes are logged. Humans fall well below the general mammalian regression for both axes.

far unverified, is that humans compensate for a relatively small nasal epithelium with more air turbulence in the nose (see below).

A second structural factor to consider is the ratio of surface area of the respiratory epithelium in the nose relative to the volume of the nose itself. Because humans have no snouts, hence small internal noses, one might expect selection to operate strongly on the surface area of epithelium we cram into the nose. This ratio may help account for a particularly well-studied aspect of human ecogeographic variation: nose shape. For decades, anthropologists armed with calipers have measured nasal dimensions of skulls from all over the world. These studies have established that modern human populations from very arid climates (either cold, or dry and hot) tend to have more protruding noses that are taller relative to their width than do populations from more humid climates with less extreme temperatures (Thomson and Buxton, 1923; Weiner, 1954; Wolpoff, 1968; Carey and Steegman, 1981; Franciscus, 1995; Franciscus and Long, 1991). In fact, there is a strong correlation ($r = 0.82$) between the nasal index (width divided by height × 100) and absolute humidity (Weiner, 1954; see also Croignier, 1981).

These trends make functional sense because one expects some correlation between nose shape and surface-area-to-volume ratio. Yokley (2009) tested this relationship in a comparison of nasal dimensions from CT scans of Americans of European and African (presumably mostly West African) descent. Although the nasal index correlates poorly with internal dimensions of the nose, European Americans have relatively narrower noses (low nasal indices) with smaller total volumes, and a higher ratio of nasal respiratory epithelium surface area to nasal volume (when the nose is not congested). In contrast, African Americans have relatively wider and more voluminous noses, with lower surface-area-to-volume ratios within the nose in the decongested state. Because the ancestors of modern European Americans evolved in cold, arid climates, they may have undergone selection to narrow the nose and hence increase the turbulence of inspired air so as to increase the nose's ability to humidify and warm air on its way into the lungs. Another way to increase respiratory epithelium surface area is through expansion of the conchae within the nose. This probably explains why cold-adapted Inuit populations from the Artic have significantly larger intranasal volumes and surface areas than populations from warmer regions, but they don't have bigger overall midfacial or maxillary sinus dimensions (Shea, 1977). Chapter 13 returns to these issues in the context of the evolution of nasal form in Neanderthals and modern humans.

A final set of structural factors that affect heat and moisture exchange between air and the nasal respiratory epithelium are shape variations that influence turbulence. Laminar flows have a velocity gradient in which flow rates approach zero near the tube's wall. According to one estimate (Hahn et al., 1993), this "boundary zone" under laminar conditions is approximately 0.25 mm thick in a human nose, enough to decrease, if not defeat, the nasal chamber's capacity for heat and moisture exchange. Human noses have several features that may increase airflow turbulence. Perhaps the most important of these is the external nose, a feature unique to the genus *Homo*. Apes, like monkeys, have broad, flat noses with no external nasal chamber, and nostrils (nares) that face anteriorly. In contrast, humans have an external nose, formed of a series of cartilages that create an empty space that projects substantially from the nasal aperture. Consequently, the lateral margins of the piriform aperture lip outward (evert) in humans rather than lie flat as they do in apes. Human nostrils also face inferiorly (although with some variation) rather than anteriorly. This configuration requires air to take a 90° turn as it enters the skull, disrupting laminar flow (Hahn et al., 1993; Churchill et al., 2004).

Another derived feature of the human nose that generates turbulence is the discontinuity in cross-sectional area between the internal and external nose. As air passes from the external to the internal nose, it goes through a valve near the plane of the piriform aperture, approximately 30–40 mm^2 in area, to enter the internal nasal chamber, whose cross-sectional area is approximately 130 mm^2 (Proctor, 1982). The nasal sill, a drop in the nasal floor cavity relative to the nasal aperture, enhances the discontinuity. Because pressure and velocity vary inversely at a given temperature (Boyle's law), the abrupt shift in area between the two nasal chambers functions as a nozzle (technically, a Venturi throat) in which air shifts from low velocity and high pressure in the external nasal cavity to high velocity and low pressure in the nasal valve. Since the air's Reynolds number increases in proportion to velocity (see above), airflow turbulence increases. A similar kind of valve inversely adjusts the velocity and turbulence of water in a garden hose.

A final factor that may influence turbulence is the position of the nasal conchae. Narrow-nosed (leptorrhine) people tend to have more medially positioned middle and inferior nasal conchae. Indeed these conchae can be so close to the nasal septum that they obstruct airflow if they become inflamed (Courtiss and Goldwyn, 1993). However, in vitro studies suggest that more medially positioned turbinates actually decrease rather than increase turbulence (Churchill et al., 2004).

Putting together the above lines of evidence, humans have several derived modifications of the nose that appear to enhance turbulence. As Chapter 12 discusses, everted nasal margins, which indicate the presence of an external nose, first appear in the genus *Homo* about 2 million years ago (Franciscus and Trinkaus, 1988). Early *Homo* also differs from earlier hominins in lacking a snout. It is therefore possible that external noses, and hence more turbulent internal noses, evolved as a compensatory modification to increase the efficiency of air humidification during inspiration or, more likely, air dehumidification during expiration in these early, snoutless hominins. Another possibility is that these modifications evolved in tandem to increase respiratory efficiency during vigorous endurance activities, such as long-distance trekking and running across the hot, dry savanna. The extent to which chimpanzees generate turbulence in their noses has yet to be studied, but their lack of an external nose suggests less turbulent airflow in the nasal cavity. This makes sense, as chimpanzees live in a warm, humid environment and have a longer snout and hence a larger internal nose relative to body size. However, nasal form among humans appears to vary in ways that accord reasonably well with predictions about the relative amount of turbulence. Humans with taller, narrower noses tend to come from more arid habitats and have more inferiorly directed nares, smaller nasal valves, more inferiorly positioned conchae, and deeper nasal sills (for review, see Churchill et al., 2004). That said, in vitro studies that have tried to measure the effect of these variations on airflow turbulence are equivocal (Churchill et al., 2004; Naftali et al., 2005).

Airflow Resistance, Exercise, and Nasal Growth

Let's return to resistance and its effect on respiration in the human pharynx. Resistance, the measure of the air's opposition to flow, is generally undesirable because it increases the pressure that the lungs and thorax must generate to push and pull air through the respiratory tract. Thus the evolution of a turbulence-generating external nose in *Homo* suggests that the benefits of increasing turbulence must have outweighed the costs. A reasonable hypothesis is that selection acted on nasal shape to favor efficient function of the respiratory epithelium to humidify inspired air and to dehumidify expired air during aerobic exercise. Big, external noses may have helped our ancestors travel long distances in the hot midday sun (Franciscus and Trinkaus, 1988)—but only up to a point, because at some threshold the costs of high resistance would outweigh the benefits of turbulent airflow. Because airway resistance is much lower in laminar than in turbulent flow, increased resistance can become a problem during vigorous exercise, which increases the need for air. A fit human needs about 0.5 L of oxygen per minute when walking at an typical speed of about 5 kph, but about 3.6 L per minute when running at a moderate speed of 15 kph (Margaria et al., 1963). Because air is 21 percent oxygen at sea level, the total volume of air that passes through the respiratory tract is about 2.5 L/min when walking and 18 L/min when jogging; even more air is needed at high altitude. To get more air, the lungs increase breath volume (from about 0.5 L to 1.5 L) and breathing rate (to as high as 40 breaths per minute in elite athletes). Such demands increase flow velocities and hence the Reynolds number, making the air more turbulent and generating much higher resistance (Courtiss et al., 1984).

Experiments in which sensors that measure resistance have been inserted into human noses indicate that the normal air-pressure resistance in the nose as one exhales is approximately 0.15–0.30 Pa/cm^3, and the flow is mostly laminar (Warren et al., 1987).[5] Airflow becomes more turbulent during exercise, and airflow resistance increases appreciably, in spite of a few responses that augment the nose's cross-sectional area, including expansion of the nostrils and vasoconstriction of the nasal mucosa (Saketkhoo et al., 1979; Olson and Strohl, 1987). Breathing solely through the nose during long periods of aerobic exercise becomes problematic, often excruciatingly uncomfortable. High pressures and resistance cause breathing to become more laborious, and the lungs have to work harder to satisfy the body's higher demand for oxygen. Because the cross-sectional area of the mouth is vastly greater than the nose, almost all humans must switch from just nasal breathing to either mouth breathing or a combination of nose and mouth (oronasal) breathing during vigorous aerobic exercise (Niinimaa, 1983; Wheatley et al., 1991).

Obligate mouth or oronasal breathing is actually quite rare in mammals. Horses and dogs ventilate almost exclusively through their noses when exercising. even between bouts of panting, which involve only shallow breaths without full ventilation (Schmidt-Nielsen et al., 1970; Taylor et al., 1971). Nonhuman mammals typically ventilate only orally when suffering from hyperthermia (Robertshaw, 2006). Chimpanzees, for example, breathe solely through their noses when sprinting but pant vigorously afterward to cool

---

5. Pressure is measured in pascals (Pa), defined as newtons per square meter.

down (Wrangham, personal communication). Obligate mouth breathing thus appears to be a uniquely human trait, probably an adaptation to minimize airflow resistance in a relatively small, snoutless nose during endurance activities (Bramble and Lieberman, 2004). In other words, if we had ape-sized snouts and noses, then we might respire primarily through our noses.

Oral breathing during exercise may have an additional thermoregulatory benefit. Muscles generate considerable heat during vigorous exercise. Oral expiration may cut down on heat recovery by the nasal muscosa, helping the body dump heat, but at the cost of increased water loss (Ferrus et al., 1984; Varene et al., 1986).

Airflow resistance also plays a vital role in the growth of the respiratory tract, again mostly affecting the nose. The Hagen-Poiseuille equation (see above) tells us that as a tube gets longer, it must increase its radius only slightly to keep airflow resistance constant (because the equation's denominator is r^4). Thus doubling the trachea's length requires only a 20 percent increase in trachea width. Although we do not know how much wider a turbulent nose must grow to maintain constant resistance, there is some evidence that resistance itself is a stimulus for normal nasal and oral growth (Moss and Salentjin, 1969). The best evidence for this response comes from studies of humans with nasal obstructions (Cooper, 1989). For example, children with enlarged adenoids usually become obligate mouth breathers. Because the internal nose in these children does not experience normal pressures from resistance, they tend to have less nasal cavity growth than normal breathers, hence smaller midfaces with narrower nasal cavities and maxillae, and reduced external noses (Linder-Aronson, 1979). This has additional effects on facial growth because the palate grows from inferior drift of the nasal floor (see Chapter 4). Obligate mouth breathers have correspondingly narrower palates and taller oral cavities than normal breathers because they depress their tongues and adopt a habitual open-mouthed (slack-jawed) posture of the mandible, so that the tongue no longer pushes laterally on the dental arch. After surgical removal of their adenoids, however, such children rapidly grow wider, longer, and deeper nasal cavities and also have dramatic catch-up growth in maxillary arch width and midfacial height and depth (Woodside et al., 1991; Zucconi et al., 1993).

There is also some evidence that airflow resistance may affect oral cavity growth. Obligate mouth breathers with narrow nasal cavities, hence narrow oral cavities, tend to compensate by developing slightly taller and deeper oral cavities (Linder-Aronson, 1979; Woodside et al., 1991). In addition, opera singers and wind instrument players—people who habitually produce high intraoral pressures—have slightly but significantly larger oral cavities than their siblings (Brattstrom et al., 1989, 1991).

Finally, no review of nasal airflow resistance would be complete without mentioning snoring. Snoring noises are caused by vibrations of the soft tissues in the pharynx such as the soft palate and the constrictor muscles. For snoring to occur, first, the tissues need to be relaxed, which is why alcohol consumption increases the problem. Second, greater airflow resistance causes more vibration. Hence people with conditions such as obstructed noses are more likely to snore because they experience increased airflow resistance. Snoring, however, is not unique to humans (as any dog owner knows).

Speech

So far, I have discussed only those functions of the pharynx, oral cavity, and nose that are essential to staying alive. Speech is not as vital, but its evolution, and the evolution of language in general, are perhaps more extensively debated than any other topic in human evolution. The debate derives partly from the inherently complex nature of speech and language and the lack of traces these capacities leave in the fossil record., This lack of data has led to broad speculation without much hypothesis testing.[6] Opinions on the evolution of speech run the gamut from those who think it arose early in human evolution, triggered by a reconfiguration of the pharynx for bipedalism, to those who argue that only *H. sapiens* is capable of articulate modern speech and language.

Our focus here is not on when or why speech evolved but on how anatomical modifications for speech production may have influenced the shape of the head and skull and how these modifications relate to other functions of the pharynx. In other words, let's consider the epistemological problem of how one might test hypotheses about the anatomical bases for speech in the first place, using the fossil record. We will begin with a review of the essential acoustic properties of speech before considering how speech sounds are produced by the larynx and modified by the pharynx, tongue, lips, nose, and other parts of the head. We'll conclude by asking to what extent the soft-tissue features that make modern human speech possible can be discerned from skeletal clues, and whether these features evolved specifically for speech or were modified to permit effective swallowing and respiration in bipedal hominins.

---

6. This is not a new problem. The subject was so contentious in the mid-nineteenth century that the French Academy of Sciences banned further publications on the evolution of language in 1866.

Making a Buzz

Vocalizations are biologically generated puffs of pressurized air that we perceive as sound waves when they strike our ears. As with the electromagnetic energy we perceive as light, a sound wave can be resolved into a "spectrum" of acoustic energy at different frequencies. The key parameters we perceive from sound waves are the frequency of the cycles, which are measured in oscillations per second, termed hertz (Hz), and the amplitude of the pressure, often measured as decibels (dB). A "pure" tone, such as the sound a tuning fork produces, has energy at one specific frequency, but the acoustic signals that make up speech are more complex and have energy distributed across a range of frequencies, mostly between 100 and 5,000 Hz. This is a fraction of the full range of human hearing, which is approximately 20 to 20,000 Hz (Moller and Pedersen, 2004).

The energy source for almost all speech waves is air forced outward from the lungs during expiration. During normal expiration, the lungs expel air mostly from passive forces, because the highly elastic tissue of the lung naturally contracts like a balloon after being stretched during inhalation (Mead et al., 1968). To some extent, the muscular wall of the thoracic cavity also contracts passively during exhalation from elastic forces. Because passive exhalation leads to diminishing pressures as lung volume decreases, exhalation during speech needs to be modulated by a variety of ribcage (intercostal) and abdominal muscles to maintain an appropriate and nearly constant pressure by opposing or augmenting the elastic contraction of the lungs.[7] At the beginning of an utterance, a speaker opposes the natural recoil of the lungs by contracting inspiratory muscles of the ribcage that expand the thoracic cavity. But at the end of a breath, the passive pressure exerted by the lungs declines below the pressure needed for phonation, requiring compensatory action of muscles that contract the thoracic and abdominal cavities.

The larynx originally evolved as a valve to seal off the trachea. We sometimes use the larynx as a seal to keep the thoracic cavity inflated, like a balloon, stiffening the thorax and abdomen when lifting heavy objects. If you open your larynx and exhale without vocalizing, air simply rushes through the opened valve of the larynx, the glottis, creating highly turbulent flow that we hear as the /h/ as in the word *head*. Such sounds, typically whispers, use about ten times the volume of air of "voiced" vocalizations and cannot be very loud

---

7. The diaphragm plays almost no part in controlling respiration during speech or singing.

or long. However, animals long ago evolved several adaptations of the larynx to phonate by converting inaudible flows of air into energy into the audible range of sounds that form the basis for vocalizations (Negus, 1949).

Some details of laryngeal anatomy help explain how the larynx produces phonations. As Figures 8.2 and 8.11 illustrate, the larynx is a series of cartilages, muscles, and ligaments at the top of the trachea. The cricoid cartilage, which serves as the base of the larynx, is shaped like a signet ring that is about twice as tall in the back as in the front. Several cartilages of the larynx have jointlike articulations with the cricoid. The largest is the thyroid cartilage, which is shaped like a mask, open to the back. The thyroid has two pairs of "horns" that stick up and down at its edges. The superior horns and the entire superior margin of the thyroid are suspended from the hyoid by the thyrohyoid ligament. The inferior horns form synovial joints with the left and rights sides of the cricoid, allowing the thyroid to rock backward and forward in the sagittal plane (a critical motion). A pair of small, pyramid-shaped cartilages, the arytenoids, also articulates with the cricoid. These cartilages sit on facets atop the posterior margin of the cricoid and can swivel around their vertical axes as well as slide mediolaterally (in the coronal plane). An even tinier pair of cartilages (the corniculate and cuneiform cartilages) sits atop each arytenoid cartilage.

The larynx has several ligaments. For speech, the most important of these are a part of a pair of thin but strong and highly elastic bands, the vocal ligaments (or vocal cords). These ligaments connect the anterior (vocal) process of each arytenoid cartilage with the inside front of the thyroid cartilage (Figure 8.11). The vocal ligaments are connected to the upper margins of a loose membrane, the conus elasticus, which lines the upper trachea. Together the vocal ligaments and their associated muscles and membranes comprise the vocal folds, also shown in Figure 8.11. As the vocal folds are drawn together in the midsagittal plane, they, along with the conus elasticus, seal off the trachea. If the vocal folds are sufficiently taut, and if air pressure from the lungs is sufficiently high, then the folds rapidly open and close as air is pushed through them. When one produces speech sounds, these rapid opening and closing movements cause a stream of puffs of pressurized air to escape from the larynx. The same principle operates, albeit more crudely, when one blows air through pursed lips to make a "raspberry" sound.

In other words, the larynx is a source of acoustic energy, not unlike the reed in a wind instrument. The frequency at which the puffs of air exit the larynx is termed the fundamental frequency (F0), which we perceive as pitch; the am-

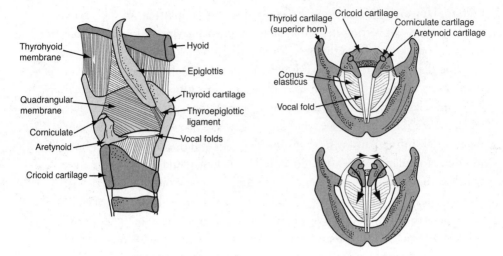

Figure 8.11. The human larynx. Left: midsagittal section. Right: superior views in abducted (top) and adducted (bottom) configurations. Not shown is the rotation of the thyroid relative to the cricoid cartilage, which helps tighten the vocal folds.

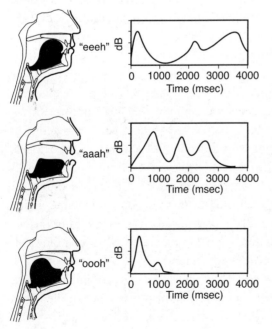

Figure 8.12. Production of sound waveforms in a modern human vocal tract. Changes in the shape of the tongue (as well as the lips and the position of the mandible) modify the harmonic peaks (formant frequencies) that differentiate speech sounds. Tongue position and formant frequencies are shown for the sounds *eeeh, aaah,* and *oooh.*

plitude of the vibrations determines the volume (Figure 8.12). We vary both the amplitude and the F0 by increasing or decreasing the pressure of the air passing through the vocal folds, and by muscular adjustments of the tension, position, and shape of the vocal folds. In general, a higher-pitched voice results from a larynx that produces higher fundamental frequencies, but the interactions between air pressure and muscular adjustments are complex, as in the case of "falsetto" or "chest" phonations (see Van den Berg, 1958). We vary the pitch by altering the tension of the vocal folds and the driving air pressure, thereby changing the frequency at which they vibrate and let through puffs of air; higher frequencies produce higher "notes."

Several key properties of phonations derive from the size of the opening between the vocal folds and their degree of tension. It follows that a number of small but critical muscles move the arytenoid and thyroid cartilages and the vocal ligament to modulate these properties. In general, the vocal muscles (whose details and actions are summarized in Table 8.2) perform two sets of functions. The first is either to tense or relax the vocal folds, thereby increasing or decreasing pitch. The one tensor of the vocal folds is the cricothyroid muscle, which rocks the thyroid forward and downward, increasing the distance between the thyroid and arytenoid cartilages (you can feel this motion with your fingers on the front of your neck when you phonate). Two muscles, the thyroarytenoid and vocalis, oppose the cricothyroid to relax the vocal folds. The latter actually inserts on the vocal folds themselves and produces

Table 8.2: Muscles of the larynx

Muscle	Origin	Insertion	Action
Cricothyroid	Anterolateral part of cricoid cartilage	Inferior margin and inferior horn of thyroid cartilage	Stretches and tenses vocal fold
Posterior cricoarytenoid	Posterior surface of cricoid cartilage	Muscular process of arytenoid cartilage	Abducts (opens) glottis
Lateral cricoarytenoid	Arch of cricoid cartilage	Muscular process of arytenoid cartilage	Adducts (closes) glottis
Thyroarytenoid	Posterior surface of thyroid cartilage	Muscular process of arytenoid process	Tenses vocal fold
Transverse and oblique arytenoids	One arytenoid cartilage	Opposite arytenoid cartilage	Adducts (closes) glottis by approximating arytenoid cartilages
Vocalis	Angle between laminae of thyroid cartilage	Vocal process of arytenoid cartilages	Lowers tension of vocal fold

minute adjustments of tension (Rodriguez et al., 1990). The vocalis and thyroarytenoid muscles are mostly active when speakers momentarily raise F0 to emphasize a word or phrase. The second set of muscular functions is the opening and closing of the glottis by drawing the vocal folds together toward the midsagittal plane (adduction) or pulling them apart (abduction). The major abductors are the posterior cricoarytenoid and the transverse and oblique arytenoids. The major adductors are the lateral cricoarytenoid and interarytenoid muscles. A single nerve, the recurrent laryngeal nerve, supplies all of these muscles (except the cricothyroid), reducing vocalizations to a whisper if it is damaged.

The length of the vocal processes of the arytenoid presents an interesting trade-off. In humans (and apes), these processes are short, permitting efficient abduction and adduction of the glottis but limiting the maximum opening of the glottis to about 50 percent of the cross-sectional area of the trachea (Negus, 1949; Kelemen, 1969). However, in many mammals the vocal processes are relatively longer, permitting the glottis to expand to about 67–75 percent of tracheal area: in horses, the vocal processes are so long they can hyperexpand the glottis to more than 100 percent of the original cross-sectional area of the trachea (Negus, 1949). Long vocal processes are useful for decreasing airflow resistance and maximizing oxygen flow in the trachea in mammals such as horses, carnivores, and ungulates that are adapted for sprinting for lengthy periods, but they increase the amount of force necessary to close and open the glottis during speech (P. Lieberman, 2006). Humans are poor sprinters but are among the most spectacular distance runners of the mammalian world, capable of running great distances at moderate speeds that do not require high rates of oxygen consumption (see Chapter 9; Carrier, 1984; Bramble and Lieberman, 2004). It is thus possible that the short arytenoids we inherited from apes played some role in the evolution of speech but also acted as a constraint on sprinting, thus favoring natural selection of variations that enhanced endurance running instead (see Chapter 9).

Filtering and Phonations

Phonations can be acoustically diverse. Speakers of English use differences in fundamental frequencies to signal emotion or grammatical information such as the end of a sentence, and some languages (e.g., Chinese) use variations in F0 to distinguish words. But there is far more to speech than phonation. Phonation produces a periodic acoustic source with energy at the fundamental frequency of phonation, F0; harmonics (integral multiples of F0, such as 2F0, 3F0, and 4F0); and a wide range of additional frequencies. We never hear these sources of acoustic energy in their entirety because the sound waves that the larynx produces are subsequently filtered by the airway through which they pass. This airway, from the larynx to lips, is called the supralaryngeal vocal tract (SVT).

To explain how the human SVT filters speech sounds, it might be useful to begin with a less complex acoustic filter in which the same acoustic principles apply. Let's consider a pipe organ, an instrument whose source of air pressure is a pump. The pressurized air generates an acoustic source through turbulent air flow that has energy distributed across a broad range of frequencies. The organ pipe sets up a sinusoidal air wave with maximum pressure at the closed end and atmospheric air pressure at the open end. The frequency, f, of this "standing wave" (see Figure 8.12) is determined by the length of the organ pipe and the velocity of sound (the equation, which is not critical for following this discussion, is $f = c/l$, where c is the velocity of the sound and l is the wavelength). Because sound waves are sinusoidal, the distance between the wave's maximum value and zero is 1/4 the length of the wave; thus the maximum length of a wave that can be sustained within the pipe is four times the length of the pipe (see Lieberman and Blumstein, 1988). Longer pipes (or vocal tracts) thus can produce standing waves with lower frequencies.

The same basic principles apply to the human SVT, which continually changes its length and shape as we talk. The frequencies at which maximum acoustic energy can pass through the SVT are termed *formants,* denoted F1, F2 and so on. These play a major role in specifying speech sounds (phonemes), as shown in Figure 8.12. Phonemes, which sound like the letters of the alphabet, are then melded together into syllables and longer segments. The length of an adult human SVT is altered either by pulling the larynx up and down or by protruding the lips (by about a centimeter in each case). More important, because the shape of the SVT is very irregular and dynamically complex (more so than any musical instrument), articulatory modifications of the SVT can generate an astonishing diversity of phonemes (for a detailed description, see P. Lieberman, 2006).

Three kinds of actions are important in creating phonemes. First, we move the tongue and other tissues to create constrictions of the SVT in different places (see Figure 8.12). Such constrictions act like Venturi throats (described above) to generate acoustic energy sources by turbulent air flow. The portion of the SVT before the turbulent source then yields a set of formants. For example, pressing the front of the tongue against the front of the palate creates the consonant /s/ (as in *hiss*) which has high-frequency formants because the SVT section before the turbulent energy source is short. Second, placing the tongue body against the midpoint of the SVT and then rapidly opening the gap will produce the momentary burst of acoustic energy of "stop" consonants such as /b/ and /t/ (as in *but*). These stop consonants are differentiated by controlling the timing of the delay between the burst of acoustic energy momentarily produced as the constriction is released and the start of phonation produced by the larynx. These delays are typically less than 25 milliseconds for consonants such as /b/ and /d/, and longer than 60 milliseconds for consonants such as /p/ and /t/.

Finally, and critically, the formant frequency patterns that differentiate vowels are produced by changes in the shape of the SVT. The SVT functions as a filter, selectively allowing maximum energy through at particular formant frequencies. Note that formant frequencies are independent of the fundamental frequency of phonation or its harmonics; they are determined by the shape of the SVT and are not simple multiples of F0. Figure 8.12 shows the formant frequencies produced by several different phonemes. In a typical 17-cm-long adult male SVT, an /i/ (as in *heel*) has a F1 of 250 Hz, a F2 of 2,300 Hz, and a F3 of 3,000 Hz; an /ae/ (as in *bat*) has a F1 of 700 Hz, a F2 of 1,800 Hz, and a F3 of 2,550 Hz. To make the /i/, we move the tongue toward the roof of the palate and open the oropharynx as wide as possible; to make the /ae/, we depress the jaw and press the tongue against the oral cavity's floor. As Figure 8.12 shows, the magnitude of energy at each peak declines for higher formant frequencies. Experimental studies show that vowel perception is primarily based on the first two or three formant frequencies (Peterson and Barney, 1952; Fant, 1960; Hillenbrand et al., 1995).[8] Put differently, speech sounds are perceived as strings of phonemes encoded into syllables, each discernible because of a distinctive set of formant frequencies.

Quantal Speech and the Human SVT

What anatomical properties make human speech special? All mammals have a larynx that generates a range of fundamental frequencies, along with a tongue

---

8. In addition, nasal-sounding speech illustrates how disruptions to the SVT affect formant frequencies. During highly nasal speech, the nasal chamber acts as a powerful filter, absorbing acoustic energy and altering or even removing certain key formant frequencies, increasing the error rate for speech perception by 30–50 percent (Bond, 1976).

and soft palate to create constrictions that alter the shape of the SVT and hence the formant frequencies generated in vocalizations. But only humans can speak. Obviously, a major basis for this ability is neuroanatomical. Chimpanzees and other mammals apparently lack much of the neural circuitry necessary to move the lips and tongue with enough speed, precision, and coordination to make the kind of rapid, endlessly recombinatorial sequences of distinct formant frequencies that make up speech. A second reason, relevant to this chapter, is the unique configuration of the human pharynx (discussed above), chiefly the rounded shape and position of the tongue along with a low larynx. To explain why the distinctively round human tongue combined with a low larynx are useful for speech, however, it is necessary to review one more acoustical property: the quantal nature of certain vowels.

Figure 8.13, from a classic paper by Peterson and Barney (1952; replicated by Hillenbrand et al., 1995), plots the first two formant frequencies (F1 and F2) of the vowels of American English for a large sample of adult men, women, and adolescents. The circles enclose 90 percent of the F1-F2 patterns for each intended vowel. The key point to note is that the range of first and second formants (F1 and F2) for a given vowel varies within the human sample but is nonetheless mostly distinct between vowels. For example, the absolute values of F1 and F2 for an /i/ (as in *bean*), and an /I/ (as in *beg*), vary considerably from speaker to speaker (particularly for F2), but an /i/ usually sounds different from an /I/. In other words, some of these vowels are typically quantal. However, when listeners heard the words that served to measure these formant frequencies in random order for different speakers, they occasionally made identification errors for most vowels except /i/ and /u/. These misidentifications occur because a person with a short SVT produces a formant frequency pattern for the word *had* that would convey the word *hid* produced by speaker with a long SVT. But because the vowels /i/ and /u/ have almost no overlap among different speakers (Nearey, 1978)—that is, they are especially quantal hence useful sounds for vocal communication.

The quantal nature of speech provides a functional explanation for the unique shape of the human SVT. Stevens's (1972) quantal theory of speech production predicts a causal relationship between the configuration of the SVT and its acoustic output. The acoustics behind the theory are complex (see P. Lieberman 2006: 279–89), but the basic point is that changing the shape of the SVT changes the formant frequencies it outputs. However, the relationship between shape changes and formant changes are nonlinear: some SVT configurations are more or less stable in terms of their effect on acoustic out-

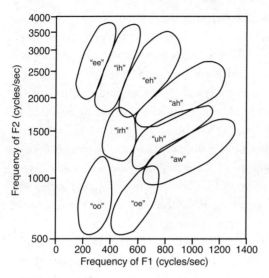

Figure 8.13. Quantal vowels plotting the first and second formant frequencies (F1 and F2) of a variety of vowels from a large sample of speakers. The vowels are quantal (distinct) in the sense that they occupy almost completely nonoverlapping positions on the graph. (From Peterson and Barney, 1952.)

put. Stevens's theory is that certain stable regions are preferable for speech because they allow the speaker to be less precise during articulation yet still produce quantal vowels such as /i/, /u/, and /a/ with a high degree of perceptibility. Put differently, they permit sloppy articulation. Furthermore, according to Stevens, an SVT needs to have two properties in order to produce stable, quantal vowels. First, the SVT should consist of two distinct tubes, each of similar length (a two-tube system). Second, one should be able to independently modify the cross-sectional area of each tube relative to that of the other by a ratio of about 10:1. The quantal theory has been subject to considerable testing in terms of both its acoustic basis and its effects on human speech (see Stevens, 1989; Lindblom, 1986, 1990; Beckman et al., 1995: Carre et al., 1995).

The human SVT has both of the properties that Stevens's acoustic model predicts. First, the adult human SVT, illustrated in Figure 8.7, clearly comprises two tubes of equal length. The horizontal tube, SVT_h, extends from our lips to the back of the oropharynx, and the vertical tube, SVT_v, extends from the soft palate to the vocal folds. These tubes are approximately equal in length in adult humans and in chimps because we have a shorter SVT_h because of having a shorter face, and a longer (taller) SVT_v because of the descent of the tongue into the pharynx, which results in a descended nonintranarial larynx. During speech, one can modify the cross-sectional area of each tube independently by about 10:1 using the rounded and highly mobile tongue (Nearey, 1978: P. Lieberman, 1984: Beckman et al., 1995). For example, when we say the vowel /i/, as in *see,* we raise the tongue to the top of the palate and move it forward toward the lips, making the cross-sectional area of the horizontal tube about ten times smaller than the vertical tube. Conversely, when we say the vowel /a/, as in *ma,* we depress and retract the tongue, making the cross-sectional area of the vertical tube about ten times smaller than the horizontal tube. As noted above, these alternate configurations maximize the differences between the formant frequencies of the two vowels. One can approximate the vowels with a nonhuman vocal tract of a different configuration, but the formant frequencies are less distinct because they are not produced with equally long tubes that have 10:1 ratios (Stevens and House, 1955; P. Lieberman et al., 1969; P. Lieberman et al., 1972b).

Why should it matter that vowels such as our *eehs* and *aahs* are acoustically distinct? The answer lies in the complex processes of normalization that our brains must perform to interpret speech sounds. The absolute value of the formant frequencies that different individuals produce varies in proportion to the length of the SVT. This means that the actual formant frequencies of the same phoneme produced by two people can differ dramatically. For example, the formant frequencies of a short person saying /ɛ/, as in *bet,* can be the same as those of a taller person saying /I/, as in *bit.* Our brain recognizes phonemes accurately, however, by standardizing the formant frequencies relative to an acoustic estimate of SVT length. We perform a comparable operation when we convert the "real" value of currencies such as dollars and euros by comparing how many euros or dollars it takes to buy a loaf of bread or a bottle of beer. We use the phoneme /i/ to estimate SVT length most accurately; /u/ also works well, but with less accuracy (Nearey, 1978; Fitch and Hauser, 1995). This estimation is possible because even though the first and second formant frequencies of an /i/ produced by a three-year-old child with a short SVT are twice those of an adult, they do not overlap with those of other vowels, regardless of SVT length (Nearey, 1978). In other words, we don't need to know the length of a speaker's SVT to identify an /i/ correctly. For this reason, some speech scientists refer to /i/ as the "supervowel," and the phenomenon may explain why /i/, /u/, and /a/ are typically the most common vowels in different languages (Greenberg, 1963; Maddieson, 1984; P. Lieberman, 2006).

Phonetic normalization may seem unlikely or unintuitive, but our brains constantly perform similar normalizations with other sensory stimuli, such as color (Hubel, 1988). When a TV is off, the screen usually appears to be a dark gray. But if we turn on the TV, we perceive deep blacks on the screen even though the TV is actually incapable of being "blacker" than its original gray. In fact, the darkest "absolute" color of the TV image is the gray of the blank

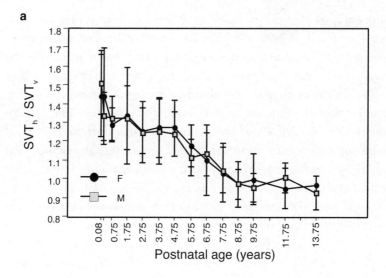

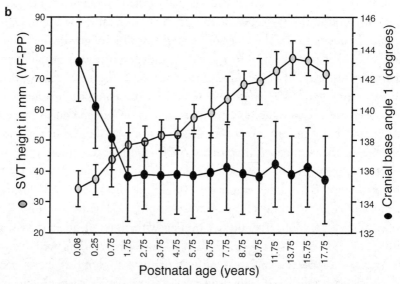

Figure 8.14. Growth of the vocal tract in modern humans. Top: the ratio of SVT_h/SVT_v, which helps make speech quantal, does not reach 1:1 until age 6. Bottom: growth of SVT_h (from the vocal folds [VF] to the palatal plane [PP])., which follows a skeletal growth trajectory, is uncorrelated with flexion of the cranial base, which occurs in a neural growth trajectory. (From Lieberman and McCarthy, 1999.)

screen, but our brains adjust to normalize perceived values so that it appears black. Acoustic signals during speech are, like colors, just relative values.

One problem with applying the quantal theory to human anatomy is the difficulty of testing the model. P. Lieberman and colleagues (1972b) modeled nonhuman SVTs that lacked 1:1 SVT_h/SVT_v proportions and found that they could not produce quantal vowels; more recently, Carre et al. (1995) used a dynamic computer model that "grew" an SVT with these proportions. However, only humans have a 1:1 SVT_h/SVT_v, and only humans speak. One way to test the functional effects of the uniquely shaped human SVT is to ask what happens to speech production and perception in humans with an SVT lacking the 1:1 SVT_h/SVT_v ratio needed to produce quantal vowels. Can such humans produce quantal speech that others can perceive? I don't know of any adult humans whose SVT proportions differ significantly from 1:1, but human ontogeny provides us with test cases: children. The rate of growth of the vertical dimensions of the pharynx appears to be constrained by the functional demands of deglutition. As discussed in Chapter 4, the hyolaryngeal complex grows inferiorly relative to the soft palate; the oral cavity grows forward at a different rate. The effect of these two different growth rates is that the ratio of SVT_h to SVT_v, shown in Figure 8.14a, doesn't reach the adult proportion of 1:1 until age 6–8 (Lieberman et al., 2001). Children with SVTs that have not reached these proportions do speak, often well, but many studies show that their formant frequencies are not as quantal, and that perception errors by listeners are higher (Buhr, 1980; P. Lieberman, 1980). Some of these problems, however, may be due to less motor control of the tongue.

Anatomical Correlates and Causes of Hyolaryngeal Descent

We have so far discussed how the uniquely low position of the larynx, combined with other features such as a short face and round tongue, improves human speech-performance capabilities. The sections above on respiration and swallowing also discussed how a low larynx increases the risk of tracheal obstruction. Hence, it is reasonable to hypothesize that the descent of the larynx conferred some adaptive benefit for speech, and that its selective advantages for improved speech perception outweighed its selective disadvantages (presumably a higher probability of death from choking). But as previously noted, the larynx is part of a suite of interrelated features with critical roles in other functions, such as respiration and deglutition, and whose configuration is integrated with other unique human features, such as a reduced snout and a neck that goes downward rather than backward from the middle of the cranium. Consequently, we need to ask if the descent of the hyolaryngeal complex and its correlated features, such as a rounder tongue, were selected primarily for speech or for other functions related to bipedalism or snout reduction and then co-opted for speech? To address this question we must consider two issues. First, are there skeletal correlates of hyoid and larynx position that allow us to infer when during human evolution the larynx descended? Second, how do changes in anatomy for bipedalism and snout reduction affect the position of the hyolaryngeal complex?

These are tough questions. Let's begin with the first, which has generated much controversy. Most of the debate has focused on whether the Neanderthals could speak like us. Although there has been speculation on the Neanderthal capacity for language since the species' discovery in the nineteenth century, the debate intensified when it took a more acoustical turn in the 1970s, thanks to Ed Crelin, Philip Lieberman, and Jeff Laitman (hereafter the "CLL group"). The CLL group proposed that the shape of the Neanderthal cranial base, in conjunction with other features in the mandible, cranium, and cervical column indicated a nonmodern SVT with a relatively high larynx (Lieberman and Crelin, 1971; P. Lieberman et al., 1972a; Laitman and Crelin, 1976; Laitman and Heimbuch, 1982; P. Lieberman, 1984). Their reasoning followed to a large extent from several differences in the anatomy of the mandible, neck, and cranial base among adult modern humans, infant humans, apes, and Neanderthals. Chief among these were the observations that modern human infants have a flat (extended) cranial base and a high, intranarial larynx (see Figure 8.7), and that the larynx descends as the cranial base flexes (George, 1978). These correlated changes were argued to be related causally by two factors. First, because the larynx is partially suspended by muscles that originate on the basioccipital, a flexed (lower) basioccipital might position the larynx lower relative to the palate. Second, basicranial flexion would shorten the anteroposterior distance between the back of the palate and the vertebral column. As the cranial base flexed, this would leave insufficient space for the larynx, requiring it to descend to keep a patent airway. Because Neanderthals had flatter cranial bases, like those of infants, the CLL group reasoned that they probably also had high larynxes, which, in combination with longer faces, would lead to $SVT_h:SVT_v$ ratios other than 1:1. If that was the case, then Neanderthals would have had difficulty articulating a full range of acoustically distinct vowels, especially vowels such as /i/ that are used for normalization. The CLL group never argued that Neanderthals couldn't speak or were stupid, but they did infer that natural selection had acted on modern human speech capabilities and therefore probably also on language capabilities.

Like many efforts to infer speech capabilities from the skeleton, the CLL group's hypothesis has problems. In the first place, the larynx is suspended below the hyoid by a half-dozen suprahyoid muscles, only some of which attach to the cranial base. We know from the above analysis of swallowing that the key function of the suprahyoid muscles is to pull the hyolaryngeal complex upward and forward to open the upper esophageal sphincter at the right moment. Further, the hyoid lies at the base of the tongue, and the attachment site of the most inferiorly positioned suprahyoid muscle, the mylohyoid, runs near the base of the mandibular corpus. Thus, for effective deglutition, the hyoid only needs to be somewhere *at the same level as* or *below* the inferior margin of the mandibular corpus. In fact, the hyoid in adult modern humans tends to be 2–3 cm below the angle of the mandible in resting position (Lieberman et al., 2001).

Another problem with the CLL group's reconstruction of SVT proportions from fossils is the assumed relationship between basicranial flexion and the position of the hyolaryngeal complex. The analysis of cranial base angulation on which the CLL group based its findings (George, 1978) incorrectly stated that basicranial flexion occurs until age 6; more recent data have shown that flexion is almost complete by age 2 (Lieberman and McCarthy, 1999). This corrected growth trajectory disproves any correlation between cranial base angle and SVT dimensions, as shown in Figure 8.14b (Lieberman et al., 2001). Flexion of the cranial base does not cause hyolaryngeal descent in humans. Another set of analyses by Laitman and colleagues (Laitman and Crelin, 1976; Laitman et al., 1979; Laitman and Heimbuch, 1982) correlated larynx position with a more complex measure of the external shape of the skull base that also included measurements of the palate. In particular, Laitman showed that the space between the back of the hard palate and the front of the foramen magnum is shorter in primates with more flexed cranial bases. This is indeed true because of the facial block described in Chapter 5. Recall that in all primates, the orientation of the back of the face is structurally integrated with the orientation of the cranial base, with little variation (a structural constraint) (McCarthy and Lieberman, 2001). Consequently, mammals with more flexed cranial bases have less space available for the pharynx. Laitman and colleagues thus inferred that Neanderthals, like infant modern humans, would have enough space for a high larynx.

But is this inference correct? Again, the hyoid in a biped must lie no higher than the base of the mandible if it is to elevate and protract the hyolaryngeal complex. As Figure 8.3 shows, the fact that the human pharynx is perpendicular to the oral cavity (thanks to our vertical necks) constrains the human larynx to be below the hyoid and the base of the tongue (in some quadrupeds, such as opossums, the larynx is in the same horizontal plane as the soft palate during feeding). But human hyoids could be higher, as is evident from studies of many children. The average anteroposterior length of the larynx in adult human is 36 mm in males and 26 mm in females (Eckel et al., 1994). Although this is too large to fit in the space occupied by the oropharynx (<40 mm in length), there is no constraint on space in the neck below the tongue. From these data one can infer that the human hyolaryngeal complex needs to be lower than it would be in a typical quadruped, but it doesn't need to be as low as it is. Indeed, monkeys and apes can assume very orthograde postures and still swallow, breathe, and vocalize.

Unfortunately, many efforts to infer hyolaryngeal position from the skeleton have also been problematic, but for different reasons. A Neanderthal hyoid from the site of Kebara in Israel has a shape quite similar to those of modern humans and different from those of apes (Arensburg et al., 1989). On this basis, Arensburg and colleagues argued, by analogy, that it must have had a modern-human-shaped SVT. However, there is no evidence that the shape of the hyoid is functionally related to a higher or lower position relative to the palate. Modern human hyoids change only slightly in shape as they grow (Miller et al., 1998), even though relative hyoid position changes considerably during this period.

A more promising approach is to try to estimate the potential range of SVT proportions based on measurements of palate and neck length. According to a study by P. Lieberman and McCarthy (2007) which estimated neck length in fossil hominins, to give a Neanderthal an SVT_v of the same length as its SVT_h would require a larynx positioned impossibly low in the chest unless Neanderthals had abnormally long necks (for which there is no evidence). This analysis also found that some very early modern humans (e.g., Skhul V) did not have 1:1 SVTs. If true, does this result mean that Neanderthals, other species of archaic *Homo,* and possibly even some early modern humans couldn't speak? Of course not. It is hard to imagine that they lacked the capacity for speech, particularly given the large size of their brains. But it may be possible that their articulation was less precise than an adult modern human's, perhaps more like that of a 4–6 year-old, lacking fully quantal *eehs* and *oohs*. Further research is needed.

One more-pessimistic point about speech, this time with regard to the tongue. Much of the complexity of speech comes from the complexity of tongue movements. This insight led Kay and coworkers (1998) to measure the size of the hypoglossal nerve (cranial nerve XII) that provides motor nerves to the tongue. If one assumes that the size of the hypoglossal canal, through which the nerve exits the skull, correlates with the number of motor neurons, then the size of this canal might "say" something about speech capabilities. Although Kay and colleagues' original analysis found modern humans (and Neanderthals) to have relatively larger hypoglossal canals than do other primates, a subsequent analysis (DeGusta et al., 1999) showed no correlation between the size of the canal and the number of neurons; in addition, when standardized for tongue size, the results for modern humans, Neanderthals, and other primates were all about the same.

Moreover, the assumption that tongue movements are more complex in speech than in oral transport and swallowing may not be correct. This problem was studied by Hiiemae and colleagues (2002), who measured the position and rate of movement of the tongue and hyoid by filming people with an X-ray machine connected to a video camera when they were chewing various foods and also when reciting the "grandfather" passage. This delightful piece of prose requires the speaker to use almost every vowel-consonant combination in the English language:

> You wish to know all about my grandfather. Well, he is nearly ninety-three years old, yet he still thinks as swiftly as ever. He dresses himself in an old black frock coat, usually several buttons missing. A long beard clings to his chin, giving those who observe him a pronounced feeling of the utmost respect. When he speaks his voice is just a bit cracked and quivers a trifle. Twice each day he plays skillfully and with zest upon our small organ. Except in the winter when the snow or ice prevents, he slowly takes a short walk in the open air each day. We have often urged him to walk more and smoke less, but he always answers, "Banana oil!" Grandfather likes to be modern in his language.

Hiiemae and colleagues found that tongue and hyoid movements during recitations of this passage were actually less complex, albeit less stereotypically patterned, than those during swallowing and oral transport, further suggesting that anatomical adaptations for feeding constrain the anatomy of the oral cavity more than those for speech do.

The evolution of speech will inevitably remain the source of much speculation. Human speech is special and derives from a number of anatomical modifications of the pharynx, but I can see no obvious way to test hypotheses about all the anatomical structures involved in speech production precisely and unambiguously without making assumptions that are also hard to test. Paradoxically, in such circumstances, the challenges of definitively rejecting null hypotheses fail to persuade skeptics yet become fodder to those whose minds are already made up. Those whose null hypothesis is that only modern humans can speak like modern humans will point to the lack of evidence to disprove the null hypothesis that fossils ancestors lack those anatomical features. The same is true of those whose null hypothesis is that fully modern speech capabilities are ancient and that they evolved well before modern *H. sapiens.*

Air Sacs

One last anatomical feature that is distinctively lacking in humans is air sacs. Air sacs are common not only to all apes but also to most primates, and they are present in many other mammals and vertebrates (see Negus, 1949; and Fitch and Hauser, 2003, for general reviews). In apes, the air sacs are bilateral outpocketings of the mucosal membrane in the laryngopharynx above the glottis (Negus, 1949). In chimpanzees and gorillas, the two spaces fuse to form an air-filled space (the hyoid recess) under the hyoid body (see Figure 8.3); they then bifurcate again and extend down the ventral surface of the larynx toward the sternum and also spread out laterally toward the armpits (Raven, 1950; Avril, 1963; Starck and Schneider, 1960; Nishimura et al., 2007). When an ape breathes out with a closed mouth and nose, the sacs fill up with air like a balloon, expanding the neck and chest. At some point in human evolution—possibly after *A. afarensis,* whose hyoid body suggests the presence of a hyoid recess (Alemseged et al., 2006)—we lost the air sacs.

The functional role of air sacs has been the subject of much debate. Negus (1949) believed that the sacs were adaptations to rebreathe air and had little to do with vocalization, but this seems unlikely, given the limited utility of inspiring carbon dioxide-rich air. Recent research has focused on the sacs' possible roles in vocalization. Hewitt et al. (2002) suggested that the air sacs reduce the tendency to hyperventilate during long sequences of repetitive calls; Fitch and Hauser (2003) suggest that they act as resonating chambers to give the appearance of a larger body. In this respect, the air sacs function in much the same way as the body of a guitar or an empty bottle amplifies sound, because the air near the hole vibrates from the "springiness" of the air inside (a structure that produces this effect is called a Helmholtz resonator). Another function of the sacs, well known in gorillas, is to act like a resonating drum when the animal thumps its chest.

In my opinion, the most interesting question is not the functional role of air sacs, but why humans lost them. My guess is that this loss had something to do with the control of speech. An empty air sac would have no effect on the production of formant frequencies by the larynx, but an inflated sac might absorb energies at certain frequencies, much as not sealing off the nasopharynx causes nasal speech sounds that have less clarity. In addition, expelling air from an inflated sac during speech could interfere with the clarity of the formant frequencies by superimposing a droning sound.

Putting All the Pipes Together

Let's conclude by drawing a deep breath and considering again the pharynx and its associated spaces in the context of swallowing, respiration, and speech. A host of major changes in the anatomy of the pharynx occurred during human evolution. One was a reduction in oral-cavity length and a shift in the shape of the tongue, which became relatively shorter and rounder. In addition, the hyolarygneal complex descended relative to most other structures in the oral cavity, leading to a nonintranarial larynx. Further, the midface as a whole became shorter, leading to reductions in nasal cavity length. Finally, an external nasal cavity was added. These anatomical changes had several major functional effects. In terms of respiration, air is probably more turbulent in the human nose than in the noses of great apes. One possibility is that greater turbulence was an adaptation to compensate for reductions in nasal cavity size that resulted from a shorter face; this change would have allowed hominins such as *Homo erectus* to humidify hot, dry air even as the relative surface area of the respiratory epithelium decreased. An alternative or additional hypothesis is that a nose that generates more turbulence, combined with obligate mouth breathing during vigorous exercise, was an adaptation to the respiratory demands of aerobic activities. Another major functional shift was for more articulate speech, thanks to the combination of a shorter oral cavity, a

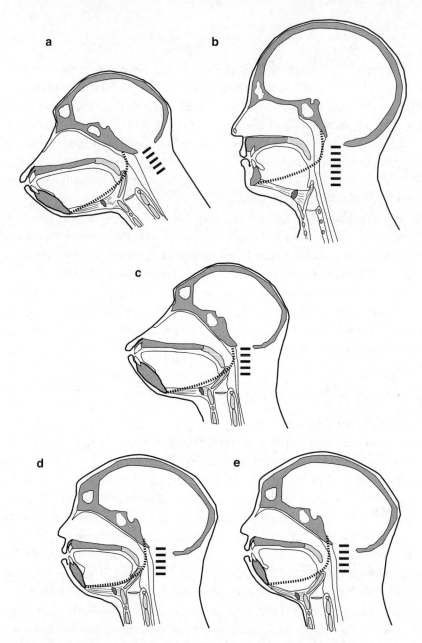

Figure 8.15. Thought experiment of different possible configurations of the oral cavity, pharynx, and vocal tract in human evolution. (a) Adult chimpanzee. (b) Modern human. (c) Hypothetical gracile australopith. Because it has a long oral cavity and a short ramus, it is likely that this hominin could have accommodated an intranarial epiglottis even though it is bipedal. (d) For *H. erectus* to have had an intranarial larynx it would have needed an relatively small tongue; but if tongue size scales isometrically with body size, then it would have had a rounder tongue, hence a descended nonintranarial larynx (e).

rounder (shorter and taller) tongue, a descended nonintranarial larynx, and a 1:1 SVT ratio. All these shifts, moreover, occurred in such a way to permit effective swallowing, albeit with a possibly higher risk of choking.

Which of these changes came first? Is the human pharynx restructured because of features related to bipedalism (a vertical neck), respiration (no snout), or merely speech? We can only speculate, as there are no hard, fossilized answers. Thus, a thought experiment might be a profitable way to conclude this discussion. Let's attempt to "design" hypothetical hominin pharynxes with alternative configurations, but following two constraints. First, we must preserve any known scaling relationships (e.g., between tongue size and body size); second, the creature must be able to breathe and swallow in accordance with the functional principles outlined above. In particular, the tongue must be appropriately shaped to push food into the oropharynx, the soft palate needs to be able to seal off the nasopharynx, the epiglottis must be able to seal the trachea, and the suprahyoid muscles must be able to protract and elevate the hyoid. Let's further assume that we can ignore the structural effects of variations in brain volume in cranial base angles, which are quite variable and poorly correlated with brain volume within hominins. (Neanderthals, for example, have brains as large as modern humans but less flexed cranial bases; see Chapter 13).

Figure 8.15 presents some hypothetical outcomes from this experiment, modified from a variety of sources (including Laitman et al., 1977). We begin with a chimpanzee and a modern human (Figure 8.15a and b). The chimpanzee has a more posteriorly oriented cervical column (oriented following Strait and Ross, 1999) that provides lots of space for the oropharynx. The hyoid lies just below the mandible in the chimp but well below the mandible in the human, and the human tongue is shorter and rounder but has approximately the same cross-sectional area because it is taller. All these differences should be familiar by now.

Figure 8.15c is a hypothetical gracile australopith whose cervical vertebrae have been reoriented into a bipedal position by moving the foramen magnum anteriorly. Just about everything is the same as in the chimpanzee, except for the posterior wall of the pharynx, which, of course, must follow the orientation of the neck. If the anatomy is right, then this is a fully operational biped with a normal intranarial epiglottis. It would be able to swallow in the same way as a chimpanzee by protracting its hyoid but not raising it much. This virtual early hominin therefore suggests that bipedalism alone does not require the hyolaryngeal complex to descend. That said, "robust" australopiths have massive, superoinferiorly tall mandibles (see Chapter 11). If the hyoid was positioned at the base of the mandible in these hominins, then it is possible that they had relatively low hyolaryngeal complexes that would have made an intranarial epiglottis unlikely, if not impossible (this hypothesis requires more research).

What about facial reduction? There has been a substantial reduction in the length of the snout in *Homo,* which changes the shape of the oral cavity and hence of the tongue. I have therefore used a computer to create two versions of a *Homo erectus*–like head in Figure 8.15d and e. In both, I shortened the snout relative to Figure 8.15c by decreasing the length of the oral cavity by 20 percent and by making the brain a little bigger. Because facial shortening makes the nasal cavity much smaller, I added an external nose to generate some turbulence (although this addition is not necessary for this exercise). A problem, however, is the tongue. In the first possibility (8.15d), I merely shortened the tongue so that it would fit in the oral cavity. This virtual hominin has an intranarial larynx with a 1:2 SVT ratio, but it also has a relatively smaller, squarer tongue than either the chimpanzee or our hypothetical australopith. This reconstructed tongue may be a problem because of the constraint noted above, that tongue volume scales isometrically with body mass across apes and modern humans (M. Clegg, personal communication). Put differently, in order for this creature to retain its high intranarial larynx, it would need a considerably smaller tongue than predicted for its body size. The alternative in Figure 8.15e has a larger, rounder tongue, about the size predicted by isometry. However, in designing this creature, I had to make the base of tongue lower, thereby dropping the hyoid (which lies at the base of the tongue) relative to the mandible, soft palate, and other structures. Such a creature would not have an intranarial epiglottis, because its epiglottis could not be long enough and still able to close the trachea.

Of course, this thought experiment is only a drawing, not a rigorous analysis. However, the exercise does suggest that although bipedalism alone does not necessitate hyolaryngeal descent, facial shortening combined with bipedalism may have contributed to the loss of an intranarial larynx, if one assumes that tongue size scales with body size. But what pressures selected for a shorter face in the first place? Hypotheses abound, and none are necessarily mutually exclusive, given the way that heads are integrated. Possibilities include changes in diet that favored smaller teeth (see Chapter 7), increased nasal turbulence for thermoregulation, and better articulation of speech sounds (i.e., a SVT_v/SVT_h that was closer to 1:1).

In short, we humans have rather odd set of tubes, with an interesting, still mysterious evolutionary history. Perhaps we would be very different creatures today if our ancestors had managed to evolve relatively smaller tongues. But as far as I can tell, the only ones of us who might have benefited would be those like poor King James, whose overly large tongue caused him such embarrassment when he drank and talked.

9

Holding Up and Moving the Head

If a man ain't got a well-balanced head, I like to see him part his hair in the middle.

JOSH BILLINGS (A.K.A. HENRY WHEELER SHAW)

You are probably relatively motionless, most likely sitting, as you read this. Yet at some point you may get up and walk or run somewhere, and in so doing you will subject your head to a variety of motions. If you walk, your head will pitch gently forward and back as your body's center of mass rises and falls. If you run, your head becomes like the top of a giant pogo stick, bouncing upward with each leap into the air and pitching forward rapidly every time a foot hits the ground. Yet during these motions you will probably manage to keep your head reasonably stable and look where you are going (with your gaze forward and slightly down) by rapidly integrating positional information about your head and body, visual data from your eyes, and contractions of the neck muscles. Unlike an amateur video, in which the image jiggles and blurs, your view of the world is generally smooth and seamless.

How we hold up and control our heads—the subject of this chapter—is an interesting problem for humans because of our unique combination of being bipedal and having a big brain. Upright posture and bipedal locomotion require several anatomical modifications in the head, the most obvious of which is that our necks point downward rather than backward from the cranium. Vertical necks help us stand and walk upright but pose some challenges for stabilizing the head during upright locomotion, especially running. Unlike most quadrupeds, whose heads are cantilevered on necks that attach to the back of the skull base, humans have short, vertical necks that attach near the center of the skull base. Because human heads and necks are not cantilevered, we cannot counter or attenuate vertical displacements of the head by flexing and extending the neck. If uncontrolled, these displacements can blur vision, potentially leading to loss of balance, hence falls, and serious injuries such as sprained ankles or broken legs. In addition, human bipeds are distinct because we walk and run with upright trunks that pitch rapidly forward and backward, especially during running, causing the neck and head to pitch forward and back. As one might expect, natural selection appears to have favored several unique structural modifications in the hominin skull, neck, and upper body to orient the eyes forward rather than upward during upright posture and locomotion and to stabilize the head during vigorous activities such as running.

This chapter has four parts. It begins with a brief review of a few essential aspects of the anatomy of the neck and its connection to the head and then discusses some novel features of the human skull that relate to bipedal posture. We'll then consider how our system of balance senses motion and relates the position of the head to visual information from the eyes. The final section explores how the head is stabilized during walking and running, and the distinct challenges these two gaits pose.

Anatomy of the Head and Neck

Darwin wrote in *The Descent of Man*: "If it be an advantage to man to have his hands and arms free and to stand firmly on his feet, of which there can be no doubt from his pre-eminent success in the battle of life, then I can see no reason why it should not have been advantageous to the progenitors of man to have been more and more erect or bipedal" (Darwin, 1871:142). Darwin was prescient despite a lack of fossil evidence. The earliest hominins—discovered more than 120 years after he penned those words—appear to have had upright postures (Galik et al., 2004; Zollikofer et al., 2005; Richmond and Jungers, 2008), suggesting that bipedalism played a key role in the divergence of the human and chimpanzee lineages (see Chapter 11). Humans, however, are not unique in the ability to stand upright. Apes frequently assume upright postures during feeding, fighting, and other behaviors (Hunt, 1992; Jablonski and Chaplin, 1993; Carrier, 2004; Thorpe et al., 2007); and other vertebrates, such as kangaroos, birds, and many dinosaurs, are bipeds of one form or another. Thus, to be precise (and to correct Plato's famous definition), humans are the only habitually striding, tailless, and featherless bipeds. This distinctive mode of posture and locomotion is made possible by several unique adaptations in the neck and head.

Let's begin our review of the cranial anatomy relevant to bipedalism with the features of the head and neck that either permit or constrain movements in three orthogonal planes. *Pitching* is flexion and extension in the sagittal plane (e.g., nodding); *yawing* is rotation around a vertical axis (e.g., shaking your head from side to side); and *rolling* is rotation around an anteroposterior axis (e.g., leaning your head to the side). These movements are typically generated around several joints, some of which are quite specialized (Figure 9.1). The first, the atlanto-occipital joint (AOJ), between the occipital condyles and the first cervical vertebra (the atlas, C1), permits mostly head pitching because of many constraining ligaments and the elongated shape of the occipital condyles and the superior articular facets of the atlas.[1] Approximately 90° of flexion and extension can occur at the AOJ in most quadrupedal mammals, but monkeys have a more limited range (about 20–30°), and this joint is extremely constrained in humans, allowing only 9–11° of flexion or extension (Graf et al., 1995a, 1995b).[2] The second joint, between the atlas and the second cervical vertebra (the axis, C2), is also specialized. The axis has an upright process (the dens) which projects behind the anterior arch of the atlas. The dens and the horizontal articular facets between C1 and C2 permit about 45° of yawing but almost no flexion and extension. The other five cervical vertebrae (C3–C7) have less distinctive shapes, with vertebral bodies and separate transverse and spinal processes for muscle and ligament attachments. The joints (termed zygapophyses) between C3 and C7 are oriented approximately 45° to the transverse plane and perpendicular to the sagittal plane, permitting a combination of flexion and extension in the sagittal plane, lateral flexion and extension, and rotation (White and Panjabi, 1978). Thus, although the upper joints of the human neck allow very restricted ranges of motion, the neck as a whole is capable of a reasonably wide range of movement (see Conley et al., 1995).

The head-neck complex has much less range of movement in humans than in most quadrupeds. Some of this immobility derives from unique anatomical features in the human head and neck. First, the upper portion of the human neck descends from the skull more or less vertically (in an upright position). In an upright human, the first few vertebrae actually angle forward and downward slightly before the neck curves posteriorly, making the anterior surface of the neck slightly convex (Figure 9.1). This curve, called a lordosis, places

---

1. Recall from Chapter 3 that the region around the foramen magnum, including the condyles, actually derives from the first few somites and evolved as a modified first cervical vertebra.

2. To observe this constraint, try to hold your neck still and see how much you can flex and extend your head at just the AOJ.

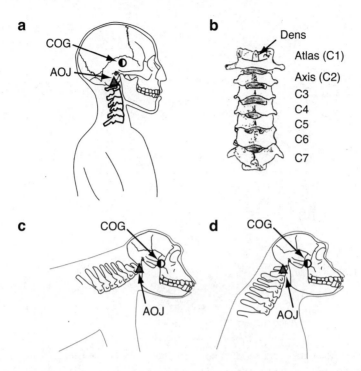

Figure 9.1. Configuration of the head and cervical spine in (a) and (b) a bipedal human; (c) a quadrupedal chimpanzee; (d) an upright chimpanzee. AOJ = atlanto-occipital joint; COG = center of gravity. The human head is nearly balanced and sits atop a short vertical spine; the chimpanzee head is much more cantilevered relative to the spine in both quadrupedal and bipedal configurations.

the head above the center of gravity of the trunk (conferring stability) and creates a spring that probably helps to attenuate forces between the neck and head. In contrast, other mammals, including apes, have cantilevered heads, in which the upper cervical vertebrae have some degree of horizontal (posteriorly directed) orientation. Even when upright, a chimp has much less of a cervical lordosis (Schultz, 1961). In addition, the foramen magnum in humans is near the center of the skull base, but in apes and all other quadrupeds the foramen magnum lies near the posterior end of the skull base. Finally, humans also differ from great apes in having low, wide shoulders. Chimps have very muscular necks (see below) with permanently hunched and relatively narrow shoulders (see Aiello and Dean, 1990; Bramble and Lieberman, 2004).

Most of the muscles that move the neck are pairs of extensors that insert posterior to the foramen magnum (see Figure 9.2). These nuchal muscles come

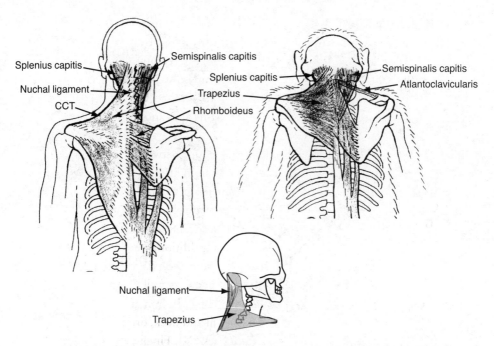

Figure 9.2. Major neck extensors in human (left, bottom) and chimpanzee (right). In humans, the head is much less connected by muscles to the thorax and the shoulder girdle, with no insertion of the rhomboid on the occipital, no atlantoclavicularis, and a much reduced trapezius and semispinalis. Humans also have a nuchal ligament, which interdigitates with the cleidocranial portion of the trapezius (CCT). Together, the nuchal ligament and CCT link the shoulder girdle with the occiput in humans. Adapted from Bramble and Lieberman (2004).

in three main layers. The deepest layer includes many short muscles (e.g., the rectus capitis and superior oblique) that arise from the atlas and axis and insert around the posterior edge of the foramen magnum. The deep nuchal muscle insertions leave a slight transverse ridge on the occipital, the inferior nuchal line. The deep nuchal muscles all have short moment arms, capable of generating fast movements but with little force. Their functional roles in humans are poorly understood because they are not amendable to EMG analysis, but it is likely that they help adjust head position, and they almost certainly provide important sensory feedback to the central nervous system. The intermediate layer of nuchal muscles (the semispinalis capitis and splenius) arises from more inferior vertebrae and insert on either side of the occipital, just above the inferior nuchal line. Finally, the superficial layer includes the trapezius and the sternocleidomastoid. These large muscles arise from the scapula and from the clavicle and sternum, respectively, and insert along a broad swath of the occipital, from one mastoid process to the other, with the trapezius in the center. The superficial nuchal insertions create a raised transverse ridge, the superior nuchal line, which is quite large in some apes and early hominins (see Chapter 11).

Figure 9.2 illustrates how the muscular connections between the head and the shoulder girdle are more substantial in apes than in humans. Apes have an especially massive upper trapezius, which helps them to produce rotational forces in the shoulder; in humans the muscle is relatively smaller, and only a very small portion of the trapezius, the cleidocranial trapezius, inserts onto the shoulder girdle (Bramble and Lieberman, 2004). In addition, the rhomboideus in apes and monkeys is a large muscle that runs from the axial (medial) margin of the scapula to the back of the occipital bone, as well as the cervical vertebrae and the upper half of the thoracic vertebrae. In humans, the rhomboideus is smaller and disconnected from the skull and neck, originating only from the upper thoracic vertebrae (and sometimes C7). Finally, humans also lack an atlantoclavicularis, a shoulder-elevating muscle present in apes that inserts on the clavicle and the transverse processes of the atlas (Aiello and Dean, 1990).

The neck musculature is dominated by extensors, but there are also a few flexors. The biggest is the longus capitis, which arises from the anterior side of the cervical vertebra and inserts on the basioccipital. In addition, several pairs of neck flexors (none of which attach to the skull) attach on the anterior side of the upper cervical vertebrae and insert either on the upper rib cage (scalenus) or more inferior cervical vertebrae (longus colli). Some of the supra- and infrahyoid muscles of the neck can also flex the head if the jaw is clenched (see Chapter 7). Finally, because the head is unbalanced, gravity also acts to flex the head at the AOJ.

Moving the head typically involves recruiting pairs of muscles. Bilateral contraction of the extensor or flexor muscles causes the head to pitch; unilateral (same-side) contraction of the extensors and flexors causes the head to roll; and contraction of flexors on one side along with extensors on the other side causes the head to yaw. Given the importance of avoiding interruption or pinching of the vital nervous and circulatory conduits that run between the head and thorax, it is not surprising that the whole complex is bound together tightly by an impressive set of ligaments. Some of the spinal ligaments also help to maintain the curves of the vertebral column, acting as springs to store and release forces in the spine.

One very different ligament, the nuchal ligament (see Figure 9.2c), is of special interest. This ligament is really a tendonlike structure that evolved independently in humans and other mammals that are well adapted for running, such as dogs and horses; nuchal ligaments are also found in animals with massive heads, such as giraffes and elephants (Bianchi, 1989; Gellman and Bertram, 2002a). However, the nuchal ligament is absent or present only as a thin plane of fascia in most other mammals, including the great apes (Vallois, 1926; Swindler and Wood, 1973; Aiello and Dean, 1990; Bramble and Lieberman, 2004). The nuchal ligament varies somewhat between species but is essentially a tendinous sheath with many elastin fibers that originates from the spinous processes of the cervical vertebrae and inserts on the midline of the occipital, below the nuchal crest. In humans, the ligament has two parts (Mercer and Bogduk, 2003). The more superficial portion is a thickened seam (a raphe), sometimes about the diameter of a pencil, which runs between the external occipital protuberance (the bump on the most posterior part of the skull) and the spinous process of the seventh cervical vertebra (C7). This tendonlike band (which does not insert onto any of the intervening cervical vertebrae) interdigitates with fibers of the upper (cranial) part of the trapezius and sometimes with the deeper semispinalis. In contrast, the deeper part of the nuchal ligament is a fascial septum that extends from the midline of the occipital below the nuchal crest to the spinous processes and interspinous ligaments of the seven cervical vertebrae. Together, the two parts of the nuchal ligament create a partially elastic connection in the midsagittal plane between the occiput, the upper neck extensors, and the shoulder girdle. Several studies on quadrupeds have demonstrated that the nuchal ligament acts like a tendon, capable of stretching and recoiling in a springlike manner (Dimery et al., 1985; Bianchi, 1989; Gellman and Bertram, 2002b). As the head pitches forward each time the animal's forefeet hit the ground, the nuchal ligament stretches and recoils, helping support the head against the effects of gravity. According to Gellman and Bertram (2000b), the ligament accounts for as much as 32 percent of the work needed to stabilize the head in a trotting horse. As discussed below, the nuchal ligament—whose presence may be detected from a sharp, everted ridge in the midsagittal plane of the occipital (the median nuchal line)—may be a derived feature that helps humans stabilize the head during running.

Bipedal Posture

Head Orientation

Let's turn now to how the head is held up. Both quadrupeds and bipeds need to orient their eyes so that they can see where they are going (usually forward), even during a wide range of head and neck movements. A study by Strait and Ross (1999) confirmed that primates tend to locomote with their orbital planes (the plane formed by the upper and lower margins of each orbit) approximately perpendicular to the ground (within 20° of earth horizontal most of the time). In turn, orbital orientation within the skull is set up mostly by angular relationships in the sagittal plane among four components of the head and neck (Figure 9.3): the orbits, the cranial base, the foramen magnum, and the neck. These components are linked in a few ways. First, the orientations of the anterior cranial base and the orbits are nearly coincident because the roof of each orbit is also part of the floor of the anterior cranial fossa. This is particularly true in monkeys and apes, whose eyes point forward rather than toward the side (more on orbital convergence in Chapter 10). If you model the orbits as a cone, the central axis (the NHA, neutral horizontal axis) is usually within 10° of the orientation of the anterior cranial base in all higher primates (McCarthy and Lieberman, 2001; see also Chapter 5). Second, the orientations of the anterior cranial base and the foramen magnum are interrelated by the cranial base angle (as indicated in Figure 9.3). The cranial base angle, however, is the most potentially variable part of the linkage between the eyes and the neck because it changes during ontogeny as the brain and face grow, and it varies significantly among species (see Chapter 5). And third, the orientation of the foramen magnum relative to that of the upper neck is partially constrained because, even though this is a mobile joint, the upper cervical vertebrae tend to be perpendicular to the plane of the foramen magnum in normal posture (Graf et al., 1995a). As noted above, the anatomy of many animals permits considerable flexion or extension at the AOJ, but these movements are more restricted in higher primates. Thus, three angles determine most of the variation in mean head orientation: neck orientation relative to earth horizontal, the angle of the foramen magnum relative to the cranial base, and the angle of the cranial base.

To illustrate the effects of these constraints on head orientation, let us consider the difficulties a chimpanzee faces when standing upright. In a common

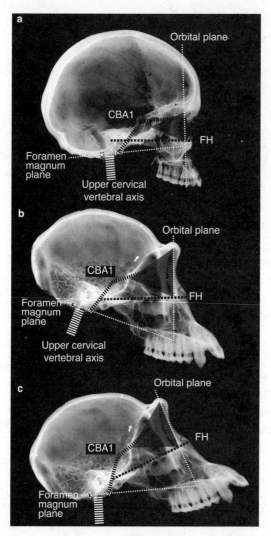

Figure 9.3. Radiographs of (a) adult human and (b) and (c) chimpanzee crania, showing the relationship between Frankfurt horizontal (FH), the orbital plane, the foramen magnum plane, the cranial base angle (CBA1), and the orientation of the upper cervical vertebrae. In the chimpanzee, the angle between the orbital and foramen magnum planes is 25° more acute than in the human; thus the chimpanzee's orbits point forward in habitual quadrupedal posture but would point upward in bipedal posture unless the chimp flexed its neck considerably.

chimpanzee, the cranial base is about 25° flatter (more extended) than in a modern human, and the foramen magnum is oriented more posteriorly. Thus (as shown in Figure 9.3b) the angle between the foramen magnum and the orbital plane is approximately 65–70° in *P. troglodytes,* but 100° in *H. sapiens* (Zollikofer et al., 2005). If a chimp were to stand like a human with an upright cervical column, then its midrange gaze would point upward (Figure 9.3c). If this standing chimp wanted to look forward without peering down its nose, it would have to flex its AOJ as much as possible and hunch its back.

A standing or walking human typically orients the head with the eyes facing forward. Viewed from the side in a radiograph (Figure 9.3), this posture places earth horizontal parallel to an empirically determined line, the Frankfurt horizontal (FH), which runs from the top of the earhole (porion) to the lowest point of the inferior rim of the orbit (orbitale) (Downs, 1952). Although the FH is sometimes useful for human studies, it is not a good marker of habitual head orientation in nonhuman primates because, as demonstrated by Strait and Ross (1999), apes and monkeys do not usually stand or walk with the FH parallel to the ground. In addition, the cranial base angle is a poor indicator of head posture for nonhuman primates: those with more vertical neck postures tend to have more flexed cranial bases, but only up to a point, as the cranial base angle and head posture are not significantly correlated when the effects of brain size are controlled for (Strait and Ross, 1999). However, because the Strait and Ross analysis included only one bipedal species with a very flexed cranial base (*H. sapiens*), one cannot conclude that a more flexed cranial base is unrelated to being bipedal. Many factors, including relative brain size and face size, affect the cranial base angle and hence the relationship between head orientation and cranial base flexion (see Chapter 5). In addition, the angle of the cranial base is only one of several components of the cranium that together determine the orientation of the head relative to the neck (the chief other one being the orientation of the foramen magnum).

Balancing

Balancing the head is a major biomechanical challenge posed by any postural mode, including bipedalism. Head balance, illustrated in Figure 9.4, is a first-class lever-load system like a seesaw. Torque on the neck side is the product of the force of the nuchal muscles, F_n, and the moment arm from the AOJ to the line of action of the nuchal muscles, r_n. Torque on the other side is the moment arm from the AOJ to the center of gravity's line of action (COG), r_{cog}, times this force, F_{cog}. Occipital condyle location can vary slightly, but in humans and most other primates the AOJ lies near the anterior margin of the foramen magnum. The other two moment arms differ more between humans and other primates, with humans being unusual in having an almost balanced head (Figure 9.4). In humans, the head's COG is located approximately in the pituitary fossa, about 1–2 cm anterior to the AOJ (Dempster and Gaughran, 1967); in contrast, a chimpanzee's COG is about 5–6 cm in front of the AOJ

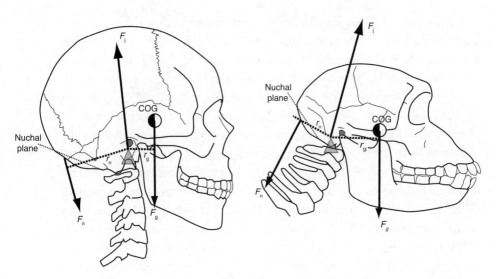

Figure 9.4. Biomechanics of head balance in human and chimpanzee heads, which differ in the orientation and length of the nuchal plane and the position of the center of gravity (COG). The human has a much shorter moment arm (r_g) from the AOJ to the vector of the head's COG (F_g) and a relatively longer moment arm (r_n) to the line of action of the head's extensors (F_n). This leads to a much higher mechanical advantage (MA) in humans (~1.75) than in chimps (~0.75). Further, the moment of inertia, I, in the pitch plane is the product of r_g^2 and the head's mass and is thus more than 10-fold lower in the human. Also shown is the predicted average orientation of the reaction force at the AOJ (F_j). Importantly, F_j in both configurations is nearly perpendicular to the plane of the foramen magnum, thus limiting potentially dangerous shearing forces on the spinal cord as it exits the skull.

(Schultz, 1942). Several factors make the human head relatively balanced. First the foramen magnum and the AOJ are positioned farther forward relative to other parts of the skull because of a relatively shorter basioccipital, a more flexed cranial base, and expansion of the brain within the posterior cranial fossa behind the foramen magnum (Biegert, 1963; Ross and Ravosa, 1993). In addition, modern humans differ from other extant primates in lacking a snout and having a shorter, more retracted face that is rotated beneath the anterior cranial fossa (see Chapter 13). These derived features all move the COG toward the AOJ.

A more centrally positioned foramen magnum combined with a relatively long and horizontally rotated nuchal plane give human head extensor muscles a high mechanical advantage (MA), meaning that they generate extensor force efficiently (albeit slowly). An even more important factor is the head's moment of inertia, *I*, which describes its resistance to changes in the rate of rotation. *I* is the head's mass times the *square* of its moment arm relative to a given axis. So small changes in the position of the COG have large effects on how much force the nuchal muscles must produce to stabilize the head. Because heads weigh about 8 percent of body mass—about 5 kg in a typical human male and 4 kg in a typical chimpanzee male—*I* is approximately 8–10 times greater in chimps. So human nuchal muscles are not only much more efficient but also have to produce much less force to stabilize the head. This difference is especially significant when the head is subject to high angular accelerations, because angular torque is the product of *I* and the angular velocity of the head (α). The latter is often 10 times higher during running than during walking (see below). The human head can thus be stabilized with at least an order of magnitude less muscular force than the chimpanzee's head.

A balanced head with a centrally located foramen magnum in which the COG of the head lies close to the AOJ is sometimes considered an adaptation for bipedalism (e.g., Broom, 1938, DuBrul, 1977), but this anthropocentric assumption is probably incorrect. Most mammals—including those with upright necks, such as giraffes, camels, and kangaroos—have COGs well in front of the AOJ (Schultz, 1942), and for more than 4 million years early hominins, such as the australopiths, had less balanced heads than modern humans despite being habitual bipeds (see Figure 9.5). Although positioning the foramen magnum near the posterior end of the skull increases the muscle force required to balance the head, the benefits of an unbalanced head are numerous. Most important, the general configuration of nonhominins with a posteriorly positioned foramen magnum permits the head to be cantilevered, with the neck oriented somewhat horizontally relative to the head. Cantilevered heads allow a much wider range of flexion and extension at the AOJ. In addition, a horizontally oriented neck can help stabilize the head in space during running. If you watch a good nonhuman runner, such as a dog, horse, or ostrich, its head seems to stay perfectly level as it runs, like a missile, even as the rest of the body rises and falls. This stability in the vertical plane occurs because the neck flexes as the body rises and then extends as the body falls (Bramble, 1989; Dunbar and Badam, 1998; Dunbar et al., 2004, 2008). Humans cannot do this because our centrally placed foramen magnum points downward rather than backward. We bounce like pogo sticks.

Another point to consider is that the anatomical shifts which lead to a more balanced head in humans pose some structural challenges to the pharynx. The trachea and esophagus run behind the face and in front of the vertebral col-

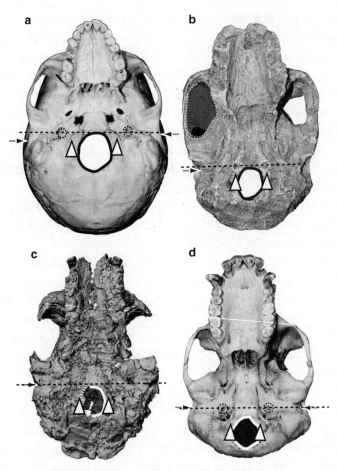

Figure 9.5. Inferior views showing the position of the foramen magnum (outlined) and occipital condyles (triangles) relative to the bicarotid line (dashed) and external occipital meatus (arrow) in (a) *H. sapiens,* (b) *Australopithecus africanus,* (c) *Sahelanthropus tchadensis,* and (d) *Pan troglodytes.* The foramen magnum and the occipital condyles are not in the center of the skull base but are more centrally located in humans than in the other bipedal hominins. In all known hominins, the foramen magnum tends to lie near the bicarotid line, but in apes such as chimpanzees, the foramen magnum and occipital condyles are even more posteriorly positioned, albeit with some variation. (For more details, see Dean and Wood, 1982b; White et al., 1994; Ahern, 2005; Schaefer, 1999; Guy et al., 2005.) Image of *Sahelanthropus* from Zollikofer et al., 2005 (courtesy M. Brunet).

umn. Quadrupeds have plenty of room for these structures, which hang below the neck. But orienting the neck upright, flexing the cranial base, and retracting the face impinge on the space behind the face for the pharynx, possibly accounting for the much smaller range of flexion allowed at the AOJ in primates, such as humans, that have more upright (orthograde) necks (Graf et al., 1995a).

Finally, being bipedal has a few additional consequences for the biomechanics of the nuchal plane and the extensor muscles that stabilize the head (illustrated in Figure 9.4). In quadrupeds, the nuchal plane slopes upward at a steep angle relative to the FH (approaching perpendicular in some species such as gorillas). Indeed, during growth, the foramen magnum in nonhuman primates drifts posteriorly, and the nuchal plane rotates vertically (see Chapter 4). In contrast, the nuchal plane in humans rotates horizontally, closer to parallel with the FH, and the foramen magnum does not drift posteriorly much. A more horizontal nuchal plane in bipedal hominins not only orients the nuchal muscles to increase their moment arm to the AOJ (see above) but also reduces shearing forces at the AOJ. When muscles act around a joint, they generate both rotational and translational forces. In this case, rotational forces extend the neck, and translational forces can generate not only compression in the nuchal plane but also shearing forces at the AOJ (see Figure 9.4). Shearing forces are worth minimizing at the foramen magnum, where the spinal cord exits the skull, and they can be reduced by rotating the nuchal plane so that the line of action of the muscles is as close as possible to perpendicular to the nuchal plane.[3] Demes (1985) showed that the orientations of the nuchal plane (and the occipital condyles of the AOJ) in bipeds and quadrupeds reduce shearing forces.

One last comment about an interesting paradox of the nuchal plane. Several researchers (Adams and Moore, 1975; Dean, 1985a,b) have pointed out that modern humans and other hominins have an expanded surface area of the nuchal plane because of a wider posterior cranial fossa and a more anteriorly located foramen magnum, but the surface area for the attachment of these muscles is smaller, presumably because a more balanced head requires less muscle force to stabilize. The result is a less crowded posterior cranial base, with smaller bony crests on which the superficial muscles insert.

---

3. The physical principle here is that the proportion of rotational to translational forces is a function of a muscle's line of action relative to the orientation of the resultant force: the closer in line to the resultant, the higher the proportion of rotational force.

Stability and Balance

The head needs to know its place, even when stationary. And when we walk, run, fall, or otherwise move about, the head needs to know how it is moving relative to the rest of the world, particularly in terms of orientation (pitch, roll, and yaw), in order to stabilize the gaze and to coordinate movements of the head with the eyes, arms, neck, and the rest of the body. Such problems confront all animals, but there are several reasons to believe they may be a special challenge for humans. First, as noted above, by becoming bipeds, our ancestors lost some of the mechanisms by which quadrupeds stabilize the head and absorb impact shocks during running. When we run, there is little we can do to keep our heads from bouncing In addition, humans are awkward bipeds with relatively long and stiff legs, upright trunks, and very little natural stability. Anyone who has played Frisbee with a dog knows that the human is much more likely to fall than the quadruped and more likely to be injured (e.g., sprain an ankle) in a fall. During most of human evolutionary history, a broken leg, a concussion, or even a sprained ankle could have been life threatening. Thus, before considering how we stabilize the head during locomotion, we'll review how the head senses changes in motion in order to maintain balance.

The Vestibular System

Changes in the position of the head are sensed by hair cells within several tiny organs, known as the vestibular system. These organs include the three semicircular ducts that lie within the bony semicircular canals, and two saclike swellings, the utricle and saccule, all of which are within the otic capsule of the inner ear within the petrous portion of the temporal bone. The other part of the inner ear, the cochlea functions to detect sound (see Chapter 10). The semicircular ducts sense only rotational changes in head orientation, and the utricle and saccule sense only linear changes in head position. Their combined shape is so complex that the entire system is often called the membranous labyrinth (which is housed in the bony labyrinth).

The vestibular system functions in much the same way as multiple gyroscopes tell a pilot how much a plane is pitching, rolling, and yawing. The duct housed in each semicircular canal detects changes in head orientation by perceiving angular accelerations in one of three perpendicular planes (Figure 9.6). In an upright human looking forward, the lateral semicircular canals (LSCs) are approximately in the transverse plane and thus detect yawing; the two pairs

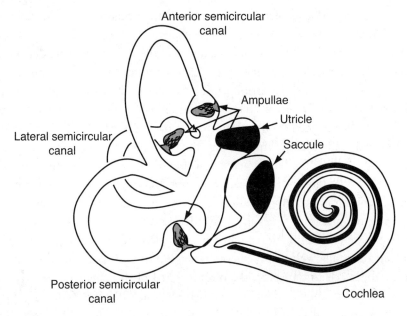

Figure 9.6. Anatomy of the membranous labyrinth, showing position and orientation of the three semicircular canals, the utricle and saccule, and the cochlea.

of vertical canals (anterior, ASC, and posterior, PSC) are each oriented at about 45° relative to the sagittal plane and together detect a combination of pitching and rolling. Note that the anterior canal on one side of the head lies approximately in the same plane as the posterior canal on the other side of the head. Thus head movements often result in different outputs from the left and right canals, so that even the most complex motion generates a unique combination of acceleration from both sides of the head in all three planes and thus can be detected and resolved.

The sensory mechanisms by which the semicircular system detects accelerations are a testament to the engineering powers of natural selection. Each semicircular duct is a hollow membranous tube that is filled with fluid (endolymph). At the base of each duct is a swelling, the ampulla, which contains bundles of hairs (stereocilia) that project into a gelatinous mass (the cupula). The fluid tends to retain its original position through inertia while the head changes orientation (much as coffee stays in a cup when you quickly rotate the cup). In this way, any pitching, rolling, and yawing motions of the head push a given ampulla against the more stationary endolymph fluid, causing the stereocilia to bend. Bending a bundle of stereocilia transduces rotational movement into a nervous signal. When changes in the rate of head rotation bend the stereocilia, they open ion channels in the cell's membrane that stimulate a short-term change in the cell's electrical potential (see Sohmer and Freeman, 2000). This depolarization causes the cells to alter the frequency with which they normally release a neurotransmitter to the nerves that synapse with each hair cell and then conduct information back to the brain. Through these pulses, hair cells provide information on the direction and rate of angular change by varying the rate of activation. Bending a hair in one direction increases the cell's rate of depolarization proportional to the amount of force applied, bending in the opposite direction decreases its rate of depolarization in the same manner (Peusner et al., 1998). Because the system works through inertia, the stereocilia function essentially as accelerometers, sensing changes in the rate of angular velocity. However, because the canal system naturally integrates the acceleration signal, the output signal of the canals to the brain is proportional to angular velocity itself (McMahon, 1984:162–63).

The utricle and saccule (which mean "little womb" and "little bag," respectively) sense linear accelerations in a similar manner. The utricle is a swelling inside the vestibule from which all the semicircular ducts arise, and the saccule projects posteriorly from the utricle (Figure 9.6). Both contain a bed of hair cells (macula) covered by a gelatinous mass that contains tiny suspended crystals, otoliths ("ear stones," made from calcium carbonate). In a standing person, the macula in the utricle is horizontally oriented, and the macula in the saccule is vertically oriented. The otoliths make the gelatinous mass denser than the surrounding endolymph fluid of the inner ear. Thus linear accelerations of the head cause the otoloth-rich gelatinous mass to move according to the force of inertia or gravity, pushing against the hair cells and stimulating the receptor cells to change their rate of neurotransmission.

Accelerations sensed by the hair cells of the semicircular ducts, the utricle, and the saccule have two major and related functions. The first is to let the brain know how the head is moving. The eighth cranial nerve carries streams of impulses from each hair cell to the brain stem; other pathways then transmit the information to a region in the parietal lobe that also receives and integrates sensory information from other parts of the body (e.g., joints, tendons, skin), leading to a conscious sense of position and movement. The second function of vestibular data is to participate in reflexes that coordinate sensory information on head position and movement with muscular coordination of the neck, eyes, and other parts of the body. These reflexes are vital because rapid positional changes of the head can cause blurry vision, disorientation, or loss of balance. Imagine that you slip on a banana skin while gazing at something important like a poisonous snake. As your head accelerates downward, the vestibular system participates in three kinds of reflexes: the *vestibulocollic reflex* (VCR) stabilizes your head relative to your trunk (so that you can continue to look at the snake without your neck snapping); the *vestibulo-occular reflex* (VOR) maintains stability of the visual field (so that your eyes' focus on the snake is not disrupted by the motion of your body); and the *vestibulospinal reflex* (VSR) tries to maintain your body's center of gravity (so you don't fall down). Additional sensory data on head and body position come to your brain from other organs, such as your eyes and specialized nerve cells (proprioceptors) that sense positional changes in various muscles, tendons, and joints.

Of the three vestibular reflexes, the VOR is particularly well understood. Because light forms an image only when projected on the retina (see Chapter 10), the eyes need to move reflexively to compensate for head movements to stabilize a given image on the retina; otherwise, vision would be blurry. For example, if the head yaws to the right (clockwise when viewed from above), then the eyes need to yaw to the left at the same rate. The VOR reflex works through a relatively simple neural circuit (summarized in Figure 9.7). When the head accelerates, hair cells in each of the semicircular ducts send a particular combination of signals to a cluster of nerve cell bodies (a ganglion) on the eighth cranial nerve; these signals are then integrated and sent on to specialized nerve cells (vestibular nuclei) in the brain stem. Half of these signals are excitatory, and half are inhibitory. When the nerves cross over to nuclei on the opposite side of the brain, they then synapse both directly with eye muscles on that side of the head, and indirectly (via another cluster of nerve cells, the oculomotor nuclei) with eye muscles on the other side of the head. The excitatory signals generate contractions, and the inhibitory signals prevent contractions. Because the VOR is driven by vestibular stimuli, these reflexes still function in the dark or when the eyes are closed.

The key limiting variables of the vestibular system are the frequency and amplitude of sensory stimuli. Under normal circumstances, the vestibular system is capable of rapidly sensing movements and quickly activating all three reflexes so that the eyes, head, and trunk can respond almost immediately (within 20–50 milliseconds) to changes in head and body position. Overly rapid changes, however, are a problem because the system saturates at high frequencies of sensory input. You can easily demonstrate this effect with a simple home VOR "experiment." Focus your eyes on some object a few feet away

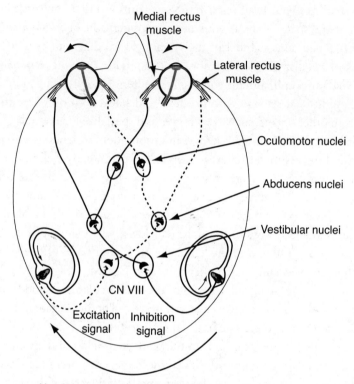

Head yaws clockwise (viewed from above)

Figure 9.7. Schematic view of the vestibulo-occular reflex (VOR) circuit. In this case the head is rotating clockwise when viewed from above, requiring the eyes to rotate counterclockwise to stay focused on the spider. Signals from the ampulla on the left side send inhibition impulses to the left medial rectus and right lateral rectus muscles, whereas signals from the right side ampulla send excitation impulses to the right medial rectus and left lateral rectus muscles. The signals travel from the inner ear to the eye muscles via the eighth cranial nerve (CN VIII) and two clusters of cell bodies in the hindbrain, the abducens nuclei and the oculomotor nuclei.

such as a piece of furniture, and then use your hand to push your head to the side. If your VOR is working correctly, then the object will not appear to jiggle or blur as your eyes remain locked on it. Next, hit your head (not too violently) while focusing on the same object. It will briefly appear to bounce and blur because the rate of head movement has exceeded the sensory limit of the VOR. Such jiggles are annoying, but can be seriously deleterious during demanding tasks such as running and figure skating, which subject the head to frequent, complex accelerations. Angular velocities above 200°/sec can cause significant losses of balance and visual acuity (Grossman et al., 1988; Viire and Demer, 1996), and the system becomes completely saturated if the head rotates faster than 350°/sec (Pulaski et al., 1981), or if the frequency of stimuli exceeds 5–10 Hz (Gauthier et al., 1984; Maas et al., 1989; Cromwell et al., 2001). Nonhuman primates and other quadrupeds typically keep angular rotations below 200°/sec (Dunbar et al., 2004). Humans, like other bipeds such as birds, appear to constrain head movements even more, probably because bipeds are inherently less stable than quadrupeds. Humans almost always manage to keep angular velocities of the head below 100°/sec, even during very bouncy activities such as running (Pozzo et al., 1990). Humans also restrict the total rotation of the horizontal semicircular canals to less than 20° during walking, running, and jumping activities (Pozzo et al., 1990) and even less during more challenging activities that require special balance (Pozzo et al., 1995).

Comparative Anatomy of the Vestibular System

To explore further the evolution of mechanisms responsible for positioning and stabilizing the human head, particularly in the context of bipedal locomotion, it is useful to review how vestibular anatomy varies among species, particularly in the position and orientation of the labyrinths, how variations affect the system's sensitivity, and how the labyrinths grow.

Position and Orientation

In humans and other primates, the orientations of the semicircular canals are similar with respect to each other and to earth horizontal. In a standing human, the lateral semicircular canals are typically tilted up about 5° relative to earth horizontal; a similar upward tilt of the LSC also characterizes most nonhuman primates and other mammals during resting postures (Fenart and Pellerin, 1988; Vidal et al., 1986; Graf et al., 1995a). In addition, most mammals, including humans, orient the maculae of the utricle with earth horizontal (Graf et al., 1995b). To what extent LSC orientation matters is questionable, as vestibular function is solely to measure changes in orientation, not orientation itself. However, humans do differ in the orientation of the face relative to the vestibular system, largely because of being bipedal. Standing humans typically orient the head with the FH roughly parallel to earth horizontal, but nonhuman primates typically hold their heads with the FH inclined upward about 20–30° to earth horizontal (Strait and Ross, 1999). This is not a case of habitual tree-gazing (or stargazing at night), but instead reflects the derived design of the human head, in which the highly flexed cranial base rotates the face downward relative to the rest of the skull (see Chapter 5).

Sensitivity

Variations in semicircular canal morphology among species have recently been scrutinized by many researchers because they provide clues to differences in perceptual sensitivity to head movements, even in fossils (see Spoor et al., 2002a, 2003, 2007b). As noted above, the vestibular system begins to decline in functional capacity at accelerations above 200°/sec and becomes saturated above 350°/sec; it also cannot detect movements that occur at frequencies much greater than 5 Hz. Broadly, four variables affect vestibular sensitivity: the lumen area and streamline length (or arc radius) of the duct, the viscosity of the endolymph fluid, and the mechanical properties of the cupula. When corrected for body size, lumen area and endolymph viscosity appear to be fairly constant among mammals (Ramprashad et al., 1984; Spoor and Thewissen, 2008), but substantial variation has been documented between species in bony canal dimensions, which reflect the arc size of the enclosed semicircular ducts. Experimental evidence shows that lengthening the radii of the ducts increases the ability to sense rapid accelerations, much like turning up the gain on an amplifier, but at the expense of noisier signals (Yang and Hullar, 2007). The relationship between duct size and sensitivity accounts for the strong negative allometry across species between body size and the radius of curvature of the semicircular ducts: the frequency of body movements declines with increasing size (big animals have longer legs with lower natural frequencies). Further, there is abundant evidence that agile or fast-moving creatures such as raptors, squirrels, and many monkeys tend to have larger arc dimensions relative to body size than sluggish, slow-moving creatures such as ducks, pigs, sloths, and lorises (Spoor et al., 2007b). Relatively enlarged semicircular ducts are also present in humans. Specifically, humans have significantly larger vertical semicircular ducts (anterior and posterior) and slightly smaller lateral semicircular ducts relative to body mass than do australopiths and other great apes (Spoor and Zonneveld, 1995, 1998) (Figure 9.8). It is therefore reasonable to conclude that humans are

Figure 9.8. Radius of the semicircular canals versus body size in nonprimate mammals, apes, modern humans, and some hominins (adapted from Spoor et al., 1994). Mammals with larger semicircular canal radii relative to body mass are more sensitive to rapid rotational movements of the head. Apes and australopiths fall below the general mammalian regression; in contrast, *H. sapiens* and *H. erectus* have relatively larger anterior and posterior semicircular canals, which sense pitching motions.

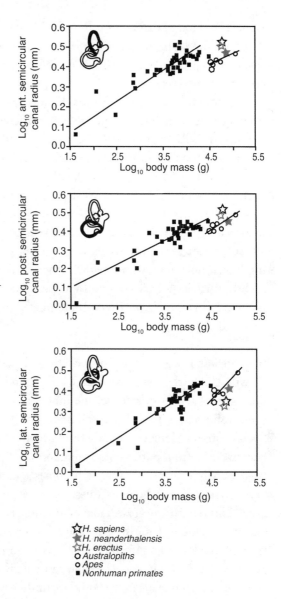

☆ *H. sapiens*
★ *H. neanderthalensis*
☆ *H. erectus*
○ *Australopiths*
○ *Apes*
■ *Nonhuman primates*

more sensitive to rapid pitching and rolling movements but slightly less sensitive to yawing movements than other apes.

If humans are more sensitive to high frequencies of head accelerations in the vertical planes (pitching and rolling) than other apes, then the functional demands of being bipedal seem to provide a likely explanation. One hypothesis is that bipedal walking introduces greater accelerations or more rapid frequencies of pitching and rolling than quadrupedal or arboreal locomotion. Spoor et al. (1994) found that bipedal australopiths had relative semicircular canal dimensions similar to those of chimpanzees and that ASC and PSC enlargement first appears in *Homo erectus* (more on this in Chapter 12). Initially, Spoor and colleagues interpreted these differences to suggest that australopiths were facultative rather than obligatory bipeds. However, mean pitch frequencies during walking in humans (1–2 Hz) are not much different from those of primates during quadrupedal locomotion (Hirasaki et al., 1999; Dunbar et al., 2004), and the rates of head pitching during walking tend to be trivial. As I argue below, and as noted by Spoor et al (2003), enlarged ASC and PSC dimensions in *H. erectus* are more likely to have evolved as adaptations to running, which generates much higher frequencies and amplitudes of pitching than walking (Grossman et al., 1988). Relatively small lateral semicircular canals in the genus *Homo* also suggest less ability to sense yawing movements, which tend to be minor during human locomotion (see below). Another twist is that Neanderthals have slightly smaller vertical (anterior and posterior) semicircular canals than do modern humans or *Homo erectus* (Hublin et al., 1996; Spoor et al., 2003).

Growth

One final and unique aspect of the labyrinth system to discuss is its mode of growth. Unlike the rest of the skull, which remodels to change in size and shape during fetal and postnatal life, the inner ear grows quickly during embryogenesis and attains its adult size and shape by 19 weeks of gestation (Jeffery and Spoor, 2004b). Moreover, the labyrinth becomes fully encapsulated a few weeks later by the petrous portion of the temporal (Scheuer and Black, 2000), which is so heavily mineralized that it becomes the densest bone of the body—probably to insulate the inner ear as much as possible from distracting vibrations. This rapid, early growth makes functional sense given that changes in the size and position of the system would require the brain to reprogram basic vestibular reflexes. Presumably, such reprogramming is either difficult or impossible, as slight errors could be fatal, particularly when being chased by a predator. Lack of remodeling in the bony labyrinth may make variations in the shape of the vestibular system particularly useful for testing systematic hypotheses.

Head Stabilization in Walking and Running

Walking and running pose quite different biomechanical problems for the head, largely because the mechanics of these two gaits differ fundamentally. Walking has been the most intensively studied even though it poses much less of a challenge for head stabilization. When one walks, the head moves up and down very slightly with the rest of the upper body, requiring some carefully tuned muscular effort (mostly neck extension) to maintain stability. When one runs, however, the head bounces, pitches, and rolls with each step, as the body leaps and falls and the torso sways from front to back and twists from side to side. In addition, as the body collides with the ground, the head pitches forward rapidly. A good way to observe these forces is to watch someone with a ponytail—a natural accelerometer—both walking and running. During a walk, the ponytail is typically quite stable. During a fast walk, the ponytail bounces a little more, particularly at heel strike. During a run, the ponytail bounces exuberantly in a figure-8 motion, simultaneously bobbing up and down and wagging from side to side, reflecting high pitching and rolling forces. But even though the runner's head also bounces up and down with the body, the head's orientation and thus the runner's gaze typically remain quite stable. The kinematic differences between the head and ponytail are evidence of unseen but critical mechanisms that stabilize the head.

Walking

Biomechanically, a walk is defined as an inverted pendular gait in which one vaults over relatively straight and stiffened legs (Figure 9.9). With each step, between heel strike and midstance, the body's center of gravity (COG) is elevated, thereby increasing the body's potential energy. Gravity then converts this energy into motion (kinetic energy) as the body's COG falls downward and forward from midstance until toe-off, at which point we begin to push the center of gravity up again and send the leg into a swing. About 70 percent of the potential energy is converted into kinetic energy with each cycle (Heglund et al., 1982). During a walk, each foot is typically on the ground for about 60 percent of a total stride cycle (which is divided into stance and swing phases), so that at least one foot is always on the ground.

To understand the head's motion during walking, it is useful to keep a few points in mind. First, the head rises and falls along with the body's COG by approximately 4–5 centimeters per stride. If walking were a purely inverted pendular gait (imagine a peg-legged person with no ankles and knees), then these movements would be more substantial, with jolting changes in direction at every heel strike. Instead, the head moves in a gradual sinusoidal wave, thanks to a number of mechanisms such as hip rotation, hip abduction, knee flexion, ankle flexion at heel strike, ankle extension at heel-off, and slight flex-

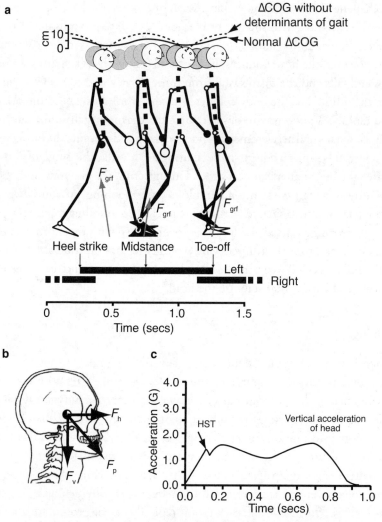

Figure 9.9. (a) Basic kinematics of the body and head during walking (adapted from Bramble and Lieberman, 2004). Left foot and arm indicated by solid shapes, right foot and arm by hollow shapes. At a moderate speed (1.4 m/sec), the center of gravity (COG) of the head rises and falls about 4–5 cm, much less than it would the case without determinants of gait (e.g., knee flexion) that smooth the head's trajectory, raising the nadir and lowering the peaks. Note also that the head typically flexes slightly at the peak of its trajectory and extends at its nadir. Approximate vectors of the ground reaction force (GRF) also shown. (b) Approximate vectors of forces acting on the head. (c) GRF of the body's interaction with the ground.

ion and extension of the trunk (backward at heel strike and forward at toe-off). These "determinants of gait" minimize and smooth vertical movements of the body's COG (Saunders et al., 1953).

During walking, the head tends to pitch in the sagittal plane; smaller rolling and yawing moments also act on the head. Pitching occurs for two reasons. The first is to stabilize the gaze to counteract the rise and fall of the body: as the body rises during the first half of stance, the head typically pitches forward slightly; and as the body falls during the second half of stance, the head pitches backward (Grossman et al., 1988; Pozzo et al., 1990). The head also pitches very slightly because of accelerations, some of which come from the ground. Every time the heel strikes the ground with a given force, the ground reacts with an equal and opposite force that sends an impulse, the heel strike transient (HST) up the leg, along the spine, and into the head (Hirasaki et al., 1999; Mulavara et al., 2002). The HST is attenuated considerably by elastic tissues in the joints and the spine (Smeathers, 1989), but it still tends to jiggle the head because, even though the trunk and neck tend to be rather vertical (usually inclined slightly forward) above the hips during walking, the head's COG is about 2 cm anterior to the AOJ. As Figure 9.9 shows, the combination of vertical and horizontal accelerations from the HST cause the head to pitch forward slightly; the head also tends to pitch forward slightly during the second half of stance phase as the body's COG falls (Mulavara et al., 2002).

A walking human's head remains fairly stable, despite the forces described above, because of a combination of passive and active mechanisms. Passive mechanisms are the intrinsic mechanical characteristics of the musculoskeletal system. These are primarily the natural inertial properties of the head and the stretch-resistant (viscoelastic) properties of the many muscles and ligaments that connect the head, neck, and trunk. Active stabilizing mechanisms mostly involve the reflex activation of muscles that extend or flex the neck. Several experiments have tried to estimate the relative contribution of passive and active types of head stabilization in the sagittal plane. These studies mostly find that although intrinsic mechanisms do play a key role, particularly in high-speed walking (Hirasaki et al., 1999), active reflexes are probably the dominant way we keep our heads still (Keshner et al., 1992, 1995; Winter et al., 1990; Peterson et al., 2001; Vasavada et al., 2002).

Figure 9.12 illustrates how rhythmic contractions of the superficial neck extensors counteract head pitching during a walk (the deep nuchal muscles, not shown, probably also make quick adjustments in position). For the most part, these muscles have very low levels of EMG activity during walking compared

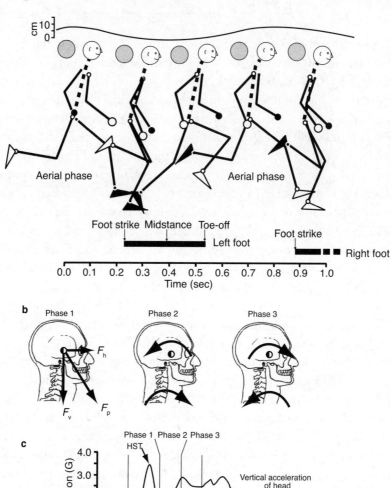

a

Aerial phase   Aerial phase

Foot strike   Midstance   Toe-off

Left foot

Foot strike

Right foot

b

Phase 1   Phase 2   Phase 3

F_h F_v F_p

c

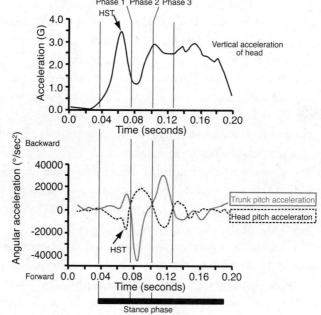

Phase 1 Phase 2 Phase 3

HST

Vertical acceleration of head

Backward

Trunk pitch acceleration

Head pitch acceleraton

HST

Forward

Stance phase

Figure 9.10. (a) Kinematics of heel-toe running at 3.5 m/sec (compare with kinematics of walking in Figure 9.9). Left foot and arm indicated by solid shapes. (b) Forces and movements during three phases that cause the head to pitch. During phase 1, the head pitches forward just after the foot strikes the ground. During phases 2 and 3, the trunk pitches forward and then backward, but reaction forces in the head and neck simultaneously pitch the head backward, then forward.

---

to running, and they mostly contract during the second half of stance, as the body's COG is falling (see also Basmajian and DeLuca, 1985). However, at very fast walking speeds (e.g., speeds at which one normally runs), these muscles have higher levels of activity, as the head is subject to much greater rates of pitching (Hirasaki et al., 1999).

Running

Most studies on head stabilization during locomotion have focused on walking, which is more common and less costly than running. For most people, the most efficient speed of walking is about 1.2–1.5 m/sec (this value is largely a function of leg length), which costs about 160 mL O_2/kg/km (Margaria et al., 1963; Rubenson et al., 2007). As we walk faster, the cost rises sharply, and we typically switch to a running gait at about 2.0–2.5 m/sec, when running becomes more economical, costing about 200–230 mL O_2/kg/km. Compared to most mammals, humans are terrible sprinters: the world's fastest humans can run at about 10 m/sec for less than 30 seconds; in contrast, a cheetah can run at 25 m/sec for about four minutes. However, we are phenomenal long-distance, aerobic endurance runners at moderate speeds (<6 m/sec), equaling or surpassing the best quadrupedal runners, such as dogs and horses, particularly in hot, arid conditions. Humans excel at long-distance running thanks to many derived anatomical and physiological specializations, many of which play no role in walking and which therefore suggest that selection for endurance-running capabilities played an important role in human evolution (Carrier, 1984; Bramble and Lieberman, 2004; see also Chapter 12).

Running is different from fast walking and has fundamentally different effects on the head. Most critically, running is an inherently bouncy gait (not unlike a biped's version of a trot), in which the legs act as springs (Figure 9.10). In contrast to walking, running has an aerial phase in which both feet are off the ground. In addition, when the foot contacts the ground (at foot strike) during a run, the hip, knee, and ankle flex so that the body's center of mass falls for the first half of the stance phase, before the leg straightens and we jump up into the air again. Leg compliance during the first half of stance converts the energy of the body's fall into elastic strain energy that is stored in the tendons and ligaments of the leg; this energy is subsequently released during the second half of stance, helping to propel the body forward and upward.

One difference between walking and running is how much the head bounces. Whereas the head rises and falls about 4–5 cm per stride in a typical walk, it experiences vertical displacements of as much as 10 cm per stride during a typical run. When a dog or an ostrich runs, its head stays remarkably still, as if the body is moving independently of the head. This stability occurs because a dog or ostrich can flex and extend its neck, which is partly horizontal and cantilevered relative to the trunk, and can also flex and extend the head, which is cantilevered relative to the neck. But humans bounce vertically when they run with upright trunks, extended legs, and a short, vertical neck. These anatomical and postural differences cause angular rotations that challenge the head's stability in several ways (see Figure 9.10b). The first challenge occurs when the foot strikes the ground after the aerial phase in running. Runners who land on their heels (heel strike) generate an impulsive ground reaction force (GRF) that is typically twice that of a walk and can approach 3 times body mass (Keller et al., 1996; Lieberman et al., 2010). The HST impulse travels rapidly up the relatively straight leg and spine to the head within 6–10 milliseconds, contributing to a peak of vertical acceleration in the head that can reach 2–4 G (G is the acceleration of gravity, 9.81 m/sec^2). Runners who land on the ball of the foot (a forefoot strike) have no appreciable impact transient (they land more gently), but the upper body still decelerates somewhat after the foot contacts the ground, causing smaller vertical accelerations in the head, about 1–2 G.

As Figure 9.10 shows, vertical accelerations such as the HST tend to pitch the head forward when there is also some degree of anterior acceleration, because the resultant of vertical and horizontal accelerations causes a forward-directed moment (a force times a moment arm). This moment can be substantial and rapid in humans right after a heel strike, causing the head's rate of pitching to increase from 0 to 100 °/sec in less than 10 milliseconds (ms); this figure would exceed 200 °/sec in less than 20–30 ms if unchecked (Lieberman et al., 2008). Because VOR performance decreases significantly above 200°/sec, such pitching is undesirable. And because 20–30 ms is too short a period for any normal reflexes, including the VCR, to control a motion (McMahon,

1984), the human head needs a self-stabilizing system to control head pitch in some way that does not rely on reflexes.[4]

An even bigger challenge to head stabilization during running (labeled phases 2 and 3 in Figure 9.10b) comes from the motion of the trunk. During a walk, the trunk is usually positioned above the hips and sways only slightly. But the much greater vector of the GRF during running, which passes well behind the trunk's center of mass, causes the trunk to pitch forward faster. In some ways, running is like a controlled, forward fall. This tendency is compounded by the fact that humans typically incline the trunk more, at about 10°, at endurance speeds (Thorstensson et al., 1984). At a moderate speed (3.5 m/sec), forward pitching of the trunk during phase 2 is typically 150–200°/sec and sometimes higher. We avoid falling down with every step because muscles of the lower back and hip, especially the gluteus maximus (the largest muscle in the human body), contract forcefully, quickly decelerating the trunk during phase 3 (Lieberman et al., 2006). Because the head and neck are connected to the trunk, some stabilizing mechanism is necessary to prevent them from pitching first forward and then backward at similarly high angular velocities, thus challenging the vestibular system.

Finally, although pitching is a major concern, the head also has a tendency to roll, usually toward the stance-side foot at heel strike and then toward the swing side of the body by midstance. Yawing motions are comparatively minor. Another issue related to head stability is the position of the upper torso and arms. During the aerial phase of running, one leg swings forward rapidly, while the opposite leg swings rapidly behind the body (see Figure 9.10). These movements generate considerable angular momentum around a vertical axis that must be counteracted to prevent the runner from moving in a zigzag path. Because the body has no foot contact with the ground to counteract these twisting forces, we swing the arms forward in opposite phase to the legs and rotate the torso in the direction of the stance foot, thereby generating nearly equal opposite angular momentum with the upper body (Hinrichs, 1990; Pontzer et al., 2009). Because the whole upper body twists, a runner must rotate the neck in the opposite direction from the trunk to keep the head pointed forward.

---

4. A muscular reflex involves the time it takes for the stimulus to be sensed and transduced (about 5–10 ms) plus the latency period between the onset of muscle stimulation and the initiation of force (usually more than 30 ms).

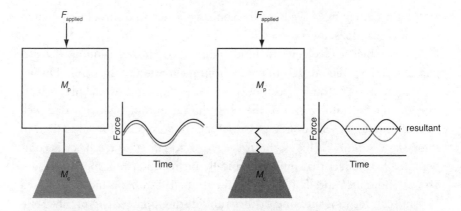

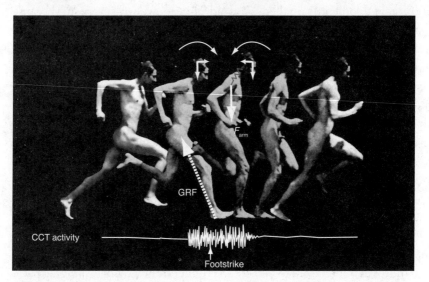

Figure 9.11. Top: mass spring model showing a primary mass (M_p) and a countermass (M_c). With a stiff linkage (left), the two masses accelerate in phase; in the elastic linkage (right), M_c accelerates out of phase with M_p, smoothing the resultant force of both masses. Bottom: schematic of how the stance-side arm acts as a mass-damper. Vertical accelerations, such as the ground reaction force (GRF) at foot strike, cause a reaction force that tends to pitch the head forward. Downward acceleration of the stance-side arm follows soon after vertical acceleration in the head. Activation of the cleidocranial trapezius (CCT) links the accelerating mass of the stance-side arm to the head via the elastic nuchal ligament.

Although running poses more substantial challenges for head stabilization than walking, we evidently manage quite well. Human runners typically keep the angular velocity of the head below 140°/sec, well within range of the VOR for integrating ocular and vestibular sensory data to stabilize gaze (Pozzo et al., 1990). In fact, humans actually control head pitching better during running in the aerobic speed range (2.5–6.5 m/sec) than during very fast walking (Grossman et al., 1988; Viire and Demer, 1996).

How do we do it? My colleagues (Dennis Bramble, Dave Raichlen, Katherine Whitcome) and I have found out that humans partly stabilize the head during running against vertical accelerations which induce pitching (phase 1 in Figure 9.10) by using the mass of the arm as a counterbalance (Lieberman et al., 2008). This unusual system is analogous to a mechanism known as a mass-damper that engineers use to control vibrations in objects like buildings (Figure 9.11). Mass dampers work to protect a primary mass (M_p) from the force of a periodic external impulse such as wind, an earthquake, or an HST. If one connects this mass to another mass (M_c) via a rigid connection, then forces applied to the primary mass will generate accelerations in both masses that are in phase (the waves are synchronized). But if the two masses are connected by a spring with some degree of elasticity, then the second mass will behave as a countermass, accelerating out of phase (out of sync) with the primary mass. As the primary mass accelerates in one direction, the countermass accelerates in the opposite direction, thereby damping the oscillation. In running, the head is the primary mass, and the stance-side arm is the countermass.

Several derived aspects of human anatomy and running kinematics enable humans to use the stance-side arm as a mass-damper for the head (Figure 9.11). Anatomically, humans differ from apes in having low, wide shoulders in which the only muscle that connects the back of the head and the shoulder girdle is a straplike portion of the trapezius, the cleidocranial trapezius (CCT) (Aiello and Dean, 1990; see also Figure 9.2). The CCT originates on the posterior occipital and inserts on the distal clavicle and the acromion process of the scapula. In contrast, apes have narrow, high, and massive shoulders in which several prominent muscles (trapezius and rhomboideus) extensively connect the shoulder girdle and occiput. This gives apes the appearance of permanently hunched shoulders. A second difference is the nuchal ligament in humans (absent in apes), which acts as the spring in the mass-damping system. Finally, each human arm constitutes about 8 percent of total body mass, almost the same as the head (5–6 kg); in contrast, the forelimb in a chimpanzee constitutes about 16 percent of body mass (about 7–8 kg), approximately twice the weight of the head (Schultz, 1942; Zihlman, 1984).

Several aspects of human kinematics (illustrated in Figure 9.11) are also relevant to the head's mass-damping system. Most important, during a run, humans habitually rotate the arms and the shoulder girdle in a direction opposite to the motion of the legs; we also habitually contract the biceps to flex the elbow so that the forearm is carried at about 90° to the upper arm. As noted above, arm and shoulder rotations help to oppose the reverse angular momentum generated by the legs during the aerial phase of running (Hinrichs, 1990), but they also position the stance-side shoulder and the linked mass of the arm posterior to the neck at heel strike, thus in line with the nuchal ligament and CCT. In addition, as shown in Figure 9.12, the CCT contracts on just the stance side before foot strike, well before any other neck extensors are activated. Therefore, just after the head starts pitching forward after foot contact, downward force from the falling stance-side arm is coupled elastically by the CCT and nuchal ligament to the back of the head. Because the arm accelerates downward after the head starts pitching forward, the inertial mass of the arm both dampens the peak vertical acceleration of the head and acts to extend the head, thereby countering its tendency to pitch forward. Another important muscle that contracts before heel strike with greater intensity on the stance side is the biceps brachii. Because this muscle crosses both the elbow and shoulder joints, it links the mass of the forearm to the head via the upper arm and shoulder. We have found that volunteers asked to run without swinging their arms normally have higher vertical accelerations in the head as well as higher rates of head pitching.

One useful feature of this system is that it is self-adjusting. By coupling the arms and head via the CCT and nuchal ligament, the inertial forces in the stance-side arm and the head (which have nearly the same mass) tend to counterbalance each other regardless of speed, because the forces that drive the mass-damping system come from the same external source—the ground—rather than muscles. Thus when running, bipedal humans can passively control head pitching at a wide range of speeds and on many different kinds of terrain.

The other major stabilization problem that confronts the human head during running is the rapid, forceful pitching (first forward, then backward) of the upright trunk. The derived anatomy of the human head-neck complex, in which the low shoulders are much less coupled than in apes to the head by muscles, also helps the head to stabilize itself passively from trunk pitching. Notably, the head and neck operate as a single segment partially independent

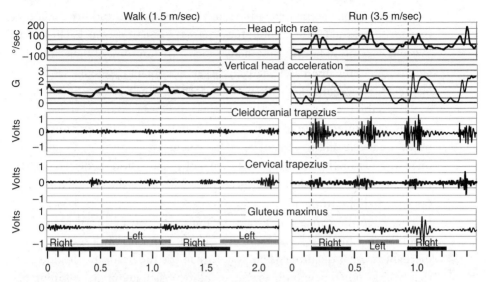

Figure 9.12. Comparison of head pitch rate, vertical head acceleration, and EMGs (from right-side muscles) of the cleidocranial trapezius, cervical trapezius, and gluteus maximus during a walk and a run. The vertical axes are the same for both gaits, and foot contacts are indicated at the bottom. Head pitch rates and vertical accelerations are much greater in the run, as are the intensity of the contractions of all three muscles. Note also that the cleidocranial trapezius fires prior to foot strike on the stance side, not the swing side.

from the trunk. Newton's third law states that for every action there is an equal and opposite reaction: so when the trunk accelerates forward and then backward (an angular acceleration), the head and neck accelerate backward and then forward at the AOJ (again, this is an angular acceleration), much as a passenger's head gets "whiplashed" in a car that suddenly lurches forward and then back. The elegance of the system is that it is self-stabilizing. The more the trunk pitches, the more the head reacts. However, this inertial damping mechanism is possible only when there are not extensive, stiff connections between the head and the upper trunk (that is, the head and neck need to flop independently of the trunk).

In other words, when hominins evolved to be bipedal, they created some challenges for head stabilization, especially during running. At some point, natural selection may have favored several derived mechanisms to keep the head stable during running. Notably, the arm functions as a mass-damping counterbalance, and the head can pitch (whiplash) backward, then forward, as the trunk pitches forward, then backward. These stabilization mechanisms may help explain several derived features of the human head-neck complex that are otherwise hard to account for in terms of function: a nuchal ligament; low and wide disconnected shoulders, which have lost most of their muscular connection to the head; and relatively larger anterior and posterior semicircular canals, which increase sensitivity to pitching accelerations (Spoor et al., 1994). To this list, we might add a more balanced head. As noted above, walking imposes minimal pitching forces on the head, and other mammals (and birds) use their necks to control head movements when running. Because humans use inertial forces to passively control pitching forces, decreasing the moment arm from the AOJ to the head's COG decreases the inertial force necessary to stabilize the head.

The unique human system for head stabilization may be a good example of evolutionary tinkering. Once hominins became bipedal, they lost many of the mechanisms that quadrupeds employ to stabilize the head. In addition, upright trunks posed new problems for stability. When endurance running became an important behavior, probably sometime after the evolution of the genus *Homo,* there might have been selection for hominins that were better at running. If so, then hominins better able to stabilize their heads, thanks to features such as the nuchal ligament and more disconnected shoulders, would have had some advantage. The likely disadvantage of the human system, however, is that it cannot operate with the anatomical configuration of our apelike arboreal ancestors. Great apes (and possibly australopiths) have a narrow upper thorax, along with massive scapular rotators, such as the trapezius and rhomboids, that connect the occiput and shoulder girdle. These rotators are advantageous for arboreal locomotion because they help pull the body up from a hanging position (Larson, 1993), but they would impede independent movements of the head and pectoral girdle that are crucial to controlling head pitch in humans during running (but not walking). Perhaps selection for stabilizing the head during running explains why the genus *Homo* lost some features of the upper body that were useful for climbing trees. We will return to this subject in Chapter 12.

Coda: Looking Ahead

Although we tend not to think of the head as a component of the postural and locomotor system, it should be evident that standing, walking, and running pose many important biomechanical challenges above the neck in bipeds. Why early hominins became bipedal is still a matter of much speculation, but it is clear that the transition, however it occurred, required some rearrangement of the skull. Most important, orienting the neck vertically meant that hominins had to reorient the orbits relative to the foramen magnum so that they could look forward rather than upward in normal posture. However, in many other respects, early hominins appear to have retained the apelike features of relatively unbalanced heads, small brains, and large faces; they probably also had extensive muscles that connected the head with the shoulder girdle.

We have much to learn about the transition to *Homo,* but by the time of *H. erectus,* the head started changing shape in several additional respects relevant to locomotion. The head became more balanced as more of the brain was positioned behind the foramen magnum and as the face became shorter. The anterior and posterior semicircular canals became relatively larger, improving the ability to sense pitching movements. And the head apparently lost some of its muscular connections to the shoulder girdle. These derived features may have improved performance during endurance running. In turn, endurance running may have helped hominins effectively hunt and scavenge for meat, behaviors that are first evident in the archaeological record starting about 2.6 mya. Chapter 12 examines in greater detail these shifts, their effects on other aspects of skull form, and their larger implications.

10

Sense and Sensitivity: Vision, Hearing, Olfaction, and Taste

This is ze nose, ze most prominent part of the human face. . . .
Through zis nose comes some of life's most rewarding sensations,
und we plan to share with you some of ze most beautiful odors
known to mankind. Unfortunately, zis same nose . . . is also responsible for bringing us some odors that are rather . . . repulsive.

Polyester, directed by John Waters, 1981

The 1981 cult film *Polyester* entertained moviegoers with the bizarre story of
a suburban housewife's tragic renaissance. To enhance the experience of the
movie, everyone in the audience was given a "scratch-'n'-sniff" card impreg-
nated with 10 numbered odors. When a particular number flashed on the
screen, audience members were supposed to scratch and then sniff the appropriate part of the card, treating themselves to a variety of scents, such as flowers, pizza, grass, and feces. Part of the fun was the audience's sharing an
unusually rich, albeit controlled, sensory experience. In a way, this experience
reflects the relative importance of the different senses in primates. In humans,
as in other primates, vision is very much the dominant sense; hearing is also
important, whereas olfaction and taste are less prominent components of our
typical sensory experience. If rats ever evolve to create and appreciate art, it is
likely that odors will play a much more prominent role in their aesthetic.

Most of the matter that enters our bodies passes through the head. Air,
water, and food enter the nose and mouth, bringing molecules we sense as taste
and smell; pressurized waves of air enter the ear as sound; and photons of light
enter the eyes. These stimuli are detected and processed by different sensory
mechanisms, and these mechanisms have had strong effects on fitness during

primate evolutionary history. As our favorite forms of entertainment attest, primates such as humans have especially acute stereoscopic and color vision, leading to the widely held presumption that vision is the dominant sense in primates. In contrast, our acoustic capabilities are average compared to most mammals, and it is generally believed that humans have below-average abilities to smell and an unexceptional sense of taste. Consensus has it that none of these senses is particularly derived in humans compared to other higher primates. Thus, there has been much research on the evolution of sensory adaptations in nonhuman primates, especially in anthropoids, but less consideration of how sensory functions have specifically influenced human craniofacial form.

But is the consensus right? There are several reasons to think about sensory mechanisms in the human head. First, our apparent lack of sensory specialization does not mean that anatomical adaptations for hearing, vision, smell, and taste are unimportant with respect to the form, function, and evolution of the human head. When impaired, these senses can have very negative consequences for fitness. Their vital roles must have a strong effect on craniofacial growth and morphology. In addition, the sensory organs, like everything else in the head, must be crammed into the small confines of the skull amid the brain, pharynx, oral cavity and other structures. Thus major changes in skull shape have had to accommodate the organs of sense without compromising their function. Finally, the consensus view that humans did not evolve any special sensory adaptations is probably wrong. Human eyes are unusually wide, colorful, and insulated, we are slightly better at hearing low frequencies than other apes, and we may have an exceptional ability to taste food flavors. The goal of this chapter is to review some basic details about how the major cranial senses develop and function and to consider how they influence the design and integration of the head. Because Chapter 9 already covered balance, we'll begin with vision, then proceed to hearing, olfaction, and taste.

Vision

If any one sense can be said to be dominant in humans, it must be vision. This dominance is a legacy of being a primate. Most animals use their eyes to detect predators, find food, and recognize each other, but primates have been especially visually oriented since our divergence from other mammals. The earliest primates appear to have been nocturnal insectivores and frugivores whose abilities to find and grasp their prey—especially cryptic prey—were aided by highly stereoscopic vision with good depth perception and relatively large eyes that are sensitive to low levels of light (R. D. Martin, 1990). Monkeys later evolved specialized adaptations for color vision to assess the nutritional qualities of fruits and leaves. In most respects, human vision does not appear to be much different from that of other diurnal higher primates, including the apes. However, the interactions between eyeball growth and skull growth have important ramifications for how the head grows in primates in general and in humans in particular. The following review thus begins with the development and anatomy of the eye and then considers how the eyes grow and function in conjunction with other parts of the head.

Ocular Function, Development and Anatomy

Eyes detect light—that is, visible electromagnetic radiation whose elementary particle is the photon. Light behaves as both a particle and wave, but for our purposes the two most important parameters of light are the frequency of the waves (wavelength), which we perceive as color, and the amplitude of the waves (intensity), which we perceive as brightness. For the eye to work, it needs a lens to admit and focus light, and a photon detector, the retina. The eye has many other tissues that increase its functionality, including the cornea, iris, pupil, and ciliary muscles, all of which help focus and filter light; the choroid plexus, which supplies blood and removes waste; the extraocular muscles that move the eye; and fluids that fill the eyeball.

The components of the eye are so elegantly integrated that opponents of Darwinian evolution point to it as evidence against natural selection. However, the eye has a well-documented evolutionary history, from a simple patch of light-sensitive pigment cells, still present in flatworms, to highly complex forms (see Fernald, 2006). The many stages by which ocular complexity evolved are reflected in how the eye develops through a series of reciprocal interactions between cells of different origins. These interactions, illustrated in Figure 10.1, begin with a pair of projections, the optic stalks, that differentiate from the forebrain (specifically, from the diencephalon, between the brain stem and the prosencephalon). Each stalk, which later becomes an optic nerve, swells at its distal end. The swellings (optic vesicles) then induce the overlying surface ectoderm of the face to differentiate and grow into a pair of thickened disks, the lens placodes. Next, each placode expands and invaginates toward the optic vesicle to form a spherical structure, the future lens. The initially hollow lens subsequently detaches from the surface ectoderm and becomes partially sur-

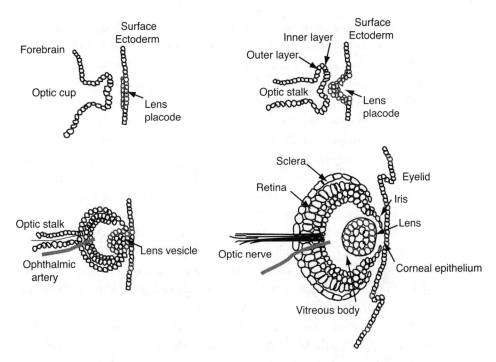

Figure 10.1. Major stages of optic development, showing how interactions between neural tissues and the lens placode of the ectoderm result in the development of the eye's major structures.

rounded, like an egg in an eggcup, by the growing optic vesicle—now called the optic cup. The lens then develops from many layers of elastic protein fibers. The inner cell layer of the optic cup becomes the light-sensitive layer of the retina; the outer layer becomes the pigmented layer of the retina. All the other tissues of the eye (including the cornea, sclera, iris, and ciliary muscles) develop from cells that invade the space between the retina and lens (for reviews, see Sperber, 2001; Edward and Kaufman, 2003).

Figure 10.2 illustrates the path of photons within the eye. The first tissue through which they pass is the cornea, a transparent, convex disc of fibrous tissue. Like the crystal on a wristwatch, the cornea protects the lens, and its curvature helps collect and bend light onto the retina. Corneal convexity is maintained by the aqueous humor, a thin, watery fluid (similar to cerebrospinal fluid) that nourishes and fills the space between the lens and the cornea. Because the cornea is bent and has a different refractive index from air, light rays from a single source hit the cornea at different angles and bend to different degrees, converging on a small point on the lens.[1] On their way to the lens, these waves pass through the iris and pupil. The iris is a tinted, fan-shaped muscle with both circular and radial muscle fibers around the pupil, an O-shaped, central sphincter. These muscle fibers function like the aperture of a camera, dilating or constricting the pupil to regulate how much light enters the eye (too much light can damage the retina). Contractions of the central, circular muscles constrict the pupil; contractions of the outer, radial fibers dilate it.

Most focusing occurs in the cornea, but the lens fine-tunes focus. The lens is a biconvex, flexible plate, typically 5 mm thick and 9 mm wide, made of transparent proteins called crystallins (their refractive index is about 1.4). The lens is suspended in the eye by a circular arrangement of many slender fibers, the zonules, that emerge from the delicate ciliary muscles.[2] Because the lens is naturally convex, when light from a distant object passes through it, the ciliary muscles relax, and natural pressure within the eye tenses the zonules, flattening the lens to reduce its refractivity. When we try to focus on a close-up object, the ciliary muscles contract to slacken the zonules, allowing the lens to return to its natural curvature and to accommodate closer focal points (Croft and Kaufman, 2006). As we age, the lens hardens, making the zonules' task more difficult.

The final leg of the light wave's journey before it hits the retina is through the vitreous humor, a clear, viscous fluid (about the consistency of egg white) that pressurizes the eyeball to help maintain its shape. The retina is the deepest of three layers that make up the wall of the eyeball. The outer layer is the sclera (which means "hard"), a thick sheet of white connective tissue that covers every part of the eyeball except the cornea. The sclera protects the eyeball, helps maintain its shape, and is visible around the iris as the white of the eye. The middle layer is the richly vascularized choroid, which supplies oxygen and removes waste from the retina. The choroid is dark and thus also helps to minimize reflections inside the eyeball.

Finally, light reaches the retina, the photoreceptive layer of the eyeball (Figure 10.3). The retina has several layers, and, oddly, photon perception takes

1. Refraction is the change in a wave's direction of travel caused by a change in velocity at the boundaries of substances with different properties (measured as the refractive index, RI: for example, RI is 1.00 for air and 1.33 for water). Refraction does not alter a wave's frequency but can alter its wavelength.

2. The eponymous name for these fibers is the zonules of Zinn, because they were first described by the eighteenth-century German anatomist and botanist Johann Gottfried Zinn (whose name also lives on botanically in the zinnia).

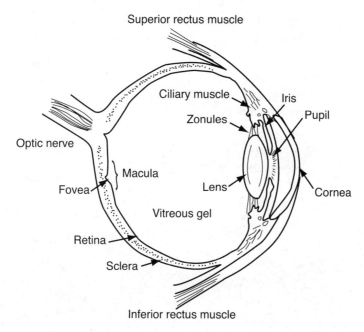

Figure 10.2. Schematic view of eyeball anatomy.

place in the outer layer (this is not the case in most invertebrates). There are two kinds of photoreceptor cells, rods and cones, both of which differentiate from the inner layer of the optic cup. The proportion of rod and cones cells varies among animals, but rod cells are more numerous (about 120 million in each human eye) and are distributed all over the retina (Ahnelt, 1998). Rods detect light intensity at low levels and contribute little to color perception (see Jacobs, 1993). Cones (of which there are about 6 million per eye in humans) are present throughout the retina but are concentrated in a small, central region (1.5 mm in diameter), the macula. A derived feature of tarsiers and anthropoids (Rohen and Castenholz, 1967), the macula gives us the sharpest vision; the center of the macula, the fovea, is a shallow pit with a particularly high density of cones from which the blood vessels are diverted away, thereby decreasing potential light refraction. The fovea works like stadium seating in a movie theater, helping pack many cones into a small space without their obstructing each other.

When a photon stimulates a rod or cone cell, the cell body sends an electrical impulse (an action potential) down its axon to a synapse with a second layer of cells, the bipolar layer, which then transmits the stimulus to the optic

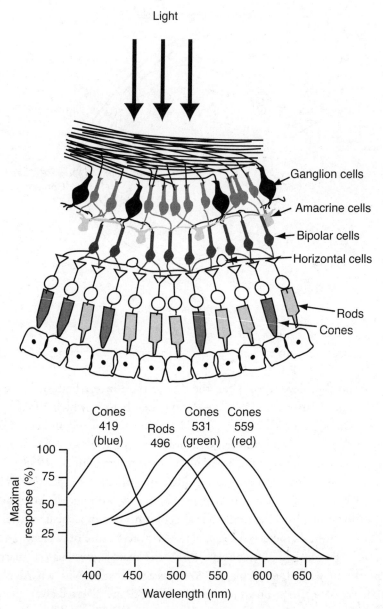

Figure 10.3. Top: Schematic view of the retina's structure. Rods and cones detect light and send information to retinal ganglion cells, which then transmit the information to the brain via the optic nerve. Several cell types (horizontal, bipolar, and amacrine cells) integrate and regulate signals produced by the rods and cones. Bottom: graph of the wavelength sensitivity of the cones and rods (numbers indicate peak sensitivity).

nerve. Between 6 and 600 neighboring rods synapse with a single bipolar cell, whereas each cone synapses with just one bipolar cell. The coalescence of many rods into a single signal accounts for their high degree of sensitivity but low acuity; in contrast, the one-to-one ratio of cones to synapses accounts for their high acuity and low sensitivity, especially in low light. The distribution of rods throughout the retina gives them a major role in peripheral vision and in sensing movement.

Rods and cones sense light via photosensitive pigments (Figure 10.3). Rods contain a single pigment, rhodopsin, which converts temporarily to vitamin A when exposed to light, triggering a nerve to fire. Rods mostly absorb light in the blue range and cannot detect wavelengths in the red range (above 600 nm), leading to monochromatic perception. Cones, in contrast, contain one of three pigments, opsins, which are sensitive to light in the short (blue), medium (green), and long (red) wavelengths. We discern different colors by integrating the relative degree of excitation from cones with different photopigments. A green object such as a leaf, for example, will excite both medium- and long-wavelength-sensitive cones.

Not all primates have color vision like that of humans. Prosimians, which are mostly nocturnal, have only two kinds of opsins in rod-dominated retinas that are well adapted for night vision and very sensitive to movement. Sometime during the Eocene (55–34 mya), monkeys apparently evolved a more cone-dominated central retina with a macula as well as a fovea. These derived features probably helped early monkeys become diurnal, and they assist all living monkeys to forage for fruits and leaves (R. D. Martin, 1990; Cartmill, 1992; Ross, 2000; Heesy and Ross, 2001). Diurnal anthropoids also have smaller corneas relative to eyeball size than prosimians (Kirk, 2004). More recently, trichromatic vision evolved in Old World monkeys. Trichromatic vision helps monkeys perceive subtle differences in the red and green wavelengths. Some researchers think that trichromatic vision evolved as an adaptation for selecting the ripest fruits (Wolf, 2002). Others researchers suggest that it evolved primarily to help monkeys select the youngest, most tender leaves during seasons when fruits are scarce (Dominy and Lucas, 2001; Dominy et al., 2003).

Human color vision appears to differ subtly from other higher primates. According to chromatic analyses of the retina as well as studies of genetic variation in the opsin genes, humans have more polymorphic photopigments and more variation in the distribution of different opsins within the fovea than other primates (Jacobs, 1996; Roorda and Williams, 1999; Verrelli and

Tishkoff, 2004). Compared to chimpanzee retinas, human retinas are more polymorphic, slightly less sensitive to the short (blue) wavelengths, and slightly more sensitive to the long (red) wavelengths (Jacobs et al., 1996). Apparently there is more variation in color perception within humans than within apes and monkeys (Alpern, 1979). Why this should be so is unclear, but it may reflect a relaxation of selection on abilities to optimize color perception so that one picks just the right fruit or leaf. This presumably also accounts for the 7–8 percent of human males who have only partial or absent color vision. One can infer that the fitness cost of reduced color perception is lower in males than females, perhaps because males engaged less in foraging.

Seeing in 3-D

Another important feature of vision is stereoscopy, the ability to perceive objects in depth. Stereoscopy derives from several aspects of how we process visual stimuli. One set of tools available to all mammals is the visual clues perceptible with a single eye about distance and relative position. Monocular cues include relative size (more distant things are smaller), clarity (more distant things are often fuzzier), and parallax (more distant objects appear to move less when we change position). Full depth perception, however, requires binocular vision, in which the fields of view of the two eyes overlap. A few details about how the brain processes visual signals help explain how binocular vision works. Stimuli from rods and cones in each portion of the retina (a receptor field) are summed up by nerve-cell bodies in the eye and sent to the optic nerve; they arrive in a region of cell bodies in the thalamus called the lateral geniculate nucleus (LGN). The LGN then sends the information to the visual cortex of the occipital lobe. Importantly, some axons in the optic nerve cross over to the opposite side of the brain, but some remain on the same side before ending up in the visual cortex. Thus each side of the visual cortex receives stimuli from each eye for the overlapping portion of the visual field. Famous experiments by Hubel and Weisel showed that cells in the visual cortex of the occipital lobe create a topographic map of the visual field for particular optical phenomena, such as edges of light (see Hubel, 1988). Ultimately, the cells integrate the similarities and differences from each eye to produce a 3-D stereoscopic picture.

The degree of overlap (the binocular field) of the eyes is an important factor in head design. As Figure 10.4 illustrates, this overlap is largely a function of three aspects of skull shape: the distance between the eyes (interorbital dis-

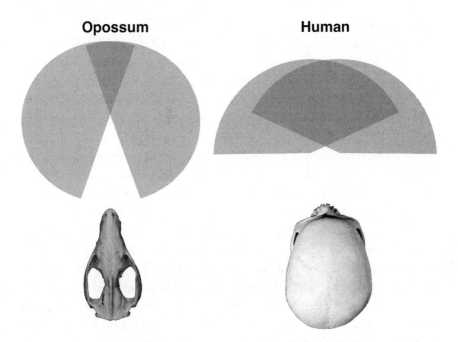

Figure 10.4. Differences in the overlap of visual fields in a nonprimate (an opossum) and a modern human. Greater orbital convergence in the human leads to a much larger degree of binocular overlap and more stereoscopic perception.

tance), the orientation of the orbits relative to the horizontal plane (frontation), and the degree to which the orbits point toward the front of the face (orbital convergence). Of these factors, convergence is the most important. Following Cartmill (1972), orbital convergence has generally been quantified as the dihedral angle (an angle formed by two planes) between the midsagittal plane and the orbital plane (the plane formed by the anterior margins of each orbit, similar to that formed by each glass in a pair of eyeglasses). This angle correlates well with binocular-field overlap (Heesy, 2004). The closer to 90° the dihedral angle is, the greater the angle of binocular visual field. Orbital convergence in humans is about 80°, causing about 140° of the visual field to be binocular (peripheral vision is not stereoscopic) (Bruce et al., 1996). This high degree of binocular vision is typical of primates (especially higher primates) and carnivores because it helps them capture prey. In contrast, most prey species have more laterally facing eyes to maximize their field of vision, but at the expense of depth perception.

As noted above, all primates have good binocular vision, reflecting the important role of predation in early primate history, and there is no evidence that ape or human stereoscopic vision is special in any way. As with many aspects of skull shape, the evolution of orbital convergence in primates is a story of changes in the scaling relationships between eyes, faces, and bodies. Across mammals and primates, orbit size scales with negative allometry (Kay and Cartmill, 1977; R. D. Martin, 1990), partly because eyeballs scale with strong negative allometry to body size (but see below). In turn, smaller-bodied animals with relatively bigger eyes and orbits have less orbital convergence because the lateral margins of the orbits cannot fit in the front of the face and are positioned more toward the back (Cartmill, 1972). However, anthropoids have more convergent orbits for a given body size than prosimians and other mammals, probably because the earliest monkeys were small, diurnal predators (Ross, 1996; Martin and Ross, 2005; Ross and Martin, 2006). Orbital convergence may also help highly social monkeys focus on facial features (Allman, 1977). Anthropoids also have more vertically oriented (frontated) orbits than prosimians, probably because of a relatively larger neocortex and more flexed cranial base (Ross and Ravosa, 1993). These factors orient the orbits more downward (ventrally) over the snout.

Fitting Eyeballs in Orbits and Faces

Eyeball size scales with negative allometry to both brain size and body size across mammals and within primates (see Figure 10.5a) (Howland et al., 2004; Kirk, 2004). Although bigger animals have relatively smaller eyes, eyeball volume in adult higher primates does not correlate strongly with either body mass or orbit volume (Figure 10.5b). During early fetal growth, the eyeball grows relatively faster than the upper face, acting like a balloon and creating tension in the many sutures of the orbit that stimulate growth (Sarnat, 1982). However, the growth rate of the skeleton surrounding the eye soon outpaces that of the eyeball, so that by birth the orbit is already much larger than the eyeball. According to Schultz (1940), the volume of a human neonatal eyeball is approximately 3.15 cm^3, slightly less than half (45 percent) the volume of the orbit; in adults, eyeball volume averages 8.2 cm^3, approximately 32 percent of the orbit's volume (24 cm^3). For comparison, mean eyeball size in adult chimps is a little smaller than in humans, about 6 cm^3, but orbital volume is about the same, approximately 24 cm^3 (Schultz, 1940). In other words, the eyeball fills a slightly larger volume of the orbit in humans (32 percent) than in chimpanzees (25 percent). These data highlight the lack of a strong rela-

Figure 10.5. Eyeball volume relative to (a) body mass and (b) orbit volume among primates. (Data from Schultz, 1940.)

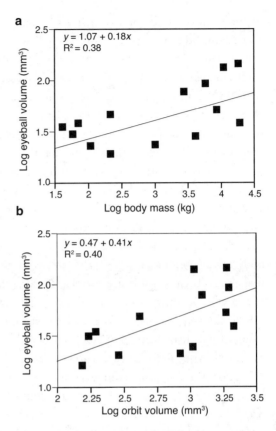

tionship between the size of the eye and the orbit in larger-bodied higher primates, such as humans and apes.

The partial independence of eyeball and orbit size in apes raises several interesting points. First, most of the cone-shaped orbit is filled with substances and structures other than the eyeball. The other contents include fat, the optic nerve, additional nerves, blood vessels, the lacrimal gland and its ducts, and seven muscles. Six of these are the extraocular muscles (the superior and inferior recti, the medial and lateral recti, and the superior and inferior obliques) that act in combinatorial pairs to rotate the eyeball. Control of eye movement includes both voluntary motor control and the vestibular ocular reflexes (VOR) that coordinate movements of the eyeballs with changes in head position (see discussion in Chapter 9). The seventh muscle in the orbit is the levator palpebris superioris, which elevates the eyelid. All the orbital muscles originate from a single tendinous ring around the optic foramen.

Another point about the poor correlation between eyeball and orbit size is made in Chapters 4 and 5: that orbits play many roles in upper facial growth in addition to housing the eyeball. Although the orbit is the primary skeletal matrix around the eyeball, the orbital walls are widely shared with other functional spaces. The roof of the orbit is the floor of the anterior cranial fossa; the orbit's medial wall is the lateral wall of the nasal cavity; and the lateral wall is part of the temporal fossa (the origin of the anterior temporalis). Indeed, six bones contribute intramembranous plates that make up the walls of the orbit: the ethmoid, sphenoid, frontal, zygomatic, maxilla, and lacrimal. These bones meet in about a dozen sutures that often vary in position, permitting growth in just about every direction (see Cartmill, 1971). It follows that many aspects of facial growth can contribute to the growth of the orbit, leading to widespread covariation between orbital size and variables such as body mass and brain volume. Among anthropoids, orbit size (usually measured as the area of the orbital aperture) correlates reasonably well with both brain volume and body mass (Aiello and Wood, 1994; Kappelman, 1996). These correlations make sense because orbit size correlates well with face size, which in turn correlates with body size and hence brain size (see Figure 4.11). Aiello and Wood (1994) showed that one can predict a species' mean body mass reasonably well from orbital area (although predictions of individual body mass have high errors). And, as is often the case with correlations, there are illuminating exceptions. For example, orbital dimensions underestimate body mass in orangutans because their upper faces are unusually shallow (Schultz, 1940). In addition, Kappelman (1996) estimated body mass from orbit size in archaic *Homo,* such as *H. heidelbergensis* and *H. neanderthalensis,* to be 80–120 kg (176–265 lbs)! These sumo-sized estimates are unsupported by more direct estimates of body mass from the postcranium (Ruff et al., 1997). Instead, as Figure 10.6 illustrates, species of archaic *Homo* had large (especially tall) faces, hence large orbits, relative to overall size (more on this in Chapter 13).

Finally, the eyeball is a fragile organ that functions with extraordinary precision in an environment that can experience dynamic loads. The retina is at particular risk. The outer, pigmented layer of the retina is firmly attached to the choroid but only loosely attached to the inner neural layer, which is largely held in place by pressure created by the vitreous humor on the inside of the eyeball. High impulsive forces from blows to the head can sometimes produce enough shearing force to separate the two layers and tear the retina.[3] Orbital

---

3. Severely nearsighted people are more susceptible to such retinal tears because their eyes are anteroposteriorly longer than average, stretching the retina and causing it to be more fragile (David et al., 1997). Until recently, myopia was very uncommon (incidence rates are less than 2 percent in hunter-

H. sapiens   **H. heidelbergensis**

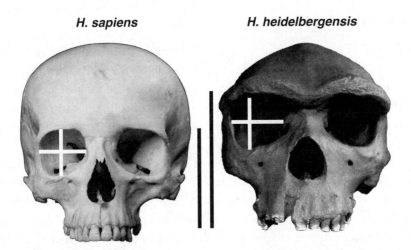

Figure 10.6. Anterior views of the face in (left) recent *H. sapiens* and (right) *H. heidelbergensis* (Kabwe), scaled to the same overall height. The face is relatively taller (solid bars) in *H. heidelbergensis,* contributing to a relatively larger (especially taller) orbit.

fat is therefore useful not only as a space filler but also as a shock absorber (Schutte et al., 2006). Another force to consider is chewing, which can generate significant stresses in the face (see Chapter 7), and which is hypothesized to account for another unique aspect of the anthropoid orbit, the postorbital septum. All mammals other than higher primates separate the orbital cavity from the temporal fossa with a sheet of fascia. Anthropoids and tarsiers, however, have replaced this soft-tissue wall with bone, the postorbital septum, which is assembled from portions of the sphenoid, frontal, and zygomatic bones. One hypothesis is that the postorbital septum evolved as an adaptation to resist deformations in the face (Cartmill, 1980; Greaves, 1985; Rosenberger, 1986). However, experiments by Ross and Hylander (1996) found little evidence that the septum resists chewing-induced strains that are effectively resisted by the postorbital bar (the pillarlike lateral wall of the orbit). A more likely explanation is that the septum evolved in early anthropoids with relatively small skulls in order to insulate the eyes from movements in the temporal fossa generated by the anterior temporalis (Cartmill, 1980; Ross, 1996).

---

gatherers), indicating that the cause is predominantly environmental (see Cordain et al., 2002, for some interesting hypotheses).

Eyebrows and Eye Color

A final word on two derived features of human eyes: eyebrows and eye color. Eyebrows are arched, thickened bands of skin above the orbits into which are set numerous, obliquely oriented hairs. Three muscles of facial expression interdigitate with the eyebrow dermis: the orbicularis oculi, the corrugator, and part of the fronto-occipitalis. These muscles, respectively, permit one to close the eyelids, wrinkle the brows, and raise the eyebrows. Eyebrows are not unique to humans, but human eyebrows are distinctive because of their high density of hair combined with the lack of neighboring hair. Why the human eyebrow is so distinctive is hard to test. One hypothesis is that the thickened dermis acts as a shock absorber to help protect the upper face from blows. This seems unlikely. Another hypothesis is that eyebrow hair, which is oriented laterally, evolved in hominins after we became furless in order to channel sweat sideways away from the eyes. Third, the eyebrows may have evolved to aid in nonverbal communication, not only in terms of their ability to move expressively, but also by providing some resistance for the muscles that insert into them. A final hypothesis is that eyebrows, attractive as they are, were favored by sexual selection. Perhaps someone will undertake the experiments necessary to test these hypotheses.

Human eyes also have distinctively shaped outlines and color. The visible outline of the human eye is wide compared to that of other primates and shows a much greater percentage of the sclera; in addition, whereas most primates have pigmented sclera that approximate the color of the nearby facial skin, humans have an unpigmented, white sclera (Kobayashi and Kohshima, 1997). Wide eyes with lots of sclera help one to change the direction of gaze quickly in the horizontal plane by moving the eyes rather than the whole head. This may be an adaptation to having a relatively narrow visual field combined with living in open habitats, a hypothesis supported by comparative evidence that animals in terrestrial habitats tend to have relatively wider eye outlines and relatively smaller irises (Kobayashi and Kohshima, 2001). In addition, plentiful visible white sclera helps other humans deduce where an individual's gaze is directed and thus may be an adaptation to communicate silently using gaze signals (Kobayashi and Kohshima, 2001).

Hearing

The second dominant sense in humans and other primates is hearing. The excellent hearing capabilities of humans are largely a legacy of being mammals, which as a group have highly derived and evolved middle and inner ears compared to reptiles. The evolution of mammalian hearing involves some famous examples of tinkering, including co-opting part of the reptilian jaw to form the middle ear. The anatomy of hearing also relates to cranial form in humans and other higher primates. Two topics are of major concern: the acoustic perception of speech and sound localization. Before considering these functions and what we know about evolution of hearing in human evolution, I first review the acoustic basis of hearing and the anatomy of development of the auditory system.

Sound Perception

Sounds are pressurized waves of molecules—usually in air but also in fluids—that we sense with our ears. Movement, such as air being pushed through the vocal folds or a foot striking the ground, generates alternating waves of higher and lower pressure. The qualities of the sounds that we perceive derive from several characteristics of these waves (see Figure 10.7). The amplitude of the waves determines the volume, which we measure in decibels (dB, the logarithmic ratio of the volume of a given sound and the minimal volume one can perceive). The frequency of the waves determines the pitch, which we measure as vibrations per second (or hertz, Hz). Most sounds generate a range of simultaneous frequencies, including harmonics (multiples of the original frequency) that together determine the tone or timbre of the sound (Chapter 8). As Figure 10.7 illustrates, most humans can hear sounds within the range of about 20–20,000 Hz (Moller and Pedersen, 2004), but the majority of sounds that we listen to, including speech, are between 100 and 7,000 Hz. Our most acute hearing occurs in the range of 2,000–4,000 Hz.

Experiments on other primates suggest that the human hearing range differs slightly from that of other monkeys and may be subtly different from that of chimpanzees. In general, Old World Monkeys, such as macaques, have hearing ranges that are shifted toward higher frequencies, in the range of 60–32,000 Hz (Heffner, 2004). Humans and monkeys hear most of the same sounds, but monkeys have more sensitivity to higher frequencies and less to extremely low frequencies. Two experiments on chimpanzees (Elder, 1935; Kojima, 1990) suggest that the chimp hearing range also may be shifted slightly toward higher frequencies than that of humans, with less sensitivity between 1,000 and 4,000 Hz (see Figure 10.7). Much has been made of these differences, but there are several reasons not to overinterpret the chimpanzee audiograms. First, the chimpanzee experiments were very limited and done using

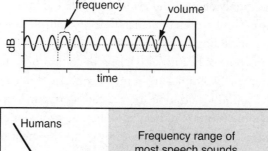

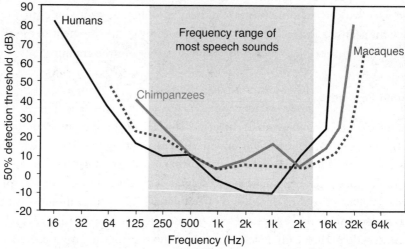

Figure 10.7. Top: illustration of two key properties of sound waves: frequency, which we perceive as pitch; and amplitude, which we perceive as volume and is measured in decibels (dB). Bottom: audiogram of human, chimpanzee, and macaque hearing (adapted from Heffner, 2004). The X axis is sound frequency; the Y axis is the volume in dB at which a sound can be detected at least 50 percent of the time. The chimpanzee data come from Kojima (1990) and Elder (1935). Humans and chimpanzees have generally similar perception within the range of most speech sounds (shaded area). As is typical, hearing in the smaller mammals is shifted toward slightly higher frequencies.

only headphones, thus not taking into account any amplifying effects of the earlobe (see below); they need to be replicated using larger samples and with loudspeakers. Second, the differences reported by Kojima (1990) are subtle. According to these data, the difference in chimpanzee and human perception of frequencies between 4,000 and 10,000 Hz is about 30 dB SPL (sound pressure level), approximately the difference in sound level between the threshold of hearing and a whisper at 5 meters. It is reasonable to conclude that chimpanzees have effectively the same range of hearing as humans for the vast range of sounds we encounter, including speech, but with slightly more sensitivity to extremely high frequencies that most humans cannot perceive.

Functional Anatomy of the Ear

Sound perception occurs as waves pass through the ear, which has three distinct portions: the outer, middle and inner ear, illustrated in Figure 10.8. Each of these components has distinct functions and embryonic origins, reflecting the ear's complex evolutionary history.

Outer Ear

The outer ear collects sounds and transmits them into the cranium. The earlobe, or pinna, which derives from swellings of ectoderm in the first and second pharyngeal arches, functions like a funnel (or an ear trumpet), trapping and reflecting sound waves into the central part of the earlobe, the concha, which narrows to become a cartilaginous tube. Many mammals, including most prosimians, have mobile pinnae that can be directed toward a sound without requiring the animal to move its head; most humans, like most monkeys and apes, have immobile pinnae. If we wish to amplify a sound, we must point our heads toward its origin or cup the hands behind the earlobes. In general, mammals whose hearing range is more sensitive to high frequencies have more pointy (tall and narrow) earlobes that decrease low-frequency reception (Rosowski, 1994). Humans, like other apes and monkeys, have relatively square pinnae, typical of mammals whose hearing is adapted to detect a wide range of frequencies, including low ones (Coleman and Ross, 2004).

The outer third of the external auditory passage (from the pinna to the tympanic membrane) is cartilaginous and lined with follicles and glands. In apes and Old World Monkeys, the ear canal becomes a bony tube lined with skin near the level of the external auditory meatus (see R. D. Martin, 1990, for details on variations). This bony part of the external ear acts as a resonating amplifier, much like a tunnel or the body of a musical instrument. When sound enters a resonating amplifier, the waves of energy bounce off the walls, increasing in amplitude at certain frequencies and decreasing in amplitude at others. These filtering characteristics are influenced by the length of the tube (longer tubes amplify lower frequencies; the peak resonance of a wave is 4 times the tube's length), and by the size of the tube's opening (smaller openings filter more frequencies) (Fant, 1960). On average, the bony tube of the external ear canal is 26 mm long, with a volume of 1 cm^3 and an opening of 115 mm^2 (Alvord and Farmer, 1997). These dimensions amplify frequencies between 2,000 and 7,000 Hz, with a peak gain (12 dB) between approximately 3,400 and 3,900 Hz (Ballachanda, 1997). This peak is above most of the pri-

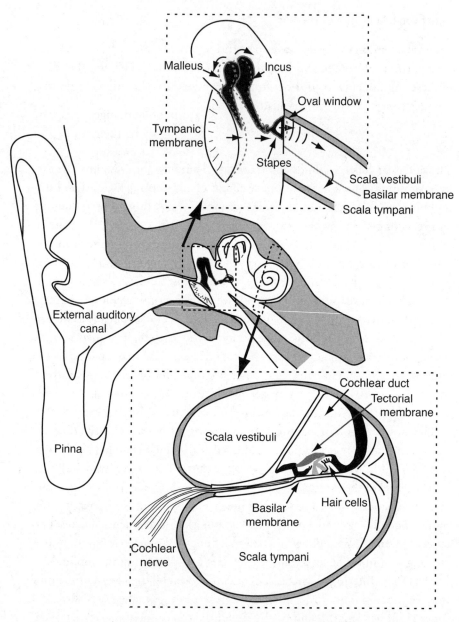

Figure 10.8. Anatomy of the outer, middle, and inner ear. Top: expanded schematic view of the ossicles, showing the bent-lever amplifier action that transmits vibrations of the tympanic membrane to the oval window. Bottom: expanded schematic cross-section of the cochlea. Sounds travel up the scala vestibuli, causing vibrations of hair cells in the inner ear.

mary formant frequencies of speech (see Chapter 8) and mostly enhances perceptions of speech sounds with middle to high frequencies, such as /s/ (as in *hiss*). Human ear canals differ from those of chimps, which are about 34 mm long and have a much smaller opening, approximately 44 mm² (M. Leakey et al., 1995; Martinez et al., 2004). The narrower ear canals of chimpanzees might account for their slightly lower perception of midrange-frequency sounds (although this difference would be offset slightly by their longer ear canals). According to experiments by Chandler (1964), decreasing the diameter of the earhole by approximately one-half results in a hearing loss of 1 dB at 1,000 Hz, 13 dB at 2,000 Hz, and 14 dB at 4,000 Hz. This sound loss is close to the audiogram levels measured in chimpanzees by Kojima (1990).

Middle Ear

Once sound waves reach the end of the outer ear, they are transmitted into the middle ear (an air-filled space within the petrous temporal) via the tympanic membrane (or eardrum). This mostly taut, fibrous membrane is stretched around the ringlike ectoympanic bone (a derivative of the angular bone of the reptilian jaw). Like a drum, the tympanic membrane bows inward and then deflects backward with every wave of pressure. The membrane presses against a chain of three middle ear ossicles: the malleus (hammer), incus (anvil), and stapes (stirrup). These tiny bones grow from the ends of the cartilage rods of the first and second pharyngeal arches and are the first bones in the skeleton to reach adult size, about 25 weeks postconception (Scheuer and Black, 2000). The ear ossicles function as a bent-lever amplifier. As illustrated in Figure 10.8, the long "handle" of the malleus is attached to the tympanic membrane. When this handle is pushed inward, it rotates the head of the malleus, which then rotates the long process of the incus medially. In turn, this movement pushes the stapes against the oval window, a membrane that separates the middle and inner ear. Since force is mass times acceleration, and pressure is force per unit area, the pistonlike action of the ossicles and the tiny size of the stapes' footplate (about one-tenth the surface area of the tympanic membrane) amplify sound energy by approximately 10-fold (Kurokawa and Goode, 1995). This amplification is critical for effective hearing because the inner ear is filled with fluid. It takes approximately 134 times more pressure to generate the same velocity of particles in the fluid of the inner ear than in air (Zwislocki, 1986).

Two other details of the middle ear are worthy of mention. First, like the outer ear, the middle ear is filled with air. Thus, for the tympanic membrane to function optimally, the pressure in the middle and the outer ear needs to be the same. Pressure equalization is made possible by the pharyngotympanic (Eustachian) tube—the remnant space left behind by the first pharyngeal pouch —that connects the pharynx and the middle ear. Although the pharyngeal opening of this tube is usually very small, actions like sneezing, swallowing and yawning help open it. Two tiny muscles also help to modulate amplification of loud sounds by the ear ossicles, preventing damage to the inner ear. First, the tensor tympani pulls medially on the handle of the malleus, lessening the tension on the tympanic membrane; second, the stapedius muscle pulls back on the stapes, damping its action.

Inner Ear

The final stage of a sound wave's journey into the head occurs in the cochlear part of the inner ear, which transduces the mechanical energy of the amplified sound waves into neural stimuli for the brain. The cochlea is a spiral-shaped tube of bone with several chambers (Figure 10.8). Sound waves from the middle ear propagate through the fluid of the cochlea every time the stapes knocks against the membrane of the oval window that separates the middle and inner ear. The lower the frequency, the further the wave propagates up the cochlea. As the wave moves up the cochlea, it causes vibrations that bend sensory hair cells within the organ of Corti in the central part of the cochlea (Figure 10.8). This complex organ has a basilar membrane from which project thousands of tiny hair cells (in four rows) that contact an overlying tectorial membrane. When the basilar membrane vibrates, the hair cells bend against the tectorial membrane, stimulating the cell nuclei to send an action potential up the cochlear nerve to the brain. Importantly, the hair cells are tuned to different frequencies, with higher frequencies received near the oval window and lower ones with increasing distance away from the oval window.

Hearing and Speech

Given the uniqueness and importance of speech in human evolution, there has been much effort to find adaptations in the human ear for speech perception. However, the acoustic range of human speech is not particularly special. If anything, natural selection appears to have geared the production of speech to the capacity of the auditory system and not the other way around.

As Chapter 8 describes, phonation occurs when air is pushed through the larynx during the production of speech. When we interpret phonations as speech rather than noise, we perceive several kinds of acoustic energy. First, energy is present at the fundamental frequency of phonation (F0), the quasi-periodic series of puffs of air that emerge from the larynx, and its harmonics (integer multiples of F0). F0, the "pitch" of a person's voice, ranges from approximately 60 Hz in adult males to 1,500 Hz in young children (Keating and Buhr, 1978). Second, we perceive the frequency patterns and spectral peaks that derive from filtering within the supralaryngeal vocal tract (SVT, the airway above the larynx). Recall that movements of the tongue and lips alter the way the SVT lets through local energy maxima at peak frequencies, known as formant frequencies (F1, F2, etc). Almost all vowel perception is based on the first three formant frequencies, which range from about 200 to 4,000 Hz (P. Lieberman, 2006). The perception of consonants depends on much the same frequency range but is enhanced by taking account of the acoustic energy generated by air turbulence, which includes frequencies as high as 7,000 Hz. Thus, the most critical frequency region for adult speech is between 700 and 3,000 Hz; even better quality perception occurs with larger bandwidths that surpass 5,000 Hz.

The frequency range of human speech sounds fall well within both the human and the chimpanzee auditory range (Figure 10.7). The only difference is that the latter may have slightly less sensitivity to frequencies between 2,000 and 4,000 Hz; chimpanzees also appear to be slightly more sensitive to frequencies around 8,000 Hz, beyond the range of most speech (Kojima, 1990). However, the chimpanzee audiogram data need to be taken with a grain of salt and considered preliminary until confirmed with larger samples and better methods of testing. If correct, they indicate that chimpanzees are able to perceive most human speech but might be slightly less able to perceive some components (notably, the second formant frequencies of some vowels) at low signal levels in the presence of environmental noise. In other words, chimps might have a harder time hearing speech at a cocktail party or perceiving whispers.

Hearing and Sound Localization

Another, more primitive function of hearing is to help animals monitor the world, especially the presence of predators. Have odd noises in a dark or scary place ever made you suspect that you might be in danger? Such noises can trigger a basic nervous response: suddenly you seem to hear more clearly, and the faintest noise causes you to turn in its direction. These instincts illustrate the importance of the ability to deduce the spatial origin of particular frequencies —from the deep thud of a woolly mammoth sneaking up behind you to the high-frequency hiss of branches rustling in the distance. Localizing the sources of sounds has obvious fitness benefits, not only for identifying danger but also for distinguishing particular sounds from a background of noise.

Most sound localization is binaural, involving comparisons of sounds between the two ears. We can do this in the horizontal plane in two ways. The first is to compare the difference in signal *amplitude* coming from each ear (the interaural level difference, ILD). If a sound is louder in the right ear than in the left, then the source is on the right side of the head. Such cues, however, work only for high frequencies in relation to the diameter of the head, which typically casts a "sound shadow." Frequencies below 500 Hz have wavelengths greater than 69 cm, more than four times the diameter of a human head. Such waves therefore pass around the human head without being diffracted and cannot be used for ILDs. Therefore high-frequency waves, above 3,000 Hz in humans, provide the best cues for sound localization based on ILDs (Hartmann, 1997).

The other way we localize sound is from the difference in *timing* of signals coming to each ear (the interaural timing difference, ITD). If a sound wave reaches the right ear before it reaches the left ear, then the sound must be coming from the right. If we know the wave's frequency, then we can calculate the angle of origin within a few degrees (Hartman, 1997). However, in contrast to ILDs, spatial perception from ITDs improves as frequencies *decrease* relative to head width. For example, when a high-frequency wave comes from the right side of the head, the right ear will perceive it first, but almost immediately thereafter, other cycles of the wave will reach the left ear, making it hard to distinguish any timing differences. In general, wavelengths less than twice the diameter of the head give the best perception for ITDs: for a human, this means frequencies below 1,500 Hz. As one would imagine, wide-headed elephants use even lower frequencies than humans to localize sounds using ILDs, and chimps use slightly higher frequencies.

In addition to these thresholds, another factor to consider is sound attenuation, the reduction in the amplitude of a waveform (measured in dB) in relation to the distance it travels. In general, obstacles such as trees, vegetation, walls, and curtains attenuate high frequencies more than low frequencies (Aylor, 1972). High-frequency attenuation accounts for why one hears just the bass sounds of a neighbor's loud stereo through a wall. Therefore, mammals are generally better off localizing sounds with the lowest-frequency sounds that the widths of their heads permit.

Finally, mammals can sense the location of sounds in the vertical plane. One way of doing this is to tilt the head, thus allowing ILD mechanisms to compare the level of sounds that reach the upper and the lower ear. Additionally, many animals use subtle differences in the spectral peaks of sounds that come from higher and lower elevations, particularly when reflected by the earlobe (Middlebrooks and Green, 1991).

The physical constraints that determine the relationship between head width and ILD and ITD function provide some clues about the evolution of acoustic perception abilities. Most importantly, animals with wider heads tend to perceive lower-frequency sounds better than do animals with narrower heads. In fact, there is a highly significant, inverse correlation ($r = -0.792$; $p < 0.0001$) between interaural distance and the highest audible frequency that a mammal can hear (Heffner, 2004). At one end of the spectrum, elephants cannot hear frequencies much higher than 10,000 Hz; at the other end, mice and fruit bats can hear frequencies higher than 90,000 Hz (Heffner and Heffner, 1980). Humans are right on the curve where we ought to be. This trend is not because of scaling effects in the inner ear: small animals with small skulls do not possess similarly scaled middle-ear bones that transduce higher frequencies. Rather the effect is most simply explained by that fact that larger animals are better able to use low-frequency ITDs to localize sound. In other words, head width suffices to explain why humans hear a slightly lower range of frequencies than chimpanzees do.

Auditory Capacities of Fossil Hominins

If it is a challenge to measure the auditory response in nonhuman primates, imagine how hard it is do so in extinct animals. It seems an impossible task, yet we do have a few clues from the fossil record. If we assume that the last common ancestor of humans and chimps had apelike hearing capacities, then at some point during human evolution there was a slight shift toward perception of lower frequencies. And, if Kojima's (1990) experimental results are reliable, then there may also have been a subtle increase in the ability to perceive frequencies between 2,000 and 4,000 Hz.

One set of clues comes from the dimensions of the bony auditory canal of the external ear. The external acoustic meatus in humans has a large cross-sectional area (115 mm^2), about twice that of chimpanzees (45 mm^2) (M. Leakey et al., 1995). Large earholes may help humans amplify mid- to high-frequency signals a little more than chimpanzees (although this deficit in chimps may be offset by slightly longer ear canals). Interestingly, the shift to a larger, human-sized earhole did not occur as brains expanded and heads got wider in the genus *Homo,* but apparently happened earlier, during australopith evolution. The few known temporal bones of hominins from the Miocene and early Pliocene for which data have been published have small, chimp-sized external acoustic meatuses (M. Leakey et al., 2001). *Au. afarensis* has a larger meatus (x = 79.7 mm^2, n = 4) but with much variability (the range is 47.7–109.3 mm^2) (Kimbel et al., 2004). All known hominins younger than *Au. afarensis* (with the sole exception of *Kenyanthropus,* a possible contemporary) have earhole diameters in the modern human range (M. Leakey et al., 2001).

The length of the bony auditory canal also decreased during human evolution before brains started to expand, but in a slightly different pattern. If one uses tympanic length as a proxy for estimating auditory canal length, then early hominins such as *Au. afarensis* and the robust australopiths have longer, chimp-sized ear canals; in contrast, *Au. africanus* and early *Homo* have shorter tympanic bones, within the modern human range (Dean and Wood, 1982b; Kimbel et al., 2004). As noted above, longer auditory canals may amplify low- to midrange-frequency sounds slightly, but these variations in length partly reflect the degree to which the lateral cranial base is expanded from pneumatization and cresting. Shortening of the bony portion of the auditory canal in *Homo* may also have arisen from the anterior advancement of the foramen magnum between the ears. Finally, there is some preliminary evidence that the cochlea is absolutely and relatively larger in *Homo* than in *Australopithecus* (Spoor, 1993), which might imply a shift toward lower-frequency perception. These measurements, however, need further study.

In spite of their size and fragility, a few fossil hominin ear ossicles have been found, sometimes because they get trapped by sediment in the middle ear or fall through the oval window of the inner ear. Among Pliocene hominins, there is an incus (SK 848) from Swartkrans that belongs to *Au. robustus* (Rak and Clarke, 1979), and a stapes (Stw 151) from Sterkfontein that belongs to *Au. africanus* (Moggi-Cecchi and Collard, 2002). Each of these ossicles differs in its own way from those of modern humans. The *Au. africanus* stapes is small and chimplike; the *Au. robustus* incus is a little more like that of humans but has a curiously inflated process where it articulates with the malleus. It is difficult to know how to interpret the potential acoustic effects of these differences given the challenges of reliably modeling the biomechanics of sound transmission in the middle ear (Rosowski, 1994). For example, audiogram data from living primates correlate weakly with biomechanical models of sound transmission that incorporate ossicle lever-arm dimensions and the cross-sectional areas of the oval window and tympanic membrane (Coleman and Ross, 2004). Thus, ossicle dimensions may be more useful sources of information about taxonomy and phylogeny than about audition.

A number of ear ossicles have been found in more recent fossil hominins, including some archaic *Homo* fossils dated to about 350,000 years ago from the site of Atapuerca in Spain (Martinez et al., 2004), as well as several Neanderthals (Arensburg et al., 1981, 1996; Masali et al., 1992). Not surprisingly, these ossicles generally resemble those of modern humans. Their morphological similarities, along with other similarities in ear canal dimensions, suggest that completely modern human hearing capacities are at least as old as the Middle Pleistocene (see Martinez et al., 2004).

Putting the evidence together, it is reasonable to hypothesize that the shift toward a modern-human-like range of hearing might have been a two-stage process. First, earholes became larger sometime during the evolution of *Australopithecus,* prior to the evolution of *Homo,* augmenting the sensitivity of the auditory system to frequencies above 2,000 Hz. This shift may reflect a transition from forest to savanna habitats, in which the absence of trees would have decreased the attenuation of high-frequency sounds, thus making shadowing effects of the head at high frequencies (ILDs) more useful for sound localization. In other words, as hominins ranged into more open habitats, perhaps there was some selective pressure to take more advantage of ILDs to localize sounds from potential predators. Later, as heads got wider in *Homo* because of bigger brains, there was a slight additional shift toward the perception of low-frequency sounds, because wider heads are more able to use ITDs. The anatomical bases for this shift are not yet clear but may involve increases in cochlea length. Importantly, neither of these shifts would have had much effect on the ability to perceive speech. The derived features of human hearing, it would seem, are all about sound localization.

Taste and Smell

Almost all the air, water, and food that enter our bodies pass through the face. Long ago, animals evolved chemosensory mechanisms to glean information about the environment from the small molecules that these incoming substances contain. Smell and taste have critical defensive roles, alerting us to predators and poisons; they also provide much information about what is in our environment, and they reward us with sensory pleasures. Most animals also find their mates using chemosensory cues. Not surprisingly, these fundamental functions are processed to a large extent in "deep," primitive parts of the brain, including the limbic system (see Chapter 6). Hence a whiff of freshly baked bread or a lover's body can trigger primal emotions such as hunger or lust. Humans also recruit many other parts of the brain to analyze chemo-sensory stimuli, associating them with memory, words, and feelings. An arche-typal example, now cliché, is recounted by Marcel Proust in *The Remembrance of Things Past*. The narrator sips a spoonful of tea in which he has soaked a madeleine cookie: "No sooner had the warm liquid mixed with the crumbs touched my palate than a shudder ran through me and I stopped, intent upon the extraordinary thing that was happening to me. An exquisite pleasure had invaded my senses, something isolated, detached, with no suggestion of its ori-gin. . . . Whence did it come? What did it mean? . . . And suddenly the mem-ory revealed itself. The taste was that of the little piece of madeleine which on Sunday mornings . . . my Aunt Léonie used to give me. . . . Immediately the old grey house upon the street, where her room was, rose up like a stage set . . . from my cup of tea" (Proust 1913–27:48–51).

We all have such "Proustian" experiences (though rarely from tea and madeleines). Yet, it is typically thought that humans, like other higher pri-mates, have an average sense of taste and a below-average sense of smell com-pared to most mammals. The standard scenario is that as our eyes expanded and moved toward the center of the face, the interorbital space that houses the olfactory portion of the nose decreased in size, and we lost most of the genes that encode the receptors for smell. In addition, Old World monkeys have lost a functional vomeronasal organ that most other mammals use to detect pheromones. That said, several aspects of human olfaction suggest that hu-mans have much greater olfactory capabilities than we have given ourselves credit for, particularly in detecting odors from the mouth that contribute to flavor perception. Our olfactory and gustatory deprivation may derive, in part, from our uniquely fastidious habits. Adult humans may be the only mammals who do not regularly put all manner of objects in their mouths and who rarely stick their noses into things other than flowers, glasses of wine, and other pleas-antly smelly objects.

How We Sniff Smells

Olfaction detects odorants, small and light molecules (with molecular weights below 294) that are at least partially soluble in water and fat and have a high vapor pressure and a low charge. Odorants, which arrive in inspired air or are released from liquids and solids in the mouth, can travel to the olfactory receptors in the upper nose via two routes (Figure 10.9). The most direct route is orthonasal, whereby molecules travel in air through the nostrils into the nasal vestibule, through the nasal valve into the internal nasal cavity, and then up to the top of the nasal cavity. But to get there, the air has to bypass a pair of turbinates (or conchae). Recall from Chapters 4 and 8 that the turbinates consist of three pairs of scroll-like processes that project off the lateral walls of the nasal cavity. All are covered with a highly vascularized, mucous-rich epithelium to create three narrow passageways, meatuses, through which air can flow. The lower two meatuses are entirely respiratory in function, and act to humidify, warm, and filter air before it enters the rest of the pharynx, and then to recover moisture from cooled air during expiration (Hillenius, 1992). Olfaction occurs in a part of the superior meatus, the olfactory epithelium (see below). During normal nasal inspiration, only about 10 percent of the air drawn in with each sniff reaches the olfactory region (Hahn et al., 1993).

The other route is retronasal, in which odorants (typically in food or drink) pass through the oral cavity, go up behind the soft palate, and then traverse the nasopharynx to the superior meatus. If you laugh while drinking and get liquid up your nose, the liquid is traveling retronasally (but not all the way up to the top of the nasal cavity). The retronasal route is longer and more circuitous and thus not used for normal sniffing, but this pathway plays a vital role in conveying odors from foods and liquids in the mouth. Between bites and swallows, exhalation of air in the pharynx pushes odorants into the nasal cavity, where some proportion of them are sensed as odors. Foods therefore lose most of their flavor when one is deprived of smell, as when congestion of the nasal cavity blocks the passage of retronasal airflow.

Once an odorant reaches the upper nasal cavity, it may trigger the sensation of a smell if it binds to an olfactory receptor in the olfactory epithelium. This specialized patch of sensory epithelium in the superior meatus of the nose contains millions of olfactory neurons. Each neuron has a single dendrite that extends to the surface of the epithelium. Several cilia project from each dendrite and lie, bathed in mucus, on the surface, waiting for passing odorants to dissolve in the mucus and bind to them. The mucus layer, about 60 microns thick, contains other proteins that help transport and bind odorants (Ohloff, 1994). Each olfactory neuron has only one of about a thousand different types of specialized olfactory receptors (ORs), which bind more strongly to certain odorant molecules and less well to others (Zhao et al., 1998). Because of this relatively broad responsiveness, many odorants can be detected by more than one type of OR (Buck, 2000).

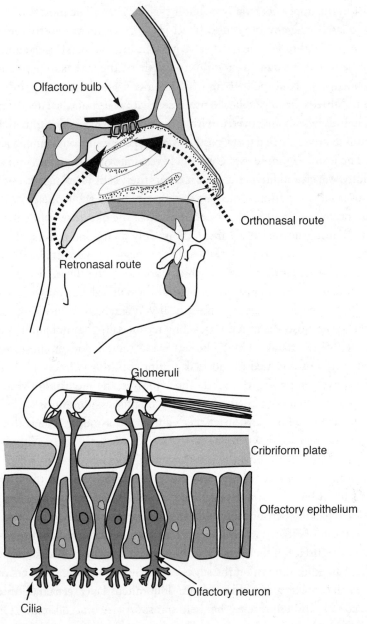

Figure 10.9. Parasagittal views of olfactory anatomy. Odorants can travel to the olfactory neurons in the olfactory epithelium via either the orthonasal or the retronasal route.

When an odorant binds to an OR, it triggers a series of impulses that are carried by the olfactory neuron's axon to the olfactory bulbs. The olfactory bulbs sit over the sievelike cribriform plate of the ethmoid bone, through which the axons pass to the olfactory epithelium (see Figure 10.9). Cribriform plate area is therefore used as a proxy for the number of olfactory axons in different species. In the olfactory bulbs, the axons from approximately 10–100 olfactory neurons, all of which have the same type of olfactory receptor, synapse in spherical structures called glomeruli. Each glomerulus in turn connects to several dozen mitral cells. There is thus a high degree of convergence (approximately 1,000:1) of olfactory neurons onto mitral cells (Hellman and Chess, 2002), which is believed to aid in the high sensitivity of smell. In addition, the intensity of a smell is a function of the concentration of odorants in the air, their ability to diffuse in gas, their solubility in mucus, and factors of airflow, such as rate and turbulence (see below) that influence how odorants bind to ORs (Keyhani et al., 1997). The output axons from the mitral cells form the lateral olfactory tract, commonly called the first cranial nerve, but actually a tract of the central nervous system.

Another key factor in smell is perception, the ability to recognize an odor and discriminate among odors. Olfactory perception depends on the formation of an activity pattern in the glomeruli due to the differential activation of ORs. This pattern is conveyed from the olfactory bulbs to the olfactory cortex (in the cerebrum) and on to the orbitofrontal cortex of the frontal lobe. In addition, signals go to the amygdala and other areas of the limbic system that contribute to properties of emotion and memory (including Proustian ones).

How Well Do We Smell?

Humans, like other primates, are frequently categorized as *microsmats* (animals with poor olfactory function) as opposed to *macrosmats*. Much of the evidence to support this idea is anecdotal, or based on humans' poor performance compared to animals such as rats and dogs in specialized tasks, like following a scent trail. Bloodhounds can follow a human's scent trail hours after it was left and even distinguish among different individuals with astonishing reliability (although they have a hard time with monozygotic twins) (Harvey and Harvey, 2003; Harvey et al., 2006). Bloodhounds, however, are obviously exceptional macrosmats, and comparing them with humans is not very fair. Most of the quantitative evidence that humans are microsmats comes from two sources. First, although the human olfactory bulb is about the same size as olfactory bulbs of many other species, it is small relative to body mass, about 30 percent of the size predicted for a primate with our brain size (Stephan et al., 1988). Second, genetic data show that we have lost most of the genes that code for olfactory receptors (ORs). Recall that the perception of an odor derives fundamentally from the combinatorial binding of the odor to different degrees with different ORs. Each OR is encoded by one of a large, ancient, and extraordinarily diverse family of genes that has 1,000–2,000 variants. Each variant responds to different degrees to a subset of different odor molecules. Mice have approximately 1,300 OR genes, of which about 200 have become nonfunctional (pseudogenes) (Zhang and Firestein, 2002). In contrast, humans have about 1,000 OR genes, of which approximately 700 are nonfunctional (Glusman et al., 2001)! Another difference is the number of olfactory neurons. Whereas humans have about 5–10 million olfactory neurons, rodents have 15–20 million, and dogs have as many as 200 million (Buck, 2004). These data suggest that humans should be able to detect almost an order of magnitude fewer odorants than a mouse.

Genetic evidence also suggests that primates differ from other mammals in their olfactory capabilities, with humans being the most microsmatic. Whereas 20 percent of OR genes are nonfunctional in rodents, about 27 percent of these genes are nonfunctional in Old World monkeys, about 50 percent are nonfunctional in apes, and about 70 percent are nonfunctional in humans (Rouquier et al., 2000). By extrapolation, humans probably cannot detect odors as well as chimpanzees on account of having maybe 200 fewer functional OR genes. In addition, although both humans and chimps have a few (albeit different) OR genes under directional selection, more OR genes appear to be under stabilizing selection in chimps (Gimelbrant et al., 2004; Gilad et al., 2005). Nevertheless, both species have polymorphisms (functional versus nonfunctional variants), suggesting that we are both still in the process of losing ORs.

To most biologists, these differences make sense. Primates live in trees above the ground, and they mostly use vision, touch, and intelligence to find and select ripe fruits and tender leaves. Unlike dogs and rodents, monkeys generally don't go about with their noses to the ground intently smelling everything. This nose-in the-air attitude is even more characteristic of apes than monkeys, and bipedalism has further distanced the human nose from the muck and filth of the ground, where many redolent smells lie.

But there are several reasons to question the dogma that primates in general and humans in particular are among the most microsmatic of mammals (Shepherd, 2004). One line of evidence comes from some ingenious perception experiments on macaques and squirrel monkeys by Laska and colleagues (e.g., Laska et al., 2000; Laska and Seibt, 2002). These studies found that monkeys can detect odorants such as alcohols at concentrations of 1 part per million or even 10 parts per billion! Such performance capabilities match those of rats and even overlap with values reported for dogs. Shepherd (2004) also notes that rats and dogs may have large numbers of olfactory receptor neurons to compensate for the drawbacks they face from constantly sticking their noses in filth. When dogs and rats sniff deeply from the ground, they vacuum up not only odorants but also bacteria, molds, and other pathogens that cause inflammatory infections (rhinitis). Thus some features of the rat and dog olfactory system might have evolved to compensate for frequent, even chronic congestion. One can remove up to 80 percent of the glomuleri in the rat olfactory bulb without causing any significant loss of the ability to detect odors (Bisulco and Slotnick, 2003). How well could humans perceive odors if we stuck our noses into things other than objects that we usually want to smell, such as flowers, wines, and cheeses? Fortunately, inquiring minds often want to know such things. Noam Sobel and colleagues tested the abilities of humans to follow complex zigzag trails of chocolate aroma laid down precisely on a large grass field (Porter et al., 2007). The blindfolded subjects had to follow the scent like a dog, on all fours and with their noses to the ground. Amazingly, two-thirds of the subjects were able to complete the task after three different trials. Laska et al. (2000) have found other odorants that humans can perceive better than dogs or rats.

Shepherd (2004) has also suggested that human olfactory perception may be especially geared toward retronasal smells, which are normally associated with foods and liquids. Humans have impressive abilities to discriminate among subtle variations in flavor. A fact not widely appreciated is that flavor is largely due to smell activated by the retronasal route, from the back of the mouth through the nasopharynx to the nasal cavity. Several derived aspects of human cranial anatomy may help augment these capabilities. First, humans have a less occluded retronasal pathway than other mammals, including chimpanzees. As Chapter 8 discussed, modern humans have a shorter oral cavity and a uniquely descended, nonintranarial larynx in which there is no contact between the epiglottis and the soft palate. Thus, when humans chew or drink,

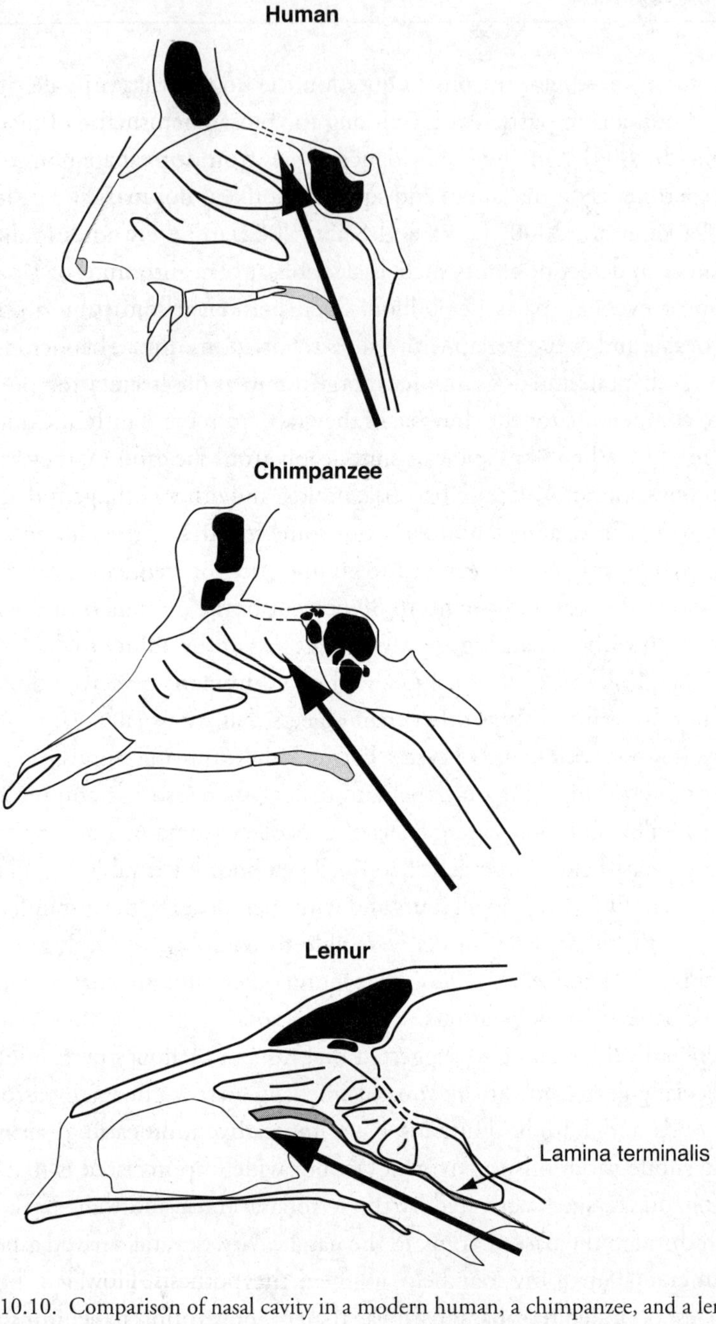

Figure 10.10. Comparison of nasal cavity in a modern human, a chimpanzee, and a lemur (adapted from Wegner, 1955). In both the human and the chimpanzee, there is a direct retronasal pathway by which odorants can travel to the olfactory epithelium. In the lemur, this pathway is blocked by a shelf of bone, the lamina terminalis.

volatile odorants can travel through a much shorter, more open passageway in order to rise above the soft palate and reach the nasal epithelium via the internal nasal cavity. In addition, humans share with other great apes a very open configuration of the two ethmoturbinates at the posterior margin of the nasal cavity. In many nonprimate mammals and prosimians, there is a shelf of bone (the lamina transversa) that divides the ethmoturbinates from the rest of the nasal cavity, effectively sealing off the olfactory epithelium from retronasal flow. As Figure 10.10 illustrates, tarsiers and higher primates have lost this wall, and in humans, as in chimpanzees, gorillas, and orangutans, the posterior portion of the nasal cavity is relatively open, creating a vertically oriented passageway for air from the oropharynx to the superior meatus (Wegner, 1955; Cave, 1973).

Another factor enhancing human olfaction may be turbulence. During normal respiration, airflow in the human nasal cavity is generally laminar, but during a vigorous sniff, airflow rates in the nose more than double, and the flow becomes increasingly turbulent (Hahn et al., 1993; Keyhani et al., 1995). As noted in Chapter 8, humans have especially turbulent nasal cavities during orthonasal breathing because of several derived features, including a nasal vestibule and a nasal valve; turbulence also increases during vigorous oral breathing. Therefore, one possibility is that humans perceive orthonasal odors with sensitivities similar to those of other mammals, despite having fewer olfactory receptors, because of more turbulent airflow patterns in the nose. As outlined in Chapter 8, there is a tradeoff between laminar versus turbulent flow in terms of velocity gradients and resistance. Laminar flows generate much less resistance and thus require less force, but they are characterized by a velocity gradient that creates an immobile boundary zone against the wall of the tube. Thus, one might hypothesize two different airflow strategies for getting more odorants to bind to ORs. The first is via low-energy laminar flow, in which velocities and pressures are low; alternatively, increasing turbulence would eliminate the boundary between the epithelium and an odorant, but at the expense of shorter durations for reception to occur and the need to generate more force. Until we know more about how odorants are drawn across the nasal epithelium, it is hard to evaluate this hypothesis. However, humans and rats have apparently equal levels of odorant absorption in the epithelium even though airflow in the olfactory region is much more turbulent in humans than rats for both orthonasal and retronasal pathways (Zhao et al., 2004, 2006). Perhaps laminar flows in rats allow more time for odorants to bind with receptors, whereas turbulent flows in humans place odorants and ORs in closer but more transient proximity. Moreover, when humans smell, they tend to maximize either the rate or the duration of flow (their sniffs are either stronger or longer) (Sobel et al., 2000). Clearly, the matter deserves more study in humans, rats, and other primates.

One last factor to consider is that many aspects of human olfactory perception are also cognitive. To perceive and analyze smells, many parts of the brain are recruited, including the limbic system, the cerebrum, and many areas of neocortex in the temporal and frontal lobes (see Small and Prescott, 2005). While noncortical structures mostly function to perceive and react to odors, the neocortex helps us to compare and distinguish them in more complex ways and to associate them with events, images, words, and feelings.[4] As noted by Shepherd (2004, 2006), the increased role of higher brain mechanisms in more encephalized mammals during olfaction may therefore compensate for their having fewer olfactory neurons and receptors. Indeed, the co-option of cognitive abilities to aid in olfaction, combined with enhanced retronasal odor detection and the invention of cooking, may account for the rich, complex pleasures that humans get from good food. Maybe evolution favored some aspects of human gustatory hedonism.

Pheromones and the Vomeronasal Organ

Another source of chemosensory stimuli is pheromones, nonvolatile molecules that are similar to odorants and commonly present in secretions such as urine, saliva, and sweat. In many creatures, including most mammals, pheromones play important roles in intraspecies communication, helping animals find and arouse mates, mark territories, signal alarm, and so on. Pheromones are classically believed to be detected in mammals and reptiles by the vomeronasal organ (VNO, or Jacobson's organ). The VNO, which develops from the olfactory placode (see Chapter 3), lies at the base of the vomer below the cartilaginous nasal septum (Keverne, 1999). Among primates, the VNO occurs in prosimians but is present only in a vestigial form in Old World monkeys and apes (Dennis et al., 2004). In humans, the VNO persists as a tiny but distinct tubular structure of incompletely differentiated cells; these cells lose their axonal connection to the brain during fetal growth (Kjaer and Fischer-Hansen, 1996;

---

4. For an integrated literary and neurological analysis that applies these processes to Proust's famous madeleine passage, see Shepherd-Barr and Shepherd (1998).

Smith and Bhatnagar, 2000). In addition, all pheromone receptor genes known to be expressed in the mouse VNO appear to be either nonfunctional or unexpressed in humans (Rodriguez and Mombaerts, 2002).

Although the human VNO appears to be vestigial, there is nevertheless some evidence that humans make and detect pheromones. Humans produce a number of molecules, primarily in apocrine sweat glands (the armpit may be the richest source), which resemble nonprimate pheromones in terms of their chemistry and behavioral effects (see Wysocki and Preti, 2004). One controversial hypothesis is that the olfactory epithelium in higher primates including humans has subsumed some of the role of detecting and transmitting pheromone signals.

Taste, Both Good and Bad

Taste, the chemosensory cousin of smell, helps to assess the chemical composition of what we are eating or mouthing. Taste perceptions help us to avoid sour or acidic substances that may be poisonous or spoiled, and they participate in our cravings for salt, sugar, and certain amino acids. Ironically, recent humans have become victims of these cravings, which surely have some evolutionary basis. Presumably we hunger after the sensory perceptions of salty, sweet, and protein-rich foods because our ancestors couldn't get enough foods with these qualities. Now, because many humans have a hard time suppressing the same cravings when faced with abundance, they are at risk for diseases such as obesity and diabetes. Many of us also develop perverse addictions to sour, bitter foods that can be harmful in excess.

Taste derives from specialized chemoreceptors within taste buds. Taste buds are aggregations of chemosensory cells that differentiate from epithelial cells in the oral cavity (see Northcutt, 2004). Most taste buds lie on the tongue, but some are also present on the epiglottis, the soft palate, and the walls of the oropharynx. Most taste buds live in small, raised projections known as papillae that cover the surface of the tongue; some buds lie in grooves or folds. As illustrated in Figure 10.11, the chemoreceptor cells in taste buds are arranged like segments of an orange below the surface of the bud. A pore at the bud's top admits molecules that subsequently contact tiny, wormlike projections (microvilli) from various receptor cells. When a molecule such as a salt ion interacts with a receptor cell, the cell's membrane then depolarizes, admitting calcium ions. These ions, in turn, stimulate the release of neurotransmitters at the cell's synapse with a sensory nerve, sending a neural impulse by one of three

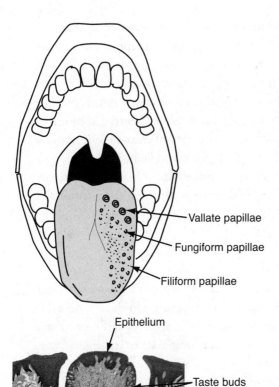

Figure 10.11. Top: taste buds are located in different kinds of projecting structures, papillae, that are distributed on the tongue's surface. Middle: histological section through the surface of the tongue showing taste buds within the papillae. Bottom: schematic view of a taste bud.

Vallate papillae

Fungiform papillae

Filiform papillae

Epithelium

Taste buds

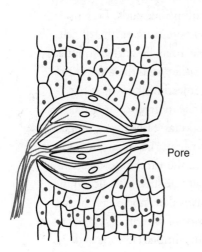

Pore

cranial nerves (VII, IX, or X) to the brain stem. From there, the pathway to perception goes to the somatosensory thalamus and neocortex, with additional connections to limbic regions (hypothalamus, amygdala, and insula) which lead to the "affective" components of taste and associated reflexes such as salivation, hunger, and disgust. Other information goes to the somatosensory cortex, which generates a conscious perception of taste. The complex combination of multiple taste and olfactory perceptions creates the perception of flavor. As noted above, most flavor is olfactory. An apple and onion can taste somewhat similar to a human deprived of olfaction.

There are only five kinds of taste bud receptors: salt, sour, sweet, bitter and savory (also called umami). Each taste receptor works differently (see Nelson, 1998). Some, such as salt-sensing taste buds, are stimulated by influx of the ion through the cell membrane; others, such as sweet- or bitter-sensing taste buds, work when a molecule such as glucose binds to a specialized receptor on the cell's surface. It is sometimes thought that the tongue has a "taste map." In fact, taste buds with different receptors are present all over the tongue. However, some areas of the tongue may be more responsive to certain tastes than other areas. Buds that are rich in salt-detecting receptors tend to be concentrated near the front of the tongue; bitter receptors are more common in buds at the sides and the back; the anterior sides of the tongue in humans may be particularly sensitive to sweet and salt or sour (Jung et al., 2004).

We know little about the evolution of taste in primates or the extent to which human taste differs from other mammals, including chimps. Some clues are beginning to come from genetic data. Bitter, sweet, and savory taste receptors are encoded by different families of genes, each with numerous variants (Behrens and Meyerhof, 2006). For the most part, humans and chimps appear to have generally similar capabilities (e.g., Hellekant and Ninomiya, 1994). However, there are some differences. For example, both chimpanzees and humans have nonfunctional polymorphisms for a receptor that senses a particular kind of bitter compound, PTC (phenylthiocarbamide). In other words, some humans and some chimps can't taste PTC. Although these polymorphisms were thought to be ancient, predating the split of the two species (Fisher et al., 1939), Wooding and colleagues (2006) showed that the molecular basis for nontaster PTC alleles differs in the two species, and therefore the variants evolved independently. Other taste-receptor variants present in humans may yield clues about selection. Some individuals (so-called "supertasters" or "nontasters") have unusually high or low percentages of receptors for certain bitter tastes, with resulting differences in taste sensitivity (e.g., Prescott et al., 2001).

Evolution of Smell and Taste in Hominins

It is a challenge to infer how smelling and tasting capabilities changed during human evolution. On the one hand, there is some genetic and morphological evidence to suggest a loss of olfactory function in primates in general and humans in particular. We have fewer functional olfactory genes, relatively smaller olfactory bulbs, and generally smaller noses. However, the performance data reviewed above raise the hypothesis that humans are much better at detecting odors and flavors than is commonly appreciated, especially via retronasal pathways. Several derived features may contribute to these abilities: a shorter oral cavity, the loss of an intranarial larynx, and a smaller nasopharynx with fewer ethmoturbinates. In addition, expansion of the neocortex probably plays an important role in our capabilities for perceiving smell and taste.

It is therefore reasonable to hypothesize that there has been some evolution of hominin chemosensory capabilities. The first step was probably the origin of bipedalism, which removed the nose spatially from regular contact with the odor-rich but filthy forest floor. As in chimpanzees, smell and taste probably played important roles in early hominin foraging and evaluating the environment. Any major shift, if it occurred, probably happened during the evolution of the genus *Homo* because of two types of craniofacial changes. First, the shortening of the nasopharynx and oral cavity, combined with a more turbulent nose (e.g., from an external nasal vestibule) and possibly a lower, nonintranarial larynx, may have opened up the retronasal pathways. Second, expansion of the neocortex, including connections between the somatosensory, language, and memory regions, may have improved the ability to analyze and perceive odors, hence flavors. Thus, we may be more sensitive to perceiving flavors, perhaps as a by-product of other evolutionary shifts (e.g., for thermoregulation and cognition), but perhaps in part from some degree of selection on chemosensory perception as well. Put differently, we may have evolved to perceive flavors better. This hypothesis, however, needs to be tested. A good start would be to test more carefully how well humans, chimpanzees, and other primates discriminate among various subtle aromas and flavors.

If humans do have an elaborated ability to perceive flavors, then this derived trait occurred in the only species ever to have invented cuisine. As discussed in Chapter 7, the mechanical processing of food certainly became more common by 2.6 mya, and there is debate over when cooking first evolved. Wrangham and colleagues suggest that cooking evolved about 2 million years ago (Wrangham et al., 1999; Wrangham and Conklin-Brittain, 2003); others think it evolved more recently, maybe 250,000 years ago (e.g., Brace et al., 1991). Regardless, it is tempting to speculate that mechanical processing and cooking, the latter of which releases particularly intense and complex aromas, helped drive some of our chemosensory capabilities. Alternatively, if cooking is a more recent invention, then it is possible that an enhanced retronasal pathway combined with a bigger neocortex predisposed humans toward a greater appreciation of cooked food when it was invented. Either way, evolution has helped some of us to enjoy fine cuisine.

11

Early Hominin Heads

It does not represent an ape-like man, a caricature of precocious hominin failure, but a creature well advanced beyond modern anthropoids in just those characters, facial and cerebral, which are to be anticipated in an extinct link between man and his simian ancestor.

RAYMOND DART, *"Australopithecus africanus:* The Man Ape of Southern Africa," 1925

This is the first of three chapters that focus on the hominin fossil record. This chapter examines the first hominins and the subsequent radiation of australopiths. Chapters 12 and 13 look at the origins of the genus *Homo* and of *H. sapiens.*

These chapters have several goals. The first is to summarize what we know about major changes in hominin craniofacial anatomy. This is a daunting task because, after more than a century of concerted effort, we have thousands of specimens from at least a dozen species that span millions of years and several continents. Our lineage did not evolve, as once was supposed, through a simple, unilinear progression of transformations from apelike creatures into modern humans with just a few intervening intermediate "stages." Instead, the hominin fossil record is diverse, moderately speciose, and with much variation both within and between species. Coping with this variation is challenging, especially when trying to determine how many species there were and their relationships to each other. That said, too much focus on subtle variations can make one can lose sight of the forest for the trees. Although this chapter reviews australopith diversity, how many species there were, and how they might be related, what follows relies necessarily on some generalizations.

A second goal is to speculate about the developmental bases for major shifts that occurred in the head during hominin evolution. One hypothesis is that many of the ontogenetic transformations necessary to generate morphological differences among closely related species may be fairly simple. As highlighted in Chapters 1–5, the ways in which heads are integrated, developmentally and functionally, permit considerable evolvability via tinkering. A few developmental changes can generate a broad suite of integrated shifts. For example, processes that flex the cranial base influence many aspects of skull morphology (e.g., facial position, neurocranial shape, and brow ridge length), which are subsequently accommodated by further epigenetic interactions during growth and development. This is not to say that the transformations we observe, such as flexed cranial bases, originated from single genetic mutations that altered just one developmental process. Rather, seemingly small types of tinkering can have widespread, substantial effects on the phenotype.

A third goal is to speculate about the functional consequences of some major transformations that we observe in the skull and, by extension, the head. Even if a limited number of developmental shifts underlie many major craniofacial transformations in hominin evolution, the substantial differences evident among humans, apes, and various fossil hominins are evidence of much evolutionary change. What were the selective pressures that drove these changes? Addressing this question is tricky because, as is often noted, it is difficult to discern the extent to which particular novel features are products of natural selection, favored because they improved organisms' abilities to survive and reproduce. Views on how to recognize adaptations differ widely. Some biologists believe that the concept should be applied only when there is good evidence that a feature is novel, heritable and improves fitness. Unfortunately, we rarely have enough data to apply such criteria to most features. But that doesn't mean that natural selection hasn't operated strongly on the hominin head since our divergence with the other great apes.

Given the critical importance of the head's many functions, and the strong negative consequences of impaired or suboptimal performance in tasks such as hearing, seeing, or chewing, I suspect that the head is a treasure trove of adaptations. But one needs to test hypotheses about particular adaptations against two criteria. First, a derived feature must be shown to improve performance relative to its ancestral condition and in the context of the environmental pressures acting on the organisms in question. Second, one needs to demonstrate that the feature is not simply a by-product of some other change that was itself the real adaptation. As an example, were smaller canines in early hominins an adaptation to accommodate larger cheek teeth, an adaptation to permit more forceful chewing, or a by-product of less competition between males? The head is so profoundly integrated that most components of the skull perform multiple functions. Only by considering a given feature in the context of when and where it first arose, and the many functions in which the skull participates (e.g., locomotion, vision, respiration, speech, chewing, and so on), can we evaluate alternative hypotheses of adaptation.

A final goal is to propose some hypotheses about why the human head looks the way it does. My thesis is that when we step back from the many changes and seemingly confusing variations evident in the fossil record, we can see that human evolution involved only a few major phases or "episodes" of tinkering, all contingent on previous events. This chapter examines two such episodes: the origins of the hominin lineage and the subsequent diversification of australopiths. As far as we can tell, the earliest hominin heads differed from those of great apes in a few crucial ways to accommodate being bipedal and to be slightly better able to chew lower-quality, more mechanically demanding foods. The subsequent australopith radiation was an elaboration of the same trends. Australopith heads vary considerably in the extent to which their teeth, faces, and jaws were adapted for masticating lots of mechanically demanding foods, perhaps driven by varied use of open habitats.

Two caveats before proceeding. The first is that, inevitably, some of what I write will be out of date by the time you read this because new discoveries, both fossil and molecular, are announced regularly. In spite of the hyperbole that occasionally accompanies these discoveries, human evolutionary biology is a sufficiently mature and predictive field that most of them will fill in gaps in our knowledge rather than require complete reinterpretations of fundamental ideas and theories. The second caveat is that I cannot possibly justice here to the hominin fossil record. There are too many fossil specimens, and they vary too much. The good news is that many detailed analyses of these fossils have been published, and I have tried to cite key references for those who wish delve further into particular morphological features, specimens, or species. In addition, although the details of each specimen of each species are important, I am more interested in considering the major transitions that occurred in human evolution. This goal necessarily requires a focus on the major features of the best-known species based on the more complete specimens, and thus we must breeze by some poorly known species and pass over many interesting variations.

Evidence of the Last Common Ancestor and the Earliest Hominins

The first transition in hominin evolution was the divergence of the human and chimpanzee lineages. Unfortunately our knowledge of the last common ancestor (LCA) of chimpanzees and humans remains hypothetical and the subject of much conjecture, but we have a limited sample of fossils more than 4 million years old that are likely early hominins and are currently assigned to three genera: *Sahelanthropus, Orrorin,* and *Ardipithecus.* We can thus attempt to draw some inferences about the first hominins, how they evolved from the LCA, and maybe even why.

The Unknown LCA

The LCA of chimpanzees and humans is unknown, and the currently available fossil record of chimpanzee evolution amounts to a few teeth that are about 500,000 years old (McBrearty and Jablonski, 2005). However, several sources of data permit some conjectures about the LCA. As Chapter 1 notes (see Figure 1.1), molecular data indicate that humans and chimpanzees shared a common ancestor sometime in the Miocene. This divergence is dated to 5–10 million years ago (Ruvolo, 1997; Takahata and Satta, 1997; Gagneux et al., 1999; Steiper and Young, 2006; Patterson et al., 2006; Barton, 2006; Wakeley, 2008). The close relationship of chimpanzees and humans is significant in light of the many morphological similarities between gorillas and chimpanzees. To a considerable extent, gorillas and chimpanzees are scaled versions of each other (B. T. Shea, 1983; 1986; Penin et al., 2002). That is, if a chimpanzee were to grow to a gorilla's size, its skeleton would resemble that of a gorilla (though not exactly). Unless these and other similarities, such as knuckle walking, evolved independently, then the LCA of gorillas and chimpanzees and the LCA of humans and chimpanzees must have been somewhat like a chimpanzee, enough to be classified in the genus *Pan* (or *Gorilla,* which amounts to nearly the same thing) (Pilbeam, 1996). Such convergence cannot be ruled out, but is improbable.

Further evidence about the LCA comes from the paucity of unique derived features evident in chimpanzees. As one might expect, chimps have some likely derived features relative to other apes (e.g., their molars have slightly thin enamel crowns and wide occlusal basins), but despite concerted efforts, researchers have identified few derived features of soft-tissue anatomy or the skeleton that distinguish chimpanzees from other extant great apes in terms which cannot be explained by the effects of scaling (Groves, 1986; Gibbs et al., 2000, 2002). In addition, the two species of chimpanzees (*P. troglodytes* and *P. paniscus*) are very similar (paleontologists would have a difficult time distinguishing them), despite having diverged about a million years (Won and Hey, 2005; Becquet et al., 2007). Many of their differences can be explained by heterochrony (bonobos are paedomorphic chimpanzees) (see Chapter 5 and Lieberman et al., 2007a).

The argument that the LCA of humans and chimps resembled an extant African great ape has been challenged on the basis of the skeleton of *Ardipithecus ramidus* (Lovejoy et al., 2009a–d; White et al., 2009; Suwa et al., 2009a). These researchers argue that *Ar. ramidus,* which is dated to 4.4 mya, has a number of derived hominin features but also many primitive features not shared with chimpanzees or gorillas. In the skull, these features include reduced subnasal prognathism and the lack of a marked supratoral sulcus. More importantly, the species' skeleton is interpreted to be that of an early biped that lacked locomotor adaptations seen in chimps and gorillas either for knuckle walking or for suspensory climbing and hanging with a vertical (orthograde) trunk. However, the *Ardipithecus* skeleton postdates the chimp-human divergence by at least 2.5 million years, it has many chimplike features, and functional interpretations of its anatomy will undergo much scrutiny and reinterpretation. Further, the hypothesis that the LCA was a monkeylike ape that did not climb trees in an orthograde manner requires a degree of convergence in the human, chimpanzee, gorilla, and orangutan lineages that many find improbable.

Regardless of just what *Ardipithecus* turns out to be, and what it suggests about the LCA, it is important to emphasize that no one thinks that the LCA of chimpanzees and humans was exactly like a living chimpanzee, or that chimpanzees are "living fossils" with no history of evolutionary change of their own. For instance, tooth crown anatomy differs in several ways between chimpanzees, gorillas, and late Miocene apes, suggesting that some aspects of chimpanzee dental shape may be derived (an inference complicated by our poor understanding of the relationships among Miocene apes). But even if there was some convergence between chimps and gorillas, the balance of evidence suggests that whatever changes occurred within the chimpanzee lineage since their divergence from hominins in terms of skeletal morphology were not as considerable as those that occurred in the human lineage. This degree of relative stasis in the chimpanzee lineage should not be surprising, because lack of change is a common characteristic of phenotypic evolution, especially in hard tissues that tend to fossilize (Gould, 2002). Consider gibbons. There are about a dozen species of gibbons, most of which diverged from each other about 8–10 million years ago (Takacs et al., 2005; Chatterjee, 2006). Because modern gibbon species differ from each other only subtly, we can infer that they have not changed much over the last 8–10 million years. Ultimately, we need more ape fossils from the late Miocene to definitively test hypotheses about the LCA of humans and chimpanzees.

Sahelanthropus tchadensis

The oldest known fossil species to consider is *Sahelanthropus tchadensis,* discovered in Chad in 2001 by a team led by Michel Brunet (Brunet et al., 2002, Zollikofer et al., 2005; Brunet et al., 2005). The material assigned to *S. tchadensis* is at least 6 million years old, probably between 6.8 and 7.2 million years old (Vignaud et al., 2002; Lebatard et al., 2008), and thus close to the time range estimated from genetic data for the divergence of the chimpanzee and human lineages. Importantly, these fossils were found 2,500 km west of the Rift Valley of eastern Africa, where all other early hominins have so far been found. *Sahelanthropus* comes from sediments that contain more than a thousand fossils. Many of these fossils are aquatic, such as fish, crocodiles, and hippos, but they also include species that inhabit gallery forests, such as monkeys, and other species typical of more open habitats, such as giraffes, pigs, and bovids. It is difficult to be precise about the paleoenvironment in which *Sahelanthropus* lived, but a reasonable reconstruction would be a densely wooded habitat (a gallery forest) close to a lake (Brunet et al., 2005).

The major fossil attributed to *S. tchadensis* is a nearly complete cranium, TM 266–01–060–1, nicknamed Toumaï (a name given to babies born just before the dry season begins, which means "hope of life" in the local Goran language). As of 2010, the species is also represented by four mandibular fragments and several isolated teeth (a right M^3, a right I^1, a right deciduous canine, and a right P^3). Although Toumaï's cranium is fractured in several places and partially distorted, a high-resolution CT scan of the skull was virtually reconstructed to recreate its antemortem shape, shown in Figure 11.1 (Zollikofer et al., 2005). As detailed below, Toumaï's cranium has many primitive features, including a small endocranial volume (360–370 cm^3). But it also has some derived hominin features, including a relatively vertical face with a short premaxillary region; a more anteriorly positioned foramen magnum than that of chimps, with a horizontally rotated nuchal plane; and a large supraorbital torus (Guy et al., 2005). Compared with those of great apes, the molars have lower and more rounded cusps, with enamel crowns intermediate in thickness between those of chimpanzees and those of early australopiths. The canines are also smaller and less daggerlike than in male chimpanzees, and they lack evidence of a C-P_3 honing complex between the upper canine and lower first

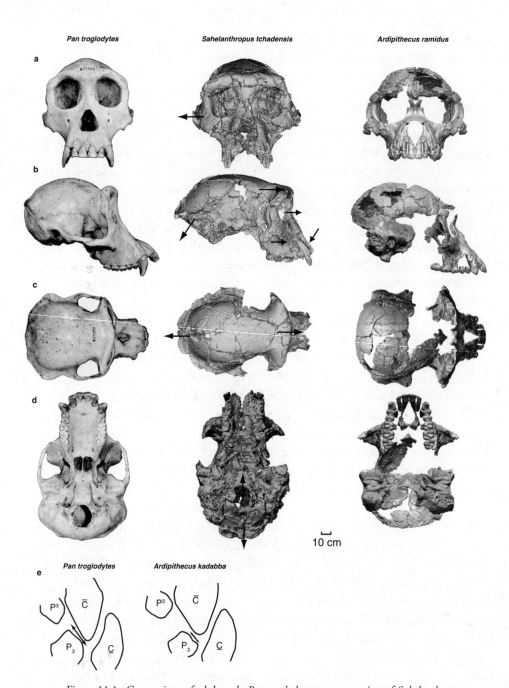

Figure 11.1. Comparison of adult male *Pan troglodytes,* reconstruction of *Sahelanthropus tchadensis,* and *Ardipithecus ramidus* in (a) anterior, (b) lateral, (c) superior, and (d) inferior views. Arrows on *Sahelanthropus* highlight major differences in shape from *Pan.* (e) Lateral

view of reduction in the C-P$_3$ honing complex in (left) a *P. troglodytes* female compared with that of (right) *Ar. kadabba* (adapted from Haile-Selassie et al., 2004). Arrow shows approximate region of honing between the canine and premolar. Renderings of *Sahelanthropus* reconstruction from Zollikofer et al. (2005: Figure 2), courtesy M. Brunet. Renderings of *Ardipithecus* reconstruction © 2009 G. Suwa and T. White (from Suwa et al. 2009a, Figure 2a–d).

---

premolar (Brunet et al., 2005). This honing complex, typical of chimpanzees but absent in *Sahelanthropus,* occurs when the distal margin of the upper canine sharpens (hones) itself against the mesial margin of the P$_3$. Honing leaves a facet on the premolar and a ridge on the back of the canine. Toumaï's large brow ridges suggest he was a male.

Orrorin tugenensis

Thirteen fossils are assigned to *Orrorin tugenensis* (Senut et al., 2001). The fossils, which come from the Tugen Hills near Lake Baringo in Kenya, are about 6 million years old (for more information on their discovery, see Gibbons, 2006). The fossils include two mandibular fragments, six isolated teeth, and some postcranial material, including much of a femur. The associated fauna suggest that the site was a wooded habitat near a lake (Pickford and Senut, 2001).

Without more information and more fossils, it is difficult to say much about the *Orrorin* material. The femur has features that suggest bipedalism (Pickford et al., 2002; Galik et al., 2004; Richmond and Jungers, 2008). So far, the craniodental material has not been described in detail.

Ardipithecus ramidus and Ardipithecus kadabba

Fossils currently assigned to the genus *Ardipithecus* come from Ethiopia. The initial discovery was made in 1992 at Aramis in the Middle Awash region of Ethiopia in sediments dated to 4.4 mya (White et al., 1994). Subsequent fieldwork yielded more than 100 hominin fossils from at least three dozen individuals (White et al., 2009). These include a spectacular partial skeleton, ARA-VP-6/500, that was found in a highly fragile and fragmentary state and required extraordinary efforts to recover and reassemble. Additional, older fossils from the Middle Awash, dated to 5.2–5.8 mya, were initially considered a subspecies of *Ar. ramidus* (Haile-Selassie, 2001) but were later assigned to a new species, *Ar. kadabba* (Haile-Selassie et al., 2004). Additional *Ar. ramidus* fossils have been discovered at Gona, in the Afar region of Ethiopia, in sediments 4.3–4.5 million years old (Semaw et al., 2005). The paleoenvironmental contexts of these different sites suggest woodland habitats (Su et al., 2009; WoldeGabriel et al., 2009; Louchart et al., 2009).

The cranial material assigned to the older *Ardipithecus* species, *Ar. kadabba,* includes one partial mandible, plus teeth from a minimum of 10 individuals (Haile-Selassie et al., 2009). The teeth are of particular interest. The canines are similar in size to those of female chimpanzees, but the crowns are less triangular and more diamond-shaped, and there is evidence for a partial C-P$_3$ honing complex in the older teeth (Haile-Selassie et al., 2004, 2009). The molar crowns are intermediate in size and thickness between those of chimpanzees and gracile australopiths (see below), and some have low crowns typical of later hominins.

More is known about the *Ar. ramidus* head from a partial skull (shown in Figure 11.1) reconstructed from many fragments of one individual (ARA-VP-6/500) and from a partial temporal and occipital fragment of a second individual (ARA-VP-1/500) (Suwa et al., 2009a). In a number of ways, the cranium looks like a smaller version of the *Sahelanthropus* cranium, with a diminutive endocranial volume (ECV) of 280–350 cm^3 and a short basioccipital, giving it an anteriorly placed foramen magnum that appears to be oriented inferiorly as one expects for a biped. The posterior occipital is missing, but overall the vault is long, low, and somewhat like a chimp's, with traces of a compound temporonuchal crest. As in *Sahelanthropus,* the *Ar. ramidus* face has a modest degree of upper and midfacial projection, creating somewhat less subnasal prognathism than is seen in *P. troglodytes* or early australopiths. In addition, as in *Sahelanthropus,* there is no marked sulcus between the frontal squama and supraorbital torus, but the superoinferior thickness of the torus is much less than in Toumaï.

The dentition of *Ar. ramidus* is known from more than 100 teeth from at least 14 individuals from Aramis (Suwa et al., 2009b) and 7 from Gona (Semaw et al., 2005). The incisors, both upper and lower, are relatively small compared to those of *Pan, Sahelanthropus,* and *Ar. kadabba.* As in *Ar. kadabba,* the lower canines are about the size of a female chimpanzee's, but the upper canines are considerably smaller (like those of *Sahelanthropus* and *Au. afarensis*), with a diamond-shaped lateral profile with elevated shoulders and lack of any honing with the mesiobuccal face of the P$_3$ (see below). The postcanine teeth (especially the second and third molars) are slightly larger than in *P. troglodytes,* with low, bunodont cusps and enamel crown thickness measurements intermediate between *Pan* and *Australopithecus.*

The published descriptions of the ARA-VP-6/500 skeleton include many features that are proposed to challenge the conventional wisdom of what one expects to see in a 4.4-million-year-old hominin (Lovejoy et al., 2009a–d). For one, the skeleton is probably a female's but has an estimated body mass of 50 kg, which is large relative to both female chimps and australopiths (which tend to be about 30–40 kg); in addition, sexual dimorphism in *Ar. ramidus* canine size is estimated to be low (Suwa et al., 2009b). These inferences are likely to be debated. Even more interesting is the postcranium: the heavily reconstructed pelvis has some features suggestive of bipedalism (short, laterally facing iliac blades), but the skeleton also has a host of other primitive features, such as a very divergent big toe, long curved phalanges, and equally long forelimbs and hindlimbs. Lovejoy et al. (2009d) claim that these features are inconsistent with descent from a chimpanzeelike knuckle walker and orthograde climber. But there will be much discussion and many alternative interpretations of this fossil, and a definitive evaluation of the species' functional morphology and phylogenetic relationships is premature at this point.

Developmental Bases for the Early Hominin Divergence

What sorts of developmental changes occurred to transform the unknown LCA into the earliest hominins? As outlined above, the currently published fossil evidence consists of just two partially complete crania (one of *Sahelanthropus* and another of *Ardipithecus*), some teeth, and a few additional mandibular and cranial fragments. Because the crania of *Sahelanthropus* and *Ardipithecus* have a high degree of resemblance, the two taxa can be considered together. However, we know little about variation in these species, and this simplifying assumption needs to be tested with more fossils. We also lack enough information on key features, such as the angle of the cranial base. Nevertheless, the available information suggests that a modest number of developmental transformations are needed to modify the ontogeny of a somewhat chimplike LCA to transform it into an early hominin.

Tinkering with Early Hominin Dentition

Dentally, the earliest hominins are primitive in some respects but differ importantly from extant African great apes and Miocene apes in others (White et al., 1994; Brunet et al., 2002; Haile-Selassie et al., 2004; Brunet et al., 2005; Semaw et al., 2005; Haile-Selassie et al., 2009; Suwa et al., 2009b). A single I^1 attributed to *Sahelanthropus* is chimpanzeelike in size and shape, as may also be the case for a single worn I_2 from *Ar. kadabba;* incisors from *Ar. ramidus* are smaller. A major difference is the canines, which tend to be smaller (especially in males), more symmetrical, and less dagger-shaped in hominins than in chimpanzees (see Figure 11.1). There is also less space (diastema) between the premolars and canines, and a reduction of honing (wear) between the distal margin of the upper canine and the mesial margin of the P_3. Finally, the cheek tooth crowns (mostly the molars) are slightly larger and flatter (see Figure 11.7). As an example, the M_1 crown averages 11.2 ± 0.5 mm in mesiodistal length and 9.8 ± 0.6 mm in buccolingual width in *P. troglodytes*. It is approximately 10 percent longer and 15 percent wider in the one known *Sahelanthropus* M_1 but less than 10 percent larger in *Ar. ramidus* (Brunet et al., 2002, 2005; Suwa et al., 2009b). Enamel thickness is a continuous trait with lots of variation and hard to encapsulate from a few measurements (see Suwa and Kono, 2005). With this caveat in mind, the published values for maximum radial thickness of crown enamel in *Sahelanthropus* molars range from 1.2 to 1.9 mm (Brunet et al., 2002, 2005); those of *Ardipithecus* range from 1.1 to 2.0 mm, with thicker enamel on the functional than on the nonfunctional cusps (White et al., 1994; Semaw et al., 2005; Haile-Selassie et al., 2009; Suwa et al., 2009b). Thus enamel thickness in early hominin molar crowns is intermediate between that of chimps and australopiths (see G. T. Schwartz, 2000).

These modest dental changes would not require massive developmental shifts. We'll focus more on the developmental bases for variations in anterior and posterior tooth size and thickness below in the context of the australopiths, but a few points are worth raising here. First, canine height varies considerably among extant apes because of a combination of differences in the rate and duration of formation: apes with bigger canines can grow them faster or for longer (G. T. Schwartz et al., 2001). As yet, we don't know to what extent canine reduction in early hominins occurred as a result of shorter growth, slower growth, or some combination of the two processes, but comparisons between apes and australopiths (reviewed below) suggest that the primary developmental cause was a reduction in the duration of crown formation (Dean, 2009). Regardless of the mechanism, shorter and smaller early hominin canines, combined with slight expansion of the cheek teeth, probably reduced the degree to which their canines projected beyond the postcanine occlusal plane. Smaller, less projecting, and more dagger-shaped canines account for the loss of the C-P$_3$ honing complex along with more wear on the tips than on the sides of these teeth (for details, see Haile-Selassie et al., 2004, 2009; Suwa et al., 2009b).

Larger-crowned molars mostly arise from more mitosis along the inner enamel epithelium during the bell stage of crown development, which also leads to a larger population of ameloblasts (see Chapter 3). Crown size can be increased by thicker enamel, either through a longer duration of crown growth or through elevated daily enamel secretion rates by amelobasts. We currently lack detailed histological studies of early hominin teeth, but comparisons of chimp and early australopith teeth (more details below) suggest that the earliest hominins differ from great apes in enamel thickness primarily because of faster daily secretion rates (Dean et al., 2001; LaCruz et al., 2008). It is reasonable to speculate that a similar mechanism accounts for the slightly greater size (megadonty) of the earliest hominin molars.

Tinkering with the Early Hominin Cranium

How cranial growth differed between the LCA and early hominins such as *Sahelanthropus* and *Ardipithecus* is hard to deduce, let alone test. Some clues come from a few derived differences (highlighted in Figure 11.1) that make the basicranium of the African great apes different from that of either *Sahelanthropus* or *Ardipithecus*. In the early hominin crania, the foramen magnum is positioned more anteriorly relative to more lateral structures in the temporal bone, the basioccipital is shorter, the petrous temporals are oriented more coronally, the nuchal plane is more horizontally oriented, and the nuchal crest is downwardly lipped (the latter two features can be evaluated only in *Sahelanthropus*). The reconstructed vaults of *Ardipithecus* and especially *Sahelanthropus* are low and flat, and in both, the upper and middle face project more relative to the neurocranium. This projection sets up a relatively long brow ridge lacking a deep sulcus between the supraorbital torus and the frontal squama (which is especially horizontal in the *Sahelanthropus* reconstruction). This projection also creates a more vertical face in which the maxillary root of the zygomatic arch is positioned a little further forward relative to the molars, so that projection of the premaxilla relative to the nasal aperture (lower facial prognathism) is reduced compared to chimps. In many other respects, *Sahelanthropus* and *Ardipithecus* cranial anatomy resembles that of chimpanzees (Guy et al., 2005; Suwa et al., 2009a).

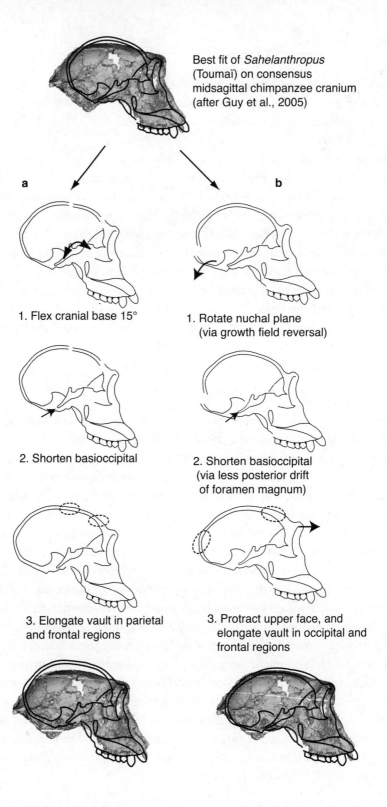

Best fit of *Sahelanthropus* (Toumaï) on consensus midsagittal chimpanzee cranium (after Guy et al., 2005)

a

1. Flex cranial base 15°

2. Shorten basioccipital

3. Elongate vault in parietal and frontal regions

b

1. Rotate nuchal plane (via growth field reversal)

2. Shorten basioccipital (via less posterior drift of foramen magnum)

3. Protract upper face, and elongate vault in occipital and frontal regions

Figure 11.2. Alternative hypothetical transformations of a male *Pan troglodytes* cranium into the reconstruction of *Sahelanthropus tchadensis.* Left: transformation based on flexing the cranial base. Right: transformation based on rotation and extension of the nuchal plane, plus more facial projection. It is possible that the actual transformation involved some combination of both derived patterns (see discussion in text). *Sahelanthropus* lateral view from Zollikofer et al. (2005), courtesy M. Brunet.

What ontogenetic differences would generate these derived features? One longstanding hypothesis has been that the earliest hominins have more flexed cranial bases than apes. As previously discussed (see Chapters 5 and 9), the cranial base is about 25° more flexed in modern humans than in common chimpanzees, rotating the human face downward relative to the rest of the skull and helping orient the eyes forward when the neck is vertical. Human cranial base flexion occurs in part because the human face stays short and retracted and thus does not stimulate cranial base extension, which occurs in chimpanzees throughout the entire period of postnatal craniofacial growth; bonobos have slightly less flexed cranial bases than chimps, probably reflecting their relatively shorter, paedomorphic faces. As discussed below, cranial base flexion is quite variable in hominins but is often (although not always) more flexed than in chimpanzees (Spoor, 1997; Lieberman et al., 2000b). However, more than cranial base flexion would be needed to transform an apelike cranium into that of *Sahelanthropus,* and it is possible that the cranial base of *Sahelanthropus* might not be more flexed than a chimpanzee's. Although internal structures of the cranial base are not preserved, several clues support this hypothesis. First, although a more flexed cranial base would rotate *Sahelanthropus's* foramen magnum closer to the center of skull, it would also rotate the face more ventrally because of the facial block (see Chapters 4 and 5), decreasing upper facial projection and tending to shorten rather than lengthen the suprarobital torus. *Sahelanthropus* has a somewhat vertical face, but its brow ridge is long for a chimpanzee-sized ape. As the thought experiment shown on the left of Figure 11.2 suggests, flexing the cranial base is probably not sufficient to adequately "transform" a chimpanzee cranium into a Toumaï-shaped cranium: the nuchal plane is the wrong orientation, the upper face is too retracted, and the brow ridges are too short.

The right side of Figure 11.2 presents an alternative hypothetical transformation from a chimplike LCA to a hominin such as *Sahelanthropus* that does not involve much cranial base flexion. It requires three developmental shifts, the first of which is a growth-field reversal in the occipital behind the foramen magnum. In nonhuman primates, the nuchal plane rotates vertically and the foramen magnum drifts posteriorly relative to the cranial base because of bone resorption on the external and deposition on the internal aspect of the occipital (for details, see Chapter 4). Humans have the opposite pattern, and the nuchal plane of *Sahelanthropus* raises the possibility that this reversal occurred very early in hominin evolution. Second, the basioccipital elongates and the foramen magnum migrates more posteriorly in nonhuman primates than in humans from resorption on the posterior and deposition on the anterior margin of the foramen (again, see Chapter 4). We lack histological data for any early hominins, but judging from the horizontal orientation of the nuchal plane in *Sahelanthropus* and the anterior position of the foramen magnum in *Sahelanthropus* and *Ardipithecus,* both changes appear to have occurred in these species. Finally, the upper face in modern humans and nonhuman primates grows forward relative to the basicranium and neurocranium from drift because the external surface of the upper face is a depository growth field. It seems probable that *Sahelanthropus* and *Ardipithecus* had long brow ridges and somewhat vertical faces because they had more anterior upper facial and midfacial growth than chimpanzees do. This drift also pulled forward the lateral margin of the orbits and the maxillary root of the zygomatic; both shifts augment the masseter's mechanical advantage (see below). Toumaï's brow ridges are especially large and would have made his face rather menacing, perhaps a useful trait when competing with other *Sahelanthropus* males.

Of course, the above hypothesis is just a thought experiment and needs to be tested with data not yet available (especially intact cranial bases). Furthermore, it is possible that *Sahelanthropus* has some degree of cranial base flexion plus facial projection and nuchal plane rotation, and it has been hypothesized that the *Ar. ramidus* cranium has a more flexed cranial base than *P. troglodytes* (Suwa et al., 2009a). Whatever the case, the important point is that one would need to modify the pattern of development of an African great ape in just a few ways to transform the hypothetical LCA into an early hominin such as *Sahelanthropus* or *Ardipithecus*. In fact, it might not have been necessary to flex the cranial base. This possibility raises the conjectural hypothesis that increased cranial base flexion may be a later derived feature in human evolution that occurred more than once as relative brain size began to increase and faces became shorter and more vertical.

The Australopith Radiation

The fossil record increases appreciably starting about 4 mya. Fossil finds, mostly from eastern and southern Africa, document a modest radiation of hominins between 4.2 and 1.3 mya. Features of some major fossils are summarized in Table 11.1. There are differences among the australopiths, but their similarities are also striking. For the most part, they were variations on a theme: small-brained, habitual bipeds with big cheek teeth. Nevertheless, their variety raises a number of questions. Why were the australopiths so diverse? How much were their variations driven by adaptations for different diets? In what other ways were these hominins different from the LCA and earlier hominins?

At least nine species of australopiths have been proposed, of which only four are reasonably well known: *Au. afarensis, Au. africanus, Au. boisei,* and *Au. robustus.* All share some primitive features with chimps and gorillas, but they also share many derived similarities with the earliest hominins, notably bipedalism, smaller canines, and larger-crowned, more thickly enameled cheek teeth. Many of their derived craniodental traits involve specializations for generating and resisting high chewing forces. These masticatory features include larger, flatter cheek teeth; robust mandibles; and large, heavily buttressed faces, often with extra crests for muscle attachments. These and other related specializations vary so substantially, however, that is useful to divide the australopiths into "gracile" and "robust" groups based on the extent of their apparent adaptations for masticating hard or tough foods. Some researchers (e.g., Wood, 1991; Strait and Grine, 2004) think that the robust australopiths belong to a distinct lineage and classify them in a different genus, *Paranthropus.* Given uncertainty over this hypothesis, I lump them in the single genus *Australopithecus.*

Gracile *Australopithecus*

Paleontologists have proposed as many as seven species of gracile australopiths (see Figure 11.3): *Au. anamensis, Au. afarensis, Au. africanus, Au. garhi, Au. bahrelghazali, Au. sediba,* and *Au. (Kenyanthropus) platyops.* The last four are known from very small sample sizes, and the last three might not be distinct species.[1] Moreover, most differences among the australopiths are

---

1. A problem with small sample sizes, especially singletons, is that one cannot estimate confidence intervals of variation. For a discussion of this problem, see R. J. Smith (2005).

Table 11.1: Major cranial remains of early hominins (excluding mandibles and isolated teeth)

Fossil	Location	Species	Age (mya)	ECV (cm³)ᵃ	Elements found
TM 266–01–060–1	Chad	*Sahelanthropus tchadensis*	6.8–7.2	360–70	Cranium
ARA VPÑ1/500	Middle Awash, Ethiopia	*Ardipithecus ramidus*	4.4		Partial cranium
ARA VPÑ1/600	Middle Awash, Ethiopia	*Ar. ramidus*	4.4	280–350	Partial skeleton
KNM-KP 29281	Kanapoi, Kenya	*Australopithecus anamensis*	4.1		Mandible and temporal fragments
KNM-KP 29283	Kanapoi, Kenya	*Au. anamensis*	4.1		Maxilla
BEL VP 1/1	Belohdelie, Ethiopia	*Au. afarensis*	3.9		Partial cranium (frontal)
AL 58–22 (base fragments)	Hadar, Ethiopia	*Au. afarensis*	3.0–3.4		Partial cranium (face and cranial
AL 162–28	Hadar, Ethiopia	*Au. afarensis*	3.0–3.4		Partial vault
AL 168–28	Hadar, Ethiopia	*Au. afarensis*	3.0–3.4	400	Partial cranium
AL 200–1	Hadar, Ethiopia	*Au. afarensis*	3.0–3.4		Maxilla
AL 288–1	Hadar, Ethiopia	*Au. afarensis*	3.0–3.4		Very partial skull (and postcranium)
AL 333–105	Hadar, Ethiopia	*Au. afarensis*	3.0–3.4	400	Partial cranium (juvenile)
AL 333–45	Hadar, Ethiopia	*Au. afarensis*	3.0–3.4	492	Partial cranium
AL 333–43/86	Hadar, Ethiopia	*Au. afarensis*	3.0–3.4	400	Maxilla and mandible
AL 444–2	Hadar, Ethiopia	*Au. afarensis*	3.0–3.4	550	Skull (male)
AL 822–1	Hadar, Ethiopia	*Au. afarensis*	3.0–3.4		Skull (female)
Dikika 1–1	Dikika, Ethiopia	*Au. afarensis*	3.3–3.4	275–300	Partial skull (and postcranium)
KNM-ER 2602	Lake Turkana, Kenya	*Au. afarensis*	3.3		Partial calvarium
Taung 1	Taung, South Africa	*Au. africanus*	2.0–2.5	440	Partial skull (juvenile)
MLD 1	Makapansgat, South Africa	*Au. africanus*	2.5–3.0	510	Partial calvarium
MLD 37/38	Makapansgat, South Africa	*Au. africanus*	2.5–3.0	435	Partial cranium
Sts 5	Sterkfontein, South Africa	*Au. africanus*	2.15	485	Cranium (no teeth)
Sts 17	Sterkfontein, South Africa	*Au. africanus*			Partial cranium
Sts 19/58	Sterkfontein, South Africa	*Au. africanus*	2–3	436	Partial cranium
Sts 52	Sterkfontein, South Africa	*Au. africanus*	2–3		Maxilla, mandible
Sts 71	Sterkfontein, South Africa	*Au. africanus*	2–3	428	Partial cranium
Stw 13	Sterkfontein, South Africa	*Au. africanus*	2–3		Partial cranium

Specimen	Locality	Species		ECV	Description
Stw 505	Sterkfontein, South Africa	*Au. africanus*	2–3	550	Cranium
Stw 573	Sterkfontein, South Africa	*Au. africanus*	2–3		Skull (and postcranium)
TM 1512	Sterkfontein, South Africa	*Au. africanus*	2–3		Right maxilla
KNM-WT 40000	Lake Turkana, Kenya	*Au. (Kenyanthropus) platyops*	3.5	400–450	Partial cranium
Bou VP 12/130	Bouri, Ethiopia	*Au. garhi*	2.5	450	Partial cranium
KNM-WT 17000	West Turkana, Kenya	*Au. aethiopicus*	2.5	410	Partial cranium
OH 5	Olduvai, Tanzania	*Au. boisei*	1.79–1.85	520	Partial cranium
KNM-ER 406	Lake Turkana, Kenya	*Au. boisei*	1.6	500	Partial cranium
KNM-ER 407	Lake Turkana, Kenya	*Au. boisei*	1.85	510	Calvarium
KNM-ER 732	Lake Turkana, Kenya	*Au. boisei*	1.60	500	Partial cranium
KNM-ER 13750	Lake Turkana, Kenya	*Au boisei*	1.87	545	Cranium
KNM-ER 23000	Lake Turkana, Kenya	*Au. boisei*	1.87	490	Partial cranium
KNM-WT 17400	Lake Turkana, Kenya	*Au. boisei*	1.7	400	Partial cranium
KNM-CH1	Chesowanja, Kenya	*Au. boisei*	>1.42		Skull
Konso (KHA-10–525)	Konso, Ethiopia	*Au. boisei*	1.4	545	Partial cranium (juvenile)
Omo L338y-6	Omo, Ethiopia	*Au. boisei*	2.4	427	Partial cranium
Omo 323–896	Omo, Ethiopia	*Au. boisei*	2.2		Partial cranium
TM 1517	Kromdraai, South Africa	*Au. robustus*	2.0		Skull
DNH7	Drimolen, South Africa	*Au. robustus*	1.5–2.0		Crushed calotte (juvenile)
SK 54	Swartkrans, South Africa	*Au. robustus*	1.5–2.0	500	Partial cranium juvenile
SK 859	Swartkrans, South Africa	*Au. robustus*	1.5–2.0	450	Right half of endocranial cast
SK 1585	Swartkrans, South Africa	*Au. robustus*	1.5–2.0	530	Partial cranium (face)
SK 17	Swartkrans, South Africa	*Au. robustus*	1.5–2.0		Partial cranium
SK 46	Swartkrans, South Africa	*Au. robustus*	1.5–2.0		Cranium
SK 48	Swartkrans, South Africa	*Au. robustus*	1.5–2.0		Partial cranium and base
SK 52/SKW 18	Swartkrans, South Africa	*Au. robustus*	1.5–2.0		Partial cranial vault (juvenile)
SK 54	Swartkrans, South Africa	*Au. robustus?*	1.5–2.0		Crushed cranium (juvenile)
SK 47	Swartkrans, South Africa	*Au. robustus*	1.5–2.0		Partial cranium (face and part of base)
SK 83	Swartkrans, South Africa	*Au. robustus*	1.5–2.0		

aECV estimates from Holloway et al. (2004).

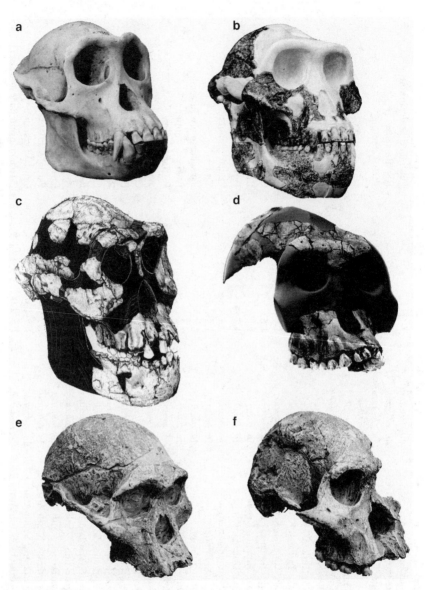

Figure 11.3. Three-quarters views of relatively complete gracile australopith specimens: (a) *Pan troglodytes* (male); (b) *Au. afarensis* (Hadar reconstruction); (c) *Au. afarensis* (AL. 444–2); (d) *Au. garhi* (BOU-VP-12/130); (e) *Au. africanus* (Sts 5); (f) *Au. africanus* (Sts 71). Image of AL. 444–2 courtesy W. Kimbel; Image of BOU-VP-12/130 © 1999 David L. Brill.

craniodental, with limited evidence for modest differences in postcranial anatomy (see Green et al., 2007). In general, female australopiths were chimplike in body size, weighing approximately 25–35 kg; males were probably 50 percent more massive, averaging 40–50 kg (McHenry, 1992), larger than most male chimpanzees measured in the wild (Pusey et al., 2005).

Australopithecus anamensis includes material from Allia Bay and Kanapoi, Kenya, and Asa Issie and Fejej, Ethiopia (M. Leakey et al., 1995; M. Leakey et al., 1998; Ward et al., 2001; White et al., 2006). Mandibular fragments from Lothagam and Tabarin, Kenya, dated to between 4–5 mya might also be included in the species. The sites listed above were probably woodland habitats, which varied in their degree of openness. Craniodentally, the species is represented by a temporal bone fragment, two partial maxillae, several mandibular fragments, and a good number of teeth. Much of the material is chimpanzeelike. The temporal bone has a small external auditory meatus that is elliptical in shape, and a flat TMJ with a horizontal tympanic plate and a flat preglenoid plane. The upper jaw has a procumbent premaxilla with a convex nasoalveolar clivus. The mandibular symphysis is deep and sloping, and there is a slight diastema between the canines and incisors. In terms of the teeth, the incisors are broad and spatulate, a little smaller than in chimpanzees. The canines are intermediate in size and shape between *Ar. ramidus* and *Au. afarensis* (they are less symmetrical than in *Au. afarensis* but more so than in *Ardipithecus* and *Sahelanthropus*); they have long, robust roots. The premolars are oval, with unequally sized buccal and lingual cusps. The upper molar crowns are broader mesially than distally. The most distinctive derived feature of *Au. anamensis* is the size and thickness of the cheek teeth: the crowns are larger, flatter and more bulging than in *Ar. ramidus,* with slightly thicker enamel (maximum radial thickness ranges from 1.3–2.3 mm) (White et al., 2006).

Australopithecus afarensis is one of the best-sampled species of early hominins, with fossils from sites in Ethiopia (Behlodelhie, Dikika, Fejej, Hadar, Maka, and White Sands), Kenya (West and East Turkana), and Tanzania (Laetoli), ranging in age from 3 to 4 million years old. *Au. afarensis* is in many ways a slightly more derived version of *Au. anamensis,* and a good argument can be made that the two species represent ends of an evolutionary continuum (Kimbel et al., 2006; White et al., 2006). For a thorough review of *Au. afarensis*'s context and anatomy, see Kimbel et al. (1984, 2004) and Kimbel and Delezene (2009). Not surprisingly, much about *Au. afarensis* is chimpanzeelike. The upper incisors are rather spatulate and large (although smaller than in *Pan*), with I^1 much larger than I^2; there is frequently a slight diastema between the incisors and canines; the premaxilla is long and convex, with a smooth transition between the nasal cavity floor and the nasoalveolar clivus; the palate is narrow, usually shallow, and with parallel tooth rows; the mastoid processes are large and very pneumatized; the mandibular fossa is shallow, with little development of an articular eminence. The brow ridge, which resembles that of chimpanzees in many ways, is coronally oriented and gets thicker along the lateral margins. So far, no *Au. afarensis* crania are complete enough to measure the cranial base angle, but a reasonable hypothesis is that CBA1 is relatively unflexed, as in chimpanzees (see below).

The major derived features in *Au. afarensis* mostly relate to being bipedal and to masticatory features related to chewing mechanically demanding diets. Correlates of the former include a foramen magnum that is shifted forward relative to other parts of the cranial base, as in *Ar. ramidus* and *S. tchadensis;* the nuchal plane is low and tends to be more horizontally oriented than in chimpanzees (albeit a little convex as in *Pan*); the basioccipital is short, with more coronally oriented petrous temporals; and the face is slightly less prognathic than in *Pan,* largely because of more upper facial projection (Kimbel et al., 2004). Consequently, the angle between the plane of the foramen magnum and the orbital plane is approximately 90°, directing the gaze forward in upright posture. Many other derived features of *Au. afarensis* appear to be related to the combination of larger cheek teeth, smaller canines, and more bite force generation and resistance. Most importantly, the molar crowns are substantially larger, flatter and with thicker enamel than in *Pan* or any earlier hominins (White et al., 2000). The canines have some degree of oblique wear, but they are smaller and more symmetrical than those of chimpanzees or any of the earliest hominins (Haile-Selassie et al., 2004). The lower first premolars (P_3) are often unicuspid and somewhat oval, as in *Pan;* but some are biscupid, and they tend to be bigger than those of chimpanzees. There is evidence for hypertrophy of the jaw adductors, including the presence of sagittal and temporonuchal crests in some male specimens for the posterior temporalis, and a large, thickened zygomatic arch.

A few other derived features of *Au. afarensis* merit attention. First, ECV in *Au. afarensis* can be estimated reliably from four adult skulls (Holloway et al., 2004): AL 444–2 (550 cm³), AL 162–28 (400 cm³), AL 333–105 (400 cm³), and AL 333–45 (485 cm³), yielding an average of 458 cm³ (see Figure 11.5). *Au. afarensis* ECVs are thus slightly but significantly larger than those of the known *Sahelanthropus* and *Ardipithecus* crania (both less than 400 cm³) and chimpanzees, in which males and females average about 400 cm³ and 370 cm³, respectively (Tobias, 1971). Larger brains account for a variety of subtle features related to a larger and slightly rounder neurocranium than that of *Pan* or *Sahelanthropus.* The size of the external auditory meatus varies considerably but is generally larger than those of *Pan* and earlier hominins (Kimbel et al., 2004). Finally, most *Au. afarensis* crania have an occipitomarginal sinus venous drainage pattern (see Chapter 6), which is rare in hominins and great apes but common in robust australopiths (Kimbel, 1984; Falk, 1986).

Australopithecus bahrelghazali is known from a single mandible fragment and a maxillary premolar found in Chad, dated biochronologically to approximately 3.5 mya (Brunet et al., 1995; Lebatard et al., 2008). The proposed species has thicker enameled and larger cheek teeth with more symmetrical premolars than those of *Pan, A. anamensis,* and the other earliest hominins— all features present in *Au. afarensis.* The mandible has a vertically oriented symphysis with a reduced superior transverse torus, which may indicate a species-level distinction (Guy et al., 2008).

Australopithecus africanus is represented by a large sample of crania, jaws, and teeth from at least four South African sites (Taung, Makapansgat, Sterkfontein, and Gladysvale) dated to 2–3 mya. Two partial skeletons from the South African site of Malapa, dated to 1.8–2.0 mya, have been proposed as a new species, *Au. sediba,* but they probably also belong in the *africanus* hypodigm. *Au. africanus* has a number of similarities to *Au. afarensis* but also some differences. These include a more vertical face that has more features probably related to generating and resisting high chewing forces, including "anterior nasal pillars" above the premolars that frame the lateral margins of the nasal aperture, more anteriorly positioned zygomatic arches (bringing the masseter forward), and a relatively wide mandibular corpus. In addition, *Au. africanus* tends to have smaller anterior teeth and larger-crowned, more thickly enameled postcanine teeth than *Au. afarensis.* ECV estimates range from 400 cm^3 to 560 cm^3, with a mean of 461 cm^3 (Holloway et al., 2004; see also Falk et al., 2000). CBA1, measured in one specimen, Sts 5, is 147° (Spoor, 1997), thus intermediate between the mean CBA1 in chimps (159°) and in modern humans (137°).

Australopithecus garhi, proposed by Asfaw et al. (1999), is primarily represented by a partial cranium (BOU-VP-12/130), along with some other material from the site of Bouri, Ethiopia. For the most part, the Bouri cranium resembles *Au. afarensis* in having a small ECV (450 cm^3); small sagittal crests; a prognathic face; a projecting premaxilla with procumbent incisors; and a shallow palate. What is surprising (*garhi* means surprise in the Afar language) is that the *Au. garhi* cranium also possesses large anterior and posterior teeth, including enormous (but slightly oval) premolars and large molars in the robust australopith size range (Asfaw et al., 1999).

One final proposed species to note is *Au. (K.) platyops,* discovered by Meave Leakey and colleagues on the west side of Lake Turkana, Kenya, and dated to 3.2–3.5 mya (M. Leakey et al., 2001). The single cranium (KNM-WT 40000), which is highly distorted, resembles *Au. anamensis* and *Au. afarensis* in many respects, with moderately sized molars, a small external acoustic meatus, and a small brain (probably 400–450 cm^3). But the cranium also has an apparently flat, vertically oriented, tall midface and subnasal clivus, and the zygomatic process of the maxilla is positioned above the premolars rather than the molars. These derived features bear some similarities to *H. rudolfensis* (see below). Some researchers think that the fossil, which is substantially deformed, is a distorted cranium of *Au. afarensis* (White, 2003).

Robust *Australopithecus*

The term *robust* means "strongly built" and has come to be informally synonymous with the three more megadont australopith species: *Au. boisei, Au. robustus,* and *Au. aethiopicus,* illustrated in Figure 11.4. These species are often put into a distinct genus, *Paranthropus,* by those who believe they constitute a distinct clade. In a way, the term *robust* is unfortunate, because only certain aspects of the skull are strongly built, mostly in males rather than females.

Australopithecus boisei, first proposed as *Zinjanthropus boisei* (L. Leakey, 1959), comes from several East African sites, including Olduvai and Peninj (Tanzania); Omo Shungura and Konso Gardula (Ethiopia); Chesowanja, Koobi Fora, and West Turkana (Kenya); and possibly Meleme (Malawi) (see Wood and Constantino, 2007). The sites span a period from 2.3 to 1.3 mya. Much of *Au. boisei*'s anatomy is dominated by the combination of small incisors and canines with enormous premolars and molars that have large crowns, hyperthick enamel, and flat occlusal surfaces. The *Au. boisei* face is massive and wide, especially in the midface, because of a large, platelike infraorbital region combined with a widely flared zygomatic arch. Because these arches are wide and positioned anteriorly relative to other facial structures, the face is slightly dished in lateral view and diamond-shaped in frontal view (see Figure 11.4). Endocranial volume ranges from 400 to 550 cm^3, averaging 492 cm^3. The upper face projects substantially in front of the neurocranium, setting up a massive, flat brow ridge with much postorbital constriction relative to the

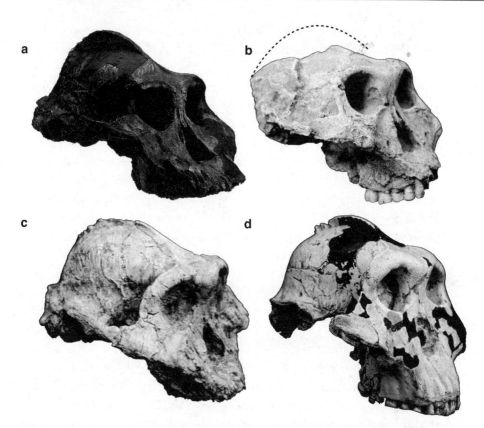

Figure 11.4. Three-quarters views of several relatively complete robust australopith crania: (a) *Au. aethiopicus* (KNM-WT 17000); (b) *Au. robustus* (SK 48); (c) *Au. boisei* (KNM-ER 406); (d) *Au. boisei* (OH 5). Note that SK 48 is partially deformed.

width of the face. A cranial base that is as flexed as in modern humans—CBA1 in OH 5 is 135°—contributes to *Au. boisei*'s facial orthognathism and may also help account for a more anteriorly positioned foramen magnum. *Au. boisei* males have large cranial crests (sagittal and compound temporonuchal) for the origin of an obviously massive temporalis muscle. The zygomatic arches are wide and thick and arise very anteriorly from the maxilla (above the premolars). The palate is thick, and the subnasal clivus is wide and somewhat horizontally oriented, with a smooth entrance into the nasal cavity. The TMJ has a deep, wide mandibular fossa, a pronounced articular eminence, and a short preglenoid plane. The mandibles are strikingly robust, with big condyles, a deep and buttressed symphysis, and a wide and tall corpus whose cross-sectional area is two to three times larger than expected for the species' body mass (Hylander, 1988). Like *Au. afarensis, Au. boisei* has an occipitomarginal venous drainage pattern.

Australopithecus aethiopicus is an interesting comparison with *Au. boisei* (see Figure 11.4). The species is known primarily from one cranium (KNM-WT 17000) and a mandible (KNM-WT 16005) from West Turkana, Kenya (Walker et al., 1986); and from some jaws and teeth from Omo, Ethiopia, including the type specimen Omo 18–1967–18 (Arambourg and Coppens, 1968). The material is mostly dated between 2.3 and 2.5 mya. *Au. aethiopicus* has large premolars (less molarized than in *Au. boisei*) and large molars, but probably had larger anterior teeth, as judged by the size of the alveolar crest around the canine and incisor tooth roots. Its cranium is also less derived and more like that of *Au. afarensis,* with a very unflexed cranial base in the chimpanzee range (CBA1 of WT 17000 is 156°), a more prognathic face, a more procumbent premaxilla, a smaller brain (ECV of the one known cranium is 410 cm^3), and enormous sagittal and compound temporonuchal crests for the posterior temporalis muscles. The TMJ is also more chimplike, with a shallow mandibular fossa, a flat preglenoid plane, and a low articular eminence.

Australopithecus robustus fossils come from the sites of Kromdraai, Swartkrans, Drimolen, and Gondolin in South Africa, all dated to between 2.0 and 1.5 mya. The species generally has small incisors and canines but large-crowned, thickly enameled premolars and molars—although not quite as large as those of *Au. boisei* (Grine and Martin, 1988). As in *Au. boisei,* the *robustus* face is large, flat, and vertical, with a thick palate. A reliable ECV, of 530 cm^3, is known from only one adult fossil, SK 1585, which may be slightly larger than the species' mean brain size, judging from other partially complete or immature fossils. The cranial base in *Au. robustus* is probably quite flexed. The combination of a big face, large chewing muscles, and a modest-sized brain results in ectocranial crests and strong postorbital constriction behind a variably thick supraorbital torus. The TMJ resembles that of *Au. boisei* in being deep, short, and wide, with little preglenoid plane. The mandible is also massively robust.

Despite its many similarities to *Au. boisei, Au. robustus* differs in several respects. Among them, the malar surface below each orbit is not flat and visorlike, as is typically the case in *Au. boisei,* but has a triangular depression because the maxillary portion is recessed relative to the zygomatic portion and the nasal margins (Rak, 1983). Also, the nasoalveolar clivus in *Au. robustus* tends to be more depressed relative to the anterior nasal pillars below the nose, giving the clivus a more gutterlike shape.

Developmental Bases of the Australopith Radiation

Australopith diversity raises many questions. How many species were there, and how were they related to one another? What shifts occurred to transform a basal hominin, something like *Sahelanthropus* or *Ardipithecus,* into an early australopith? And what developmental shifts transformed earlier, more gracile species such as *Au. anamensis* and *Au. afarensis* into more robust species such as *Au. boisei* and *Au. robustus?* The goal here is not to dwell on all the detailed features that help define each species and elucidate their phylogenetic relationships but to consider major developmental and functional shifts that account for the broad range of variation within this group of hominins. This kind of broad approach, which looks at the entire range of australopith variation, underscores the importance of several general trends related to tooth size, and adaptations for producing large bite forces and resisting the high strains these bites generated. In fact, one can make the argument that four interrelated suites of transformations suggestive of evolutionary tinkering account for a large proportion of the observed variation: (1) slight increases in brain size; (2) an increase in posterior tooth size combined with a decrease in anterior tooth size; (3) variations in the degree and nature of facial prognathism and projection; and (4) variations in facial size and other features related to producing and resisting masticatory forces.

Endocranial volumes

Compared to later hominins, australopiths have relatively small, apelike ECVs, ranging from 387 cm^3 to 560 cm^3 (see Figure 11.5). The smallest-brained australopiths, many of which are older than 2.5 mya, fall within the adult chimpanzee range of variation (282–500 cm^3) (Tobias, 1971). A number of larger-brained crania, such as AL 444–2 (550 cm^3, *Au. afarensis*), Stw 505 (550 cm^3, *Au. africanus*), SK 1585 (530 cm^3, *Au. robustus*), and KGA10–525 (545 cm^3, *Au. boisei*), fall significantly outside the chimpanzee range but are within the gorilla range (350–752 cm^3), although comparisons must allow for the fact that gorillas have much larger body masses.

As discussed in Chapter 6, we do not know to what extent body mass variation accounts for the variation in australopith ECVs, or whether some of these increases derive from the presence of more neurons or other developmental factors. Regardless, variation in ECV influences craniofacial shape considerably, with larger-brained australopith crania tending to have more rounded

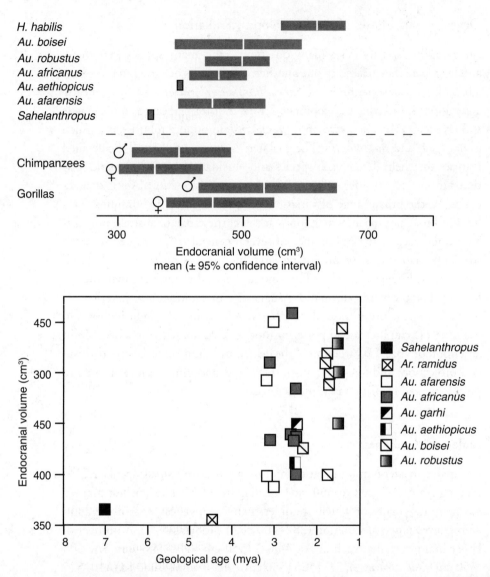

Figure 11.5. Endocranial volume (ECV) in chimpanzees, gorillas and australopiths. Top: by species. Bottom: over time.

cranial vaults (see Kimbel et al., 2004: 26–34), more flexed cranial bases, and sometimes wider cranial bases. As discussed below, these factors in turn interact with facial variations to affect brow ridge shape, facial orientation, and other aspects of cranial shape.

Tinkering with Teeth

If there is one quintessential feature of the australopiths, it is their large, thickly enameled cheek teeth combined with a general reduction in the anterior teeth—first the canines and then the incisors. In the interests of brevity, I focus here on just the permanent teeth and on the major trends in tooth size and shape from earlier, more gracile australopiths to later, more robust forms. Readers interested in more details about tooth size and shape variations should consult more extensive reviews (e.g., White et al., 1981; Blumenberg, 1985; Wood et al., 1988; Aiello and Dean, 1990; Wood, 1991; Suwa et al., 1994, 1996; Lockwood et al., 2000; Teaford and Ungar, 2000; Ward et al., 2001; Scott and Lockwood, 2004; White et al., 2006).

Let's start with the front teeth (Figure 11.6). Early gracile australopiths such as *Au. anamensis* and *Au. afarensis* have incisors that are significantly smaller than those of chimps, both absolutely and relative to cranial size; but like chimpanzees, they tend to have large and spatulate maxillary (upper) central incisors. Geologically younger australopiths, including *Au. boisei* and *Au. robustus,* have even smaller incisors, especially uppers. Important exceptions with larger incisors include *Au. garhi* (Asfaw et al., 1999) and probably *Au. aethiopicus; Au. (K.) platyops* also has large incisors (as judged by the roots) in which the centrals and laterals are about the same size (M. Leakey et al., 2001). Canine reduction is another trend continued from earlier hominins, whose canines are already smaller, less dimorphic, more symmetrical, and less dagger-shaped than those of great apes—and therefore without a C-P$_3$ honing complex (see above; Haile-Selassie et al., 2004). Despite much sexual size dimorphism, one can characterize *Au. afarensis* and *Au. anamensis* canines as diamond-shaped, even smaller, and less projecting beyond the occlusal plane than those of great apes or the earliest hominins (see Figures 11.1 and 11.6). In later australopiths and continuing in *Homo,* the canines became more symmetrical, incisor-shaped, and less projecting. As australopith canines became more incisiform, they also have wear patterns more typical of incisors. In apes, and to a lesser extent the earliest hominins, the upper and lower canines wear along their mesial and distal margins, respectively, but more incisiform australopith canines wear primarily at the crown tips.

As the front teeth became smaller in australopiths, the back teeth expanded in several ways. One trend is that the premolars became larger and more molarlike (Figure 11.7). For example, the P$_3$ in apes tends to be asymmetrically oval-shaped, with a large, conical buccal cusp (protoconid). In *Au. ana-*

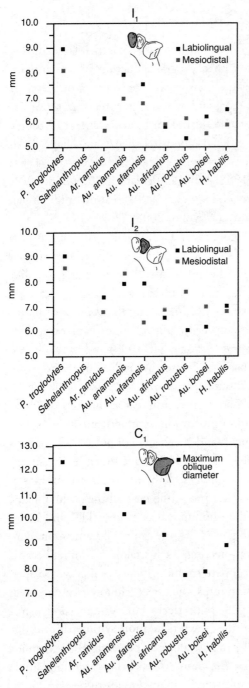

Figure 11.6. Mean labiolingual and mesiodistal crown dimensions of the anterior dentition in *P. troglodytes* and early hominins. (Data from White et al., 1981; Wood et al., 1988; Tobias, 1991; Wood, 1991; Brown and Walker, 1993; Suwa et al., 1994, 1996; Lockwood et al., 2000; Ward et al., 2001; Brunet et al., 2002, 2005; Kimbel et al., 2004; Scott and Lockwood, 2004; White et al., 2006; Haile-Selassie et al., 2009, and Suwa et al., 2009b.)

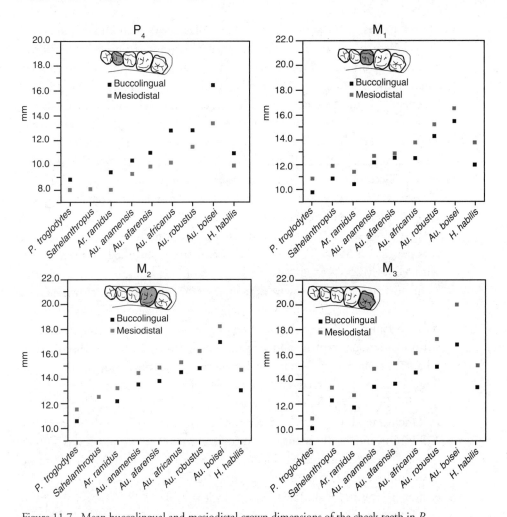

Figure 11.7. Mean buccolingual and mesiodistal crown dimensions of the cheek teeth in *P. troglodytes* and early hominins. (Data from White et al., 1981; Wood et al., 1988; Tobias, 1991; Wood, 1991; Brown and Walker, 1993; Suwa et al., 1994; Suwa et al., 1996; Lockwood et al., 2000; Ward et al., 2001; Brunet et al., 2002, 2005; Kimbel et al., 2004; Scott and Lockwood, 2004; White et al., 2006; Haile-Selassie et al., 2009, and Suwa et al., 2009b.)

mensis and *Au. afarensis,* the P_3 is slightly larger and more symmetrical, and in *Au. afarensis* it often has an expanded lingual cusp (metaconid) that is more anteriorly placed so that the crown is slightly rounder. This trend continues in later australopiths, such as *Au. africanus,* and reaches its extreme in species such as *Au. boisei* and *Au. robustus,* in which the P_3s are as large and square as chimp molars. In addition, the other premolars, both mandibular and maxillary, are larger, squarer (from buccolingual widening) and more "molariform" in australopiths than in the earliest hominins and *Pan,* with extreme enlargement in the robust species. Finally, the molars differ dramatically and follow a generally equivalent trend (despite much variation) with substantial size increases in robust australopiths as a whole (see Figure 11.6).[2] As an example, the average crown area of M_2 in *Au. afarensis* is 193 mm^2, about 60 percent larger than in chimps (120 mm^2); M_2 crown area in *Au. boisei* is approximately 306 mm^2, about 60 percent larger than in *Au. afarensis* (Wood, 1991). Crown expansion is more exaggerated in the distal teeth, partly from increases in length but especially from increases in width. In both chimps and humans, the first molars tend to be largest, and the third molars tend to be smallest (M1 > M2 > M3), but in robust australopiths, the trend is reversed, with M3 > M2 > M1.

As postcanine tooth crowns expanded in the australopiths, crown enamel also became thicker. It is difficult to characterize enamel thickness in simple terms because it is a continuous variable, with considerable variation both within individual crowns and between teeth (for example, as discussed in Chapter 7, enamel tends to be thicker on the "functional" side of cusps—the lingual side of upper teeth and the buccal side of lower teeth). At the risk of oversimplifying, the maximum enamel thickness of the cervical wall in the first molars tends to be just under 1 mm in chimpanzees (0.58–0.98 mm) and 1–3 mm in modern humans (Schwartz et al., 2000; Suwa and Kono, 2005). Similar measures of enamel thickness range from 1.3 to 2.9 mm in gracile australopiths such as *Au. anamensis, Au. afarensis,* and *Au. africanus* (Ward et al., 2001; White et al., 1994; G. T. Schwartz et al., 1998; White et al., 2006) and from 1.3 to 3.5 mm in *Au. robustus* (G. T. Schwartz et al., 1998). The enamel on *Au. boisei* teeth is even thicker (Grine and Martin, 1988). Thicker-enameled, larger crowns in the more megadont robust australopiths are typically accompanied by increased complexity of the crown surface, such as a shelf of enamel around the base of the protocone or protoconid (termed, respectively, a *protostyle* or *protostylid),* supernumerary cusps, and more intricate patterns of fissures between the cusps. This complexity tends to wear away soon after the tooth erupts into occlusion.

2. These generalizations ignore many interesting variations. For example, some fossils assigned to *Au. africanus* from the western part of the Sterkfontein site (Stw) have comparatively large molars, whereas some molars assigned to *Au. robustus* from the site of Kromdraai are comparatively small (Grine, 1988).

Detailed histological studies are beginning to elucidate the developmental bases for anterior tooth reduction and postcanine crown expansion in australopiths. There are three basic ways to generate variations in tooth size and thickness (for reviews, see Macho and Wood, 1995; G. T. Schwartz et al., 2000; Dean, 2006; T. Smith et al., 2007a). First, duration of tooth crown growth can be altered (longer for larger or thicker; shorter for smaller or thinner). For example, apes secrete enamel at an average rate of roughly 4μm/day, so enamel that is 0.5 mm thicker could be achieved by extending the period of crown formation by approximately four months (125 days) (Dean et al., 2001). Second, the rates at which secretory ameloblasts synthesize enamel matrix can be varied (faster for larger or thicker; slower for smaller or thinner). Third, the developing enamel-dentine junction (EDJ) can vary in size (and shape) because of variations in the rate of mitosis and hence the number of active cells in the inner enamel epithelium as it grows. This happens mostly during the bell stage of crown development (see Chapter 3).

Although there are important differences in EDJ morphology among apes, early hominins, and australopiths, histological studies suggest that the front teeth of robust australopiths are smaller than those of apes because they have shorter crown-formation times. For example, I^1 crown formation lasts approximately 3.5 years in chimps and gracile australopiths but only 2.5 years in robust australopiths (Beynon and Dean, 1988; Dean et al., 2001). Further, chimpanzees grow their permanent canines in about 6–7 years, but in australopiths this period of growth appears to be reduced to 3–4 years (Schwartz and Dean, 2001; Dean et al., 2001).

Cheek teeth are a different story. In spite of their size differences, postcanine crowns take about the same time to form enamel in great apes and australopiths, both robust and gracile. The growth period for a first molar crown, for example, is about 2.5 years in some early hominins and African apes but closer to 3.0 years in modern humans (Beynon and Wood, 1987; Beynon and Dean, 1988; Dean, 2006; Reid and Dean, 2006). Apparently one way that big-toothed australopiths grew larger teeth with much thicker enamel was by increasing daily secretion rates. According to LaCruz et al. (2008), the average rate of cuspal enamel secretion in chimps and modern humans is about 4.2 μm/day, approximately 4.6 μm/day in *Au. afarensis,* and 5.8 μm/day in *Au. boisei.* But because such differences would account for crowns that are at most 25–40 percent bigger or thicker enameled, a significant proportion of tooth size increases in megadont australopiths must have come from more growth of the EDJ at the bell stage. Increased numbers of differentiated cells at the EDJ, together with faster differentiation at the inner enamel epithelium, not only expands the tooth crown area that will eventually be capped with enamel but also increases the number of secretory amelobasts. This increase sets up a more acutely angled secretory cell front, apparent in the orientation of the striae of Retzius (see Shellis, 1984). Support for this hypothesis comes from LaCruz et al. (2006), who found that molars in *Au. robustus* have more acutely angled Retzius lines and higher enamel extension rates near the cervical margin than those in *Au. africanus.*

We have yet to identify what genetic mechanisms underlie the processes that cause variations in australopith megadonty. In baboons, crown size (specifically crown width) correlates slightly with body size, but enamel thickness is independent of body size (Hlusko et al., 2004, 2006), and body size probably did not vary much between australopith species. This, in light of the different mechanisms reviewed above, suggests that selection acting on australopith tooth size was multifactorial. Some degree of postcanine megadonty likely occurred from changes in the size of growth fields along the jaw specified by patterning genes (Tucker and Sharpe, 2004). Other postcanine size increases could hypothetically result from particular genes that are active within developing tooth germs that lead to faster ameloblast secretion rates (e.g., Jernvall et al., 2000; Salazar-Ciudad and Jernvall, 2005). Finally, within ape and presumably also hominin species, males have bigger canine crowns than females because of longer durations of crown formation (Schwartz and Dean, 2001), but differences between species suggest that variations in canine crown size would have been made possible by changes to processes that regulate either the duration or the rate of crown formation (Dean, 2009). Moreover, we cannot rule out the possibility that some degree of megadonty in the permanent dentition derived from epigenetic factors during growth, as young individuals chewed tough diets with their deciduous teeth while growing their permanent crowns (see Chapter 7).

Bigger tooth crowns also come with additional features, including bigger, thicker, more robust, and more complex root systems. Chimp lower premolars typically have two roots, as in the earliest hominins and *Au. afarensis* and *Au. anamensis* (Wood, 1991; Ward et al., 2001; White et al., 2006). In South Africa, some *Au. africanus* and *Au. robustus* specimens also have double-rooted premolars, but in many cases these become thickened and fused into a single root (known as a Tomes' root) that still preserves a large groove running down its length, thus resembling the single rooted premolars of the genus *Homo* (Wood et al., 1988). In eastern Africa, however, the premolar roots of *Au. boisei* are not only stouter but also more spread out than those of earlier hominins, with more divergence of the individual roots (Wood et al., 1988). *Au. boisei* also has significantly more robust roots in the molar teeth (Wood et al., 1988). Root size and complexity is functionally important because bigger roots can distribute masticatory forces over a greater surface area and therefore lower stresses (see below). Root development is also integrated with the growth of the alveolar process, and it would be interesting to test the correlation among various hominin species between the height and width of the alveolar crest and the size of the tooth roots. Finally, tooth size (both crown and root) relates to many other aspects of skull form in the mandible and maxilla, including facial position and shape, the TMJ, and the zygomatic arches. These correlates of big cheek teeth and small anterior teeth are treated below.

Facial Projection and Prognathism

Australopiths vary in how the face is hafted onto the rest of the skull. Here it is critical to distinguish between *projection* and *prognathism*. As discussed in Chapter 5, facial projection describes the forward growth of the face relative to the cranial base, either from displacement relative to the middle cranial fossa at the face's posterior margin or from drift along the face's anterior surface. In contrast, prognathism describes the orientation of the facial profile relative to the rest of the skull, which is influenced to a large extent (though not solely) by variation in the cranial base angle. Because of the facial block (Figure 5.8), a more extended CBA rotates the whole face dorsally (up and out), leading to a more prognathic face and contributing to a longer supraorbital torus; a more flexed CBA rotates the face more ventrally (down and in), leading to a less prognathic face and contributing to a shorter supraorbital torus. Additional factors influence facial prognathism and projection, especially brain size, anterior tooth size, and the relative size of the midface. Brain size also affects upper facial projection, because larger brains can elongate the anterior cranial fossa relative to the front of the anterior cranial base, leading to a more vertical, curved frontal squama and reduced supraorbital structures. Larger, more procumbent incisors and canines augment lower facial prognathism because larger front teeth have larger root cross-sections, enlarging the alveolar process of the premaxillary region. Consequently, hominins with large, procumbent anterior teeth tend to have more convex, projecting lower faces below the nose. Finally, variations in the anterior drift of the zygomatic region and the nasal margins influence the advancement of the midface relative to the cranial base and neurocranium and hence the degree of midfacial projection.

Australopiths vary considerably in facial orientation and projection because of differing combinations of the above factors. Although we lack enough relatively completely crania with intact cranial bases to be definitive, the available evidence suggests three general combinations within the genus as a whole, illustrated in Figure 11.8. First, *Au. afarensis* retains a fairly primitive configuration, partly reminiscent of chimpanzees and of *Sahelanthropus* (see above), but with a few differences. One important primitive feature may be the cranial base angle. The most complete *Au. afarensis* skull, AL 444–2, has a partly reconstructed cranial base, and the foramen caecum is not preserved (Kimbel et al., 2004). Nonetheless, one can estimate from various clues (e.g., the position of nasion, the orientation of the presphenoid plane, and the shape of the deep olfactory pit) that this skull's CBA1 is somewhere between 152 and 162°, probably about 155°. This crude estimate, if correct, reinforces the general impression that the *Au. afarensis* cranial base angle was not significantly different from that in chimpanzees and helps account for its chimplike degree of prognathism. As in chimps, facial prognathism in *Au. afarensis* is enhanced by the species' large, procumbent anterior teeth, generating a convex, prognathic subnasal region.

Facial profiles, however, differ in *Au. afarensis* and chimps because of additional factors. First, *Au. afarensis* has a slightly larger ECV, averaging 455 cm³; ECV in AL 444–2 is 550 cm³ (Holloway et al., 2004). This encephalization, though modest, leads to a more rounded and anteriorly expanded anterior cranial fossa (see Figures 11.3 and 11.8), thus generating less upper facial projection relative to the neurocranium and slightly smaller supraorbital structures in which there is no supratoral sulcus behind the brow ridges (see Chapter 5). In addition, faces project in *Au. afarensis* more than in chimps, particularly in the midface, leading to a slightly less concave, more vertical face in midsagittal profile (Kimbel et al., 2004). So far, no complete *Au. anamensis* crania have been discovered, but the available evidence suggests that *Au. anamensis* and *Au. afarensis* probably share a similar facial configuration, with the former perhaps being slightly more primitive (M. Leakey et al., 1995; Ward et al., 2001; Kimbel et al., 2006).

The other australopith species manifest three different configurations, all variations on the primitive condition in *Au. afarensis* (and presumably *Au. anamensis*). One group, which is represented by *Au. aethiopicus* and might also include *Au. garhi*, basically resembles *Au. afarensis* in having a very extended, albeit short, cranial base, combined with a relatively small brain and presumably large incisors (Figure 11.8). The KNM-WT 17000 cranium, attributed to *Au. aethiopicus,* has a more extended CBA of 156°, an extremely large and prognathic face, and an ECV of 410 cm³. We have no measure of CBA1 in *Au. garhi,* but

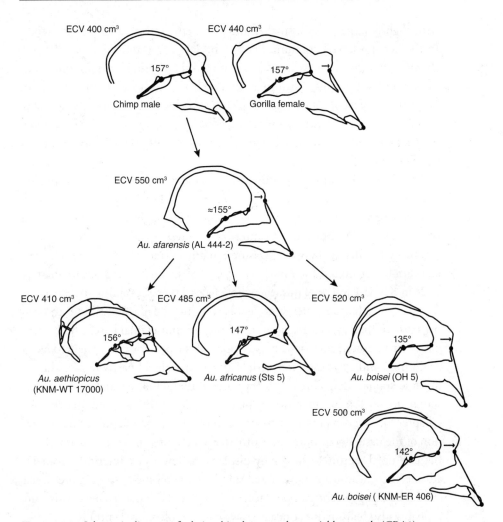

Figure 11.8. Schematic diagram of relationships between the cranial base angle (CBA1), endocranial volume (ECV), and facial projection in African great apes and different australopith crania. In general, crania with more prognathic faces have more extended cranial bases, and more vertically oriented faces occur in crania with more flexed cranial bases. Midsagittal sections of australopiths adapted from Kimbel et al. (2004).

the type specimen (BOU-VP-12/130) has a fairly prognathic face combined with an ECV of 450 cm³ (Asfaw et al., 1999). Both species, however, differ from *Au. afarensis* in other respects related to megadonty (see below).

A second derived australopith configuration is evident in *Au. africanus,* which has a somewhat more flexed cranial base than *Au. afarensis,* combined with slightly larger ECVs, smaller anterior teeth, and expanded cheek teeth. In the best-preserved cranium, Sts 5, CBA1 is 147°, and ECV is 485 cm³. *Au. africanus* thus has less facial prognathism than earlier australopiths, both overall and subnasally, but also lacks much of the upper or middle facial projection typical of robust australopiths (see below). Consequently, *Au. africanus* faces tend to have a somewhat reduced development of the supraorbital region and slightly more vertical faces than other gracile australopiths (see Figure 11.8).

The third configuration to consider is evident in *Au. boisei* and probably also in *Au. robustus* to a lesser extent. As Figures 11.5 and 11.8 show, both species have very projecting upper and middle faces, giving the masseter and anterior temporalis muscles very anteriorly located origins relative to the TMJ (see below). CBA1 cannot be measured in any currently known *Au. robustus* cranium but is quite flexed in *Au. boisei:* 135° in OH 5, and approximately 142° in KNM-ER 406. Thus, the whole face in these hominins appears to be rotated ventrally from a flexed cranial base, while the massive middle and upper parts of the face are displaced from the anterior cranial base. The combination of a flexed cranial base, hypertrophy of the posterior face, marked facial projection, and small anterior teeth leads to vertical and partly retracted faces in these crania. McCollum (1999) proposed that nasal septum growth in response to postcanine megadonty played a role in this reconfiguration (see Chapter 5). The vomer in robust australopiths inserts on the premaxillary portion of the nasal floor, more anteriorly than in chimps and early australopiths. Because nasal septum growth may displace the lower face inferiorly, vomeral advancement leads to a flatter nasal sill. When combined with a more flexed cranial base, a projecting upper face, and a taller, more massive face, this shift probably also helps redirect facial growth as a whole more ventrally.

The alternative configurations summarized above are not phylogenetic reconstructions, in part because characteristics such as ECV, CBA1, and facial projection are not simple or independent. As Chapter 5 outlines, many factors affect CBA1, including brain size, cranial base length, and facial length. Further, it is possible that *Au. boisei* and *Au. robustus* independently evolved somewhat similar faces. Both australopiths apparently combined slight increases in relative brain size (which contribute to a more flexed cranial base) with increased upper facial projection, which improves the face's efficiency in generating and resisting masticatory forces (see below). Other aspects of facial shape noted below suggest that these two species evolved big faces in slightly different ways.

Robusticity of Other Features Related to Mastication

Prognathism and projection are but two aspects of variation in australopith skulls associated with the demands of producing high masticatory forces and then resisting the strains these forces generate. Additional features to consider include the width and height of the face; how the face and jaws are buttressed to withstand high, repetitive chewing strains; and the ways in which the muscles of mastication attach to the skull. As discussed in Chapter 7, the basic principle of generating high force is either to augment a muscle's cross-sectional area or to boost its mechanical advantage (MA) by increasing the ratio of the muscle's lever and load arms; the basic principle of resisting strain is to add mass in the plane of deformation. Later we'll consider how these principles apply to australopith craniofacial function, but first let's review briefly how different australopiths accommodated large teeth and big masticatory muscles, and how they grew large, buttressed faces. Again, many of these variations fall on a continuum between more gracile, early species such as *Au. afarensis* and later, more robust species such as *Au. boisei.* Some variations also reflect alternative ways to achieve the same general effect. Readers interested in more details of australopith craniofacial variation should consult more comprehensive reviews by Rak (1983, 1988), Kimbel et al. (1984), Bilsborough and Wood (1988), Aiello and Dean (1990), and Kimbel et al. (2004).

Facial Masks

One obvious source of variation among australopith faces when viewed from the front is the size and shape of the roughly hexagonal facial "mask," illustrated in Figure 11.9. The top and bottom of the mask are the supraorbital torus and the alveolar process, respectively; the lateral margins of the orbits and zygomatic arches form the sides. Because faces can grow in a limited number of ways (reviewed in Chapter 4), we can infer that a few growth trends dominate the variation we observe. First, the robust australopith species grew taller faces primarily in the bottom half of the face. A large percentage of this extra growth must have been from more deposition along the diagonally oriented zygomaticomaxillary suture (thus simultaneously widening the midface) and from taller alveolar processes. (There are no significant differences in orbit height among the australopiths: all are between 30 and 40mm). Some increases in midfacial height may also have occurred from more inferior drift of the nasal cavity (see above). The combined effect is that a greater proportion of the face in the robust australopiths is *below* rather than *above* the zy-

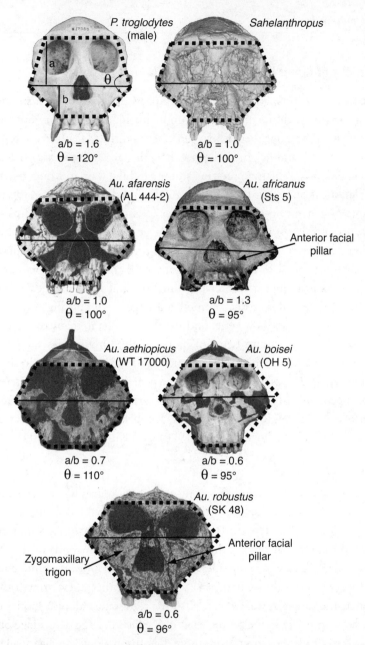

Figure 11.9. Anterior view of face in *P. troglodytes, S. tchadensis, Au. africanus, Au. africanus, Au. aethiopicus, Au. boisei,* and *Au. robustus,* highlighting variations in the proportion and angle of the facial masks and other key features (anterior nasal pillars and the zygomaxillary trigon). Ratio a/b describes the proportion of the face above the maximum width of the zygomatics; θ is the angle between the lateral margin of the orbits and the lateral margin of the zygomaxillary crest. Image of *Sahelanthropus* from Zollikofer et al. (2005), courtesy M. Brunet; Image of AL. 444–2 from Kimbel et al. (2004), courtesy W. Kimbel.

gomatic arches, placing the zygomatic arches higher relative to the occlusal plane. Taller midfaces in robust australopiths are accompanied by wider midfaces. We can infer that facial widening occurred largely from sutural deposition in the zygomaticomaxillary suture and from lateral drift of the zygomatic arches—both of which displace the zygomatic arches more laterally relative to the tooth row. Some facial widening in *Au. boisei* also must have occurred from sutural growth in the midsagittal plane, giving this species a slightly wider nose, palate, and interorbital region (see Kimbel et al., 2004).

Pillars, Gutters, and Convexities

Other aspects of large, robust faces in various australopiths may have grown slightly differently. In *Au. africanus* and *Au. robustus,* the lateral margins of the nose are slightly rounded and columnar because of the presence of structures called anterior nasal pillars, which rise above the premolars (Figure 11.9). Further, the maxillary portion of the infraorbital region in the two South African species did not drift anteriorly as much as the zygomatic portion, creating a steplike zygomaticomaxillary suture with a depressed "trigon" between the anterior nasal pillars and the zygomatic arches (Rak, 1983). In contrast, the infraorbital surface is smoother in East African species such as *Au. afarensis, Au. aethiopicus,* and *Au. boisei.* These differences, summarized in Figure 11.9, suggest that the midface became wider and more anteriorly positioned in slightly different ways (maybe independently) in the South and East African robust species.

Additional differences are evident in terms of facial flatness. The primitive condition, seen in the African great apes and *Au. afarensis,* is a face that is slightly concave in the midsagittal plane and convex in the transverse plane (see Figures 11.3, 11.4). Midsagittal concavity results from a procumbent premaxillary region combined with an extended cranial base and limited upper facial projection (see above). Transverse convexity comes from the combination of a receding zygomatic and a prognathic lower face. Thus, within the australopiths, we see several trends in facial flatness. As the anterior dentition became smaller and less procumbent, faces became flatter and less concave in the midsagittal plane (compare *Au. afarensis* and *Au. aethiopicus* with *Au. robustus* and *Au. boisei*). As the zygomatic bone drifted forward, the transverse contour of the face went from being convex in *Au. afarensis* to concave in *Au. aethiopicus* (which combines a prognathic face with zygomatics that are swung forward) to nearly flat in *Au. boisei* and *Au. robustus.* In addition, swinging the zygomatic arch forward via drift, combined with more cranial base flexion that ventrally rotates the whole face, leads to variations in the position of the zygomatic process of the maxilla. The anterior end of the zygomatic arch arises above the middle of the dental arcade (M^1 to M^1/P^4) in chimps and *Au. afarensis,* above M^1/P^4 to P^4 in *Au. africanus* (see Figure 11.3), and above the premolars in the robust australopiths (see Figure 11.4).

Ectocranial Crests

Big teeth come with big chewing muscles, including the temporalis. As discussed in Chapter 7, the anterior, middle, and posterior fibers of this fan-shaped muscle differ functionally: the more horizontally oriented posterior fibers generate bite force more effectively in the anterior teeth, and the more vertically oriented anterior fibers generate more force on the postcanine teeth. In addition, large temporalis muscles on small-brained heads often lead to the formation of ectocranial crests to accommodate the muscle's origin. These factors account for much of the variation in cresting evident among apes and australopiths, illustrated in Figure 11.10. In particular, species with large canines or incisors combined with small ECVs tend to have extensive development of the sagittal crest toward the posterior end of the vault. This crest often continues laterally above the nuchal region, toward the supramastoid portion of the temporal (a compound temporonuchal crest). Such cresting is most extreme in gorillas and in KNM-WT 17000 (*Au. aethiopicus*); crests associated with a large posterior temporalis are also present to a lesser extent in male chimps and *Au. afarensis,* both of which have relatively large incisors. Species with large cheek teeth, such as *Au. boisei,* tend to have a large middle and anterior temporalis that leads to the development of more anteriorly placed sagittal crests that extend onto the frontal bone, again with relatively smaller crests on bigger-brained individuals and females. The clear standout in terms of cresting is KNM-WT 17000, which has a tiny ECV (410 cm^3) combined with large anterior *and* posterior teeth. This cranium has an enormous sagittal crest that extends from somewhere on the frontal bone all the way to the occipital bone, where it participates in a large compound temporonuchal crest.

Another source of variation relating to temporalis size is the encroachment of the anterior temporalis onto the supraorbital torus. In chimps and early gracile australopiths, the temporal lines, which mark the margin of the temporalis origin on the vault, tend to converge relatively posteriorly on the vault and then diverge toward the front of the skull (Figure 11.10). This pattern reflects a relatively large posterior temporalis. In robust australopiths with big molars, the insertion area of the substantial anterior temporalis is concomi-

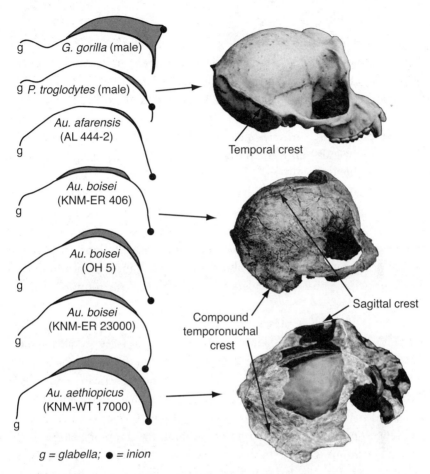

Figure 11.10. Ectocranial crests in chimpanzees, gorillas, and several australopiths (modified from Kimbel et al., 2004, Figure 3.6.5). Variations in the height and position of the sagittal crest and the development of the temporal crests, which merge with the nuchal crests in australopiths, indicate relative size of the different parts of the temporalis muscle.

tantly larger and extended anteriorly on the vault. As noted by Rak (1983), *Au. boisei* represents the extreme condition in which the temporalis inserts along most of the posterior margin of the supraorbital torus. The combination of wide faces, small brains, laterally positioned zygomatic arches, large temporalis muscles, and ventrally rotated faces creates a marked degree of postorbital constriction and a wider, rounder temporal foramen in the more robust australopiths than in the more gracile species.

Massive Mandibles

Because mandibles tend to preserve well, there are dozens of fossil australopith mandibles, allowing researchers to study their variation in detail (Johanson and White, 1979; Johanson et al., 1982; Hylander, 1988; Wood, 1991;
Lockwood et al., 2000; White et al., 2000; Ward et al., 2001; Taylor, 2006;
Rak et al., 2007). As with the cranium, we will not review australopith mandibular variation comprehensively but instead focus on the developmental bases
for some key trends that relate to the masticatory system's overall size, shape
and robusticity. In considering these trends, remember from Chapter 4 that
variations in mandible size and shape derive in part from variations in the size
of the mesenchymal condensations from which the bone grows. In addition,
during fetal and postnatal ontogeny, the mandible essentially grows forward
and downward from bone deposition in the condyles and along the posterior
margin of the ramus; the corpus also grows forward and outward from drift
(the V principle); and the alveolar crest grows around the tooth roots above
the corpus.

As Figure 11.11 illustrates, several major trends stand out in a comparison
of gracile and robust species. First, in robust species, as faces get taller and occlusal planes descend further below the TMJ, the mandibular ramus becomes
relatively taller. Presumably a large proportion of this increase in ramus height
derives from more growth within the condyle (see Chapters 4 and 7). A second trend is that the mandibular ramus in robust species is anteroposteriorly longer, possibly from less resorption along the anterior margin of the
ramus. The ramus is longer in this respect in *Au. afarensis* than in great apes,
and even longer in *Au. boisei* and *Au. robustus*. Third, as the major jaw adductor muscles become relatively larger in the more robust species, the muscles induce larger origin and insertion sites. In these species, the coronoid
process, the insertion of the temporalis, is relatively higher and deeper, with a
smaller notch between the coronoid and the condylar process; the gonial region of the mandible, where much of the masseter and medial pterygoids insert, also appears to be relatively larger and thicker. A fourth trend is that the
alveolar process is taller and wider in more megadont species, reflecting longer,
wider tooth roots. Fifth, the symphyseal region and the corpus also become
thicker and taller in the more robust forms, with relatively more development
of the superior than the inferior transverse torus. Sixth, the orientation of the
symphysis is highly variable, but in general its midsagittal contour becomes
more vertical and less rounded in more robust forms. Finally, the shape of the

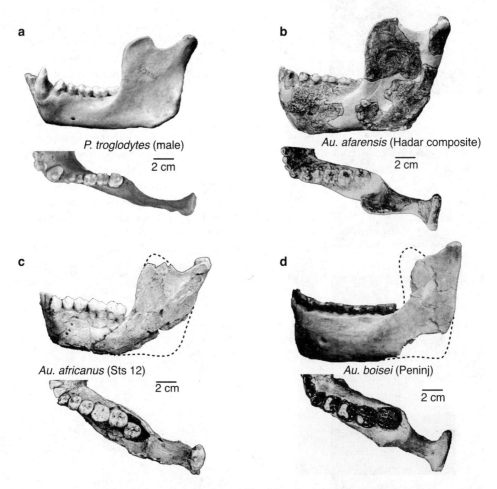

Figure 11.11. Lateral and superior views of mandibles: (a) male chimpanzee (courtesy Peabody Museum, Harvard University); (b) *Au. afarensis* (a composite of several specimens from the AL. 333 site); (c) *Au. africanus* (Sts 12, likely female); (d) *Au. boisei* (Peninj). The more megadont species, especially *Au. boisei,* tend to have taller rami and more robust, well-buttressed mandibular corpora with expanded alveolar crests.

jaw when viewed from above is more U-shaped (as in apes) in earlier, gracile species because of the combination of large front teeth (hence, a wide anterior region of the corpus) and a narrow intercondylar distance; as the anterior teeth become smaller and more crowded and the condyles move laterally in the more robust species, the shape of the tooth row becomes more V-shaped.

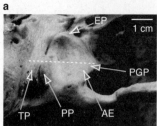

a

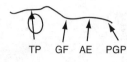

P. troglodytes (male)

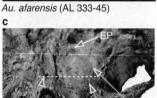

b

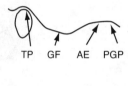

Au. afarensis (AL 333-45)

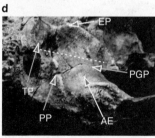

c

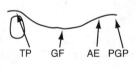

Au. africanus (Sts 19)

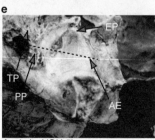

d

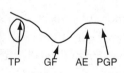

Au. aethiopicus (KNM-WT 17000)

e

Au. boisei (OH 5)

Figure 11.12. Left: oblique views of the temporal region in a male chimpanzee and representative australopiths, highlighting several variations in the anatomy of the TMJ. TP, tympanic plate; PP, postglenoid process; GF, glenoid fossa; EP, entoglenoid process; AE, articular eminence; PGP, preglenoid plane. Right: contour maps through the middle of the glenoid fossa (location of contour indicated by dashed line at left). In robust australopiths, the glenoid fossa is relatively shorter and more concave and has less of a preglenoid plane.

Glenoid Fossae

The glenoid fossa varies considerably among the australopiths (see Figure 11.12). At the primitive end of the spectrum, the configuration in *Au. anamensis, Au. afarensis,* and *Au. aethiopicus* (and also probably *Kenyanthropus*) generally resembles that in chimpanzees, gorillas, and early hominins such as *Sahelanthropus.* These species typically have a shallow glenoid fossa with a long preglenoid plane; a slightly tubular, relatively horizontal tympanic plate; a strong, laterally positioned postglenoid process; and a large entoglenoid process (M. Leakey et al., 1995; Ward et al., 2001; Kimbel et al., 2004). At the other end of the spectrum is *Au. boisei,* which looks as if the whole glenoid region was compressed from front to back, leading to a deep, short glenoid fossa, with a short, more convex, and less coronally oriented articular eminence. *Au. boisei* also has a more vertical tympanic plate that essentially replaces the postglenoid process as the posterior wall of the glenoid fossa. *Au. africanus* and *Au. robustus* have glenoid fossae that are intermediate between the two extremes.

Functional Shifts in Early Hominin Evolution

We now turn to function. Of the many differences evident between the hypothetical LCA, the earliest hominins, and various australopith species, the majority would have improved two kinds of performance: being bipedal, and chewing foods that require generating and resisting more force.

Locomotion

If any fundamental difference in function between apes and early hominins is paramount, it is bipedalism. Bipedal posture and walking pose several challenges to the head: to orient the eyes so that they look forward; to stabilize the head, and hence the gaze, against angular accelerations; and to reduce shearing forces at the atlanto-occipital joint (AOJ) (see Chapter 9). Many derived hominin features, evident in the crania of early hominins such as *Sahelanthropus* and *Ardipithecus* and later in the australopiths, would have improved performance in these functions during bipedal but not quadrupedal locomotion.

When primates stand or locomote, they generally look forward, holding their heads with the plane of the orbits (the line connecting the top and bottom of the orbital margins) usually within 20° of earth horizontal (Strait and Ross, 1999). In addition, the AOJ in higher primates in general, and in humans especially, is tightly constrained: the upper cervical vertebrae are habit-

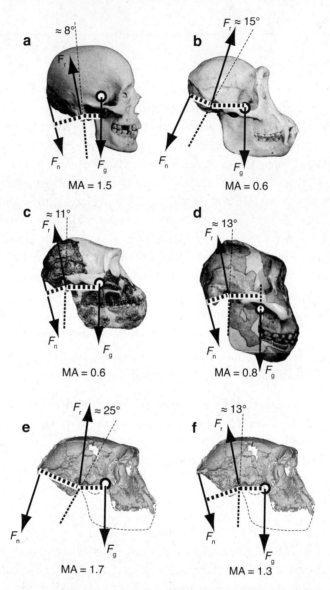

Figure 11.13. Postural biomechanics in (a) *H. sapiens;* (b) *P. troglodytes* (male); (c) *Au. afarensis* (Hadar composite); (d) *Au. boisei* (OH 5 with the Peninj mandible); (e) *Sahelanthropus* (reconstructed as a quadruped); (f) *Sahelanthropus* (reconstructed as a biped). I have plotted the approximate vector of the nuchal muscles (F_n, oriented toward C7), the head's center of gravity (F_g, calculated from area centroid in lateral view), their moment arms, and the estimated resultant force (F_r) and shearing angle at the atlanto-occipital joint. Reconstructing *Sahelanthropus* as a quadruped leads to high shearing forces. Also shown are the estimated mechanical advantages (MA) for the superficial neck extensors, which also would have been high in a quadrupedal *Sahelanthropus*. Image of *Sahelanthropus* from Zollikofer et al. (2005), courtesy M. Brunet.

ually oriented close to 90° to the plane of the foramen magnum, with the joint permitting only 10–12° of flexion and extension in humans (Graf et al., 1995a,b). These constraints match some key differences between ape and human cranial anatomy. In chimpanzees, the upper cervical vertebral column points backward and downward from the back of the skull base, but in humans, the upper cervical vertebral column curves slightly forward (there is a lordosis) from near the center of the skull base. Consequently, the angle between the foramen magnum and the orbital plane averages 64° ± 6° in chimpanzees, but 103° ± 7° in modern humans. All known hominin crania match the human configuration, indicative of upright posture (Table 11.2). This angle is approximately 95° in the *Sahelanthropus* reconstruction (Zollikofer et al., 2005), and the *Ar. ramidus* reconstruction also has an inferiorly oriented foramen magnum (Suwa et al., 2009a). In australopiths, the angle varies between 82° and 100°.

As discussed above, the change in the orientation of the foramen magnum to accommodate upright posture may have occurred in early hominins through alterations to the pattern of resorption and deposition in the nuchal portion of the occipital. Such a shift not only reorients the foramen magnum more perpendicularly to the orbits but also lowers shearing forces on the AOJ. As Figure 11.13 illustrates, shearing forces are caused by a joint reaction force at the AOJ that is not perpendicular ("normal") to the plane of the foramen magnum. Primate skull design suggests there has been strong selection to minimize shearing strains at the AOJ, which can endanger the brain stem and spinal cord if improperly countered (Demes, 1985). It is difficult to estimate AOJ shearing forces precisely because one must estimate the orientation of the

Table 11.2: Angle between foramen magnum (FM) and orbital plane (OP)

Species (specimen)	FM-OP angle (degrees)
Pan troglodytes (male and female)	64 ± 6
Sahelanthropus (TM 266–01–060–1)	95
Australopithecus afarensis (Hadar reconstruction)	93
Au. africanus (Sts 5)	85
Au. boisei (OH 5)	88
Au. boisei (KNM-ER 406)	89 (estimate)
Au. aethiopicus (KNM-WT 17000)	82 (estimate)
Homo habilis (KNM-ER 1813)	97
H. sapiens (male and female)	103 ± 7

nuchal muscle resultant (done here by pointing the resultant at the estimated location of the seventh cervical vertebra), the center of mass of the head (using the skull's area centroid), and the resultant of these two vectors when the head is in equilibrium. Imprecise estimates, illustrated in Figure 11.13, show that horizontally rotating the nuchal plane in early hominins, including the *Sahelanthropus* reconstruction, would have kept shearing at the AOJ low by orienting the resultant force approximately within the range observed in modern humans and chimpanzees, between roughly 5° and 15°. If hominins such as *Sahelanthropus* had been quadrupedal, this configuration would have subjected the AOJ, where the spinal cord exits the skull, to much higher shearing forces. Such shearing might have been dangerous for a quadrupedal hominin during activities like jumping that generate high forces in the head.

Reorienting the nuchal plane also changes the orientation of the nuchal muscles' line of action, thus altering their mechanical advantage (MA). Admittedly imprecise estimates (also shown in Figure 11.13) suggest that the MA of this system is typically close to 0.6 in chimpanzees and approximately 1.2–1.5 in modern humans. Humans have much more balanced heads than chimps because the center of mass of the head is closer to the AOJ and the nuchal plane is relatively long. Early hominins generally fit the modern human configuration, but with MAs of approximately 1 or lower if they were bipedal. If Toumaï's head was oriented like a quadruped's, then his MA would be about 1.25, but if it was oriented like a biped's, his MA would be lower, about 0.8. Australopiths such as Sts 5 (*Au. africanus*) and KNM-ER 406 (*Au. boisei*) have MAs of approximately 0.8 to 1.0. All nonhuman tetrapods tend to have low MAs because of tradeoffs. Although a high MA means that the nuchal muscles have to generate less force to balance the head, such economy occurs at the expense of slower angular velocities and a decrease in the head's range of movement at the AOJ (see Chapter 9).

The slightly lower MA of the nuchal muscles in *Sahelanthropus* and some big-faced robust australopiths accounts for an interesting feature also present in some great apes such as gorillas: downward "lipping" of the nuchal crest (see Figure 11.4). Such lipping suggests large, forceful muscles that have to compensate for short moment arms. If the nuchal muscles in fossils such as Toumaï, AL 444–2 (*Au. afarensis*), KNM-WT 17000 (*Au. aethiopicus*), and OH 5 (*Au. boisei*) had higher, more efficient MAs, then perhaps their nuchal crests would have been smaller and less downwardly "pulled."

In short, the functional consequences of the derived configuration in hominins, in which the foramen magnum and nuchal plane are more horizontally rotated, make sense in terms of selection for improved performance as bipeds. A slightly more anterior position of the foramen magnum would have improved the nuchal muscles' MA, but this is a less critical adaptation. Thus, based solely on cranial evidence, we can infer that early hominins such as *Sahelanthropus* were bipeds. Other postcranial evidence is necessary to assess to what extent and how the earliest hominins were bipedal.

Diet

The second suite of derived anatomical differences that dominate early hominin and australopith heads is related to mastication. Compared to great apes, early hominins such as *Sahelanthropus* and *Ardipithecus* had slightly wider, thicker-enameled postcanine crowns, especially in the molars. In addition, their canines were smaller and their faces slightly more orthognathic and protracted. For the most part, these trends continue in the australopiths, albeit with many variations. The robust australopiths are the most derived, with enormous, thick-crowned premolars and molars, along with dozens of other features throughout the skull that must have increased the ability to generate massive bite forces, possibly to break down more mechanically demanding foods, and to resist the high, repetitive strains these forces generated.

Occlusion

Chewing is first and foremost an interaction between food and teeth, so a good place to start evaluating differences in dietary adaptations among early hominids is dental variation.

Incisors/ Wide, spatulate incisors tend to be an adaptation for frugivory, leading to a good correlation in anthropoids between diet and the length of the incisor row relative to body mass (Hylander, 1979; Kay and Ungar, 1997). Monkeys with relatively wider incisor rows tend to eat larger, tougher fruits with thick husks and flesh that adheres to large hard seeds (think of mangoes); monkeys with relatively narrower incisors tend to eat foods such as berries, seeds, and leaves, which require less incisal preparation. Chimpanzees and orangutans have especially wide incisor rows for their body size; gorillas fall on the anthropoid line. Assuming that this relationship is useful for predicting hominin diets, then the one upper incisor attributed to *Sahelanthropus*—which is wide, spatulate, and chimpanzee-sized—appears to have been well adapted for incising the sorts of tough fruits that chimpanzees often consume.

Younger species, such as *Au. anamensis, Au. afarensis, Au. aethiopicus,* and to some extent *Au. africanus,* fall, like gorillas, on the anthropoid line, suggesting that they used their incisors a lot. *Au. boisei* and *Au robustus,* like modern humans, have relatively narrow incisors that were less adapted for ingesting diets that required much incisal preparation. These inferences are supported by microwear studies (Ryan and Johanson, 1989).

Canine Size. Canine reduction is an important trend in early hominin evolution. As noted above, *Sahelanthropus* and *Ardipithecus* (especially *Ar. ramidus*) have smaller, less conical canines than chimpanzees (especially in terms of crown height and mesiodistal crown length), with a loss of the C-P_3 honing complex. Canine size is further reduced in australopiths. Because long, well-honed canines are useful, especially in males for fighting and threat displays, it is sometimes thought that canine diminution reflects less competition between males (Lovejoy, 2009). However, canine dimorphism is actually a poor predictor of male competition among primates (Plavcan, 2000). An alternative explanation, proposed by Hylander and Vinyard (2006), is that early hominin canine reduction is a by-product of other adaptations for generating high chewing forces on the molars and premolars. As discussed in Chapter 7, one effective way to increase postcanine bite force is to advance the masseter's origin on the cranium by displacing the zygomatic arch forward. Anterior positioning of this region increases the masseter's lever arm but limits gape (particularly in a muscle such as the masseter, with a high angle of pennation). As one might expect, long canines appear to be incompatible with small gapes, as is evident by the close relationship among mammals in general, and primates in particular, between maximum gape and the degree to which the canines project beyond the occlusal plane (Hylander and Vinyard, 2006). Australopiths and humans thus resemble mammals such as horses, which have large, anteriorly placed masseters, can't open their mouths very wide, and have short canines. Another hypothesis is that long, projecting canines minimize the ability to generate transverse movements when chewing on the postcanine teeth, especially in mammals with short mandibles.[3]

3. Transverse chewing would also have been facilitated in robust australopiths by their unique entoglenoid process, which turns posteriorly at the medial portion of the articular eminence, overlapping the medial end of the tympanic plate. This turn would have facilitated rotation of the condyle around a vertical axis, permitting a wide range of transverse movements (DuBrul, 1977; Aiello and Dean, 1990).

Postcanine Teeth. One of the strongest trends in early hominin evolution is expansion of the premolars combined with larger (especially wider) molars with thicker enamel crowns. On average, the permanent molars of *Ardipithecus* and *Sahelanthropus* are just a little longer and wider than those of chimpanzees (data from Brunet et al., 2002, 2005; Haile-Selassie, 2001; Haile-Selassie et al., 2004; Semaw et al., 2005; Suwa et al., 2009b). The summed occlusal area of the mandibular dentition (P_3–M_3) averages about 450–500 mm^2 in chimpanzees, between 760 and 850 mm^2 in gracile australopiths, and 970–1,228 mm^2 in robust australopiths (data from Wood, 1991). Put differently, occlusal area is more than 2.5 times larger relative to body mass in *Au. boisei* than in a chimpanzee, and 1.6 times larger than in *Au. afarensis* (McHenry, 1988). In addition, hominin postcanine teeth tend to be flatter and less bulbous than those of African great apes, in part because they have slightly thicker enamel crowns, nearly three times thicker in some robust australopiths. In general, the gracile australopiths have more occlusal relief than the robust species (Grine, 1981; Kay, 1985).

We can make several inferences about occlusal function and diet from these differences, keeping in mind that natural selection often promotes adaptations for fallback diet items during seasons of scarcity rather than preferred food items during periods of abundance (for examples, see Grant 1991; Laden and Wrangham, 2005; Marshall and Wrangham, 2007). Flatter, more bulbous teeth are less capable of shearing tough, fibrous foods such as leaves and meat that are highly pliant. Large occlusal surface areas not only increase the rate of food breakdown but also create more breakage sites, accommodating a wider range of food particle sizes and shapes, from large, soft fruits to small, hard items such as seeds (Lucas, 2004). Thicker-enameled crowns also help to resist wear damage (Lucas et al., 2008a,b), permitting hominins to chew more abrasive food items that wear teeth down more rapidly (Ungar and M'Kirera, 2003). Putting all the evidence together, there is a clear trend in early hominin evolution toward teeth better adapted to grinding hard, abrasive foods in which the lower teeth are not only pulled upward but also dragged medially against the upper teeth with high bite forces in the vertical and transverse planes (Lucas, 2004). Although fruits were probably preferred foods, early hominins apparently underwent selection for teeth that were better adapted for eating seeds and underground storage organs, such as tubers. These inferences are supported by microwear studies (R. S. Scott et al., 2005; Grine et al., 2006; Ungar, 2004, but see Ungar et al., 2008).

Bite Force Generation

Harder, tougher foods require either more chews per unit item or more force per chew. One cannot estimate bite forces precisely, but we can get a reasonable indication of masticatory-force generation capabilities from examining variations in the mechanical advantage of the major jaw elevator muscles in different planes and from combining these data with estimates of muscle cross-sectional areas to estimate torques.

Figure 11.14 compares lateral and anterior views of a few representative early hominin crania (*Sahelanthropus, Au. afarensis,* and *Au. boisei*) with that of an ape. I have chosen a female gorilla rather than a chimpanzee for the comparison because gorillas tend to eat a tougher diet (but keep in mind that female gorillas weigh 70–80 kg, about 50 percent more than male australopiths). These figures provide only rough estimates of the MA of the temporalis and masseter muscles for a chew on the second molars (in which the load arm is the distance from the TMJ to the M^2). Lever arms, the perpendicular distance from the TMJ to the muscle resultant, are harder to estimate. The masseter resultant is estimated from the angle of the mandible (the intersection of the posterior border of the ramus and the inferior border of the corpus, a point that has to be estimated in some cases) to the midpoint of the masseter's insertion on the zygomatic arch; the temporalis resultant is estimated from the coronoid process to the highest point on the temporal line. These resultants are, frankly, educated guesses and do not reflect the variation that undoubtedly existed, but they nonetheless permit some inferences. The first is that the *Sahelanthropus* reconstruction could probably generate large bite forces more efficiently than a female gorilla. Several factors contribute to higher MAs in this hominin, including a more projecting face (see Chapter 5), which positions the zygomatic arch more anteriorly relative to the TMJ, increasing the MA of the masseter for generating adductor force by approximately 33 percent. *Sahelanthropus* also has a wider midface relative to the TMJ, increasing the MA of the masseter in the transverse plane. A reasonable inference is that *Sahelanthropus* was comparatively efficient at generating high vertical and transverse forces.

The trend toward higher MAs continues in the australopiths, especially in *Au. boisei*. In part, MAs in these species are higher because of a more anteriorly positioned and laterally flared zygomatic arch. Because the zygomatic root of the maxilla is advanced relative to the TMJ, these species not only have more room for additional masseter muscle fibers but also orient the masseter's resultant more horizontally, increasing its moment arm (perpendicular distance) rel-

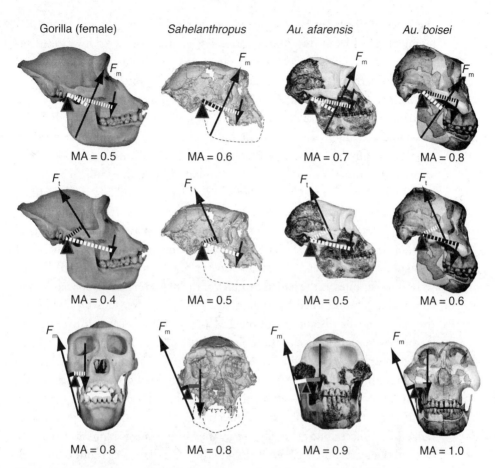

Figure 11.14. Comparisons of mechanical advantages (MA) for the temporalis (F_t) and masseter (F_m) muscles in the sagittal plane, and for the masseter in the transverse plane in *G. gorilla* (female); *Sahelanthropus; Au. afarensis* (Hadar composite); and *Au. boisei* (OH 5 with the Peninj mandible). Image of *Sahelanthropus* from Zollikofer et al. (2005), courtesy M. Brunet.

ative to the TMJ (but limiting gape, as noted above). Widening the zygomatic arch increases the masseter's moment arm in the transverse plane and increases the space available for a larger temporalis cross-sectional area (see below).

Another aspect of australopith facial shape to consider is the height of the jaw and the position of the occlusal plane, which is further from the TMJ in the more robust forms. Because the resultants of the masseter and medial pterygoid are oblique, a taller (lower) mandible augments the moment arm for these two muscles (as is evident in Figure 11.14) and potentially increases the space available for large muscle cross-sectional areas.[4] A final factor that augments the MA of the adductor muscles is a more retracted lower face relative to the TMJ, which decreases the load arm length. As noted above, facial retraction derives in part from cranial base flexion, which is considerable in *Au. boisei* and moderate in *Au. africanus* and possibly *Au. robustus.*

MA is an important parameter, but to estimate how much torque the muscles could actually exert (torque is a force times a moment arm), we need estimates of muscle forces, which are proportional to muscle cross-sectional areas. This estimation was done for a number of hominin skulls in a creative analysis by Demes and Creel (1988), which I have expanded and modified slightly. Demes and Creel estimated temporalis force from the muscle's cross-sectional area in the temporal fossa; they estimated masseter force as 80 percent of temporalis force (based on regression data from macaques and orangutans). They then estimated adductor bite force *equivalents* (BFEs) for second molar bites as the added product of the masseter and temporalis cross-sectional areas multiplied by each muscle's lever arm (see above), divided by the load arm (the distance from the TMJ to the M^2). Because bite forces are applied over a tooth's occlusal surface, these estimated BFEs are most appropriately graphed relative to M^2 area, as in Figure 11.15a. Note that males typically have higher BFEs than females and that the robust australopiths could produce very high BFEs—about three times that of a modern human's and about twice that of a male chimpanzee's. Gracile australopiths yield a wide range of estimates but generally are in the chimp range. Note also that the slope of the regression between BFEs and the M^2 surface area is linear and well correlated. Thus, although more megadont australopiths could produce much higher bite forces, the pressures these chews generated were approximately the same. In other words, more megadont hominins were not necessarily chewing harder food that required higher occlusal stresses to fracture particles but instead were chewing food that required larger occlusal surface areas. In all likelihood, their diet was tougher and lower in quality, perhaps because it contained more fiber (see below).

What about anterior bite forces? As outlined in Chapter 7, the horizontally oriented fibers of the posterior temporalis are major contributors to incisal and canine bite forces. We can therefore assess anterior bite forces in part from the relative size of the origin of the posterior temporalis at the back of the vault,

---

4. A lower occlusal plane, however, increases the anteroposterior translation required to ensure vertical movement of the lower teeth during occlusion (see Chapter 7).

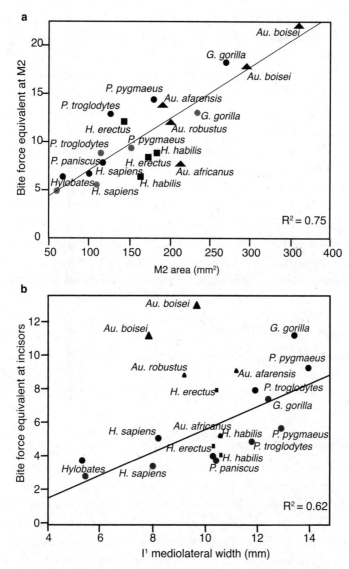

Figure 11.15. Estimated bite force equivalents for (a) chews on the second molars plotted against second molar (M2) area, and (b) incisal chews plotted against the mesiodistal length of the first incisor in extant hominoids (circles), australopiths (triangles), and fossil *Homo* (squares). Black symbols indicate males; gray symbols indicate females. Molar bite forces vary enormously, but bite pressures (force per unit area) were probably similar across hominin species. In contrast, incisor bite pressures may have been exceptionally high in robust australopiths. (Adapted from Demes and Creel, 1988.)

particularly along the temporonuchal crest, which creates space for a band of horizontally oriented fibers that run above the mastoid process to the coronoid process. The size of this compartment is especially large in gorillas and male chimps but is also comparatively large in other species with large anterior dentitions, including *Au. afarensis* and *Au. aethiopicus.* Australopiths with smaller incisors, such as *Au. africanus, Au. boisei,* and *Au. robustus,* tend to have less developed crests for the insertion of the posterior temporalis. However, the more orthognathic faces of many of these species would have reduced the load arm of the muscles, increasing their mechanical advantage. In addition, because the robust australopiths also had mesiodistally narrower incisors, anterior bite forces were distributed over smaller areas, thus increasing bite stresses produced (see Figure 11.15b). Thus, although we typically think of megadont australopiths as mostly selected for producing high masticatory forces on the postcanine teeth, they might also have been able to produce remarkably high anterior bite pressures, more than twice those of gracile australopiths or extant apes. Interestingly, a relatively large posterior compartment of the temporalis may leave a trace on the coronoid process of the mandible, whose size and shape are affected by the size of the temporalis insertion and the orientation of its resultant (Herring, 1993). The coronoid process in *Au. afarensis* resembles that of gorillas in its relative height, anteroposterior length, and posterior orientation, presumably because of a large posterior temporalis (Rak et al., 2007).

Masticatory Strain Resistance

Substantial masticatory forces generate high, repetitive strains that need to be resisted. The basic principle by which bones resist strain—adding mass in the planes of deformation—explains much variation evident in australopith jaws and faces. Chewing hard, tough foods generates high magnitudes of strain in the mandible in the form of sagittal bending, wishboning (lateral transverse bending), and twisting (see Chapter 7). Sagittal bending is resisted by making the mandibular corpus taller, especially by adding more bone to the base of the corpus. Wishboning is resisted by making the corpus mediolaterally wider and the symphysis anteroposteriorly deeper. Because bone resists compression better than tension, wishboning is best resisted by adding bone to the posterior margin of the symphysis (Hylander, 1988). Great apes achieve this reinforcement with a deep shelf of bone, the inferior transverse torus, at the base of the symphysis; australopiths also have an inferior transverse torus, and some more-robust mandibles have a second, extensive shelf of alveolar bone, a superior transverse torus, above it. Finally, twisting is resisted by making the corpus and the symphysis taller and wider, thus increasing the polar moment of inertia (J).

Resistance to all three types of strain is evident in all australopith mandibles, which, compared with those of great apes, tend to have thick, wide corpora, and tall, deep symphyses, often with pronounced transverse tori (see Figure 11.11). Both the width and height of the mandibular corpus relative to jaw length are significantly greater in gracile australopiths than in extant apes and even greater in robust australopiths such as *Au. boisei* and *Au. robustus* (see Hylander, 1988). Increases in width, which help counteract both wishboning and twisting, are especially marked. The robusticity index of the corpus at M_1 (width/height 100) averages 48 in chimps but 56 in *Au. afarensis* and 68 in *Au. robustus* (Chamberlain and Wood, 1985; Daegling and Grine, 1991). Mastication also generates high compressive strains in the condyles, which grow upward and backward in response to compression (Fuentes et al., 2002; see Chapter 7). Thus it is not surprising that australopiths, especially robust forms, also tend to have relatively large condylar processes, more like those of gorillas than those of chimps. Larger condyles presumably lower stresses in the TMJ by distributing forces over a larger surface area.

We know less about strain magnitudes and patterns in the face, particularly in nonprognathic faces, but we can make a few generalizations. First, as discussed in Chapter 7, strains tend to follow a gradient, with higher strain magnitudes in the alveolar process around the tooth roots and the palate, and lower strain magnitudes in the upper face; the zygomatic arches also experience high strain magnitudes. Second, to a large extent the primate face is bent or sheared in the sagittal plane during mastication and incision, and twisted in the coronal plane around an anteroposterior axis during mastication. The face may also be subject to vertical loads, often compressive, that are asymmetrical during unilateral mastication. We see many structures in the australopith face that would improve resistance to these sorts of strains. Most important, making the face taller and wider not only resists vertical bending and shearing but also resists twisting in the coronal plane. This sort of expansion is evident in early hominins such as *Sahelanthropus, Ar. ramidus,* and *Au. afarensis,* which have faces that are chimplike in many respects but also relatively wider and taller, especially in the infraorbital region, presumably from additional growth in the zygomaticomaxillary suture. Hominins such as *Au. afarensis* retain some degree of convexity in the sagittal plane that helps resist bending in the sagittal plane, especially from large incisal bite forces (Demes, 1987). In spite of variation in prognathism, more robust forms such as *Au. aethiopicus, Au. boisei,* and *Au. robustus* have much wider, flatter faces that would have offered substantial resistance to bending, shearing, and twisting (although calculating resistance is hard without better information on the loads applied). Palate thickness is also greater in australopiths, especially robust forms. And some species, notably *Au. africanus* and *Au. robustus,* also have anterior nasal pillars alongside the nasal aperture above the premolars. These pillars were probably adaptations to buttress the face against compressive loads from bending and shearing during chews on the premolars, which are expanded, thickly enameled, and very molarlike (Rak, 1983; Strait et al., 2009).

Mastication also generates strains elsewhere in the skull. Although strains in the vault bones were probably minor (Lieberman, 1996), strains in the sutures between the vault bones were probably more substantial (see Herring and Teng, 2000; Lieberman et al., 2004a). Interestingly, several crania of *Au. boisei* have a unique configuration of the posterior vault in which the parietal overlaps both the mastoid portion of the temporal bone and the superolateral corner of the occipital (Kimbel and Rak, 1985). In addition, high chewing forces produce compression in the TMJ. Although compressive joint reaction forces are probably dissipated primarily by the large condylar processes and the articular disk for posterior chews when the condyles are retracted below the glenoid fossa, they might be more problematic for anterior chews in which the condyle is protracted forward below the preglenoid plane. Early hominins, such as *Sahelanthropus* and *Ar. ramidus,* and australopiths with large anterior teeth, such as *Au. anamensis, Au. afarensis,* and *Au. aethiopicus,* tend to have long, shallow preglenoid planes similar to those of chimps and gorillas. The TMJ in later species with reduced anterior dentition is deeper and more compressed from front to back (more like that of *Homo*).

Adaptations for Open Habitats

The australopiths generally resemble the African great apes in many other aspects of craniofacial anatomy. Their semicircular canals and hence their system of balance were not different from those of apes (Spoor et al., 1994), but they lacked a cantilevered neck-head complex. There is no evidence of any special features of their vision or olfaction. Their rate of ontogeny was similar to that of apes, or possibly just a little slower (Bromage and Dean, 1985; Beynon and Dean, 1988; B. H. Smith, 1994; Dean, 2006). And as Chapter 8 surmises, gracile australopiths probably had a primitive pharynx with an intranarial larynx (the robust australopith configuration is harder to reconstruct). The one hyoid so far known from *Au. afarensis* (Alemseged et al., 2006) suggests that they retained the primitive configuration of air sacs.

One derived feature evident in some australopiths might have been helpful for living in open habitats: hearing. As Chapter 10 discussed, the cross-sectional size and length of the external auditory canal influence how certain frequencies are amplified before they hit the eardrum. Long canals amplify low to midrange frequencies slightly, and larger cross-sectional areas amplify midrange to high frequencies. Early hominins such as *Sahelanthropus, Au. anamensis,* and *Au. (K.) platyops* have small earholes, in the chimpanzee size range (25–45 mm²), whereas later hominins have larger earholes, more or less in the modern human size range (115 mm²) (M. Leakey et al., 2001). Average earhole diameter in *Au. afarensis* is intermediate (79.7 mm²) but with much variability (47.7–109.3 mm², *n* = 8) (Kimbel et al., 2004). This variation suggests that some *Au. afarensis* individuals might been more sensitive to frequencies above 2,000 Hz. Such frequencies are useful for detecting sound location via interaural level differences (ILDs). Because ILDs are more useful in open habitats than in heavily forested environments, where trees and bushes tend to attenuate high frequencies, increases in earhole size might reflect selection for these hominins to localize sounds from potential predators (or each other) as they ranged into more open habitats.

Another challenge that early hominins would have faced in open habitats is thermoregulation. In this respect, being bipedal gives a slight advantage by exposing less of the body's surface area to solar radiation (Wheeler, 1991). That said, there is no evidence that australopiths had any special features for cooling the brain (see Chapter 6). In addition, we lack evidence to determine how much fur australopiths had and whether they had few or many eccrine glands for cooling via sweating. One can only speculate that the earliest hominins and the australopiths, like most mammals, were not frequently active in open, hot habitats during the middle of the day.

Summary and Final Speculations

What were the heads of early hominins and australopiths like, and what selective forces drove their evolution? As far as we can tell, the first hominins were not dissimilar from the African great apes in many aspects of their craniofacial anatomy, with the major exceptions of one suite of adaptations for being bipedal and another for chewing harder and tougher food. These transformations required only a few small changes in craniofacial development. If we could observe *Sahelanthropus* and *Ardipithecus,* we might conclude they were mostly frugivores whose diet also included an occasionally important component of tough, fibrous fallback foods, such as shoots, piths, and underground storage organs (Walker, 1981; Laden and Wrangham, 2005). Fallback foods—usually less desirable items with higher fiber contents than preferred foods—are important targets of natural selection because the ability to process them effectively has major fitness consequences during periods of food scarcity and high competition. These foods may have been infrequent components of the diet in early hominins but were probably more important to the australopiths, who became increasingly specialized for eating not only large quantities of tough foods but also hard foods, such as seeds, that are not typically available to apes in forests (see Jolly, 1970). A reliance on fallback foods might have been seasonal or possibly even occasional for some of these hominins.

Given paleoenvironmental evidence that the earliest hominins lived in woodland habitats, one can imagine that selection for bipedalism first occurred to enable these animals to exploit resources in more open habitats or to travel more efficiently between widely spaced woodland habitats as forests became increasingly fractionated in parts of Africa at the end of the Miocene. In this context, it is relevant that hominins possibly evolved from a knuckle-walking LCA. Knuckle walking is a comparatively inefficient form of locomotion, costing chimpanzees considerably more energy than is used by human bipeds (Sockol et al., 2007). Extending the hips and knees and reorienting the neck to walk upright would have helped hominins traverse long distances between patches of trees more efficiently than chimps, without losing the ability to climbing trees well. Bipedalism may also have helped early hominins use upright postures to forage for hard-to-reach foods (Hunt, 1991; Thorpe et al., 2007).

In short, as Darwin (1871) inferred, it is reasonable to speculate that bipedalism was one of the most important factors that set the human lineage on a different evolutionary path from the apes. Thanks to some fortuitous tinkering, early hominins were able to occupy a new, more open ecological niche, mostly by becoming bipedal and also by generating and resisting higher chewing forces. These adaptations led to a radiation of different forms that vary mostly in their degree of megadonty. But the australopiths set the stage for additional, major transformations that occurred in the genus *Homo.* These transformations, at least those from the neck up, are the focus of the next chapter.

12

Ecce Early *Homo*

Your marked postorbital constriction
Would clearly justify prediction
That had you lived to breed your kind
They would have had the childish mind
That feeds upon the comic strips
And reads with movements of the lips.

E. A. HOOTON, "Ode to Homo some-jerktensis,"
Up from the Ape, 1946

In many ways, the origin of the human genus was as transformative as the divergence of the hominin lineage from the African apes, if not more so. Over the course of the transition from *Australopithecus* to *H. erectus,* brain size increased substantially (partly because of increases in body size); the cheek teeth shrank; the face lost its snout and became more vertical and narrower; the nose added an external vestibule; the head as a whole became more rounded, balanced, and sensitive to pitching; and sexual dimorphism decreased. These and other shifts (many of them postcranial) reflect a series of profound behavioral changes in which we can begin to see the emergence of the hunter-gatherer way of life. Among other things, the changes suggest that early *Homo* (defined here broadly as *H. habilis* and early *H. erectus*) were not just foragers but also social, diurnal omnivores (foragers and carnivores), relying more on meat and less on tough, fibrous foods. Thermoregulatory and locomotor adaptations for trekking and running long distances in the heat also suggest that early *Homo* was more active in open, hot habitats than australopiths were. And in early *Homo* we see evidence for cognitive advances and behavioral innovations such as making stone tools, processing food, and possibly communicating more with speech. These changes, which did not all happen at once, set the stage for less transformative but no less important events, including the origin of our own species, *H. sapiens.*

Before reviewing species of early *Homo* and their developmental and functional differences, it might be useful to consider some relevant epistemological problems. One is that by the time of *H. erectus,* the hominin fossil record becomes better sampled and covers a much wider geographic range than for earlier hominins. In addition, the fossil record of early *Homo* comes with an archaeological counterpart, providing direct evidence of behavior. Consequently, we can make inferences and test hypotheses about the evolution of early *Homo* with more confidence than we can with earlier hominins. But as the number of fossils ascribed to putative species increases and the fossils derive from a wider geographic range, the amount of variation we perceive goes up, and it becomes less obvious where to draw boundaries between species.

H. erectus typifies this problem. Fossils assigned to *H. erectus* span about 1.8 million years and come from many parts of the Old World. Although *H. erectus* crania generally resemble each other in many respects, such as the overall shape of the vault, they can also vary considerably in other features, such as brain volume. The largest *H. erectus* has an ECV about 70 percent bigger than that of the smallest (this difference is only 40 percent in *Au. afarensis*). We have difficulties sorting the variation, in part because it is unclear how much variation can occur within a single species, we don't know how to model variation through time, and we don't understand what some of the variation means. Some experts sort *H. erectus*–like fossils into more than one species, depending on traits such as the degree to which the brow ridges are slightly arched or barlike, whether the maximum breadth of the vault lies a little higher or lower relative to the mastoid processes, or whether there is a slight degree of keeling on the vault. How do we know that these variations provide taxonomically useful and reliable information? How do we ensure that taxonomic hypotheses are based on testable scientific methods?

The most common way to test taxonomic hypotheses is to compare variation in hypothetical fossil hominin taxa with extant primate species (e.g., chimpanzees or gorillas). Patterns of variation in known, related species are assumed to provide a benchmark for how much variation to expect in fossil species. But this assumption is complicated by the extent to which variation can be ascribed to parameters such as sexual dimorphism, geography, geological age, and so on. Further, it is possible that fossil hominin taxa had patterns of variation that differ from those observed in living primate species. So these analyses are rarely definitive.

Cladistic studies complement taxonomic analyses by assessing the evolutionary relationships among putative taxa. Cladistics makes use of the simple logic that taxa sharing the same derived (novel) features most likely share a common ancestor. But cladistic analyses of morphological data can have a hard time coping with continuously variable features (e.g., features that vary in size or degree of curvature), they sometimes make tenuous assumptions about the independence and homology of morphological characters, and they have difficulty in managing the presence of convergence (independent evolution) and character conflict in most data sets.[1] Character conflict is evident when different characters imply alternative trees. Lots of character conflict means that slight changes to cladistic data sets yield different, incompatible trees with nearly equal likelihoods of being correct, limiting confidence in the results of any analysis. A final problem to note is that cladistic logic applies only to taxa that diverge cleanly into distinct lineages. Substantial hybridization between distinct species can confound the results of any cladistic analysis.

These systematic problems are not insignificant, but our major goal is here is to summarize the general variation and trends in early members of the genus *Homo,* to ask what developmental shifts were responsible for the major transformations evident, and to hypothesize how natural selection might have been operating on function and behavior, spurring on evolutionary change. Key questions include when and why *Homo* evolved a long period of childhood, when and to what extent *Homo* became a carnivore, why and how species of *Homo* have such big brains, when language and speech capabilities evolved, and when and why hominins switched to thermoregulating predominantly by sweating. In fact, these and other functional and behavioral issues raise an additional problem: how do we cope with the degree and quantity of change in the transition from *Australopithecus* to *Homo?* Much happened during this transition, suggesting intense levels of natural selection. Yet because heads are so integrated, both functionally and developmentally, it is difficult to identify bona fide adaptations. Were big brow ridges adaptations for sexual display or merely by-products (spandrels) of the way that large faces project forward from the cranial base? Is an orthognathic face an adaptation for increasing nasal turbulence, improving the balance of the head, or facilitating articulate

---

1. One doesn't really inherit morphological features, but instead the genes that help regulate the processes by which they grow. Thus it is easier to assess the homology (similarity due to common ancestry) of DNA base pairs than of morphological features and to test their independence.

speech, or is it a by-product of having smaller teeth? With so much change to consider, it is little wonder that there has been so much speculation about the origins of the human genus.

As in Chapter 11, we will begin by reviewing the cast of characters and then make some hypotheses about the developmental bases for the transformations we observe between them. We'll finish by considering the possible functional consequences and selective forces behind these shifts.

Species of Early *Homo*

Our knowledge of the origins of the human genus is frustratingly murky. The fossil record is sparse and inadequate between 3 and 2 million years ago, when the genus probably first evolved. The oldest well-dated fossil reliably attributed to *Homo* is a maxilla (AL 666–1, shown in Figure 12.1) from Hadar, Ethiopia, dated to 2.3 mya (Kimbel et al., 1997). Interestingly, and perhaps not coincidentally, early *Homo* fossils first appear after the oldest dated stone tools from 2.6 mya (Semaw et al., 2003). But the crop of fossils we have from between 1.9 and 1.5 mya includes much confusing variation, leading to a lack of consensus on how to sort them out. So far, three species are generally recognized: *H. habilis, H. rudolfensis,* and *H. erectus.* One of them, *H. rudolfensis,* may belong in the genus *Australopithecus* (or possibly even *Kenyanthropus*). And, paradoxically, the one we think we know the best, early *H. erectus,* is puzzlingly variable, enough to lead some experts to believe it comprises more than one species.

Homo habilis

The first thing to know about this confusing species is that ever since *H. habilis* was proposed, it has tended to accommodate fossils that do not belong in *Australopithecus* but which are not obviously *H. erectus* either. This state of affairs began in 1960, when Louis and Mary Leakey discovered in Bed I of Olduvai Gorge a partial skull and some hand bones of a hominin, Olduvai Hominid (OH) 7, which did not belong to the more robust *Au. boisei* species (then called *Zinjanthropus*) they had previously discovered. The discovery of OH 7 was soon followed by finds of additional partial crania (e.g., OH 13 and OH 16). All had smaller teeth than the two australopith species then known, *Au. boisei* and *Au. africanus,* and brain volumes that were slightly larger (600–700 cm³) but not as large as those of known *H. erectus* crania. Accordingly, Louis Leakey, Phillip Tobias, and John Napier proposed the species *H. habilis* in 1964 (L. S. B. Leakey et al., 1964). Despite much criticism, recognition of the species has endured, in part because of more discoveries from Olduvai Gorge (e.g., OH 24, OH 62); from Koobi Fora, Kenya (e.g., KNM-ER 1813) (Wood, 1991); and from Members G and H of the Omo Shungura Formation, Ethiopia (e.g., L894–1) (Boaz and Howell, 1977; Coppens, 1980; Suwa et al., 1996). As noted above, the oldest securely dated fossil that might be attributed to *H. habilis* is a maxilla from Hadar, Ethiopia (AL 666–1), dated to 2.33 mya (Kimbel et al., 1997); the youngest is a maxilla from Kenya (KNM-ER 42703), dated to 1.44 mya (Spoor et al., 2007a). Figure 12.1 illustrates these two maxillae, along with two partial crania attributed to *H. habilis.* For a summary of major fossils attributed to *H. habilis,* see Table 12.1.

A definitive description of *H. habilis* is hard to achieve. For the most part, fossils attributed to *H. habilis* have ECVs between 500 cm³ and 700 cm³ (the smallest is KNM-ER 1813 at 509 cm³, and the largest is OH 7, an estimated 697 cm³) with a mean of 609 cm³ (Holloway et al., 2004). The few reasonably complete *H. habilis* crania (e.g., KNM-ER 1813, OH 24) have fairly vertical faces, with a distinct though modest supraorbital torus and a sulcus between the torus and the frontal squama. Unlike australopiths, the *H. habilis* midface is narrower than the upper face, with a nearly vertical malar (infraorbital) region. An everted nasal margin in some crania suggests the presence of an external nasal vestibule. The palate is thin and shallow, with a short premaxillary region and minimal alveolar prognathism. The foramen magnum is more centrally located than in australopiths (Dean and Wood, 1982b), and the nuchal plane in a few crania, such as KNM-ER 1813, has a median nuchal line, suggesting the presence of a nuchal ligament (see Chapter 9). Temporals assigned to *H. habilis* have a short and deep TMJ with a gradually sloped articular eminence. The mandibles are quite variable in size, but the smaller ones tend to be less robust than in *Australopithecus,* with a reduced inferior transverse torus.

The teeth of *H. habilis* tend to have a little more occlusal relief than in australopiths (they are cuspier), but they are not much smaller than in *Australopithecus* and much bigger than in *H. erectus* (Figure 12.2). Average postcanine tooth area in *H. habilis* is 478 mm², only 8 percent smaller than in *Au. africanus* (518 mm²) and about 4 percent larger than in *Au. afarensis* (460 mm²)

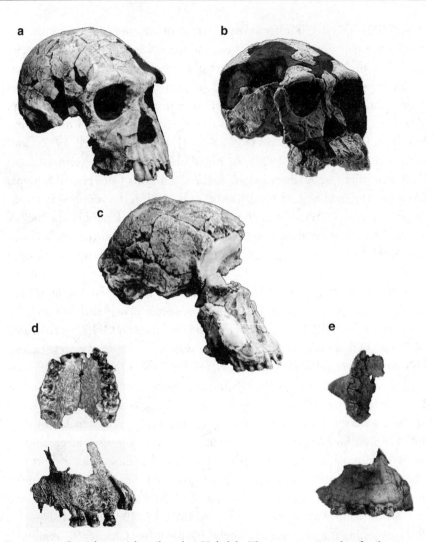

Figure 12.1. Cranial material attributed to *H. habilis*. The two most complete fossils are (a) KNM-ER 1813 and (b) OH 24, both likely females; (c) KNM-ER 1805 (which lacks much of its face) is a likely male. Other important fossils include two maxillae: (d) AL. 666–1, and (e) KNM-ER 42703 maxilla. Image of KNM-ER 42703 from Spoor et al. (2007a), courtesy Kenya National Museum; image of AL. 666–1 courtesy W. Kimbel. Images not to scale.

(McHenry and Coffing, 2000). That said, the dentition of *H. habilis* is distinctive in having buccolingually narrower, nearly square premolars and molars (Wood, 1991). Further, habiline M_3 and M_2 crown areas tend to be equal in size (whereas in australopiths, M_3 is bigger than M_2).

Table 12.1: Major cranial fossils attributed to *H. habilis* and *H. rudolfensis*

Fossil	Location	Age (millions of years)	Elements found
		H. habilis	
OH 7	Olduvai Gorge, Tanzania	1.7–1.8	Mandible, parietals, dentition (also hand bones)
OH 13	Olduvai Gorge, Tanzania	1.6–1.8	Fragmented skull
OH 16	Olduvai Gorge, Tanzania	1.7–1.8	Fragmented cranium, dentition
OH 24	Olduvai Gorge, Tanzania	1.7–1.8	Fragmented cranium
OH 62	Olduvai Gorge, Tanzania	1.7–1.8	Partial skeleton including cranial and mandibular fragments
OH 65	Olduvai Gorge, Tanzania	1.7–1.8	Maxilla
KNM-ER 1813	Turkana Basin, Kenya	1.65–1.8	Cranium
KNM-ER 1805	Turkana Basin, Kenya	1.8	Cranium and mandible fragments
KNM-ER 3735	Turkana Basin, Kenya	1.7–1.8	Fragmented cranium
KNM-ER 42703	Turkana Basin, Kenya	1.44	Maxilla
Omo L-894	Omo, Ethiopia	1.7–1.9	Fragmented cranium
		H. rudolfensis	
KNM-ER 1470	Turkana Basin, Kenya	1.7–1.8	Partial cranium
KNM-ER 1590	Turkana Basin, Kenya	1.8–1.9	Fragmented cranium
KNM-ER 1802	Turkana Basin, Kenya	1.7–1.8	Mandible
KNM-ER 3732	Turkana Basin, Kenya	1.7–1.8	Partial cranium
UR 501	Uraha, Malawi	2.3–2.5	Mandible
Omo 75–14	Omo, Ethiopia	1.9–2.1	Maxilla

Body size is a frustrating issue in *H. habilis*. Only a few postcrania are associated with the species, including one incomplete and highly fragmentary skeleton, OH 62 (Johanson et al., 1987), and another, even more partial fragmentary skeleton, KNM-ER 3735 (Leakey and Walker, 1985). Both partial skeletons are of small individuals (approximately 30 kg), with relatively long arms and short legs (Haeusler and McHenry, 2004, 2007). These rather scrappy specimens, along with estimates of body size from orbits (Aiello and Wood, 1994), suggest that the encephalization quotient (EQ) of *H. habilis* falls in the australopith range, encouraging some researchers to reassign the species to the genus *Australopithecus* (Wood and Collard, 1999a,b). More postcrania associated with crania are needed.

Almost everything we know about *H. habilis* comes from East Africa, but a few fossils from South Africa need to be mentioned. One of these is an enig-

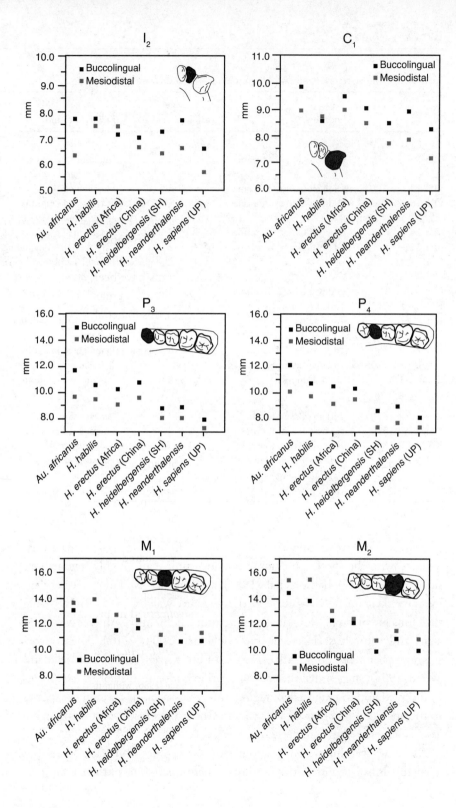

Figure 12.2. Tooth crown dimensions in mm in the genus *Homo* by species (with *H. erectus* from Africa and Asia separated). Tooth sizes for *H. heidelbergensis* are from the Sima de los Huesos (SH) site, discussed in Chapter 13. Human tooth sizes come from an Upper Paleolithic (UP) sample. Data from Weidenreich (1937); Frayer (1977); Tobias (1991); Wood (1991); Brown and Walker (1993); Martinón-Torres et al. (2008).

matic partial cranium from Sterkfontein, Stw 53, from either Member 4 or Member 5. Although initially ascribed to *H. habilis* (Hughes and Tobias, 1977), its big teeth and wide midface suggest that it is more likely an australopith (Kuman and Clarke, 2000; Schwartz and Tattersall, 2003). In addition, more than half a dozen fossils from Swartkrans have been proposed to belong to *H. habilis,* including a partial cranium, SK 847; a mandible, SK 15; and various isolated teeth (Grine, 1993). Much of this material is hard to assess, but several scholars have made the case that SK 15 and SK 847 should be attributed to *H. erectus sensu lato (s.l.)* (Rightmire, 1990; Grine et al., 1993; Smith and Grine, 2008).

Homo rudolfensis

H. rudolfensis is a puzzling species. Of the few fossils assigned to the species (see Table 12.1), almost all were originally attributed to *H. habilis* and then reassigned to *H. rudolfensis* for two reasons.[2] First, if they were included in *H. habilis,* this species would have been more variable than extremely sexually dimorphic species such as gorillas or orangutans, which seems unlikely. Second, the fossils assigned to *H. rudolfensis* have a number of anatomical differences from those of *H. habilis.* Many of these features are shared with australopiths and are difficult to account for as sexual dimorphism (Wood, 1985; Stringer, 1986; Lieberman et al., 1988; Wood, 1991; Kramer et al., 1995; Grine et al., 1996). I thus denote the species as *H. (Au.) rudolfensis.* The proposed species includes a few partial crania from Koobi Fora, Kenya (KNM-ER 1470, 1590, 1802, 3732), dated to between 1.8 and 1.9 mya. *H. (Au.) rudolfensis* might

2. The species name *rudolfensis* was first proposed by Alexeev (1986) within the genus *Pithecanthropus.* Although Alexeev chose KNM-ER 1470 as the type specimen, the proposed taxon included other fossils clearly belonging to *H. erectus.* According to the rules of zoological nomenclature, the name *rudolfensis* still applies to whatever species KNM-ER 1470 belongs to, assuming it is not an invalid species. Because this was the first alternative proposed for the fossil, the name still applies, but in a different context. To add to the confusion, Alexeev used the former name of Lake Turkana, Lake Rudolf. Because the lake's name was changed in 1975, a better name would have been *H. turkanaensis.*

also include a maxilla dated to 2.0 mya from Omo, Ethiopia (Omo 75–14) and some material from Uraha, Malawi (UR 501), which could be as old as 2.3–2.5 mya (Bromage et al., 1995). Another possible early contender is a fragment of temporal bone, KNM-BC 1, from the Chemeron formation in Lake Baringo, Kenya. Whether the fossil belongs to *Homo* is debated (Hill et al., 1992; Sherwood et al., 2002). Its provenance is uncertain, but it could be as old as 2.4 mya (Deino and Hill, 2002).

Figure 12.3 shows some relatively complete fossils attributed to *H. (Au.) rudolfensis*. The most distinct feature is brain size, which can be estimated reliably in KNM-ER 1470 (752 cm^3) and perhaps a little less so in KNM-ER 1590 (825 cm^3) and KNM-ER 3732 (750 cm^3), yielding an average ECV of 775 cm^3 (Holloway et al., 2004). Relative brain size, however, is hard to calculate because there are no postcranial fossils reliably attributed to the species. Enlarged brains, a derived feature, stand in contrast to the relatively primitive face of *H. (Au.) rudolfensis*, which is flat, and wide—especially in the midfacial region below the orbits—and slightly prognathic, but not as prognathic as that of *Australopithecus* (Kimbel, 2009). In addition, the TMJ in *H. (Au.) rudolfensis* is shallower (hence more primitive) than that of *H. habilis*. Unlike *H. habilis*, *rudolfensis* has a less distinct supratoral sulcus behind the torus, making the brow ridges in these bigger crania look less robust than in the smaller habilines. KNM-ER 1470 might have had an everted nasal margin, but because this region is damaged, it is hard to be sure. The mandibular corpus appears to be rather large. And, importantly, the species' teeth are big. The postcanine teeth are actually larger on average than those of *Au. africanus*, and unlike in *H. habilis*, the M$_3$ appears to be larger than M$_2$ (Wood, 1991); the upper incisors and canines also appear to have been large.

H. erectus (sensu lato)

H. erectus is a fascinating but frustrating species to consider. On the one hand, *H. erectus* is often regarded as the first hominin with a modern body, bigger brain, smaller teeth, and vertical, snoutless face. In addition, *H. erectus* appears to have been the first hominin to disperse out of Africa into much of the Old World. But the closer we look, the more problems we see with this standard story. Although the fossils generally assigned to *H. erectus* share features such as a long, low vault, and big brow ridges, the sample as a whole subsumes a whopping degree of variation. This variation calls into question the null hypothesis, generally accepted since 1950, that they belong to one species unless

a   b

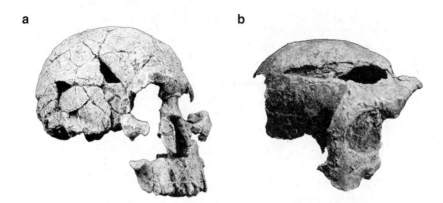

Figure 12.3. Two partial crania attributed to *H. rudolfensis:* (a) KNM-ER 1470 (the type specimen); (b) KNM-ER 3732. The face in KNM-ER 1470 is not attached to the rest of the cranium, so its degree of prognathism is uncertain. Images not to scale.

data prove otherwise. But the variation is not easy to sort, and there is no consensus on whether or how to divide the species up. Further, the lines between *H. erectus* and what follows (e.g., *H. heidelbergensis*) are blurry. Paleontologists disagree not only on how many species there are but also on how to answer the question in the first place.

Because we cannot resolve all these taxonomic issues, its best to treat *H. erectus* in the *sensu lato*. That is, although *H. erectus sensu lato* may comprise more than one species, the variants from Africa, Georgia, and Asia are not so different that much information is lost by lumping them together. Table 12.2 summarizes the major cranial material attributed to *H. erectus s.l.*

The Context of H. erectus

A little history helps make sense of *H. erectus.* The species was discovered by Eugène Dubois (1858–1940), a Dutch doctor who joined the Dutch East Indian Army so that he could travel to Java to pursue his dream of finding the "missing link" between apes and humans (see Shipman, 2001). In 1891, along the banks of the Solo River near the town of Trinil, Dubois' workers found several fossils, including a large-brained, big-browed skullcap (Trinil 2). Dubois eventually named the species *Pithecanthropus erectus* (Dubois, 1894). Dubois' finds were later followed by discoveries of similar fossils in Java, and in China at Zhoukoudian Cave (Locality 1). These finds were originally separated into different species, sometimes in different genera (the Peking fossils, for example, were assigned to *Sinanthropus pekinensis,* and fossils from Sangiran were

Table 12.2: Major cranial material attributed to *Homo erectus s.l.*

Fossil	Location	Age (millions of years)	Elements found
		Africa	
KNM-ER 730	Turkana Basin, Kenya	1.5–1.6	Fragmented skull
KNM-ER 820	Turkana Basin, Kenya	1.5–1.6	Mandible
KNM-ER 992	Turkana Basin, Kenya	1.5–1.6	Mandible
KNM-ER 2598	Turkana Basin, Kenya	1.9	Occipital
KNM-ER 3733	Turkana Basin, Kenya	1.8	Cranium
KNM-ER 3883	Turkana Basin, Kenya	1.6–1.7	Partial cranium
KNM-WT 15000	Turkana Basin, Kenya	1.5–1.6	Skeleton
KNM-ER 42700	Turkana Basin, Kenya	1.5–1.6	Partial cranium
SK 847	Swartkrans, South Africa	1.5–2.0	Partial skull
SK 15	Swartkrans, South Africa	1–1.7	Mandible
OH 9	Olduvai Gorge, Tanzania	1.4	Calvarium
OH 12	Olduvai Gorge, Tanzania	0.9–1.2	Fragmented cranium
OH 22,23	Olduvai Gorge, Tanzania	0.8–0.9	Mandibles
KGA 10–1	Konso, Ethiopia	1.4	Mandible and M3
Omo L996–17	Omo, Ethiopia	1.3–1.4	Parietal and temporal fragments
Melka Konturé	Ethiopia	0.8–1.2	Cranial fragment
Bouri	Ethiopia	1	Cranium
Buia	Eritrea	1	Skull
Sidi Abderarahman	Morocco	0.5–0.6	Fragmented mandible
Ternifine	Algeria	0.7–0.8	Fragmented skull
Thomas Quarries fragments	Morocco	0.5–0.6	Mandibular and cranial
		Europe	
D211	Dmanisi, Georgia	1.7–1.8	Mandible
D2600	Dmanisi, Georgia	1.7–1.8	Mandible
D2280	Dmanisi, Georgia	1.7–1.8	Cranium
D2282	Dmanisi, Georgia	1.7–1.8	Cranium linked to D211 mandible
D2700/D2735	Dmanisi, Georgia	1.7–1.8	Skull
D2735	Dmanisi, Georgia	1.7–1.8	Mandible
D3444/D3900	Dmanisi, Georgia	1.7–1.8	Skull
		Asia	
Modjokerto	Java	1.8	Partial vault
Sangiran	Java	0.7–1.6	Teeth, mandibles, partial crania
Trinil 2	Java	0.9–1.2	Cranial vault
Samgbungmacan	Java	0.5–0.8	Cranial vaults
Ngandong 1–12	Java	0.03–0.1	Cranial vaults
Gongwangling (Lantian)	China	1	Partial cranium

Table 12.2:   Continued

Fossil	Location	Age (millions of years)	Elements found
Yunxian 1,2	China	0.4–0.8	Crania
Chenjiawo	China	0.5–0.6	Mandible
Yuanmou	China	0.5–0.8	Teeth
Zhoukoudian	China	0.4–0.75	Partial skulls (>14 individuals)
Hexian	China	0.15–0.3	Cranial vault
Dali*	China	0.18–0.23	Cranium
Xujiayao*	China	0.1–0.13	Skull fragments (many individuals)
Maba*	China	0.1–0.15	Partial cranium
Jinnuishan*	China	0.2–0.3	Skeleton
Narmada	India	Middle Pleistocene	Vault

*Fossil often attributed to archaic *Homo* sp. (see Chapter 13).

assigned to *Meganthropus paleojavanicus*). The ensuing confusion was resolved by an outsider to hominin paleontology, Ernst Mayr, who sensibly condensed (synonymized) the various taxa into just one species, *H. erectus* (Mayr, 1944, 1951). Since then, the null hypothesis has generally been that *H. erectus* is one species unless proved otherwise (Le Gros Clark, 1964; Howell, 1978; Howells, 1980; Rightmire, 1988; Wolpoff, 1999), even as dozens of new discoveries from Asia and Africa have swelled its ranks (Table 12.2; Figure 12.4). In the early 1960s, Louis Leakey discovered several *H. erectus* fossils at Olduvai Gorge, substantially expanding the species' geographic and temporal range and establishing that *H. erectus* was the first species to migrate out of Africa. Other notable finds in Africa include a number of early fossils from Lake Turkana that may extend back as far as 1.9 mya, including a nearly complete skeleton of a juvenile (KNM-WT 15000), dated to 1.55 mya, from Nariokotome on the west side of Lake Turkana (Walker et al., 1986; Walker and Leakey, 1993). The Nariokotome boy has come to epitomize our image of *H. erectus*. This "strapping youth" was about 9 years old when he died but would have been nearly 6 feet tall and 65 kg when fully grown, with a postcranial skeleton much like that of modern humans. Additional African fossils assigned to *H. erectus* include some mandibles and other fragments from Ternifine, Algeria (Arambourg, 1963), a mandible from Thomas Quarry, Morocco (Ennouchi, 1969; Sausse, 1975), and a partial cranium from Salé,

Figure 12.4. Map of major *H. erectus s.l.* sites. See Table 12.2 for dates and information on other sites.

Morocco (Hublin, 1985). A handful of fossils from Swartkrans in South Africa, including a partial cranium, SK 847, probably also belong in *H. erectus* (Grine et al., 1993).

Recent discoveries have added some unexpected twists to the *H. erectus* story. Of particular import is an assemblage of fossils from Dmanisi, Georgia, dated to 1.77 mya. As of 2008, at least five crania (four with matching mandibles) and several partial skeletons have been found at the site. These fossils generally resemble *H. erectus* material from both Africa and Europe, but they are smaller: brain sizes range from 600 to 775 cm³, and body sizes range from 40 to 50 kg (Lordkipanidze et al., 2005, 2006, 2007; Rightmire et al., 2006). These size differences and other variations (see below) have led some researchers to classify the Dmanisi fossils as a different species, *H. geor-*

gicus (J. Schwartz, 2000; Vekua et al., 2002). Another source of complexity comes from geological ages. Dmanisi is a very old site; moreover, several *H. erectus* fossils from the Javanese sites of Sangiran and Mojokerto, long thought to be about one million years old, were redated to 1.6–1.8 mya (Swisher et al., 1994). And redating of another group of fossils from Ngandong, Indonesia, originally thought to be about 250,000 years old, indicates that they are less than 100,000 years old and possibly only 53–27,000 years old (Swisher et al. 1996, 1997; Yokoyama et al., 2008). If these dates stand the test of time, then some *H. erectus* fossils in Asia are nearly as old as the oldest *H. erectus* fossils in Africa, but the species also apparently persisted in Asia well after the origin of *H. sapiens.*

A final complication to note is that the oldest dated fossils reliably attributed to *H. habilis,* the putative ancestor of *H. erectus,* are the same age as the oldest *H. erectus* fossils, dating to about 1.8–1.9 mya. Thus, unless *H. erectus* evolved almost immediately from *H. habilis* and then spread around the world in less than 100,000 years, we are missing important information about the presumably habiline ancestry of *H. erectus.* As noted above, this hypothesis accords with the evidence of one fossil, a maxilla from Hadar dated to 2.3 mya (AL 666), which might be *H. habilis* (Kimbel et al., 1997).

General Anatomy of H. erectus s.l.

Before discussing variation in *H. erectus sensu lato,* let's review the species' cranial anatomy, as illustrated in Figure 12.5 (see also Andrews, 1984; Wood, 1984; Rightmire, 1990; Antón, 2002, 2003; Antón et al., 2007). Perhaps most characteristic is the shape of the cranial vault, which is typically long, low, and broadest near the base, with two ridges of bone at either end. The ridge at the back is the occipital or transverse torus, a horizontal shelf of bone between the nuchal plane and the posterior part of the occipital, which are at nearly 90° to each other in the sagittal plane. The most superior point of the nuchal lines in the midsagittal plane (a point called inion) is coincident with the most posterior point on the skull (opisthocranion). Some *H. erectus* crania also have an angular torus, an enlargement of the posterior portion of the temporal lines as they approach the mastoid angle of the parietal. At the other end of the vault is the supraorbital torus, a continuous shelf of bone above the two orbits, separated from the low-rising frontal squama by either a flat shelf or a slight sulcus (or "gutter"). Because the *H. erectus* face projects forward from the vault, there is always a considerable degree of postorbital constriction between the cranial vault and supraorbital torus. The vault in *H. erectus*

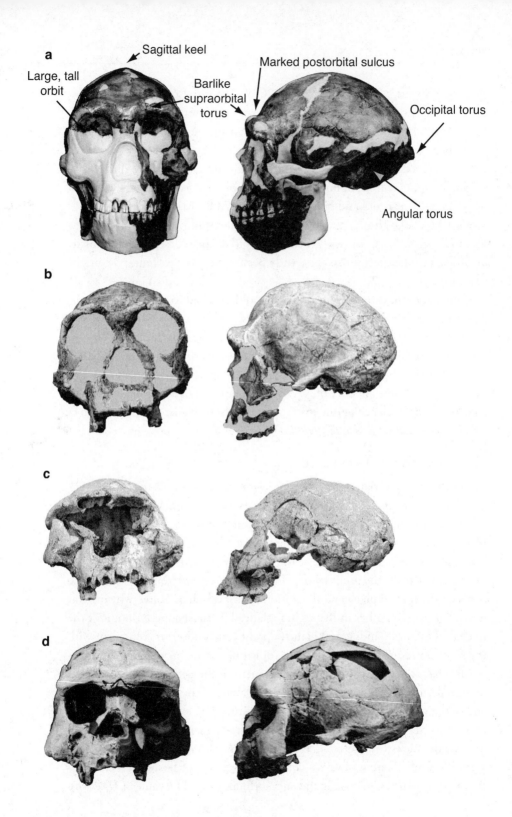

a

Sagittal keel

Large, tall orbit

Barlike supraorbital torus

Marked postorbital sulcus

Occipital torus

Angular torus

b

c

d

Figure 12.5. Cranial anatomy in *H. erectus s.l.,* highlighting several derived features in fossils from Africa, Europe, and Asia: (a) reconstructed composite from Zhoukoudian; (b) KNM-ER 3733, an early African example; (c) Dmanisi 2280, courtesy D. Lordkipanidze; (d) Sangiran 17. Images not to scale.

tends to be thick, in large part from an expanded diploe, but this feature is very variable (Lieberman, 1996). Another common but variable feature is the sagittal keel, a slight mounding of bone on either side of the sagittal suture and the midline of frontal (former location of the metopic suture).

ECVs in *H. erectus s.l.* (including specimens from Dmanisi) range from approximately 600 to 1200 cm³, with a general tendency for larger ECVs over time (Figure 12.6). As expected from the way that the vault grows around the brain (see Chapters 5 and 6), bigger-brained individuals tend to have more curved vaults and less marked postorbital constriction. The posterior cranial base in *H. erectus,* as in *H. habilis,* is short, and with petrous temporals that are rotated inward. The cranial base is not well enough preserved in most *H. erectus* crania to measure the cranial base angle, but there are enough measurements to show that the cranial base is moderately flexed. In one small cranium from Indonesia, Sangiran 4, CBA1 is 141° (Baba et al., 2003); in KNM-ER 42700, it is 138° (Spoor et al., 2007a), and in KNM-ER 3733, CBA1 is 145°. In OH 9, the angle between the presphenoid plane and the basioccipital clivus (known as CBA4) is estimated at 97° (range 93°–104°), significantly below the mean (111°) for this angle in modern humans (Maier and Nkini, 1984; Ross and Henneberg, 1995; Lieberman and McCarthy, 1999). Based on Mauer and Nkini's reconstruction, CBA1 in OH 9 is 133°–141°. Putting together all the evidence, the average CBA1 for *H. erectus* is about 141°, at the upper end of the modern human range and intermediate between the species average for *H. sapiens* and archaic *Homo,* such as *H. neanderthalensis,* in which the base is more extended (see Chapter 13).

There are only few well-preserved faces of *H. erectus,* but these indicate that the face was generally orthognathic, except for a slight degree of alveolar prognathism below the nasal sill. As in *H. habilis,* the upper face around the orbits is the widest part of the face. The infraorbital region is vertical or sometimes slopes slightly backward and is recessed relative to the projecting margin of the nasal cavity (piriform) aperture. Sometimes the lateral margins of the nasal aperture have a pillarlike configuration (paranasal maxillary pillars). *H. erectus* faces vary considerably in size. Some fossils, such as OH 12 and the Dmanisi

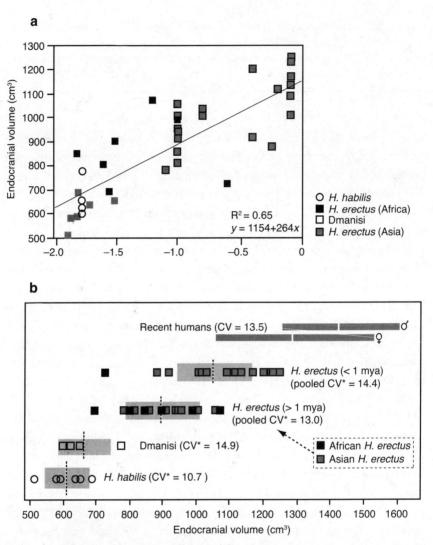

Figure 12.6. Endocranial volume (ECV) in *H. habilis* and *H. erectus s.l.* (a) over time and (b) by species and region. Gray bars in (b) indicate ± 1 standard deviation; dotted lines indicate species or population mean. Coefficients of variation (the standard deviation divided by the mean) are corrected for sample-size effects as CV*= CV · (1 + 1/4n).

crania, have diminutive faces, but others, such as Sangiran 17, have massive faces that are both superionferiorly tall and anteroposteriorly long (see Rightmire, 1990; Antón, 2004). As faces get bigger in *Homo,* so does the size of the supraorbital torus (Lieberman, 2000; Antón et al., 2007). Mandibles are notoriously variable, but *H. erectus* mandibles are more robust than those of modern humans, with a posteriorly sloping symphysis that lacks a chin. The cheek teeth are much smaller than those of *H. habilis* (average postcanine area is about 370 mm^2 in *H. erectus,* compared to 478 mm^2 in *H. habilis*) and only a little larger than those of modern humans. Many *H. erectus* molars have a mesial marginal ridge and accessory cusps; sometimes, the P^{4}s have two roots.

Variation in H. erectus s.l.

Given the long temporal span and wide geographic range of *H. erectus s.l.,* one expects considerable variation. However, the variation we observe is uncomfortably great by any standards, and not just as a consequence of time or space. This problem is amply demonstrated by the variation in endocranial volume. As Figure 12.6a shows, there is a general trend toward bigger brains over time, with some smaller specimens (600–900 cm^3) from Dmanisi and East Turkana at the older end of the spectrum, and larger-brained skulls (>1000 cm^3) in Asia that tend to be younger (Figure 12.6a). Within Asia and Africa, this trend is weaker. One simple way to assess variation is with the coefficient of variation (CV), the standard deviation of the sample standardized by the mean.[3] As illustrated in Figure 12.6b, the CV of brain size for pooled sex samples is 9.6 in chimpanzees, 11 in gorillas, and 12–14 in recent humans (Tobias, 1971). The CV* for ECV in *H. habilis* is a reasonable 10.7, but the CV* for *H. erectus s.l.* is 18.6 for the entire sample and 13.0 for fossils at least one million years old. If we further break down the sample, the CV* is still high: 14.9 in Dmanisi, 16.4 in the East African sample, and 13.3 in the Asian sample. We can draw two inferences from this variability. First, *H. erectus s.l.* is extremely variable in brain volume, even more so than gorillas. Some of this variation comes from the species' temporal span, but much of the variation within populations of *H. erectus s.l.,* if it was a single species, must result from males' being substantially bigger than females. Another possibility to consider is that *H. erectus s.l.* samples more than one species.

3. CVs need to be corrected for sample size (*n*), computed here as CV* = CV · (1 + 4/*n*) (Sokal and Rohlf, 1981).

If *H. erectus s.l.* actually comprises more than one species, then we would expect to see some geographical patterning of the variation. Yet despite efforts to find patterns, they are difficult to discern. One hypothesis is that the early African sample is a different species, *H. ergaster* (Andrews, 1984; Wood, 1991, 1994; Martinez and Arsuaga, 1997; Wood and Richmond, 2000), which also might include the Dmanisi sample (Gabunia et al., 2000b).[4] This diagnosis is based on several features thought to distinguish African and Asian samples, including a thinner vault, absence of keeling, and a less prominent angular torus. However, the hypothesis has not received much support, because several East African fossils do have keeling (e.g., KNM-ER 3733 and KNM-ER 42700), and others are thick, with large tori (e.g., OH 9, Daka). Further, vault thickness not only is highly variable but also scales with size (see Chapter 9; see also Gauld, 1996; Antón et al., 2007). So many of these distinctions may simply be size-related. Another possibility is that the Georgian material is a separate species, *H. georgicus,* which perhaps has closer affinities to *H. habilis.* Again, the primary support for this hypothesis is size. The largest-brained skull from Dmanisi (D2282) has an ECV of 775 cm^3 and must be a male, but it is smaller than all but two *H. erectus s.l.* crania from East Africa (KNM-ER 42700 and OH 12). In terms of shape, the Dmanisi crania generally resemble other *H. erectus* crania (Gabunia and Vekua, 1995; Rosas and Bermudez de Castro, 1998; Gabunia et al., 2000a,b; Vekua et al., 2002), and the scale with *H. erectus* crania from East Africa and East Asia and not with *H. habilis* (Lieberman and Bar-Yosef, 2005; Spoor et al., 2007a; Baab, 2008). Finally, although there are some subtle features that tend to be more prevalent or different in Asia than in Africa (e.g., a fissure between the tympanic and mastoid processes, a narrower glenoid fossa, a more frequent incidence of marginal ridges and accessory cusps on the molars), it is hard to distinguish fossils from the two regions using any scheme (see Antón, 2003; Terhune et al., 2007).

In short, it is possible that *H. erectus s.l.* comprises more than one species, but we cannot easily reject the hypothesis that it is a single species. In any event, the ramifications of splitting or lumping *H. erectus* are probably not profound, given what follows in the fossil record.

4. The name *H. ergaster* was originally intended to sort out variation in *H. habilis* (Groves and Mazak, 1975), and the species was proposed without comparison to *H. erectus.* The type specimen chosen for the new species, KNM-ER 992, is almost certainly an early African *H. erectus.* Because of the rules of zoological nomenclature, *H. ergaster* has priority for any species that includes KNM-ER 992.

Early *Homo:* Developmental Transformations

What developmental transformations underlie the origins and evolution of the head in early *Homo?* Although one could characterize this transition superficially by comparing species from before and after (e.g., *Au. afarensis* and *H. erectus*), this comparison would overlook evidence that the transition from *Australopithecus* to *Homo* was probably complex and involved at least one transitional, habiline stage. But until we have more fossil evidence, many aspects of this possible evolutionary sequence are conjectural.

The first transition to consider is from *Australopithecus* to the earliest species of *Homo,* which is probably *H. habilis* or something like it (but see above). Although cladistic analyses differ as to whether *H. habilis* evolved from *Au. africanus* or from something more like *Au. afarensis* (e.g., Lieberman et al., 1996; Skelton and McHenry, 1998; Strait et al., 1997; Strait and Grine, 2004), it is reasonable to assume that *H. habilis* evolved from a gracile australopith, keeping in mind that an unknown species might be the real ancestor. Further, I compare *H. habilis* primarily to *A. africanus* rather than to *A. afarensis,* because the former is younger and a little more derived. These assumptions don't really change the big picture. Compared to a generic gracile australopith, *H. habilis* grew a slightly larger brain on a shorter posterior cranial base, resulting in a slightly more rounded cranial vault and a more centrally located center of gravity. *H. habilis* also has a less prognathic face with relatively less lateral and anterior growth. Overall, these shifts give the impression of a slightly less robust masticatory system in *H. habilis* than in gracile australopiths.

The second, related transition to examine is the origin of *H. erectus,* which is complicated by uncertainty over whether *H. habilis* is really the ancestor of *H. erectus,* and whether *H. erectus s.l.* actually comprises more than one species. Again, we'll look at the big picture and assume that the answers to these two questions are "yes" and "no," respectively. If so, then in *H. erectus s.l.* we see a continuation of some trends that characterized the evolution of *H. habilis,* but with some additional derived features. Most important, most (but not all) *H. erectus* individuals grew bigger brains than *H. habilis,* with increases over geological time so that ECVs in *H. erectus* average 1,000 cm^3 by 1 mya. In addition, the face in *H. erectus,* especially in males, grew larger than in *H. habilis.* But like earlier hominins. *H. erectus* must have had considerable sexual dimorphism in body mass, which would help account for much of the variability in overall cranial size and robusticity, including cranial superstructures such as tori and keels. This pattern of robusticity persisted until recently, after the origins of *H. sapiens.*

One can only speculate on the developmental mechanisms behind these general shifts, using basic principles of craniofacial growth and development (reviewed in Chapters 2–5), along with some simple geometric morphometric comparisons of relatively complete fossil crania (from CT scans using methods outlined in Chapter 5). For practical reasons, we'll ignore *H. (Au.) rudolfensis,* which, despite its large brain size, could be an australopith and is unlikely to be the ancestor of *H. erectus* (Lieberman et al., 1996; Wood and Collard, 1999a; Strait and Grine, 2004).

Brain Size and Basicranial and Neurocranial Growth in Early *Homo*

ECV ranges from 400 to 560 cm^3 in *Au. africanus* (mean 462 cm^3, n = 8); from 509 to 687 cm^3 in *H. habilis* (mean 609 cm^3, n = 6); and from 600 to 900 cm^3 in *H. erectus* crania older than 1.5 million years (mean 737 cm^3, n = 8). So absolute brain size increased over both transitions. We do not yet know the developmental bases for these increases, and endocasts of early *Homo* crania are not well enough preserved to permit reliable inferences about the extent to which they involved general scaling of the brain as a whole or increases in particular regions of the neocortex (Holloway et al., 2004).

One important question to consider is the extent to which brain expansion in early *Homo* is a function of differences in body mass. As noted above, the limited postcranial evidence available for *H. habilis* (notably, the OH 62 skeleton) suggests that female body size in this species was australopith-like, leading to an estimated EQ of approximately 3.0 to 3.3. This is at the upper end of the range for australopiths, whose EQs range from 2.0 to 3.1 (Aiello and Dean, 1990; McHenry and Coffing, 2000; see Chapter 6). In the Dmanisi population, EQ estimates range from 2.6 to 3.1 (Lordkipanidze, et al., 2007), about the same as for the *H. erectus* boy from Nariokotome, whose adult EQ is estimated to be near 3.2. Therefore, brains in early *Homo* are, at most, only slightly larger relative to body size than in *Australopithecus.*

How did increased absolute brain size in early *Homo* affect the growth of the head as a whole? As Chapters 4 and 5 discussed, the brain, cranial vault, and cranial base grow as an integrated unit, partly driven by interactions between the volume of the brain and the dimensions of the cranial base. Given these principles, there are two possible explanations (not mutually exclusive) for the more spherical shape of the cranium in *H. habilis.* The first is a reorganization of brain structure, such as expansion of the parietal association cortex, which would lead to a rounder brain (see Holloway, 1983; Deacon, 1998). Data to test this hypothesis are currently scarce. The second is the spatial packing effect of growing a relatively larger brain on a relatively shorter cranial base (Ross and Ravosa, 1993; Lieberman et al., 2008a). This hypothesis is also difficult to test because there are no complete habiline cranial bases, but the best-preserved example known so far from Olduvai Gorge (OH 24) has a short basioccipital (30 percent shorter than in *Au. africanus*) combined with more coronally oriented petrous temporals.[5] A shorter posterior cranial base—resulting from less growth in the spheno-occipital synchondrosis or less posterior drift of the foramen magnum—positions the foramen magnum in *H. habilis* more centrally in the cranium than in gracile australopiths (Dean and Wood, 1982b). Further, a modestly larger brain combined with a shorter posterior cranial base leads to a relative expansion and general rounding of the neurocranium and possibly more coronally oriented temporals (see Chapter 4). These shifts are evident in Figure 12.7a, which depicts a thin-plate spline (TPS) transformation of 3-D landmarks from CT scans of an *Au. africanus* cranium, Sts 5 (the reference), into the landmark configuration of KNM-ER 1813 (the target) after correcting for size (see Chapter 5 for an explanation of the methods). Even a small-headed habiline like KNM-ER 1813 (whose ECV is only 510 cm^3) has a slightly more rounded frontal than the gracile australopith. When viewed from the back, the parietal walls in *H. habilis* are also more vertical than those in *Australopithecus* (e.g., the most lateral point on the parietal, euryon, is higher). No *H. habilis* crania are well enough preserved to measure either the length or angle of the cranial base, but there are suggestions that CBA1 was probably about as flexed as in *Au. africanus* and maybe more flexed than in *Au. afarensis* (see discussion in Chapter 11).

The developmental bases for basicranial and cranial vault differences between *H. habilis* and *H. erectus* (see Figure 12.7b) are mostly a continuation of trends we see in *H. habilis*. First, absolute ECVs in *H. erectus* tend to be even larger. Second, as in *H. habilis*, *H. erectus* skulls have a short basioccipital and coronally oriented petrous pyramids. Thus, to a slight extent, the posterior cranial base in *H. erectus* appears to be "rolled up" (spatially packed) from a bigger brain, causing the sagittal and coronal contours of the cranial

5. OH 24 is a worrying specimen, as its cranial vault was highly damaged and fragmented, but the cranial base is in pretty good shape. The OH 24 basioccipital is 16.1 mm, 31 percent shorter than the mean for three *Au. africanus* specimens (Sts 5, Sts 19, Mld 37/38).

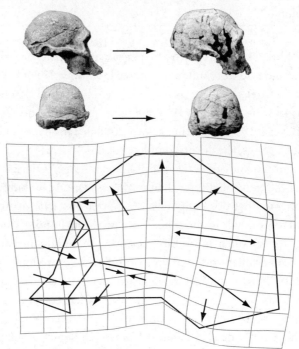

a

Sts 5 (reference); KNM-ER 1813 (target)

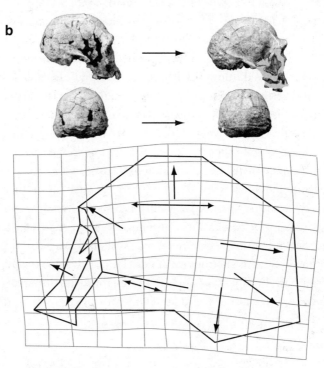

b

KNM-ER 1813 (reference); KNM-ER 3733 (target)

Figure 12.7. Thin-plate spline transformations of size corrected 3-D landmarks (data from CT scans). (a) Transformation from Sts 5, an *Au. africanus* female (reference) into KNM-ER 1813, a *H. habilis* female (target). (b) Transformation from KNM-ER 1813 (reference) into KNM-ER 3733, a likely male *H. erectus* (target). Arrows indicate major changes in the landmark configurations (see text for discussion).

vault to become a little rounder, particularly in larger-brained fossils (Antón et al., 2007). To what extent cranial base angulation participates in spatial packing is uncertain, in part because of variations in relative brain and face size. The cranial base angle in one big-faced *H. erectus* cranium, OH 9, is less flexed than in *H. sapiens;* but another relatively smaller skull from Java, Sambungmacan 4, has a cranial base as flexed as those of modern humans (Ross and Henneberg, 1995; Baba et al., 2003).

Another growth difference between *H. habilis* and *H. erectus* is that some neurocranial expansion in the latter occurs at the posterior end of the vault, probably through more deposition in the lambdoid suture and from more posterior drift of the nuchal plane (and maybe in the occipitomastoid suture). This extra growth pushes out the posterior end of the vault, giving *H. erectus* its characteristically long, low occipital, in which the most posterior point on the vault (opisthocranion) is nearly coincident with the meeting point of the superior nuchal lines (inion). The elongated occipital squama and nuchal plane help make the vault look lemon-shaped in lateral view, an impression exaggerated by superstructures such as the occipital torus (see below). Finally, bigger ECVs cause the neurocranium and basicranium in *H. erectus* to grow a little wider than in *H. habilis,* particularly in the posterior cranial fossa. More lateral growth displaces the TMJs more laterally. Because the palate in *H. erectus* is approximately the same width as in *H. habilis* (ca. 40 mm), the tympanic portion of the temporal needs to be rotated about 30° around a vertical axis in order to align the long axis of the mandibular ramus slightly inward from the TMJ so that the lower and upper tooth rows are parallel and similar in width; TMJ rotation in *H. erectus* sets up a 90° angle between the petrous and tympanic portions of the temporal (Figure 12.8; see also Terhune et al., 2007).

Facial Size, Shape, and Orientation in Early *Homo*

Early *Homo,* like gracile australopiths, has a projecting upper face relative to the neurocranium, and hence substantial brow ridges and some degree of pos-

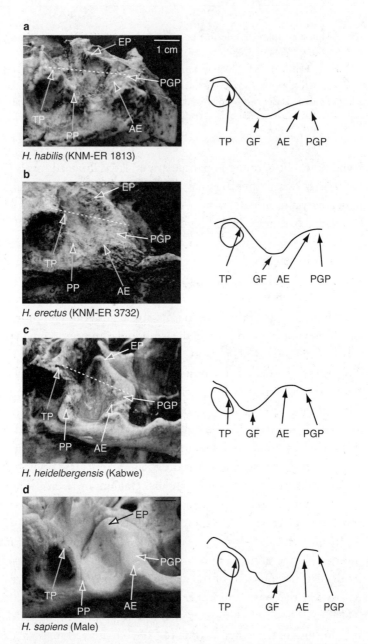

Figure 12.8. Left: oblique views of the temporal region: (a) *H. habilis* (KNM-ER 1813);
(b) *H. erectus* (KNM-ER 3732); (c) *H. heidelbergensis* (Kabwe); (d) *H. sapiens* (recent male).
TP, tympanic plate; PP, postglenoid process; GF, glenoid fossa; EP, entoglenoid process; AE,
articular eminence; PGP, preglenoid plane. Right: contour maps through the middle of the
glenoid fossa (location of contour indicated by dashed line at left).

torbital constriction. However, one can see evidence for two major shifts in facial growth during the evolution of early *Homo.* First, between gracile australopiths and *H. habilis,* the face became more orthognathic as a whole and the midface mediolaterally narrower. Second, in *H. erectus* there was a general trend toward a superoinferiorly taller face, albeit with much variation from sexual dimorphism and other factors.

In terms of prognathism, *H. habilis* is more orthognathic than *Au. afarensis* or *Au. africanus* (see Figure 12.7). As reviewed in Chapters 4 and 5, more orthognathic faces can derive from a more flexed cranial base, from less anterior displacement of the maxillary complex, or from a shift in growth fields in which the anterior surface of the face is covered by a resorptive rather than a depository growth field, as is the case in humans but not in nonhuman primates (Duterloo and Enlow, 1970). We currently lack histological data on facial growth fields in early *Homo;* further, we have no reliable measurements of CBA1 in *H. habilis* and only indirect estimates for most *H. erectus* crania. That said, we can infer tentatively that CBA1 is unlikely to be much more flexed in *H. habilis* than in *Au. africanus* (see above). This inference raises the hypothesis that habiline orthognathism derives from a relatively shorter, more ventrally rotated maxillary complex. Such a change might have occurred from a growth field shift in which the anterior surface of the face in *H. habilis* is more resorptive than in gracile australopiths, which retain the primitive pattern of chimpanzees (Bromage, 1989). If so, then early *Homo* may lack a snout because of a shift in growth field regulation on the anterior surface of the maxilla and zygomatic (a hypothesis ripe for testing). Note that the TPS in Figure 12.7a suggests a more posteriorly positioned facial surface relative to the rest of the face and cranial base.

H. habilis lacks the characteristic australopith facial configuration in which the midface is wider than the upper face. Instead, the *Homo* midface is typically narrower than the upper face. This slimming down presumably occurs from less lateral bone deposition in the zygomaticomaxillary suture or less lateral drift of the zygomatic. Together these changes lead to more vertically oriented lateral orbit margins and a squarer (less triangular) malar region.

The face in *H. erectus* essentially exhibits a continuation of the trends we see in *H. habilis* but is relatively bigger in some individuals, especially from more vertical growth (see Figure 12.7b). This increase in facial size among probable males may partly reflect the lack of any well-preserved faces of male *H. habilis* individuals. With this caveat in mind, when one compares most *H. erectus* faces (e.g., KNM-ER 3733, Sangiran 17) with the face of KNM-ER

1813, just about everything in the former is superoinferiorly taller in proportion to size, including the supraorbital torus (thicker), the shape of the orbits (taller and bigger), and the nose (taller). Presumably taller faces in *H. erectus* are partly a scaling consequence of having bigger bodies, but they could also result partly from selection for relatively bigger faces. In either case, bigger-faced *H. erectus* individuals, presumably males, have more massive brow ridges and other features that create an overall pattern of marked craniofacial robusticity compared to recent humans (see below).

Tooth Size and Masticatory Robusticity in Early *Homo*

Many important developmental differences between *Australopithecus* and *Homo* involve a reduction in the size of the entire masticatory system, beginning with the teeth. When Leakey and colleagues proposed *H. habilis* in 1964, they considered smaller postcanine teeth to be one of the species' defining features. In fact, as Figure 12.2 depicts, dental reduction did not occur in *H. habilis*. Tooth size in *H. habilis* is very variable, but the species' postcanine tooth area (the summed products of the mesiodistal and buccolingual dimensions of the P_4, M_1, and M_2 crowns) averages 487 mm^2, only a little smaller than in *Au. africanus* (511 mm^2), and much bigger than in early *H. erectus*, whose postcanine area averages 377 mm^2—just a little larger than in *H. sapiens* (334 mm^2) (McHenry and Coffing, 2000). That said, there is some limited evidence that *H. habilis* postcanines might have slightly thinner enamel crowns, perhaps caused by slower daily secretion rates. Enamel crown thickness in molars is notoriously variable both within and between teeth, but in *Au. africanus*, maximum thickness averages 2.2 (± 0.4) mm on the occlusal surface and 2.3 (± 0.5) mm on the cervical surface, with a total range of 1.3–2.9 mm (G. T. Schwartz, 2000; see also Chapter 11). Of the few *H. habilis* molars from East Africa that have so far been measured from natural fractures, maximum thickness averages 1.4 (± 0.4) mm on the occlusal surface and 1.6 (± 0.3) mm on the cervical surface, with a total range of 1.2–2.2 mm (Beynon and Wood, 1986). These thickness measurements need to be taken with several grains of salt given high levels of variability, small sample sizes, and the difficulties of attributing isolated teeth to species, but they do accord with the general observation that crown surfaces tend to show more relief in *H. habilis* than in gracile australopiths. Further, according to histological studies by Lacruz et al. (2008), the average daily secretion rate of molar cuspal enamel declines from approximately 5.5 μm/day in *Au africanus* to 4.7 in *H. habilis* and 4.2 in *H. erectus*. More data are needed to test this hypothesis.

It is worth emphasizing that really major reductions in crown size in early *Homo* occurred between *H. habilis* and *H. erectus,* especially in the third molar and in the mesiodistal lengths of the other postcanine teeth (Wood, 1991). Crown reduction is complicated by much intraspecific variation but must have occurred from reductions in the size of the enamel dentine junction (EDJ), presumably because of fewer cell divisions in the inner enamel epithelium during the bell stage of crown development (see also Chapter 11). Tooth roots, especially premolars, also tend to be simpler in *H. erectus* (Wood et al., 1988). In spite of smaller crowns, however, the overall schedule of tooth growth and eruption in early *H. erectus* was relatively rapid and similar to that of apes (Dean et al., 2001).

As one might expect from its dentition, *H. habilis* is intermediate between gracile australopiths and *H. erectus* in terms of overall masticatory robusticity. There is no single way to measure this, but one useful assessment is the size of the mandible relative to an estimate of body mass. When corrected for body mass, measures of mandibular robusticity (e.g., the height and thickness of the symphysis and corpus) along with measures of crown size in *H. habilis* are more similar to those of australopiths than to those of *H. erectus,* which are much less robust (Chamberlain and Wood, 1985; Wood and Aiello, 1998; Wood and Collard, 1999a,b). We do not know to what extent these differences result from diminished growth in response to loading or from other developmental mechanisms, such as smaller condensation sizes during embryonic growth. Regardless, thinner, smaller molars and premolars in early *Homo,* especially *H. erectus,* do appear to correspond to a decrease in the overall size of the adductor muscles or a shift to smaller, fast-twitch fibers, as discussed in Chapter 7 (Stedman et al., 2004). Smaller or weaker jaw muscles may also explain other derived features of *H. erectus,* such as relatively narrower mandibular condyles and TMJs and less robust zygomatic arches.

Sexual Dimorphism and Overall Robusticity

Although the masticatory system as a whole is relatively smaller and less robust in early *Homo* than in *Australopithecus,* one should not consider early *Homo* to be gracile. As noted above, many early *Homo* crania, especially some fossils attributed to *H. erectus,* have classic features associated with general craniofacial robusticity, including thick cranial vaults and cranial superstructures such as supraorbital tori, angular tori, and keels. We don't know the developmental bases for all these features, but many of them correlate with body size, which is highly variable.

One obvious source of this variation is sexual dimorphism. Although *H. erectus* has traditionally been considered to have low, levels of sexual dimorphism, this view needs to be reconsidered given recent discoveries from East Africa and Georgia. Of the four East African crania attributed to *H. erectus* that are more than 1.5 million years old, one, KNM-ER 42700 is clearly a small, gracile female; it has an ECV of 691 cm³. The three others (KNM-ER 3733, KNM-ER 3883, and KNM-WT 15000) are larger, more robust, and likely to be males, with an average ECV of 850 cm³ (range 804–900 cm³). Assuming that KNM-ER 42700 is an average female, then ECV dimorphism in the species was about 20 percent, less than in australopiths but far greater than in modern humans, which averages about 8 percent (Tobias, 1971).[6] A similar degree of dimorphism characterizes other geographical and chronological populations of *H. erectus*. For example, the Dmanisi fossils include a group of smaller crania (D2700, D2280, D3444) with an average ECV of 625 cm³, and a larger cranium (D2282) with an ECV of 775 cm³, yielding an estimated dimorphism of 24 percent (see Figure 12.9). Similarly, dividing the Sangiran crania into larger (S3, S12, S17) and smaller (S2, S4, S10) groups leads to an estimated dimorphism of 17 percent.[7] Comparable or even higher levels of dimorphism are evident from a comparison of the one male *H. erectus* cranium from East Africa that is about one million years old—Daka, with an ECV of 995 cm³ (Asfaw et al., 2002)—with two roughly contemporary putative females, OH 12 and KNM-OL 45500—both with ECVs estimated to be less than 800 cm³ (Holloway et al., 2004; Potts et al., 2004).

It is difficult to be definitive, but *H. habilis* also appears to display considerable variability in size and robusticity (Wood, 1991, 1993). For example, of the fossils reliably attributed to *H. habilis,* larger crania, such as KNM-ER

---

6. As Chapter 11 notes, australopiths probably had moderate to high levels of dimorphism, with a body mass in males of approximately 45 kg, about 40–50 percent higher than in females, who averaged 30 kg (McHenry, 1996). Given that brain size scales with negative allometry to body size, dimorphism of ECV values tends to be a little lower, with male australopiths such as AL 444–2 having ECVs that may be approximately 30–40 percent larger than those of females such as AL 288–1.

7. Sangiran 4 has a modest ECV (908 cm³), but the specimen is robust, with a strong parietal keel, a massive occipital torus, and cresting in the mastoid region; it might be a male.

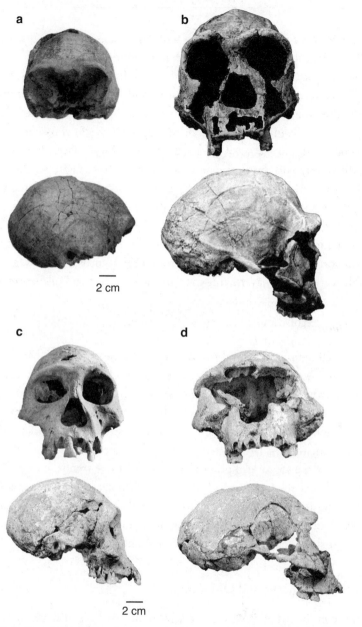

Figure 12.9. Sexual dimorphism in *H. erectus s.l.* Top: comparison of likely female (a, KNM ER 42700) and male (b, KNM-ER 3733) crania from the Turkana Basin. Bottom: comparison of likely female (c, D2700) and male (d, D2282) crania from Dmanisi. Image of KNM-ER 42700 from Spoor et al. (2007a) (courtesy Kenya National Museum); image of Dmanisi crania courtesy G. P. Rightmire and D. Lordkipanidize.

1805, tend to be much more robust, with larger cranial superstructures than smaller crania, such as KNM-ER 1813 (see Figure 12.1).

Appreciating the substantial levels of sexual dimorphism in early *Homo,* as well as size variation between populations and over time, helps us interpret some of the variability in cranial superstructures, especially in *H. erectus.* As the comparisons in Figure 12.9 make clear, bigger *H. erectus* crania are generally more robust, in part because of bigger brains or bigger faces, both of which scale differently with body size and have different effects on growth. For example, face size scales with brow ridge size and other features related to mastication (see Chapter 5); other aspects of cranial vault growth, such as vault thickness, scale with brain size or body size (see Chapter 6). There are few *H. erectus* crania with intact faces, but Antón and colleagues (2007) find a significant correlation between ECV and supraorbital torus thickness, and other variable features, such as the thickness of the posterior end of the vault.

Early *Homo:* Functional and Behavioral Changes

What were the functional and behavioral implications of cranial variations in early *Homo?* And can we make inferences about selective pressures during the transition from *Australopithecus* to *Homo,* and from earliest *Homo* to *H. erectus?* To summarize what little we know, the first of these transitions mostly involved a slight increase in brain size, a shorter posterior cranial base, less facial prognathism, and less lateral growth of the midface. In the second transition to *H. erectus,* we see an elaboration of many of these trends, combined with much smaller cheek teeth and a scaling up of face size, which presumably corresponds to an increase in body size, particularly in males. Together these changes point to a combination of cognitive, dietary, thermoregulatory, and locomotor shifts that suggest selection for a new way of life.

Cognition, Language, and Ontogeny

Given the problems of estimating relative brain size, it is difficult to ascertain how much of the endocranial volume expansion evident in early *Homo* is a byproduct of larger bodies. Regardless, there are other lines of evidence for some sort of cognitive shift from *Australopithecus* to earliest *Homo,* and from *H. habilis* to *H. erectus.*

The most direct source of information about cognition is the archaeological record. The first known archaeological sites, from approximately 2.6 mya, are mostly just scatters of simple stone tools, often associated with animal bone fragments (Semaw et al., 2003; Roche et al., 1999). This earliest stone tool industry, the Oldowan, was made by using one stone, a hammerstone, to knock flakes off another fine-grained stone, called a core (Figure 12.10). The resulting products, sharp flakes and a chopper made from the remnant core, were used to cut meat off bones, to cut wood, and to extract marrow from bones (see Schick and Toth, 1993). Bones found in the earliest sites have cut marks and percussion marks, which are evidence of defleshing and marrow extraction, respectively (de Heinzelin et al., 1999; Semaw et al., 1997; Semaw, 2000). We cannot attribute these first stone tools to any species, but their appearance just prior to the oldest dated fossils of early *Homo* seems unlikely to be mere coincidence. Further, in spite of their crudeness, Oldowan tools indicate a level of planning and complexity beyond the capabilities of chimpanzees (Toth et al., 1993). And, in spite of disagreement whether the animal bones in these sites were acquired by scavenging or hunting, their presence is a signal that hominins were incorporating meat and animal tissues in their diet (see below).

A second technological transition occurred about 1.7 mya, with the invention of more sophisticated flaked-tool industries known as the Acheulian and the Developed Oldowan (Dominguez-Rodrigo, 2001). These industries included the production of symmetrical, bifacial tools; the major difference between them is that only the Acheulian includes handaxes (Figure 12.10), sometimes found in sites by the thousands. It's a reasonable inference that this new technology is related to the evolution of *H. erectus*. Bifacial tools such as handaxes, although crude by our standards, require more steps to fabricate than Oldowan tools, and they come in more stereotypical forms, often very symmetrical. These toolmakers must have had a clear idea, a mental template, of what they were trying to make. In addition, animal bones found in sites less than 1.7 million years old often come from relatively complete carcasses that were probably acquired by hunting, not scavenging (Potts, 1988; Bunn, 2001; Dominguez-Rodrigo, 2002; Dominguez-Rodrigo et al., 2006).

In other respects, we have little additional evidence to compare cognitive complexity in early *Homo* with that in australopiths beyond the data on relative and absolute brain size (see above). One recurrent subject of speculation is whether larger brains, the presence of stone tools, and hunting are correlated with other cognitively complex capabilities such as language. It seems probable that *H. erectus* had some form of language, but whether they had modern speech capabilities is hard to test. As discussed in Chapter 8, one can

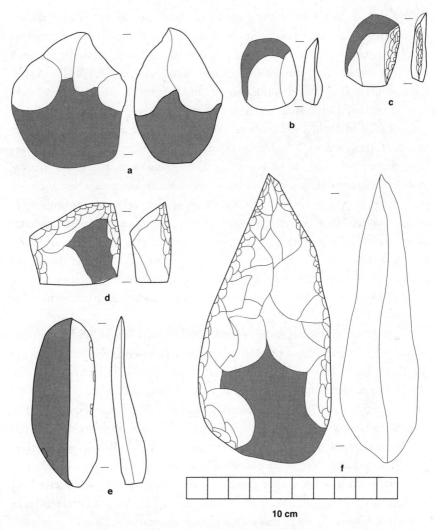

10 cm

Figure 12.10. Lower Paleolithic lithic technology showing different views of representative tools: (a) Oldowan chopper; (b) flake; (c) retouched flake or simple scraper; (d) complex scraper; (e) flake blade, (f) Acheulian handax. Drawings courtesy Dr. J. Shea.

reconstruct early *Homo* with a primitive (intranarial) larynx, in which the epiglottis and soft palate contact only if the tongue was unusually small relative to body mass. It is therefore reasonable to infer that *H. erectus* had a somewhat modern configuration of the pharynx, with a rounded tongue and a partially descended larynx; it probably also lacked air sacs. These derived features would have improved speech capabilities. But because *H. erectus* has such a long face, the ratio of vertical to horizontal portions of the vocal tract was unlikely to be 1:1, suggesting speech that was not as quantal, hence somewhat less articulate and perceptible as modern human speech.

Another indirect clue about cognition comes from ontogeny. Modern humans have a long, extended period of childhood, which delays the onset of reproduction but has benefits, including extending a period that facilitates many kinds of learning (see Bogin, 2001). Paleoanthropologists once assumed from tooth eruption patterns that a long, delayed ontogeny evolved in early *Homo,* with M1 eruption (which coincides with 95 percent completion of brain growth) occurring at about 6 years of age. However, studies by Dean and colleagues (2001) on incremental structures in enamel and dentine (see Chapter 2) indicate that the rate and timing of tooth crown formation and eruption in *H. erectus* fossils were more like those of australopiths than those of modern humans. According to their estimates, M1 eruption in *H. erectus* occurred at approximately 4.4 years of age, about 1.6 years earlier than in *H. sapiens.* We'll return to this subject and the relationship between the rate of ontogeny and the rate of brain growth in Chapter 13.

Diet

The transition from *Australopithecus* to *Homo* involved a shift in diet. Before considering what early *Homo* ate and how they chewed it, it is useful to think about dietary needs, many of which were driven by increases in brain and body size. Across primates, daily energetic expenditure, DEE (in kcal) = $93.3W^{0.75}$, where W is body mass in kg (Leonard and Robertson, 1997; Key and Ross, 1999). Larger primates need absolutely more but relatively fewer calories per day. Additional costs to consider are those of reproduction. Gestation elevates DEE on average by 25 percent, and lactation increases it by 37 to 41 percent (Oftedahl, 1984; Coehlo, 1986). Applying these equations to hominins (Table 12.3), with slight adjustments for body mass, suggests that a typical 45 kg australopith male would have needed 1,610 kcal per day; a 30 kg australopith female would have needed 1,175 kcal/day when not reproducing, 1,469 kcal/day

Table 12.3:   Estimated daily energetic expenditures (DEE) of chimpanzees, australopiths, and early *Homo*

Species	Male body mass (kg)	Male DEE (kcal)	Female body mass (kg)	Female DEE (kcal)	Female DEE gestating (kcal)	Female DEE lactating (kcal)
Pan troglodytes	40	1,332	30	1,175	1,469	1,671
Australopithecus afarensis	45	1,610	30	1,175	1,469	1,671
Homo. erectus (small)	50	1,700	40	1,332	1,665	1,874
H. erectus (large)	63	2,087	52	1,807	2,269	2,487
Old World human foragers (nonequestrian)	70	2,259	57	1,931	2,414	2,665

Adapted from Aiello and Key (2002) and Marlowe (2005).

when gestating, and 1,671 kcal/day when lactating (Aiello and Key, 2002). Larger bodies and larger brains would increase DEE in *H. erectus,* especially in reproducing females. In smaller *H. erectus* morphs (such as those at Dmanisi), lactating mothers' costs would have increased approximately 12 percent to 1,874 kcal/day, but once *H. erectus* females attained modern human sizes, about 50 kg, then their DEE costs when lactating would have been 2,487 kcal/day, a substantial 48 percent increase compared to those of australopith mothers.

What kinds of foods did *H. erectus* individuals, especially females, eat to supply their higher daily energy needs? Many mammals solve this problem by taking advantage of the fact that metabolism scales to body mass to the power of 0.75. Because bigger mammals need fewer calories per unit of body mass, they can afford longer gut passage times (hence more fermentation and absorption) and thus process larger quantities of higher-fiber foods (Gaulin, 1979; Lambert, 1998; du Toit and Yetman, 2005). This relationship, known as the Jarman-Bell principle, explains why smaller-bodied apes such as gibbons tend to feed almost exclusively on higher-quality resources, such as fruit, whereas larger-bodied apes such as gorillas have a much higher component of fiber in their diets.

There are several lines of evidence suggesting that early *Homo,* especially *H. erectus,* did not follow the Jarman-Bell principle. The first of these is tooth size. As noted above, compared to australopiths, *H. habilis* had similar-sized crowns that perhaps were a little thinner, but *H. erectus* certainly had both thinner crowns and a 22 percent decrease in postcanine tooth area. As Chapter 7 discusses, small, thin postcanine crowns are less effective than large, thickly enameled ones at breaking down hard foods that require repetitive, high forces (Kay, 1985; L. B. Martin, 1985; G. T. Schwartz, 2000; Lucas, 2004). Further, occlusal surface area correlates with the rate of food breakdown, because larger teeth have more breakage sites, accommodate a greater range of food particle shapes, and perform better at comminuting small, hard items such as seeds. To generate a similar fracture, tooth size scales with food toughness to the power of 1/3 (Lucas, 2004). Hence, teeth that are 20 percent smaller in surface area suggest a 50 percent reduction in food toughness. Therefore the smaller, thinner postcanine teeth of early *H. erectus* imply that they were chewing a higher-quality diet that was neither as tough nor as hard as what australopiths ate. These inferences are supported by analyses of microwear on *H. erectus* teeth (Teaford and Ungar, 2000; Ungar et al., 2006). Microwear analyses of *H. habilis* teeth also suggest a more versatile diet than that of *Australopithecus,* with less reliance on fallback foods (Ungar and Scott, 2009).

Additional sources of information about the diet of early *Homo* come from assessments of craniofacial biomechanics. Here, the evidence is equivocal, starting with comparisons of the efficiency of bite force production. Figure 12.11 gives a simplistic example by comparing the mechanical advantage of the temporalis and the masseter in the sagittal and frontal planes for a composite *Au. afarensis* skull, a *H. habilis* (KNM-ER 1813, probably a female), a reconstruction of a likely male *H. erectus* from Zhoukoudian (Tattersall and Sawyer, 1996), and a recent human male. Although the zygomatic arch is less advanced in *Homo,* the face is also less prognathic, leading to slightly more efficient vertical force generation for the masseter and slightly less for the temporalis. The combination of a wide face with a more medially placed TMJ also increases the MA of the masseter in the coronal plane. Thus there is no substantial decrease in the efficiency of masticatory force generation in early *Homo* compared to that of gracile australopiths. Note, however, that estimates of MA indicate only how efficiently muscles *could* produce force, not how much force they actually produced. To determine this parameter, we need to return to estimates of bite force production in hominins that use data from cross-sectional areas of the muscles and the lengths of their lever and load arms (see Chapter 7). Estimates by Demes and Creel (1988), shown in Figure 11.15, suggest that *H. habilis* and *H. erectus* could produce about as much pressure during incision and mastication as gracile australopiths, such as Sts

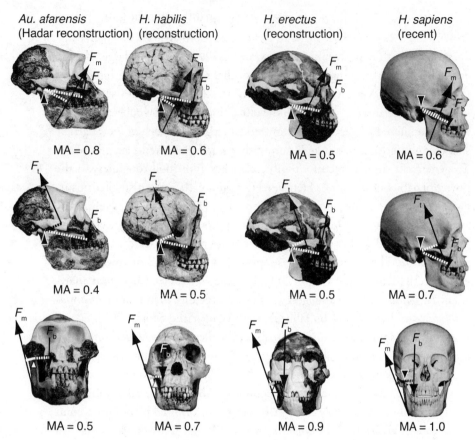

Figure 12.11. Comparison of mechanical advantages (MA) for the temporalis (F_t) and masseter (F_m) muscles in the sagittal plane, and for the masseter in the transverse plane (bottom) in *Au. afarensis* (Hadar composite), *H. habilis* (composite of KNM-ER 1813 and OH 13), *H. erectus* (Zhoukoudian composite), and a recent *H. sapiens*. Bite force efficiency in *Homo* decreases in the sagittal plane for the temporalis but increases for the masseter; in addition, recent *Homo* is more efficient at generating transverse bite forces. Keep in mind that MA measures only the efficiency of torque generation, not the magnitude of bite forces actually generated. *H. habilis* composite courtesy G. Sawyer and V. Deak.

5. Again, we should be cautious about these estimates, because they require a number of assumptions based on incomplete evidence and do not take into account muscle fiber type. Humans differ from apes in having a derived, in-activated version of the *MYH16* gene, which evolved approximately 2.4 mya and reduces the size of fast-twitch muscle fibers in the jaw adductors (Rowler-son et al., 1983; Stedman et al., 2004). This suggests less bite force capability in early *Homo* than in *Australopithecus*.

A further indicator of diet is the extent to which different regions of the face and jaws would have coped with high or repetitive masticatory stresses. More robust jaws with relatively larger cross-sectional and second moment areas can better resist the substantial bending and twisting strains that unilateral mastication generates (reviewed in Chapter 7). Although comparisons of mandibular cross-sectional shape between primate species correlate poorly with differences in the mechanical properties of their diets (Daegling, 2002), the robusticity of the mandibular corpus (breadth relative to height) is about 15 percent lower ($p < 0.01$) in *H. erectus* than in *H. habilis* or gracile australopiths, which are about equally robust (Chamberlain and Wood, 1985; Wood and Collard, 1999a,b). In addition, the *H. erectus* face has other features that give the impression of being less capable of resisting the high, frequent bite forces associated with a biomechanically challenging diet. These include less robust zygomatic arches and less mass in the coronal and sagittal planes of the face that would counter twisting, bending, and shearing.

Evidence for a dietary shift in early *Homo* raises the question of what they were eating that was softer, less tough, and required less work to chew. The period in question was characterized by considerable climatic variability, with an overall trend toward expansion of grassland habitats, particularly by 1.7–1.9 mya (Behrensmeyer et al., 1997; Potts, 1998; Reed, 1997; Bobe et al., 2002; Reed and Russak, 2009). One plentiful source of nutrients in the open habitats in which many early *Homo* fossils and archaeological sites are found, such as the Turkana Basin and Olduvai Gorge, is meat. Thanks to the archaeological record, we know that early *Homo* did have access to meat and other animal tissues, such as marrow and brain, which are rich in calories, proteins, and fats but low in fiber (more on this below). These habitats also offer a number of other high-quality foods, including underground storage organs (USOs), such as tubers. The latter were probably gathered by australopiths, and they must have been as important for early *Homo* as they are for modern foragers (Hatley and Kappelman, 1980; Hawkes et al., 1997; Ungar, 2004; Laden and Wrangham, 2005). Because most USOs are tough and fibrous when raw, it seems reasonable to hypothesize that early *Homo* consumed USOs using mechanical processing techniques such as pounding; further, some paleoanthropologists have suggested that early *Homo* cooked USOs, which would have increased their energy returns (Wrangham et al., 1999; Wrangham, 2006, 2009).

A final set of dietary variables to consider are social ones. Early *Homo* and australopith mothers probably had different diets, but the former may also have obtained more resources from increased levels of food sharing or more provisioning by males (Isaac, 1978). Although these behaviors are invisible in the fossil record, they are considered below.

Viewed together, the craniodental evidence suggests that the dietary transition from *Australopithecus* to *Homo* occurred in two stages. First, one can hypothesize that *H. habilis* was able to process a diet generally similar to that of gracile australopiths. But perhaps this species was not as dependent as *Australopithecus* on low-quality fallback foods and instead relied more on scavenged meat or processed (e.g., pounded) USOs when fruits and other high-quality gathered or extracted foods were scarce. This strategy was certainly elaborated on and expanded by *H. erectus,* which must have consumed a generally higher-quality diet that was less fibrous, tough, and hard. That said, we should be under no illusions about *H. erectus* cuisine. By any standard, *H. erectus* probably ate a much more mechanically demanding diet than we eat today.

Locomotion, Respiration, and Thermoregulation

To dine in the Early Stone Age, one first had to hunt or forage for dinner. Thus, if the transition from *Australopithecus* to *Homo* involved a shift in diet and energetics, it must have also involved some changes in locomotion and related physiological processes, such as respiration and thermoregulation. Before considering how early *Homo* hunted and gathered, let's review the craniofacial evidence for locomotor, respiratory, and thermoregulatory differences between early *Homo* and *Australopithecus.*

We'll begin with the nose. In some crania of *H. habilis* (e.g., KNM-ER 1813, KNM-ER 1805) and in most early *H. erectus* crania (e.g., D2700, KNM-ER 3733, KNM-WT 15000) the nose is derived relative to *Australopithecus* in two important respects: the margins of the piriform aperture are sharp and everted, and a distinct nasal sill separates the nasal cavity from the face (see Figure 12.5). In addition, the nasal bones in early *H. erectus* tend to be convex, and they are tilted up in the sagittal plane relative to the nasal process of the frontal. As Franciscus and Trinkaus (1988) insightfully pointed out, these features are markers of an external nasal vestibule, a derived feature of *Homo* that helps generate turbulent airflow during inspiration. Turbulent flow occurs partly because the external nose points the nares downward, thus creating a right angle between the outside world and the nasal vestibule; additional causes of turbulence include the nasal sill and the discontinuity created by the nasal valve between the internal and external nose (see Chapter 8).

The one obvious benefit to increasing nasal turbulence in early *Homo* would have been to humidify air during rapid, forceful respiration. Although turbulence generates much more airflow resistance and thus requires the lungs to work harder to pump air, turbulent flows lack a boundary zone along the surface of the epithelium where heat and moisture exchange occur. In addition, the vortices characteristic of turbulent flow bring a greater volume of air into contact with the mucosal lining of the respiratory epithelium. Therefore, more turbulent noses are more effective at humidifying air during vigorous activities that require higher airflow rates. This would have been an advantage for early *Homo* individuals hunting or gathering in the midday sun, helping bring air to the requisite 75–80 percent humidity during inspiration, and helping to recover some of that water during expiration. External noses, however, are more an adaptation for walking than for running. When running, humans must switch to oral breathing (or possibly oronasal breathing) because resistance in the nose becomes too high for effective respiration (see Chapter 8). We lack good enough models of airflow in fossil hominins to deduce whether early *Homo* was an obligate mouth breather when running. One major advantage of obligate mouth breathing is to dump heat, although at the cost of more water loss.

In short, the most likely explanation for selection to increase nasal turbulence in early *Homo* is to improve respiratory function during aerobic activity in hot, arid conditions. External noses probably helped early *Homo* engage in strenuous activities, such as trekking, during the heat of midday, when most mammals rest and seek shade to avoid hyperthermia. If this hypothesis is correct, then we should also see evidence for commensurate thermoregulatory adaptations, especially sweating and the loss of fur. Unfortunately, elaboration of eccrine glands and the loss of terminal hair over most of the body do not leave a trace in the fossil record, and their genetic basis is currently unknown. At this point, we can only speculate that *H. habilis* or *H. erectus* was more furless and sweaty than extant apes and australopiths.

As discussed in Chapter 8, there is some evidence that early *Homo* had additional mechanisms to keep the brain cool. According to Falk (2004, 2007), the frequency of emissary veins in the parietal and mastoid bones increases from approximately 20–40 percent in *Australopithecus* to greater than 70 percent in *H. habilis* and *H. erectus* (the Dmanisi sample appears to be intermediate). Emissary veins normally transport blood from the brain to the scalp, where it can be cooled by evapotranspiration. Cooled blood in the scalp helps lower core body temperature and has also been proposed to cool the brain directly during conditions of hyperthermia through reverse flow back into the diploe (Cabanac and Brinnel, 1985; Zenker and Kubic, 1996).

If the above speculations are correct, then early *Homo* was probably trekking more than australopiths in the heat. At some point, hominins also evolved the ability to run long distances at impressive speeds under aerobic capacity in such conditions, and there are several lines of evidence that these locomotor capabilities were present to some extent in early *Homo*. Musculoskeletal adaptations for endurance running are mostly postcranial (see Bramble and Lieberman, 2004), but at least three cranial features would have improved the ability to stabilize the head during running. Perhaps the most definitive is the radius of the anterior and posterior semicircular canals relative to body mass, which reflect the sensitivity of the vestibular system to rapid pitching movements (Spoor et al., 2007b). Spoor and colleagues (1994) found that the radii of the anterior and posterior semicircular canals are about 25 percent larger, and thus much bigger relative to body mass, in early *H. erectus* fossils than in australopiths or chimpanzees (see Figure 9.8). So far, no bony labyrinths from east African *H. habilis* crania have been measured, but Spoor et al. (1994) did study two ears that might be from early *Homo* from South Africa: Stw 53, which is probably an australopith but is sometimes attributed to *H. habilis;* and SK 847, which is probably *H. erectus*. SK 847 has canal dimensions identical to those of other *H. erectus* fossils and modern humans; Stw 53 is a puzzle because it resembles humans in having relatively large anterior and posterior semicircular canals compared to australopiths but also has relatively large lateral semicircular canals compared to any ape. Although the vestibular system needs further study in *H. habilis,* more sensitive anterior and posterior semicircular canals in *H. erectus* would have improved gaze stability during jerky movements that rapidly accelerate the head in the sagittal plane. As discussed in Chapter 9, the best functional explanation for this derived feature is to help stabilize gaze during running.

Another cranial feature related to running performance in hominins is head balance (see Figure 12.12). Several morphological factors, described above, position the head's center of gravity more centrally in *Homo* than in *Australopithecus,* including a shorter posterior cranial base, more expansion of the posterior cranial fossa behind the foramen magnum (especially in *H. erectus*), and less facial prognathism. As Chapter 9 reviewed, we can best assess head balance by its effect on the moment of inertia, I, the head's mass times the *square* of its moment arm between the atlanto-occipital joint (AOJ), and the vertical vector of the head's center of gravity (estimated from the skull's area centroid). Although calculations of head mass from the skull are imprecise, estimates suggest that I is about three times greater in representative crania of *H. habilis* and *H. erectus* than in crania of gracile australopiths. Figure 12.12 also shows that the MA of the head extensors in *Homo* is more than twice that in *Australopithecus.*

Au. afarensis (Hadar reconstruction)

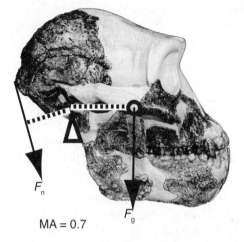

F_n

MA = 0.7

F_g

H. habilis (KNM-ER 1813 & OH 13)

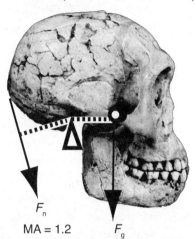

F_n

MA = 1.2

F_g

H. erectus (KNM-WT 15000)

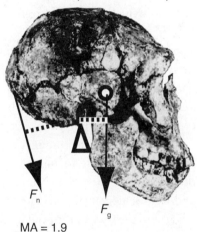

F_n

F_g

MA = 1.9

H. erectus (Zhoukoudian reconstruction)

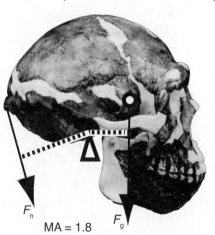

F_n

MA = 1.8

F_g

Figure 12.12. Biomechanics of head balance in *Homo,* showing the mechanical advantage (MA) for the superficial head extensors and estimates of the moment of inertia (*I,* calculated as head mass times the square of the moment arm of the center of gravity). F_n is the estimated nuchal muscle resultant; F_g is the force of the mass of the head. The center of gravity (circles) is estimated from the area centroid of the skull, and the mass of the head is estimated as 8 percent of total body mass. Estimates of *I* are much lower in *Homo* (0.003–0.005 kg · m^2) than in *Australopithecus* (~0.013–0.016 kg · m^2). *H. habilis* composite courtesy G. Sawyer and V. Deak.

As with the semicircular canals, a significantly lower I in early *Homo* compared to *Australopithecus* is more beneficial for running than walking. To stabilize the head in the sagittal plane, the flexors and extensors of the head need to counter the head torque, the product of I and α (angular velocity). Angular velocities are much lower, and hence much less of a challenge to head stability, in walking than in running. Quadrupeds have very unbalanced heads but stabilize the head by flexing and extending the neck and head, both of which are strongly cantilevered relative to the body (see Chapter 9). Early hominins lost this stabilization system by orienting the neck vertically and moving the foramen magnum more centrally in the skull base. Thus the hominin head bounces up and down during running like a person on a pogo stick. Presumably this jiggling was no more than a nuisance until hominins started to run regularly over long distances, at which point reduced head stability, hence reduced gaze stability, probably increased the likelihood of injurious falls (not a trivial problem for bipeds, who are intrinsically less stable than quadrupeds). Under such circumstances, any reduction in head pitching would decrease the amount of counter torque necessary to achieve head stability. (See Chapter 9 for a description of these mechanisms, many of which are usually recruited passively.)

A final, related source of evidence for head stabilization in *Homo* during running is the nuchal ligament (NL), a derived feature in humans that is absent in apes. This ligament is a two-part structure (see Figure 9.2). The superior portion is a tendon that interdigitates with the trapezius's cranial fibers and runs from the external occipital protuberance to the spinous process of the seventh cervical vertebra. The deeper portion is a fascial septum that runs from the midline of the occipital below the nuchal crest to the spinous processes and interspinous ligaments of the cervical vertebrae. The NL acts as an elastic element that, in conjunction with the cleidocranial portion of the trapezius (CCT), connects the mass of each arm to the back of the head in the midsagittal plane. Because the CCT contracts prior to foot strike during running, and because the trailing arm accelerates downward and out of phase with head accelerations, the NL helps the mass of the arm dampen vertical accelerations in the head during running that contribute to pitching movements. Crucially, the presence of an NL may be detectable in the skull because it leaves a sharp, everted ridge on the midline of the occipital, known as the median nuchal line (Figure 12.13). Median nuchal lines that are everted, hence indicative of a NL, are absent in apes and australopiths; they first appear in crania

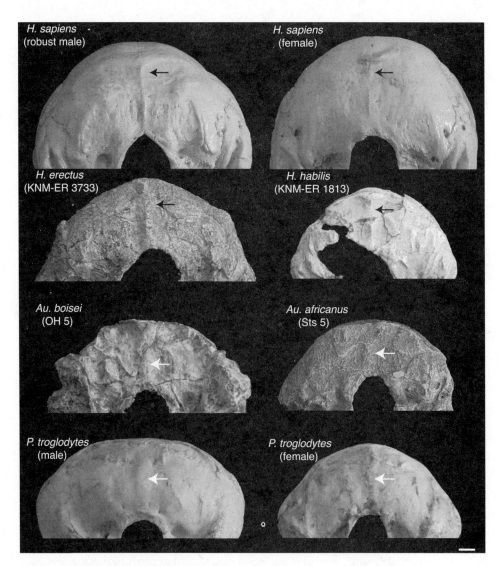

Figure 12.13. Inferior view of the nuchal plane of the occipital. The median nuchal line, which is present in *Homo* but not in *Pan* or *Australopithecus,* is evident as an everted ridge (dark arrows) in the midline of the nuchal plane below the superior nuchal lines. This ridge differs from a rounded ridge (white arrows) sometimes present in early hominins and apes that represents the margin between the two sides of the trapezius. Scale bar 1 cm.

attributed to *H. habilis* (e.g., KNM-ER 1813) and early *H. erectus* (e.g., KNM-ER 3733). (The median nuchal line is not to be confused with a low, convex line that sometimes forms in apes between the left and right pairs of nuchal muscles.)

Speculations on the Origins of Early *Homo*

The transition from *Australopithecus* to *Homo* involved considerable change, and attempts to synthesize what happened during this period are necessarily conjectural, especially given the inadequate fossil record between 3 and 2 mya. That said, the balance of evidence suggests that the emergence of the genus *Homo* was essentially about energy. Because of a peculiar set of circumstances, species of early *Homo* found new, successful ways to extract energy from the environment and to use these added calories to shift to a new reproductive strategy. This energetic shift involved at least two steps, with *H. habilis* as an intermediate between australopiths and *H. erectus sensu lato*.

H. habilis is an enigmatic mosaic. Although we cannot say for sure that *H. habilis* is geologically older and ancestral to *H. erectus,* that seems a reasonable guess. *H. habilis* shares many ancestral features with gracile australopiths, including large cheek teeth and a small body with relatively long arms and short legs (evidence for this is limited). Some experts suggest that *H. habilis* is not sufficiently derived to warrant inclusion in the genus *Homo* (Wood and Collard, 1999a,b). But it also shares some significant derived traits with *H. erectus,* including a short and deep TMJ; a more orthognathic, narrower face; an external nose; a more balanced head; and perhaps a slightly larger brain relative to body size. These features, plus the archaeological record, hint at a small but important shift in diet and behavior. Perhaps the species was less reliant on hard and tough fibrous foods and instead included more animal tissue in the diet.

The climatic context in which these changes occur leads to the hypothesis that *H. habilis* found a new niche to exploit during seasons when preferred foods such as fruits were scarce (Lieberman et al., 2008b, 2009). Between 2.5 and 1.9 mya, East Africa became more arid, and the savannas expanded (Trauth et al., 2005; Reed and Russak, 2009). Hominins were therefore living in habitats that were less forested and more intensely seasonal, favoring alternative ways to cope during periods of food scarcity. The robust australopiths, which evolved at about the same time, clearly focused on low-quality "fallback" foods, which accounts for their big, thick cheek teeth, massive faces, and other enlarged masticatory features. In contrast, *H. habilis* may have invented several new strategies, some of them social (such as food sharing and provisioning, which are discussed below) but many of them dietary. First, by 2.6 mya, hominins (which we cannot prove were early *Homo*) regularly started to acquire meat, marrow, and other animal tissues. Perhaps this shift in diet mostly occurred during periods of fruit scarcity. In addition, early *Homo* may have foraged more widely for a greater diversity of foods. And there is evidence that early *Homo* engaged in more extra-oral processing of foods using stone tools. Stone tools were used not only to process meat but probably also to mechanically process low-quality, fibrous foods such as USOs. Techniques such as pounding (an ideal use for a spheroid) help break down cell walls, reduce toughness, and increase the quantity of calories available from a given unit of food.

If these inferences are correct, then *H. habilis* was an australopith-like hominin that accomplished a slight dietary grade shift, becoming more of a specialist of high-quality resources. This shift accounts not only for its derived masticatory features but also for some other novelties, such as external noses and more balanced heads that might have played a role in becoming a carnivore (discussed below). And, as is so often the case in evolution, one change makes others possible. Specifically, more energy from diet, especially from proteins and fats in animal tissue, may have released constraints on selection for additional aspects of the phenotype, including brain and body size. As emphasized above, bigger brains and bodies are costly. One can pay for them by eating more low-quality food, or by switching trophic levels and eating higher-quality food. Maybe early *Homo* did the latter, eventually leading to the emergence of *H. erectus,* a bigger-bodied, wide-ranging, tool-dependent, nonarboreal hominin that was reliant on hunting in addition to gathering. Viewed in this light, *H. erectus* may be the product of several selection gradients that started in *H. habilis*. First, *H. erectus* has smaller cheek teeth with thinner enamel. Second, *H. erectus* has more features indicative of intense endurance capabilities. And third, both brain and body size increased in *H. erectus,* although not necessarily rapidly, or all at once. The end result is a much more humanlike hominin.

Foraging, Scavenging, and Hunting

How might novel features evident in early *Homo* have helped these hominins acquire high-quality foods, including meat? One part of the equation is foraging, which almost certainly supplied the majority of calories and nutrients for early *Homo,* as it does for modern hunter-gatherers. Chimpanzees typically travel 2–3 km/day, but foragers in more open habitats have to travel much farther, typically 10–15 km per day (Pontzer and Wrangham, 2006; Marlowe, 2005). In addition, whereas chimpanzees collect almost all their foods, hunter-gatherers extract a significant portion of their resources by digging, processing, and other means. We don't know daily travel ranges for *Australopithecus,* but it is reasonable to hypothesize that they increased in *Homo.* Most of the evidence for this increase is postcranial, such as longer and more robust legs. There are also several cranial features, including an external nose and a more balanced head, which suggest that *H. habilis* and *H. erectus* would have had improved abilities to trek long distances in hot, arid conditions. Tooth size reductions in *H. erectus* also indicate food processing, which is consistent with the development of extractive technology.

Acquiring meat is a much greater challenge than foraging. I would assume that early *Homo,* like modern hunter-gatherers, took advantage of every opportunity and used multiple strategies to get meat, by either scavenging or hunting. Although the relative importance of hunting versus scavenging in the Early Stone Age is much debated, there is good evidence that by 1.7 mya, *H. erectus* was successfully hunting medium-sized to large animals (Potts, 1998; Dominguez-Rodrigo, 2002). But evidence for carnivory raises a question: How could small, weaponless, hominins hunt or scavenge effectively? Both tasks are, by nature, dangerous, difficult, and highly competitive. Meat is an ephemeral, highly valued resource. As soon as carnivores abandon a carcass, scavengers such as hyenas, vultures, and jackals compete for the remains, leaving little to eat within hours (Van Valkenburgh, 2001). Thus scavengers require speed to reach carcasses before other scavengers; they need weapons, such as claws and fangs, to compete for carcasses with other scavengers; and they must have natural defenses, such as fur or speed, to protect them during the inevitable fights that ensue. Similarly, most hunting carnivores use speed and power to overcome their prey and to compete with each other. In these respects, early *Homo* seems ill suited to either scavenging or hunting. Bipedal hominins cannot gallop, and thus are intrinsically slow and unsteady when running. The fastest humans today can run at only 10 m/sec for less than 30 seconds, but lions and other carnivores can gallop at speeds of 20 m/sec or more for up to 4 minutes. Hominins could not have run fast enough or with sufficient agility to catch most prey, such as gazelle or wildebeest, which can run and turn at speeds of 15–20 m/sec (Garland, 1983). Another serious problem for bipedal hominin hunters or scavengers is the lack of natural weaponry and defenses such as fangs, claws, and heavy paws. What match would any *H. habilis* have been for a hyena or a saber-toothed tiger?

A classic narrative of human evolution is that early humans were able to enter the carnivore guild by using technology to compensate for their lack of speed, power, and natural weaponry. Darwin was an early proponent of the brains-over-brawn hypothesis: "The slight corporeal strength of man, his little speed, his want of natural weapons, etc, are more than counterbalanced, firstly by his intellectual powers, through which he has, whilst remaining in a barbarous state, formed for himself weapons and tools, etc, and secondly by his social qualities which lead him to give aid to his fellow-men and to receive it in return" (Darwin 1871:157). However, according to the archaeological record, the projectile weaponry needed to hunt did not exist during the Early Stone Age. The spear point was not invented until about 300,000 years ago, and the bow and arrow was invented less than 100,000 years ago (J. J. Shea, 2006). The best weapons available to early *Homo* hunters were clubs and untipped spears, neither of which are very lethal. To kill an animal with an untipped spear, one must attack at very close range (a few meters) because the blunt point is ineffective at penetrating a hide. Further, most spears cause death not from the wound itself but from lacerations to internal organs caused by the jagged edges of the spearhead. So, early *Homo* hunters who tried to kill animals at short range with untipped spears had low chances of killing their prey and risked life-threatening injuries. One kick from a wildebeest or an encounter with its horns would have significantly reduced an early hunter's fitness, selecting against such risky behaviors.

An alternative hypothesis is that early *Homo* entered the carnivore guild not by outsprinting prey and outfighting competitors but by becoming the first specialized diurnal carnivore (actually, omnivore). Several adaptations improved the ability of hominin hunters to run and walk long distances in the heat of the midday sun, when other mammals seek shade. There was probably no one strategy by which early *Homo* hunted or scavenged, but the ability to run would have been crucial to both hunting and scavenging, as it is for most social carnivores.

Running would have helped hominin hunters in many ways, such as allowing them to get close to prey, but it would have been especially important in persistence hunting (Carrier, 1984). During persistence hunts, a group of hunters typically chases an animal in the midday heat; when it gallops away to hide and cool itself, they follow and track it, intermittently walking and running. The faster they track the prey, the less time it has to restore its core body temperature, and eventually, usually within 10–15 km, the animal collapses from hyperthermia (Liebenberg, 2006). The basis for this kind of hunting lies in human endurance capabilities, which exceed those of other mammals, especially in the heat. Reasonably fit humans are unique in being able to run 10–20 km or more in hot conditions at speeds (2.5–6 m/sec that exceed the trot-gallop transition of most quadrupeds (Bramble and Lieberman, 2004). Because quadrupeds, including hyenas and dogs, cannot simultaneously gallop and pant (Bramble and Jenkins, 1993), a human hunter armed with nothing more lethal than a club or an untipped spear can safely run a large mammal such as a kudu into hyperthermia by chasing the prey above its trot-gallop transition speed. In fact, hunters prefer to chase large animals because they overheat faster: heat production scales to body mass to the power of 3, but heat loss in nonsweating animals scales closer to isometry (Schmidt-Nielsen, 1990). Persistence hunting was practiced until recently by Bushmen in the Kalahari and by hunter-gatherers in other hot, open habitats (see Lieberman et al., 2007b for a review). In addition, persistence hunting is very effective (more than 50 percent of Bushman hunts are successful), requires almost no technology, and is remarkably safe because a hunter never needs to get close to the prey until it is already dying. This form of hunting requires only the ability to walk and run long distances, the ability to keep cool by sweating rather than panting (hence the need for lots of sweat glands and no fur), and access to water for hydration.

There are several reasons to suspect that *H. erectus* was an effective long-distance runner able to engage in this kind of hunting. As noted above, *H. erectus* has several cranial features that would have improved the ability to run (e.g., enlarged anterior and posterior semicircular canals, a more balanced head, a nuchal ligament). In addition, other adaptations for running are evident in the *H. erectus* postcranium (reviewed in Bramble and Lieberman, 2004). The ability to trek long distances and engage in endurance running are probably related: both are parts of a subsistence system that involved covering large distances quickly. Certainly, hunters who ran down an animal would have walked home with the meat.

Running might also have helped early hominins scavenge. Vultures circling over carcasses would tip off hominins to the location of potential scavenging opportunities, but because carcasses are quickly eaten, it would be advantageous to reach them as quickly as possible. This sort of scavenging may have been particularly effective in hot conditions. Hyenas are good long-distance runners, able to run 10–15 km a day (Van Valkenburgh, 2001), but only when it is cool at dawn, at dusk, and at night, because they cannot thermoregulate when galloping by sweating or panting. So early hominins would have had an advantage over hyenas when it was hot. Additionally or alternatively, hominins might have used their thermoregulatory advantages to patrol more wooded habitats near streams, where scavengeable carcasses last longer (Blumenschine, 1987).

Energy and Early *Homo*

To reiterate, early *Homo* appears to have exploited a new niche as the climate changed at the end of the Pliocene and the early Pleistocene. *H. erectus* elaborated on a trend that possibly started in *H. habilis* to became a diurnally active omnivore with a diet rich in high-quality, low-fiber foods, including animal tissues and USOs, many of which were processed using simple stone tools. Acquiring this diet benefited from or favored new adaptations for thermoregulation, such as turbulent external noses, sweating, and the loss of fur, and new adaptations for locomotion, such as more balanced heads, more sensitive vestibular systems, and long, tendon-filled legs. The new diet also favored other changes in the dentition, TMJ, and face, reviewed above.

This new energetically intensive and presumably successful way of life is consequential because it led to a release of constraints on body size and brain size. It is probably not a coincidence that bigger, more costly bodies and brains come *after*, not *before*, evidence for shifts in diet, locomotion, tool making, and thermoregulation. But once the selection gradient for increases in brain and body size started, it must have been a self-reinforcing process. A major source of positive feedback is that big, costly brains and bodies would have favored lower interbirth intervals (IBIs), which would improve fitness but further reinforce the need to obtain lots of energy by hunting, processing food, and so on. Chimpanzee IBIs tend to be about 5–6 years (Goodall, 1986; Brewer-Marsden et al., 2006), but modern human foragers in nonequestrian societies are able to wean their offspring earlier, shortening the IBI to 3–4 years (Marlowe, 2005). A shorter IBI makes sense energetically, because lactating is so expensive that a 50 kg hominin female with a chimplike IBI of 5 years would not only have fewer total offspring (hence lower fitness), but would also spend a whopping extra 1.47 Mcal per offspring—a 45 percent increase (Aiello and Key, 2002). But if *H. erectus* females had an IBI close to that of human foragers, approximately 3 years, then their cost per infant would have been about 20 percent less than that in australopiths, in spite of higher daily costs. We can therefore hypothesize that natural selection would have pushed early *Homo* toward shorter IBIs (a hypothesis that requires testing). And the evidence presented above indicate that *H. erectus* achieved this shift in part by relying more on high-quality foods such as meat and marrow, and by processing foods to increase their energy return.

Another means of sustaining a more energetically intensive reproductive strategy with shorter IBIs in hominins was to adopt a number of behaviors that are absent or rare in ape social groups. We can only make conjectures about the prevalence of these sorts of behaviors, but lactating mothers and their dependent offspring would have benefited significantly from provisioning by males, which in turn would benefit a male's fitness if he was provisioning his own offspring. Further, because hunting is dangerous and difficult with infants or toddlers in tow, a shift to hunting probably involved a sexual division of labor in which males did most of the hunting and females foraged for more predictable resources. Finally, the hunter-gatherer strategy relies heavily on various kinds of cooperation within groups, ranging from food sharing to defense from predators.

In short, there is reason to hypothesize that the momentous transition from *Australopithecus* to *Homo* involved, at heart, a shift in diet and energetics. Through a combination of trekking, running, making stone tools, and cooperating socially, early *Homo* found a way to get access to high-quality resources such as processed food, meat, and marrow, becoming the first active, diurnal social omnivore. Once early *Homo* started on this new path, it opened the door for a new, hunter-gatherer way of life that favored selection for larger bodies, larger brains, and shorter interbirth intervals. These shifts, in turn, set the stage for later transitions in human evolution such as the further expansions in brain size and slower ontogenies we see in archaic and modern humans.

13

The Evolution of the Head
in *Homo sapiens*

You interest me very much, Mr. Holmes. I had hardly expected
so dolichocephalic a skull or such well-marked supra-orbital development. Would you have any objection to my running my
finger along your parietal fissure? A cast of your skull, sir, until
the original is available, would be an ornament to any anthropological museum. It is not my intention to be fulsome, but I con-
fess that I covet your skull.

A. CONAN DOYLE, *The Hound of the Baskervilles,* 1902

Finally, we turn to the evolution of the head in humans, both archaic and modern. Humans are naturally inclined to consider our species' origin a major transformation. Yet most of the changes we can observe above the neck between *H.
erectus* and *H. sapiens* appear to have involved minor tinkering. Species of archaic *Homo,* notably *H. heidelbergensis* and *H. neanderthalensis,* differ from early
H. erectus primarily in having slightly larger brains and faces. In turn, the heads
of *H. sapiens* differ from those of archaic *Homo* in just a few major ways: we
have smaller faces that are almost entirely retracted underneath the anterior cranial fossa; the cranial base is more flexed; the cranial vault is very spherical; and
we have short oral cavities, with short, rounded tongues. We also tend to think
of humans as uniquely gracile, with absent or diminutive brow ridges and thin-
walled cranial vaults; but most features related to gracility are not only highly
variable but also comparatively recent, postdating the origins of *H. sapiens.* Further, many of the features that differentiate *H. sapiens* skulls from different parts
of the world are also recent, variable, and subtle.

Even though evolutionary changes in the head of late *Homo* may be slight,
they have had profound implications for anyone reading this book, and they

highlight some common problems with thinking about our species' origin. Perhaps the biggest challenge is to avoid the burden of preconceived biases and assumptions about what it is to be human. Such assumptions—sometimes explicit, but often implicit—color how we define species of *Homo,* how we test hypotheses about their evolutionary relationships, and how we evaluate their behavioral and functional capabilities. To begin with, what do we mean by the term *human?* To some scholars, large brains, intelligence, and the capacity to reason are quintessentially human traits, and thus all fossil taxa with large brains must be a kind of human. According to such (grade-based) logic, our species, *H. sapiens,* is variable and polytypic, and includes groups such as the Neanderthals (which would then be a subspecies, *H. sapiens neanderthalensis*). Others adopt a more evolutionary (clade-based) approach. Thus, evidence that the Neanderthals are a distinct lineage and not direct ancestors of modern *H. sapiens* suggests that they belong to a separate species, *H. neanderthalensis,* regardless of how smart they were.

Another, related source of debate arises from disagreement over what null hypotheses to adopt for important "human" traits. Consider language and complex cognition. Should the null hypothesis be that only *H. sapiens* has sophisticated language and cognitive capabilities unless proven otherwise? Or should we assume that all big-brained hominins have modern capabilities unless proven otherwise? Because it is difficult to disprove a null hypothesis about traits such as language that do not leave clear fossil or archaeological traces, the choice of null hypothesis is a major source of bias.

Although our epistemological problems are difficult to resolve, we are fortunate to have considerable information—fossil, genetic, archaeological, and linguistic—with which to make inferences and test hypotheses about the evolution of *H. sapiens.* Arguably, there are few speciation events for which we have so much diverse data. Many disagreements about human origins stem not from a lack of information but from different assumptions about the relevance of the available information to the questions we ask. For example, do morphological features of the skull, such as the shape of the brow ridges, the size of the mastoid processes, or the curvature of the infraorbital region constitute good markers of phylogenetic relatedness? And to what extent does absence of evidence in the archaeological record for typically "modern" behaviors, such as the production of symbols, indicate an absence of abstract thinking capabilities?

I would argue that the best strategy is to be epistemologically cautious about any phylogenetic and behavioral hypotheses. Although skulls contain much data about phylogeny and function, this information is often obfuscated by processes of integration that make it hard to identify independent features that were selected or that reflect actual units of inheritance. Further, many, if not most, variations in craniofacial shape among recent humans may simply reflect the effects of drift and other random, unselected evolutionary changes (Roseman and Weaver, 2007). Until we better understand the relationship between genotype and phenotype, we can adopt several strategies to mitigate the effects of ignorance. First, we should avoid picking unjustifiable null hypotheses. There is no logical reason to assume a priori that an important trait such as language is either unique to modern humans or shared by all large-brained hominins unless proven otherwise. Second, we should evaluate whether skeletal features contain phylogenetic or functional information before assuming that they do (or don't). For a given feature, this means knowing something about its developmental bases, how it co-varies with other features, and how it affects performance. And third, we need to be careful about variation. An important fact about human evolution is that more than 85 percent of our species' genetic variation exists within any given population (Lewontin, 1972). Similarly, about 80–90 percent of the craniometric variation we can measure is also present within any given population (Relethford, 1994; González-José et al., 2004; Roseman and Weaver, 2004), and most variation among recent human skulls appears to reflect the effects of neutral evolutionary change rather than natural selection (Roseman, 2004; Roseman and Weaver, 2004, 2007). Indeed, many aspects of skull shape that differ among populations tend to be quite recent in origin, so that most human skulls dated between roughly 150,000 and 10,000 years ago are more similar to other like-aged skulls than they are to those of historic populations, regardless of geographic location (Howells, 1989, 1995; Lahr, 1996). This fact means that we cannot use most cranial variations to reconstruct reliable phylogenies beyond 10,000 years ago in any one region, and it raises questions about whether similar features can be used to make reliable inferences over much greater spans of time.

It is not possible to review here all the many detailed differences in the skull between *H. erectus* and *H. sapiens*. In order to focus on the big picture, we will begin with a review of the fossil record and the species involved, then consider some developmental shifts that may underlie the major transformations we observe, and conclude by speculating on their functional and behavioral implications and how natural selection might have operated.

As previously, the term *human* is used here to mean *H. sapiens*. Where different interpretations of the word could lead to confusion between *H. sapiens* and other species of archaic *Homo,* such as *H. neanderthalensis* and *H. heidelbergensis,* I employ the term *modern human.*

Archaic *Homo* Species and the Muddle in the Middle

In some ways, what happened during the last million years of human evolution is fairly straightforward (see Figure 13.1). Once early *Homo* (probably *H. erectus s.l.*) moved into the Old World, its descendants continued to evolve in different regions, mostly undergoing increases in brain and face size. Thus, by one million years ago, there were big-brained and big-faced *H. erectus*–like hominins in Africa, western Asia, East Asia, South Asia, and probably Europe. Apparently *H. erectus* persisted in Asia for a long time with relatively little change. In contrast, the descendants of *H. erectus* in Europe and Africa underwent modest evolutionary transformations, eventually giving rise to several species. One, the Neanderthals (*H. neanderthalensis*), lived primarily in Europe and went extinct approximately 30,000 years ago (30 kya). The other, *H. sapiens,* evolved in Africa between 300 and 200 kya. *H. sapiens* dispersed into the Middle East starting about 100 kya, and then into Europe and Asia between 60 and 40 kya. These modern humans replaced all other species of *Homo,* apparently with very low levels of interbreeding with Neanderthals (Green et al., 2010).

Many details of this general scenario, however, remain obscure and unresolved. One major source of confusion and debate is the taxonomy of fossils in the genus *Homo* that are neither early *H. erectus* nor modern *H. sapiens.* One extreme view, favored by "lumpers," is to call all these things *H. sapiens,* making only grade-based distinctions between more "archaic" and "modern" *H. sapiens* (Wolpoff et al., 1984; Wolpoff, 1999; Bräuer, 2008). This scheme, however, is refuted by genetic and paleontological evidence (reviewed below) that Neanderthals are a distinct lineage and that all modern humans trace their ancestry to Africa within the last 200,000 years. Most paleoanthropologists thus separate fossils that are clearly *H. erectus* from fossils that are younger than about 500,000 years or have brains larger than about 1,000 cm³. Of the two archaic species most commonly recognized, the older one is *H. heidelbergensis,* which is found in both Africa and in Europe (Rightmire, 1998, 2008); the younger species is *H. neanderthalensis,* which is found mostly in Europe and western Asia (see below). A few scholars recognize even more species, including *H. antecessor, H. rhodesiensis, H. mauritanicus, H. helmei,* and *H. soloensis* (for example, see Bermúdez de Castro et al., 1997; Lahr and Foley, 1998; Tat-

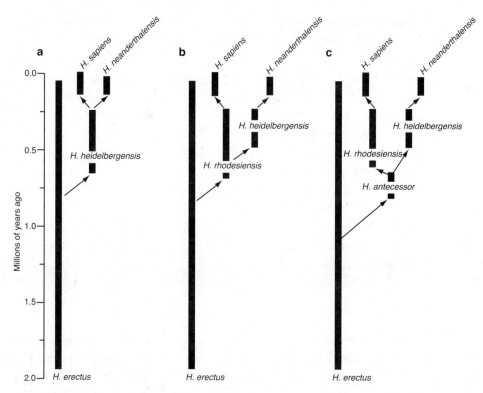

Figure 13.1. Alternative phylogenies for the genus *Homo* (modified from Rightmire, 2008).

tersall and Schwartz, 2000; Hublin, 2001; Schwartz and Tattersall, 2005). Of these, *H. rhodesiensis* is hypothesized to be an African species of archaic *Homo* ancestral to *H. sapiens* (see Figure 13.1b). Identifying *H. rhodesiensis* as a separate species makes sense if one thinks that Neanderthals evolved in Europe from *H. heidelbergensis* as part of a lineage that was long separate (more than 600 kya) from that of *H. sapiens*. *H. antecessor* is hypothesized to be a descendant of *H. erectus* that is ancestral to both *H. sapiens* and *H. neanderthalensis* (see Figure 13.1c). Some of these taxa might be bona fide species, but it is difficult to reject the null hypothesis of fewer species based on analyses of variation or the presence and absence of shared derived characters. Another species to add is *H. floresiensis,* a hypothesized dwarfed species of *Homo* from the island of Flores, Indonesia (see below).

As with *H. erectus,* we simply lack the knowledge to test hypotheses about taxonomy at the fine scale we would like. A sensible solution is to recognize that there were probably several lineages of "archaic" *Homo,* all variations of the same basic plan. These taxa share with *H. erectus* many features such as long, low vaults, vertical but projecting faces, thick cranial vaults, and marked brow ridges; they differ from *H. erectus* primarily in having relatively larger brains and faces and brow ridges that arch more over each orbit, and in lacking a few characteristic features of *H. erectus,* such as sagittal keels and long nuchal planes. With the exception of the Neanderthals and modern humans, both of which have suites of distinctive features, there is no obvious way to group these fossils into different species on the basis of craniofacial variations. One possible approach is to classify them as populations of indeterminate taxonomic status in the genus *Homo:* "archaic" *Homo* spp. Alternatively, one can consider them to be a single, widespread, and diverse species, *H. heidelbergensis* (Rightmire, 2008). Given the difficulties of resolving this taxonomy with much confidence, I will consider these fossils in the taxon *H. heidelbergensis sensu lato (s.l).*

Late *Homo* in Africa

The fossil record in Africa, which includes a wide range of variation, can be divided into three broad groups (see Table 13.1). At the more primitive, older end are fossils such as OH 9, Buia, and Daka that date from 800 kya or earlier and have endocranial volumes less than 1,100 cm^3 (Figure 13.2a). These Lower Pleistocene crania fit within the general hypodigm of *H. erectus s.l.,* although some, such as Buia, lack a few features common to many *H. erectus* crania, such as a sagittal keel (Abbate et al., 1998).

A second, Middle Pleistocene group of fossils includes crania such as Bodo, Kabwe (Broken Hill), and Eyasi (see Table 12.1 and Figure 13.2b) that date to between 750 and 300 kya, or possibly even more recently in some cases (see Dominguez-Rodrigo et al., 2008). These crania tend to have ECVs greater than 1,200 cm^3 (an exception is Salé, with an ECV of 880 cm^3, which might be a female). Contemporary postcranial remains are rare but are generally large and robust (Pearson, 2000). This group of fossils tends to be classified as *H. heidelbergensis* (Rightmire, 1998, 2008; Hublin, 2001) but are sometimes classified as *H. rhodesiensis* if they are considered distinct from archaic *Homo* in Europe (Bermúdez de Castro et al., 1997; Hublin, 2001). The justification for distinguishing them from *H. erectus* is twofold. First, they show a reduction of some unique derived features of *H. erectus,* such as sagittal keels and occipital tori. In addition, they share a number of derived features, some of which stem from having larger brains (at least in males). These features include a relatively longer occipital scale (occipital squama length relative to nuchal plane

Figure 13.2. Examples of African late *Homo* crania: (a) Olduvai Hominid (OH) 9 (*H. erectus*); (b) Kabwe (*H. heidelbergensis s.l.*); (c) Bodo (*H. heidelbergensis s.l.*). Images not to scale.

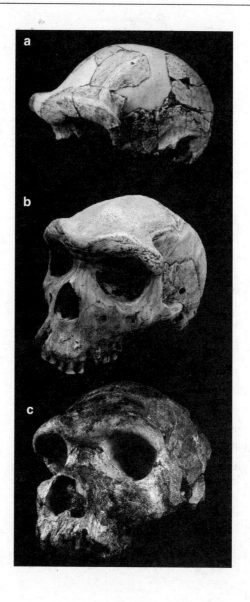

length), more vertically oriented parietals, a more rounded occipital, and less postorbital constriction with broader temporal squamae. According to Rightmire (2008), *H. heidelbergensis* also differs from *H. erectus* in a few discrete traits, such as thinner tympanic plates and a more vertically oriented lateral margin of the nasal aperture.

The final group of African fossils to consider includes crania younger than 350 kya, which are "modern" in various ways (see Figure 13.8). These fossils have ECVs in the recent human range, and their cranial vaults are rounder

Table 13.1: Major cranial fossils attributed to late *Homo* in Africa from the last million years

Fossil(s)	Location	Species	Date (kya)	ECV (cm³)	Elements found
Olduvai 12	Tanzania	H. erectus	900–1200	725–750	Partial cranium
Buia	Eritrea	H. erectus	ca. 1000	750–800	Cranium
Daka	Ethiopia	H. erectus	1000	995	Cranium
KNM-OL 45500	Kenya	H. erectus	900–970		Partial cranium
Olduvai 22, 23	Tanzania	H. erectus	800–900		Mandibles
Ternifine	Algeria	H. erectus	700–800		Fragmented skull
Thomas Quarry	Morocco	H. erectus	400–600		Cranial fragment, mandible
Bodo	Ethiopia	H. heidelbergensis s.l.	600	1250	Partial cranium
Elandsfontein	South Africa	H. heidelbergensis s.l.	500–1000	1225	Partial skull
Kabwe (Broken Hill)	Zambia	H. heidelbergensis s.l.	300–600	1280	Cranium
Ndutu	Tanzania	H. heidelbergensis s.l.	400	1100	Partial cranium
Salé	Morocco	H. heidelbergensis s.l.	400	880	Cranium
Sidi Abderrahman	Morocco	H. heidelbergensis s.l.	400		Mandible
Eyasi	Tanzania	H. heidelbergensis s.l.	200–400		Partial cranium
Cave of Hearths	South Africa	H. heidelbergensis s.l.	250–350		Mandible
Laetoli 18 (Ngaloba)	Tanzania	H. heidelbergensis s.l.	200–300	1367	Partial cranium
Florisbad	South Africa	H. heidelbergensis s.l.	150–300		Partial cranium
Omo I	Ethiopia	H. sapiens	195		Partial skull
Omo II	Ethiopia	H. sapiens	ca 195	1435	Partial cranium
Herto 16/1	Ethiopia	H. sapiens	160	1450	Cranium
Herto 16/2	Ethiopia	H. sapiens	160		Partial cranium
Herto 16/5	Ethiopia	H. sapiens	160		Cranium (child)
Singa	Sudan	H. sapiens	140–150	1550	Partial cranium
Jebel Irhoud I	Morocco	H. sapiens	150–200	1305	Partial cranium
Jebel Irhoud II	Morocco	H. sapiens	150–200	1450	Partial cranium
Jebel Irhoud III	Morocco	H. sapiens	150–200		Mandible
Eliye Springs	Tanzania	H. sapiens	<200		Partial cranium
Dar es Soltane 2	Morocco	H. sapiens	<130		Cranial fragment, mandible
Klasies River	South Africa	H. sapiens	65–125		Partial crania and mandibles
Aduma	Ethiopia	H. sapiens	80–105		Cranial fragment
Bouri	Ethiopia	H. sapiens	80–105		Partial cranium
Border Cave	South Africa	H. sapiens	<80	1510	Partial cranium, mandible
Hofmeyr	South Africa	H. sapiens	33–39		Skull

than those of more archaic fossils, but they are more robust than most recent humans, especially in the brow ridges. Examples include fossils from Omo, Herto, Jebel Irhoud, Laetoli, and Florisbad. Some, such as Omo II, look more archaic and possibly do not belong in *H. sapiens.*[1] But for reasons discussed below, I think that most if not all of these fossils are early *H. sapiens.*

One question posed by late *Homo* fossils in Africa is whether the differences we see reflect considerable variation within a single but diverse species, across several species, or over a gradually evolving lineage. All three views are tenable but hard to test. On the one hand, we have only a few dozen, poorly dated fossils sampling more than one million years of evolution across an enormous continent. Given gaps in the fossil record, we may lack the resolution necessary to demarcate real species. On the other hand, we know that *H. sapiens* originated in Africa, and the trend we see toward bigger brains, more spherical skulls, and more retracted faces might represent different stages of evolution within a lineage (see Bräuer, 2008). If so, then it will be difficult to determine where to draw the line between species. A further complication is that the Middle Pleistocene was an era of major and repeated climatic fluctuations. It is possible, maybe likely, that there was more than one species of *Homo* in Africa for a while and that dispersals between Africa and Eurasia led to complex interchanges between taxa in the two continents (see Hublin, 2000; Burroughs, 2007).

In the end, it is difficult to make secure taxonomic distinctions in the African fossil record, particularly between *H. erectus* and *H. heidelbergensis s.l.* Few disagree that *H. erectus* continued to evolve in Africa, eventually leading to the speciation of *H. sapiens* sometime after 300 kya. Between *H. erectus* and *H. sapiens,* there are plenty of fossils loosely assigned to *H. heidelbergensis* or *H. rhodesiensis,* which may or may not have been evolutionarily separate from similar hominins in western Asia and Europe. That said, it is important to stress that *H. erectus* and *H. heidelbergensis s.l.* differ from each other less than either differs from *H. sapiens* (see below).

Late *Homo* in Europe and Western Asia

As discussed in Chapter 12, *H. erectus s.l.,* or something like it, was present in Georgia at 1.77 mya. In addition, there are traces of hominins in the Middle

---

1. Omo II is much less modern in shape than Omo I. Although the two specimens were found in the same general region, they may not be the same age (McDougall et al., 2005).

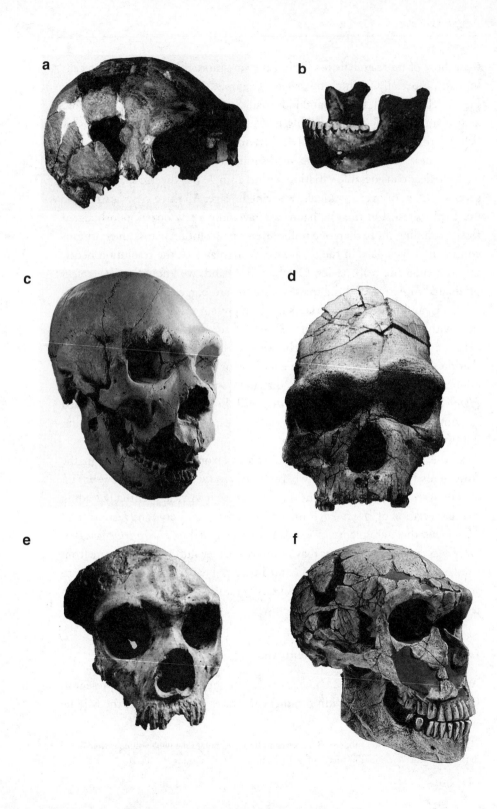

Figure 13.3. Examples of European late *Homo:* (a) Ceprano (*H. erectus s.l.*); (b) Mauer mandible (*H. heidelbergensis*); (c) Sima de los Huesos 5 (*H. heidelbergensis*); (d) Arago 21 (*H. heidelbergensis*); (e) Forbes Quarry (*H. neanderthalensis*); (f) La Ferrassie 1 (*H. neanderthalensis*). Images not to scale. Image of SH-5 (Javier Trueba, Madrid Scientific Films) courtesy J. L. Arsuaga; image of Ceprano courtesy G. Manzi.

East more than 1 million years old at sites such as 'Ubeidiya (Tobias, 1966). But, for the most part, the fossil trail in Europe and western Asia goes cold until about 1.2 mya.[2] Table 13.2 summarizes key fossils from Europe in several phases, the oldest of which dates to between 1.2 and 0.78 mya, when Europe experienced several warm interglacial periods (interspersed with some cold snaps). The few known hominins from this time (see Figure 13.3a) include a partial cranium from Ceprano, Italy (Ascenzi et al., 1996; Manzi et al., 2001), and fragments of about half a dozen individuals from some limestone caves in the Sierra de Atapuerca, Spain, near Burgos (Carbonell et al., 1995; Bermúdez de Castro et al., 1997; Carbonell et al., 1999). Most of these fossils come from the cave of Gran Dolina, but the oldest is from the cave of Sima del Elefante (Carbonell et al., 2008).

These fossils generally resemble contemporaneous *H. erectus s.l.* crania from Africa. The Ceprano cranium is moderately large (1,165 cm^3) with a low vault; a robust, double-arched supraorbital torus; a very angled occiput with a torus; and no sagittal keel. The Gran Dolina material, although fragmentary (possibly because of cannibalism), includes a double-arched supraorbital torus from a robust adult, a partial juvenile cranium, and a small mandible. The fossils' discoverers have proposed that they are a distinct species, *H. antecessor,* ancestral to both Neanderthals and modern humans (Bermúdez de Castro et al., 1997; Carbonell et al., 2005). This hypothesis is not widely accepted because the primary evidence for shared derived features with *H. sapiens* is from an immature maxillary fragment. In addition, the material is not vastly different from other archaic *Homo* fossils in Africa (see above). It is probably sensible to set the name *H. antecessor* aside until there is better material to test the hypothesis.

Europe became glacial and inhospitable for hominins between 770 and 625 kya (with a brief warming spell around 650 kya), which probably accounts for their absence during this period. However, a second, much larger group

2. Archaeological sites such as Pirro Nord in Italy suggest that even older European fossils may be found (Arzarello et al., 2006).

Table 13.2: Major cranial fossils attributed to late *Homo* in Europe

Fossil(s)	Location	Species	Date (kya)	ECV (cm³)	Elements found
Sima del Elefante	Spain	H. erectus/antecessor	1200–1400		Mandible
Ceprano	Italy	H. erectus/antecessor	500–900	1185	Partial cranium
Gran Dolina	Spain	H. erectus/antecessor	800–900		Fragments
Sima de los Huesos 4	Spain	H. heidelbergensis	530–600	1390	Partial skull
Sima de los Huesos 5	Spain	H. heidelbergensis	530–600	1125	Partial skull
Steinheim	Germany	H. heidelbergensis	250–350	900	Cranium
Swanscombe	England	H. heidelbergensis	400	1250	Partial cranium
Petralona	Greece	H. heidelbergensis	200–700	1220	Cranium
Apidima	Greece	H. heidelbergensis			Crania
Arago 21	France	H. heidelbergensis	350–600	1166	Partial cranium
Vértesszőlős	Hungary	H. heidelbergensis	180–400		Occipital
Mauer	Germany	H. heidelbergensis			Mandible
Bilzingsleben	Germany	H. heidelbergensis	300–400		Partial crania
Reilingen	Germany	H. heidelbergensis	125–250	1430?	Partial cranium
Ehringsdorf	Germany	H. neanderthalensis	200–205	1450	Partial crania
Biache-St-Vaast	France	H. neanderthalensis	160–190		Partial cranium
Le Lazaret	France	H. neanderthalensis	160–200	1250	Partial skull (child)
Saccopastore 1	Italy	H. neanderthalensis	122–129	1245	Cranium
Saccopastore 2	Italy	H. neanderthalensis	122–129	1300	Partial cranium
Krapina	Croatia	H. neanderthalensis	120–140	1450	Many skull fragments
Pech de L'Azé	France	H. neanderthalensis	>103	1150	Partial skull (juvenile)
La Ferrassie 1	France	H. neanderthalensis	60–75	1640	Skull (plus skeleton)
Regourdou	France	H. neanderthalensis	65–75		Mandible (plus partial skeleton)
La Quina 5	France	H. neanderthalensis	65	1350	Partial skull (and skeleton)
La Quina 18	France	H. neanderthalensis	65	1200	Cranium (child)

Guattari	Italy	*H. neanderthalensis*	50–60	1550	Partial cranium
La Chapelle aux Saints	France	*H. neanderthalensis*	47–56	1625	Skull (plus skeleton)
Roc de Marsal	France	*H. neanderthalensis*	>50	1260	Skull, partial skeleton (infant)
Feldhofer	Germany	*H. neanderthalensis*	40?	1525	Partial cranium
Engis 2	Belgium	*H. neanderthalensis*	40–60	1362	Partial cranium (child)
Devil's Tower	Gibraltar	*H. neanderthalensis*	50?	1400	Partial skull (child)
Forbes Quarry	Gibraltar	*H. neanderthalensis*	50?	1200	Partial cranium
Spy 1	Belgium	*H. neanderthalensis*	50?	1553	Partial skull (and skeleton)
Spy 2	Belgium	*H. neanderthalensis*	50?	1305	Partial skull (and skeleton)
Le Moustier	France	*H. neanderthalensis*	40	1565	Partial skull (and skeleton)
St.-Césaire	France	*H. neanderthalensis*	36		Partial cranium
Vindija	Croatia	*H. neanderthalensis*	33–42		Mandibles and cranial fragments
Arcy-sur-Cure	France	*H. neanderthalensis*	34		Temporal fragment
Zafarraya	Spain	*H. neanderthalensis*	28		Mandible
Peştera cu Oase	Romania	*H. sapiens*	34–35		Mandible and partial crania
Mladeč 1	Czech Republic	*H. sapiens*	31	1390–1650	Multiple skeletons
Cro Magnon	France	*H. sapiens*	27	1600–1730	Multiple skeletons
Poedmostí	Czech Republic	*H. sapiens*	>30	1250–1580	Multiple skeletons
Dolni Vĕstonice	Czech Republic	*H. sapiens*	26	1285–1547	Multiple skeletons
Brno	Czech Republic	*H. sapiens*	23–26	1304–1600	Multiple skeletons
Grimaldi	Italy/France	*H. sapiens*	25–30	1375–1880	Multiple skeletons
Lagar Velho	Portugal	*H. sapiens*	24		Partial skull and skeleton (child)
Arene Candide	France	*H. sapiens*	23	1414–1520	Multiple skeletons
Abri Pataud	France	*H. sapiens*	22	1380	Multiple skeletons
Oberkassel	Germany	*H. sapiens*	12	1279–1500	Multiple skeletons
Chancelade	France	*H. sapiens*	12	1710	Skeleton

(or set of groups) shows up between 600 and 300 kya at sites throughout Europe (see Table 13.2 and Figure 13.3b). The most spectacular sample comes from the Sima de los Huesos ("Pit of Bones"), deep within the Atapuerca cave system. The remains of more than 30 individuals were dumped into the pit, possibly as a form of burial (Arsuaga et al., 1997). The fossils probably date to 530–600 kya (Bischoff et al., 2003, 2007). The Sima de los Huesos fossils, which constitute the most spectacular assemblage of nonmodern hominin fossils ever found, include two well-preserved crania (SH 4 and SH 5). Table 13.2 lists other important archaic *Homo* remains from this general period in Europe, including a slightly squashed cranium from Steinheim, Germany; a partial cranial vault from Swanscombe, England; a well-preserved cranium from Petralona, Greece; and some fragmentary fossils from Arago, France (dating to 350–600 kya), including the partial cranium Arago 21 (see Figure 13.3b).

In many ways, the archaic *Homo* crania from the Middle Pleistocene of Europe resemble those of contemporary African fossils, such as Bodo and Kabwe. Most of them are robustly built and large, with ECVs between 1,100 and 1,400 cm^3; long, low vaults; big, projecting faces characterized by double-arched supraorbital tori, large orbits, and wide interorbitals; and somewhat vertically oriented parietals. They also differ from *H. erectus* in the same ways that the African archaic crania do. It is logical to attribute both groups of hominins to the species *H. heidelbergensis,* whose type specimen is a mandible from Mauer, Germany. However, the relationship between the European and African samples is uncertain, especially with regard to the Neanderthals (see below). Some *H. heidelbergensis* crania have features that later typify the Neanderthals. As examples, the Swanscombe occipital has a suprainiac fossa, and the Reilingen vault has an occipital bun and a suprainiac fossa (see below). Other possible shared derived features include a slight inflation of the midface around the nasal aperture (hence the absence of a canine fossa), broad nasal bones, a long occipital squama above inion, a high position of glabella, and a series of dental traits (Arsuaga et al., 1993; Hublin, 1995, 2001; Martinón-Torres et al., 2007). If these are homologous and phylogenetically informative characters, then *H. heidelbergensis* and *H. neanderthalensis* are part of a distinct European lineage. But the case for continuity is clouded because some features (e.g., midfacial inflation) are quite variable in their expression in the European record and also present in some African archaic *Homo* crania (e.g., Kabwe, Bodo). Moreover, temperate Europe underwent cycles of intense glaciations during the Middle Pleistocene. During glacial periods, populations of hominins would either have gone extinct or been pushed south toward the

Mediterranean basin and southwest Asia. Because African hominins may have been periodically pushed northward, archaic *Homo* in Europe and Africa probably interacted at times (Hublin, 2001; see below).

H. neanderthalensis

Of all the extinct species of fossil hominins, the Neanderthals loom largest in the collective imagination, partly because the history of research on Neanderthals is deeply woven into the intellectual history of research on human evolution (see Trinkaus and Shipman, 1993). The first Neanderthals were discovered between 1829 and 1856, before Darwin's *Origin of Species* was published. Since the species was first formally proposed in 1863,[3] hundreds of sites have been excavated, yielding a rich fossil record that includes dozens of complete or partially complete skulls (Table 13.2). Despite much disagreement, few question that the Neanderthals are a taxon of archaic *Homo,* with a characteristic suite of features that by now should be very familiar: long, low cranial vaults; large ECVs (in the modern human range); large, projecting, double-arched supraorbital tori; massive faces with large orbits; and robust, chinless mandibles. However, Neanderthal skulls also have many distinctive features, including an expanded nasal cavity in which the nasal aperture is enlarged relative to the rest of the midface (see Figure 13.4). Like other archaic *Homo,* Neanderthals have long faces, but their zygomatic arches are positioned more posteriorly, usually emerging above M^2 rather than M^1 (Trinkaus, 1987). This "zygomatic retreat," combined with a wider, expanded nasal cavity, gives the infraorbital region a slightly more parasagittal orientation (Rak, 1986). In addition, Neanderthals have an expanded temporal around the mastoid region, creating the impression of small mastoid processes; they also have a wide digastric fossa and a projecting juxtamastoid eminence (a bony crest behind and below the mastoid process) (Hublin, 1978; Santa Luca, 1978). The back of the occiput in Neanderthals typically projects posteriorly in a "bun" (or chignon) above the nuchal plane, as if someone had tried to punch through the occiput from the inside. There is always a horizontal depression (suprainiac fossa) above the posteriormost point on the midline of the occipital. Neanderthals also have a slightly elongate, heart-shaped foramen magnum. In Ne-

---

3. The Irish anatomist William King proposed the species *H. neanderthalensis* in 1863, after the type specimen from the Neander valley. At the time, the German word for valley, *Thal,* included an *h.* In 1901, German spelling reform eliminated the *h,* leading to differences of opinion as to whether the English vernacular for the species should be *Neanderthal* or *Neandertal.*

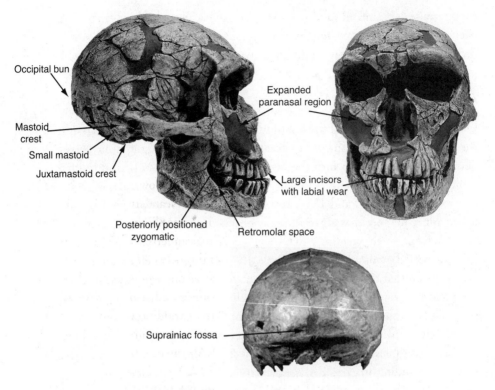

Figure 13.4. Major derived features of *H. neanderthalensis*. Anterior and lateral views are of La Ferrassie 1; posterior view is of La Quina 5.

anderthals, the anterior and posterior semicircular canals tend to be relatively small, with the latter situated lower relative to the lateral semicircular canal (Spoor et al., 2003). Finally, the Neanderthal mandible has a retromolar space between the M_3 and the ascending ramus, and Neanderthals typically have anteroposteriorly thick incisors, which were often heavily worn on the front (labial) surface. Neanderthals also tend to exhibit other dental features, such as asymmetric P_4s, that are present at much higher frequencies than in populations from other parts of the world (Bailey, 2005).

Although the above features vary, they usually come as an ensemble, making it easy to identify a Neanderthal. One can thus conclude with confidence that the species was part of a geographically and temporally restricted lineage. Most of the known Neanderthals fossils come from Europe, but some have also been found in the Levant (the eastern end of the Mediterranean), and a few occur as far to the east as Uzbekistan and possibly the Altai mountains of

Siberia (Krause et al., 2007). The oldest definitive Neanderthal fossils (e.g., Biache-Saint-Vaast, Ehringsdorf, Le Lazaret) postdate180 Ka (Klein, 2009). But the heyday of Neanderthals in Europe—the age of most fossils—was between 130 and 40 kya. Over this time, Europe experienced repeated cold periods that pushed the Neanderthals southward toward the Mediterranean. They appeared in the Levant during one such cold snap, between 80 and 60 kya (Bar-Yosef, 2000). The last known Neanderthal fossil is from Zafarraya, at the southern tip of the Iberian Peninsula, dated to 28–30 kya (Hublin et al., 1995). It looks as if the Neanderthals went extinct at the edge of Europe about 10,000 years after the first *H. sapiens* arrived (see below).

For years there was disagreement over whether the Neanderthals were our ancestors or closely related cousins. The answer is primarily the latter. There is abundant fossil and genetic evidence that the Neanderthals went extinct with little contribution to the modern human lineage. In terms of the fossil data, many early *H. sapiens* in Europe are robust (as one would expect), but none has any convincing *derived* Neanderthal features (see below). Likewise, Neanderthal fossils lack derived *H. sapiens* features. In addition mitochondrial (thus maternally inherited) DNA sequences have been obtained from more than a dozen Neanderthal fossils, and nuclear DNA sequences from three Neanderthal fossils. The mtDNA sequences indicate that *H. neanderthalensis* and *H. sapiens* genomes diverged approximately 660,000 (± 140) years ago (Green et al., 2008)[4]. Ancient nuclear DNA yield similar results. Most of the these data come from a 38,000-year-old fossil from Vindija, Croatia, that was analyzed independently by two groups (Noonan et al., 2006; Green et al., 2006). After correcting for contamination by recent modern human DNA (Wall and Kim, 2007), both data sets suggest that the average coalescence time between *H. neanderthalensis* and *H. sapiens* was about 700 kya and estimate that the population ancestral to both species split about 370 kya.[5] The nuclear DNA, moreover, indicates that there was a small degree (1–4 percent) of genetic admixture between Neanderthals and non-African humans (but not the in other

4. Keep in mind that this is a coalescence date (when the genomes diverged), not a speciation date (when the species or populations actually diverged). The average time of genetic coalescence is always older than the time of speciation or population divergence because of segregation within the ancestral population: many genes will segregate in a population before a species splits.

5. One of the original analyses of the Vindija DNA (Green et al., 2006) estimated a more recent time of divergence with possible interbreeding, because it did not correct for contamination in long sequences, which are less likely to survive intact than short ancient DNA sequences (Wall and Kim, 2007).

direction), presumably after humans started to disperse from Africa (Green et al., 2010). Analyses of phenotypic variation that model rates of random change over time come up with remarkably similar estimates for the split between *H. neanderthalensis* and *H. sapiens:* 311–435 kya, with a 95 percent confidence interval of 182–592 kya (Weaver et al., 2008). Putting all the data together, it is reasonable to conclude that *H. neanderthalensis* and *H. sapiens* last shared a common ancestor in the Middle Pleistocene and probably split between 500 and 200 kya. Any hybridization that occurred was not substantial.

What is most controversial about the Neanderthals is not their phylogenetic relationship to us, but what they were like behaviorally and cognitively. Over the years, the pendulum of opinion has swung between two extreme views. In the nineteenth and early twentieth centuries, Neanderthals were typically considered to have been savage brutes: stooped, unkempt, and incapable of sophisticated modern thought or speech. But after World War II, many paleoanthropologists reclassified Neanderthals as a subspecies of humans (*H. sapiens neanderthalensis*) and proposed that a well-coiffed Neanderthal in a suit and hat could pass unnoticed on the subway. A major basis for this change of opinion was a reappraisal of ECV, which averages 1,488 cm³ in Neanderthals, with a range of 1,172–1,740 cm³ (Holloway et al., 2004). The mean of these ECVs is greater than the recent human mean of 1,330 cm³ (Tobias, 1971). Given the high cost of growing and maintaining brain tissue (see Chapter 6), large brains must have been related to selection for intelligence. That some Neanderthals had brains slightly larger than most modern humans is best explained as a scaling effect of body mass, estimated to average 76 kg in Neanderthals (Ruff et al., 1997). Large bodies are a common adaptation to cold temperatures because they lower the ratio of surface area to body mass (body mass scales to the power of 3, but surface area scales to the power of 2). Early modern humans from Ice Age Europe also have larger brains and bodies than modern human populations.[6]

The archaeological record provides mixed evidence that is difficult to interpret about Neanderthal cognitive capacities. On the one hand, the Neanderthals were impressively adapted to harsh conditions in Ice Age Europe, able to make sophisticated tools and weapons and to hunt large, dangerous ani-

---

6. Neanderthals also had relatively shorter distal limbs (calves and forearms), which lower the ratio of surface area to body mass (Trinkaus, 1981; see also Weaver and Steudel-Numbers, 2005). Early modern humans in Europe (Gravettians from 24 kya) had more tropical body proportions, with longer distal arms and legs, but their descendants also evolved shorter distal limbs, presumably to help them cope with intensely cold winters (Holliday, 2002).

mals. When they moved southward into the Levant about 80 kya, they apparently displaced (maybe with some interbreeding) early *H. sapiens* who were living there and were also making Middle Paleolithic tools (Bar-Yosef, 2000). About 50 kya, however, *H. sapiens* invented a new industry—and a new way of life—known as the Upper Paleolithic, and spread once again into Europe. For reasons that we may never fully understand, the Neanderthals went extinct as modern humans from Africa thrived in Europe. Why *H sapiens* survived and Neanderthals did not may never be solved. Hypotheses include superior language and cognitive capacities, faster rates of reproduction, and more diverse strategies for hunting and gathering (for reviews, see Stringer and Gamble, 1993; Hublin, 2000; Klein, 2003; Mellars, 2004; Trinkaus, 2007).

Late *Homo* in Asia

Western, eastern, and southern Asia have different fossil records (summarized in Tables 13.3 and 13.4). Western Asia, especially the Levant, connects Africa with Eurasia and has long had a complex mixture of fauna and flora from both continents. The earliest hominins in Western Asia were probably *H. erectus* and possibly also *H. heidelbergensis.* By approximately 100 kya, early *H. sapiens,* which must have dispersed northward out of Africa, were in the region (see below). But from 80 to 50 kya, the region was populated by *H. neanderthalensis,* which presumably moved in from southeastern Europe or Anatolia during a colder period. Because *H. sapiens* was apparently absent during this period (some dates, however, are controversial), one can surmise that Neanderthals temporarily displaced the modern humans. However, by 40–60 kya, *H. sapiens* populations dispersed again out of Africa, this time with a more sophisticated Upper Paleolithic technology, and the Neanderthals disappeared, as they did elsewhere.

Most of the hominins from East and southern Asia older than about 60 kya probably belong in *H. erectus s.l.* (Table 13.4). Middle and Upper Pleistocene fossils attributed to *H. erectus* in Asia include not only the Ngandong fossils from Indonesia, which are probably younger than 100 kya, but also some older Chinese fossils, such as Hexian and Yunxian, that have many of the typical derived features of *H. erectus* (Wu and Poirier, 1995; Antón, 2003). By now, these features should be familiar: long, low vaults of moderate volume (800–1,200 cm^3); marked brow ridges; low frontals; the frequent presence of sagittal keeling and angular tori; a low-angled occiput with a transverse torus; and large, orthognathic faces with some alveolar prognathism. A few fossils dated

Table 13.3: Major cranial fossils attributed to late *Homo* in western Asia

Fossil/site	Location	Species	Date (kya)	ECV (cm³)	Elements found
Zuttiyeh	Israel	H. heidelbergensis s.l.	>250		Partial cranium
Tabun 2	Israel	H. sapiens	180–140		Mandible
Skhul 1	Israel	H. sapiens	90–150		Partial skull
Skhul 4	Israel	H. sapiens	90–150	1550	Partial skull
Skhul 5	Israel	H. sapiens	90–150	1520	Partial skull
Skhul 9	Israel	H. sapiens	90–150	1590	Partial skull
Qafzeh 6	Israel	H. sapiens	90–100	1568	Cranium (male)
Qafzeh 9	Israel	H. sapiens	90–100	1531	Skull (female)
Qafzeh 11	Israel	H. sapiens	90–100	1280	Skull (child)
Teshik Tash	Uzbekistan	H. neanderthalensis	70	1525	Partial skull (child)
Tabun 1	Israel	H. neanderthalensis	70–80	1271	Partial skull (and skeleton)
Kebara 2	Israel	H. neanderthalensis	60		Mandible (and partial skeleton)
Shanidar 1	Iraq	H. neanderthalensis	50–70	1600	Cranium
Shanidar 2	Iraq	H. neanderthalensis	50–70		Partial cranium
Shanidar 5	Iraq	H. neanderthalensis	50–70	1550	Partial cranium
Amud I	Israel	H. neanderthalensis	50–70	1740	Skull (and partial skeleton)
Amud 7	Israel	H. neanderthalensis	50–70		Partial skull (and skeleton) of infant
Dederiyeh 1	Syria	H. neanderthalensis	50	1096	Partial skeleton (infant)
Dederiyeh 2	Syria	H. neanderthalensis	50	1089	Partial skeleton (infant)
Nazlet Khater 1	Egypt	H. sapiens	37–38	1420	Partial skeleton

Note: Many of these sites have multiple burials (Skhul has 14, Qafzeh more than 15, Amud more than 15, and Shanidar 9).

Table 13.4: Major cranial fossils attributed to late *Homo* in East Asia and Australasia

Fossil(s)	Location	Species	Date (kya)	ECV (cm³)	Elements found
Sangiran 2	Java, Indonesia	*H. erectus*	1000–1600	813	Partial cranium (vault)
Sangiran 3	Java	*H. erectus*	1000–1600	950	Partial cranium (vault)
Sangiran 4	Java	*H. erectus*	1000–1600	908	Partial cranium
Sangiran 10	Java	*H. erectus*	1000–1600	855	Partial cranium (vault)
Sangiran 12	Java	*H. erectus*	1000–1600	1059	Partial cranium (vault)
Sangiran 17	Java	*H. erectus*	1000–1600	1004	Partial cranium
Trinil 2	Java	*H. erectus*	900–1200	940	Cranial vault
Sangbungmacan 1	Java	*H. erectus*	50–100	1035	Cranial vault
Sangbungmacan 3	Java	*H. erectus*	50–100	917	Cranial vault
Sangbungmacan 4	Java	*H. erectus*	50–100	1006	Partial cranium
Ngawi 1	Java	*H. erectus*	50–100	1000	Partial cranium
Ngandong 1 (Solo I)	Java	*H. erectus*	27–125	1172	Partial cranium (vault)
Ngandong 3 (Solo III)	Java	*H. erectus*	27–125		Partial cranium
Ngandong 6 (Solo V)	Java	*H. erectus*	27–125	1251	Partial cranium
Ngandong 7 (Solo 6)	Java	*H. erectus*	27–125	1013	Partial cranium
Ngandong 12 (Solo IX)	Java	*H. erectus*	27–125	1135	Partial cranium
Ngandong 13 (Solo X)	Java	*H. erectus*	27–125	1231	Partial cranium
Ngandong 14 (Solo XI)	Java	*H. erectus*	27–125	1090	Partial cranium
Wadjak 1	Java	*H. sapiens*	<10	1550	Cranium (plus partial skeleton)
Liang Bua 1	Flores, Indonesia	*H. floresiensis*	13–95	417	Skull (plus partial skeleton)
Narmada	India	*H. erectus**	Middle Pleistocene	1260	Vault
Gongwangling (Lantian)	China	*H. erectus*	1000	780	Partial cranium
Yunxian 1	China	*H. erectus*	400–800	1200	Partial cranium
Yunxian 2	China	*H. erectus*	400–800		Partial cranium
Chenjiawo	China	*H. erectus*	500–600		Mandible
Yuanmou	China	*H. erectus*	500–800		Teeth
Zhoukoudian II (III,L)	China	*H. erectus*	600–780	1030	Partial cranium
Zhoukoudian III	China	*H. erectus*	600–780	915	Partial cranium
Zhoukoudian V	China	*H. erectus*	400–500	1140	Partial cranium

(*continued*)

Table 13.4: Continued

Fossil(s)	Location	Species	Date (kya)	ECV (cm³)	Elements found
Zhoukoudian X (I,L)	China	*H. erectus*	600–780	1225	Partial cranium
Zhoukoudian XI	China	*H. erectus*	600–780	1015	Partial cranium
Zhoukoudian XII	China	*H. erectus*	600–780	1030	Partial cranium
Hexian	China	*H. erectus*	150–300	1120	Cranial vault
Dali	China	*H. erectus**	180–230		Cranium
Jinnuishan	China	*H. erectus**	200–300	1300	Skull (plus skeleton)
Xujiayao	China	*H. erectus**	100–130		Skull fragments (many individuals)
Maba	China	*H. erectus**	100–150		Partial cranium
Liujiang	China	*H. sapiens*	10–60	1480	Cranium
Tianyuandong	China	*H. sapiens*	35		Mandible (and partial skeleton)
Zhoukoudian Upper Cave 101	China	*H. sapiens*	10–30	1500	Skull
Zhoukoudian Upper Cave 102	China	*H. sapiens*	10–30	1380	Skull
Zhoukoudian Upper Cave 103	China	*H. sapiens*	10–30	1290	Skull
Minatogawa 1	Japan	*H. sapiens*	16–18	1390	Skull (plus skeleton)
Minatogawa 2	Japan	*H. sapiens*	16–18	1170	Skull (plus skeleton)
Minatogawa 4	Japan	*H. sapiens*	16–18	1090	Skull (plus skeleton)
Niah Cave	Borneo, Indonesia	*H. sapiens*	40		Cranium
Talgai	Australia	*H. sapiens*	17–24		Partial cranium
Mungo 1	Australia	*H. sapiens*	15–50		Partial cranium
Mungo 3	Australia	*H. sapiens*	~12		Partial skeleton
Keilor	Australia	*H. sapiens*	~12	1497	Partial cranium
Kow Swamp 1	Australia	*H. sapiens*	9–15		Partial skeleton
Kow Swamp 2	Australia	*H. sapiens*	13		Partial skeleton
Coobol Creek	Australia	*H. sapiens*	~12	1444	Partial skeleton
Nacurrie	Australia	*H. sapiens*	11–12		Skeleton
Willandra Lakes 50	Australia	*H. sapiens*	12–18	1540	Partial cranium

*Fossil often attributed to archaic *Homo* sp.

Data from Holloway et al. (2004); Schwartz and Tattersall (2003); Antón (2003).

to between 280 and 130 kya are considered by some to be *H. heidelbergensis* (or some other species of archaic *Homo*) rather than *H. erectus:* Dali, Jinnuishan, and Mapa (see Figure 13.5). As noted above, a clear distinction between *H. erectus* and archaic *Homo* is difficult to draw under the best of circumstances, and each of these crania has some classic *H. erectus* features, such as slight keeling and angular tori. Another fossil to note is a poorly dated partial cranium from Narmada, India, probably from the Middle Pleistocene, which falls in the same general category (Badam et al., 1989; Kennedy et al., 1991; Cameron et al., 2004).

And then there is *H. floresiensis.* In 2003, a team of Australian and Indonesian researchers uncovered a partial skeleton (nicknamed "the Hobbit" by the media) from the cave of Liang Bua, on the island of Flores in the Indonesian archipelago (Brown et al., 2004; Morwood et al., 2004). Further excavations recovered the remains of at least six individuals dated to between 95 and 17 kya (Morwood et al., 2005). The fossils are of tiny people, about a meter tall and weighing 25–30 kg. The ECV of the one skull (LB1) is 417 cm^3, the size of a chimpanzee's. The cranium has an odd mix of features (Figure 13.6). The face is vertical but projecting, with a relatively large, double-arched supraorbital torus; the frontal squama is somewhat sloping; the occipital is rounded, the vault bones are relatively thick; and the temporals are heavily pneumatized. The teeth, while slightly large, are typical of the genus *Homo.* The two mandibles have many primitive features, including the lack of a chin, superior and inferior transverse tori in the symphysis, and two partially fused (Tomes') roots in the lower premolars. The P^{4}s are rotated 90° in the jaw (presumably from crowding). Many postcranial features are also primitive relative to *H. erectus,* including relatively short legs, long feet, narrow shoulders, australopith-like wrists, and a flared pelvis (Brown et al., 2004; Morwood et al., 2005; Larson et al., 2007; Tocheri et al., 2007; Jungers et al., 2009).

There is much controversy over these finds. The discoverers originally proposed that the Liang Bua finds represent a dwarfed species of *Homo,* possibly *H. erectus.* Occasionally, isolated species on small islands undergo a phenomenon known as insular dwarfism or gigantism, in which large species become small and small species become large.[7] The island of Flores has examples of both, including giant rats, enormous varanid lizards (Komodo dragons), and

7. Dwarfism may occur when limited resources favor smaller-bodied animals because of competition for food (smaller animals need less food) or when the lack of predators removes any selection pressure for larger bodies.

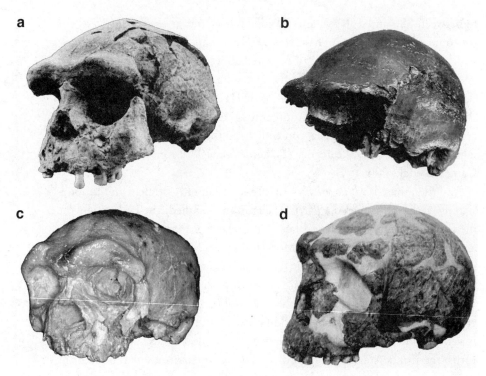

Figure 13.5. Some crania of Asian late *Homo:* (a) Sangiran 17 (*H. erectus*); (b) Ngandong 6 (*H. erectus*); (c) Dali (*H erectus s.l.* or *H. heidelbergensis s.l.*); (d) Jinnuishan (*H erectus s.l.* or *H. heidelbergensis s.l.*). Photos are not to scale. Photo of Dali courtesy M. Wolpoff; photo of Jinnuishan courtesy Lu Zune.

dwarf elephants. Further, archaeological finds on the island have been proposed to be as old as 800,000 years, suggesting a long period of habitation (Morwood et al., 1998). The hypothesis is that hominins were also subject to the same kinds of selective pressures, and several studies suggest that the LB1 endocast and cranium most closely resemble crania of early *Homo* (especially *H. erectus*), after correcting for the effects of size and scaling (Falk et al., 2005; Gordon et al., 2008; Baab and McNulty, 2009). Analyses of the postcranium, moreover, reveal that the limb proportions, shoulder, wrist, and foot of *H. floresiensis* are actually more primitive than those of *H. erectus* and more closely resemble those of *H. habilis* (Larson et al., 2007; Tocheri et al., 2007; Jungers et al., 2009). These resemblances raise a surprising hypothesis: that *H. floresiensis* is the descendant of a habiline species of *Homo* that left Africa very early, got all the way to Indonesia, and remained isolated and unchanged for mil-

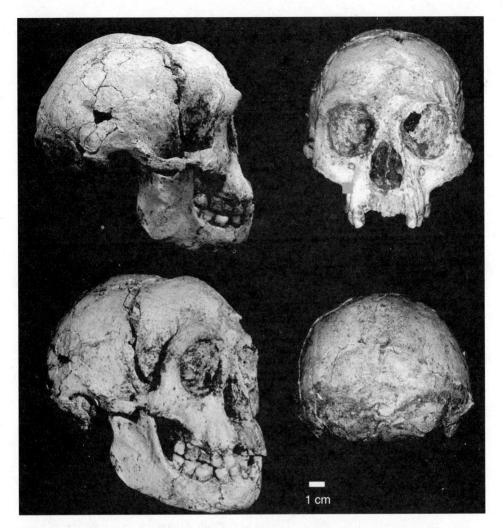

Figure 13.6. Liang Bua 1 skull (*H. floresiensis*): (a) lateral, (b) anterior, (c) three-quarters, and (d) posterior views. From Brown et al. (2004) (courtesy P. Brown).

lions of years. Alternatively, early *H. erectus,* the possible ancestor of *H. flore-siensis,* may have been more habiline than previously recognized.

Although it seems likely that *H. floresiensis* is a new species, some scholars are skeptical for many reasons, not the least of which is that LB1's brain is too small to fit typical scaling relationships between brain and body size. Across species, brain mass scales to body mass to the power of 0.75; among closely related species the scaling exponent is 0.2–0.4, and within species the scaling exponent is 0.25 (Martin et al., 2006). Thus if LB1 were a dwarfed modern human with a 50 percent smaller body mass, such an allometry would predict its brain size to be reduced by 16 percent, to approximately 1,100 cm³; if it were a dwarfed *H. erectus* descended from the Dmanisi population, then its brain size would be expected to be 500–650 cm³. Another way of putting it is that with a coefficient of allometry of 0.25, a Dmanisi hominin's body mass would have to be reduced to 6 kg to achieve the brain size seen in LB1 (Martin et al., 2006). However, analyses of brain size reduction in dwarfed species of hippos from the Mediterranean indicate that endemic dwarfing in mammals can lead to a higher scaling relationship (of 0.47) between brain and body size (Weston and Lister, 2009). Such a coefficient of allometry could accommodate dwarfing of a Dmanisi-sized *H. erectus s.l.*

An alternative hypothesis is that *H. floresiensis* is descended from *H. habilis*, which has a smaller, more primitive body and a smaller brain than *H. erectus* (Jungers et al., 2009). Even then, however, brain size would have had to undergo some dwarfing without body size decreases (the smallest habiline ECV is 510 cm³). Accordingly, several researchers have suggested that the Flores material represents some form of microcephalic pathology, which often includes dwarfism (Jacob et al., 2006); Laron's syndrome (Hershkovitz et al., 2007); or endemic cretinism (Obendorf et al., 2008). But these contentions are disputed by analyses of endocasts (Falk et al., 2007), crania (Argue et al., 2006), and postcrania (Larson et al. 2007; Tochieri et al., 2007).

Archaic *Homo* Teeth

There is much dental variation in late *Homo*, with *H. erectus* and *H. heidelbergensis* postcanine crowns tending to be larger than those of *H. neanderthalensis* and *H. sapiens*. The most dentally distinctive species of late *Homo* is not *H. sapiens* but *H. neanderthalensis*. Although Neanderthals retain shovel-shaped incisors (a primitive feature also present in *H. erectus*), they have high (albeit very variable) frequencies of several distinctive traits that provide extra relief on the crowns, such as tubercles emerging from the lingual cingulum of I^1 and accessory cusps in the P^3, P_4, and M^1 crowns (for details see Bailey, 2006, 2007). Neanderthal tooth roots also tend to be relatively long, especially in the incisors (Bailey, 2005), and the molars are often taurodont (with elongated pulp chambers that bifurcate near the distal end of the root). Cuspal enamel is slightly thinner in *H. neanderthalensis* than in *H. sapiens* (T. Smith et al., 2007b; Olejniczak et al., 2008).

Origin and Dispersals of *H. sapiens*

Before we consider our own species, the only hominin species left extant, it may help to clarify some terminological baggage left over from the troubled history of research on recent human origins. As noted above, there has been a lengthy debate over where to draw the line between *H. sapiens* and other proposed species of late *Homo,* including *H. neanderthalensis.* Before World War II, most experts considered archaic and modern humans to be distinct species, and some classified different human "races" as subspecies (Carl Linnaeus in 1758 distinguished four human subspecies: *Homo sapiens afer, H. s. asiaticus, H. s. americanus,* and *H. s. europaeus*). After the war, there was a healthy reaction to these scientifically unsupportable, sometimes racist views. Many postwar anthropologists preferred to recognize *H. sapiens* as a highly variable, polytypic and ancient species that included Neanderthals and other archaic taxa (e.g., Brace, 1967). According to this scheme, all modern people are *H. sapiens sapiens* or "anatomically modern" humans, whereas equally encephalized Neanderthals and other extinct, "archaic" humans were classified as subspecies (e.g., *H. sapiens neanderthalensis*). Recent genetic and developmental evidence, however, requires us to abandon this scheme. Accordingly, the term "archaic *Homo sapiens*" should be discarded (unless you believe in multiregional evolution), and the term "anatomically modern *Homo sapiens*" is superfluous (unless you believe that some *H. sapiens* were not "anatomically modern"). That said, humans have become less robust and less variable since the end of the Ice Age (Howells, 1973, 1989, 1995; Lahr, 1996; Lahr and Wright, 1996). It is therefore sometimes useful to make a distinction between early, more robust *H. sapiens* and recent, more gracile *H. sapiens.*

Derived Anatomical Features of *H. sapiens*

Before considering the developmental and functional shifts that make the head of *H. sapiens* different from the heads of other hominin species, it is useful to first define the anatomical differences between *H. sapiens* and other species of late *Homo.* This task involves some circular logic: to determine which aspects of modern human cranial anatomy are unique derived features, we first have to define which crania to attribute to *H. sapiens,* which requires making assumptions about which features are derived in the first place. In spite of this circularity (addressed later using developmental data), there is general consensus on the major derived features that characterize the *H. sapiens* skull (Day and Stringer, 1982; Stringer et al., 1984; Groves, 1989; Tattersall, 1992; Lieberman, 1995; Lahr, 1996; Lieberman et al., 2002, 2004b; Schwartz and Tattersall, 2005; Tattersall and Schwartz, 2008). These features, illustrated in Figure 13.7, boil down to two major differences from archaic *Homo*. First, *H. sapiens* has a very spherical (globular) cranial vault that is tall, wide, and short. Neurocranial sphericity leads to an integrated suite of derived features of the cranial vault, including parietals that are long, wide and very curved in both the sagittal and coronal planes; an occipital region that is long and does not project very far posteriorly; and a frontal squama that rises steeply above the orbits.

The second major and distinctive difference is that the *H. sapiens* face is mostly retracted underneath the anterior cranial fossa, in large part because the face is smaller (especially anteroposteriorly shorter) and the cranial base is more flexed. Whereas much of the face in archaic *Homo* crania projects in front of the anterior cranial fossa, almost all the face in *H. sapiens* lies *below* the anterior cranial fossa (Figure 13.7). Facial retraction leads to a related suite of differences, the most obvious of which are that *H. sapiens* typically have a weak, noncontinuous supraorbital torus, with distinct medial and lateral portions; a distinct canine fossa (a depression on the anterior surface of the maxilla just lateral to the bulge around the canine's root); a short space for the oropharynx between the back of the palate and the foramen magnum; and orbital cavities that are rectangular (wider than tall). Another important unique derived feature is the chin. Only modern humans have an upside-down T-shaped projection at the base of the mandibular symphysis, below the alveolar process (Lam et al., 1996; Schwartz and Tattersall, 2000). Schwartz and Tattersall (2005: 507) have proposed some additional derived features that would further restrict *H. sapiens* to mostly recent, nonrobust crania.

Although the modern human skull has a broadly accepted suite of derived features, there is much variability. As one might expect (it is statistically inevitable), some modern human crania fall outside the 95 percent confidence intervals of recent *H. sapiens* variation for classically diagnostic features in the neurocranium and brow ridges (as defined by Day and Stringer, 1982; and Stringer et al., 1984); conversely, a few skulls often attributed to archaic *Homo* fall within the 95 percent range of *H. sapiens* variation for some of these features (Kidder et al., 1992). Thus some researchers still consider *H. sapiens* a morphologically diverse species with "archaic" and "anatomically modern" grades (e.g., Wolpoff et al., 2001). However, for the two most fundamental, unique derived differences between *H. sapiens* and archaic *Homo*—facial re-

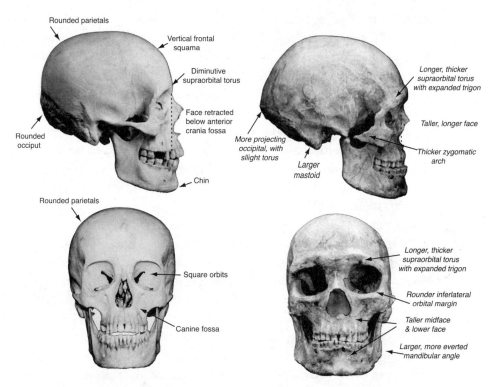

Figure 13.7. Derived features of *H. sapiens* and differences between a recent, gracile skull (left, a male from Italy), and an older, more robust Pleistocene skull (right, Zhoukoudian Upper Cave 101). Also indicated (in italics) are a number of features in the Pleistocene skull associated with robusticity. Images courtesy of the Peabody Museum, Harvard University.

traction and overall neurocranial sphericity—no known fossils of archaic *Homo* fall within the modern human range (Lieberman et al., 2002).

Much disagreement about where to draw a line between *H. sapiens* and other species of *Homo* stems from the question of how to cope with varying degrees of robusticity. The term *robusticity* can have alternative meanings but generally describes a skull, such as the one illustrated in Figure 13.7 (right), that is overall large and thick-walled, with marked cranial superstructures. As described by Lahr (1996), robust cranial features typically include thick supra-orbital tori in both the central (supraciliary) and lateral (trigone) portions, zygomaxillary tuberosities (a bulge or inferior projection on the infraorbital region of the malar, usually at the junction of the maxilla and the zygomatic bone), strong median nuchal lines (from the insertion of the nuchal ligament along the midline of the nuchal plane), and an occipital torus (a horizontal bar of bone between the top of the nuchal plane and the occipital squama). Additional features of skull robusticity include a pronounced chin, rounding of the inferolateral margin of the orbit, midsagittal keeling of the frontal or parietal bones, thick zygomatic arches, large mastoid processes, and extra ridges posterior and superior to the mastoid processes. Few populations from the last 10,000 years are extremely robust in this sense (exceptions include some from Tierra del Fuego, Patagonia, and Australia). But for most of its history, *H. sapiens* had high, albeit variable, levels of robusticity. In other words, robusticity is an ancestral (nonderived) characteristic of *H. sapiens* with little phylogenetic meaning.

What causes craniofacial robusticity is not entirely known, but a proportion is size related. In general, human crania have become more than 10 percent smaller over the last 10,000–12,000 years (Kidder et al., 1992; Lieberman, 1998), and skulls that are larger—overall and in terms of tooth size, jaw size, and upper facial size—are significantly more robust (Lahr and Wright, 1996). Robusticity is also more exaggerated and common in skulls that are narrow (dolichocephalic), probably because narrow cranial bases lead to more posterior elongation of the back of the skull and anterior elongation of the face (Lieberman et al., 2000a). Not all aspects of robusticity, however, can be accounted for by size. It is reasonable to hypothesize that some component of robusticity derives from hormonal mechanisms, which is supported by the fact that human males tend to be more robust than females even after taking size into account (see Bernal et al., 2006). Experiments and comparative studies also show that some degree of robusticity derives from more mechanical loading from chewing harder, less processed diets (Corruccini, 1999; Lieberman et al., 2004b; O'Connor et al., 2005; see also Chapter 7).

Fossil Evidence

Armed with features to diagnose modern *H. sapiens,* both gracile and robust, we can now consider the fossil evidence for when and where our species evolved. Major fossils, their dates, and ECVs (when known) are listed by region in Tables 13.1 to 13.4.

Africa

The earliest known *H. sapiens* fossils include the Omo I cranium, from the Kibish formation in Ethiopia (Day, 1969; R. E. F. Leakey et al., 1969), which

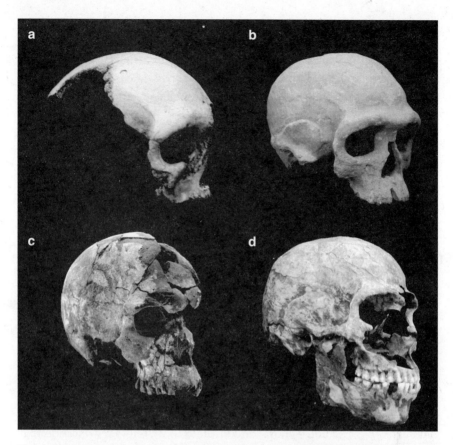

Figure 13.8. Some early *H. sapiens* crania from Africa and southwest Asia: (a) Florisbad; (b) Jebel Irhoud 1 (reconstruction); (c) Herto (BOU-VP-16/1); (d) Skhul V (a cast with the reconstructed face removed). Note that all these crania are robust (as in Figure 13.7.) but have globular vaults and retracted faces. Images not to scale. Image of BOU-VP-16/1 © 2003 David L. Brill.

may be as old as 195 kya (McDougall et al., 2005); three crania from Herto, Ethiopia, dated to 150–160 kya (White et al., 2003); a partial cranium from Singa, Sudan, dated to 133 kya (McDermott et al., 1996); a partial cranium from Laetoli, Tanzania (LH 18, Ngaloba) that probably dates to about 129–108 kya (M. D. Leakey and Harris, 1987); crania from Aduma and Bouri, Ethiopia, both from 105–80 kya (Haile-Selassie et al., 2004); and about 30 fragments from multiple individuals from the caves at Klasies River, South Africa, dated to between 120 and 90 kya (Singer and Wymer, 1982; Rightmire and Deacon, 1991; Grine et al., 1998; Deacon and Deacon, 1999). Other

African fossils that are quite robust but probably qualify as *H. sapiens* based on overall sphericity and facial retraction include some crania from Jebel Irhoud, Morocco, traditionally dated to 160 kya (Grün and Stringer, 1991; Hublin, 2001; Smith et al., 2007c); and some less securely dated fossils from Eliye Springs, Kenya, and Florisbad, South Africa (see Figure 13.8).

Western Asia

If modern humans evolved in Africa, then the next oldest place they would presumably occur is western Asia (see Figure 13.9 and Table 13.3), which connects Africa and Eurasia. And so it is. The earliest known *H. sapiens* in this region are from three sites in Israel dated to between 130 and 90 kya (Bar-Yosef, 2000). They include a mandible with a chin from layer C of Tabun Cave (McKown and Keith, 1939; Quam and Smith, 1998); nine burials from Skhul Cave (McKown and Keith, 1939); and another series of burials from Qafzeh Cave (Vandermeersch, 1981; Tillier, 1999). As noted above, these modern humans precede the Neanderthals, who showed up in the region about 80–50 kya.

Europe

The oldest known European *H. sapiens* fossils are from the Romanian site of Peştera cu Oase, dated to about 35 kya (Trinkaus et al., 2003a,b). The abundance of *H. sapiens* fossils in Europe increases from 30 kya onward: the famous skeletons from Cro-Magnon, France, are dated to about 28 kya (Henri-Gambier, 2002). As noted above (and in Table 13.2), only a few Neanderthal fossils have been found in Europe that postdate the arrival of *H. sapiens,* suggesting that Neanderthals were outcompeted either directly or indirectly. It has been argued that a few early European *H. sapiens* fossils bear traces of interbreeding (hybridization) with Neanderthals (Trinkaus, 2005, 2007; Zilhão, 2006).[8] The cranial evidence for these claims comes primarily from robust fossils that have derived features of *H. sapiens* (e.g., chins; spherical cranial vaults; small, divided brow ridges) but also have somewhat large faces and some archaic dental features. Although minor hybridization occurred between the two species, large faces and teeth and high levels of robusticity are also characteristic of other early *H. sapiens* from Africa and elsewhere, and thus are

8. One hypothesized case of hybridization is the skeleton of a 4-year-old boy from Lagar Velho in Portugal, dated to about 24 kya, which has relatively robust and short distal limb bones (Duarte et al., 1999; Trinkaus and Zilhão, 2002). Most of the preserved features of the skull fall within the juvenile *H. sapiens* range of variation, including overall shape (Zollikofer et al., 2002), the bony labyrinth (Spoor et al., 2002b), the chin (Trinkaus, 2002), and the teeth (Hillson and Trinkaus, 2002).

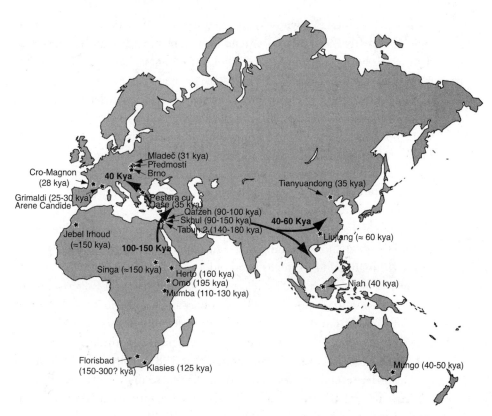

Figure 13.9. Dates and locations of some of the oldest *H. sapiens* fossils from different parts of the Old World and likely ages for dispersals out of Africa.

not necessarily shared-derived features between Neanderthals and *H. sapiens.* Further, whatever interbreeding occurred, it cannot have been substantial, given other genetic and skeletal evidence (see above).

Asia

The fossil record of modern human origins in Asia is scanty, but the oldest currently known Asian *H. sapiens* fossil is a mandible and partial skeleton from Tianyuan Cave in China, dated to between 42 and 38 kya (Shang et al., 2007). Other early material includes some skulls from Zhoukoudian Upper Cave that date to 30–10 kya; a cranium from Niah Cave, Sarawak (Malaysia), that may be 40 kya (Brothwell, 1960; Barker et al., 2007); fossils from Sri Lanka dated to 33–28 kya (Kennedy et al., 1987); and a handful of Chinese fossils, such as Liujiang, whose dates are uncertain (Wu, 1992). As in Europe and Africa, early Asian *H. sapiens* tend to be robust, with slightly larger teeth than recent humans, leading some scholars to suggest that these modern humans interbred with or evolved from *H. erectus* or archaic *Homo sp.* Interbreeding cannot be ruled out, but, as noted above, *H. erectus* may have persisted in Indonesia until quite recently, possibly less than 70 to 40 kya (Swisher et al., 1997; Yokoyama et al., 2008). The most likely scenario is that early, somewhat robust *H. sapiens* arrived in Asia about 60 kya and then mostly replaced other hominins, just as they replaced *H. neanderthalensis* in Europe.

Australia

The oldest archaeological sites in Australia are close to 45 kya (Bowler et al., 2003; Habgood and Franklin, 2008), and the oldest Australian fossils so far found (Mungo 1 and 3) might be as old as 40 kya but are probably slightly younger (Bowler et al., 2003; Gillespie, 1997; Gillespie and Roberts, 2000). These earliest Australians are much less robust and archaic-looking than some younger crania from the sites of Kow Swamp, Coobool Creek, and Keilor (all between 14 and 12 kya). Despite claims of regional continuity between *H. erectus* and *H. sapiens* (e.g., Thorne and Wolpoff, 1981; Wolpoff, 1999; Wolpoff et al., 2001), several studies have shown that these more recent fossils look superficially archaic because of artificial cranial deformation (Brown, 1981, 1992; Antón and Weinstein, 1999; Durband, 2008).

Recent Changes, Variation, and Neutral Evolution

One complication in interpreting early *H. sapiens* fossils is that our species' heads have changed noticeably since the Pleistocene in several respects. First, there has been a general decline in overall body size and head size: many facial dimensions have decreased by 10–30 percent (Kidder et al., 1992; Lahr, 1996; Lieberman, 1998), postcanine crown size has shrunk by about 12 percent (Brace et al., 1991), and brain size has declined by approximately 11 percent (Ruff et al., 1997). Second, size decreases correspond to a concomitant decline in robusticity, which is moderately correlated with overall size (Lahr and Wright, 1996), and which may also reflect differences in climate and the effects of less mechanical loading, such as smaller jaws, zygomatic arches, and regions of insertion for the masticatory muscle attachments (see Chapter 7).

In addition to shrinking and becoming more gracile, modern humans have also undergone some regional differentiation since the Pleistocene. One should be wary of exaggerating regional differences, because approximately 80–90 percent of all modern human craniometric variation (like genetic variation) occurs *within* rather than *between* populations (Relethford, 1994; Roseman and Weaver, 2004). Nonetheless, there are statistically significant differences between population means for a few aspects of skull shape, mostly in the face (Howells, 1989; Lahr, 1996; see also Gonzáles-José et al., 2004). In general, East Asians have faces that are somewhat flatter and taller than in other populations; Europeans tend to have a more projecting upper nasal region, along with more retracted and narrower jaws; Australo-Melanesians tend to have a relatively narrow frontal combined with a broad, slightly projecting face below; Africans are highly diverse but typically have a narrow cranial base (biporionic breadth) combined with a narrow midface, a convex frontal squama, a small glabellar region, and a short face; Native Americans, also very diverse, tend to have a short, broad cranial base and vault, a wide lower face, and a prominent nasal saddle. A key fact is that most of these differences are less than 10,000 years old (Howells, 1989). Figure 13.10 summarizes a cluster analysis of principal components of variation, based on 57 measurements taken on a large sample of males from 28 modern human groups and 22 premodern samples. Note that Holocene crania tend to cluster together in regional groups, and almost all the fossils 10,000 years old or older cluster together, regardless of their geographic location. In other words, the cranial features that make people from various parts of the world look different are not very ancient.

Another important insight is that most recent changes do not strongly signal the effects of natural selection. Instead, several lines of evidence suggest that processes of neutral evolution, such as genetic drift and random gene flow, can explain most phenotypic variation among recent human skulls. First, the proportion of phenotypic variance in the skull within rather than between human populations is almost the same as the proportion of genetic variance within rather than between populations (Relethford, 1994; 2002; Roseman and Weaver, 2004; Roseman, 2004). For most traits, the percentage of total variation explained by intrapopulation variation is less than 10 percent, lower than observed in most other mammals (Lynch, 1990). Moreover, under conditions of neutral evolution, one expects lower levels of variance to occur from drift and the effects of small population size. Intriguingly, measures of intragroup variation in the cranium are higher in Africa and become lower with increasing geographic distance from Africa (Manica et al., 2007), as is clearly the case in the genotype (see Wells, 2002; Jobling et al., 2004; Pearson, 2004; Prugnolle et al., 2005). Another line of evidence reflecting low levels of selec-

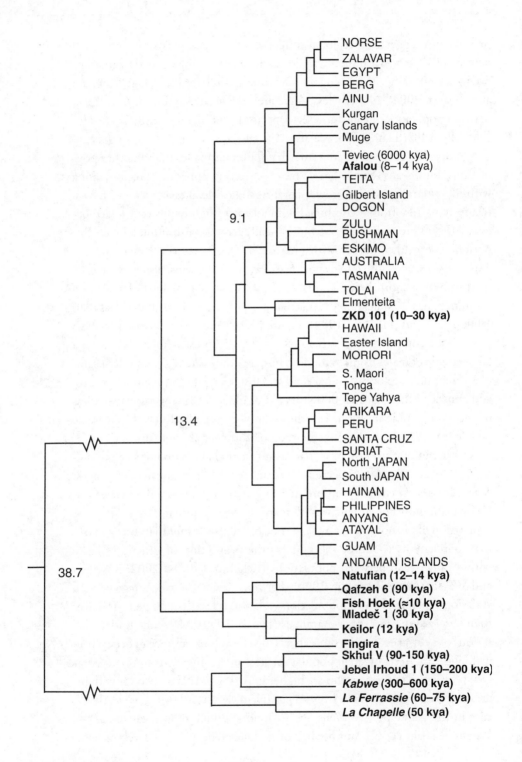

Figure 13.10. Overall shape similarities among skulls from different parts of the world (adapted from Howells, 1989: Figure 16). The graph is a cluster analysis of size-corrected measurements of shape calculated from 57 measurements taken on more than 2,500 crania. Recent populations are in capitals, Holocene archaeological populations are in title case, and Pleistocene skulls (along with approximate dates) are in boldface. Note that almost all Pleistocene crania are more similar to each other than they are to any crania from the last 10,000 years, regardless of geographic location.

---

tion on craniofacial variation is the high correspondence between measures of genetic and craniofacial difference among populations (after correcting for lower heritabilities of phenotypic than genotypic traits), indicating that phenotypic variance maps on well to genetic variance (Harvati and Weaver, 2006; González-José et al., 2004). And finally, estimates of the divergence time between Neanderthals and modern humans using phenotypic data, when applied to models of neutral evolutionary change, correspond well to molecular estimates (Weaver et al., 2007; Weaver et al., 2008).

Developmental Transformations in Late *Homo*

What developmental shifts underlie the craniofacial transformations we observe from *H. erectus s.l.* to archaic *Homo,* and then from archaic *Homo* to *H. sapiens?* A theme often repeated above is that the differences between *H. erectus* and various taxa of archaic *Homo* (e.g., *H. neanderthalensis* and *H. heidelbergensis*) are subtler than the differences between archaic *Homo* and *H. sapiens.* Whereas the former involved mostly increases in brain and face size, the latter occurred primarily from decreases in face size combined with elongation and flexion of the anterior cranial base, particularly in the middle cranial fossa. Later, after *H. sapiens* evolved, there were further decreases in face size and other aspects of robusticity.

Let's focus on what appear to be three key or representative transformations: from *H. erectus s.l.* to *H. heidelbergensis s.l;* from *H. heidelbergensis s.l.* to *H. neanderthalensis;* and from *H. heidelbergensis s.l.* to early *H. sapiens.* All three cases assume that archaic *Homo* crania from the Middle Pleistocene of Africa and Europe fall within a single, diverse taxon, *H. heidelbergensis s.l.,* rather than multiple species (e.g., *H. rhodesiensis, H. antecessor*). As in Chapters 11 and 12, several kinds of data, including geometric morphometric comparisons from CT scans of well-preserved and relatively complete crania, are integrated to make some inferences about the major developmental shifts underlying these transformations.

The Transition from *H. erectus s.l.* to *H. heidelbergensis s.l.*

The essential plan of the skull in *H. erectus* and *H. heidelbergensis s.l.* is very similar. Both have a long, low cranial vault and a large but vertical (orthognathic) face that projects well in front of the anterior cranial fossa. But there are differences, some of which are evident in the thin-plate spline analysis shown in Figure 13.11 (see Chapter 5 for a description of this method). The spline shows the transformations necessary to warp an early *H. erectus* cranium (KNM-ER 3733) into an *H. heidelbergensis* cranium from Africa, Kabwe (also known as Broken Hill). Both are probably males. The transformation is relatively subtle, particularly after correction for the effects of size. One shift is the posterior cranial fossa, which is more flexed relative to the rest of the skull, and in which the upper occipital scale is relatively longer. In addition, the face of *H. heidelbergensis* is relatively longer and taller, and the cranial base is slightly more extended.

Brain Size and Vault Shape

Figure 13.12 plots ECV over time in *H. erectus* and *H. heidelbergensis*. There is a significant temporal trend toward larger ECVs within *H. erectus* (about 100 cm³ more brain every 500,000 years), and ECVs in *H. heidelbergensis* are slightly but significantly larger than in *H. erectus* (Rightmire, 2004). Before considering how larger ECVs account for shape differences (see below), we need to ask how brains got larger in *H. erectus*. One way to address this question is to ask whether bigger brains are the result of bigger bodies. In East Africa, there are only a few postcrania attributable to *H. erectus s.l.* (KNM-ER 736, KNM-ER 737, KNM-ER 1808, KNM-WT 15000, OH 28, and OH 34). These fossils yield estimates of adult body mass that range from 51 to 68 kg, averaging 58 kg (Ruff and Walker, 1993), similar to body mass estimates for the Zhoukoudian specimens. In Dmanisi, body masses are smaller, about 40–50 kg. When combined with ECV data, EQs for *H. erectus s.l.* remain between 3.0 and 3.8 (Ruff et al., 1997; Lordkipanidze et al., 2007). Thus much of the variation in brain size within *H. erectus s.l.* appears to be the result of scaling (see Figure 13.12b).

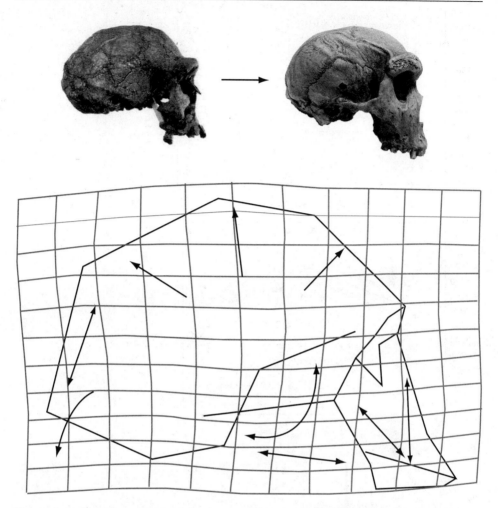

Figure 13.11. Thin-plate spline transformation of an *H. erectus* reference cranium (KNM-ER 3733) into an *H. heidelbergensis* target (average of Kabwe and Petralona). Arrows indicate major differences in growth.

There are few crania with associated postcranial elements for *H. heidelbergensis* (mostly from the Sima de los Huesos). Body mass estimates for *H. heidelbergensis* range from 65 to 71 kg, and mean ECV is approximately 1,200 cm^3, yielding an estimated average EQ for the species of between 4.3 and 4.7 (see Figure 13.12c; for data; see Ruff et al., 1997; Arsuaga et al., 1997; Holloway et al., 2004). One obtains similar results by plotting ECVs against

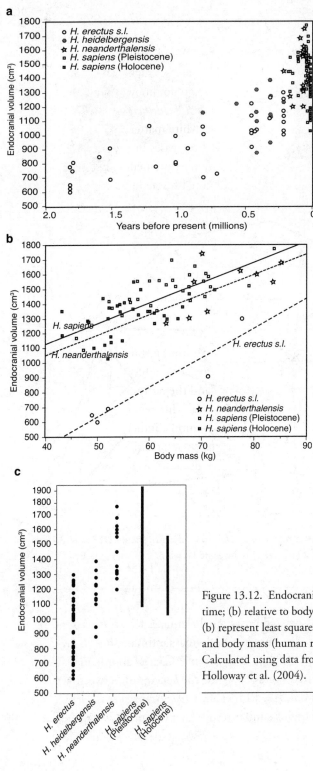

Figure 13.12. Endocranial volume in late *Homo:* (a) over time; (b) relative to body mass; and (c) by species. Lines in (b) represent least squares regression between brain size and body mass (human regression is shown by solid line). Calculated using data from Ruff et al. (1997) and Holloway et al. (2004).

orbital area, a useful proxy for body mass (Aiello and Wood, 1994; Rightmire, 2004, 2008). Thus body mass in *H. heidelbergensis* increases slightly compared to that in *H. erectus,* and brains get even bigger. We do not know what factors make brains both absolutely and relatively bigger in *H. heidelbergensis.* A likely possibility is more neurons, hence relatively more white matter (see Chapter 6). It is also possible that there were proportional increases in some parts of the brain, but this hypothesis is not supported by measurements of the relative size of the posterior, middle, and anterior cranial fossae in *H. erectus s.l.* and *H. heidelbergensis* (Bastir et al., 2008).

Bigger brains lead to some changes in vault shape, evident in Figure 13.11. *H. erectus s.l.* and *H. heidelbergensis* crania share the same basic elongated shape (long and low, widest just above the mastoid processes, low frontal angles, and prominent brow ridges), but relatively and absolutely bigger brains contribute to several derived features in *H. heidelbergensis* cranial vaults that make them intermediate between *H. erectus* and *H. sapiens.* First, brain expansion widens the anterior cranial fossa, causing significantly less postorbital constriction (measured as minimum frontal breadth relative to maximum biorbital breadth) in *H. heidelbergensis* (Rightmire, 2008). Second, brain expansion makes the parietals slightly more rounded in the coronal plane, and the occiput significantly more rounded in the sagittal and coronal planes. This difference is evident in the occipital scale index (inion-opisthion/lambda-inion), indicating that the occipital squama is relatively longer and the nuchal plane relatively shorter in *H. heidelbergensis* (Rightmire, 2008). Bigger brains also lead to taller vaults that are relatively higher than in *H. erectus s.l.* but not as globular as in *H. sapiens* (see below).

Face Size and Shape

Apart from the Dmanisi sample, KNM-ER 3733, KNM-WT 15000, and Sangiran 17, there are few complete *H. erectus* faces. These crania and partial upper faces, such as Sangiran 27, suggest that the face in *H. erectus s.l* and *H. heidelbergensis,* although orthognathic, was generally large, especially in males. Figure 13.13a, which plots a simple measure of overall face size (a geometric mean of five dimensions: orbit height, palate length, palate breadth, upper facial breadth, and facial height), indicates that *H. heidelbergensis* faces are on average 11 percent larger than those of *H. erectus* (including the Dmanisi sample). Because faces grow in a skeletal growth trajectory in tandem with the postcranium (see Chapter 2), relatively bigger faces may derive in part from

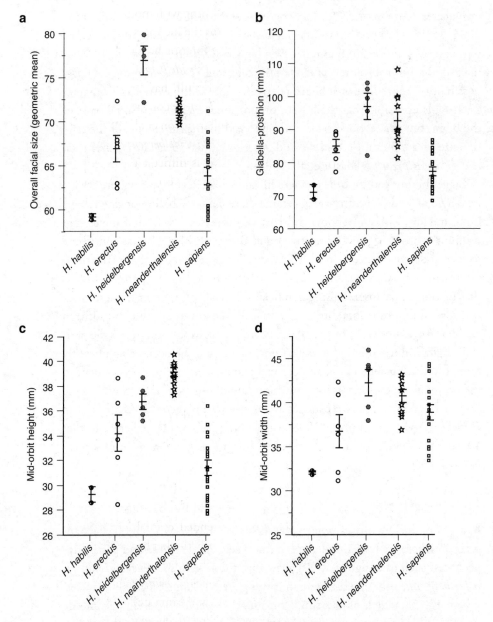

Figure 13.13. Facial size and dimensions in late *Homo:* (a) overall facial size (the geometric mean of orbit height, orbit width, palate length, palate breadth, upper facial breadth, and facial height); (b) facial height; (c) midorbit height; (d) midorbit width. Bar indicates mean ± 1 standard deviation. The *H. sapiens* sample consists entirely of Pleistocene fossils (Holocene samples are even smaller).

relatively bigger bodies (a hypothesis that needs testing with more postcrania). Figure 13.13 also shows that a large proportion of the increases in face size in *H. heidelbergensis* come from increases in superoinferior height and anteroposterior length but not in mediolateral width. *H. heidelbergensis* faces, especially likely males such as Petralona and Broken Hill, have tall orbits and tall infraorbital (malar) regions, but facial width is more constrained in skulls than facial height and length (see Chapter 5; Hallgrímsson et al., 2007a). The internal nasal cavity is also capacious in many *H. heidelbergensis* crania (see below). Despite much variation in mandibles, it is difficult to discern any major shifts in mandibular morphology or growth between *H. erectus s.l.* and *H. heidelbergensis*.

Spatial Packing and Cranial Base Angulation

Chapter 5 reviews how interactions between brain size and face size influence cranial shape. One effect is that relatively larger brains tend to increase cranial base flexion, but relatively longer faces tend to decrease cranial base flexion. CBA1 can be measured accurately in only a few fossils (see Table 13.5), but the combination of these interactions apparently leads to cranial bases that are moderately flexed in *H. erectus:* the average CBA1 is 141°, above the *H. sapiens* mean (134°) and at the upper 95 percent of the modern human range (127°–141°). In most *H. heidelbergensis* crania, CBA1 is between 140° and 145°, thus more extended than in most *H. erectus* and *H. sapiens* crania, possibly because of relatively larger faces (see above).

Another zone of interaction between the cranial base and face is the supraorbital torus. As explained in Chapter 5, the anteroposterior length of the brow ridge is partly influenced by the anteroposterior length of the face relative to the length of the anterior cranial base, and by the angle of the cranial base (relatively longer faces and more extended cranial bases lead to relatively longer brow ridges). In addition, vertically thicker brow ridges scale with relatively taller faces. The combination of these factors apparently accounts for the marked brow ridge development in *H. heidelbergensis*, especially in likely males such as Petralona, Kabwe, and Sima de los Huesos 5, which have long and tall faces combined with relatively extended cranial bases. However, the brow ridges in *H. heidelbergensis* tend to arch more over each orbit than they do in *H. erectus*, in which the torus is shaped more like a single beam.

Table 13.5: Cranial base angle, endocranial volume, and encephalization quotient for late *Homo* species

Fossil(s)	ECV (cm³)	CBA1 (degrees)	Body Mass (kg)	EQ	References (for CBA)
H. erectus					
Sangbungmacan 4	1006	141			Baba et al., 2003
KNM-ER 42700	691	138			Spoor et al., 2007a
KNM-ER 3733	848	145			This study
OH 9	1067	133–142			Maier and Nkini, 1984
Species average	*970*	*141*	*58*	*3.0–3.8*	
H. heidelbergensis					
Petralona	1220	140–145			This study
Kabwe	1280	145			This study
Bodo	1225	139			This study
SH 5	1125	145			This study
Species average	*1263*	*142.5*	*68*	*4.3–4.7*	
H. neanderthalensis					
Forbes Quarry	1200	150			This study
Guattari	1500	150			This study
Species average	*1500*	*150*	*76*	*4.8*	
H. sapiens (early)					
Singa	1550	140			Spoor, personal communication
Eliye Springs	1212	127			Bräuer et al., 2004
Skhul IV	1550	133			This study
Skhul V	1520	135			This study
Oberkassel I	1500	131			This study
Abri Pataud 1	1380	135			This study
Cro Magnon 1	1730	135			This study
Species Average	*1495*	*134*	*64*	*5.3*	
H. sapiens (recent)	1350	134	58.2	5.3	Lieberman and McCarthy, 1999

ECV and body mass data from Holloway et al. (2004) and Ruff et al. (1997).

The Transition from *H. heidelbergensis* to *H. neanderthalensis*

Neanderthals are a distinct European and western Asian species of archaic *Homo* (see above). Many experts posit that *H. neanderthalensis* evolved exclusively in Europe from *H. heidelbergensis*. This lineage would account for the presence of some Neanderthal features in European fossils dating to earlier than 350 kya (e.g., Arsuaga et al., 1997; Dean et al., 1998; Hublin, 1998). An alternative hypothesis is that *H. sapiens* and Neanderthals evolved from a single common ancestor, *H. heidelbergensis,* that inhabited both Africa and Europe and remained a single, albeit diverse, species, thanks to population movements between the two continents during the Pleistocene (Hublin, 2001; Rightmire, 2008). Regardless, the morphological and hence developmental differences between *H. heidelbergensis* and *H. neanderthalensis* are not extensive. Figure 13.14 plots a thin-plate spine transformation from an average of two *H. heidelbergensis* crania (Kabwe and Petralona) into that of Guattari, a relatively complete Neanderthal from Italy. The TPS (which is based on CT scans) does not capture small-scale differences in morphology, such as the juxtamastoid crests, the suprainiac fossa, or the teeth (see above); nonetheless, it shows that only a few subtle developmental shifts are necessary to transform a typical skull of *H. heidelbergensis* into an *H. neanderthalensis.* These shifts include growing a bigger (especially longer) brain, particularly in the posterior cranial fossa, and having less anterior migration of the zygomatic region.

Brain Size and Shape

H. neanderthalensis has large ECVs, averaging approximately 1,500 cm^3 and with a range of 1,172–1,740 cm^3 (Holloway et al., 2004), well above the average for *H. heidelbergensis* (1,263 cm^3) and at the upper end of the *H. sapiens* range. Mean estimated body mass in *H. neanderthalensis* is 76 ± 1.4 kg, larger than in *H. sapiens,* yielding a slightly lower EQ of 4.8 (Ruff et al., 1997). Given the interactions between brain expansion and neurocranial and basicranial growth, large brains in Neanderthals lead to a number of predictable differences in vault shape from that of *H. heidelbergensis,* evident in Figure 13.14. Although the basic shape of the vault in both species is long and low, expansion of the anterior cranial fossa in Neanderthals typically causes the frontal to be a little more curved in the sagittal plane and mediolaterally wider behind the upper face (thus with less postorbital constriction). The maximum width of the parietals is a little higher in Neanderthals (it tends to be just above the mastoids in *H. heidelbergensis*). The posterior part of the vault is also enlarged in several ways, including more rotation of the occipital behind the foramen magnum, leading to a more convex nuchal plane. Similarly, the occipital in Neanderthals appears to be "pushed out" above the tentorium cerebelli, whose insertion creates a distinct internal occipital protuberance (Figure 13.15; see also Chapter 4), elongating the upper scale of the occipital (called lambdoid flattening). Several studies suggest that occipital bunning results from posterior expansion of a relatively large neocortex (Trinkaus and LeMay,

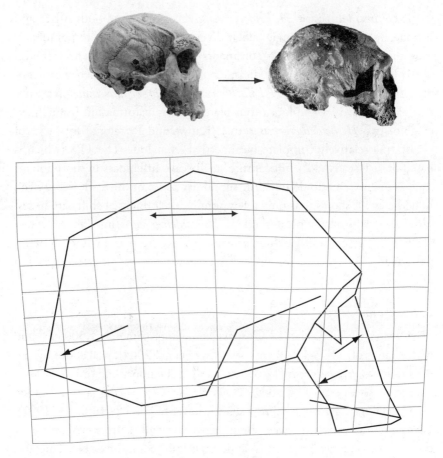

Figure 13.14. Thin-plate spline transformation of *H. heidelbergensis* reference cranium (average of Kabwe and Petralona) into *H. neanderthalensis* target (Monté Circeo cranium). Arrows indicate major shifts in relative growth.

1982; Lieberman et al., 2000a; Gunz and Harvati, 2007). It thus should not be surprising that some large-brained *H. sapiens* with narrow cranial bases also have occipital bunlike projections (Lieberman et al., 2000a).

Facial Shape

As Figures 13.4 and 13.14 depict, Neanderthals have a distinctively project-ing midface. At the risk of glossing over the variation evident in this species, the Neanderthal face is generally characterized by a tall, wide, and everted

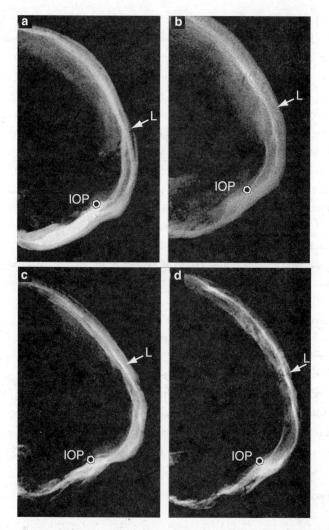

Figure 13.15. Lateral radiographs (from Lieberman et al., 2000a) of the occipital region in two Pleistocene *H. sapiens* crania from Europe: (a) Abri Patuad, and (b) Cro-Magnon I; and two *H. neanderthalensis* crania, (c) La Chapelle aux Saints, and (d) La Ferrassie I. The *H. sapiens* crania have slight occipital buns ("hemibuns"), but in the Neanderthals, the occipital buns are much more marked and project further posteriorly relative to the internal occipital protruberance (IOP), where the tentorium cerebelli inserts on the occipital, and below the intersection of the parietals and occipital bone (lambda, L).

nasal aperture in combination with infraorbital (zygomaxillary) regions that are tall, wide, and "swept back," in the sense that they are oriented slightly less in the coronal plane than in *H. erectus* and in many (but not all) *H. heidelbergensis* faces (Rak, 1986). As noted by Trinkaus (2003), the distinctive morphology of the Neanderthal midface is not due to more midfacial prognathism or projection than in *H. heidelbergensis*. Instead, one can deduce that three processes would have contributed to the Neanderthal configuration. First, the Neanderthal internal nose grew to be very wide, especially in the upper part (see also Schwartz and Tattersall, 1996; Franciscus, 1999). Second, the lateral margin of the midface (the zygomatic process of the maxilla) did not migrate as far forward (perhaps from less drift) in *H. neanderthalensis* as in *H. erectus* or *H. heidelbergensis* (Trinkaus, 1987). Consequently, the zygomatic root in Neanderthals typically arises above M2–M3, rather than M1–M2, as is most often the case in *H. sapiens* and *H. heidelbergensis*. Finally, the paranasal sinuses in Neanderthals appear to be inflated, resulting in a smooth, visorlike contour of the midface.

Zygomatic retreat in the Neanderthal face is matched by changes in the mandible (see Figure 13.4), which also reflects a more posteriorly positioned masseter relative to the postcanine teeth. Notably, *H. neanderthalensis* mandibles are as long overall as those of *H. heidelbergensis,* but the ramus is about 10 percent anteroposteriorly shorter, and the postcanine crowns are mesiodistally shorter. The combination of these two reductions leads to the frequent presence of a retromolar space between the M3 and the anterior margin of the ramus (Franciscus and Trinkaus, 1995).

The Transition from *H. heidelbergensis* to *H. sapiens*

The origin of *H. sapiens* marks a significant shift in craniofacial architecture from earlier taxa of *Homo*. Because heads are so integrated, it is a challenge to unravel all the developmental causes underlying this transformation, but several major factors—all interrelated—appear to be especially important (see Figure 13.16): relatively bigger brains that differ in shape, relatively smaller faces, more retracted faces, and more cranial base flexion. A number of other, smaller developmental differences are also evident.

Brain Size and Proportions

As summarized in Figure 13.12 and Table 13.5, ECVs are significantly larger in Pleistocene *H. sapiens* (1,495 cm^3) than in recent modern humans (1,350 cm^3),

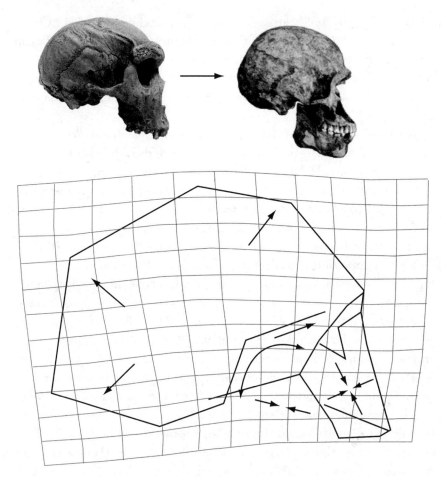

Figure 13.16. Thin-plate spline transformation of *H. heidelbergensis* reference cranium (Kabwe) into Pleistocene *H. sapiens* target (average of Skhul V and Cro-Magnon I). Arrows indicate major shifts in relative growth.

not significantly different from Neanderthals, and about 200 cm³ larger than the mean for *H. heidelbergensis* (Holloway et al., 2004). Some of this variation reflects differences in body mass. According to Ruff and colleagues (1997), the average body mass of recent modern humans (excluding the current epidemic of obesity) is 58.2 kg, yielding an EQ of 5.3; during the Late Pleistocene, body mass in *H. sapiens* was slightly bigger, averaging between 63 and 67 kg, yielding an equivalent EQ of 5.3 (see Table 13.5). As noted above, estimates of body mass for the slightly smaller-brained *H. heidelbergensis* range from 65 to

71 kg (Ruff et al., 1997; Arsuaga et al., 1997), yielding estimated EQs between 4.3 and 4.7. Put together, these data indicate a shift in relative brain size in *H. sapiens,* both early and recent, in which brain size is slightly but significantly larger relative to body size than in archaic *Homo.* This grade shift is evident in the higher intercept of the slope between brain size and body size in Figure 13.12b.

What processes made *H. sapiens* brains relatively larger is unknown, but we have several clues. First, as discussed in Chapter 6, the volume of the frontal lobe is proportionately the same relative to brain size in humans and apes (despite increases in the prefrontal cortex), but the relative size of the temporal lobe is as much as 25 percent greater in humans (Semendeferi, 2001; Rilling and Seligman, 2002; Schenker et al., 2005). Temporal expansion is probably due to more white matter in the lateral and inferior parts of the lobe (Rilling, 2006). It is impossible to be definitive about when the temporal lobe became bigger, but several analyses show that the middle cranial fossa, which houses the temporal lobes, is about 20 percent wider and longer in *H. sapiens* than in other species of *Homo* (Lieberman et al., 2002, 2004b; Bastir et al., 2008; see also Maier and Nkini, 1984; Seidler et al., 1997; Baba et al., 2003). These analyses raise the possibility that a unique derived feature of *H. sapiens* is a relatively larger temporal lobe, possibly in regions implicated in language. Human temporal lobes, however, are 200–240 cm^3, approximately 16 percent of total brain volume (Semendeferi, 2001), so a bigger temporal lobe can account for only a portion of modern human brain-size increases. Comparisons of endocasts also suggest that parts of the parietal lobe may be relatively larger in *H. sapiens* (Bruner et al., 2003; Holloway et al., 2003). More data are needed to figure out what exactly is enlarged and by how much in the modern human parietal lobe, not to mention other lobes, relative to those of apes and other hominins. Because the temporal and parietal lobes are above and to the side of the central portion of the cranial base, their expansion might contribute to a more flexed cranial base (see below).

Face Size

Another distinctive aspect of the *H. sapiens* head is a smaller face. Figure 13.13 compares facial size and dimensions in recent and early *H. sapiens, H. heidelbergensis,* and *H. neanderthalensis.* There is much variation, particularly in Pleistocene *H. sapiens* (in which there are some very large, robust males), but note that overall facial size, quantified as a geometric mean, is 15–25 percent smaller in *H. sapiens* than in taxa of archaic *Homo,* both absolutely and relative to brain size. This diminution results mostly from the combination of an anteroposteriorly and superinferiorly shorter face (Lieberman, 1998; Bastir et al., 2007). A smaller face gives modern humans less projecting faces, more rectangular orbits (similarly wide but not as tall), and vertically shorter infraorbital and subnasal regions (see Figure 13.16).

The developmental bases of facial diminution in *H. sapiens* are unknown. Probably there are several contributing factors. One might be smaller condensations of mesenchyme (e.g., maxillary and mandibular arches, the frontonasal prominence) that initially form the embryonic face (see Chapter 3). In addition, modern human faces may grow less in sutures and in growth fields along the posterior margin of the face through diminished response to hormones such as growth hormone or from other mechanisms that influence osteoblast activity (see Chapter 2). The premaxillary suture in *H. sapiens,* for example, apparently fuses earlier in ontogeny than it does in *H. neanderthalensis* (Maureille and Bar, 1999). In addition, modern humans may have higher rates of resorption along the anterior surface of the face. Finally, changes in diet must have had some epigenetic effect, as evinced by substantial (approximately 6 percent) decreases in facial length over the last few hundred years, as foods have become softer and more processed (see Chapter 7).

Cranial Base Angle

As Chapter 5 discussed, CBA1 is an important influence on overall cranial shape. CBA1, in turn, is affected by several factors, including brain volume relative to cranial base length, and face length relative to anterior cranial base length (Ross and Ravosa, 1993; Lieberman et al., 2008a). Both factors conspire to flex the *H. sapiens* cranial base. Modern humans have not only shorter faces but also slightly larger brains, particularly in the middle cranial fossa around the spheno-occipital synchondrosis, the major site of cranial base flexion (see Chapter 4). CBA1 averages $134 \pm 3.1°$ in recent *H. sapiens* (Lieberman and McCarthy, 1999). Most Pleistocene *Homo* fossils have damaged cranial bases, but we have enough good fossils (see Table 13.5 and Figure 13.17) to think that the cranial base was as flexed in Pleistocene *H. sapiens* as in recent humans (the average is 133°), but about 10°–15° more flexed than in *H. heidelbergensis s.l.* and *H. neanderthalensis.*

Facial Retraction

In all other hominins, a substantial portion of the face projects in front of the anterior cranial fossa and the frontal lobe, but in modern humans the face is

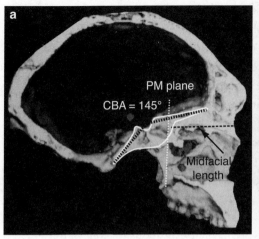
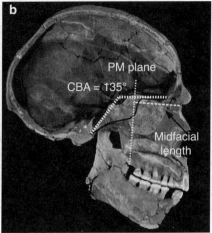

Figure 13.17. Effects of cranial base angle (CBA1), facial length, and position of the posterior maxillary (PM) plane on facial projection and orientation in two representative crania: (a) Kabwe (*H. heidelbergensis s.l.*), and (b) Skhul V (*H. sapiens*). As discussed in the text, Kabwe's cranial base is approximately 10° more extended, and the face is relatively longer; both features contribute to a greater degree of facial projection, longer brow ridges, and a less vertical frontal squama. CT scan of Kabwe courtesy Natural History Museum, U.K.; CT scan of Skhul V courtesy Peabody Museum, Harvard University.

tucked almost entirely below the frontal lobe. Several interrelated factors, illustrated in Figure 13.17 for a comparison of Kabwe and Skhul V (see also Figure 5.9), lead to facial retraction. The first is an anteroposteriorly shorter face. Because the face grows inferiorly from the anterior cranial base and displaces forward from the middle cranial fossa, a longer face will project more in front of the anterior cranial fossa, bringing with it the outer table of the frontal (see Chapters 4 and 5). Facial projection correlates strongly with brow ridge length, which explains why many early *H. sapiens* crania (e.g., Skhul V) with anteroposteriorly longer faces compared to recent humans also have longer brow ridges. A second factor is a more flexed cranial base (itself affected by facial length and brain size). As outlined in Chapter 5, the roof of the orbits is the floor of the anterior cranial fossa, and the back of the face is always perpendicular to the neutral horizontal axis (NHA) of the orbits (Enlow, 1990). These constraints cause the face to rotate as a "block" along with the anterior cranial base; thus a more flexed CBA1 rotates the entire face ventrally relative to the rest of the skull, tucking more of the face beneath the anterior cranial fossa and shortening the length of the pharynx behind the palate (Mc-

Carthy and Lieberman, 2001). The average anteroposterior length of the oropharynx behind the palate (approximated by the distance from basion to staphylion) is approximately 1 cm shorter in *H. sapiens* than in archaic *Homo.*

Neurocranial Sphericity

The vault in both recent and Pleistocene modern humans is significantly more rounded than in archaic *Homo* (Lieberman et al., 2002). Two factors that might influence vault sphericity are relatively larger temporal and parietal lobes (see above). Because the vault grows around the brain largely in response to forces exerted by the brain itself, a more globular brain leads to more globular vault. Also, as Chapter 5 reviewed, vault sphericity is partly influenced by the volume of the brain relative to the size and shape of the cranial base platform. Brain size alone is not responsible for the more flexed cranial base in *H. sapiens* (otherwise Neanderthals would have an equally flexed CBA1), but the more flexed cranial base in modern humans combined with a large brain do lead to a rounder, more spherical vault (Lieberman et al., 2004b).

Chin

The most distinctive derived feature of the *H. sapiens* mandible is the presence of a chin, defined as an upside-down-T-shaped protuberance along the anterior, lower margin of the symphysis. The inverted T creates two hollows on either side (mental fossae), which are present prior to birth (Buschang et al., 1992; Schwartz and Tattersall, 2000). The functional basis of the chin is unclear, but its developmental origin stems from the unique presence in modern humans of a resorptive growth field on the anterior margin of the symphysis below the alveolar crest (Duterloo and Enlow, 1970; Bromage, 1990). In all other extant primates and presumably hominins, this field is depository (although this hypothesis needs testing; see Figure 4.14).

Other Features

Modern humans have a few additional distinct features whose developmental bases are poorly understood. Among them is a distinct canine fossa—a vertical depression on the maxilla above the canine root and between the nasal aperture and the malar surface. Canine fossa depth is independent of variation in facial retraction and sphericity and may be associated with more retracted zygomatic bones or a smaller maxillary sinus (Lieberman et al., 2002). For additional proposed derived features of *H. sapiens,* see Tattersall and Schwartz (2008).

Late *Homo:* Functional and Behavioral Changes

How might the heads of species of late *Homo* differed functionally? And how might any differences in function relate to differences in behavior? As reviewed above, there is a trend within *H. erectus s.l.* toward bigger brains and faces, presumably from bigger bodies; we also see in *H. erectus s.l.* an increase in overall cranial robusticity (mostly in males) combined with a trend toward smaller teeth. *H. heidelbergensis* and *H. neanderthalensis* crania are generally *H. erectus*–like but with a slight increase in face size and relative brain size. In contrast, *H. sapiens* represents a shift in craniofacial architecture, with a retracted, smaller face, and a more spherical cranial vault. It is hard to be definitive, but these and other derived features in *H. sapiens* suggest some functional differences in cognition, diet, respiration, locomotion, and speech capabilities. Not all of these changes were adaptations, but some may be related to a suite of behaviors argued to distinguish modern humans from other hominins (a loose category of traits often termed *behavioral modernity*). These behaviors potentially include eating a more diverse diet; a greater reliance on cooked or otherwise processed food; formation of more complex social networks, with longer-distance exchange of goods; and perhaps greater creativity. It is reasonable to hypothesize (but hard to prove) that these differences helped give modern humans a competitive advantage over other taxa of late *Homo*.

Cognition, Ontogeny, and Language

Few would question the inference that late *Homo* got smarter than early *Homo*. But was this transition gradual or punctuated, and was *H. sapiens* cognitively different from other species of late *Homo*? One line of evidence is brain size. As reviewed above, *H. heidelbergensis s.l., H. neanderthalensis,* and *H. sapiens* have absolutely and relatively bigger brains than *H. erectus*. Brain tissue is metabolically costly, so the benefits of a large brain must have outweighed the costs to synthesize and maintain additional brain tissue. An adult *H. habilis* would have needed 150–200 kcal/day to feed a 650 g brain, and an early *H. erectus* adult would have needed 230–380 kcal/day to feed a 800 g brain; a Neanderthal or *H. sapiens* brain costs about 400–650 kcal a day (see Chapter 6). Brains also require high-quality nutrients, such as essential fatty acids, that are costly to synthesize or difficult to acquire. And large infant brains pose serious dangers during birth.

Did the larger brains of late *Homo* take longer to grow, and how are rates of brain growth related to cognitive abilities? One line of thinking is that bigger brains slow down the rate of ontogeny because they require more energy and scarce resources. This hypothesis predicts that big-brained species of late *Homo* do not deviate much from the strong correlation documented among extant primate species between brain size and ontogenetic rates (B. H. Smith, 1989; see Chapter 6). If so, then Neanderthals would have erupted their M1s at about 6 years of age, as modern humans do (M1 eruption coincides with 95 percent completion of brain growth). Another hypothesis is that the metabolic and reproductive costs of slowing ontogeny are outweighed by the cognitive benefits of growing a bigger brain over a longer period. A more prolonged neural ontogeny might lower the cost of development per unit of time (Hublin and Coqueugniot, 2006) and extend the period of development during which neuronal synapses are wired, potentially making possible more complex cognition (Parker, 2005; R. K. Walker et al., 2006a,b). According to this line of reasoning, some of the proposed cognitive differences among species of *Homo* may derive from variations in the rate and duration of ontogeny, with modern humans taking longer to grow their brains than Neanderthals. Keep in mind, however, that the brain structure continues to change through adolescence, long after it finishes increasing in size (see Kandel et al., 2000).

Different lines of evidence yield conflicting results about how long it took the brain to attain adult size in different species of *Homo*. The controversy starts with *H. erectus*. One estimate comes from an analysis of an infant *H. erectus* cranium from Mojokerto that may date to 1.8 mya. Coqueugniot and colleagues (2004) estimated Mojokerto to have died approximately a year after birth, based on cranial indicators of maturation such as the degree of suture closure. Leaving aside the problem of accurately estimating the fossil's age of death (published estimates range from 0.5 to 6 years), Mojokerto's ECV is 663 cm^3, approximately 72 percent the average adult ECV of Asian *H. erectus* (921 cm^3). If Mojokerto died at the age of 1.0, then it had completed proportionately more brain growth than a similarly aged modern human, in which ECV averages approximately 50 percent of adult volume (Figure 13.18a).

Proportional measures of brain growth may not be a reliable indicator of how long it took the brain to grow, because of errors in estimating neonatal brain size and because of ontogenetic changes in growth rates (Leigh, 2004, 2006; DeSilva and Lesnik, 2006). If one graphs *absolute* brain size against postnatal age (Figure 13.18b), then the Mojokerto child falls on an inter-

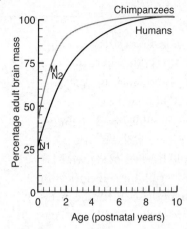

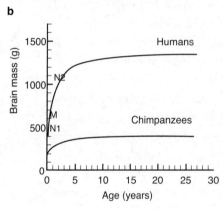

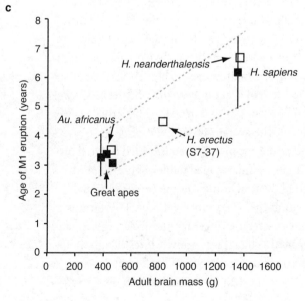

Figure 13.18. Endocranial volume (ECV) ontogeny in modern humans, chimpanzees, and late *Homo*: (a) as percentage of adult ECV relative to age; (b) absolute ECV relative to age; (c) age of M1 eruption relative to brain mass (lines indicate recent modern human range of eruption variation). The Modjokerto infant (M), here assumed to be 1 year old, falls closer to the chimpanzee line when percentage adult ECV is plotted against age but closer to the modern human line when absolute ECV is plotted relative to age. Neanderthal infants from Mesmaiskaya (N1, perinatal), and Dederiyeh (N2, 1.5–2 years old by modern human standards) fall close to the modern human line for both absolute and relative volume. (Data from Coqueugniot et al., 2004; Leigh, 2004; Ponce de León et al., 2008.) As (c) shows, the estimates of M1 eruption in australopiths and one *H. erectus* (from Sangiran) are consistent with the relationship between the age of M1 eruption and adult brain size. (Data from Aiello and Dean, 1990; Dean et al., 2001.) The one Neanderthal estimate is a juvenile from the site of La Chaise (Macchiarelli et al., 2006); however, other estimates of Neanderthal tooth eruption suggest a faster rate of dental ontogeny (T. Smith et al., 2007b).

mediate trajectory of brain growth predictable from its brain size: more rapid than in chimpanzees and australopiths, but slightly slower than in modern humans. A secure estimate of this fossil's age of death is necessary to resolve the issue, but other studies of *H. erectus* fossils suggest that this species had a rate of brain ontogeny that was faster than that of modern humans, in line with a smaller brain. Dean and colleagues (2001) used incremental markings to estimate the age of M1 eruption (hence the age its brain reached 95 percent adult size) in an *H. erectus* tooth from Java (S7–37) at close to 4.5 years. This estimate falls close to the best-fit line in mammals between ECV and the age of brain completion (Figure 13.18c). Another fossil that indicates an intermediate rate of ontogeny in *H. erectus* is the Nariokotome boy (KNM-WT 15000), who was probably about 8 or 9 years old when he died, but whose age by modern human standards would be about 10–12 years based on tooth eruption and 13–15 years based on skeletal maturation (Dean and Smith, 2009). Finally, neonatal brain size in *H. erectus* is estimated to be between 270 and 320 g (Simpson et al., 2008; DeSilva and Lesnik, 2008), intermediate between modern humans and chimps, whose brains at birth are approximately 30 percent and 40 percent of adult brain size, respectively (DeSilva and Lesnik, 2006).

The ontogenetic trajectory of Neanderthals, whose adult brain was as large as that of an early *H. sapiens,* is also unresolved. Given that neonatal brain size in Neanderthals was approximately 400 cm^3 (Ponce de León et al., 2008), within the recent human range of 380–420 cm^3, did Neanderthals also have a slow, modern human–like rate of brain growth? Direct evidence from dental aging provides conflicting results, in part because of much variation in dental maturation rates within modern humans and presumably also in Neanderthals. One study of incremental growth lines in Neanderthal teeth, from the French site of La Chaise, estimated M1 eruption (hence the age of 95 percent adult brain completion) at 6.7 years, at the high end of the *H. sapiens* range (Macchiarelli et al., 2006). But analysis of a Neanderthal from the Belgian site of Sclayn suggests a relatively rapid rate of ontogeny, with an age of M2 eruption at 8 years, at the low end of the *H. sapiens* range of maturation rates (T. Smith et al., 2007b). The latter finding accords with other analyses of enamel crown formation times (Dean et al., 1986; Stringer et al., 1990; Tompkins, 1996; Guatelli-Steinberg et al., 2005; Ramirez Rozzi and Bermudez de Castro, 2004). The balance of evidence suggests that the ontogenetic trajectory of Neanderthals was probably at the low end of the modern human range of variation, but how much they were shifted to the end of the range may be unresolvable given intraspecies variation for M1 eruption and other markers of developmental age. As Figure 13.18c shows, age at M1 eruption ranges from 4.7 to 7.0 years in modern humans (Liversidge, 2003) and from 2.7 to 4.1 years in chimps (Conroy and Mahoney, 1991; Zihlman et al., 2004).

Another source of data on cognition is brain shape. As reviewed above, there are some indications that *H. sapiens* brains may differ in several ways from those of other species of late *Homo,* including a larger prefrontal cortex and relatively larger temporal and parietal lobes (Lieberman et al., 2002; Bruner et al., 2003; Bastir et al., 2008). These expansions require further study, and efforts to make inferences about cognitive changes from data on brain shape are tentative at best. But if correct, they raise some interesting hypotheses. As Chapter 6 discussed, the temporal lobe participates in many aspects of complex cognitive function, among them organizing sensory inputs and memory (e.g., for words, images, sounds, and smells). The temporal lobes thus play key roles in language and other tasks that involve sensory perception and processing (Kandel et al., 2000). The prefrontal cortex is important for a number of related, abstract cognitive functions, such as language, reasoning, planning, and interpreting social interactions (Wood and Grafman, 2003). One can tentatively hypothesize that selection for language or related cognitive abilities played some role in the evolution of *H. sapiens.*

Different challenges hamper interpretations of the archaeological record, our only "direct" source of evidence on cognition. Because hominins invent tool industries, not vice versa, one expects technological changes to follow rather than coincide with speciation events if species differed in cognitive abilities. But absence of evidence for cognitive skills is hardly evidence for absence of those capabilities. Just as farming and computers weren't invented by humans who suddenly evolved novel cognitive capacities, one cannot infer that the invention of stone-tool industries such as the Upper Paleolithic occurred because of some evolutionary shift in brain function (for alternative views, see Klein, 1995; Bar-Yosef, 1998). Even so, correlations between archaeological transitions and the first appearance of species of *Homo* merit some consideration. The first of these correlations, the origins of *Homo* by 2.3 mya and of the Oldowan by 2.6 mya is too poorly sampled to be reliable (see Chapter 12), but the oldest dated *H. erectus* fossils (ca. 1.9 mya) do appear slightly before the first Acheulian and the Developed Oldowan (ca. 1.7 mya). Although Acheulian assemblages are not abundant in East or Southeast Asia, *H. erectus* in Asia did have the ability to make handaxes (Yamei et al., 2000).

Early *H. heidelbergensis s.l.* is found with both the Acheulian and Developed Oldowan tool industries, and thus this species' origins cannot be associated with any major change in technology. But *H. heidelbergensis* apparently invented prepared-core technologies that were the basis for the Middle Paleolithic/ Middle Stone Age sometime between 300 and 500 kya, probably in Africa. In this kind of technology, the toolmaker first shapes the surface of a stone core in order to knock off thin flakes of predetermined size and shape (see Figure 13.19). Making prepared cores requires much planning and skill, but they help hunters to make thin, triangular spearheads, which are far more lethal than untipped spears (J. J. Shea, 2006). Prepared cores also facilitate the fabrication of other types of tools (e.g., scrapers and burins). In addition, archaeological evidence suggests that Middle Paleolithic/Middle Stone Age hominins engaged in more long-distance exchanges of raw material (McBrearty and Brooks, 2000) and that they regularly used hearths to control fire for cooking (James, 1989; Rowlett, 2000). One hypothesis to consider is that *H. heidelbergensis* invented prepared core technologies in Africa and that some populations subsequently dispersed into Europe, where they evolved into the Neanderthals (perhaps interbreeding with local populations of archaic *Homo*). Such a scenario accords with archaeological evidence for Levallois technology (an early technique for making prepared cores), which first appears in Eurasia between 400 and 300 kya (Santonja and Villa, 2006), and with genetic evidence for a shared common ancestor of *H. neanderthalensis* and *H. sapiens* about 600–300 kya (see above).

By 250 kya, MSA/MP stone-tool industries were common to all large-brained species of *Homo* in Africa, Europe, and western Asia. There is evidence, however, for changes in Africa within the MSA that may correspond uniquely with the evolution of *H. sapiens*. According to McBrearty and Brooks (2000), there is a gradual, albeit patchy, accrual in the African MSA of many quintessentially "modern" human behaviors such as long-distance exchange, shellfishing, the systematic processing of pigments, and the fabrication of new kinds of tools, including bone points (see also Henshilwood, 2006). These innovations, which are less prevalent in the MP of Eurasia, suggest a modern, more creative type of thinking that has led to, and still generates, an accelerating pace of innovation. One of these innovations was the Upper Paleolithic (UP), which appears to have been invented by *H. sapiens* about 50–40 kya in either the Levant or North Africa (Bar-Yosef, 2000). UP stone tool–making technology is based on a new way of making cores that are prismatic in shape,

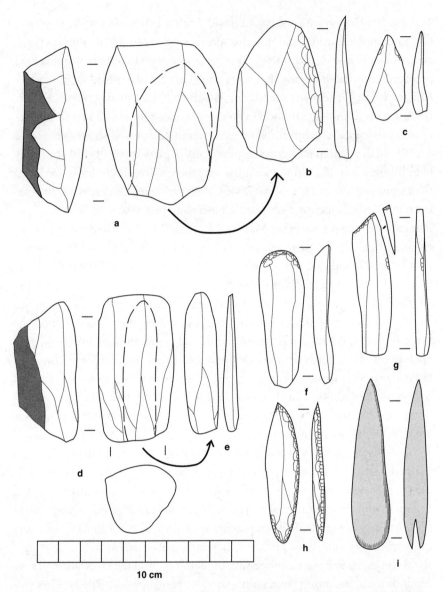

Figure 13.19. Comparison of Middle and Upper Paleolithic technology and tools showing different views. Middle Paleolithic tools shown include a Levallois core (a), which could be used to create the blank for a sidescraper flake (b) or a Levallois point (c). Upper Paleolithic tools were made from prismatic blade cores (d) to produce blades (e) that could be fashioned into tools such as endscrapers (f), burins (g), and points (h); Upper Paleolithic toolmakers also fashioned split-based points from antlers (i). Drawings courtesy J. Shea.

facilitating the mass production of long, sharp stone blades with lots of edge that were then transformed into many distinctive shapes (e.g., points, burins, scrapers, and knives). UP hunter-gatherers also made a wide range of tools from bone, ivory, and antler; they often produced art; and they hunted and gathered a much broader range of resources, including small animals, fish, and birds (Stiner et al., 1999, 2000; Mellars, 2004; Klein, 2009).

Once invented, the UP—and its *H. sapiens* makers—quickly spread throughout the Old World. UP industries first appear in western Asia sometime between 50 and 40 kya; from there they spread to Europe about 40–35 kya and to North Asia by 35 kya. Whether *H. neanderthalensis* was able to make UP tools is open to question. According to some archaeologists, the last Neanderthals in Western Europe (35–29 kya) made a kind of imitation UP industry, the Chatelperronian. But there are differences of opinion over what the Chatelperronian is and who made it (Hublin et al., 1996; Bar-Yosef, 2006; Zilhão et al., 2006; Mellars, et al., 2007).

Altogether, multiple lines of evidence suggest what one might expect: as brains got bigger in *Homo,* cognitive abilities became more sophisticated. Although early *H. erectus* never seems to have progressed beyond a chopper-chopping tool industry, larger-brained taxa of late *Homo,* especially *H. sapiens* and *H. neanderthalensis,* made extremely sophisticated prepared-core tools and presumably were cognitively complex. But only *H. sapiens* seems to have had the ability or proclivity to invent new technologies and cultural styles such as the UP. To what extent this shift was made possible by novel neural adaptations (Klein, 1995) or by accidents of culture and environment akin to those responsible for the invention of agriculture will endure as a topic of much speculation.

Speech Abilities in Late *Homo*

Chapter 8 reviewed the functional anatomy of speech and the elusive evidence for the evolution of speech capabilities, but several issues relevant to this perennially contentious topic are worth revisiting, with two caveats. First, speech is just one component of language, so an ability to produce slightly more articulate speech does not necessarily imply greater linguistic capacities (or vice versa). Second, we must be wary of unfalsifiable null hypotheses. Some scholars assume that language and fully modern speech capabilities are unique to *H. sapiens* unless proved otherwise; others assume that large-brained hominins such as Neanderthals possessed modern human–like language abilities unless proved otherwise. Because neither speech nor the tissues that produce it fossilize, both null hypotheses are impossible to reject. That said, it is reasonable to suppose that species of late *Homo* such as *H. neanderthalensis* had some sort of complex language, given their large brains and the sophisticated tools that they produced.

With these caveats in mind, what can we infer from skulls about the speech (not language) capabilities of late *Homo*? As discussed in Chapter 8, a key attribute of articulate speech in humans is that errors of production or perception are minimized by using the tongue to modify independently the cross-sectional areas of the vertical and horizontal portions of the supralaryngeal vocal tract (SVT_v and SVT_h). When the SVT_v and SVT_h are equal in length, a speaker can produce more phonetically discrete (quantal) vowels with greater accuracy and with less need for precision (Peterson and Barney, 1952; Stevens, 1972, 1989; P. Lieberman, 2006). This configuration is made possible in modern humans by a short, round tongue that positions the hyoid and larynx low in the neck, setting up a nonintranarial larynx (the soft palate and epiglottis do not make contact in the resting position). It follows that a key parameter for assessing speech capabilities in different species of *Homo* is the ratio of SVT_v to SVT_h. Estimating this ratio, however, is easier said than done. Although SVT_h (the distance from the lips to the back of the oropharynx) can be measured in fossils from the length of the palate and the oropharynx behind the palate (and adding 1 cm for the lips), SVT_v (the distance from the soft palate to the vocal folds) is determined mostly by soft tissues and cannot be estimated precisely from skeletal features such as the angle of the cranial base, hyoid shape, or mandibular dimensions (Lieberman and McCarthy, 1999). As illustrated in Figure 8.15, it is possible to reconstruct a *H. erectus s.l.* head with a high or low larynx, depending on whether tongue size scales isometrically with body mass.

In spite of uncertainties about larynx position and SVT dimensions in archaic *Homo,* one can hypothesize that two derived aspects of *H. sapiens* might have improved performance capabilities for producing more quantal vowels. First, because the modern human face is shorter, the modern human oral cavity (palate length) is on average 57 ± 5.1 mm, shorter by about 10 mm than in *H. heidelbergensis* (68.3 ± 5.1 mm) and *H. neanderthalensis* (62.3 ± 6.5 mm). Second, the *H. sapiens* cranial base is approximately 10–15° more flexed, thus rotating the whole face, including the back of the palate, ventrally underneath the anterior cranial base, and shortening the pharyngeal space behind the palate by about one centimeter (McCarthy and Lieberman, 2001).

These two shifts combine to shorten by approximately 10 percent the skeletal length of the SVT_h (prosthion-basion) in Pleistocene *H. sapiens* (10.5 cm) compared to *H. neanderthalensis* (11.7 cm) or *H. heidelbergensis* (11.8 cm). Because the lips add 1 cm to SVT_h, we can infer that to have a modern SVT (with an SVT_h/SVT_v ratio of 1:1), an archaic *Homo* would have needed an STV_v of about 13 cm, 2–3 cm longer than an average adult modern human's. Such an SVT would yield a rich baritone voice but presents some anatomical problems. Whereas the hyoid in an average modern human lies 2–3 cm below the lower margin of the mandible, about the level of the third and fourth cervical vertebrae (Haralabakis et al., 1993; Lieberman et al., 2001), an archaic *Homo* hyoid would have had to be 2 cm lower to have a 1:1 ratio of SVT_h/SVT_v. Because the vertical dimensions of the cervical vertebrae in archaic *Homo* and *H. sapiens* are similar (Stewart, 1962; Trinkaus, 1983; Gómez-Olivencia et al., 2007) we can infer that a modern SVT configuration in an archaic *Homo* would have required the hyoid to be at about the level of the fourth to the sixth cervical vertebrae. The larynx is suspended another 2 cm below the hyoid, so this configuration would position the larynx extremely low, perhaps within the thoracic cavity. Such a position could prevent infrahyoid muscles such as the sternothyroid from depressing the hyolaryngeal complex, thus making swallowing either impossible or different from all other mammals.

A longer SVT_h, does not rule out the possibility that archaic *Homo* could speak or had sophisticated language, but it does suggest slightly less articulate (quantal) speech, perhaps comparable to a 4–6-year-old modern human's. Given constraints on larynx position in the neck, combined with the risks of asphyxiation that presumably increase with greater separation of the epiglottis and soft palate, one can merely speculate that the derived, shorter face of modern humans might have been selected for speech performance. To test this hypothesis more completely, however, we need better indicators of the potential range of hyoid and larynx position in fossil crania and their effects on speech production (e.g., Lieberman and McCarthy, 2007). Further, our understanding of speech evolution is likely to improve as we learn more about the neurological and genetic bases for language. So far, most attention has focused on one gene, *Foxp2* (a transcription factor). *Foxp2* appears to have been selected for language ability: the gene is derived in humans relative to chimps, and both humans and mice with a mutated form of the gene have difficulties with the motor control of vocalization (Enard et al., 2002; Enard et al., 2009). Interestingly, Neanderthals might have a humanlike version of *Foxp2* (Krause et al., 2007). We can look forward to a better understanding of language evolution as we discover and study other genes involved in the behavior.

Chewing and Cooking

Masticatory evolution in late *Homo* presents a slight paradox: although there is a modest trend toward more efficient bite force production, the ability to generate and withstand high bite forces declines. Figure 13.20 compares the mechanical advantage (MA) of the temporalis and masseter muscles in the sagittal plane for representative skulls of *H. erectus*, *H. heidelbergensis* (a composite), *H. neanderthalensis,* and Pleistocene *H. sapiens,* along with species averages (males only) computed from large samples. The estimates show that MAs have remained broadly similar over the Pleistocene, with slightly higher MAs in *H. sapiens* because of anteroposteriorly shorter faces that decrease the length of the masseter and temporalis load arms. The biggest differences occur after the Pleistocene, as faces become smaller, yielding significantly higher MAs in recent humans. Narrower faces also give *H. sapiens,* especially recent humans, a slightly higher MA for the masseter in the transverse plane (not shown).

A higher MA in recent *H. sapiens* doesn't necessarily mean higher bite forces. As described in Chapter 7, one can estimate the maximum force a chewing muscle generates from estimates of its principal cross-sectional area (the muscle's cross sectional area corrected for fiber angle, giving the total number of muscle fibers), its MA, and an average maximum force production of 35 N/cm^2 (Zajac, 1989). As shown by O'Connor et al. (2005), summing the force produced by the major adductors allows one to estimate maximum bite forces not only in humans and other primates but also in fossil skulls. Such estimates are summarized for modern humans and other species of late *Homo* in Table 13.6 for incisal bites and first molar chews. The estimates indicate that maximum masticatory forces and pressures (bite force/molar area) in recent humans are approximately 20–25 percent lower than in archaic *Homo,* but that bite forces in Pleistocene modern humans were about as high as in archaic *Homo.* In other words, maximum bite force production in humans has declined in concert with reductions in cranial size (O'Connor et al., 2005), primarily over the last few thousand years, as overall craniofacial robusticity has decreased. This inference corresponds with an approximate 1 percent decrease in molar crown area every 1,000 years during the Pleistocene (Brace et al., 1991).

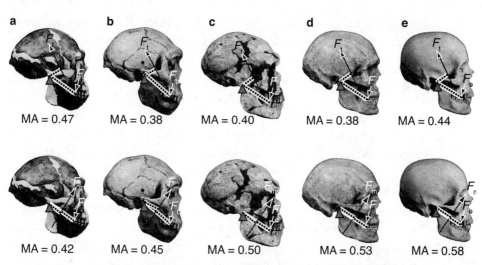

Figure 13.20. Mechanical advantages (MA) of the temporalis (top) and masseter (bottom) muscles in sagittal view for representative late *Homo* crania: (a) Zhoukoudian reconstruction (*H. erectus*); (b) Kabwe (with the Mauer mandible, both *H. heidelbergensis*); (c) La Ferrassie 1 (*H. neanderthalensis*); (d) Zhoukoudian 101 (Pleistocene *H. sapiens*); (e) recent Holocene *H. sapiens*. MAs do not differ greatly among these specimens but tend to be relatively high in recent modern humans because of facial shortening. However, other factors influence bite force production capability, and estimated absolute bite forces are lower in recent humans because of smaller muscle cross-sectional areas.

The counterpart of bite force production is the face's ability to resist masticatory stresses, which can be high in magnitude and very repetitive (recall that chimps spend about half the day feeding). As noted in Chapter 7, the human face—both archaic and modern—is likely subjected to twisting stresses around an anteroposterior axis, combined with shearing stresses in the coro-

Table 13.6: Maximum bite force estimates in late *Homo* (males only)

Taxon	Sample size	M1 force (N)	I1 force (N)	M1 force/molar area (N/cm²)
H. sapiens (recent)	11	1041	616	3.8
H. sapiens (Pleistocene)	6	1365	553	4.8
H. neanderthalensis	4	1275	550	4.7
H. erectus s.l.	3	1213	745	4.3
H. heidelbergensis	3	1106	545	4.0

nal and sagittal planes. In addition, the strains that the face experiences are almost certainly highest near the occlusal plane and at sites of muscle attachments (e.g., the zygomatic arch), and then diminish away from these regions, with only tiny magnitudes in the upper face and the braincase. Both archaic and modern human faces probably resist such stresses and strains in a similar manner, by having a tall, wide, and flat face in which most of the face's mass is distributed in the coronal plane (see Ross, 2001). In this respect, the larger (mostly taller) face of archaic *Homo* would have been better adapted than the smaller (mostly shorter) face of both early *H. erectus* and *H. sapiens* to withstand more or higher-magnitude strains from twisting and shearing. For example, the superoinferior height of the malar (infraorbital) region is approximately 30 percent smaller in Pleistocene *H. sapiens* than in archaic *Homo,* with further decreases occurring in the last 12,000 years (see Figures 13.7 and 7.21).

The role of anterior bite forces in the Neanderthal face merits special attention. Neanderthals have large incisors with robust roots, and the crowns are typically worn heavily on the labial (front) side as if they were used as tools to grip skins and other objects (Coon, 1962; Brace, 1962; Trinkaus, 1983; Smith and Paquette, 1989). Spencer and Demes (1993) found that the likely vector of masseter force in Neanderthals has an especially oblique (anterior) orientation, increasing the muscle's moment arm and hence its efficiency for producing high incisal bite forces. Other studies that have attempted to model bite forces by incorporating estimates of muscle cross-sectional areas have found that maximum Neanderthal incisal bite force magnitudes, although high, were probably not higher than those of other archaic *Homo* or robust *H. sapiens* (Antón, 1994; O'Connor et al., 2005; see also Table 13.6). It follows that heavy wear on Neanderthal incisors, combined with frequent signs of degenerative joint disease in their TMJs, probably resulted from frequent, repetitive loading rather than from unusually high magnitudes of loading. Evidence that Neanderthals used their anterior teeth heavily accords with the unique, projecting shape of the Neanderthal midface (Demes, 1987). Whatever the cause, the convexity of the Neanderthal midface in the sagittal plane (see above) would have helped to resist sagittal shearing and bending forces from incisal bites; in addition, the somewhat tubular shape of the paranasal region would have helped resist twisting forces around an anteroposterior axis generated by unilateral chews on the canines, which are somewhat laterally positioned in Neanderthals.

Although archaic *Homo* taxa, including Neanderthals, may have had greater resistance than *H. sapiens* to high or repetitive chewing forces, it is a challenge to distinguish adaptations from spandrels in terms of facial resistance to masticatory loading. In Neanderthals, the expanded midface would have conferred increased resistance to anterior dental loading, but it is also possible that the Neanderthal midface, with its especially large nasal cavity, evolved via random drift or as an adaptation for thermoregulation in cold, arid climates (see below). More generally, larger faces in archaic *Homo* than in earlier *H. erectus* or in more recent *H. sapiens* may derive, at least partially, from scaling effects of body size (Ruff et al., 1997). Moreover, some proportion of recent facial diminution is probably environmental, because masticating softer, more processed foods generates less growth in the lower face (Carlson, 1976; Carlson and Van Gerven, 1977; Corruccini, 1999; Lieberman, et al., 2004a).

Regardless of what factors actually selected for smaller faces, it should be clear that size reduction in the *H. sapiens* face was made possible by decreasing the adaptive value of having a large face. More specifically, it is reasonable to hypothesize that cooking and other food-processing technologies released constraints on the need to have a tall, deep and wide face to resist high, frequent strains. As Chapter 7 reviews, cooking renders most foods considerably softer and more tender; further, processing techniques such as pounding and grinding reduce food particle size and toughness. As a result, humans who cook and process their food generate much less loading both in terms of the magnitude and number of cycles. But when and to what extent hominins began to cook is unclear. One possibility is that cooking was invented by early *H. erectus* (Wrangham et al., 1999). There is no convincing archaeological evidence, however, that cooking is so ancient. Undisputable traces of fire in hominin sites extend back only to 750 kya (Goren-Inbar et al., 2004), and the oldest definitive evidence of cooking comes from the Middle Paleolithic (James, 1989; Brace, 1995; Rowlett, 2000; Goldberg et al., 2001). Even though one cannot rule out the hypothesis that *H. erectus* cooked, certainly cooking intensified following the evolution of *H. heidelbergensis* and evidently became so important that it changed our biology in numerous and profound ways, ultimately making humans dependent on the behavior (Wrangham, 2009).

Cooking and food production also correspond to another trend in late *Homo*, dental reduction. Cooking would have substantially reduced any selective advantage for having large, thickly enameled tooth crowns. Recall from Chapter 7 that tooth crown area scales to food toughness to the power of $1/3$ (Lucas, 2004; see also Chapter 7). According to predictions of fracture scaling, the approximately 20 percent reduction of postcanine crown area in *H.*

erectus relative to *H. habilis* would have been made possible by a 50 percent reduction in food toughness. In addition, the 12 percent reduction in tooth size in *H. sapiens* compared to *H. erectus* would correspond to a further 30 percent reduction in toughness. Why these reductions occurred, however, is not so clear. Although chewing softer and more tender food reduces the need for big, thickly enameled teeth, the absence of selection for large teeth doesn't actually explain why teeth got smaller. One attempt to explain this phenomenon is the probable mutation effect (PME) hypothesis (Brace, 1964; Wolpoff, 1975), which posits that a relaxation of selection for large teeth allowed mutations to accrue that cause less enamel growth. A problem with the PME hypothesis is that absence of selection tends to increase variability rather than cause directional change (see also Calcagno and Gibson, 1988). A more plausible explanation is that smaller teeth were favored as a byproduct of an anteroposteriorly shorter face (see above) to prevent dental crowding and other problems of malocclusion. Alternatively or additionally, smaller permanent tooth crowns may derive from integrative mechanisms that help teeth fit properly into jaws (see Chapter 7).

Of course, cooking and food processing have other important benefits. Cooked food is more nutritious, keeps longer, and has fewer parasites and pathogens than raw food (Wrangham, 2009). Cooked, processed food also improves the ability to supplement infant diets. A typical, well-fed human mother in today's world of abundance cannot produce enough energy from breast milk to meet an infant's metabolic demands after 8–10 months, requiring her to supplement with solid food (Kuzawa, 1998). Further, human deciduous molars don't start erupting until after one year (the mean age at eruption of M1 is 16 months). So cooking and mechanically processing foods were probably necessary to enable human infants to grow and thrive, to lower the risk of gastrointestinal infections, and to help mothers lower interbirth intervals.

Given that a preference for cooked food would benefit one's fitness, it is tempting to make some speculations about why humans can derive so much sensory pleasure from the flavor of cooked food. One reason that cooked food often tastes better is that heat increases the number and complexity of odorants, which contribute to approximately 80 percent of perceived flavor via retronasal pathways. As argued in Chapter 10, humans may be uniquely adept at appreciating flavor because of several derived features, including a nonintranarial larynx, a vertically oriented pharynx, an expanded retronasal space, and a large neocortex (see also Shepherd, 2006). None of these features in and of themselves would necessarily have been selected for appreciating flavors, but they may have played some evolutionary role because hominins who were better able to perceive odorants might have differentially preferred cooked food and thus had higher fitness.

Respiration and Thermoregulation

A few aspects of cranial shape changes in late *Homo* suggest modified respiratory and thermoregulatory capabilities. Chapters 8 and 12 discussed how the expansion of the external nose and the absence of a snout in early *Homo* likely increased airflow turbulence, enhancing the ability to humidify and then dehumidify water during vigorous activity in hot, dry conditions. These features are further derived in different ways in late *Homo*, especially *H. neanderthalensis* and *H. sapiens.*

   H. neanderthalensis and *H. heidelbergensis,* like *H. erectus s.l.,* retain an anteroposteriorly long, tall, and vertical midface with a wide piriform aperture (Trinkaus, 2003). In Neanderthals, this breadth, almost 1 cm greater than in *H. sapiens,* partly stems from a prognathic face (Holton and Franciscus, 2008). In Neanderthals and in some *H. heidelbergensis,* the upper part of the nasal cavity is also wider, and the lateral margins of the nose (the frontal processes of the maxilla) are markedly everted. In addition, the inferior margins of the nasal bones in Neanderthals (and many *H. heidelbergensis*) are both absolutely and relatively wide (Franciscus, 1999). Together, these factors produce a more rectangular (rather than pear-shaped) piriform aperture and contribute to the slightly more parasagittal orientation of the paranasal and infraorbital regions (Rak, 1986; Trinkaus, 1987). Archaic *Homo* in general had large internal nasal cavities, and Neanderthals probably also had large external noses.

   Given the lack of well-preserved nasal cavities in fossil crania, it is impossible to be definitive about the functional significance of expanded noses in archaic *Homo,* especially in Neanderthals. One likely effect would have been to increase the relative surface area of the respiratory epithelium. As reviewed in Chapter 8, the surface area of the respiratory epithelium (both nasal and tracheal) relative to body mass is much lower in humans than in other mammals. This reduction may help humans dump heat during exhalation, but it also leads to greater vapor loss during exhalation and to less effective humidification and thermoregulation of inspired air. Recovering moisture during exhalation was probably a more important challenge for hominins in hot, arid habitats in Africa; humidifying and warming inspired air would have been more problematic in cold, arid habitats such as Europe, especially in the winter. Thus we can hypothesize that the tall, long, and wide (hence voluminous) nasal cavities of different taxa of archaic *Homo* would have been adaptive in more than one way. First, long nasal cavities would have increased the surface area of the nasal epithelium relative to body mass, primarily by lengthening the lower and middle meatuses. Second, wider nasal cavities would have permitted larger and more projecting inferior, middle and upper nasal conchae, thus increasing the relative surface area of the nasal epithelium within the inner nose (a hypothesis that remains difficult to test without preserved turbinates in fossil crania). Finally, because the piriform aperture is the approximate location of the nasal valve, a smaller nasal aperture might have helped increase turbulence during inspiration, thus increasing convective contact between air and the mucosa (Churchill et al., 2004; Yokley, 2006).

Several studies of recent populations lend some support to these hypotheses. People from the Artic tend to have larger nasal cavities than people from temperate or tropical regions (Shea, 1977), and Americans of European descent have significantly longer, narrower, and more projecting noses than Americans of African descent, giving them larger surface-area-to-volume ratios in the nasal cavity when the nasal mucosa are decongested (Yokley, 2009). Thus European noses may be more turbulent during inspiration and hence better adapted to warming and humidifying cold, dry air, especially during vigorous activity. Further research is needed on this topic, but we can tentatively conclude that archaic human noses were well adapted for vigorous activity levels, with Neanderthals being especially well adapted for the additional challenges of warming and humidifying inspired air in extremely cold conditions.

The *H. sapiens* pharynx is derived in a different way. As described above, the modern human face is shorter both superoinferiorly and anteroposteriorly, which means that the nasopharynx is also smaller. The breadth and height of the piriform aperture vary among human populations, but average nasal breadth is about 1 cm narrower in *H. sapiens* than in archaic *Homo* (Franciscus, 1995). Nasopharynx length in recent humans is also about 10–12 percent shorter than in archaic *Homo* (Lieberman, 1998). Consequently, if one crudely models the nasal cavity as a prism (nasal height × nasal breadth × nasal length × 0.5) and ignores differences in turbinate size, then total nasopharynx volume in recent and Pleistocene *H. sapiens* are approximately 29.1 ± 2.8 cm^3 and 32.8 ± 7.8 cm^3, respectively, about 50 percent smaller than in *H. neanderthalensis* and *H. heidelbergensis,* which have nasal volumes of about 52–63 cm^3 (analyses based on data from Franciscus, 1995). In addition, because of cranial base flexion and the facial block, the length of the oropharynx behind the palate in modern humans is shorter by about 1 cm. Altogether, if neck length has not changed much in the genus *Homo,* then the pharynx in *H. sapiens* is significantly shorter overall (possibly by about 12–15 percent) as well as narrower (about 20–25 percent) relative to body mass than in archaic *Homo.* Note, however, dimensions of the nasal cavity as a whole do not accurately predict the surface area and volume of the actual airways within the nose (Yokley, 2009).

Given their many functional roles and position in the middle of the highly integrated face, it would be unwise to assume that a smaller internal nose and a shorter pharynx were selected in *H. sapiens* as adaptations for respiration and thermoregulation, but we can deduce that the derived features of the human airway possibly affected these functions in several ways. First, during "normal" nasal respiration (i.e., not during vigorous exercise), airflow resistance in conditions of laminar flow might have been slightly higher in modern humans than in archaic *Homo* because of a shorter, narrower pharynx. The basis for this hypothesis is the Hagen-Poiseuille equation, which calculates the resistance of laminar flow, R, as

$$R = 8nl/\pi r^4$$

where n is the viscosity of the gas, l is the length of the tube, and r is the radius of the tube (see Chapter 8). Although shortening the pharynx would tend to decrease resistance, this effect would be more than offset by the effects of a narrower pharynx scaled to the power of 4. How much higher resistance might be in a modern than in an archaic nasopharynx is hard to quantify, but the difference would be most important during exercise. As the body's demand for oxygen goes up, flow rates and pressures increase severalfold, generating more turbulent flow with far higher resistance. More turbulent flow has important implications for heat and moisture exchange. During inspiration, the epithelial lining of the nasal cavity and pharynx (which is highly vascularized and covered by mucosa) functions to humidify and warm air to approximately 75–80 percent humidity and 37°C, respectively; during expiration, the same tissues recapture some of this heat and moisture. More turbulent flow lacks a boundary zone and causes circulating vortices of air to come into greater contact with the nasal epithelium, leading to more heat and moisture exchange. So the more constricted modern human nasopharynx might have been slightly more turbulent, with more resistance. During nasal breathing, this additional turbulence might have increased the efficiency of heat and moisture exchange.

And during very vigorous exercise, the modern human configuration might have increased the need for oral breathing, which increases the ability to dump heat but decreases the ability to conserve water.

A second effect of the narrower modern human nose, particularly among recent humans of European descent, would be to narrow the nasal valve, which is located near the junction between the external and internal nose. As described in Chapter 8, the nasal valve probably functions like a Venturi throat, increasing the turbulence of inspired air and thus increasing convective contact between inspired air and the nasal mucosa, which permits the nose to warm and humidify inspired air. So far, efforts to test this hypothesis have been inconclusive (Churchill et al., 2004).

Locomotion

A few derived aspects of the head in late *Homo* probably affected locomotor performance. As in *H. erectus,* heads are more balanced in archaic *Homo* than in earlier hominins, with high MAs for the superficial neck extensors. Even though some archaic *Homo* skulls have enormous faces, tending to move the center of gravity (COG) anteriorly, large brains, combined with less globular (more lemon-shaped) cranial vaults, help compensate for this added mass by positioning a substantial portion of the brain's mass behind the atlanto-occipital joint (AOJ). Even so, the head in *H. sapiens* is significantly more balanced than in any other species of *Homo,* largely because of a smaller, less projecting face (see Figure 13.21). The moment arm from the AOJ to the COM averages 2.5 cm ± 0.5 cm in Pleistocene *H. sapiens* skulls, approximately 40–80 percent shorter than in other species of *Homo,* including *H. heidelbergensis* (mean =3.6 cm), *H. neanderthalensis* (mean = 4.3 cm) and *H. erectus s.l.* (mean =4.5 cm). Recent modern humans with smaller faces have even more balanced heads, with moment arms of approximately 2.0 cm. More balanced heads correlate with higher MAs for the superficial neck extensors (approximately 1.8 in Pleistocene *H. sapiens* but 1.3 in archaic *Homo* and *H. erectus*).

As reviewed in Chapter 9, a nearly balanced head is unusual in mammals. Other primates, including apes, have more cantilevered heads that enable a wide range of motion and permit independent counterrotations of the trunk, neck, and head in order to stabilize the head, and hence gaze, during locomotion. When hominins initially became bipedal, the AOJ was rotated under the cranium (see Chapter 11). A vertical, centrally located neck decreases the energy cost of stabilizing the head but this configuration also restricts head

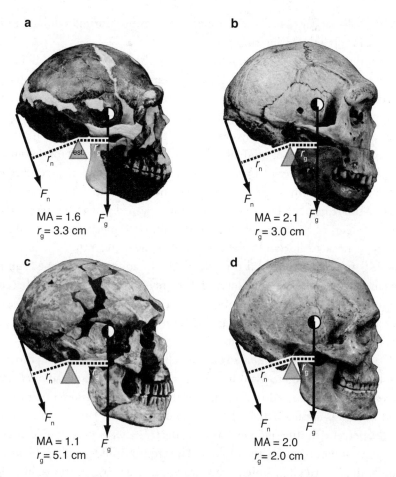

Figure 13.21. Biomechanics of head balance in late *Homo* in (a) Zhoukoudian reconstruction (*H. erectus*); (b) Kabwe (with the Mauer mandible, both *H. heidelbergensis*); (c) La Ferrassie 1 (*H. neanderthalensis*); (d) Zhoukoudian 101 (Pleistocene *H. sapiens*). Estimated force resultant and moment arm for the nuchal muscles are indicated by F_n and r_n; estimated gravitational force and moment arm for the head (measured to the area centroid of the skull, marked by a circle) are indicated by F_g and r_g; triangle shows location of atlanto-occipital joint (estimated in *H. erectus*). Below each skull are estimates of mechanical advantage (MA) for the superficial head extensors and r_g. Because of a smaller, more retracted face, r_g is much smaller in the *H. sapiens* skull, leading to a much lower moment of inertia in the pitch plane.

mobility and poses challenges for head stabilization. The human head is essentially an eccentric mass atop a series of upright supports, poorly able to control vertical oscillations and highly subject to forces that tend to pitch and roll the head every time the body collides with the ground and when the trunk pitches, rolls, and yaws. These movements are minor during walking but increase considerably during running.

Chapter 9 reviewed several novel, passive mechanisms for head stabilization that evolved in the genus *Homo*. First, the human head is partially stabilized by recruiting the mass of the arms as dampers through their linkage to the head via the cleidocranial trapezius and nuchal ligament. Second, because the head and the trunk are largely decoupled, forward acceleration of the trunk causes backward acceleration of the head and neck (a sort of stabilizing whiplash), thus countering the effects of forward pitch of the trunk. Likewise, as the gluteus maximus and erector spinae muscles contract forcefully to decelerate the trunk, the head and neck are whiplashed forward, thus countering the effects of backward pitch of the trunk. Both mechanisms require the head and neck to flex and extend somewhat independently (i.e., to be decoupled) from the trunk and pectoral girdle. In addition, both mechanisms benefit exponentially (to the power of 2) from a more balanced head because the torque exerted by pitching of the head equals $I \cdot \alpha$, where I is the head's moment of inertia (the square of the head's moment arm times its mass), and α is the head's angular acceleration.

Many features in late *Homo* suggest excellent endurance-running capabilities, but there is reason to hypothesize that the more balanced head of *H. sapiens* would have increased the stability of the head. If we assume that overall head mass was approximately the same in all species of late *Homo,* then the 40–80 percent shorter moment arm of the *H. sapiens* head would reduce pitching torques by at least twofold. Such torques would probably be insubstantial during walking, which generates low angular accelerations, but high during running. In such conditions, even slight reductions in facial projection and prognathism that move the head's COG closer to the AOJ would significantly reduce extensor forces needed to maintain a stable head. The extent to which *H. sapiens* might have been better able to stabilize the head than archaic *Homo* is hard to evaluate. We asked volunteers to clench a 4 kg weight in the mouth while running on a treadmill to increase the head's moment of inertia. The extra weight typically doubled head pitch accelerations, although not beyond the threshold of the vestibular-ocular reflexes, because the extensor muscles of the neck fire with more intensity. And it is possible that, with training, humans could tune their passive head-stabilization systems (described above) to cope passively with the extra mass. Regardless, a two- to threefold higher I probably explains why the insertion of the nuchal ligament on the occipital, the median nuchal line, tends to be much more accentuated in archaic *Homo* than in *H. sapiens* (Figure 12.13). The insertion of the nuchal ligament is also usually much more rugose on larger, robust human skulls (Lahr and Wright, 1996).

The issue of head balance in *H. neanderthalensis* merits special consideration. Like other archaic *Homo,* Neanderthals have large, heavy faces, which would have led to a two- to threefold higher moment of inertia in the pitch plane than modern humans experience. Although the occipital bun would have been a slight counterbalance to the mass of the face, one can calculate (by virtually eliminating the bun) that the bun shortened the moment arm by less than 5 percent. In addition, the anterior and posterior semicircular canals, both of which sense pitch accelerations, are smaller relative to body mass in Neanderthals than in other taxa of *Homo* (Spoor et al., 2003). Because the relative size of the semicircular canals influences the ability to sense rapid, jolting accelerations of the head during locomotion (Spoor et al., 2007b), it follows that Neanderthals might have been less adept at stabilizing their gaze during running. This trend may reflect the effects of selection, as the Neanderthals also evolved in cool, semiarctic conditions that would not have been conducive to methods of persistence hunting common in the tropics, in which humans drive their prey into hyperthermia by forcing animals to gallop long distances in the heat (Liebenberg, 2006; Lieberman et al., 2009).

Coda

Paradoxically, the evolutionary transformation marked by the origin of *H. sapiens* appears to be modest and subtle compared to earlier transformations that led to the initial divergence of hominins from the other apes and the shift from *Australopithecus* to *Homo.*

As *H. erectus* thrived and dispersed throughout Africa and parts of Eurasia because of the success of the hunting and gathering way of life, the species continued to evolve slightly, mostly in terms of brain size. *H. erectus* persisted for more than 1 million years in Asia but evolved in Africa and western Eurasia into *H. heidelbergensis* by approximately 600,000 years ago. This diverse species (which probably oversimplifies the complex phylogenetic relationships among taxa of archaic *Homo*) was only subtly different from *H. erectus:* primarily more encephalized, but perhaps also with a slightly larger body and face. One can surmise that bigger (and maybe more complexly structured) brains permitted some cognitive advances in *H. heidelbergensis,* including the invention of prepared core technologies, notably the Levallois technique, in order to make more effective hunting weapons and tools for processing. *H. heidelbergensis* probably had other innovations, perhaps including the more extensive use of cooking and more advanced linguistic capacities, which helped the species survive the climatic challenges of the Middle Pleistocene.

A final pair of speciation events occurred more recently. One of these was the evolution of *H. neanderthalensis* from *H. heidelbergensis s.l.* The Neanderthals were part of a distinct European lineage dating back to at least 500–350 kya, but they may also have been related to populations of *H. heidelbergensis s.l.* in Africa that segregated after the invention of the Levallois technique, some of them dispersing into western Asia and Europe. Neanderthals are thus a highly derived variant of *H. heidelbergensis s.l.* that became well adapted to the extremely variable and often harsh conditions of temperate habitats. Derived Neanderthal features include bigger brains, larger bodies, enlarged nasal cavities, and large faces well designed to resist high strains generated on the anterior dentition.

The other major speciation event—our own—occurred at about the same time in Africa. Precisely how *H. sapiens* diverged from *H. heidelbergensis s.l.* is unknown, but the evidence reviewed above suggests that several factors might have been important. The first was a reorganization of the brain involving the expansion of several parts. These expansions affected the overall shape of the vault and possibly contributed to some subtle but important cognitive shifts, including a greater use of symbolic behaviors; more extended interactions with larger, more dispersed social groups (manifested in more long-distance transport of raw materials); increased exploitation of new food resources, such as fish and small vertebrates; and perhaps a greater division of labor (for a range of views, see Klein, 1995, 2009; Mellars, 2004; McBrearty and Brooks, 2000; Kuhn and Stiner, 2006). At the same time, innovations such as cooking and other food processing techniques would have relaxed selection for maintaining a big face and perhaps interacted synergistically with selection for an overall smaller face in *H. sapiens*. A shorter face would have improved capabilities for language (a more quantal vocal tract), thermoregulation (a more turbulent nasopharynx), and locomotion (a more balanced head). The end result was the distinctively shaped, rounded modern human head with a smaller and especially shorter face retracted beneath the frontal lobe.

One can only speculate, but these various derived features might have helped *H. sapiens* become more effective hunter-gatherers, leading to population growth, range expansion, and an accelerated pace of technological and cultural innovation (factors that reinforce each other via positive feedback).

Since then, the combination of complex cognitive abilities and high population densities (resulting from high birth rates and low mortality rates) has fueled ever more rapid change and invention. One landmark event occurred around 50,000 years ago, when *H. sapiens* invented the Upper Paleolithic and subsequently dispersed throughout the Old World, mostly replacing the Neanderthals in Europe and *H. erectus* (or its descendants) in Asia. The next most transformative set of events was the invention of agriculture (which occurred at multiple places and times).

In short, a paradox of human evolution is that the transformations that led to the origin of *H. sapiens* appear to have been minor tinkering events, at least as far as the head is concerned. Although modern human heads differ considerably from those of chimpanzees and gorillas, we are basically a slightly more encephalized, smaller-faced version of *H. erectus*. There were probably additional differences compared to other species of late *Homo* in terms of brain organization and function, but these changes were probably subtle compared to the substantial differences that distinguish humans from the great apes. In this respect, the nature of the differences we see between *H. sapiens* and other hominins returns us to several questions discussed in Chapter 1. Notably, how is that skulls and heads can vary so much yet still evolve and function? How can seemingly small shifts have such major effects? And what role did contingent events play in the evolution of the human head? Now that we have reviewed in depth how human heads grow and function, and the fossil evolution for evolutionary changes that occurred since the divergence from the chimpanzee lineage, it is time to reconsider briefly in Chapter 14 the roles of complexity, tinkering, and contingency in the evolution of the human head.

14

Final Thoughts and Speculations

Once you were apes, and even now man is more of an ape
than any ape.

FRIEDRICH NIETZSCHE, *Thus Spoke Zarathustra* (1883–1885)

Who hasn't locked eyes with another mammal—a chimpanzee, a dog, or even
a cow—and recognized some sort of kinship? We may not entirely understand
each other's thoughts or perceptions, but it is difficult to escape the conclusion that we have much in common, including a shared sense of wonderment
about each other. Some of this feeling may be psychological projection (I want
to think that my dog is thinking about me), but our ability to connect at some
deep level with other mammals stems also from the fact that biologically we
are not very different. Our heads have almost entirely the same anatomical
structures, and they perform the same basic functions of respiration, balance,
vision, cognition, and so on. We see, hear, smell, taste, and feel the world in
mostly the same way, with only slight differences in the range of colors, frequencies, and odorants we perceive. Our craniofacial differences stem primarily from variations in the relative sizes of the head's components and the way
they are arranged architecturally. Evolution has modified the same basic design in numerous ways and to great effect, but deep down we are all tinkered
versions of each other.

Yet these differences matter. Even though chimpanzees look at human be-
ings with some intelligence, they cannot think about us in as sophisticated a
way as we can think about them. A major basis for this difference, of course,
is brain size and structure. But humans are not just apes with bigger, more
complex brains. Evolution transformed our ancestors from apes to modern
humans over many stages, leaving numerous distinctive traces in our cranio-

facial anatomy: a downward-pointing foramen magnum; a horizontally oriented nuchal plane; small, but relatively thickly enameled molars and premolars; small canines and incisors; no snout; a vertical, small, and retracted face; and a short, wide, and spherically shaped cranium. These and other transformations raise numerous questions, three of which have been the central topics of this book. First, what happened in human evolution to make our heads the way they are? Second, why did these transformations occur? And finally, how do humans heads manage to function as they do—to let us think, breathe, thermoregulate, speak, chew, swallow, see, hear, balance, taste, and smell—yet also be so evolvable?

In spite of various ongoing debates, answering what happened in the evolution of the human head is probably the easiest question to address. More than a century of concerted research has yielded a rich body of evidence showing that many of the key transformations we observe when we compare ourselves to other great apes occurred in three major stages, each of which was contingent on previous events. Morphologically, the first hominins shared many features with the African great apes, but they probably were bipeds of a sort, and they had slightly larger and more thickly enameled molars, along with slightly smaller canines. Eventually, early hominins were successful enough outside woodland habitats to make possible a radiation of australopith species. These diverse hominins were excellent bipeds but also good at climbing trees; they probably preferred fruit, but they were able to eat a broad diet, including various types of low-quality, mechanically demanding foods such as tubers. One group, the robust australopiths, evolved between three and two million years ago and were especially well adapted to processing low-quality foods that became more abundant as woodland habitats shrank at the expense of more open habitats. At approximately the same time, the genus *Homo* evolved to exploit a higher-quality diet, eventually leading to the evolution of *H. erectus*. This important species had a bigger brain, a more vertical face, smaller teeth, an external nose, a more balanced head, and various other features that helped it trek and run in the midday sun in order to hunt and gather. With *H. erectus* we see the beginnings of a recognizably human foraging way of life that made possible major increases in relative brain size and also helped the species to thrive and disperse out of Africa into parts of Eurasia. The descendants of *H. erectus* varied considerably, and included several large-brained species of late *Homo*, including *H. neanderthalensis* and *H. sapiens*. Frankly, our own species is only subtly different from our recent cousins, primarily in having a smaller, more retracted face, and a more globular braincase.

These transformations raise many fundamental and challenging questions about *why* these events occurred. As Ernst Mayr (1988) emphasized, to answer such questions, it is essential to distinguish between proximate and ultimate mechanisms of causation. The proximate causes of human craniofacial uniqueness are still murky, but their bases lie to a great extent in evolution's marvelous ability to tinker. Heads first evolved about 500 million years ago, probably to make early vertebrates better predators (Gans and Northcutt, 1983). A central premise of this book is that the way in which vertebrate heads evolved gave them special qualities that considerably enhanced their subsequent evolvability. As Chapters 2–5 reviewed, the head develops and grows in a complex way that causes its many key modules—organs and functional spaces—to use the same developmental pathways and to share the same common skeletal framework. Almost every wall of bone in the head participates in more than one functional matrix. Further, these walls of bone interact with their neighboring tissues in a dynamic, highly interactive way, with many levels and modes of epigenesis. So even though the eyes, ears, noses, teeth, and the brain are all packed in together, they become highly integrated by mutually accommodating each other's growth. As examples, the brain and the face push in different ways on the cranial base, and both the eyes and nasopharynx affect the same wall of bone that separates them. These epigenetic mechanisms, which lead to high levels of correlation and covariation throughout the skull, are vital to permit heads to function effectively during ontogeny even as the brain, body, and face all grow at different rates and to different degrees. Importantly, these mechanisms also accommodate considerable variation. Heads can function well despite considerable differences in brain size, tooth size, nasopharynx shape, and so on.

An emergent property of integration and modularity is evolvability, especially in the head, where levels of epigenetic integration are so high. Chapters 11–13 reviewed the evidence that many of the same mechanisms outlined in Chapters 2–5, which permit the head to accommodate variations in the size and shape of various modules, such as the brain, face and eyes, during growth, also accommodate a reasonable degree of variation between species in terms of the same modules. We can only speculate, but it is reasonable to hypothesize that near the end of the Pliocene, early *Homo* diets became higher in quality and more reliable, releasing constraints on variation in brain size. At that point, individuals with slightly larger brains probably had slightly higher fitness. Regardless of the developmental mechanisms that generated bigger brains (e.g., genes that regulate neuron proliferation), the heads of these early *Homo* individuals were able to accommodate a bigger brain without any compromise in function, using mostly the same mechanisms that accommodate brain growth during ontogeny: synchondroseal growth, drift, and sutural growth in the posterior, middle, and anterior cranial fossae; sutural expansion in the cranial vault; and more flexion of the cranial base.

Similarly, we can assume that comparable integrative mechanisms enabled hominins to accommodate other shifts that were advantageous or occurred by chance, such as bigger teeth, shorter faces, more vertical necks, and smaller oral cavities. Paradoxically, the mechanisms that enable heads to be so complex also permit heads to be so evolvable. The transformations we see in the hominin head may seem remarkable, but they appear modest compared to those we see during the evolution of other lineages, such as whales or elephants. Further, in spite of the obvious differences evident between human, chimp, and hippo heads, we are actually just variants of one another, with the same basic organs, growth centers, and mechanisms of growth. Thanks to the way vertebrate heads initially evolved and their basic mechanisms of growth and development, a little tinkering can go a long way.

Despite significant recent advances in our understanding of craniofacial patterning, morphogenesis, and integration, we have far to go before we understand the genetic and cellular mechanisms that underlie how heads grow and vary. Someday, maybe soon, we will identify which genes changed (and when) during human evolution to make brains bigger, teeth smaller, noses more external, foreheads sweaty, and so on. Understanding these mechanisms will not only provide us with useful clinical insights but will also help us learn more about how, when, and why many major transformations occurred. Interpreting these developmental shifts will be especially illuminating when examined in the context of the fossil record. Humans didn't just spring from chimpanzees. Instead, our evolutionary history involved many intermediate transformations with particular ecological contexts. Considered in this light, the fossil and archaeological records of human evolution will become more rather than less interesting as our knowledge of genomics and the relationship between genotype and phenotype improves.

Integrating developmental and paleontological evidence brings up the issue of more ultimate mechanisms of causation necessary to understand why the human head looks the way it does. Here there is much basis for speculation, including the extent to which natural selection is an engine of change. Over the past few decades, we have become increasingly aware that mechanisms other than natural selection can be potent agents of evolutionary change. Most important, there is compelling evidence that most population-level variation stems from random (nonselected) processes such as genetic drift and founder effects (Lewontin, 1972). As with genetic variation, modern human heads vary far more within than between populations, and most of the differences we do perceive between populations derive primarily from random genetic shifts (Relethford, 1994, 2002; Roseman and Weaver, 2007; Weaver, et al., 2008; Manica et al., 2007).

Another factor to consider is the difficulty of testing whether novel features are true adaptations, selected because they improved an organism's fitness. To be sure, natural selection does promote adaptations, but mechanisms of integration, such as pleiotropy (in which a gene has multiple effects) and linkage (in which selection on one gene indirectly "selects" for neighboring genes) are also frequent. Many of the changes we observe in the human head that are commonly assumed to be adaptations may have been byproducts of evolutionary change elsewhere. My favorite example is the extraordinary human ability to perceive flavor, which I hypothesize is a byproduct of a nonintranarial epiglottis, a short oral cavity, and an expanded neocortex. It is likely that none of these features were selected for improved flavor perception. That said, one could argue that an enhanced ability to perceive flavors may have favored the human proclivity for cooked foods, a proclivity that has undoubtedly led to many energetic and health benefits.

One should be cautious about invoking hypotheses selection without testing them (see Rose and Lauder, 1996), but there are nonetheless many reasons to believe that, over the long term, natural selection was a dominant force behind many of the major transformations we observe in the evolution of the human head. As formulated by Darwin (1859), natural selection is basically an emergent property of three key factors intrinsic to all biological systems: heritability, variation, and differential reproductive success. Of these, fitness is impossible to assess in the fossil record, but there is no question that the head participates in dozens of key functions, reviewed in Chapters 6–10, with profound fitness consequences. Some of these functions, such as vision, hearing, and balance, are not particularly acute in humans (although there are some differences between us and apes), but that does not make them any less vital. Until recently, hominins with impaired vision, hearing, or balance probably had significantly lower chances of surviving and reproducing. Other functions, such as cognition, speech, thermoregulation, mastication and deglutition, are no less crucial and have clearly undergone much more dramatic changes since our divergence from the African apes. The performance effects of these changes suggest very intense levels of selection. And, as I argued above, high levels of selection combined with a high capacity for evolvability have made the head a very dynamic locus for evolutionary change in the human lineage.

Major Selective Forces in the Evolution of the Human Head

I'd like to conclude with some speculations about the major selective forces responsible for key transformations in the evolution of the human head. Although selection was undoubtedly complex, I think one can make the case that three major suites of adaptive shifts were dominant factors in making our heads (and bodies) so different from those of other apes.

Nonforest Habitats

The first of these selective forces was to survive better in habitats other than forests. This selection occurred over a long period but perhaps was most crucial when the hominin lineage diverged from the other apes—which mostly live in dense forest habitats and have regular access to ripe fruit. Several lines of evidence indicate that the first hominins were probably like chimpanzees in many (although not all) morphological respects, but that they lived in habitats that were somewhat marginal for apes that ate mostly fruit. If so, then one can hazard that they would have faced more intense periodic shortages of preferred foods (fruits). It is under such conditions that natural selection can be most intense (Grant, 1991; Marshall and Wrangham, 2007), and comparative studies of apes, along with the glimpses provided by fossil remains of *Sahelanthropus, Ardipithecus,* and *Orrorin,* allow us to infer that some of these hominins had slightly larger, thicker-enameled cheek teeth, which helped them break down mechanically demanding fallback foods that were necessary to eat when fruits were scarce. In addition, early hominins might have been better at standing and walking bipedally, which would have helped them in a variety of tasks, such as foraging, wading across rivers, and saving energy when trekking between increasingly distant patches of food (see Hunt, 1991; Thorpe et al., 2007; Sockol et al., 2007). It is reasonable to hypothesize that the earliest hominins were selected for being better at surviving in more open habitats than chimps.

Selection for more open habitats probably intensified during the early and middle Pliocene (roughly 5 to 3 million years ago), as temperature and rainfall patterns continued to change in Africa (see Behrensmeyer et al., 1997; Potts, 1998; Trauth et al., 2005). As the forests shrank, finding and eating food became a major challenge for early hominins. Over time, the australopiths appear to have become more adapted to chewing lower-quality, mechanically demanding diets, thanks to thicker, larger molars and premolars, along with jaws and faces better able to generate and resist large, repetitive chewing forces. In addition, their skulls, like their postcrania, indicate that they probably retained some adaptations for climbing trees but also that they probably were economical bipedal walkers, well adapted for trekking longer distances to acquire food. Enlargement of their external auditory canals testifies to improved abilities to perceive frequencies above the 2,000 Hz range, useful for detecting the location of predators in more open habitats (see Chapter 10). But it is important not to gloss over the many primitive similarities that australopiths share with the African great apes.

Another bout of selection for open habitats occurred at the end of the Pliocene (between 3 and 2 million years ago), as the world continued to cool, causing dramatic climatic changes in Africa, including much variability. As the forests and woodlands shrank and the savannas expanded, one "solution" was favored in the evolution of the robust australopiths. These hominins were essentially intensified australopiths, even more extremely adapted than their more gracile forebears to chewing low-quality foods. The other "solution," discussed in Chapter 12, was the evolution of early *Homo*, which became adept at exploiting a wide range of habitats including open ones (see Potts, 1996). Although the transition from *Australopithecus* to *Homo* occurred gradually, we see at first in *H. habilis* and then in *H. erectus* evidence for several kinds of improved performance capabilities for acquiring high-quality foods in open habitats. Such habitats offer fewer fruits but contain an abundance of grazing ungulates. In particular, early *Homo* evolved a more turbulent nasopharynx, better at coping with being active in hot, arid habitats; a more balanced head and sensitive vestibular system, better able to handle the mechanical demands of running; and smaller teeth, more specialized for eating high-quality food. There is good evidence that *H. erectus* occupied a new and successful niche, that of a diurnal social omnivore (part carnivore, part forager). This was such a successful niche that *H. erectus s.l.* probably expanded quickly, dispersing into parts of Eurasia.

Energy Acquisition and Use

Climate change was a key driving force for much of human evolution, but the origins of *Homo,* especially *H. erectus,* involved more than just adaptations to new habitats. At some point, members of the human genus became better at acquiring and using energy differently than other primates. As outlined in Chapter 12, it is reasonable to speculate that this shift was partly if not largely a byproduct of early *Homo*'s successes as a trekker (perhaps a trend that began in *Australopithecus*), as an opportunistic forager, and most importantly as a carnivore. *H. habilis* apparently had access to meat, but *H. erectus* certainly became an excellent hunter, thanks to a suite of derived features. These features, some of which may be adaptations for running and hunting, and others of which may have been exaptations of other functions, were important because they enabled early members of the genus *Homo* to range widely to acquire reliably plenty of high-quality food. This shift in ranging and diet not only permitted selection to make teeth smaller but also made more energy available to hominins, setting in motion other selective forces (see Pontzer and Kamilar, 2009). Most important, more energy probably allowed females to wean infants earlier, lower their interbirth intervals, and have more babies that survived to become adults. In addition, more energy allowed individuals, both males and females, to grow and fuel bigger bodies and bigger brains. In turn, bigger bodies and brains would have ramped up the selective pressure to lower interbirth intervals (Aiello and Key, 2002; Chapter 12).

Once one commits to the strategy of acquiring lots of energy in order to spend it on reproduction, then natural selection can promote a positive feedback loop (Van Valen, 1973). Other factors that improved the ability of *H. erectus* and then *H. heidelbergensis* to acquire high-energy resources included food processing, cooking, the invention of stone tools and projectile technologies, and probably also behaviors that are hard to document in the fossil or archaeological records, such as cooperative hunting, food sharing, language, and tracking prey (Isaac, 1978; Liebenberg, 1990). In turn, these behaviors led to higher fitness, which helped promote further advances—many of them cognitive or behavioral—in ways to exploit and use energy. I don't think it is coincidental that *H. erectus s.l.* dispersed rapidly out of Africa and into Eurasia. This new way of life persisted (albeit with many modifications and elaborations, such as the Middle and Upper Paleolithic) until the invention of agriculture. And, of course, the positive feedback loop has never ended. Once humans invented agriculture, farmers were able to wean their babies earlier and increase fertility, leading to a population explosion that is still accelerating following the Industrial Revolution. Ever since *H. erectus,* our genus has been relentlessly dependent on acquiring more and more energy.

Brains, Language, and Other Complex Cognitive Functions

The final, obvious, and related selective pressure evident in the evolution of the human head was for increased brain size and changes in brain structure. So many researchers have considered this selective force in such depth that only a few points merit brief comment here. The first is that big brains are obviously advantageous because they help make possible a variety of complex cognitive functions, such as language. In addition, the same mechanisms that accommodate brain growth during ontogeny probably required only minimal tinkering to accommodate selection for bigger brains during hominin evolution. However, brain tissue is terribly costly, and voluminous brains pose all sorts of anatomical and physiological challenges for thermoregulation, protection, birthing, locomotion, and so on (see Chapter 6). Thus it makes sense that substantial increases in brain size (both absolute and relative) didn't occur until after 2 million years ago, when other factors, such as hunting, released energetic constraints on their expansion. In addition, selection for bigger brains was clearly complex and multiphasic. Endocranial volumes were a little smaller in *Sahelanthropus* and *Ardipithecus* than in most chimps, and australopiths have only modestly larger brains. The first substantial increases in brain size evident in the genus *Homo* were probably driven mostly by increases in body size. Really major increases in relative brain size began approximately one million years ago and accelerated with the evolution of *H. heidelbergensis.* In addition, as previously noted, modern humans really aren't much more encephalized than other species of late *Homo.* And, of course, there is far more to brain function than size alone (topics for other books by other authors).

One last point to consider is that selection for intelligence, which partly derives from a bigger brain, leads to other, related selection pressures. Darwin clearly articulated this idea in 1871 in *The Descent of Man,* when he suggested that tool making (itself spurred on by bipedalism) led to selection for a bigger brain and other capabilities, such as language. This topic has engendered perhaps more disagreement than any other in human evolution (an impressive feat), but not without reason, because language is so important yet leaves almost no physical traces. Most of the biological bases for language cannot be studied from skulls, with the very limited exception of speech. Absence of evidence leaves much room for speculation. Here is mine: As I outlined in Chapters 8 and 13, I would be surprised if *H. erectus* and other species of late *Homo* did not have some impressive speech capabilities, far beyond those of chimpanzees. As brains got bigger, so probably did language abilities. But there is reason to believe that some derived aspects of *H. sapiens* head shape, notably our unusually short oral cavities, were selected for because they modestly improved our ability to speak with greater clarity. Perhaps someday we will be able to clone a Neanderthal and then have a drink together at the bar to discuss the issue more fully.

Heading Off

One final comment. Although I have filled many hundreds of pages on how human heads develop, function, and evolved, it should be abundantly evident that the more we discover about how and why our heads are they way they are, the more we recognize how much we don't know. As science progresses, we can look forward to many new discoveries—fossil, functional, genetic, and developmental—that will disprove or alter much of what I have written. That is the nature of scientific inquiry. But the fun and frustration of studying a system as complex as the head is the benefit of using integrated rather than reductionist approaches. It is sometimes rewarding to focus a problem down to its simplest element and then solve it. But heads defy many efforts to simplify because they are, by nature, complex and highly integrated systems. Future advances will come from considering these different lines of evidence using as integrated an approach as possible.

Glossary

Acheulian (or Acheulean): Lower Paleolithic industry that includes handaxes.

adaptation: A feature favored by natural selection because it improves an organism's ability to survive and reproduce.

allometry: Scaling relationship between size and shape.

ameloblast: Cell that secretes dental enamel.

anlage: A primordium of cells from which an organ develops; bones develop from a mesenchymal anlage.

balancing side: Side opposite the side of occlusion during unilateral mastication.

basion: Landmark in the midline of the skull at the most anterior point on the foramen magnum in the midline plane (in the occipital bone).

blood brain barrier (BBB): System that separates blood from cerebrospinal fluid (CSF) in the central nervous system. Lining cells restrict diffusion across the barrier to very small molecules, such as oxygen and hormones; other molecules need to be transported actively across the BBB.

branchial arches: In embryonic development, paired streams of cells that migrate bilaterally around the pharynx from alongside the hindbrain toward the midline of the head, forming most of the face and neck.

branchial pouches: Spaces between the branchial arches on the endodermal side of developing face and neck.

bregma: Landmark in midline of the skull at the intersection of the parietal bones and the frontal bone (also at the intersection of the sagittal and coronal sutures).

caudal: Directional term meaning "toward the tail."

cementoblast: Cell that secretes cementum along the surface of tooth roots, helping anchor the tooth to the periodontal ligament.

cerebrospinal fluid (CSF): A clear fluid that surrounds the brain and spinal cord in the subarachnoid space and fills the ventricular system around and inside the brain. Produced in the choroid plexus.

chondroblast: Cell that synthesizes cartilage.

chondrocranium: Embryonic precursor to basicranium (cranial base).

cladistics: Method of determining evolutionary relationships among taxa. If taxa such as species evolve though change over time (what Darwin termed descent with modification), then more closely related species should share derived features.

coefficient of variation (CV): Normalized measure of dispersion, defined as the standard deviation divided by the mean. CV* denotes a CV corrected for the effects of small sample sizes (see main text).

cranial: Directional term meaning "toward the head."

cranial base angle (CBA): Angle that summarizes the orientation of the endocranial fossae relative to each other in the midsagittal plan. The most common measure is CBA1 from basion (the anterior margin of the foramen magnum) to sella (the center of the hypophyseal fossa) to the foramen caecum (the inferiormost point on the anterior cranial base).

craniosynostosis: Malformed or premature fusion of suture(s) in cranial vault.

decibel: A logarithmic measurement of the amplitude of sound relative to the threshold of perception of an average human.

dental lamina: Epithelial cells from branchial arch ectoderm that fold into the front and back of each mandibular arch, giving rise to ectodermal component of teeth.

desmocranium: Embryonic precursor to neurocranium (cranial vault).

development: Changes in shape or function during ontogeny.

displacement: Movement of one structure relative to another either from growth in one of the structures (primary displacement) or from growth in an intermediary structure (secondary displacement).

dorsal: Directional term meaning "toward the posterior or back."

drift: Movement of a wall of bone relative to other parts of the skeleton via deposition on one surface and resorption on the opposing surface.

dura mater: Tough outermost of the three layers of the membranes that surround the brain and spinal cord.

ectocranium: Periosteal membrane on outside of cranial vault.

ectoderm: In embryonic development, one of three germ layers arising from gastrulation. Ectodermal cells contribute to neural crest, skin, central nervous system, and many other parts of the body.

ectomeninx: Membrane surrounding the brain and spinal cord that gives rise to meningeal membranes.

enamel dentine junction (EDJ): Boundary between the epithelial cells of the enamel organ and the mesenchymal cells of the dental papilla, which underlies the tooth crown. The size and shape of the EDJ influence tooth crown shape.

endochondral bone growth: Process by which bones grow, involving formation and then replacement of a cartilage precursor, with elongation occurring in growth plates. The cranial base and most of the postcranium grow endochondrally.

endocranium: Periosteal membrane on inside of cranial vault.

endoderm: One of three germ layers following gastrulation. Endodermal cells contribute to the lining of the gastrointestinal tract (from mouth to anus), and of the digestive glands, lungs, bladder and other parts of the body.

epiblast: One layer of cells in the bilaminar germ disc of the embryo (the other being the hypoblast) that during gastrulation gives rise to the ectoderm, mesoderm, and endoderm.

epigenetic: Term describing interaction between a cell's genome and stimuli external to the cell. In cell biology, the term has also come to refer to heritable changes in a cell's genomic function that do not alter DNA sequences.

epithelial cells: Lining cells that have many functions, including induction of neighboring tissues, secretion, absorption, protection, and transcellular transport.

ethmomaxillary complex: Region of the middle face around the nasal cavity, including the ethmoid, maxilla, and palatine.

evolvability: A taxon's ability to change over time, first though generating diversity and then through natural selection.

falx cerebri: Vertical fold of dural membranes, from midline of the tentorium cerebelli to the midsagittal top of the cranium and the ethmoid.

foramen caecum: Most anterior and inferior point on the cranial base.

form: The description of an object's morphology, including both shape and size.

formant frequency: In speech, an amplitude peak in the frequency spectrum of a sound (F1, F2, etc). Variations in formant frequencies change a sound's tone.

frequency: The number of occurrences of a repeating event per unit time (the duration of a cycle).

frontonasal prominence: Region of the embryonic head, induced by the prosencephalic organizing center, that forms most of the upper face.

functional matrix hypothesis (FMH): Theory of craniofacial modularity developed by M. Moss that considers the head as a series of functional matrices, genetically or functionally determined soft tissues and the spaces they occupy, each enclosed by a functional cranial component that derives its shape principally from its associated matrix.

fundamental frequency: A source frequency. In speech, F0 is the frequency at which puffs of air exit the larynx, which we perceive as pitch.

gastrulation: Key transformation during early development, which forms the embryonic germ layers and many fundamental aspects of body plan.

geometric morphometrics: Technique of using landmarks to quantify size and shape. Two or more objects with the same set of landmarks may be compared by transposing them into the same coordinate system, scaling them to the same size, and rotating them to minimize the summed square of the landmark differences.

genotype: Genetic constitution of a cell or organism.

glabella: Landmark in the midline of the skull; a slightly elevated point between the two brow ridges and above the nose.

gracile: Characteristic of being lightly built (opposite of robust).

growth: Changes in size (as opposed to shape) during ontogeny.

Haversian remodeling: Process in which old bone tissue is replaced with new tissue. The sequence usually involves the activity of osteoclasts, which tunnel out old bone, followed by osteoblasts, which lay down new bone in cylindrical lamellae around a central channel. These are termed secondary osteons or Haversian systems.

heterochrony: Pattern of changes in relative timing and rate of developmental processes that leads to morphological change.

Holocene: Current geological epoch, which began approximately 12,000 years ago, after the Pleistocene.

homeobox genes: Family of transcription factor genes that share a short sequence (about 180 DNA base pairs) known as the homeodomain. These genes help define segment position and identity.

hominin: Term for any organism more closely related to a human than to a chimpanzee. Because humans and chimps are actually more closely related to each other than either species is to gorillas, those species more closely related to humans than chimps belong in the tribe Hominini.

hypophyseal: Term referring to basicranial cartilages that form around hypophyseal (pituitary) gland.

induction: Initiation of a process by which one cell affects the function or differentiation of another cell.

integration: The way in which different components (modules) of a system are combined into a whole. Morphological integration quantifies patterns of correlation and covariation among modules; functional integration quantifies interactions among components that affect performance; and developmental integration quantifies interactions among components during growth and development.

intramembranous bone growth: The process by which most cranial bones grow, involving formation of bone (without a cartilage precursor) within a periosteal membrane.

lambda: Landmark in the midline of the skull at the intersection of the occipital bone and the two parietal bones.

lamellar bone: The most common type of bone, consisting of thin, platelike layers of bone, often with collagen aligned at different angles between layers.

laminar flow: A pattern of flow in which gas or fluid moves in parallel layers with little turbulence.

laryngopharynx: Bottom portion of the pharynx, between the epiglottis and the inlet (or aditus) of the larynx.

Later Stone Age (LSA): Period in African prehistory roughly contemporaneous with the Upper Paleolithic (see below), during which a wide variety of stone and bone tools were produced.

lever arm: A kind of moment arm, defined as the distance between the center of rotation and the point where force is applied (also known as an in-lever). Torque is the product of the force applied and the moment arm.

ligament: Fibrous connective tissue between two bones.

load arm: A kind of moment arm defined as the distance between the center of rotation and the application of force (also known as an out-lever).

Lower Paleolithic: Earliest subdivision of the Paleolithic or Old Stone Age, which includes the Oldowan, Acheulian, and the Developed Oldowan industries.

mandibular prominences: Paired derivatives of first branchial arch, which contribute to mandible and temporomandibular joint.

mastication: Process of breaking down (comminuting) food on the postcanine teeth.

maxillary prominences: Paired derivatives of first branchial arch, which contribute to maxilla and palate.

mechanical advantage (MA): Ratio of lever arm (in-lever) to load arm (out-lever) that determines the trade-off between the efficiency of torque generation and the velocity of angular movement. A higher MA is more efficient but slower.

mechanotransduction: A mechanism by which cells convert a mechanical stimulus (e.g., light, sound, strain) into chemical or neural activity.

Meckel's cartilage: Cartilaginous rod, derived from first branchial arch, at the center of the embryonic mandible, around which intramembranous bone forms.

mesenchyme: Stem cell line that develops into connective tissues as well as blood cells, the lymphatic system, smooth muscle, and the circulatory system. Mesenchyme derives from either mesoderm or neural crest tissue.

mesethmoid: Precursor to many parts of ethmoid; persists as nasal septum.

mesoderm: One of three germ layers following gastrulation. Mesodermal cells contribute to the musculoskeletal, urinary, reproductive, and circulatory systems, among others.

Middle Paleolithic (MP): Middle subdivision of the Paleolithic or Old Stone Age. Often characterized by prepared core technologies.

Middle Stone Age (MSA): Term for the equivalent of the Middle Paleolithic in Africa.

Miocene: Geological epoch from 23 million to 5 million years ago.

modeling: Process by which a bone adds new tissue, thus growing in size and possibly also changing shape. Modeling differs from remodeling, in which a bone changes shape without changing size or has tissue turnover.

modularity: The concept that organisms comprise semi-independent units (modules). Modules can be defined at many levels of structure, from genes to organs and systems.

moment of inertia: Measure of an object's resistance to angular acceleration, I. For a point mass, $I = mr^2$, where m is the mass and r is the moment arm (the perpendicular distance between the center of mass and its point of rotation).

morphogenesis: Processes that generate the position, shapes, and cell types of tissues and organs.

nasopharynx: Upper portion of pharynx, above the soft palate and behind the two openings to the nasal cavity (choanae or posterior nares).

neural crest: Migratory stem cells that derive from neural folds and give rise to much of the head.

neural-growth trajectory: Rate of growth characteristic of the central nervous system (CNS). In humans, the CNS grows most rapidly during the first two years of life, more slowly until about age 6 or 7 years, and then almost completely ceases.

neutral horizontal axis of orbits (NHA): Anatomical line that describes the orientation of the orbits, modeled as cones in lateral view (in a radiograph). The NHA goes from the superoinferior midpoint between the superior orbital fissures and the inferior rims of the optic canals to the superoinferior midpoint between the lower and upper orbital rims.

notochord: Stiffened rod of mesoderm that emerges after gastrulation and forms the central axis of the embryo.

odontoblast: Cell that secretes dentine.

Oldowan: Earliest Lower Paleolithic industry, characterized by fabrication of simple flakes, bifaces, unifacial choppers, and other tool forms.

ontogeny: Sequence of events by which an organism develops from a fertilized egg to its mature form.

opisthion: Landmark in the midline of the skull at the most posterior point on the foramen magnum (in the occipital bone).

opisthocranion: Landmark in the midline of the skull at the most posterior point on the occipital bone relative to glabella (see above).

oropharynx: Middle portion of the pharynx, between the epiglottis and the soft palate and behind the oral cavity.

osteoblast: Cell that secretes bone and regulates bone metabolism.

osteoclast: Cell that resorbs bone tissue.

osteocyte: Mature osteoblast that is surrounded by bone tissue within a space (lacuna) and communicates with other osteocytes via tubelike processes known as canaliculi.

osteoid: Unmineralized bone matrix, mostly collagen, but also containing other proteins.

osteon: A cylindrical arrangement of bone layers (lamellae) surrounding a central channel (Haversian channel).

parachordal (or postchordal) region: Region of cranial base dorsal or lateral to the notochord (which terminates below the hypophyseal fossa).

paraxial mesoderm: Mesoderm that differentiates on either side of the notochord and gives rise to somites and to part of branchial arches.

perichondrium: Membrane of mesenchymal cells that surrounds a cartilage (see above), often in a developing bone.

periosteum: Membrane of mesenchymal cells that surrounds a bone.

pharynx: Passageway for food and air that runs from the mouth and nasal cavity to the esophagus and trachea.

phenotype: Any observable trait of a cell or organism, including anatomy, physiology, development, and behavior.

phonation: The process by which the vocal folds produce sounds, which are then modified by the vocal tract.

pitch: Motion of the head in the sagittal plane (e.g., nodding).

placode: In embryonic development, outpocketing of thickened epithelial tissue derived from neural ectoderm, associated with sensory organs. Often plays an inductive role in development of other structures.

Pleistocene: Geological epoch, commonly known as the Ice Age, from either 2.6 or 1.8 million years ago (date was recently changed) to 12,000 years ago.

Pliocene: Geological epoch from 5 million years ago to either 2.6 or 1.8 million years ago (date was recently changed to 2.6 mya).

postchordal: Basicranial cartilages that form behind (caudal to) the notochord.

posterior maxillary plane (PM plane): Line that describes the midline back of the face in lateral view in a radiograph. Defined as the midsagittal projection of a line from the maxillary tuberosities to the anterior poles of the middle cranial fossa (MCF).

prechordal: Basicranial cartilages that form in front of (rostral to) the notochord.

prognathism: Measure of the anteroposterior protrusion of the lower face relative to the upper face.

projection: Measure of the anteroposterior protrusion of the face relative to the cranial base and neurocranium.

prosencephalic organizing center: Inductive center (derived from primitive-streak mesoderm) in front of the forebrain that induces differentiation of cells that contribute to upper face.

Rathke's pouch: A depression in the roof of the developing mouth (stomodeum) cranial to the buccopharyngeal membrane, which later gives rise to the anterior pituitary gland.

remodeling: Process by which a bone changes shape and or composition. The most common kind of remodeling is Haversian remodeling (see above).

resistance: A force that opposes motion.

resonance: The tendency of a system to oscillate with higher amplitudes at some frequencies than at others; these frequencies are called resonant frequencies.

rhombencephalic organizing center: Inductive center by the hindbrain that induces differentiation of cells that contribute to the branchial arches.

rhombencephalon: The hindbrain; derives from segments of the embryonic brain known as rhombomeres.

robusticity: Term referring to strongly built anatomy: in the skull, refers to being large and thick-walled, with marked superstructures.

roll: Motion of the head in the coronal plane (e.g., leaning the head to one side or the other).

rostral: Directional term meaning "toward the face or snout."

sclerotome: Precursor skeletal tissue of paraxial mesoderm tissue that condenses next to the notochord to form vertebral bodies in the skeleton and occipital cartilages in the cranium.

second moment of area: resistance of a beam's cross-section to bending around an axis (also known as the area moment of inertia). Using calculus, defined as $I_x = \int y^2 \cdot dA$, where A is an area and y is its moment arm to the axis of bending, x. An algebraic formula is provided in Chapter 6.

sella: Landmark in midline of the skull at the center of the pituitary (hypophyseal) fossa.

sexual dimorphism: Difference in size or shape for a given feature between males and females in a population or species.

shape: All the geometrical information for an object independent of size (scale), location, and orientation.

shearing: Deformation within a given plane.

sinus: A cavity within an anatomical structure.

size: A measure of quantity (e.g., mass, volume, length).

skeletal growth trajectory: Rate of postnatal growth characteristic of the skeleton. In humans, the rate of skeletal growth has an S-shaped curve, with early rapid growth followed by a period of slower, steady growth; an adolescent growth spurt; and cessation of growth between 17 and 20 years of age.

somites: Segments of mesoderm around the notochord that form the axial skeleton and part of the occipital bone.

sphincter: A structure, usually a circular muscle, that constricts an orifice or tubular passageway.

stiffness: Resistance of an elastic material to strain (deformation) from an applied force.

stomodeum: In embryonic development, shallow depression around which the mouth forms at the cranial junction of the ectoderm and endoderm.

strain: A dimensionless measure of deformation per unit length.

stress: A measure of force per unit area, usually measured in pascals (Pa), equivalent to newtons (N) per square meter.

superstructures: Features of the skull, such as crests, processes, ridges, eminences, or tuberosities, associated with attachment of muscles, ligaments, or tendons.

synchondrosis: A cartilaginous joint or endochondral growth site.

telencephalon: The cerebrum or forebrain.

tendon: A tough band of fibrous connective tissue that connects muscle to bone and resists tension, sometimes storing elastic energy.

tentorium cerebelli: Semihorizontal fold of dural membranes that supports the neocortex; attaches to petrous crests of the temporals and the lesser wings of the sphenoid and extends around rim of the posterior cranial fossa.

thin-plate spline: Graphical representation of the transformation of one form into another, analogous to the bending of a thin sheet of metal.

torque: The tendency of a force to rotate an object around an axis, $\tau = r \cdot F$, where r is the moment arm and F is the applied force.

toughness: The resistance of a material to fracture.

trabecular bone: Macrostructural category of bone (also known as cancellous or spongy bone) consisting of a lattice of tiny spicules or struts (also called trabecula).

turbulence: A nonlaminar flow of fluid or gas characterized by chaotic patterns, with high variation in pressure, velocity, and direction.

Upper Paleolithic (UP): Most recent subdivision of the Paleolithic or Old Stone Age. Often characterized by blade core technologies, bone tools, and cultural variations in time and geography.

ventral: Directional term, meaning "toward the anterior or front."

vestibulo-ocular reflex (VOR): A reflex in which an image is stabilized on the retina by moving the eye in the opposite direction and at an equal angular rate to a movement of the head.

vomeronasal organ: A chemosensory organ at the base of the nasal septum; vestigial in humans.

Wolff's law: The idea (but not really a law) that bone responds dynamically to its mechanical environment. The idea was originally applied to the orientation of trabecular struts in a joint relation to orientations of compression and tension.

working side: The side of the head engaged in occlusion during unilateral mastication.

Wormian bones: Irregularly shaped, small extra bones that occasionally form between standard bones of the cranial vault.

woven bone: A disorganized type of bone in which collagen fibers are randomly oriented. Woven bone is often formed quickly during rapid growth or injury repair.

yaw: Motion of the head in the transverse plane (e.g., turning to the side).

References

Abbate, E., A. Albianelli, A. Azzaroli, M. Benvenuti, B. Tesfa-mariam, P. Bruni, N. Cipriani, et al.1998. A one-million-year-old *Homo* cranium from the Danakil (Afar) Depression of Eritrea. *Nature* 393:458–60.

Abeles, M. 1991. *Corticonics: Neural Circuits of the Cerebral Cortex.* Cambridge: Cambridge University Press.

Ackermann, R. R. 2005. Ontogenetic integration in the hominoid face. *Journal of Human Evolution* 48:175–97.

Ackermann, R. R., and G. E. Krovitz. 2002. Common patterns of facial ontogeny in the hominid lineage. *Anatomical Record* 269:142–47.

Adamopoulos, G., G. Tassopoulos, E. Ferekydis, M. Bosinakou, and T. E. Kontozoglou. 1994. Nasomaxillary skeletal dimensions complex in patients with osseous nasal septum deformities. *Journal of Otolaryngology* 23:84–87.

Adams, L. M., and W. J. Moore. 1975. Biomechanical appraisal of some skeletal features associated with head balance and posture in the Hominoidea. *Acta Anatomica* (Basel) 92:580–84.

Aganastopolou, S., D. D. Karamaliki, and M. N. Spyropoulos. 1988. Observations on the growth and orientation of the anterior cranial base in the human foetus. *European Journal of Orthodontics* 10:143–48.

Ahern, J. C. M. 2005. Foramen magnum position variation in *Pan troglodytes*. *American Journal of Physical Anthropology* 127:267–76.

Ahnelt, P. K. 1998. The photoreceptor mosaic. *Eye* 12:531–40.

Aiello, L., and C. Dean. 1990. *An Introduction to Human Evolutionary Anatomy.* London: Academic Press.

Aiello, L. C., and B. Wood. 1994. Cranial variables as predictors of hominine body mass. *American Journal of Physical Anthropology* 95:409–26.

Aiello, L. C., and P. Wheeler. 1995. The expensive-tissue hypothesis: The brain and the digestive system in human and primate evolution. *Current Anthropology* 36:199–221.

Aiello, L. C., and C. Key. 2002. The energetic consequences of being a *Homo erectus* female. *American Journal of Human Biology* 14:551–65.

Alba, D. M. 2002. Shape and stage in heterochronic models. In *Human Evolution through Developmental Change,* ed. N. Minugh-Purvis and K. J. McNamara, 28–50. Baltimore, MD: Johns Hopkins University Press.

Alberius, P., and H. Friede 1992. Skull growth. In *Bone Growth,* ed. B. K. Hall, 129–53. Boca Raton, FL: CRC Press.

Alemseged, Z., F. Spoor, W. H. Kimbel, R. Bobe, D. Geraads, D. Reed, and J. G. Wynn. 2006. A juvenile early hominin skeleton from Dikika, Ethiopia. *Nature* 443:296–301.

Alexeev, V. P. 1986. *The Origin of the Human Race.* Moscow: Progress Publishers.

Allman, J. 1977. Evolution of the visual system in the early primates. *Progress in Psychobiology and Physiological Psychology* 7:1–53.

Allman, J. M., A. Hakeem, J. M. Erwin, E. Nimchinsky, and P. Hof. 2001. The anterior cingulate cortex: The evolution of an interface between emotion and cognition. *Annals of the New York Academy of Sciences* 935:107–17.

Allman, J. M., K. K. Watson, N. A. Tetreault, and A. Y. Hakeem. 2005. Intuition and autism: A possible role for Von Economo neurons. *Trends in Cognitive Sciences* 9:367–73.

Alpern, M. 1979. Lack of uniformity in colour matching. *Journal of Physiology* 288:85–105.

Alvord, L. S., and B. L. Farmer. 1997. Anatomy and orientation of the human external ear. *Journal of the American Academy of Audiology* 8:383–90.

Amis, T. C., N. O'Neill, T. Van der Touw, and A. Brancatisano. 1996. Control of epiglottic position in dogs: Role of negative upper airway pressure. *Respiratory Physiology* 105:187–94.

Andrews, P. 1984. An alternative interpretation of the characters used to define *H. erectus. Courier Forschungsinstut Senckenberg* 69:167–75.

Andrews, P., and L. Martin. 1991. Hominoid dietary evolution. *Philosophical Transactions of the Royal Society of London, Series B* 334:199–209.

Antón, S. C. 1989. Intentional cranial vault deformation and induced changes in the cranial base and face. *American Journal of Physical Anthropology* 79:253–67.

———. 1994. Mechanical and other perspectives on Neandertal craniofacial morphology. In *Integrative Paths to the Past: Palaeoanthropological Advances in Honor of F. Clark Howell,* ed. R. S. Coruccinni and R. L. Ciochon, 677–95. Inglewood Cliffs, NJ: Prentice-Hall.

———. 2002. Evolutionary significance of cranial variation in Asian *Homo erectus. American Journal of Physical Anthropology* 118:301–23.

———. 2003. Natural history of *Homo erectus. Yearbook of Physical Anthropology:* 46:126–70.

———. 2004. The face of Olduvai Hominid 12. *Journal of Human Evolution* 46:337–47.

Antón, S. C., C. R. Jaslow, and S. M. Swartz. 1992. Sutural complexity in artificially deformed human crania. *Journal of Morphology* 214:321–32.

Antón, S. C., and K. J. Weinstein. 1999. Artificial cranial deformation and fossil Australians revisited. *Journal of Human Evolution* 36:195–209.

Antón, S. C., F. Spoor, C. Fellmann, and C. C. Swisher III. 2007. Defining *Homo erectus:* Size considered. In *Handbook of Paleoanthropology,* ed. W. Henke, H. Rothe, and I. Tattersall, 1655–93. Berlin: Springer.

Arambourg, C. 1963. Le gisement de Ternifine, II: *L'Atlanthropus mauritanicus. Archives de l'Institut de Paléontologie Humaine, mémoire* 32:37–190.

Arambourg, C., and Y. Coppens. 1968. Decouverte d'un Australopithecien nouveau dans les gisements de l'Omo (Ethiopia). *South African Journal of Science* 64: 58–59.

Ardran, G. M., F. H. Kemp, and J. Lind. 1958. A cineradiographic study of bottle feeding. *British Journal of Radiology* 31:11–22.

Arensburg, B., M. Harell and H. Nathan. 1981. The human middle ear ossicles: Morphometry, and taxonomic implications. *Journal of Human Evolution* 10: 197–205.

Arensburg, B., A.-M. Tillier, B. Vandermeersch, H. Duday, L. A. Schepartz, and Y. Rak. 1989. A Middle Palaeolithic human hyoid bone. *Nature* 338:758–60.

Arensburg, B., I. Pap, and A.-M. Tillier. 1996. The Subalyuk 2 middle ear stapes. *International Journal of Osteoarchaeology* 6:185–88.

Argue, D., D. Donlon, C. Groves, and R. Wright. 2006. *Homo floresiensis:* Microcephalic, pygmoid, *Australopithecus,* or *Homo? Journal of Human Evolution* 51:360–74.

Arsuaga, J. L., I. Martínez, A. Gracia, J. M. Carretero, and E. Carbonell. 1993. Three new human skulls from the Sima de los Huesos Middle Pleistocene site in Sierra de Atapuerca, Spain. *Nature* 362:534–37.

Arsuaga, J. L., I. Martínez, A. Gracia, J. M. Carretero, C. Lorenzo, and N. García. 1997. Sima de los Huesos (Sierra de Atapuerca, Spain): The site. *Journal of Human Evolution* 33:109–27.

Arzarello, M., F. Marcolini, G. Pavia, M. Pavia, C, Petronio, M. Petrucci, L. Rook, and R. Sardella. 2006. Evidence of earliest human occurrence in Europe: The site of Pirro Nord (southern Italy). *Naturwissenschaften* 94:107–12.

Ascenzi, A., I. Biddittu, P. F. Cassoli, A. G. Segre, and E. Segre Naldini.1996. A calvarium of late *Homo erectus* from Ceprano, Italy. *Journal of Human Evolution* 31: 409–23.

Asfaw B., T. D. White, C. O. Lovejoy, G. Suwa, and S. Simpson. 1999. *Australopithecus garhi:* A new species of early hominid from Ethiopia. *Science* 284:629–35.

Asfaw, B., W. H. Gilbert, Y. Beyene, W. K. Hart, P. Renne, G. WoldeGabriel, E. Vrba, and T. D. White. 2002. Remains of *H. erectus* from Bouri, Middle Awash, Ethiopia. *Nature* 416:317–20.

Ashton, E. H., and S. Zuckerman. 1956. Age changes in the position of the foramen magnum in hominoids. *Journal of the Zoological Society of London* 126:315–25.

Atchley, W. R., and B. K. Hall. 1991. A model for development and evolution of complex morphological structures. *Biology Review of the Cambridge Philosophical Society* 66:101–57.

Avril, C. 1963. Kehlkopf und Kehlsack des Schimpansen, *Pan troglodytes*. *Gegenbaurs morphologisches Jahrbuch* 105:74–129.

Aylor, D. 1972 Noise reduction by vegetation and ground. *Journal of the Acoustical Society of America* 51:197–205.

Baab, K. L. 2008. A re-evaluation of the taxonomic affinities of the early *Homo* cranium KNM-ER 42700. *Journal of Human Evolution* 5:741–46.

Baab, K. L., and K. P. McNulty. 2009. Size, shape, and asymmetry in fossil hominins: The status of the LB1 cranium based on 3D morphometric analyses. *Journal of Human Evolution* 57:608–22.

Baba, H., F. Aziz, Y. Kaifu, G. Suwa, R. T. Kono, and T. Jacob. 2003. *Homo erectus* calvarium from the Pleistocene of Java. *Science* 299:1384–88.

Babineau, T. A., and J. H. Kronman. 1969. A cephalometric evaluation of the cranial base in microcephaly. *Angle Orthodontics* 39:57–63.

Bach-Peterson, S., and I. Kjaer. 1993. Ossification of lateral components in the prenatal cranial base. *Journal of Craniofacial Genetics and Developmental Biology* 13:76–82.

Bach-Petersen, S., I. Kjaer, and B. Fischer-Hansen. 1994. Prenatal development of the human osseous temporomandibular region. *Journal of Craniofacial Genetics and Developmental Biology* 14:135–43.

Badam, G. L., V. D. Misra, J. N. Pal, and J. N. Pandey. 1989. A preliminary study of Pleistocene fossils from the Middle Son Valley, Madhya Pradesh. *Man and Environment* 13:41–47.

Bailey, S. E. 2005. Diagnostic dental differences between Neandertals and Upper Paleolithic modern humans: Getting to the root of the matter. In *Current Trends in Dental Morphology Research,* ed. E. Zadzinska, 201–10. Lodz, Poland: University of Lodz Press.

———. 2006. Beyond shovel shaped incisors: Neandertal dental morphology in a comparative context. *Periodicum Biologorum* 108:253–67.

———. 2007. The evolution of non-metric dental variation in Europe. *Mitteilungen der Gesellschaft für Urgeschichte* 15:9–30.

Baker, M. A., and L. W. Chapman. 1977. Rapid brain cooling in exercising dogs. *Science* 195:781–83.

Bakker, A. D., K. Soejima, J. Klein-Nulend, and E. H. Burger. 2001. The production of nitric oxide and prostaglandin E(2) by primary bone cells is shear stress dependent. *Journal of Biomechanics* 34:671–77.

Bakker, A. D., M. Joldersma, J. Klein-Nulend, and E. H. Burger. 2003. Interactive effects of PTH and mechanical stress on nitric oxide and PGE2 production by primary mouse osteoblastic cells. *American Journal of Physiology—Endocrinology and Metabolism* 285:E608–E613.

Ballachanda, B. B. 1997. Theoretical and applied external ear acoustics. *Journal of the American Academy of Audiology* 6:411–20.

Bamford, O., V. Taciak, and I. H. Gewolb. 1992. The relationship between rhythmic swallowing and breathing during suckle feeding in term neonates. *Pediatric Research* 31:619–24.

Bamiou, D. E., S. Worth, P. Phelps, T. Sirimanna, and K. Rajput. 2001. Eighth nerve aplasia and hypoplasia in cochlear implant candidates: The clinical perspective. *Otology and Neurotology* 22:492–96.

Barker, G., H. Barton, M. Bird, P. Daly, I. Datan, A. Dykes, L. Farr,et al. 2007. The "human revolution" in lowland tropical Southeast Asia: The antiquity and behavior of anatomically modern humans at Niah Cave (Sarawak, Borneo). *Journal of Human Evolution* 52:243–61.

Barteczko, K., and M. Jacob. 1999. Comparative study of the shape, course, and disintegration of the rostral notochord in some vertebrates, especially humans. *Anatomy and Embryology* 200:345–66.

Barton, N. H. 2006. Evolutionary biology: How did the human species form? *Current Biology* 16:647–50.

Barton, R. A., and P. H. Harvey. 2000. Mosaic evolution of brain structure in mammals. *Nature* 405:1055–58.

Bar-Yosef, O. 1998. The chronology of the Middle Paleolithic of the Levant. In *Neandertals and Modern Humans in Western Asia,* ed. T. Akazawa, K. Aoki, and O. Bar-Yosef, 39–56. New York: Plenum.

———. 2000. The Middle and Upper Paleolithic in Southwest Asia and neighboring regions. In *The Geography of Neandertals and Modern Humans in Europe and the Greater Mediterranean,* ed. O. Bar-Yosef and D. Pilbeam, 107–56. Cambridge, MA: Peabody Museum Bulletin 8.

———. 2006. Neanderthals and modern humans: A different interpretation. In *Neanderthals and Modern Humans Meet,* ed. N. J. Conard, 165–87. Tübingen: Tübingen Publications in Prehistory, Kerns Verlag.

Basmajian, J. V., and C. J. DeLuca. 1985. *Muscles Alive: Their Functions Revealed by Electromyography.* Baltimore, MD: Williams and Wilkins.

Bass, S. L., L. Saxon, R. M. Daly, C. H. Turner, A. G. Robling, E. Seeman, and S. Stuckey. 2002. The effect of mechanical loading on the size and shape of bone in pre-, peri-, and postpubertal girls: A study in tennis players. *Journal of Bone and Mineral Research* 17:2274–80.

Bastir, M., and A. Rosas. 2004a. Facial heights: Evolutionary relevance of postnatal ontogeny for facial orientation and skull morphology in humans and chimpanzees. *Journal of Human Evolution* 47:359–81.

———. 2004b. Comparative ontogeny in humans and chimpanzees: Similarities, differences and paradoxes in postnatal growth and development of the skull. *Annals of Anatomy* 186:503–9.

———. 2005. Hierarchical nature of morphological integration and modularity in the human posterior face. *American Journal of Physical Anthropology* 128:26–34.

———. 2006. Correlated variation between the lateral basicranium and the face: A geometric morphometric study in different human groups. *Archives of Oral Biology* 51:814–24.

Bastir, M., P. O'Higgins, and A. Rosas. 2007. Facial ontogeny in Neanderthals and modern humans. *Proceedings of the Royal Society of London, Series B* 274:1125–32.

Bastir, M., A. Rosas, D. E. Lieberman, and P. O'Higgins. 2008. Middle cranial fossa anatomy and the origin of modern humans. *Anatomical Record* 291:130–40.

Baume, L. J. 1951. The postnatal growth of the maxilla in *Macaca mulatta*. *Journal of Dental Research* 30:501–2.

———. 1957. A biologist looks at the sella point. *Transactions of the European Orthodontic Society* 33:150–59.

Beckman, M. E., T.-P. Jung, S. Lee, K. de Jong, A. K. Krishnamurthy, S. C. Ahalt, K. B. Cohen, and M. Collins. 1995. Variability in the production of quantal vowels revisited. *Journal of the Acoustical Society of America* 97:471–90.

Becquet, C., N. Patterson, A. C. Stone, M. Przeworski, and D. Reich. 2007. Genetic structure of chimpanzee populations. *PLoS Genetics* 3: E66.

Beecher, R. M., R. S. Corruccini, and M. Freeman 1983. Craniofacial correlates of dietary consistency in a nonhuman primate. *Journal of Craniofacial Genetics and Developmental Biology* 3:193–202.

Begley, S. 2007. The new science of human evolution. *Newsweek,* 149:53–58.

Behe, M. 1996. Darwin's Black Box: The Biochemical Challenge to Evolution. New York: The Free Press.

Behrens, M., and W. Meyerhof. 2006. Bitter taste receptors and human bitter taste perception. *Cellular and Molecular Life Sciences* 63:1501–9.

Behrensmeyer, A. K., N. E. Todd, R. Potts, and G. E. McBrinn. 1997. Late Pliocene faunal turnover in the Turkana Basin, Kenya and Ethiopia. *Science* 278:1589–94.

Behrents, R. G., D. S. Carlson, and T. Abdelnour. 1978. *In vivo* analysis of bone strain about the sagittal suture in *Macaca mulatta* during masticatory movements. *Journal of Dental Research* 57:904–8.

Behrents, R. G., and E. F. Harris. 1991. The premaxillary-maxillary suture and orthodontic mechanotherapy. *American Journal of Orthodontics and Dentofacial Orthopedics* 99:1–6.

Berge, C., and X. Penin. 2004. Ontogenetic allometry, heterochrony, and inter-specific differences in the skull of African apes, using tridimensional Procrustes analysis. *American Journal of Physical Anthropology* 124:124–38.

Berger, L. R., D. J. de Ruiter, S. E. Churchill, P. Schmid, K. J. Carlson, P. H. G. M. Dirks, and J. M. Kibii. 2010. *Australopithecus sediba:* A new species of *Homo*-like australopith from South Africa. *Science* 328:195–204.

Bergsneider, M. 2001. Evolving concepts of cerebrospinal fluid physiology. *Neurosurgery Clinics of North America* 12:631–8.

Berkovitz, B. K. B., B. J. Moxham, and H. N. Newman. 1995. *The Periodontal Ligament in Health and Disease.* 2nd ed. London: Mosby-Wolfe.

Bermúdez de Castro, J. M., J. L. Arsuaga, E. Carbonell, A. Rosas, I. Martínez, and M. Mosquera. 1997. A hominid from the lower Pleistocene of Atapuerca, Spain. *Science* 276:1392–95.

Bernal, V., S. I. Perez, and P. N. Gonzalez. 2006. Variation and causal factors of craniofacial robusticity in Patagonian hunter-gatherers from the late Holocene. *American Jounral of Human Biology* 18:748–65.

Bertram, J. E. A., and S. M. Swartz. 1991. The "law of bone transformation": A case of crying Wolff? *Biological Review* 66:245–73.

Beynon, A. D., and B. A. Wood. 1986. Variations in enamel thickness and structure in East African hominids. *American Journal of Physical Anthropology* 70:177–93.

———. 1987. Patterns and rates of enamel growth in the molar teeth of early hominids. *Nature* 326:493–96.

Beynon, A. D., and M. C. Dean. 1988. Distinct dental development patterns in early fossil hominids. *Nature* 335:509–14.

Bhatty, N., A. H. Gilani, and S. A. Nagra. 2000. Nutritional value of mung bean (*Vigna radiata*) as affected by cooking and supplementation. *Archivos latino-americanos de nutrición* 50:374–79.

Bianchi, M. 1989. The thickness, shape and arrangement of elastic fibres within the nuchal ligament from various animal species. *Anatomischer Anzeiger* 169: 53–66.

Biegert, J. 1957. Der Formandel des Primateschïndels und seine Bezeihungen zur ontogenetischen Entwicklung und den phylogenetischen Spezialisationen der Kopforgane. *Morphologisches Jahrbuch* 98:77–199.

Biegert, J. 1963. The evaluation of characters of the skull, hands and feet for primate taxonomy. In *Classification and Human Evolution,* ed. S. L. Washburn, 116–45. Chicago: Aldine de Gruyter.

Biewener, A. A. 1991. Musculoskeletal design in relation to body size. *Journal of Biomechanics* 24:19–29.

———. 1992. *In vivo* measurement of bone strain and tendon force. In *Biomechanics: Structures and Systems; A Practical Approach,* ed. A. A. Biewener, 123–47. Oxford: Irl Press.

Biewener, A. A., and J. E. A. Bertram. 1993. Skeletal strain patterns in relation to exercise training during growth. *Journal of Experimental Biology* 185:51–69.

Biewener, A. A., G. W. Soghikian, and A. W. Crompton. 1985. Regulation of respiratory airflow during panting and feeding in the dog. *Respiration Physiology* 61:185–95.

Biewener, A. A., and C. R. Taylor. 1986. Bone strain: A determinant of gait and speed? *Journal of Experimental Biology* 123:383–400.

Biewener, A. A., S. M. Swartz, and J. E. A. Bertram. 1986. Bone modeling during growth: Dynamic strain equilibrium in the chick tibiotarsus. *Calcified Tissue International* 39:390–95.

Biewener, A. A., N. L. Fazzalari, D. D. Konieczynski, and R. V. Baudinette. 1996. Adaptive changes in trabecular architecture in relation to functional strain patterns and disuse. *Bone* 19:1–8.

Bikle, D. D., T. Sakata, and B. P. Halloran. 2003. The impact of skeletal unloading on bone formation. *Gravitational and Space Biology Bulletin* 16:45–54.

Bilsborough, A., and B. A. Wood. 1988. Cranial morphometry of early hominids: Facial region. *American Journal of Physical Anthropology* 76:61–86.

Bischoff, J. L., D. D. Shamp, A. Aramburu, J. L. Arsuaga, E. Carbonell, and J. M. Bermúdez de Castro. 2003. The Sima de los Huesos hominids date to beyond U/Th equilibrium (>350 kyr) and perhaps to 400–500 kyr: New radiometric dates. *Journal of Archaeological Science* 30:275–80.

Bischoff, J. L., R. W. Williams, R. J. Rosenbauer, A. Aramburu, J. L. Arsuaga, N. Garcia, and G. Cuenca-Bescós. 2007. High-resolution U-series dates from the Sima de los Huesos hominids yields 600 kyrs: Implications for the evolution of the early Neanderthal lineage. *Journal of Archaeological Science* 34:763–70.

Bishnoi, S., and N. Khetarpaul. 1994. Protein digestability of vegetables and field peas (*Pisum sativum*): Varietal differences and effect of domestic processing and cooking methods. *Plant Foods for Human Nutrition* 46:71–6.

Bisulco, S., and B. Slotnick. 2003. Olfactory discrimination of short chain fatty acids in rats with large bilateral lesions of the olfactory bulbs. *Chemical Senses* 28: 361–70.

Björk, A. 1955. Cranial base development. *American Journal of Orthodontics* 41: 198–225.

Blaney, S. P. A. 1986. An allometric study of the frontal sinus in *Gorilla, Pan* and *Pongo. Folia Primatologica* 47:81–96.

———. 1990. Why paranasal sinuses? *Journal of Laryngology and Otology* 104: 690–93.

Blaxter, K. 1989. *Energy metabolism in animals and man.* Cambridge: Cambridge University Press.

Blumenberg, B. 1985. Biometrical studies upon hominoid teeth: The coefficient of variation, sexual dimorphism and questions of phylogenetic relationship. *Biosystems* 18:149–84.

Blumenschine, R. J. 1987. Characteristics of an early hominid. scavenging niche. *Current Anthropology* 28:383–407.

Boaz, N. T., and R. L. Ciochon. 2004. Headstrong hominids. *Natural History* 113: 29–34.

Boaz, N. T., and F. C. Howell. 1977. A gracile hominid cranium from upper Member G of the Shungura Formation, Ethiopia. *American Journal of Physical Anthropology* 46:93–108.

Bobe R., A. K. Behrensmeyer, and R. E. Chapman. 2002. Faunal change, environmental variability and late Pliocene hominin evolution. *Journal of Human Evolution* 42:475–97.

Boesch, C., and H. Boesch. 1990. Tool use and tool making in wild chimpanzees. *Folia Primatologica* 54:86–99.

Bogin, B. 2001. *The Growth of Humanity.* New York: Wiley-Liss.

Bolk, L. 1910. On the slope of the foramen magnum in primates. *Koninklijke Nederlandse Akademie van Wetenschappen* 12:525–34.

———. 1926. *Das Problem der Menschwerdung.* Jena, Germany: Gustav Fischer.

Bond, Z. S. 1976. Identification of vowels excerpted from neutral and nasal contexts. *Journal of the Acoustical Society of America* 59:1229–32.

Bookstein, F. L. 1991. *Morphometric Tools for Landmark Data: Geometry and Biology.* Cambridge: Cambridge University Press.

Bookstein, F. L., P. Gunz, P. Mitteroecker, H. Prossinger, K. Schaefer, and H. Seidler. 2003. Cranial integration in *Homo:* Singular warps analysis of the midsagittal plane in ontogeny and evolution. *Journal of Human Evolution* 44:167–87.

Bosma, J. F. 1986. *Anatomy of the Infant Head.* Baltimore, MD: Johns Hopkins University Press.

———. 1992. Pharyngeal swallow: Basic mechanisms, development, and impairments. *Advances in Otolaryngology—Head and Neck Surgery* 6:225–75.

Bouvier, M., and W. L. Hylander. 1982. The effect of dietary consistency on morphology of the mandibular condylar cartilage in young macaques (*Macaca mulatta*). *Progress in Clinical and Biological Research* 101:569–79.

———. 1996. The mechanical or metabolic function of secondary osteonal bone in the monkey *Macaca fascicularis.* *Archives of Oral Biology* 41:941–50.

Bowler, J. M., H. Johnston, J. M. Olley, J. R. Prescott, R. G. Roberts, W. Shawcross, and N. A. Spooner. 2003. New ages for human occupation and climatic change at Lake Mungo, Australia. *Nature* 421:837–40.

Brace, C. L. 1962. Cultural factors in the evolution of the human dentition. In *Culture and the Evolution of Man,* ed. M. F. A. Montagu, 343–54. New York: Oxford University Press.

————. 1964. The probable mutation effect. *American Naturalist* 98:453–55.

————. 1967. *The Stages of Human Evolution.* 1st ed. Englewood Cliffs, NJ: Prentice-Hall.

————. 1995. *The Stages of Human Evolution.* 5th ed. Englewood Cliffs, NJ: Prentice-Hall.

Brace, C. L., S. L. Smith, and K. D. Hunt. 1991. What big teeth you had, Grandma! Human tooth size, past and present. In *Advances in Dental Anthropology,* ed. M. A. Kelley and C. S. Larsen, 33–57. New York: Wiley-Liss.

Brain, C. K., and A. Sillen. 1988. Evidence from the Swartkrans cave for the earliest use of fire. *Nature* 336:464–66.

Bramble, D. M. 1989. Axial-appendicular dynamics and the integration of breathing and gait in mammals. *American Zoologist* 29:171–86.

Bramble, D. M., and F. A. J. Jenkins Jr. 1993. Mammalian locomotor-respiratory integration: Implications for diaphragmatic and pulmonary design. *Science* 262: 235–40.

Bramble, D. M., and D. E. Lieberman. 2004. Endurance running and the evolution of *Homo. Nature* 432:345–52.

Brattstrom, V., L. Odenrick, and E. Kvam. 1989. Dentofacial morphology in children playing musical wind instruments: A longitudinal study. *European Journal of Orthodontics* 11:179–85.

Brattstrom, V., L. Odenrick, and R. Leanderson. 1991. Dentofacial morphology in professional opera singers. *Acta Odontologica Scandinavica* 49:147–51.

Bräuer, G. 2008. The origin of modern anatomy: By speciation or intraspecific evolution? *Evolutionary Anthropology* 17:22–37.

Bräuer, G., and R. Leakey. 1986. The ES-11693 cranium from Eliye Springs, West Turkana, Kenya. *Journal of Human Evolution* 15:289–312.

Bräuer, G., C. Groden, F. Gröning, A. Kroll, K. Kupczik, E. Mbua, A. Pommert, and T. Schiemann. 2004. Virtual study of the endocranial morphology of the matrix-filled cranium from Eliye Springs, Kenya. *Anatomical Record* 276A:113–33.

Brewer-Marsden, S., D. Marsden, and M. Emery Thompson. 2006. Demographic and female life history parameters of free-ranging chimpanzees at the Chimpanzee Rehabilitation Project, River Gambia National Park. *International Journal of Primatology* 27:391–410.

Briggs, L. C. 1955. *The Stone Age Races of Northwest Africa.*Cambridge, MA: American School of Prehistoric Research Bulletin, no 18.

Bromage, T. G. 1989. Ontogeny of the early hominid face. *Journal of Human Evolution* 18:751–73.

————. 1990. On ontogenetic and phylogenetic origins of the human chin. *American Journal of Physical Anthropology* 83:263–5.

————. 1991. Enamel incremental periodicity in the pig-tailed macaque: A polychrome fluorescent labeling study of dental hard tissues. *American Journal of Physical Anthropology* 86:205–14.

Bromage, T. G., and M. C. Dean. 1985. Re-evaluation of the age at death on immature fossil hominids. *Nature* 317:525–27.

Bromage, T. G., F. Schrenk, and F. W. Zonneveld. 1995. Paleoanthropology of the Malawi Rift: An early hominid mandible from the Chiwondo Beds, northern Malawi. *Journal of Human Evolution* 28:71–105.

Broom, R. 1938. The Pleistocene anthropoid apes of South Africa. *Nature* 142:377–79.

Brothwell, D R. 1960. Upper Pleistocene human skull from Niah Caves, Sarawak. *Sarawak Museum Journal* 9:323–49.

Brown, B., and A. Walker. 1993. The dentition. In *The Nariokotome Homo erectus Skeleton*, ed. A. Walker and R. E. F. Leakey, 161–92. Cambridge, MA: Harvard University Press.

Brown, P. 1981. Artificial cranial deformation: A component in the variation in Pleistocene Australian Aboriginal crania. *Archaeology in Oceania* 16:156–67.

———. 1992. Recent human evolution in East Asia and Australasia. *Philosophical Transactions of the Royal Society London, Series B* 337:235–42.

Brown, P., S. K. Pinkerton, and W. Lambert. 1979. Thickness of cranial vault in Australian Aboriginals. *Archaeology and Physical Anthropology in Oceania* 14:54–71.

Brown, W. A., T. I. Molleson, and S. Chinn. 1984. Enlargement of the frontal sinus. *Annals of Human Biology* 11:221–26.

Brown, P., T. Sutkina, M. J. Morwood, R. P. Soejono, Jatmiko, E. W. Saptomo, and R. A. Due. 2004. A new small-bodied hominin from the Late Pleistocene of Flores, Indonesia. *Nature* 431:1055–61.

Bruce, V., P. R. Green, M. A. Georgeson. 1996. *Visual Perception: Physiology, Psychology and Ecology.* 3rd ed. Hove, U.K.: Taylor and Francis.

Bruner, E. 2004. Geometric morphometrics and paleoneurology: Brain shape evolution in the genus *Homo*. *Journal of Human Evolution* 47:279–303.

Bruner, E., G. Manzi, and J. L. Arsuaga. 2003. Encephalization and allometric trajectories in. the genus *Homo*: Evidence from the Neandertal and and modern lineages. *Proceedings of the National Academy of Sciences* 100:15335–40.

Brunet, M., A. Beauvilain, Y. Coppens, E. Heintz, A. H. E. Moutaye, and D. Pilbeam. 1995. The first australopithecine 2500 kilometres west of the Rift Valley. *Nature* 378:273–75.

Brunet, M., F. Guy, D. Pilbeam, H. T. Mackay, A. Likius, A. Djimboumalbaye, et al. 2002. A new hominid from the upper Miocene of Chad, central Africa. *Nature* 418:145–51.

Brunet, M., F. Guy, D. Pilbeam, D. E. Lieberman, A. Likius, H. T. Mackaye, M. S.

Ponce de León, C. P. Zollikofer, and P. Vignaud. 2005. New material of the earliest hominid from the Upper Miocene of Chad. *Nature* 434:752–55.

Buck, L. B. 2000. The molecular architecture of odor and pheromone sensing in mammals. *Cell* 100:611–18.

———. 2004. Olfactory receptors and odor coding in mammals. *Nutrition Reviews* 62: S184–8.

Buhr, R. D. 1980. The development of vowels in an infant. *Journal of Speech and Hearing Research* 23:75–94.

Bull, J. W. D. 1969. Tentorium cerebelli. *Proceedings of the Royal Society of Medicine* 62:1301–10.

Bunn, H. T. 2001. Hunting, power scavenging, and butchering by Hadza foragers and by Plio-Pleistocene *Homo.* In *Meat-Eating and Human Evolution,* ed. C. B. Stanford, and H. T. Bunn, 199–218. Oxford: Oxford University Press.

Bürglin, T. R. 1994. A comprehensive classification of homeobox genes. In *Guidebook to the Homeobox Genes,* ed. D. Duboule, 25–71. Oxford: Oxford University Press.

Burke, A. C., C. E. Nelson, B. A. Morgan, and C. J. Tabin. 1995. Hox genes and the evolution of vertebrate axial morphology. *Development* 121:333–46.

Burroughs, William J. 2007. *Climate Change in Prehistory: A Multidisciplinary Approach.* 2nd ed. Cambridge: Cambridge University Press.

Buschang, P. H., K. Julien, R. Sachdeva, and A. Demirjian. 1992. Childhood and pubertal growth changes of the human symphysis. *Angle Orthodontist* 62:203–10.

Bush, E., and J. Allman. 2004. The scaling of frontal cortex in primates and carnivores. *Proceedings of the National Academy of Sciences USA* 101:3962–66.

Bütow, K.-W. 1990. Craniofacial growth disturbance after skull base and associated suture synostoses in the newborn Chacma baboon: A preliminary report. *Cleft Palate Journal* 27:241–52.

Buxhoeveden, D. P., and M. F. Casanova 2002. The minicolumn and evolution of the brain. *Brain, Behavior and Evolution* 60:125–51.

Cabanac, M. 1993. Selective brain cooling in humans: "Fancy" or fact? *Federation of American Societies for Experimental Biology Journal* 17:1143–6.

Cabanac, M., and M. Caputa. 1979. Natural selective cooling of the human brain: Evidence of its occurrence and magnitude. *Journal of Physiology* 286:255–64.

Cabanac, M., and Brinnel, H. 1985. Blood flow in the emissary veins of the human head during hyperthermia. *European Journal of Applied Physiology* 54:172–76.

———. 1988. Beards, baldness, and sweat secretion. *European Journal of Applied Physiology and Occupational Physiology* 58:39–46.

Cai, J., G. L. Hoff, P. C. Dew, V. J. Guillory, and J. Manning. 2005. Perinatal periods of risk: Analysis of fetal-infant mortality rates in Kansas City, Missouri. *Maternal and Child Health Journal* 9:199–205.

Calcagno, J. M., and K. R. Gibson. 1988. Human dental reduction: Natural selection or the probable mutation effect? *American Journal of Physical Anthropology* 77:505–17.

Cameron, D., R. Patnaik, and A. Sahni. 2004. The phylogenetic significance of the Middle Pleistocene Narmada hominin cranium from central India. *International Journal of Osteoarchaeology* 14:419–47.

Cameron, J. 1930. Craniometric memoirs, no. II: The human and comparative anatomy of Cameron's craniofacial axis. *Journal of Anatomy* 64:324–36.

Canalis, E., M. Centrella, W. Burch, and T. L. McCarthy. 1989. Insulin-like growth factor I mediates selective anabolic effects of parathyroid hormone in bone cultures. *Journal of Clinical Investigations* 83:60–65.

Carbonell, E., J. M. Bermúdez de Castro, J. L. Arsuaga, J. C. Díez, A. Rosas, G. Cuenca-Bescos, R. Sala, M. Mosquera, and X. P. Rodriguez. 1995. Lower Pleistocene Hominids and Artifacts from Atapuerca-TD6 (Spain). *Science* 269:826–32.

Carbonell, E., M. Mosquera, X. P. Rodríguez, and R. Sala. 1999. Out of Africa: The Dispersal of the Earliest Technical Systems Reconsidered. *Journal of Anthropological Archaeology* 18:119–36.

Carbonell, E., J. M. Bermúdez de Castro, J. L. Arsuaga, E. Allue, M. Bastir, A. Benito, I. Cáceres, et al. 2005. An Early Pleistocene hominin mandible from Atapuerca-TD6, Spain. *Proceedings of the National Academy of Sciences USA* 102:5674–78.

Carbonell, E, J. M. Bermúdez de Castro, J. M. Parés, A. Pérez-González, G. Cuenca-Bescós, A. Ollé, M. Mosquera, et al. 2008. The first hominin of Europe. *Nature* 452:465–69.

Carey, J. W., and A. T. Steegman Jr. 1981. Human nasal protrusion, latitude and climate. *American Journal of Physical Anthropology* 56:313–19.

Carlson, D. S. 1976. Temporal variation in prehistoric Nubian crania. *American Journal of Physical Anthropology* 45:467–84.

Carlson, D. S., and D. P. Van Gerven. 1977. Masticatory function and post-Pleistocene evolution in Nubia. *American Journal of Physical Anthropology* 46:495–506.

Carlsson, G. E., and G. Persson. 1967. Morphological changes in the mandible after extraction and wearing of dentures: A longitudinal, clinical and x-ray cephalometric study covering 5 years. *Odontology Reviews* 18:27–54.

Carre, R., B. Lindblom, and P. MacNeilage. 1995. Acoustic factors in the evolution of the human vocal tract. *Comptes rendus de l'Académie des Sciences de Paris, série 2b* 320:471–76.

Carrier, D. R. 1984. The energetic paradox of human running and hominid evolution. *Current Anthropology* 25:483–96.

———. 2004. The running-fighting dichotomy and the evolution of aggression in hominids. In *From Biped to Strider: The Emergence of Modern Human Walking, Running, and Resourch Transport,* ed. J. Meldrum and C. Hilton, 135–62. New York: Kluwer/Plenum Press.

Carroll, L. 1885. *Alice's Adventures in Wonderland.* London: Macmillan.

Carroll, S. B. 2005. Endless Forms Most Beautiful: The New Science of Evo Devo and the Making of the Animal Kingdom. New York: Norton.

Carroll, S. B., J. K. Grenier, and S. D. Weatherbee. 2001. *From DNA to Diversity: Molecular Genetics and the Evolution of Animal Design.* Malden, MA: Blackwell Scientific.

Carter, D. R., and G. S. Beaupré. 2001. Skeletal Function and Form: Mechanobiology of Skeletal Development, Aging, and Regeneration. New York: Cambridge University Press.

Carter, R. L. 1999. *Mapping the Mind.* Berkeley: University of California Press.

Cartmill, M. 1971. Ethmoid component in the orbit of primates. *Nature* 232:566–67.

———. 1972. Arboreal adaptations and the origin of the order Primates. In *The Functional and Evolutionary Biology of Primates,* ed. R. Tuttle, 97–122. Chicago: Aldine.

———. 1980. Morphology, function and evolution of the anthropoid postorbital septum. In *Evolutionary Biology of the New World Monkeys and Continental Drift,* ed. R. L. Ciochon and A. B. Chiarelli, 243–74. New York: Plenum.

———. 1992. New views on primate origins. *Evolutionary Anthropology* 1:105–11.

Cave, A. J. E. 1973. The primate nasal fossa. *Biological Journal of the Linnaean Society* 5:377–87.

Cave, A. J. E., and R. W. Haines. 1940. The paranasal sinuses of the anthropoid apes. *Journal of Anatomy* 72:493–523.

Cerny, R., P. Lwigale, R. Ericsson, D. Meulemans, H. H. Epperlein, and M. Bronner-Fraser. 2004. Developmental origins and evolution of jaws: New interpretation of "maxillary" and "mandibular." *Developmental Biology* 276:225–36.

Chai, Y., X. Jiang, Y. Ito, J. Han, and D. H. Rowditch. 2000. Fate of the mammalian cranial neural crest during tooth and mandibular morphogenesis. *Development* 127:1671–79.

Chamberlain, A. T., and B. A. Wood. 1985. A reappraisal of variation in hominid mandibular corpus dimensions. *American Journal of Physical Anthropology* 66:399–405.

Chandler, J. R. 1964. Partial occlusion of the external auditory meatus: Its effect upon air and bone conduction hearing acuity. *Laryngoscope* 74:22–54.

Changizi, M. A. 2001. Principles underlying mammalian neocortical scaling. *Biological Cybernetics* 84:207–15.

Changizi, M. A., and S. Shimojo. 2005. Parcellation and area-area connectivity as a function of neocortex size. *Brain Behaviour and Evolution* 66:88–98.

Chatterjee, H. 2006. Phylogeny and biogeography of gibbons: A dispersal-vicariance analysis. *International Journal of Primatology* 27:699–712.

Chen, F., and W. Li. 2001. Genomic divergences between humans and other hominoids and the effective population size of the common ancestor of humans and chimpanzees. *American Journal of Human Genetics* 68:444–56.

Chernoff, B., and P. M. Magwene. 1999. Morphological integration: Forty years later. In *Morphological Integration,* ed. E. C. Olsen, and R. L. Miller, 319–53. Chicago: University of Chicago Press.

Cheverud, J. M. 1982. Phenotypic, genetic, and environmental integration in the cranium. *Evolution* 36:499–516.

———. 1988. A comparison of genetic and phenotypic correlations. *Evolution* 42: 958–968.

———. 1989. A comparative analysis of morphological variation patterns in the papionins. *Evolution* 43:1737–47.

———. 1995. Morphological integration in the saddle-back tamarin (*Saguinus fuscicollis*) cranium. *American Naturalist* 145:63–89.

———. 1996. Developmental integration and the evolution of pleiotropy. *American Zoologist* 36:44–50.

Cheverud, J. M., L. A. Kohn, L. W. Konigsberg, and S. R. Leigh. 1992. Effects of fronto-occipital artificial cranial vault modification on the cranial base and face. *American Journal of Physical Anthropology* 88:323–45.

Chimpanzee Sequencing and Analysis Consortium. 2005. Initial sequence of the chimpanzee genome and comparison with the human genome. *Nature* 437: 69–87.

Chizhikov, V. V., and K. J. Millen. 2004. Mechanisms of roof plate formation in the vertebrate CNS. *Nature Reviews: Neuroscience* 5:808–12.

Chu, B. C., A. Narita, K. Aoki, T. Yoshida, T. Warabi, and K. Miyasaka. 2000. Flow volume in the common carotid artery detected by color duplex sonography: An approach to the normal value and predictability of cerebral blood flow. *Radiation Medicine* 18:239–44.

Churchill, S. E., L. L. Shackelford, J. N. Georgi, and M. T. Black. 2004. Morphological variation and airflow dynamics in the human nose. *American Journal of Human Biology* 16:625–38.

Ciochon, R. L., R. A. Nisbett, and R. S. Corruccini. 1997. Dietary consistency and craniofacial development related to masticatory function in minipigs. *Journal of Craniofacial Genetics and Developmental Biology* 17:96–102.

Clark, D. D., and L. Sokoloff. 1999. Circulation and energy metabolism of the brain. In *Basic Neurochemistry: Molecular, Cellular And Medical Aspects,* 6th ed., ed. G. J. Siegel, B. W. Agranoff, R. W. Albers, S. K. Fisher, and M. D. Uhler, 637–70. Philadelphia: Lippincott-Raven.

Clemens, T. L., and S. D. Chernausek. 2004. Genetic strategies for elucidating insulin-like growth factor action in bone. *Growth Hormone and IGF Research* 14:195–99.

Cobb, S. N., and P. O'Higgins. 2004. Hominins do not share a common postnatal facial ontogenetic shape trajectory. *Journal of Experimental Zoology,* 302B: 302–21.

Coehlo, A. M. J. 1986. Time and energy budgets. In *Comparative Primate Biology,* ed. G. Mitchell, and J. Erwin, 141–66. New York: Alan R. Liss, Inc.

Cohen, M. M., Jr. 2000a. Sutural biology. In *Craniosynostosis: Diagnosis, Evaluation and Management,* ed. M. M. Cohen, and R. E. MacLean, 11–23. New York: Oxford University Press.

———. 2000b. Merging the old skeletal biology with the new, I: Intramembranous ossification, endochondral ossification, ectopic bone, secondary cartilage, and pathologic considerations. *Journal of Craniofacial Genetics and Developmental Biology* 20:84–93.

Cole, P. 1982. Modification of inspired air. In *The nose: Upper airway physiology and the atmospheric environment,* ed. D. F. Proctor and I. Andersen, 351–75. Amsterdam: Elsevier Biomedical Press.

Coleman, M. N., and C. F. Ross. 2004. Primate auditory diversity and its influence on hearing perfomance. *Anatomical Record* 281A: 1123–37.

Compston, J. E. 2001. Sex steroids and bone. *Physiological Reviews* 81:419–47.

Conan Doyle, A. 1902. *The Hound of the Baskervilles.* New York: Grosset and Dunlap.

Conklin-Brittain, N. L., R. W. Wrangham, and C. C. Smith. 2002. A two-stage model of increased dietary quality in early hominid evolution: The role of fiber. In *Human Diet: Its Origin and Evolution,* ed. P. Ungar and M. Teaford, 61–76. Westport, CT: Bergin and Garvey.

Conley, M. S., R. A. Meyer, J. J. Bloomberg, D. L. Feeback, and G. A. Dudley. 1995. Noninvasive analysis of human neck muscle function. *Spine* 20:2505–12.

Conroy, G. C., and C. J. Mahoney. 1991. Mixed longitudinal study of dental emergence in the chimpanzee, *Pan troglodytes* (Primates, Pongidae). *American Journal of Physical Anthropology* 86:243–54.

Cook, I. J., W. J. Dodds, R. A. Dantas, B. T. Massey, M. K. Kem, I. M. Lang, et al. 1989. Opening mechanisms of the human upper esophageal sphincter. *Dysphagia* 4:8–15.

Coon, C. S. 1962. *The Origins of Races.* New York: Alfred A. Knopf.

Cooper, B. C. 1989. Nasorespiratory function and orofacial development. *Otolaryngologic Clinics of North America* 22:413–41.

Copp, A. J., N. D. E. Greene, and J. N. Murdoch. 2003. The genetic basis of mammalian neurulation. *Nature Reviews: Genetics* 4:784–93.

Coppens, Y. 1980. The differences between *Australopithecus* and *Homo:* Preliminary conclusions from the Omo Research Expedition's studies. In *Current Argument on Early Man,* ed. L. G. Königsson, 207–25. Oxford: Pergamon Press.

Coqueugniot, H., J.–J. Hublin, F. Veillon, F. Houet, and T. Jacob. 2004. Early brain growth in *Homo erectus* and implications for cognitive ability. *Nature* 431: 299–302.

Cordain, L, S. B. Eaton, J. Brand Miller, S. Lindeberg, and C. Jensen. 2002. An evolutionary analysis of the aetiology and pathogenesis of juvenile-onset myopia. *Acta Ophthalmologica Scandinavica* 80:125–35.

Corruccini, R. S. 1984. An epidemiologic transition in dental occlusion in world populations. *American Journal of Orthodontics and Dentofacial Orthopaedics* 86: 419–26.

———. 1990. Australian aboriginal tooth succession, interproximal attrition, and Begg's theory. *American Journal of Orthodontics and Dentofacial Orthopaedics* 97:349–57.

———. 1999. How Anthropology Informs the Orthodontic Diagnosis of Malocclusion's Causes. Lewiston, NY: Mellen Press.

Corruccini, R. S., and R. M. Beecher. 1982. Occlusal variation related to soft diet in a nonhuman primate. *Science* 218:74–6.

———. 1984. Occlusofacial morphological integration lowered in baboons raised on soft diet. *Journal of Craniofacial Genetics and Developmental Biology* 4:135–42.

Corruccini, R. S., A. M. Henderson, and S. S. Kaul. 1985. Bite-force variation related to occlusal variation in rural and urban Punjabis (North India). *Archives of Oral Biology* 30:65–69.

Corruccini, R. S., G. C. Townsend, L. C. Richards, and T. Brown. 1990. Genetic and environmental determinants of dental occlusal variation in twins of different nationalities. *Human Biology* 62:353–67.

Corti, O., C. Hampe, F. Darios, P. Ibanez, M. Ruberg, and A. Brice. 2005. Parkinson's disease: From causes to mechanisms. *Comptes rendus biologiques* 328:131–42.

Cottrill, C. P., C. W. Archer, and L. Wolpert. 1987. Cell sorting and aggregate formation in micromass culture. *Developmental Biology* 122:503–15.

Coulombre, A. J. 1965. The eye. In *Organogenesis,* ed. R. L. Dehaan and J. Ursprung, 219–15. New York: Holt, Reinhart and Winston.

Coulombre, A. J., J. L. Coulombre, and H. Mehta. 1962. The skeleton of the eye, I: Conjunctival papillae and scleral ossicles. *Developmental Biology* 5:382–401.

Couly, G., A. Grappin-Botton, P. Coltey, and N. M. Le Duarin. 1996. The regeneration of the cephalic neural crest, a problem revisited: The regenerating cells originate from the contralateral or from the anterior and posterior neural fold. *Development* 122:3393–407.

Count, E. W. 1947. Brain and body weight in man: Their antecedents in growth and evolution. *Annals of the New York Academy of Sciences* 46:993–1122.

Courtiss, E. H., and R. M. Goldwyn. 1993. The effects of nasal surgery on airflow. *Plastic Reconstructive Surgery* 72:9–21.

Courtiss, E. H., T. J. Gargan, and G. B. Courtiss. 1984. Nasal physiology. *Annals of Plastic Surgery* 13:214–23.

Cowin, S. C., S. Weinbaum, and Y. Zeng. 1995. A case for bone canaliculi as the anatomical site for strain generated potentials. *Journal of Biomechanics* 28:1281–7.

Cowin, S. C., and M. L. Moss. 2001. Mechanosensory mechanisms in bone. In *Bone Biomechanics Handbook,* 2nd ed., ed. S. C. Cowin, 1–29. Boca Raton, FL: CRC Press.

Craven, A. H. 1956. Growth in the width of the head of the *Macaca* rhesus monkey as revealed by vital staining. *American Journal of Orthodontics* 42:341–62.

Cray, J. Jr., R. S. Meindl, C. C. Sherwood, and C O. Lovejoy. 2008. Ectocranial suture closure in *Pan troglodytes* and *Gorilla gorilla:* Pattern and phylogeny. *American Journal of Physical Anthropology* 136:394–99.

Croft, M. A., and P. L. Kaufman. 2006. Accommodation and presbyopia: The ciliary neuromuscular view. *Ophthalmology Clinics of North America* 19:13–24.

Croignier, E. 1981. The influence of climate on the physical diversity of European and Mediterranean populations. *Journal of Human Evolution* 10:611–14.

Crompton, A. W. 1981. The origin of mammalian occlusion. In *Orthodontics,* ed. H. G. Barrer, 3–18. Philadelphia: University of Pennsylvania Press.

————. 1995. Masticatory function in nonmammalian cynodonts and early mammals. In *Functional Morphology in Vertebrate Paleontology,* ed. J. Thomason, 55–75. Cambridge: Cambridge University Press.

Crompton, A. W., R. Z. German, and A. J. Thexton. 1997. Mechanisms of swallowing and airway protection in infant mammals (*Sus domesticus* and *Macaca fascicularis*). *Journal of Zoology, London* 241:89–102.

Crompton, A. W., D. E. Lieberman, and S. Aboelela. 2006. Tooth orientation during occlusion and the functional significance of condylar translation in primates and herbivores. In *Amniote Paleobiology: Perspectives on the Evolution of Mammals, Birds, and Reptiles,* ed. M. Carrano, T. Gaudlin, R. Blob, and J. Wible, 367–88. Chicago: University of Chicago Press.

Crompton, A. W., D. E. Lieberman, T. Owerkowicz, R. V. Baudinette, and J. Skinner. 2008. Motor control of masticatory movements in the southern hairy-nosed wombat (*Lasiorhinus latifrons*). In *Primate Craniofacial Function and Biology,* ed. C. Vinyard, M. J. Ravosa and C. Wall, 83–111. New York: Springer.

Cromwell, R. L., R. A. Newton, and L. G. Carlton. 2001. Horizontal plane head stabilization during locomotor tasks. *Journal of Motor Behavior* 33:49–58.

Currey, J. D. 2002. *Bones: Structure and Mechanics.* Princeton, NJ: Princeton University Press.

Dabelow, A. 1929. Über Korrelationen in der phylogenetischen Entwicklung der Schädelform, I. *Morphologisches Jahrbuch* 63:1–49.

Daegling, D. J. 1989. Biomechanics of cross-sectional size and shape in the hominoid mandibular corpus. *American Journal of Physical Anthropology* 80:91–106.

———. 1993. Functional morphology of the human chin. *Evolutionary Anthropology* 1:170–77.

———. 2002. Bone geometry in cercopithecoid mandibles. *Archives of Oral Biology* 47:315–25.

Daegling, D. J., and F. E. Grine. 1991. Compact bone distribution and biomechanics of early hominid mandibles. *American Journal of Physical Anthropology* 86: 321–39.

Daegling, D. J., and W. L. Hylander. 1997. Occlusal forces and mandibular bone strain: Is the primate jaw "overdesigned"? *Journal of Human Evolution* 1997 33:705–17.

Daegling, D. J., M. J. Ravosa, K. R. Johnson, and W. L. Hylander. 1992. Influence of teeth, alveoli, and periodontal ligaments on torsional rigidity in human mandibles. *American Journal of Physical Anthropology* 89:59–72.

Damasio, H., T. Grabowski, R. Frank, A. M. Galaburda, and A. R. Damasio. 1994. The return of Phineas Gage: Clues about the brain from the skull of a famous patient. *Science* 264:1102–5.

Dart, R. 1922. The misuse of the term 'visceral.' *Journal of Anatomy* 56:167–76.

———. 1925. *Australopithecus africanus:* The Man Ape of Southern Africa. *Nature* 115:195–99.

Darwin, C. R. 1859. *On the Origin of Species.* London: John Murray.

———. 1871. The Descent of Man, and Selection in Relation to Sex. London: John Murray.

Dattani, M. T., and I. C. Robinson. 2000. The molecular basis for developmental disorders of the pituitary gland in man. *Clinical Genetics* 57:337–46.

David, T., S. Smye, T. James, and T. Dabbs. 1997. Time-dependent stress and displacement of the eye wall tissue of the human eye. *Medical Engineering and Physics* 19:131–9.

Davidovitch, Z. 1991. Tooth movement. *Critical Reviews in Oral Biology and Medicine* 2:411–50.

Day, M. H. 1969. Omo human skeletal remains. *Nature* 222:1135–38.

Day, M. H., and C. B. Stringer. 1982. A reconsideration of the Omo Kibish remains and the erectus-sapiens transition. In *Homo erectus et la Place de l'Homme de Tautavel parmi les Hominidés Fossiles,* ed. M. A. de Lumley, 814–46. Nice: Premier Congrès International de Paléontologie Humaine.

de Beer, G. R. 1937. *The Development of the Vertebrate Skull.* Oxford: Oxford University Press.

———. 1958. *Embryos and Ancestors.* 3rd ed. Oxford: Oxford University Press.

de Heinzelin, J., J. D. Clark, T. D. White, W. Hart, P. Renne, G. WoldeGabriel, Y. Beyene, and E. S. Vrba. 1999. Environment and behavior of 2.5-million-year-old Bouri hominids. *Science* 284:625–29.

de Ricqlès, A., F. J. Meunier, J. Castanet, and H. Francillon-Vieillot. 1991. Comparative microstructure of bone. In *Bone,* vol 3, ed. B. K. Hall, 1–77. Boca Raton, FL: CRC Press.

de Winter, W., and C. E. Oxnard. 2001. Evolutionary radiations and convergences in the structural organization of mammalian brains. *Nature* 409:710–14.

Deacon, H. J., and J. Deacon. 1999. *Human Beginnings in South Africa: Uncovering the Secrets of the Stone Age.* Cape Town: David Philip Publishers.

Deacon, T. W. 1998. The Symbolic Species: The Co-evolution of Language and the Brain. New York: W. W. Norton.

Dean, D., J.-J. Hublin, R. Holloway, and R. Ziegler. 1998. On the phylogenetic position of the pre-Neandertal specimen from Reilingen, Germany. *American Journal of Physical Anthropology* 34:485–508.

Dean, M. C. 1985a. Comparative myology of the hominoid cranial base, I: The muscular relationships and bony attachments of the digastric muscle. *Folia Primatologica* (Basel). 43:234–48.

———. 1985b. Comparative myology of the hominoid cranial base, II: The muscles of the prevertebral and upper pharyngeal region. *Folia Primatologica* (Basel). 44:40–51.

———. 1987. Growth layers and incremental markings in hard tissues: A review of the literature and some preliminary observations about enamel structure in *Paranthropus boisei. Journal of Human Evolution* 16:157–72.

———. 1988. Growth processes in the cranial base of hominoids and their bearing on morphological similarities that exist in the cranial base of *Homo* and *Paranthropus.* In *Evolutionary History of the "Robust" Australopithecines,* ed. F. E. Grine, 107–12. New York: Aldine de Gruyter.

———. 1995. Variation in the developing root cone angle of the permanent mandibular teeth of modern man and certain fossil hominoids. *Journal of Anatomy* 98:121–31.

———. 2006. Tooth microstructure tracks the pace of human life-history evolution. *Proceedings of the Royal Society of London, Series B: Biological Sciences* 273: 2799–808.

———. 2009. Growth in tooth height and extension rates in modern human and fossil hominin canines and molars. In *Frontiers of Oral Biology: Interdisciplinary Dental Morphology,* ed. T. Koppe, G. Meyer and K. W. Alt, 68–73. Basel: Karger.

Dean, M. C., and B. H. Smith. 2009. Growth and development of the Nariokotome youth, KNM-WT 15000. In *The First Humans: Origin and Early Evolution of the Genus* Homo, ed. F. E. Grine, J. G. Fleagle, and R. E. F. Leakey, 101–20. New York: Springer.

Dean, M. C., and B. A. Wood. 1981. Developing pongid dentition and its use for ageing individual crania in comparative cross-sectional growth studies. *Folia Primatologica* 36:111–27.

———. 1982a. Metrical analysis of the basicranium of extant hominoids and *Australopithecus*. *American Journal of Physical Anthropology* 54:63–71.

———. 1982b. Basicranial anatomy of Plio-Pleistocene hominids from East and South Africa. *American Journal of Physical Anthropology* 59:157–74.

———. 1984. Phylogeny, neoteny and growth of the cranial base in hominoids. *Folia Primatologica* 43:234–48.

Dean, M. C., C. B. Stringer, and T. G. Bromage. 1986. Age at death of the Neanderthal child from Devil's Tower, Gibraltar, and the implications for studies of general growth and development in Neanderthals. *American Journal of Physical Anthropology.* 70:301–9.

Dean, M. C., M. G. Leakey, D. J. Reid, F. Schrenk, G. T. Schwartz, C. Stringer, and A. Walker. 2001. Growth processes in teeth distinguish modern humans from *Homo erectus* and earlier hominins. *Nature* 414:628–31.

Dechow, P. C., and W. L. Hylander. 2000. Elastic properties and masticatory bone stress in the macaque mandible. *American Journal of Physical Anthropology* 112: 553–74.

DeGusta, D., W. H. Gilbert, and S. P. Turner. 1999. Hypoglossal canal size and hominid speech. *Proceedings of the National Academy of Sciences USA* 96:1800–1804.

Deino, A. L., and A. Hill. 2002. 40Ar/(39)Ar dating of Chemeron Formation strata encompassing the site of hominid KNM-BC 1, Tugen Hills, Kenya. *Journal of Human Evolution* 42:141–51.

Dejaeger, E., W. Pelemans, E. Ponette, and G. Vantrappen. 1994. Effect of body position on deglutition. *Digestive Diseases and Sciences* 39:762–5.

Demes, B. 1982. The resistance of primate skulls against mechanical stresses. *Journal of Human Evolution* 11:687–91.

———. 1984. Mechanical stresses at the primate skull base caused by the temporomandibular joint force. In *Food Acquisition and Processing in Primates,* ed. D. J. Chivers, B. A. Wood, and A. Bilsborough, 407–13. New York: Plenum.

———. 1985. Biomechanics of the primate skull base. *Advances in Anatomy, Embryology, and Cell Biology* 94:1–59.

———. 1987. Another look at an old face: Biomechanics of the Neanderthal facial skeleton reconsidered. *Journal of Human Evolution* 16:297–303.

Demes, B., and N. Creel. 1988. Bite force, diet and cranial morphology in fossil hominids. *Journal of Human Evolution* 17:657–70.

Dempster, W. T., and G. R. L. Gaughran. 1967. Properties of body segments based on size and weight. *American Journal of Anatomy* 120:33–54.

Denis, D., M. Righini, C. Scheiner, F. Volot, L. Boubli, J. Vola, and J.-B. Saracco.

1992. Ocular growth in the fetus, 1: Comparative study of the axial length and biometric parameters in the fetus. *Ophthalmologica* 206:924–33.

Denis, D., F. Faure, F. Volot, C. Scheiner, L. Boubli, X. Dézard, and J.-B. Saracco. 1993. Ocular growth in the fetus, 2: Comparative study of the growth of the globe and the orbit and the parameters of fetal growth. *Ophthalmologica* 207:125–32.

Dennis, J. C., T. D. Smith, K. P. Bhatnagar, C. J. Bonar, A. M. Burrows, and E. E. Morrison. 2004. Expression of neuron-specific markers by the vomeronasal neuroepithelium in six species of primates. *Anatomical Record* 281A:1190–200.

Depew, M. J., J. K. Liu, J. E. Long, R. Presley, J. J. Meneses, R. A. Pedersen, and J. L. Rubenstein. 1999. Dlx5 regulates regional development of the branchial arches and sensory capsules. *Development* 126:3831–46.

Depew, M. J., T. Lufkin, and J. R. L. Rubenstein. 2002. Specification of jaw subdivision by *Dlx* genes. *Science* 298:381–85.

Depew, M. J., C. A. Simpson, M. Morasso, and J. L. Rubenstein. 2005. Reassessing the Dlx code: The genetic regulation of branchial arch skeletal pattern and development. *Journal of Anatomy* 207:501–61.

DeSilva, J., and J. Lesnik. 2006. Chimpanzee neonatal brain size: Implications for brain growth in *Homo erectus. Journal of Human Evolution* 51:207–12.

———. 2008. Brain size at birth throughout human evolution: A new method for estimating neonatal brain size in hominins. *Journal of Human Evolution* 55: 1064–74.

Destexhe, A., and E. Marder. 2004. Plasticity in single neuron and circuit computations. *Nature* 431:789–95.

Devlin, M. J., and D. E. Lieberman. 2007. Variation in estradiol level affects cortical bone growth in response to mechanical loading in sheep. *Journal of Experimental Biology* 210:602–13.

Dibbets, J. M. H. 1990. Mandibular rotation and enlargement. *American Journal of Orthodontics and Dentofacial Orthopedics* 98:29–32.

Diewert, V. M. 1983. A morphometric analysis of craniofacial growth and changes in spatial relations during secondary palate development in human embryos and fetuses. *American Journal of Anatomy* 167:495–522.

———. 1985. Development of human craniofacial morphology during the late embryonic and early fetal periods. *American Journal of Orthodontics and Dentofacial Orthopaedics* 88:64–76.

Dill, D. B., A. V. Bock, and H. T. Edwards. 1933. Mechanisms for dissipating heat in man and dog. *American Journal of Physiology* 104:36–43.

Dimery, N. J., R. M. Alexander, and K. A. Deyst. 1985. Mechanics of the ligamentum nuchae of some artiodactyls. *Journal of Zoology, London* 206:341–51.

Dixon, A. D., D. A. N. Hoyte, and O. Rönning. 1997. *Fundaments of Craniofacial Growth.* Boca Raton, FL: CRC Press.

Dobson, S. D., and E. Trinkaus. 2002. Cross-sectional geometry and morphology of the mandibular symphysis in Middle and Late Pleistocene *Homo*. *Journal of Human Evolution* 43:67–87.

Dodd, J. S., J. A. Raleigh, and T. S. Gross. 1999. Osteocyte hypoxia: A novel mechanotransduction pathway. *American Journal of Physiology* 48: C598–C602.

Doden, E., R. Halves. 1984. On the functional morphology of the human petrous bone. *American Journal of Anatomy* 169:451–62.

Dominguez-Rodrigo, M. 2001. A study of carnivore competition in riparian and open habitats of modern savannas and its implications for hominid behavioural modelling. *Journal of Human Evolution* 40: 77–98.

———. 2002. Hunting and scavenging by early humans: The state of the debate. *Journal of World Prehistory* 16:1–54.

Dominguez-Rodrigo, M., T. R. Pickering, S. Semaw, and M. J. Rogers. 2006. Cut-marked bones from Pliocene archaeological sites at Gona, Afar, Ethiopia: Implications for the function of the world's oldest stone tools. *Journal of Human Evolution* 48:109–21.

Domínguez-Rodrigo, M., A. Mabulla, L. Luque, J. W. Thompson, J. Rink, P. Bushozi, F. Díez-Martin, and L. Alcala. 2008. A new archaic *Homo sapiens* fossil from Lake Eyasi, Tanzania. *Journal of Human Evolution* 54:899–903.

Dominy, N. J., and P. W. Lucas. 2001. Ecological importance of trichromatic vision to primates. *Nature* 410:363–6.

Dominy, N. J., J. C. Svenning, and W. H. Li. 2003. Historical contingency in the evolution of primate color vision. *Journal of Human Evolution* 44:25–45.

Donahue, S. W., C. R. Jacobs, and H. J. Donahue. 2001. Flow-induced calcium oscillations in rat osteoblasts are age, loading frequency, and shear stress dependent. *American Journal of Physiology* 81: C1635–C1641.

Doran, D. M., A. McNeilage, D. Greer, C. Bocian, P. Mehlman, and N. Shah. 2002. Western lowland gorilla diet and resource availability: New evidence, cross-site comparisons, and reflections on indirect sampling methods. *American Journal of Primatology* 58:91–116.

Downs, W. R. 1952. The role of cephalometrics in orthodontic case analysis. *American Journal of Orthodontics* 38:162–82.

Dryden, I. L., and K. V. Mardia. 1998. *Statistical Shape Analysis*. New York: John Wiley.

Duarte, C., J. Maurício, P. B. Petit, P. Souto, E. Trinkaus, H. van der Plicht, and J. Zilhão. The early Upper Paleolithic human skeleton from the Abrigo do Lagar Velho (Portugal) and modern human emergence in Iberia. *Proceedings of the National Academy of Sciences* 96:7604–9.

Dubois, E. 1894. *Pithecanthropus erectus:* Eine menschanähnliche Übergangsform aus Java. Batavia: Landsdruckerei.

DuBrul, E. L. 1977. Early hominid feeding mechanisms. *American Journal of Physical Anthropology* 47:305–20.

DuBrul, E. L., and D. M. Laskin. 1961. Preadaptive potentialities of the mammalian skull: An experiment in growth and form. *American Journal of Anatomy* 109: 117–32.

DuBrul, E. L., and H. Sicher. 1954. *The adaptive chin.* Springfield, IL: Charles Thomas.

Ducy, P., M. Starbuck, M. Priemel, J. Shen, G. Pinero, V. Geoffroy, M. Amling and G. Karsenty. 1999. A Cbfa1-dependent genetic pathway controls bone formation beyond embryonic development. *Genes and Development* 13:1025–1036.

Dunbar, D. C., and G. L. Badam. 1998. Development of posture and locomotion in free-ranging primates. *Neuroscience and Biobehavioral Reviews* 22:541–46.

Dunbar, D. C., G. L. Badam, B. Hallgrímsson, and S. Vieilledent. 2004. Stabilization and mobility of the head and trunk in wild monkeys during terrestrial and flat-surface walks and gallops. *Journal of Experimental Biology* 207:1027–42.

Dunbar, D. C., J. M. Macpherson, R. W. Simmons, and A. Zarcades. 2008. Stabilization and mobility of the head, neck and trunk in horses during overground locomotion: Comparisons with humans and other primates. *Journal of Experimental Biology* 211:889–907.

Durband, A. C. 2008. Artificial cranial deformation in Pleistocene Australians: The Coobool Creek sample. *Journal of Human Evolution* 54:795–813.

Duterloo, H. S., and D. H. Enlow. 1970. A comparative study of cranial growth in *Homo* and *Macaca. American Journal of Anatomy* 127:357–68.

du Toit, J. T., and C. A. Yetman. 2005. Effects of body size on the diurnal activity budgets of African browsing ruminants. *Oecologia* 143:317–25.

Eble, G. J. 2002. Multivariate approaches to development and evolution. In *Human Evolution through Developmental Change,* ed. N. Minugh-Purvis and K. J. McNamara, 51–78. Baltimore, MD: Johns Hopkins University Press.

Eckel, H. E., C. Sittel, P. Zorowka, and A. Jerke. 1994. Dimensions of the laryngeal framework in adults. *Surgical and Radiologic Anatomy* 16:31–36.

Edward, D. P., and L. M. Kaufman. 2003. Anatomy, development, and physiology of the visual system. *Pediatric Clinics of North America* 50:1–23.

Ekberg, O., and S. V. Sigurjonsson. 1982. Movement of the epiglottis during deglutition: A cineradiographic study. *Gastrointestinal Radiology* 7:101–7.

Elder, J. H. 1935. The upper limit of hearing in the chimpanzee. *American Journal of Physiology* 112:109–15.

Elia, M. 1992. Organ and tissue contribution to metabolic weight. In *Energy Metabolism: Tissue Determinants and Cellular Corollaries,* ed. J. M. Kinney, and H. N. Tucker, 61–79. New York: Raven Press.

Elliot Smith, G. 1912. Address to the Anthropological Section. *British Association for the Advancement of Science, Sect. H, Dundee* 191:575–98.

Ellis, E., G. S Throckmorton, and D. P. Sinn. 1996. Bite forces before and after surgical correction of mandibular prognathism. *Journal of Oral and Maxillofacial Surgery* 54:176–81.

Ellison, P. T. 2003. *On Fertile Ground: A Natural History of Human Reproduction.* Cambridge, MA: Harvard University Press.

Enard, W., M. Przeworski, S. E. Fisher, C. S. L. Lai, V. Wiebe, T. Kitano, A. P. Monaco, and S. Pääbo. 2002. Molecular evolution of FOXP2, a gene involved in speech and language *Nature* 418:869–72.

Enard, W., S. Gehre, K. Hammerschmidt, S. Hölter, T. Blass, M. Somel, M. Brückner, et al. 2009. A humanized version of FOXP2 affects cortico-basal ganglia circuits in mice. *Cell* 137:961–71.

Endo, B. 1966. Experimental studies on the mechanical significance of the form of the human facial skeleton. *Journal of the Faculty of Sciences of the University of Tokyo,* 3:1–106.

English, J. D., P. H. Buschang, and G. S. Throckmorton. 2002. Does malocclusion affect masticatory performance? *Angle Orthodontist* 72:21–27.

Englyst, H. N., and J. H. Cummings. 1986. Digestion of the carbohydrates of banana (*Musa paradisiaca sapientum*) in the human small intestine. *American Journal of Clinical Nutrition* 44:42–50.

Engström, C., S. Kiliaridis, and B. Thilander. 1986. The relationship between masticatory muscle function and cranial morphology, II: A histological study in the growing rat fed a soft diet. *European Journal of Orthodontics* 8:271–79.

Enlow, D. H. 1963. *Principles of of Bone Remodeling.* Springfield, IL: Charles C. Thomas.

———. 1966. A comparative study of facial growth in *Homo* and *Macaca. American Journal of Physical Anthropology* 24:293–308.

———. 1977. The remodeling of bone. *Yearbook of Physical Anthropology* 19–34.

———. 1990. *Facial Growth.* 3rd ed. Philadelphia: Saunders.

Enlow, D. H., and M. Azuma. 1975. Functional growth boundaries in the human and mammalian face. In *Morphogenesis and Malformation of Face and Brain,* ed. D. Bergsma, 217–30. New York: Alan R. Liss.

Enlow, D. H., and S. Bang. 1956. Growth and remodelling of the human maxilla. *American Journal of Orthodontics* 51:446–64.

Enlow, D. H., and D. B. Harris. 1964. A study of the postnatal growth of the human mandible. *American Journal of Orthodontics* 50:25–25.

Ennis, L. M. 1937. Roentgenographic variations of the maxillary sinus and the nutrient canals of the maxilla and mandible. *International Journal of Orthodontics and Oral Surgery* 23:173–93.

Ennouchi, E. 1969. Découverte d'un Pithécanthropien au Maroc. *Comptes rendus de l'Académie des Sciences de Paris, série D,* 269: 763–65.

Evans, P. D., S. L. Gilbert, N. Mekel-Bobrov, E. J. Vallender, J. R.,Anderson, L. M. Vaez-Azizi, S. A. Tishkoff, R. R. Hudson, and B. T. Lahn. 2005. Microcephalin, a gene regulating brain size, continues to evolve adaptively in humans. *Science* 309:1717–22.

Falk, D. 1986. Evolution of cranial blood drainage in hominids: Enlarged occipital/marginal sinuses and emissary foramina. *American Journal of Physical Anthropology* 70:311–24.

———. 1990. Brain evolution in *Homo:* The "radiator" theory. *Behavioral and Brain Sciences* 13:333–81.

———. 2004. *Braindance.* 2nd ed. Gainesville: University Press of Florida.

———. 2007. Constraints on brain size: The radiator hypothesis. In *The Evolution of Primate Nervous Systems,* ed. Todd M. Preuss and Jon H. Kaas, 347–54. New York: Elsevier Press.

Falk, D., L. Konigsberg, R. Criss Helmcamp, J. Cheverud, M. Vannier, and C. Hildebodt. 1989. Endocranial suture closure in rhesus macaques (*Macaca mulatta*). *American Journal of Physical Anthropology* 80:417–28.

Falk, D., J. C. Redmond, J. G. Guyer, G. C. Conroy, W. Reicheis, G. W. Weber, and H. Seidler. 2000. Early hominin brain evolution: A new look at old endocasts. *Journal of Human Evolution* 38:695–717.

Falk, D., C. Hildebolt, K. Smith, M. J. Morwood, T. Sutikna, P. Brown, Jatmiko, E. W. Saptomo, B. Brunsden, and F. Prior. 2005. The brain of LB1, *Homo floresiensis. Science* 308:242–45.

Falk, D., C. Hildebolt, K. Smith, M. J. Morwood, T. Sutikna, Jatmiko, E. W. Saptomo, H. Imhof, H. Seidler, and F. Prior. 2007. Brain shape in human microcephalics and *Homo floresiensis. Proceedings of the National Academy of Sciences USA.* 104:2513–8.

Fant, G. G. 1960. *Acoustic theory of speech production.* The Hague: Mouton.

Fenart, R., and C. Pellerin. 1988. The vestibular orientation method: Its application in the determination of an average human skull type. *International Journal of Anthropology* 3:241–49.

Fernald, R. D. 2006. Casting a genetic light on the evolution of eyes. *Science* 313:1914–18.

Ferrus, L., D. Commenges, J. Gire, and P. Varene. 1984. Respiratory water loss as a function of ventilatory or environmental factors. *Respiration Physiology* 56:11–20.

Findlay, D. M., and P. M. Sexton. 2005. Calcitonin. *Growth Factors* 22:217–24.

Finlay, B. L., and R. B. Darlington. 1995. Linked regularities in the development and evolution of mammalian brains. *Science* 268:1578–84.

Fisher, R. A., E. B. Ford, and J. Huxley. 1939. Taste-testing the anthropoid apes. *Nature* 144:750.

Fitch, W. T., and J. Giedd. 1999. Morphology and development of the human vocal tract: A study using MRI. *Journal of the Acoustical Society of America* 60:718–24.

Fitch, W. T., and M. Hauser 1995. Vocal production in nonhuman primates: Acoustics, physiology and functional constraints on "honest" advertisement. *American Journal of Primatology* 37:191–219.

———. 2003. Unpacking "Honesty": Vertebrate Vocal Production and the Evolution of Acoustic Signals. In *Acoustic Communication,* ed. A. Simmons, R. R. Fay, and A. N. Popper, 65–137. New York: Springer-Verlag.

Fitch, W. T., and D. Reby. 2001. The descended larynx is not uniquely human. *Proceedings of the Royal Society (London) Biological Sciences* 268:1669–75.

Flügel, C., and J. W. Rohen. 1991. The craniofacial proportions and laryngeal position in monkeys and man of different ages: A morphometric study based on CT-scans and radiographs. *Mechanisms of Ageing and Development* 61:65–83.

Foley, R. A., and M. M. Lahr. 1997. Mode 3 technologies and the evolution of modern humans. *Cambridge Archaeological Journal* 7:3–36.

Fondon, J. W. III, and H. R. Garner. 2004. Molecular origins of rapid and continuous morphological evolution. *Proceedings of the National Academy of Sciences USA* 101:18058–63.

Ford, E. H. R. 1958. Growth of the human cranial base. *American Journal of Orthodontics* 44:498–506.

Forrester, D. J., N. K. Carstens, and D. B. Shurtleff. 1966. Craniofacial configuration of hydrocephalic children. *Journal of the American Dental Association* 72:1399–404.

Forsberg, H., F. Crozet, amd N. A. Brown. 1998. Waves of mouse Lunatic fringe expression, in four hour cycles at two hour intervals, precede somite boundary formation. *Current Biology* 8:1027–1030.

Fortelius, M. 1985. Ungulate cheek teeth: Developmental, functional and evolutionary interrelationships. *Acta Zoologica Fennica* 180:1–76.

Frahm, H. D., H. Stephan, and M. Stephan. 1982. Comparison of brain structure volumes in Insectivora and Primates, I: Neocortex. *Journal für Hirnforschung* 23:375–89.

Franciscus, R. G. 1995. Later Pleistocene nasofacial variation in Western Eurasia and Africa and modern human origins. PhD diss., University of New Mexico.

———. 1999. Neandertal nasal structures and upper respiratory tract "specialization." *Proceedings of the National Academy of Sciences USA* 96:1805–9.

Franciscus, R. G., and E. Trinkaus. 1988. Nasal morphology and the emergence of *Homo erectus. American Journal of Physical Anthropology* 75:517–27.

———. 1995. Determinants of retromolar space presence in Pleistocene *Homo* mandibles. *Journal of Human Evolution* 28:577–95.

Franciscus, R. G., and J. C. Long. 1991. Variation in human nasal height and breadth. *American Journal of Physical Anthropology* 85:419–27.

Frayer, D. 1977. Metric changes in the Upper Paleolithic and Mesolithic. *American Journal of Physical Anthropology* 46:109–20.

Friede, H. 1981. Normal development and growth of the human neurocranium and cranial base. *Scandinavian Journal of Plastic and Reconstructive Surgery* 115: 163–69.

Frost, H. M. 1986. *The Intermediary Organization of the Skeleton.* Boca Raton, FL: CRC Press.

———. 1987. Bone "mass" and the "mechanostat": A proposal. *Anatomical Record* 219:1–9.

———. 1990. Skeletal structural adaptations to mechanical usage (SATMU). *Anatomical Record* 226:403–22.

———. 1999. Why do bone strength and "mass" in aging adults become unresponsive to vigorous exercise? Insights of the Utah paradigm. *Journal of Bone and Mineral Metabolism* 17:90–97.

Fuentes, M. A., L. A. Opperman, L. L. Bellinger, D. S. Carlson, and R. J. Hinton. 2002. Regulation of cell proliferation in rat mandibular condylar cartilage in explant culture by insulin-like growth factor-1 and fibroblast growth factor-2. *Archives of Oral Biology* 47:643–54.

Gabunia, L., and A. Vekua. 1995. A Plio-Pleistocene hominid from Dmanisi, East Georgia, Caucasus. *Nature* 373:509–12.

Gabunia, L., A. Vekua, and D. Lordkipanidze. 2000a. The environmental context of early human occupation of Georgia. *Journal of Human Evolution* 38:785–802.

Gabunia, L., A. Vekua, D. Lordkipanidze, C. C. Swisher III, R. Ferring, A. Justus, M. Nioradze, et al. 2000b. Earliest Pleistocene hominid cranial remains from Dmanisi, Republic of Georgia: Taxonomy, geological setting, and age. *Science* 288:1019–1025.

Gagneux, P., C. Wills, U. Gerloff, D. Tautz, P. A. Morin, C. Boesch, B. Fruth, G. Hohmann, O. A. Ryder, and D. S. Woodruff. 1999. Mitochondrial sequences show diverse evolutionary histories of African hominoids. *Proceedings of the National Academy of Sciences USA* 96:5077–82.

Galik, K., B. Senut, M. Pickford, D. Gommery, J. Treil, A. J. Kuperavage, and R. B. Eckhardt. 2004. External and internal morphology of the Bar 1002'00 *Orrorin tugenensis* femur. *Science* 307:1450–53.

Gans, C., and R. G. Northcutt. 1983. Neural crest and the origin of the vertebrates: A new head. *Science* 220:268–74.

Garland, T. 1983. The relation between maximal running speed and body mass in terrestrial mammals. *Journal of Zoology* 199:1557–70.

Gasser, R. F. 1967. Early formation of the basicranium in man. In *Development of the Basicranium,* ed. J. F. Bosma, 29–43. Bethesda, MD: National Institutes of Health.

Gauld, S. C. 1996. Allometric patterns of cranial bone thickness in fossil hominids. *American Journal of Physical Anthropology* 100:411–26.

Gaulin, S. J. C. 1979. A Jarman-Bell Model of Primate Feeding Niches. *Human Ecology* 7:1–20.

Gauthier, G. B., J. P. Piron, J. P. Roll, E. Marchetti, and B. Martin. 1984. High-frequency vestibulo-ocular reflex activation through forced head rotation in man. *Aviation, Space, and Environmental Medicine* 55:1–7.

Gellman, K. S., and J. E. A. Bertram. 2002a. The equine nuchal ligament, 1: Structural and material properties. *Journal of Veterinary Orthopedic Traumatology* 15: 1–6.

———. 2002b. The equine nuchal ligament, 2: The role of the head and neck in locomotion. *Journal of Veterinary Orthopedic Traumatology* 15:7–14.

Geoffroy Sainte-Hilaire, E. 1836. Considerations sur les singes, les plus voisins de l'homme. *Comptes rendus de l'Académie des Sciences de Paris* 2:92–95.

George, S. L. 1978. A longitudinal and cross-sectional analysis of the growth of the postnatal cranial base angle. *American Journal of Physical Anthropology* 49: 171–78.

Gerber, H. P., T. H. Vu, A. M. Ryan, J. Kowalski, Z. Web, and N. Ferrara. 1999. VEGF couples hypertrophic cartilage remodeling, ossification and angiogenesis during endochondral bone formation. *Nature Medicine* 5:623–28.

Gerhart, J., and M. Kirschner. 1997. *Cells, Embryos, and Evolution.* Oxford: Blackwell Science.

German, R. Z., A. W. Crompton, and A. J. Thexton. 1998. The coordination and interaction between respiration and deglutition in young pigs. *Journal of Comparative Physiology [A]* 182:539–47.

Gibbons, A. 2006. The First Human: The Race to Discover Our Earliest Ancestors. New York: Doubleday.

Gibbs, S., M. Collard, and B. A. Wood. 2000. Soft-tissue characters in higher primate phylogenetics. *Proceedings of the National Academy of Sciences* 97:11130–32.

———. 2002. Soft tissue anatomy of the extant hominoids: A review and phylogenetic analysis. *Journal of Anatomy* 200:3–49.

Gibson, K. R., D. Rumbaugh, and M. Beran. 2001. Bigger is better: Primate brain size in relationship to cognition. In *Evolutionary Anatomy of the Primate Cerebral Cortex,* ed. D. Falk and K. R. Gibson, 79–97. Cambridge: Cambridge University Press.

Giedd, J. N. 2008. The teen brain: Insights from neuroimaging. *Journal of Adolescent Health* 42:335–43.

Gilad, Y, O. Man, and G. Glusman. 2005. A comparison of the human and chimpanzee olfactory receptor gene repertoires. *Genome Research.* 15:224–23.

Gilbert, S. F. 1989. *Embryology: Constructing the Organism.* Sunderland, MA: Sinauer Associates.

———. 2006. *Developmental Biology.* 8th ed. Sunderland, MA: Sinnauer Associates.

Gilbert, S. F., and J. A. Bolker. 2000. Homologies of process: Modular elements of embryonic construction. In *The Character Concept in Evolutionary Biology*, ed. G. P. Wagner, 435–54. New Haven, CT: Yale University Press.

Gilbert, S. L., W. B. Dobyns, and B. T. Lahn. 2005. Genetic links between brain development and brain evolution. *Nature Reviews: Genetics* 6:581–90.

Giles, W. B., C. L. Philips, and D. R. Joondeph. 1981. Growth in the basicranial synchondroses of adolescent *Macaca macaca. Anatomical Record* 199:259–66.

Gillespie, R. 1997. Burnt and unburnt carbon: Dating charcoal and burnt bone from the Willandra Lakes, Australia. *Radiocarbon* 39:225–36.

Gillespie, R., and R. G. Roberts. 2000. On the reliability of age estimates for human remains at Lake Mungo. *Journal of Human Evolution* 38:727–30.

Gimelbrant, A. A., H. Skaletsky, and A. Chess. 2004. Selective pressures on the olfactory receptor repertoire since the human-chimpanzee divergence. *Proceedings of the National Academy of Sciences USA* 101:9019–22.

Gionhaku, N., and A. A. Lowe. 1989. Relationship between jaw muscle volume and craniofacial form. *Journal of Dental Research* 68:805–9.

Giraud-Guille, M. M. 1988. Twisted plywood architecture of collagen fibrils in human compact bone osteons. *Calcified Tissue International* 42:167–80.

Glusman, G., I. Yanai, I. Rubin, and D. Lancet. 2001. The complete human olfactory subgenome. *Genome Research* 11:685–702.

Godfrey, L. R., and M. R. Sutherland. 1996. The paradox of peramorphic paedomorphosis: Heterochrony in human evolution. *American Journal of Physical Anthropology* 99:17–42.

Goldberg, P., S. Weiner, O. Bar-Yosef, Q. Xu, and J. Liu. 2001. Site formation processes at Zhoukoudian, China. *Journal of Human Evolution* 41:483–530.

Goldman, P. S., and T. W. Galkin. 1978. Prenatal removal of frontal association cortex in the fetal rhesus monkey: Anatomical and functional consequences in postnatal life. *Brain Research* 152:451–85.

Goldman-Rakic, P. S. 1980. Morphological consequences of prenatal injury to the primate brain. *Progress in Brain Research* 53:1–19.

Gomez, C., E. M. Özbudak, J. Wunderlich, D. Baumann, J. Lewis, and O. Pourquié. 2008. Control of segment number in vertebrate embryos. *Nature* 454:335–39.

Gómez-Olivencia A., J. M. Carretero, J. L. Arsuaga, L. Rodríguez-García, R. García-González, and I. Martínez. 2007. Metric and morphological study of the upper cervical spine from the Sima de los Huesos site (Sierra de Atapuerca, Burgos, Spain). *Journal of Human Evolution* 53:6–25.

González-José, R., S. Van Der Molen, S. González-Pérez, and M. Hernández. 2004. Patterns of phenotypic covariation and correlation in modern humans as viewed from morphological integration. *American Journal of Physical Anthropology* 123:69–77.

Goodall, J. 1986. *The Chimpanzees of Gombe: Patterns of Behavior.* Cambridge, MA: Harvard University Press.

Gordon, A. D., L. Nevell, and B. Wood. 2008. The *Homo floresiensis* cranium (LB1): Size, scaling, and early *Homo* affinities. *Proceedings of the National Academy of Sciences USA* 105:4650–5.

Goren-Inbar, N., N. Alperson, M. E. Kislev, O. Simchoni, Y. Melamed, A. Ben-Nun, and E. Werker. 2004. Evidence of hominin control of fire at Gesher Benot Ya'aqov, Israel. *Science* 304:725–27.

Goret-Nicaise, M., and A. Dhem. 1982. Presence of chondroid tissue in the symphyseal region of the growing human mandible. *Acta Anatomica (Basel)* 113:189–95.

Gorlin, R. J., M. M. Cohen, and R. C. M. Hennekam. 2001. *Syndromes of the Head and Neck*. Oxford: Oxford University Press.

Gould, S. J. 1977. *Ontogeny and Phylogeny*. Cambridge, MA: Belknap Press.

———. 2002. *The Structure of Evolutionary Theory*. Cambridge, MA: Harvard University Press.

Gould, S. J., and R. C. Lewontin. 1979. The spandrels of San Marco and the Panglossian paradigm: A critique of the adaptationist programme. *Proceedings of the Royal Society of London, Series B* 205:581–98.

Graf, W., C. de Waele, and P. P. Vidal. 1995a. Functional anatomy of the head-neck movement system of quadrupedal and bipedal mammals. *Journal of Anatomy* 186:55–74.

Graf, W., C. de Waele, P. P. Vidal, D. H. Wang, and C. Evinger. 1995b. The orientation of the cervical vertebral column in unrestrained awake animals, II: Movement strategies. *Brain, Behavior and Evolution* 45:209–31.

Grant, Peter R. 1991. Natural selection and Darwin's finches. *Scientific American* 265:82–87.

Greaves, W. S. 1978. The jaw lever system in ungulates: A new model. *Journal of the Zoological Society of London* 184:271–85.

———. 1985. The mammalian post-orbital bar as a torsion-resisting helical strut. *Journal of the Zoological Society of London* 207:125–36.

———. 2000. Location of the vector of jaw muscle force in mammals. *Journal of Morphology* 243:293–9.

Green, D. J., A. D. Gordon, and B. G. Richmond. 2007. Limb-size proportions in *Australopithecus afarensis* and *Australopithecus africanus*. *Journal of Human Evolution* 52:187–200.

Green, R. E., J. Krause, S. E. Ptak, A. W. Briggs, M. T. Ronan, J. J. Simons, L. Du, et al. 2006. Analysis of one million base pairs of Neanderthal DNA. *Nature* 444:330–36.

Green, R. E., A. S. Malaspinas, J. Krause, A. W. Briggs, P. L. Johnson, C. Uhler, M. Meyer, et al. 2008. A complete Neandertal mitochondrial genome sequence determined by high-throughput sequencing. *Cell* 134:416–26.

Green, R. E., J. Krause, A. W. Briggs, T. Maricic, U. Stenzel, M. Kircher, N. Patterson, et al. 2010. A draft sequence of the Neanderthal genome. *Science* 328:710–22.

Greenberg, J. 1963. *Universals of Language.* Cambridge, MA: MIT Press.

Grine, F. E. 1981. Trophic differences between gracile and robust australopithecines. *South African Journal of Science* 77:203–30.

———. 1988. New craniodental fossils of *Paranthropus* from the Swartkrans formation and their significance in "robust" australopithecine evolution. In *Evolutionary History of the "Robust" Australopithecines,* ed. F. E. Grine, 223–43. New York: Aldine de Gruyter.

———. 1993. Description and preliminary analysis of new hominid craniodental materials from the Swartkrans formation. *Transvaal Museum Monograph* 8: 167–94.

Grine, F. E., and L. B. Martin. 1988. Enamel thickness and development in *Australopithecus* and *Paranthropus.* In *Evolutionary History of the "Robust" Australopithecines,* ed, F. E. Grine, 3–42. New York: Aldine de Gruyter.

Grine, F. E., B. Demes, W. L. Jungers, and T. M. Cole III. 1993. Taxonomic affinity of the early *Homo* cranium from Swartkrans, South Africa. *American Journal of Physiology* 92:411–26.

Grine, F. E., W. L. Jungers, and J. Schulz. 1996. Phenetic affinities among early *Homo* crania from East and South Africa. *Journal of Human Evolution* 30:189–225.

Grine, F. E., O. M. Pearson, R. G. Klein, and G. P. Rightmire. 1998. Additional human fossils from Klasies River Mouth, South Africa. *Journal of Human Evolution* 35:95–107.

Grine, F. E., P. S. Ungar, M. F. Teaford, and S. El-Zaatari. 2006. Molar microwear in *Praeanthropus afarensis:* Evidence for dietary stasis through time and under diverse paleoecological conditions. *Journal of Human Evolution* 51:297–319.

Gross, M. D., G. Arbel, and I. Hershkovitz. 2001. Three-dimensional finite element analysis of the facial skeleton on simulated occlusal loading. *Journal of Oral Rehabilitation* 2001 28:684–94.

Grossman, G. E., R. J. Leigh, L. A. Abel, D. J. Lanska, and S. E. Thurston. 1988. Frequency and velocity of rotational head perturbations during locomotion. *Experimental Brain Research* 70:470–76.

Groves, C. P. 1986. Systematics of the great apes. In *Comparative Primate Biology,* vol. 1, *Systematics, Evolution and Anatomy,* ed. D. Swinger and J. Erwin, 187–217. New York: Alan R Lisss.

———. 1989. *A Theory of Human and Primate Evolution.* Oxford: Oxford University Press.

Groves, C. P., and V. Mazak. 1975. An approach to the taxonomy of the Hominidae: Gracile Villafranchian hominids of Africa. *Casopis pro Mineralogii a Geologii* 20: 225–47.

Grün, R., and C. B. Stringer. 1991. Electron spin resonance dating and the evolution of modern humans. *Archaeometry* 33:153–99.

Grymer, L. F., and C. Bosch. 1997. The nasal septum and the development of the midface: A longitudinal study of a pair of monozygotic twins. *Rhinology* 35:6–10.

Guatelli-Steinberg, D., D. J. Reid, andand T. A. Bishop. 2005. Anterior tooth growth periods in Neandertals were comparable to those in modern humans. *Proceedings of the National Academy of Sciences, USA* 102:14197–202.

Gunz, P., and K. Harvati. 2007. The Neanderthal "chignon": Variation, integration and homology. *Journal of Human Evolution* 52:262–74.

Gurdjian, E. S., H. R. Lissner, V. R. Hodgson, and L. M. Patrick L. M. 1964. Mechanism of head injury. *Clinical Neurosurgery* 12:112–28.

Guy, F., D. E. Lieberman, D. Pilbeam, M. Ponce de León, A. Likius, H. T. Mackaye, P. Vignaud, C. Zollikofer, and M. Brunet. 2005. Morphological affinities of the *Sahelanthropus tchadensis* (Late Miocene hominid from Chad) cranium. *Proceedings of the National Academy of Sciences* 102:18836–41.

Guy, F., H. T. Mackaye, A. Likius, P. Vignaud, M. Schmittbuhl, and M. Brunet 2008. Symphyseal shape variation in extant and fossil hominoids, and the symphysis of *Australopithecus bahrelghazali*. *Journal of Human Evolution* 55:37–47.

Habgood, P. J., and N. R. Franklin. 2008. The revolution that didn't arrive: A review of Pleistocene Sahul. *Journal of Human Evolution* 55:187–222.

Haeckel, E. 1866. Generalle Morphologie der Organismen: Allgemeine grundzüge der organischen Formen-Wissenschaft, mechanische begründet durch die von Charles Darwin reformierte Descendenz-Theorie. Berlin: Georg Reimer.

Haeusler, M., and H. M. McHenry. 2004. Body proportions of *Homo habilis* reviewed. *Journal of Human Evolution* 46:433–65.

———. 2007. Evolutionary reversals of limb proportions in early hominids? Evidence from KNM-ER 3735 (*Homo habilis*). *Journal of Human Evolution* 53:383–405.

Hahn, I., P. W. Scherer, and M. M. Mozell. 1993. Velocity profiles measured for airflow through a large-scale model of the human nasal cavity. *Journal of Applied Physiology* 75:2273–87.

Haile-Selassie, Y. 2001. Late Miocene hominids from the Middle Awash, Ethiopia. *Nature* 412:178–81.

Haile-Selassie, Y., G. Suwa, and T. D. White. 2004. Late Miocene teeth from Middle Awash, Ethiopia, and early hominid dental evolution. *Science* 303:1503–5.

———. 2009. Hominidae. In *Ardipithecus kadabba: Late Miocene Evidence from the Middle Awash, Ethiopia,* ed. Y. Haile-Selassie and G. WoldeGabriel, 159–236. Berkeley: University of California Press.

Halder, G., P. Callaerts, and W. J. Gehring. 1995. Induction of ectopic eyes by targeted expression of the eyeless gene in *Drosophila*. *Science* 267:1788–92.

Hales, J. R., F. R. Stephens, A. A. Fawcett, R. A. Westerman, J. D. Vaughan, D. A. Richards, and C. R. Richards. 1986. Lowered skin blood flow and erythrocyte sphering in collapsed fun-runners. *Lancet* 1:1495–96.

Hall, B. K. 1978. *Developmental and Cellular Skeletal Biology.* New York: Academic Press.

———. 1987. Tissue interactions in the development and evolution of the vertebrate head. In *Development and Evolution of the Neural Crest,* ed. P. F. A. Maderson, 215–19. London: Wiley and Sons.

———. 1998. Germ layers and the germ-layer theory revisited: Primary and secondary germ layers, neural crest as a fourth germ layer, homology, and the demise of the germ-layer theory. *Ecology and Evolutionary Biology* 30:121–86.

———. 1999. *The Neural Crest in Development and Evolution.* New York: Springer-Verlag.

———. 2003. Unlocking the black box between genotype and phenotype: Cell condensations as morphogenetic (modular) units. *Biology and Philosophy* 18: 219–47.

———. 2005. Bones and Cartilage: Developmental and Evolutionary Skeletal Biology. Amsterdam: Elsevier.

Hall, B. K., and S. Hörstadius. 1988. *The Neural Crest.* Oxford: Oxford University Press.

Hall, B. K., and T. Miyake. 1995. How do embryos measure time? in *Evolutionary Change and Heterochrony,* ed. K. J. McNamara, 2–20. New York: Wiley.

Hallgrímsson, B., C. J. Dorval, M. L. Zelditch, and R. Z. German. 2004. Craniofacial variability and morphological integration in mice susceptible to cleft lip and palate. *Journal of Anatomy* 205:501–17.

Hallgrímsson, B., J. J. Brown, A. F. Ford-Hutchinson, H. D. Sheets, M. L. Zelditch, and F. R. Jirik. 2006. The brachymorph mouse and the developmental-genetic basis for canalization and morphological integration. *Evolution and Development* 8:61–73.

Hallgrímsson, B., D. E. Lieberman, W. Lie, A. F. Ford-Hutchinson, and F. R. Jirik. 2007a. Epigenetic interactions and the structure of phenotypic variation in the skull. *Evolution and Development* 9:76–91.

Hallgrímsson, B., D. E. Lieberman, N. M. Young, T. Parsons, and S. Wat. 2007b. Evolution of covariance in the mammalian skull. *Novartis Foundation Symposium* 284:164–8.

Hanken, J., and B. K. Hall. 1993. *The Skull,* vols. 1–3. Chicago: University of Chicago Press.

Hannam, A. G., and W. W. Wood. 1989. Relationships between the size and spatial morphology of human masseter and medial pterygoid muscles, the craniofacial skeleton, and jaw biomechanics. *American Journal of Physical Anthropology* 80: 429–45.

Haralabakis, N. B., M. Toutountzakis, and S. Yiagtzis. 1993. The hyoid bone position in adult individuals with open bite and normal occlusion. *European Journal of Orthodontics* 5:249–63.

Harlow, J. M. 1848. Passage of an iron bar through the head. *Boston Medical Surgery Journal* 39:389–93.

Harris, E. F. 1993. Size and form of the cranial base in isolated cleft lip and palate. *The Cleft Palate Craniofacial Journal.* 30:170–4.

Harris, E. F., and M. G. Johnston. 1991. Heritability of craniometric and occlusal variables: A longitudinal sib analysis. *American Journal of Orthodontics and Dentofacial Orthopaedics* 99:258–68.

Harris, E. F., and R. J. Smith. 1980. A study of occlusion and arch widths in families. *American Journal of Orthodontics* 78:155–63.

Harris, E. F., R. H. Potter, and J. Lin. 2001. Secular trend in tooth size in urban Chinese assessed from two-generation family data. *American Journal of Physical Anthropology* 115:312–18.

Hartmann, W. M. 1997. *Signals, Sound, and Sensation.* New York: American Institute of Physics Press.

Hartwig-Scherer, S. 1993. Body weight prediction in early fossil hominids: Towards a taxon-"independent" approach. *American Journal of Anthropology* 92:17–36.

Harvati, K., and T. D. Weaver. 2006. Human cranial anatomy and the differential preservation of population history and climate signatures. *Anatomical Record,* 288A:1225–33.

Harvey, L. M., and J. W. Harvey. 2003. Reliability of bloodhounds in criminal investigations. *Journal of Forensic Science* 48:811–16.

Harvey L. M., S. J. Harvey, M. Hom., A. Perna, and J. Salib. 2006. The use of bloodhounds in determining the impact of genetics and the environment on the expression of human odor type. *Journal of Forensic Sciences* 51:1109–14.

Hatley, T., and J. Kappelman. 1980. Bears, pigs, and Plio-Pleistocene hominids: A case for the exploitation of belowground food resources. *Human Ecology* 8: 371–87.

Hatta, K., A. W. Puschel, and C. B. Kimmel. 1994. Midline signaling in the primordium of the zebrafish anterior central nervous system. *Proceedings of the National Academy of Sciences* 91:2061–65.

Haug, H. 1987. Brain sizes, surfaces, and neuronal sizes of the cortex cerebri: A stereological investigation of man and his variability and a comparison with some mammals (primates, whales, marsupials, insectivores, and one elephant). *American Journal of Anatomy* 180:126–42.

Hauser, G., and G. F. DeStefano. 1989. *Epigenetic Variants of the Human Skull.* Stuttgart: E. Schweizerbart'sche Verlag.

Hawkes, K., J. F. O'Connell, and N. G. Blurton Jones. 1997. Hadza women's time allocation, offspring provisioning and the evolution of long postmenopausal life spans. *Current Anthropology* 38:551–77.

Heaney, R. P., and C. M. Weaver. 2005. Newer perspectives on calcium nutrition and bone quality. *Journal of the American College of Nutrition* 24:574S–81S.

Heath, M. R.1982. The effect of maximum biting force and bone loss upon masticatory function and dietary selection of the elderly. *International Dental Journal* 32:345–56.

Heesy, C. P. 2004. On the relationship between orbit orientation and binocular visual field overlap in mammals. *Anatomical Record* 281A: 1104–10.

Heesy, C. P., and C. F. Ross. 2001. Evolution of activity patterns and chromatic vision in primates: Morphometrics, genetics and cladistics. *Journal of Human Evolution* 40:111–49.

Heffner, R. S. 2004. Primate hearing from a mammalian perspective. *Anatomical Record* 281A: 1111–22.

Heffner, R., and H. Heffner. 1980. Hearing in the elephant (*Elephas maximus*). *Science* 208:518–20.

Heglund, N. C., M. A. Fedak, C. R. Taylor, and G. A. Cavagna. 1982. Energetics and mechanics of terrestrial locomotion, IV: Total mechanical energy changes as a function of speed and body size in birds and mammals. *Journal of Experimental Biology* 97:57–66.

Hellekant, G., and Y. Ninomiya. 1994. Bitter taste in single chorda tympani taste fibers from chimpanzee. *Physiology and Behavior* 56:1185–88.

Hellman, A., and A. Chess. 2002. Olfactory axons: A remarkable convergence. *Current Biology* 12:R849–51.

Henri-Gambier, D. 2002. Les fossiles de Cro-Magnon (Les Eyzies-de-Tayac, Dordogne): Nouvelles données sur leur position chronologique et leur attribution culturelle. *Bulletins et mémoires de la Société Anthropologique de Paris* 14:89–112.

Henshilwood, C. S. 2006. Modern humans and symbolic behaviour: Evidence from Blombos Cave, South Africa. In *Origins,* ed. G. Blundell, 78–83. Cape Town: Double Storey.

Herring, S. W. 1993. Epigenetic and functional influences on skull growth. In *The Skull,* ed. J. Hanken and B. K. Hall, 237–71. Chicago: University of Chicago Press.

Herring, S. W., and T. C. Lakars. 1982. Craniofacial development in the absence of muscle contraction. *Journal of Craniofacial Genetics and Developmental Biology* 1:341–57.

Herring, S. W., and R. J. Mucci. 1991. In vivo strain in cranial sutures: The zygomatic arch. *Journal of Morphology* 207:225–39.

Herring, S. W., and S. Teng. 2000. Strain in the braincase and its sutures during function. *American Journal of Physical Anthropology* 112:575–93.

Herring, S. W., S. Teng, X. Huang, R. J. Mucci, and J. Freeman. 1996. Patterns of bone strain in the zygomatic arch. *Anatomical Record* 246A:446–57.

Herring, S. W., K. L. Rafferty, Z. J. Liu, and C. D. Marshall. 2001. Jaw muscles and the skull in mammals: The biomechanics of mastication. *Comparative Biochemistry and Physiology, Part A: Molecular and Integrative Physiology* 131:207–19.

Hershkovitz, I., L. Kornreich, and Z. Laron. 2007. Comparative skeletal features between *Homo floresiensis* and patients with primary growth hormone insensitivity (Laron syndrome). *American Journal of Physical Anthropology* 109:211–27.

Hewitt, G., A. MacLarnon and K. E. Jones. 2002. The functions of laryngeal air sacs in primates: A new hypothesis. *Folia Primatologica* 73:70–94.

Hiiemae, K. M. 2000. Feeding in mammals. In *Feeding: Form, Function, and Evolution in Tetrapod Vertebrates,* ed. K. Schwenk, 411–48. San Diego, CA: Academic Press.

Hiiemae, K. M., and A. W. Crompton. 1985. Mastication, food transport, and swallowing. In *Functional Vertebrate Morphology,* ed. M. Hildebrand, D. M. Bramble, K. F. Liem, and D. B. Wake, 262–90. Cambridge, MA: Harvard University Press.

Hiiemae, K. M., and R. F. Kay. 1972. Trends in the evolution of primate mastication. *Nature* 240:486–87.

Hiiemae, K. M., and J. B. Palmer. 1999. Food transport and bolus formation during complete feeding sequences on foods of different initial consistency. *Dysphagia* 14:31–42.

Hiiemae, K. M., J. B. Palmer, S. W. Medicis, J. Hegener, B. S. Jackson, and D. E. Lieberman. 2002. Hyoid and tongue surface movements in speaking and eating. *Archives of Oral Biology* 47:11–27.

Hiley, P. G. 1976. The thermoregulatory responses of the galago (*Galago crassicaudatus*), the baboon (*Papio cynocephalus*) and the chimpanzee (*Pan satyrus*) to heat stress. *Journal of Physiology* 254:657–67.

Hill, A., S. Ward, A, Deino, G. Curtis, and R. Drake. 1992. Earliest *Homo. Nature* 355:719–22.

Hill, R. S., and C. A. Walsh. 2005. Molecular insights into human brain evolution. *Nature* 437:64–67.

Hillenbrand, J. L., A. Getty, M. J. Clark, and K. Wheeler. 1995. Acoustic characteristics of American English vowels. *Journal of the Acoustical Society of America* 97:3099–3311.

Hillenius, W. J. 1992. The evolution of nasal turbinates and mammalian endothermy. *Paleobiology* 18:17–29.

Hillson, S. W., and E. Trinkaus. 2002. Comparative dental crown metrics. In *Portrait of the Artist as a Child: The Gravettian Human Skeleton from the Abrigo do Lagar Velho and Its Archeological Context,* ed. J. Zilhão and E. Trinkaus, 356–64. Lisbon: Instituto Português de Arqueologia.

Hinrichs, R. 1990. Upper extremity function in distance running. In *Biomechanics of Distance Running,* ed. P. R. Cavanagh, 107–31. Champaign, IL: Human Kinetics.

Hinton, R. J. 1983. Relationships between mandibular joint size and craniofacial size in human groups. *Archives of Oral Biology* 28:37–43.

———. 1993. Effect of dietary consistency on matrix synthesis and composition in the rat condylar cartilage. *Acta Anatomica (Basel)* 147:97–104.

Hirabayashi, M., M. Motoyoshi, T. Ishimaru, K. Kasai, and S. Namura. 2002. Stresses in mandibular cortical bone during mastication: Biomechanical considerations using a three-dimensional finite element method. *Journal of Oral Science* 44: 1–6.

Hirasaki, E., S. T. Moore, T. Raphan, and B. Cohen. 1999. Effects of walking velocity on vertical head and body movements during locomotion. *Experimental Brain Research* 127:117–30.

Hlusko, L. J., G. Suwa, R. Kono, and M.C. Mahaney. 2004. Genetics and the evolution of primate enamel thickness: A baboon model. *American Journal of Physical Anthropology* 124:223–33.

Hlusko, L. J., L. R. Lease, and M. C. Mahaney. 2006. The evolution of genetically correlated traits: Tooth size and body size in baboons. *American Journal of Physical Anthropology* 131:420–7.

Hodgkin, A. L. 1954. A note on conduction velocity. *Journal of Physiology* 125: 221–24.

Hofer, H. 1960. Studien zum Problem des Gestaltwandels des Schädels der Säugetiere, insbesondere der Primaten, I: Die medianen Kruemmungen des Schädels und ihre Erfassung nach Landzert. *Zeitschrift für Morphologie und Anthropologie* 50: 299–316.

———. 1969. On the evolution of the craniocerebral topography in primates. *Annals of the New York Academy of Sciences* 162:341–56.

Hofer, H., and W. Spatz. 1963. Studien zum Problem des Gestaltwandels des Schädels der Säugetiere, insbesondere der Primaten, II: Über die Kyphosen fetaler und neonater Primatenschädel. *Zeitschrift für Morphologie und Anthropologie* 53:29–52.

Hofman, M. A. 1989. On the evolution and geometry of the brain in mammals. *Progress in Neurobiology* 32:137–58.

Hohl, T. H. 1983. Masticatory muscle transposition in primates: Effects on craniofacial growth. *Journal of Oral and Maxillofacial Surgery* 11:149–56.

Holbourn, A. H. S. 1943. Mechanics of head injuries. *Lancet* 2:438–41.

Holcombe, S. J., C. J. Cornelisse, C. Berney, and N. E. Robinson. 2002. Electromyographic activity of the hyoepiglotticus muscle and control of epiglottis position in horses. *American Journal of Veterinary Research* 63:1617–21.

Holland, L. Z. 2000. Body-plan evolution in the Bilateria: Early antero-posterior patterning and the deuterostome-protostome dichotomy. *Current Opinion in Genetics and Development* 10:434–42.

Holliday, T. W. 2002. Body size and postcranial robusticity of European Upper Paleolithic hominins. *Journal of Human Evolution* 43:513–28.

Holloway, R. L. 1983. Human paleontological evidence relevant to language behavior. *Human Neurobiology* 2:105–14.

Holloway, R. L., and D. G. Post. 1982. The relativity of relative brain measures and hominid mosaic evolution. In *Primate Brain Evolution,* ed. E. Armstrong and D. Falk, 57–76. New York: Plenum Press.

Holloway, R. L., D. C. Broadfield, and M. S. Yuan. 2003. Morphology and histology of the chimpanzee primary visual striate cortex indicate that brain reorganization predated brain expansion in early hominid evolution. *Anatomical Record* 273A: 594–602.

Holloway, R. L., D. C. Broadfield, M. S. Yuan, J. H. Schwartz, and I. Tattersall. 2004. *The Human Fossil Record,* vol. 3, *Brain Endocasts: The Paleoneurological Evidence.* New York: Wiley.

Holton, N. E., and R. G. Franciscus. 2008. The paradox of a wide nasal aperture in cold-adapted Neandertals: A causal assessment. *Journal of Human Evolution* 55: 942–51.

Hooton, E. A. 1946. *Up from the Ape.* New York: Macmillan.

Howell, F. C. 1978. *Hominidae.* In *Evolution of African Mammals,* ed. V. J. Maglio and H. B. S. Cooke, 154–248. Cambridge, MA: Harvard University Press.

Howells, W. W. 1973. *Cranial Variation in Man.* Cambridge, MA: Papers of the Peabody Museum of Archaeology and Ethnology.

———. 1980. *Homo erectus:* Who, when and where; A survey. *Yearbook of Physical Anthropology* 23:1–23.

———. 1989. *Skull Shapes and the Map: Craniometric Analyses in the Dispersion of Modern* Homo. Cambridge, MA: Papers of the Peabody Museum of Archaeology and Ethnology.

———. 1995. *Who's Who in Skulls: Ethnic Identification of Crania from Measurements.* Cambridge, MA: Papers of the Peabody Museum of Archaeology and Ethnology.

Howland, H. C., S. Merola, and J. R. Basarab. 2004. The allometry and scaling of the size of vertebrate eyes. *Vision Research* 44:2043–65.

Hoyte, D. A. 1991. The cranial base in normal and abnormal growth. *Neurosurgery Clinics of North America* 2:515–37.

Hubel, D. 1988. *Eye, Brain, and Vision.* New York: W. H. Freeman.

Hublin, J.-J. 1978. Quelques caractères apomorphes du crâne néandertalien et leur interprétation phylogénétique. *Comptes rendus de l'Académie des Sciences de Paris* 287:923–26.

———. 1985. Human fossils from the North African Middle Pleistocene and the origin of *Homo sapiens.* In *Ancestors: The Hard Evidence,* ed. E. Delson, 283–88. New York: Liss.

———. 1998. Climatic changes, paleogeography, and the evolution of the Neandertals. In *Neandertals and Modern Humans in Western Asia,* ed. T. Akazawa, K. Aoki, and O. Bar-Yosef, 295–310. New York: Plenum Press.

———. 2000. Modern-nonmodern hominid interactions: A Mediterranean perspective. In *The Geography of Neandertals and Modern Humans in Europe and the*

Greater Mediterranean, ed O. Bar-Yosef and D. Pilbeam, 157–82. Cambridge, MA: Peabody Museum Bulletin.

———. 2001. Northwestern African Middle Pleistocene hominids and their bearing on the emergence of *Homo sapiens.* In *Human Roots: Africa and Asia in the Middle Pleistocene,* ed. L. Barham and K. Robson-Brown, 99–121. Bristol: Cherub.

Hublin, J.-J., and H. Coqueugniot. 2006. Absolute or proportional brain size: That is the question; A reply to Leigh's (2006) comments. *Journal of Human Evolution* 50:109–13.

Hublin, J.-J., C. Barroso Ruiz, P. Medina Lara, M. Fontugne, and J.-L. Reyss. 1995. The Mousterian site of Zafarraya (Andalucia, Spain): Dating and implications on the Palaeolithic peopling processes of Western Europe. *Comptes rendus de l'Académie des Sciences de Paris, série 2a* 321:931–37.

Hublin, J.-J., F. Spoor, M. Braun, F. Zonneveld, and S. Condemi. 1996. A late Neanderthal associated with Upper Palaeolithic artefacts. *Nature* 381:224–26.

Hughes, A. R., and P. V. Tobias. 1977. A fossil skull probably of the genus *Homo* from Sterkfontein, Transvaal. *Nature* 265:310–12.

Humphrey, L. T., M. C. Dean, and C. B. Stringer. 1999. Morphological variation in great ape and modern human mandibles. *Journal of Anatomy* 195:491–513.

Hunt, K. D. 1991. Mechanical implications of chimpanzee positional behavior. *American Journal of Physical Anthropology* 86:521–36.

———. 1992. Positional behavior of *Pan troglodytes* in the Mahale Mountains and Gombe Stream National Parks, Tanzania. *American Journal of Physical Anthropology* 87:83–105.

Hylander, W. L. 1977. The adaptive significance of Eskimo craniofacial morphology. In *Orofacial Growth and Development,* ed. A. A. Dahlberg and T. M. Graber, 129–69. The Hague: Mouton.

———. 1978. Incisal bite force direction and the. functional significance of mammalian mandibular translation. *American Journal of Physical Anthropology* 48: 1–8.

———. 1979. Experimental analysis of temporomandibular joint reaction force in macaques. *American Journal of Physical Anthropology* 51:433–56.

———. 1984. Stress and strain in the mandibular symphysis of primates: A test of competing hypotheses. *American Journal of Physical Anthropology* 64:1–46.

———. 1985. Mandibular function and biomechanical stress and scaling. *American Zoologist* 25:315–30.

———. 1986. *In vivo* bone strain as an indicator of masticatory bite force in *Macaca fascicularis. Archives of Oral Biology* 31:149–57.

———. 1988. Implications of *in vivo* experiments for interpreting the functional significance of "robust" australopithecine jaws. In *Evolutionary History of the "Robust" Australopithecines,* ed. F. Grine, 55–83. New York: Aldine de Gruyter.

————. 2006. Functional anatomy and biomechanics of the masticatory apparatus. In *Temporomandibular Joint Disorders,* ed. C. S. Greene, D. M. Laskin, and W. L. Hylander, 3–34. Hanover Park, IL: Quintessence Publishers.

Hylander, W. L., and R. Bays. 1978. Bone strain in the subcondylar region of the mandible in *Macaca fascicularis* and *Macaca mulatta. American Journal of Physical Anthropology* 48:408–15.

Hylander, W. L., and K. R. Johnson. 1994. Jaw muscle function and wishboning of the mandible during mastication in macaques and baboons. *American Journal of Physical Anthropology* 94:523–47.

Hylander, W. L., and C. J. Vinyard. 2006. The evolutionary significance of canine reduction in hominins: Functional links between jaw mechanics and canine size. *American Journal of Physical Anthropology Supplement* 42:107.

Hylander, W. L., K. R. Johnson, and A. W. Crompton. 1987. Loading patterns and jaw movement during mastication in *Macaca fascicularis:* A bone strain, electromyographic and cineradiographic analysis. *American Journal of Physical Anthropology* 72:287–314.

Hylander, W. L., P. G. Picq, and K. R. Johnson. 1991. Masticatory-stress hypotheses and the supraorbital region of primates. *American Journal of Physical Anthropology* 86:1–36.

Hylander, W. L., P. G. Picq, and K. R. Johnson. 1992. Bone strain and the supraorbital region of primates. In *Bone Biodynamics in Orthodontic and Orthopaedic Treatment,* ed. D. S. Carlson and S. A. Goldstein, 315–49. Ann Arbor, MI: Center for Human Growth and Development.

Hylander, W. L., M. J. Ravosa, C. F. Ross, and K. R. Johnson. 1998. Mandibular corpus strain in primates: Further evidence for a functional link between symphyseal fusion and jaw-adductor muscle force. *American Journal of Physical Anthropology* 107:257–71.

Hylander, W. L., M. J. Ravosa, C. F. Ross, C. E. Wall, and K. R. Johnson. 2000. Symphyseal fusion and jaw-adductor muscle force: An EMG study. *American Journal of Physical Anthropology* 112:469–92.

Ichim, I., M. Swain, and J. A. Kieser. 2006. Mandibular biomechanics and development of the human chin. *Journal of Dental Research* 85:638–42.

Ingervall, B., and E. Helkimo. 1978. Masticatory muscle force and facial morphology in man. *Archives of Oral Biology* 23:203–6.

Isaac, G. 1978. The food-sharing behavior of proto-human hominids. *Scientific American* 238:90–108.

Israel, H. 1973. Age factor and the pattern of change in craniofacial structures. *American Journal of Physical Anthropology* 39:111–28.

Iwasaki, K. 1989. Dynamic responses in adult and infant monkey craniums during occlusion and mastication. *Journal of the Osaka Dental University* 23:77–97.

Jablonski, N. 2006. *Skin: A Natural History.* Berkeley: University of California Press.

Jablonski, N., and G. Chaplin. 1993. Origin of habitual terrestrial bipedalism in the ancestor of the Hominidae. *Journal of Human Evolution* 24:259–80.

Jacob, F. 1977. Evolution and tinkering. *Science* 196:1161–66.

Jacob, T., E. Indriati, R. P. Soejono, K. Hsü, D. W. Frayer, R. B. Eckhardt, A. J. Kuperavage, A. Thorne, and M. Henneberg. 2006. Pygmoid Australomelanesian *Homo sapiens* skeletal remains from Liang Bua, Flores: Population affinities and pathological abnormalities. *Proceedings of the National Academy of Sciences USA* 103:13421–26.

Jacobs, G. H. 1993. The distribution and nature of colour vision among the mammals. *Biological Reviews* 68:413–71.

———. 1996. Primate photopigments and primate color vision. *Proceedings of the National Academy of Sciences USA* 93:577–81.

Jacobs, G. H., J. F. Deegan II, and J. L. Moran. 1996. ERG measurements of the spectral sensitivity of common chimpanzee (*Pan troglodytes*). *Vision Research* 36: 2587–94.

James, S. R. 1989. Hominid use of fire in the Lower and Middle Pleistocene: A review of the evidence. *Current Anthropology* 30:1–26.

Janis, C. 1983. Muscle of the masticatory apparatus in two genera of Hyrax (*Procavia* and *Heterohyrax*). *Journal of Morphology* 176:61–87.

Jaslow, C. R. 1990. Mechanical properties of cranial sutures. *Journal of Biomechanics* 23:313–21.

Jaslow, C. R., and A. A. Biewener. 1995. Strain patterns in the horncores, cranial bones and sutures of goats (*Capra hircus*) during impact loading. *Journal of Zoology* 235:193–210.

Jaworski, Z. F. G., M. Liskova-Kair, and H. K. Uhthoff. 1980. Effects of long-term immobilization on the pattern of bone loss in older dogs. *Journal of Bone and Joint Surgery* 59:204–8.

Jee, W. S. S., T. J. Wronski, E. R. Morey, and D. B. Kimmel 1983. Effects of space flight on trabecular bone in rats. *American Journal of Physiology* 244:310–14.

Jeffery, N. 2003. Brain expansion and comparative prenatal ontogeny of the non-hominoid primate cranial base. *Journal of Human Evolution* 45:263–84.

Jeffery, N., and F. Spoor. 2002. Brain size and the human cranial base: A prenatal perspective. *American Journal of Physical Anthropology* 118:324–40.

———. 2004a. Ossification and midline shape changes of the human fetal cranial base. *American Journal of Physical Anthropology* 123:78–90.

———. 2004b. Prenatal growth and development of the modern human labyrinth. *Journal of Anatomy* 204:71–92.

Jenkins, G. N. 1978. *Physiology and Biochemistry of the Mouth.* Oxford: Blackwell Scientific.

Jerison, H. 1973. *The Evolution of the Brain and Intelligence.* New York: Academic Press.

Jernvall, J. 1995. Mammalian molar cusp patterns: Developmental mechanisms of diversity. *Acta Zoologica Fennica* 198:1–61.

Jernvall, J., and H. S. Jung. 2000. Genotype, phenotype and developmental biology of molar tooth characters. *Yearbook of Physical Anthropology* 43:171–90.

Jernvall, J., and I Thesleff. 2000. Reiterative signaling and patterning during mammalian tooth morphogenesis. *Mechanisms of Development* 92:19–29.

Jernvall, J., P. Kettunen, I. Karavanova, L. B. Martin, and I. Thesleff. 1994. Evidence for the role of the enamel knot as a control center in mammalian tooth cusp formation: Non-dividing cells express growth stimulating Fgf-4 gene. *International Journal of Developmental Biology* 38:463–9.

Jernvall, J., S. V. E. Keränen, and I. Thesleff. 2000. Evolutionary modification of development in mammalian teeth: Quantifying gene expression patterns and topography. *Proceedings of the National Academy of Sciences USA* 97:14444–48.

Jiang, X., S. Iseki, R. E. Maxson, H. M. Sucov, and G. M. Morriss-Kay. 2002. Tissue origins and interactions in the mammalian skull vault. *Developmental Biology* 241:106–16.

Jinushi, H., T. Suzuki, T. Naruse, and Y. Isshiki. 1997. A dynamic study of the effect on the maxillofacial complex of the face bow: Analysis by a three-dimensional finite element method. *Bulletin of the Tokyo Dental College* 38:33–41.

Jobling, M. A., M. E. Hurles, and C. Tyler-Smith. 2004. *Human Evolutionary Genetics: Origins, Peoples and Disease.* New York: Garland Science.

Johanson, D. C., and T. D. White. 1979. A systematic assessment of early African hominids. *Science* 202:321–30.

Johanson, D. C., C. O. Lovejoy, W. H. Kimbel, T. D. White, S. C. Ward, M. E. Bush, B. M. Latimer, and Y. Coppens. 1982. Morphology of the Pliocene partial hominid skeleton (AL288–1) from the Hadar Formation, Ethiopia. *American Journal of Physical Anthropology* 57:453–99.

Johanson, D. C., F. T. Masao, G. C. Eck, T. D. White, R. C. Walter, W. H. Kimbel, B. Asfaw, P. Manega, P. Ndessokia, and G. Suwa. 1987. New partial skeleton of *Homo habilis* from Olduvai Gorge, Tanzania. *Nature* 327:205–9.

Johnson, J. I., J. A. Kirsch, R. L. Reep, and R. C. Switzer III. 1994. Phylogeny through brain traits: More characters for the analysis of mammalian evolution. *Brain, Behavior and Evolution* 43:319–47.

Johnston, M. C., and P. T. Bronsky. 2002. Craniofacial embryogenesis: Abnormal developmental mechanisms. In *Understanding Craniofacial Anomalies: The Etiopathogenesis of Craniosynostoses and Facial Clefting,* ed. M. P. Mooney and M. I. Siegel, 61–124. New York: Wiley-Liss.

Jolly, C. J. 1970. The seed eaters: A new model for hominid differentiation based on a baboon analogy. *Man* 5:5–26.

Jones, H. H., J. D. Priest, W. C. Hayes, C. C. Tichenor, and D. A. Nagel. 1977. Humeral hypertrophy in response to exercise. *Journal of Bone and Joint Surgery* 59A:204–8.

Joseph, B. K., N. W. Savage, W. G. Young, and M. J. Waters. 1994. Insulin-like growth factor-I receptor in the cell biology of the ameloblast: An immunohistochemical study on the rat incisor. *Epithelial Cell Biology* 3:47–53.

Joseph, J. T., and D. L. Cardozo. 2004. *Functional Neuroanatomy: An Interactive Text and Manual.* Hoboken, NJ: J. Wiley and Sons.

Jung, H. S., K. Akita, and J. Y. Kim. 2004. Spacing patterns on tongue surface-gustatory papilla. *International Journal of Developmental Biology* 48:157–61.

Jungers, W. L., A. B. Falsetti, and C. E. Wall. 1995. Shape, relative size, and size-adjustments in morphometrics. *Yearbook of Physical Anthropology* 38:137–61.

Jungers, W. L., W. E. Harcourt-Smith, R. E. Wunderlich, M. W. Tocheri, S. G. Larson, T. Sutikna, R. A. Due, and M. J. Morwood. 2009. The foot of *Homo floresiensis. Nature* 459:81–84.

Kahrilas, P. 1997. Esophageal motility disorders: Current concepts of dysphagia with visual examination of the gag reflex and velar movement. *Dysphagia* 12:21–23.

Kandel, E. R., J. H. Schwartz, and T. M. Jessel. 2000. *Principles of Neural Science.* 4th ed. New York: Mcgraw-Hill.

Kangas, A. T., A. R. Evans, I. Thesleff, and J. Jernvall. 2004. Nonindependence of mammalian dental characters. *Nature* 432:211–4.

Kantomaa, T., and O. Rönning. 1991. Growth of the Mandible. In *Bone,* vol. 6, ed. B. K. Hall, 157–83. Boca Raton, FL: CRC Press.

Kapoor, K., V. K. Kak, and B. Singh. 2003. Morphology and comparative anatomy of circulus arteriosus cerebri in mammals. *Anatomia, Histologia, Embryologia* 32: 347–55.

Kappelman, J. 1996. The evolution of body mass and relative brain size in fossil hominids. *Journal of Human Evolution* 30:243–76.

Karsenty, G. 1999. The genetic transformation of bone biology. *Genes and Development* 13:3037–3051.

Karsenty, G., and E. F. Wagner. 2002. Reaching a genetic and molecular understanding of skeletal development. *Developmental Cell* 2:389–406.

Kay, R. F. 1975 The functional adaptations of primate molar teeth. *American Journal of Physical Anthropology* 43:195–216.

———. 1984. On the use of anatomical features to infer foraging behavior in extinct primates. In *Adaptations for Foraging in Nonhuman Primates: Contributions to an Organismal Biology of Prosimians, Monkeys, and Apes,* ed. P. S. Rodman and J. G. H. Cant, 21–53. New York: Columbia University Press.

———. 1985. Dental evidence for the diet of *Australopithecus*. *Annual Review of Anthropology* 14:315–41.

Kay, R. F., and M. Cartmill. 1977. Cranial morphology and adaptations of *Palaechthon nacimienti* and other paramomyidae (Plesiadapoidea?, Primates) with a description of a new genus and species. *Journal of Human Evolution* 6:9–53.

Kay, R. F., and K. M. Hiiemae. 1974. Jaw movement and tooth use in recent and fossil primates. *American Journal of Physical Anthropology* 40:227–56.

Kay, R. F., and P. S. Ungar. 1997. Dental evidence for diet in some Miocene catarrhines. In *Function, Phylogeny and Fossils: Miocene Hominoids and Great Ape and Human Origins,* ed. D. R. Begun, C. Ward, and M. Rose, 131–51. New York: Plenum.

Kay, R. F., M. Cartmill, and M. Balow. 1998. The hypoglossal canal and the origin of human vocal behavior. *Proceedings of the National Academy of Sciences USA.* 95:5417–19.

Keating, P., and R. Buhr. 1978. Fundamental frequency in the speech of infants and children. *Journal of the Acoustical Society of America* 63:567–71.

Kelemen, G. 1969. Anatomy of the larynx and the anatomical basis of vocal performance. In *The Chimpanzee,* vol. 1, ed. G. H. Bourne, 35–186. Basel: Karger.

Keller, T. S., A. M. Weisberger, J. L. Ray, S. S. Hasan, R. G. Shiavi, and D. M. Spengler. 1996. Relationship between vertical ground reaction force and speed during walking, slow jogging, and running. *Clinical Biomechanics* 11:253–59.

Kennedy, K. A. R., S. U. Deraniyagala, W. J. Roertgen, J. Chiment, and T. Disotell. 1987. Upper Pleistocene fossil hominids from Sri Lanka. *American Journal of Physical Anthropology* 72:441–61.

Kennedy, K. A. R., A. Sonakia, J. Chiment, and K. K. Verma. 1991. Is the Narmada hominid an Indian *Homo erectus? American Journal of Physical Anthropology* 86:475–96.

Keranen, S. V., P. Kettunen, T. Aberg, I. Thesleff, and J. Jernvall. 1999. Gene expression patterns associated with suppression of odontogenesis in mouse and vole diastema regions. *Development Genes and Evolution* 209:495–506.

Kerley, E. R. 1965. The microscopic determination of age in human bone. *American Journal of Physical Anthropology.* 23:149–64.

Keshner, E. A., R. Cromwell, and B. W. Peterson. 1992. Head stabilization during vertical seated rotations and gait. *Gait and Posture* 1:105–8.

Keshner, E. A., G. Peng, T. Hain, and B. W. Peterson. 1995. Characteristics of head and neck stabilization in two planes of motion. In *Multisensory Control of Posture,* ed. T. Mergner and F. Hlavacka, 83–94. New York: Plenum Press.

Keverne, E. B. 1999. The vomeronasal organ. *Science* 288:716–20.

Key, C. A., and C. Ross. 1999. Sex differences in energy expenditure in non-human primates. *Proceedings of the Royal Society of London, Series B* 266:2479–85.

Keyhani, K., P. W. Scherer, and M. M. Mozell. 1995. Numerical simulation of airflow in the human nasal cavity. *Journal of Biomechanical Engineering* 117:429–41.

———. 1997. A numerical model of nasal odorant transport for the analysis of human olfaction. *Journal of Theoretical Biology* 186:279–301.

Kidder, J. H., R. L. Jantz, and F. H. Smith. 1992. In *Continuity or Replacement: Controversies in* Homo sapiens *Evolution,* ed. G. Bräuer and F. H. Smith, 157–77. Rotterdam: A. A. Balkema.

Kiliaridis, S. 1995. Masticatory muscle influence on craniofacial growth. *Acta Odontologica Scandinavica* 53:196–202.

Kiliaridis, S., C. Engström, and B. Thilander. 1986. The relationship between masticatory muscle function and cranial morphology, I: A cephalometric longitudinal analysis in the growing rat fed a soft diet. *European Journal of Orthodontics* 7:273–83.

Kiliaridis, S., C. Mejersjö, and B. Thilander. 1989. Muscle function and craniofacial morphology: A clinical study in patients with myotonic dystrophy. *European Journal of Orthodontics* 11:131–38.

Kiliaridis, S., B. Thilander, H. Kjellberg, N. Topouzelis, and A. Zafiriadis. 1999. Effect of low masticatory function on condylar growth: A morphometric study in the rat. *American Journal of Orthodontics and Dentofacial Orthopaedics* 116:121–25.

Kimbel, W. H. 1984. Variation in the pattern of cranial venous sinuses and hominid phylogeny. *American Journal of Physical Anthropology* 63:243–63.

———. 2009. The Origin of *Homo.* In *The First Humans: Origin and Early Evolution of the Genus* Homo, ed. F. E. Grine, J. G. Fleagle, and R. E. F. Leakey, 31–37. New York: Springer.

Kimbel, W. H., and L. K. Delezene. 2009. Lucy redux: A review of research on *Australopithecus afarensis. Yearbook of Physical Anthropology* 49:2–48.

Kimbel, W. H., and Y. Rak. 1985. Functional morphology of the asterionic region in extant hominoids and fossil hominids. *American Journal of Physical Anthropology* 66:31–54.

Kimbel, W. H., T. D. White, and D. C. Johanson. 1984. Cranial morphology of *Australopithecus afarensis:* A comparative study based on a composite reconstruction. *American Journal of Physical Anthropology* 64:337–88.

Kimbel, W. H., D. C. Johanson, and Y. Rak. 1997. Systematic assessment of a maxilla of *Homo* from Hadar, Ethiopia. *American Journal of Physical Anthropology* 103:235–62.

Kimbel, W. H., C. A. Lockwood, C. V. Ward, M. G. Leakey, Y. Rak, and D. C. Johanson. 2006. Was *Australopithecus anamensis* ancestral to *A. afarensis?* A case of anagenesis in the hominin fossil record. *Journal of Human Evolution* 51:134–52.

Kimbel, W. H., Y. Rak, and D. C. Johanson. 2004. *The Skull of* Australopithecus afarensis. Oxford: Oxford University Press.

Kirk, E. C. 2004. Comparative morphology of the eye in primates. *Anatomical Record* 281A: 1095–1103.

Kirkendall, D. T., and W. E. Garrett Jr. 2001. Heading in soccer: Integral skill or grounds for cognitive dysfunction? *Journal of Athletic Training* 36:328–33.

Kirschner, M. W., and J. C. Gerhart. 2005. *The Plausibility of Life.* New Haven, CT: Yale University Press.

Kjaer, I. 1989. Prenatal skeletal maturation of the human maxilla. *Journal of Craniofacial Genetics and Developmental Biology* 9:257–64.

———. 1990. Ossification of the human basicranium. *Journal of Craniofacial Genetics and Developmental Biology* 10:29–38.

Kjaer, I., and B. Fischer Hansen. 1996. The human vomeronasal organ: Prenatal developmental stages and distribution of luteinizing hormone-releasing hormone. *European Journal of Oral Sciences* 104:34–40.

Klein, R. G. 1995. Anatomy, behavior, and modern human origins. *Journal of World Prehistory* 9:167–98.

———. 2003. Whither the Neanderthals? *Science* 299:1525–7.

———. 2009. *The Human Career.* 3rd ed. Chicago: University of Chicago Press.

Klevezal, G. A. 1996. Recording Structures of Mammals: Determination of Age and Reconstruction of Life History. Rotterdam: A. A. Balkema.

Kobayashi, H., and S. Koshima. 1997. Unique morphology of the human eye. *Nature* 387:767–68.

Kobayashi, H., and S. Koshima. 2001. Unique morphology of the human eye and its adaptive meaning: Comparative studies of external morphology of the primate eye. *Journal of Human Evolution* 40:419–35.

Kodama, G. 1976a. Developmental studies of the presphenoid of the human sphenoid bone. In *Development of the Basicranium,* ed. J. F. Bosma, 141–55. Bethesda, MD: National Institutes of Health.

———. 1976b. Developmental studies of the body of the human sphenoid bone. In *Development of the Basicranium,* ed. J. F. Bosma, 156–65. Bethesda, MD: National Institutes of Health.

———. 1976c. Developmental studies of the orbitosphenoid of the human sphenoid bone. In *Development of the Basicranium,* ed. J. F. Bosma, 166–76. Bethesda, MD: National Institutes of Health.

Kohn, L. A. P., S. R. Leigh, S. C. Jacobs, and J. M. Cheverud. 1993. Effects of annular cranial vault modification on the cranial base and face. *American Journal of Physical Anthropology* 90:147–68.

Kojima, S. 1990. Comparison of auditory functions in the chimpanzee and human. *Folia Primatologica* 55:62–72.

Kollar, E. J., and G. R. Baird. 1970. Tissue interactions in embryonic mouse tooth germs, I: Reorganization of the dental epithelium during tooth germ reconstruction. *Journal of Embryology and Experimental Morphology* 24:159–71.

Komori, T., H. Yagi, S. Nomura, A. Yamaguchi, K. Sasaki, K. Deguchi, Y. Simizu, R. T. Bronson, Y. H. Gao, and M. Inada. 1997. Targeted disruption of Cbfa1 results in complete lack of bone formation owing to maturational arrest of osteoblasts. *Cell* 89:755–64.

Kongtes, G., and A. Lumsden. 1996. Rhombencephalic neural crest segmentation is preserved throughout craniofacial ontogeny. *Development* 122:3229–42.

Korioth, T. W., and A. Versluis. 1997. Modeling the mechanical behavior of the jaws and their related structures by finite element (FE) analysis. *Critical Reviews in Oral Biology and Medicine* 8:90–104.

Kouprina, N., A. Pavlicek, G. H. Mochida, G. Solomon, W. Gersch, Y. H. Yoon, R. Collura, et al. 2004. Accelerated evolution of the ASPM gene controlling brain size begins prior to human brain expansion. *PLoS Biology* 2:E126.

Kramer, A., S. M. Donnelly, J. H. Kidder, S. D. Ousley, and S. M. Olah. 1995. Craniometric variation in large-bodied hominoids: Testing the single-species hypothesis for *Homo habilis*. *Journal of Human Evolution* 29:443–62.

Krause, J., L. Orlando, D. Serre, B. Viola, K. Prüfer, M. P. Ricards, J.-J. Hublin, C. Hänni, A. P. Derevi5anko, and S. Pääbo. 2007. Neanderthals in central Asia and Siberia. *Nature* 449:902–4.

Krogman, W. M. 1969. Growth changes in the skull, face, jaws and teeth of the chimpanzee. In *The Chimpanzee,* vol. 1, ed. G. H. Bourne, 104–64. Basel: Karger.

Kuhn, S. L., and M. C. Stiner. 2006. What's a mother to do? The division of labor among Neandertals and modern humans in Eurasia. *Current Anthropology* 47: 953–81.

Kuman, K., and R. J. Clarke. 2000. Stratigraphy, artefact industries and hominid associations for Sterkfontein Member 5. *Journal of Human Evolution* 38:827–47.

Kuno, Y. 1956. *Human Perspiration.* Springfield, IL: Charles C Thomas.

Kuratani, S. 2005. Craniofacial development and the evolution of the vertebrates: The old problems on a new background. *Zoological Science* 22:1–19.

Kurokawa, H., and R. L. Goode. 1995. Sound pressure gain produced by the human middle ear. *Otolaryngology: Head and Neck Surgery* 113:349–55.

Kuykendall, K. L. 1996. Dental development in chimpanzees (*Pan troglodytes*): The timing of tooth calcification stages. *American Journal of Physical Anthropology* 99:135–57.

Kuzawa, C. W. 1998. Adipose tissue in human infancy and childhood: An evolutionary perspective. *Yearbook of Physical Anthropology* 41:177–209.

Lacruz, R. S., F. R. Rozzi, and T. G. Bromage. 2006. Variation in enamel development of South African fossil hominids. *Journal of Human Evolution* 51:580–90.

Lacruz, R. S., M. C. Dean, F. Ramirez-Rossi, and T. G. Bromage. 2008. Megadontia, striae periodicity and patterns of enamel secretion in Plio-Pleistocene fossil hominins. *Journal of Anatomy* 213:148–58.

Laden, G., and R. W. Wrangham. 2005. The rise of the hominids as an adaptive shift in fallback foods: Plant underground storage organs (USOs) and the origin of the Australopiths. *Journal of Human Evolution* 49:482–98.

Lager, H. 1958. A histological description of the cranial base in Macaca rhesus. *Transactions of the European Orthodontics Society* 34:147–56.

Lahr, M. M. 1996. The Evolution of Modern Human Diversity: A Study of Cranial Variation. Cambridge: Cambridge University Press.

Lahr, M. M., and R. A. Foley. 1998. Towards a theory of modern human origins: Geography, demography, and diversity in recent human evolution. *Yearbook of Physical Anthropology* 41:137–76.

Lahr, M. M., and R. S. V. Wright. 1996. The question of robusticity and the relationship between cranial size and shape in *Homo sapiens*. *Journal of Human Evolution* 31:157–91.

Laitman, J. T., and E. S. Crelin. 1976. Postnatal development of the basicranium and vocal tract region in man. In *Symposium on the Development of the Basicranium,* ed. J. F. Bosma, 206–19. Bethesda, MD: National Institutes of Health.

Laitman, J. T., and R. C. Heimbuch. 1982. The basicranium of Plio-Pleistocene hominids as an indicator of their upper respiratory systems. *American Journal of Physical Anthropology* 59:323–44.

Laitman, J. T., E. S. Crelin, and G. T. Conlogue. 1977. The function of the epiglottis in monkey and man. *Yale Journal of Biology and Medicine* 50:43–48.

Laitman, J. T., R. C. Heimbuch, and E. S. Crelin. 1979. The basicranium of fossil hominids as an indicator of their upper respiratory systems. *American Journal of Physical Anthropology* 51:15–34.

Lam, Y. M., O. M. Pearson, and C. M. Smith. 1996. Chin morphology and sexual dimorphism in the fossil hominid mandible sample from Klasies River Mouth. *American Journal of Physical Anthropology* 100:545–57.

Lambert, J. E. 1998. Primate digestion: Interactions among anatomy, physiology, and feeding ecology. *Evolutionary Anthropology* 7:8–20.

Lanyon, L. E., V. Armstrong, D. Ong, G. Zaman, and J. Price. 2004. Is *ERα* key to controlling bones' resistance to fracture? *Journal of Endocrinology* 182:183–91.

Larsen, W. J. 2001. *Human Embryology.* 3rd ed. Edinburgh: Churchill Livingstone.

Larson, J. E., and S. W. Herring. 1996. Movement of the epiglottis in mammals. *American Journal of Physical Anthropology* 100:71–82.

Larson, S. G. 1993. Functional morphology of the shoulder in primates. In *Postcranial Adaptation in Nonhuman Primates,* ed. D. Gebo, 45–69. DeKalb: Northern Illinois University Press.

Larson, S. G., W. L. Jungers, M. J. Morwood, T. Sutikna, Jatmiko, E. W. Saptomo, R. A. Due, and T. Djubiantono. 2007. *Homo floresiensis* and the evolution of the hominin shoulder. *Journal of Human Evolution* 53:718–31.

Laska, M., and A. Seibt. 2002. Olfactory sensitivity for aliphatic alcohols in squirrel monkeys and pigtail macaques. *Journal of Experimental Biology* 205:1633–43.

Laska, M., A. Seibt, and A. Weber. 2000. "Microsmatic" primates revisited: Olfactory sensitivity in the squirrel monkey. *Chemical Senses* 25:47–53.

Latham, R. A. 1972. The sella point and postnatal growth of the human cranial base. *American Journal of Orthodontics* 61:156–62.

Lavelle, C. L. B., R. P. Shellis, and D. F. G. Poole. 1977. *Evolutionary Changes to the Primate Skull and Dentition.* Springfield, IL: Charles C. Thomas.

Leakey, L. S. B. 1959. A new fossil skull from Olduvai. *Nature* 184:491–93.

Leakey, L. S. B., P. V. Tobias, and J. R. Napier. 1964. A new species of the genus *Homo* from Olduvai Gorge. *Nature* 202:7–9.

Leakey, M. D., and J. M. Harris. 1987. *Laetoli: A Pliocene Site in Northern Tanzania.* Oxford: Clarendon Press.

Leakey, M. G., C. S. Feibel, I. McDougall, and A. Walker. 1995. New four-million-year-old hominid species from Kanapoi and Allia Bay, Kenya. *Nature* 376: 565–71.

Leakey, M. G., C. S. Feibel, I. McDougall, C. Ward, and A. C. Walker. 1998. New specimens and confirmation of an early age for *Australopithecus anamensis. Nature:* 393:62–66.

Leakey, M. G., F. Spoor, F. H. Brown, P. N. Gathogo, C. Kiarie, L. N. Leakey, and I. McDougall. 2001. New hominin genus from eastern Africa shows diverse middle Pliocene lineages. *Nature* 410:433–40.

Leakey, R. E. F., and A. C. Walker. 1985. Further hominids from the Plio-Pleistocene of Koobi Fora, Kenya. *American Journal of Physical Anthropology* 67:135–63.

Leakey, R. E. F., K. W. Butzer, and M. H. Day. 1969. Early *Homo sapiens* remains from the Omo River region of southwest Ethiopia. *Nature.* 222:1132–38.

Lebatard, A. E., D. L. Bourlès, P Duringer, M. Jolivet, R. Braucher, J. Carcaillet, M. Schuster, et al. 2008. Cosmogenic nuclide dating of *Sahelanthropus tchadensis* and *Australopithecus bahrelghazali:* Mio-Pliocene hominids from Chad. *Proceedings of the National Academy of Sciences USA* 105:3226–31.

Le Douarin, N. M., C. Ziller, and G. F. Couly. 1993. Patterning of neural crest derivatives in the avian embryo: In vivo and in vitro studies. *Developmental Biology* 159:24–49.

Lee, S. H., O. Bédard, M. Buchtová, K. Fu, and J. M. Richman. 2004. The origin and fate of the maxillary prominence. *Developmental Biology* 276:207–24.

Lee, S. K., Y. S. Kim, H. S. Oh, K. H. Yang, E. C. Kim, and J. G. Chi. 2001. Prenatal development of the human mandible. *Anatomical Record* 263A: 314–25.

Leestma, J. E. 1988. *Forensic Neuropathology.* New York: Raven Press.

Le Gros Clark, W. E. 1964. The Fossil Evidence of Human Evolution: An Introduction to the Study of Paleoanthropology. 2nd ed. Chicago: University of Chicago Press.

Leigh, S. R. 1992. Patterns of variation in the ontogeny of primate body size dimorphism. *Journal of Human Evolution* 23:27–50.

———. 2001. The evolution of human growth. *Evolutionary Anthropology* 10: 223–36.

———. 2004. Brain growth, life history, and cognition in primate and human evolution. *American Journal of Primatology* 62:139–64.

———. 2006. Cranial ontogeny of Papio baboons (*Papio hamadryas*). *American Journal of Physical Anthropology* 130:71–84.

Leonard, W. R., and M. L. Robertson. 1997. Comparative primate energetics and hominid evolution. *American Journal of Physical Anthropology* 102:265–81.

Leutenegger, W. 1974. Functional aspects of pelvic morphology in simian primates. *Journal of Human Evolution* 3:207–22.

Lewontin, R. 1972. The apportionment of human diversity. *Evolutionary Biology* 6: 391–98.

Liebenberg, L. 1990. *The Art of Tracking: The Origin of Science.* Cape Town: David Philip Publishers.

———. 2006. Persistence hunting by modern hunter-gatherers. *Current Anthropology* 47:1017–1026.

Lieberman, D. E. 1993. Life history variables preserved in dental cementum microstructure. *Science* 261:1162–64.

———. 1994. The biological basis for seasonal increments in dental cementum and their application to archaeological research. *Journal of Archaeological Science* 21: 525–39.

———. 1995. Testing hypotheses about recent human evolution from skulls: Integrating morphology, function, development and phylogeny. *Current Anthropology* 36:159–97.

———. 1996. How and why recent humans grow thin skulls: Experimental data on systemic cortical robusticity. *American Journal of Physical Anthropology* 101: 217–36.

———. 1998. Sphenoid shortening and the evolution of modern human cranial shape. *Nature* 393:158–62.

———. 1999. Homology and hominid phylogeny: Problems and potential solutions. *Evolutionary Anthropology* 7:142–51.

———. 2000. Ontogeny, homology, and phylogeny in the Hominid craniofacial skeleton: The problem of the browridge. *In Development, Growth and Evolution,* ed. P. O'Higgins and M. Cohn, 85–122. London: Academic Press.

————. 2008. Speculations about the selective basis for modern human cranial form. *Evolutionary Anthropology* 17:22–37.

Lieberman, D. E., and O. Bar-Yosef. 2005. Apples and oranges: Morphological versus behavioral transitions in the Pleistocene. In *Interpreting the Past: Essays on Human, Primate and Mammal Evolution,* ed. D. E. Lieberman, R. J. Smith, and J. Kelley, 275–96. Boston: Brill Academic Publishers.

Lieberman, D. E., and A. W. Crompton. 2000. Why fuse the mandibular symphysis? A comparative analysis. *American Journal of Physical Anthropology* 112:517–40.

Lieberman, D. E., and B. K. Hall. 2007. The evolutionary developmental biology of tinkering: An introduction to the challenge. *Novartis Foundation Symposium* 284:1–19.

Lieberman, D. E., and R. C. McCarthy. 1999. The ontogeny of cranial base angulation in humans and chimpanzees and its implications for reconstructing pharyngeal dimensions. *Journal of Human Evolution* 36:487–517.

Lieberman, D. E., and O. M. Pearson. 2001. Trade-off between modeling and remodeling responses to loading in the mammalian limb. *Bulletin of the Museum of Comparative Zoology* 156:269–82.

Lieberman, D. E., D. R. Pilbeam, and B. A. Wood. 1988. A probabilistic approach to the problem of sexual dimorphism in *Homo habilis:* A comparison of KNM-ER 1470 and KNM-ER 1813. *Journal of Human Evolution* 17:503–11.

Lieberman, D. E., B. A. Wood, and D. R. Pilbeam. 1996. Homoplasy and early *Homo:* An analysis of the evolutionary relationships of *H. habilis sensu stricto* and *H. rudolfensis. Journal of Human Evolution* 30:97–120.

Lieberman, D. E., K. M. Mowbray, and O. M. Pearson. 2000a. Basicranial influences on overall cranial shape. *Journal of Human Evolution* 38:291–315.

Lieberman, D. E., C. F. Ross, and M. J. Ravosa. 2000b. The primate cranial base: Ontogeny, function and integration. *Yearbook of Physical Anthropology* 43:117–69.

Lieberman, D. E., R. C. McCarthy, K. M. Hiiemae, and J. B. Palmer. 2001. Ontogeny of hyoid and laryngeal descent: Implications for deglutition and vocalization in humans. *Archives of Oral Biology* 46:117–28.

Lieberman, D. E., B. M. McBratney, and G. Krovitz. 2002. The evolution and development of cranial form in *Homo Sapiens. Proceedings of the National Academy of Sciences USA* 99:1134–39.

Lieberman, D. E., O. M. Pearson, J. D. Polk, B. Demes, and A. W. Crompton. 2003. Optimization of bone growth and remodeling in response to loading in tapered mammalian limbs. *Journal of Experimental Biology* 206:3125–38.

Lieberman, D. E., G. E. Krovitz, F. W. Yates, M. Devlin, and M. St. Claire. 2004a. Effects of food processing on masticatory strain and craniofacial growth in a retrognathic face. *Journal of Human Evolution* 46:655–77.

Lieberman, D. E., G. E. Krovitz, and B. McBratney-Owen. 2004b. Testing hypotheses about tinkering in the fossil record: The case of the human skull. *Journal of*

Experimental Zoology, Part B: Molecular and Developmental Evolution 302B: 302–21.

Lieberman, D. E., J. D. Polk, and B. Demes. 2004c. Predicting long bone loading from cross-sectional geometry. *American Journal of Physical Anthropology* 123: 156–71.

Lieberman, D. E., D. A. Raichlen, H. Pontzer, D. Bramble, and E. Cutright-Smith E. 2006. The human gluteus maximus and its role in running. *Journal of Experimental Biology* 209:2143–55.

Lieberman, D. E., J. Carlo, M. Ponce de León, and C. P. Zollikofer. 2007a. A geometric morphometric analysis of heterochrony in the cranium of chimpanzees and bonobos. *Journal of Human Evolution* 52:647–62.

Lieberman, D. E., D. M. Bramble, D. A. Raichlen, and J. J. Shea. 2007b. The evolution of endurance running and the tyranny of ethnography. *Journal of Human Evolution.* 53:439–42.

Lieberman, D. E., B. Hallgrímsson, W. Liu, T. E. Parsons, and H. A. Jamniczky. 2008a. Spatial packing, cranial base angulation, and craniofacial shape variation in the mammalian skull: Testing a new model using mice. *Journal of Anatomy* 212:720–35.

Lieberman, D. E., D. M. Bramble, D. A. Raichlen, and J. J. Shea. 2008b. Brains, brawn and the evolution of endurance running capabilities. In *The First Humans: Origin of the Genus Homo,* ed. F. E. Grine, J. G. Fleagle, and R. E. Leakey, 77–92. New York: Springer.

Lieberman, D. E., D. M. Bramble, D. A. Raichlen, and K. Whitcome. 2008c. Functional, developmental and morphological integration: The case of the head and forelimb in bipedal hominins. *American Journal of Physical Anthropology Supplement* 46:140.

Lieberman, D. E., D. R. Pilbeam, and R. W. Wrangham. 2009. The transition from *Australopithecus* to *Homo.* In *Transitions in Prehistory: Essays in Honor of Ofer Bar-Yosef,* ed. J. J. Shea and D. E. Lieberman, 1–24. Oxford: Oxbow Press.

Lieberman, D. E., M. Venkadesan, W. A. Werbel, A. I. Daoud, S. D'Andrea, I. S. Davis, R. O. Mang'eni, and Y. M. Pitsiladis. 2010. Foot strike patterns and collision forces in habitually shod versus barefoot runners. *Nature* 463:531–35.

Lieberman, P. 1980. On the development of vowel production in young children. In *Child Phonology, Perception and Production,* ed. G. Yeni-Komshian, and J. Kavanagh, 113–42. New York: Academic Press.

———. 1984. *The Biology and Evolution of Language.* Cambridge, MA: Harvard University Press.

———. 2000. *Human Language and Our Reptilian Brain.* Cambridge, MA: Harvard University Press.

———. 2006. *Toward an Evolutionary Biology of Language.* Cambridge, MA: Harvard University Press.

Lieberman, P., and S. E. Blumstein. 1988. *Speech physiology, speech perception, and acoustic phonetics.* Cambridge: Cambridge University Press.

Lieberman, P., and E. S. Crelin. 1971. On the speech of Neanderthal man. *Linguistic Inquiry* 2:203–22.

Lieberman, P., and R. C. McCarthy. 2007. Tracking the evolution of language and speech. *Expedition* 49:15–20.

Lieberman, P., D. H. Klatt, and W. H. Wilson. 1969. Vocal tract limitations on the vowel repertoires of rhesus monkey and other nomhurnan primates. *Science* 164: 1185–87.

Lieberman, P., E. S. Crelin, and D. H. Klatt. 1972a. Phonetic ability and related anatomy of the newborn, adult human, Neanderthal man, and the chimpanzee. *American Anthropologist* 74:287–307.

Lieberman, P., K. S. Harris, P. Wolff, and L. H. Russell. 1972b. Newborn infant cry and nonhuman primate vocalizations. *Journal of Speech and Hearing Research* 14:718–27.

Lilliford, P. J. 1991. Texture and acceptability of plant foods. In *Feeding and the Texture of Food,* ed. J. F. V. Vincent and P. J. Lilliford, 93–121. Cambridge: Cambridge University Press.

Lindblom, B. 1986. Phonetic universals in vowel systems. In *Experimental Phonology,* ed. J. J. Ohala and J. J. Jeager, 13–44. Orlando, FL: Academic Press.

———. 1990. Explaining phonetic variation: A sketch of the H and H theory. In *Speech Production and Speech Modeling,* ed. W. J. Hardcastle and A. Marchal, 258–75. Dordrecht: Kluwer, 258–75.

Linder-Aronson, S. 1979. Naso-respiratory function and craniofacial growth. In *Naso-respiratory Function and Craniofacial Growth,* ed. J. A. McNamara Jr., 121–47. Ann Arbor, MI: Center for Human Growth and Development.

Liversidge, H. M. 2003. Worldwide variation in human dental development. In *Growth and Development in the Genus Homo,* ed. J. L. Thompson, A. Nelson, and G. Krovitz, 73–113. Cambridge: Cambridge University Press.

Lockwood, C. A., W. H. Kimbel, and D. C. Johanson. 2000. Temporal trends and metric variation in the mandibles and dentition of *Australopithecus afarensis. Journal of Human Evolution* 39:23–55.

Lockwood, C. A., J. M. Lynch, and W. H. Kimbel. 2002. Quantifying temporal bone morphology of great apes and humans: An approach using geometric morphometrics. *Journal of Anatomy* 201:447–64.

Lordkipanidze, D., A. Vekua, R. Ferring, G. P. Rightmire, J. Agousti, G. Kiladze, A. Muchelishvili, et al. 2005. The earliest toothless hominin skull. *Nature* 434: 717–18.

Lordkipanidze, D., A. Vekua, R. Ferring, G. P. Rightmire, C. P. Zollikofer, M. Ponce de León, J. Agousti J, G. Kiladze, et al. 2006. A fourth hominin skull from Dmanisi, Georgia. *Anatomical Record* 288A: 1146–57.

Lordkipanidze, D., T. Jashashvili, A. Vekua, M. A. Ponce de León, C. P. Zollikofer, G. P. Rightmire, H. Pontzer, et al. 2007. Postcranial evidence from early *Homo* from Dmanisi, Georgia. *Nature* 449:305–10.

Louchart, A., H. Wesselman, R. J. Blumenschine, L. J. Hlusko, J. K. Njau, M. T. Black, M. Asnake, and T. D. White. 2009. Taphonomic, avian, and small-vertebrate indicators of *Ardipithecus ramidus* habitat. *Science* 326:66e1–4.

Lovejoy, C. O. 2009. Reexamining human origins in the light of *Ardipithecus ramidus. Science* 326:74e1–8.

Lovejoy, C. O., S. W. Simpson, T. D. White, B. Asfaw, and G. Suwa. 2009a. Careful climbing in the Miocene: The forelimbs of *Ardipithecus ramidus* and humans are primitive. *Science* 326:70e1–8.

Lovejoy, C. O., G. Suwa, L. Spurlock, B. Asfaw, and T. D. White. 2009b. The pelvis and femur of *Ardipithecus ramidus:* The emergence of upright walking. *Science* 326:71e1–6.

Lovejoy, C. O., B. Latimer, G. Suwa, B. Asfaw, and T. D. White. 2009c. Combining prehension and propulsion: The foot of *Ardipithecus ramidus. Science* 326: 72e1–8.

Lovejoy, C. O., G. Suwa, S. W. Simpson, J. H. Matternes, and T. D. White. 2009d. The great divides: *Ardipithecus ramidus* reveals the postcrania of our last common ancestors with African apes. *Science* 326:100–106.

Lucas, P. W. 2004. *Dental Functional Morphology: How Teeth Work.* Cambridge: Cambridge University Press.

Lucas, P. W., C. R. Peters, and S. R. Arrandale. 1994. Seed-breaking forces exerted by orang-utans with their teeth in captivity and a new technique for estimating forces produced in the wild. *American Journal of Physical Anthropology* 94:365–78.

Lucas, P. W., P. J. Constantino, and B. A. Wood. 2008a. Structural and functional trends in tooth morphology within the hominid clade. *Journal of Anatomy* 212: 486–500.

Lucas, P., P. Constantino, B. Wood, and B. Lawn. 2008b. Dental enamel as a dietary indicator in mammals. *Bioessays* 30:374–85.

Lufkin, T., M. Mark, C. P. Hart, P. Dolle, M. Lemeur, and P. Chambon. 1992. Homeotic transformation of the occipital bones of the skull by ectopic expression of a homeobox gene. *Nature* 359:835–41.

Lukacs, J. R. 1989. Dental paleopathology: Methods for reconstructing dietary patterns. In *Reconstruction of Life from the Skeleton,* ed. M. R. Iscan and K. A. R. Kennedy, 261–86. New York: Alan R. Liss.

Lumsden, A. G. 1988. Spatial organization of the epithelium and the role of neural crest cells in the initiation of the mammalian tooth germ. *Development* 103, Supplement: 155–69.

Lumsden, A. G., and R. Krumlauf. 1996. Patterning of the vertebrate neuraxis. *Science* 274:1109–15.

Lund, V. J. 1988. The maxillary sinus in higher primates. *Acta Otolaryngologica* 105:163–71.

Lynch, M. 1990. The rate of morphological evolution in mammals from the standpoint of the neutral expectation. *American Naturalist* 136:727–41.

Lynnerup, N., J. G. Astrup, and B. Sejrsen. 2005. Thickness of the human cranial diploe in relation to age, sex and general body build. *Head and Face Medicine* 1:1–13.

Maas, E. F., W. P. Huebner, S. H. Seidman, and R. J. Leigh. 1989. Behavior of human horizontal vestibulo-ocular reflex in response to high-acceleration stimuli. *Brain Research* 499:153–56.

MacDonald, M. E., and B. K. Hall. 2001. Altered timing of the extracellular-matrix-mediated epithelial-mesenchymal interaction that initiates mandibular skeletogenesis in three inbred strains of mice: Development, heterochrony, and evolutionary change in morphology. *Journal of Experimental Zoology* 291:258–73.

Macchiarelli, R., L. Bondioli, A. Debénath, A. Mazurier, J.-F. Tournepiche, W. Birch, and C. Dean. 2006. How Neanderthal molar teeth grew. *Nature* 444:748–51.

Macho, G. A., and B. A. Wood. 1995. The role of time and timing in hominid dental evolution. *Evolutionary Anthropology* 4:17–31.

Maddieson, I. 1984. *Patterns of Sound.* Cambridge: Cambridge University Press.

Maier, W., and A. Nkini. 1984. Olduvai Hominid 9: New results of investigation. *Senckenbergische Naturforschende Gesellschaft* 69:123–30.

Main, R. P., and A. A. Biewener. 2004. Ontogenetic patterns of limb loading, *in vivo* bone strains, and growth in the goat radius. *Journal of Experimental Biology* 207:2577–88.

Mandler, M., and A. Neubuser. 2001. FGF signaling is necessary for the specification of odontogenic mesenchyme. *Developmental Biology* 240:548–59.

Manica, A., W. Amos, F. Balloux, and T. Hanihara. 2007. The effect of ancient population bottlenecks on human phenotypic variation. *Nature* 448:346–48.

Manzi, G., F. Mallegni, and A. Ascenzi. 2001. A cranium for the earliest Europeans: Phylogenetic position of the hominid from Ceprano, Italy. *Proceedings of the National Academy of Sciences USA* 98:10011–16.

Marazita, M. L. 2002. Genetic etiologies of craniosynostosis. In *Understanding Craniofacial Anomalies: The Etiopathogenesis of Craniosynostoses and Facial Clefting,* ed. M. P. Mooney and M. I. Siegel, 147–61. New York: Wiley-Liss.

Marder, E., and D. Bucher. 2001. Central pattern generators and the control of rhythmic movements. *Current Biology* 11:R986–96.

Margaria, R., P. Cerretelli, P. Aghemo, and G. Sassi. 1963. Energy cost of running. *Journal of Applied Physiology* 18:367–70.

Marlowe, F. 2005. Hunter-gatherers and human evolution. *Evolutionary Anthropology* 14:54–67.

Marotti, G. 1996. The structure of bone tissues and the cellular control of their deposition. *Italian Journal of Anatomy and Embryology* 101:25–79.

Marroig, G., and J. Cheverud. 2001. A comparison of phenotypic variation and covariation patterns and the role of phylogeny, ecology and ontogeny during cranial evolution of New World monkeys. *Evolution* 55:2576–600.

Marshall, A. J., and R. W. Wrangham. 2007. Evolutionary consequences of fallback foods. *International Journal of Primatology* 28:1219–35.

Martin, A. W., and F. A. Fuhrman. 1955. The relationship between summated tissue respiration and metabolic rate in the mouse and the dog. *Physiological Zoology* 28:18–34.

Martin, J. F., A. Bradley, and E. N. Olson. 1995. The paired-like homeobox gene MHox is required for early events of skeletogenesis in multiple lineages. *Genes and Development* 9:1237–49.

Martin, L. B. 1985. Significance of enamel thickness in hominoid evolution. *Nature* 314:260–63.

Martin, R. B. 2000. Toward a unifying theory of bone remodeling. *Bone* 26:1–6.

Martin, R. B., D. B. Burr, and N. A. Sharkey. 1998. *Skeletal Tissue Mechanics.* New York: Springer-Verlag.

Martin, R. D. 1981. Relative brain size and basal metabolic rate in terrestrial vertebrates. *Nature* 293:57–60.

———. 1983. *The Human Brain Evolution in an Ecological Context.* New York: American Museum of Natural History.

———. 1990. *Primate Origins and Evolution: A Phylogenetic Reconstruction.* Princeton, NJ: Princeton University Press.

Martin, R. D., and P. H. Harvey. 1985. Brain size allometry: Ontogeny and phylogeny. In *Size and Scaling in Primate Biology,* ed. W. L. Jungers, 147–73. New York: Plenum Press.

Martin, R. D., and C. F. Ross. 2005. The evolutionary and ecological context of primate vision. In *Structure, Function and Evolution of the Primate Visual System,* ed. J. Kremers, L. Silveira, and P. Martin, 1–36. New York: John Wiley.

Martin, R. D., A. M. MacLarnon, J. L. Phillips, and W. B. Dobyns. 2006. Flores hominid: New species or microcephalic dwarf? *Anatomical Record* 288A:1123–45.

Martinez, I., and J. L. Arsuaga. 1997. The temporal bones from Sima de los Huesos Middle Pleistocene site (Sierra de Atapuerca, Spain): a phylogenetic approach. *Journal of Human Evolution* 33:283–318.

Martinez, I., M. Rosa, J. L. Arsuaga, P. Jarabo, R. Quam, C. Lorenzo, A. Gracia, J.-M. Carretero, J.-M. Bermúdez de Castro, and E. Carbonell. 2004. Auditory capacities in Middle Pleistocene humans from the Sierra de Atapuerca in Spain. *Proceedings of the National Academy of Sciences USA* 101:9976–81.

Martinón-Torres, M., J. M. Bermúdez de Castro, A. Gómez-Robles, J. L. Arsuaga, E. Carbonell, D. Lordkipanidze, G. Manzil, and A. Margvelashvili. 2007. Dental evidence on the hominin dispersals during the Pleistocene. *Proceedings of the National Academy of Sciences USA* 104:13279–82.

Martinón-Torres, M, J. M. Bermúdez de Castro, A. Gómez-Robles, A. Margvelashvili, L. Prado, D. Lordkipanidze, and A Vekua. 2008. Dental remains from Dmanisi (Republic of Georgia): Morphological analysis and comparative study. *Journal of Human Evolution* 55:249–73.

Masali, M., S. Borgognini Tarli, and M. Maffei. 1992. Auditory ossicles and evolution of the primate ear: A biomechanical approach. In *Language Origin: A Multidisciplinary Approach,* ed. J. Wind, B. Chiarelli, B. Bichakjian, A. Nocentini, and A. Jonker, 67–86. Dordrecht: Kluwer Academic.

Matsuoka, T., P. E. Ahlberg, N. Kessaris, P. Iannarelli, U. Dennehy, W. D. Richardson, A. P. McMahon, and G. Koentges. 2005. Neural crest origins of the neck and shoulder. *Nature* 436:347–55.

Maureille, B. 2002. A lost Neanderthal neonate found. *Nature* 419:33–34.

Maureille, B., and D. Bar. 1999. The premaxilla in Neandertal and early modern children: Ontogeny and morphology. *Journal of Human Evolution* 37:137–52.

Mayr, E. 1944. On the concepts and terminology of vertical subspecies and species. *National Research Council Committee on Common Problems of Genetics, Paleontology and Systematics* 2:11–16.

———. 1951. Taxonomic categories in fossil hominids. *Cold Spring Harbor Symposia on Quantitative Biology* 15:109–18.

———. 1988. Towards a New Philosophy of Biology: Observations of an Evolutionist. Cambridge, MA: Harvard University Press.

McArthur, A. J., and J. L. Monteith. 1980. Air movement and heat loss from sheep, I: Boundary layer insulation of a model sheep, with and without fleece. *Proceedings of the Royal Society of London, Series B: Biological Sciences* 209:187–208.

McBratney-Owen, B., S. Iseki, S. D. Bamforth, B. R. Olsen, and G. M. Morriss-Kay. 2008. Development and tissue origins of the mammalian cranial base. *Developmental Biology* 322:121–32.

McBrearty, S., and A. Brooks. 2000. The revolution that wasn't: A new interpretation of the origin of modern human behavior. *Journal of Human Evolution* 39: 453–563.

McBrearty, S., and N. G. Jablonski. 2005. First fossil chimpanzee. *Nature* 437:105–8.

McCammon, R. 1970. *Human Growth and Development.* Springfield, OH: Thomas.

McCarthy, R. C. 2001. Anthropoid cranial base architecture and scaling relationships. *Journal of Human Evolution* 40:41–66.

———. 2004. Constraints and primate craniofacial growth and form. PhD diss., George Washington University.

McCarthy, R. C., and D. E. Lieberman. 2001. The posterior maxillary (PM) plane and anterior cranial architecture in primates. *Anatomical Record* 264A:247–60.

McCarthy, T. L., C, Ji, and M. Centrella. 2000. Links among growth factors, hormones, and nuclear factors with essential roles in bone formation. *Critical Reviews in Oral Biology and Medicine* 11:409–22.

McCollum, M. A. 1999. The robust australopithecine face: A morphogenetic perspective. *Science* 284:301–5.

———. 2000. Subnasal morphological variation in fossil hominids: A reassessment based on new observations and recent developmental findings. *American Journal of Physical Anthropology* 112:275–83.

McCollum, M A., F. E. Grine, S. C. Ward, and W. H. Kimbel. 1993. Subnasal morphological variation in extant hominoids and fossil hominids. *Journal of Human Evolution* 24:87–112.

McDaniel, M. A. 2005. Big-brained people are smarter: A meta-analysis of the relationship between *in vivo* brain volume and intelligence. *Intelligence* 33: 337–46.

McDermott, F., C. B. Stringer, R. Grün, T. Williams, V. Din, and C. J. Hawkesworth. 1996. New Late-Pleistocene uranium-thorium and ESR ages for the Singa hominid (Sudan). *Journal of Human Evolution* 31:507–16.

McDougall, I., F. H. Brown, and J. G. Fleagle. 2005. Stratigraphic placement and age of modern humans from Kibish, Ethiopia. *Nature* 433:733–36.

McGowan, D. A., P. W. Baxter, and J. James. 1993. *The Maxillary Sinus and Its Dental Implications.* Oxford: Wright.

McGrew, M. J., J. K. Dale, S. Fraboulet, and O. Pourquie. 1998. The lunatic fringe gene is a target of the molecular clock linked to somite segmentation in avian embryos. *Current Biology* 8:979–82.

McHenry, H. M. 1988. New estimates of body weight in early hominids and their significance to encephalization and megadontia in robust australopithecines. In *The Evolutionary History of the "Robust" Australopithecines,* ed. F. E. Grine, 133–48. New York: Aldine.

———. 1992. Body size and proportions in early hominids. *American Journal of Physical Anthropology* 87:407–31.

———. 1996. Sexual dimorphism in fossil hominids and its socioecological implication. In *The Archaeology of Human Ancestry: Power, Sex, and Tradition,* ed. J. Steels and S. Skennan, 91–109. London: Routledge and Kegan Paul.

McHenry, H. M., and K. Coffing. 2000. *Australopithecus* to *Homo:* Transformations in body and mind. *Annual Review of Anthropology* 29:125–46.

McKinney, M. L. 2002. Brain evolution by stretching the global mitotic clock of development. In *Human Evolution through Developmental Change,* ed. N. Minugh-Purvis and K. J. McNamara, 173–88. Baltimore, MD: Johns Hopkins University Press.

McKown, T. D., and A. Keith. 1939. The Stone Age of Mount Carmel, vol. 2, The

Fossil Human Remains from the Levallois-Mousterian. Oxford: Oxford University Press.

McMahon, T. A. 1984. *Muscles, Reflexes and Locomotion.* Princeton, NJ: Princeton University Press.

McNeilage, A. 2001. Diet and habitat use of two mountain gorilla groups in contrasting habitats in the Virungas. In *Mountain Gorillas,* ed. M. M. Robbins, P. Sicotte, amd K. J. Stewart, 265–92. Cambridge: Cambridge University Press.

Mead, J., A. Bouhuys, and D. G. Proctor. 1968. Mechanisms generating subglottic pressure. *Annals of the New York Academy of Sciences* 155:177–81.

Meikle, M. C. 2002. *Craniofacial Development, Growth and Evolution.* Bressingham: Bateson Publishing.

Meindl, R. S., and C. O. Lovejoy. 1985. Ectocranial suture closure: A revised method for the determination of skeletal age at death based on the lateral-anterior sutures. *American Journal of Physical Anthropology* 68:57–66.

Mekel-Bobrov, N., S. L. Gilbert, P. D. Evans, E. J. Vallender, J. R. Anderson, R. R. Hudson, S. A. Tishkoff, and B. T. Lahn. 2005. Ongoing adaptive evolution of ASPM, a brain size determinant in *Homo sapiens. Science* 309:1720–22.

Mellars, P. 2004. Neanderthals and the modern human colonization of Europe. *Nature* 432:461–65.

Mellars, P., B. Gravina, and C. Bronk Ramsey. 2007. Confirmation of Neanderthal/modern human interstratification at the Chatelperronian type-site. *Proceedings of the National Academy of Sciences USA* 104:3657–62.

Melsen, B. 1969. Time of closure of the spheno-occipital synchondrosis determined on dry skulls: A radiographic craniometric study. *Acta Odontologica Scandinavica* 27:73–90.

———. 1971. The postnatal growth of the cranial base in *Macaca rhesus* analyzed by the implant method. *Tandlaegebladet* 75:1320–29.

———. 2001. Tissue reaction to orthodontic tooth movement: A new paradigm. *European Journal of Orthodontics* 23:671–81.

Menton, D. N., D. J. Simmons, S.-L. Chang, and B. Y. Orr. 1984. From bone lining cell to osteocyte: An SEM study. *Anatomical Record* 209A:29–39.

Mercer, S. R., and N. Bogduk. 2003. Clinical anatomy of ligamentum nuchae. *Clinical Anatomy* 16:484–93.

Michejda, M. 1971. Ontogenetic changes of the cranial base in *Macaca mulatta.* Proceedings of the Third International Congress of Primatology, Zurich 1970. 1:215–25.

———. 1972a. The role of basicranial synchondroses in flexure processes and ontogenetic development of the skull base. *American Journal of Physical Anthropology* 37:143–50.

———. 1972b. Significance of basicranial synchondroses in nonhuman primates and man. *Proceedings of the Third Conference on Experimental Medicine and Surgery in Primates, Lyon,* 1:372–78.

Michejda, M., and D. Lamey. 1971. Flexion and metric age changes of the cranial base in the *Macaca mulatta:* Infants and juveniles. *Folia Primatologica* 34: 133–41.

Middlebrooks, J. C., and D. M. Green. 1991. Sound localization by human listeners. *Annual Review of Psychology* 42:135–59.

Middleton, F. A., and P. L. Strick. 2000. Basal ganglia and cerebellar loops: Motor and cognitive circuits. *Brain Research: Brain Research Reviews* 31:236–50.

Miles, R. 2003. Piltdown Man: The Secret Life of Charles Dawson and the World's Greatest Archaeological Hoax. Stroud, U.K.: Tempus Publishing.

Miller, A. J. 1982. Deglutition. *Physiological Reviews* 62:129–84.

———. 1991. Craniomandibular Muscles: Their Role in Form and Function. Boca Raton, FL: CRC Press.

Miller, K. W., P. L. Walker, and R. L. O'Halloran. 1998. Age and sex-related variation in hyoid bone morphology. *Journal of Forensic Sciences* 43:1138–43.

Mina, M. 2001. Regulation of mandibular growth and morphogenesis. *Critical Reviews in Oral Biology and Medicine* 12:276–300.

Mitchell, D., S. K. Maloney, C. Jessen, H. P. Laburn, P. R. Kamerman, G. Mitchell, and A. Fuller. 2002. Adaptive heterothermy and selective brain cooling in arid-zone mammals. *Comparative Biochemistry and Physiology, Part B: Biochemistry and Molecular Biology* 131:571–85.

Mitteroecker, P., P. Gunz, M. Bernhard, K. Schaefer, and F. L. Bookstein. 2004. Comparison of cranial ontogenetic trajectories among great apes and humans. *Journal of Humn Evolution* 46:679–98.

Mitteroecker, P., Gunz, P., and F. L. Bookstein. 2005. Heterochrony and geometric morphometrics: A comparison of cranial growth in *Pan paniscus* versus *Pan troglodytes. Evolution and Development* 7:244–58.

Moggi-Cecchi, J., and M. Collard. 2002. A fossil stapes from Sterkfontein, South Africa, and the hearing capabilities of early hominids. *Journal of Human Evolution* 42:259–65.

Moller, H., and C. S. Pedersen. 2004. Hearing at low and infrasonic frequencies. *Noise and Health* 6:37–57.

Moore, K. L., and T. V. N. Persaud. 2008. *The Developing Human: Clinically Oriented Embryology.* 8th ed. Philadelphia: W. L. Saunders.

Mori, S., and D. B. Burr. 1993. Increased intracortical remodeling following fatigue damage. *Bone* 16:103–9.

Morimoto, N., N. Ogihara, K. Katayama, and K. Shiota. 2008. Three-dimensional ontogenetic shape changes in the human cranium during the fetal period. *Journal of Anatomy* 212:627–35.

Morishima, A., M. M. Grumbach, E. R. Simpson, C. Fisher, and K. Qin. 1995. Aromatase deficiency in male and female siblings caused by a novel mutation and the physiological role of estrogens. *Journal of Clinical Endocrinology and Metabolism* 80:3689–98.

Morriss-Kay, G. M. 2001. Derivation of the mammalian skull vault. *Journal of Anatomy* 199:143–51.

Morriss-Kay, G. M, and A. O. Wilkie. 2005. Growth of the normal skull vault and its alteration in craniosynostosis: Insights from human genetics and experimental studies. *Journal of Anatomy* 207:637–53.

Morwood, M. J., P. B. O'Sullivan, F. Aziz, and A. Raza. 1998. Fission-track ages of stone tools and fossils on the east Indonesian island of Flores. *Nature* 392:173–76.

Morwood, M. J., R. P. Soejono, R. G. Roberts, T. Sutkina, C. S. M. Turney, K. E. Westaway, W. J. Rink, et al. 2004. Archaeology and age of a new hominin from Flores in eastern Indonesia. *Nature* 431:1087–91.

Morwood, M. J., P. Brown, Jatmiko, T. Sutikna, E. W. Saptomo, K. E. Westaway, R. A. Due, et al. 2005. Further evidence for small-bodied hominins from the Late Pleistocene of Flores, Indonesia. *Nature* 437:1012–17.

Moss, M. L. 1960. Inhibition and stimulation of fusion in the rat calvaria. *Anatomical Record* 136:457–67.

———. 1963. Morphological variations of the crista galli and medial orbital margin. *American Journal of Physical Anthropology* 21:159–64.

———. 1968. The primacy of functional matrices in orofacial growth. *Dental Practitioner* 19:63–73.

———. 1975. Functional anatomy of cranial synostosis. *Child's Brain* 1:22–33.

———. 1976. Experimental alteration of basi-synchondroseal cartilage growth in rat and mouse. In *Development of the Basicranium,* ed. J. F. Bosma, 541–69. Bethesda, MD: National Institutes of Health.

———. 1997a. The functional matrix hypothesis revisited, 1: The role of mechanotransduction. *American Journal of Orthodontics and Dentofacial Orthopaedics* 112:8–11.

———. 1997b. The functional matrix hyopothesis revisited, 2: The role of an osseous connected cellular network. *American Journal of Orthodontics and Dentofacial Orthopaedics* 112:221–26.

———. 1997c. The functional matrix hypothesis revisited, 3: The genomic thesis. *American Journal of Orthodontics and Dentofacial Orthopaedics* 112:338–42. 1997.

———. 1997d. The functional matrix hypothesis revisited, 4: The epigenetic antithesis and the resolving synthesis. *American Journal of Orthodontics and Dentofacial Orthopaedics* 112:410–14.

Moss, M. L., and L. Salentjin. 1969. The capsular matrix. *American Journal of Orthodontics* 56:474–90.

Moss, M. L., and R. W. Young. 1960. A functional approach to craniology. *American Journal of Physical Anthropology* 18:281–92.

Moss, M. L., C. B. Norback, and G. Robertson. 1956. Growth of certain human fetal cranial bones. *American Journal of Anatomy* 97:155–76.

Moyers, R. E. 1988. *Handbook of Orthodontics*. Chicago: Yearbook Medical Publishers.

Mucci, R. J. 1982. The role of attrition in the etiology of third molar impactions: Confirming the Begg hypothesis. PhD diss., University of Illinois—Chicago.

Muenke, M., and P. A. Beachy. 2000. Genetics of ventral forebrain development and holoprosencephaly. *Current Opinion in Genetics and Development* 10:262–69.

Mulavara, A. P., M. C. Verstraete, and J. J. Bloomberg. 2002. Modulation of head movement control in humans during treadmill walking. *Gait and Posture* 16: 271–82.

Murakami, Y, K. Uchida, F. M. Rijli, and S. Kuratani. 2005. Evolution of the brain developmental plan: Insights from agnathans. *Developmental Biology* 280:249–59.

Murray, P. D. F. 1936. Bones. A Study of the Development and Structure of the Vertebrate Skeleton. Cambridge: Cambridge University Press.

Mustarde, J. C. 1971. *Congenital Absence of the Eye*. Philadelphia: W. J. Saunders.

Naftali, S., M. Rosenfeld, M. Wolf, and D. Elad. 2005. The air-conditioning capacity of the human nose. *Annals of Biomedical Engineering* 33:545–53.

National Safety Council. 2009. Highlights from *Injury Facts*. www.nsc.org/news_resources/injury_and_death_statistics/Pages/HighlightsFromInjuryFacts.aspx, accessed May 2010.

Naunheim, R. S., P. V. Bayly, J. Standeven, J. S. Neubauer, L. M. Lewis, and G. M. Genin. 2003. Linear and angular head accelerations during heading of a soccer ball. *Medicine and Science in Sports and Exercise* 35:1406–12.

Neary, T. 1978. *Phonetic features for vowels*. Bloomington: Indiana University Linguistics Club.

Negus, V. E. 1949. *The Comparative Anatomy and Physiology of the Larynx*. London: William Heineman Medical Books.

Nelson, G. M. 1998. Biology of taste buds and the clinical problem of taste loss. *Anatomical Record* 253:70–78.

Neubuser, A., H. Peters, R. Balling, and G. R. Martin. 1997. Antagonistic interactions between FGF and BMP signaling pathways: A mechanism for positioning the sites of tooth formation. *Cell* 90:247–55.

Nevell, L., and B. A. Wood. 2008. Cranial base evolution within the hominin clade. *Journal of Anatomy* 212:455–68.

New, S. A. 2001. Exercise, bone and nutrition. *Proceedings of the Nutrition Society* 60:265–74.

Nielsen, I., L. A. Bravo, and A. J. Miller. 1989. Normal maxillary and mandibular growth and development in *Macaca mulatta:* A longitudinal cephalometric study from 2 to 5 years of age. *American Journal of Orthodontics and Dentofacial Orthopaedics* 96:405–15.

Niinimaa, V. 1983. Effect of nasal or oral breathing route on upper airway resistance. *Acta Oto-Laryngologica* 95:161–66.

Nimchinsky, E. A., E. Gilissen, J. M. Allman, D. P. Perl, J. M. Erwin, and P. R. Hof. 1999. A neuronal morphologic type unique to humans and great apes. *Proceedings of the National Academy of Sciences USA* 96:5268–73.

Nishida, S., N. Endo, H. Yamigiwa, T. Tanizawa, and H. Takahashi. 1999. Number of osteoprogenitor cells in human bone marrow markedly decreases after skeletal maturation. *The Journal of Bone and Mineral Research* 17:171–77.

Nishimura, T. 2005. Developmental changes in the shape of the supralaryngeal vocal tract in chimpanzees. *American Journal of Physical Anthropology* 126:193–204.

Nishimura, T., A. Mikami, J. Suzuki, and T. Matsuzawa. 2003. Descent of the larynx in chimpanzee infants. *National Academy of Sciences USA* 100:6930–33.

———. 2007. Development of the laryngeal air sac in chimpanzees. *American Journal of Primatology* 28:483–92.

Noden, D. 1983. The role of neural crest in patterning of avian cranial skeletal, connective and muscle tissues. *Developmental Biology* 96:144–65.

———. 1986. Origins and patterning of craniofacial mesenchymal tissues. *Anatomical Record* 208:1–13.

———. 1991. Cell movements and control of patterned tissue assembly during craniofacial development. *Journal of Craniofacial Genetics and Developmental Biology* 11:192–213.

Noonan, J. P., G. Coop, S. Kudaravalli, D. Smith, J. Krause, J. Alessi, F. Chen, et al. 2006. Sequencing and analysis of Neanderthal genomic DNA. *Science* 314:1113–18.

Northcutt, R. G. 1981. Evolution of the telencephalon in nonmammals. *Annual Review of Neuroscience* 4:301–50.

———. 2004. Taste buds: Development and evolution. *Brain, Behavior and Evolution* 64:198–206.

Obendorf, P. J., C. E. Oxnard, and B. J. Kefford. 2008. Are the small human-like fossils found on Flores human endemic cretins? *Proceedings of the Royal Society of London, Series B* 275:1287–96.

O'Connor, C. F., R. G. Franciscus, and N. E. Holton. 2005. Bite force production capability and efficiency in Neandertals and modern humans. *American Journal of Physical Anthropology* 127:129–51.

Oftedahl, O. T. 1984. Milk composition, milk yield and energy: Output at peak lactation; A comparative review. *Symposia of the Zoological Society of London* 51:33–55.

O'Higgins, P. 2000a. The study of morphological variation in the hominid fossil record: Biology, landmarks and geometry. *Journal of Anatomy* 197:103–20.

————. 2000b. Quantitative approaches to the study of craniofacial growth and evolution: Advances in morphometric techniques. In *Development, Growth and Evolution: Implications for the study of hominid evolution,* ed. P. O'Higgins and M. Cohn, 163–85. London: Academic Press.

O'Higgins, P., P. Chadfield, and N. Jones. 2001. Facial growth and the ontogeny of morphological variation within and between the primates *Cebus apella* and *Cercocebus torquatus. Journal of Zoology* 254:337–57.

Ohloff, G. 1994. *Scent and Fragrances.* Berlin: Springer-Verlag.

Ohlsson, C., B. A. Bengtsson, O. G. P. Isaksson, T. T. Adreassen, and M. Slottweg. 1998. Growth hormone and bone. *Endocrine Reviews* 19:55–79.

Ohtsuki, F. 1977. Developmental changes of the cranial bone thickness in the human fetal period. *American Journal of Physical Anthropology* 46:141–54.

Olasoji, H. O., and S. A. Odusanya. 2000. Comparative study of third molar impaction in rural and urban areas of South-Western Nigeria. *Odonto-stomatologie tropicale* 23:25–8.

Olejniczak, A. J., T. M. Smith, R. N. Feeney, R. Macchiarelli, A. Mazurier, L. Bondioli, A. Rosas, et al. 2008. Dental tissue proportions and enamel thickness in Neandertal and modern human molars. *Journal of Human Evolution* 55: 12–23.

Olsen, B. R., A. M. Reginato, and W. Wang. 2000. Bone development. *Annual Review of Cell Developmental Biology* 16:191–220.

Olson, E. C., and R. L. Miller. 1958. *Morphological Integration.* Chicago: University of Chicago Press.

Olson, L. G., and K. P. Strohl. 1987. The response of the nasal airway to exercise. *American Review of Respiratory Disease* 135:356–59.

Opperman, L. A. 2000. Cranial sutures as intramembranous bone growth sites. *Developmental Dynamics* 219:472–85.

O'Rahilly, R. R., and F. Muller. 2001. *Human Embryology and Teratology.* 3rd ed. New York: Wiley-Liss.

Osborn, J. W. 1981. *Dental Anatomy and Embryology.* Oxford: Blackwell Scientific Publications.

Owerkowicz, T. 2004. Respiratory innovations in monitor lizards, mammals and birds. PhD diss., Harvard University.

Pal, G. P., B. P. Tamanka, R. V. Routal, and S. S. Bhagwat. 1984. The ossification of the membranous part of the squamous occipital bone in man. *American Journal of Anatomy* 138:259–66.

Palmeirim, I., D. Henrique, D. Ish-Horowicz, and O. Pourquié. 1997. Avian hairy gene expression identifies a molecular clock linked to vertebrate segmentation and somitogenesis. *Cell* 91:639–48.

Palmer, J. B. 1998. Bolus aggregation in the oropharynx does not depend on gravity. *Archives of Physical Medicine and Rehabilitation* 79:691–96.

Palmer, J. B., K. M. Hiiemae, and J. Liu. 1997. Tongue-jaw linkages in feeding: A preliminary videofluorographic study. *Archives of Oral Biology* 42:429–41.

Parfitt, A. M. 1987. Bone and plasma calcium homeostasis. *Bone* 8:51–58.

Park, Z. Y., and D. H. Russell. 2000. Thermal denaturation: A useful technique in peptide mass mapping. *Analytical Chemistry* 72:2667–70.

Parker, S. T. 2005. Evolutionary relationships between molar eruption and cognitive development in anthropoid primates In *Human Evolution through Developmental Change,* ed. N. Minugh-Purvis and K. J. McNamara, 305–16. Baltimore, MD: Johns Hopkins University Press.

Passingham, R. E. 1973. Anatomical differences between the neocortex of man and other primates. *Brain, Behavior and Evolution* 7:337–59.

Patterson, N., D. J. Richter, S. Gnerre, E. S. Lander, and D. Reich. 2006. Genetic evidence for complex speciation of humans and chimpanzees. *Nature* 441:1103–8.

Pearson, O. M. 2000. Postcranial remains and the origin of modern humans. *Evolutionary Anthropology* 9:229–47.

———. 2004. Has the combination of genetic and fossil evidence solved the riddle of modern human origins? *Evolutionary Anthropology* 13:145–59.

Pearson, O. M., and D. E. Lieberman. 2004. The aging of Wolff's Law. *Yearbook of Physical Anthropology* 47:63–99.

Peleg, D., and J. A. Goldman. 1978. Fetal deglutition: A study of the anencephalic fetus. *European Journal of Obstetrics and Gynecology and Reproductive Biology* 8: 133–36.

Penin, X., C. Berge, and M. Baylac. 2002. Ontogenetic study of the skull in modern humans and common chimpanzees: Neotenic hypothesis reconsidered with a tridimensional Procrustes analysis. *American Journal of Physical Anthropology* 118: 50–62.

Persinger, M. A. 2001. The neuropsychiatry of paranormal experiences. *Journal of Neuropsychiatry and Clinical Neurosciences* 13:515–24.

Peters, H., and R. Balling. 1999. Teeth: Where and how to make them. *Trends in Genetics* 15:59–65.

Peterson, B. W., H. Choi, T. Hain, E. Keshner, and G. C. Peng. 2001. Dynamic and kinematic strategies for head movement control. *Annals of the New York Academy of Sciences* 942:381–93.

Peterson, G. E., and H. L. Barney. 1952. Control methods used in a study of the vowels. *Journal of the Acoustical Society of America* 24:175–84.

Peusner, K. D., G. Gamkrelidze, and C. Giaume. 1998. Potassium currents and excitability in second-order auditory and vestibular neurons. *Journal of Neuroscience Research* 53:511–20.

Pickford, M., and B. Senut. 2001. "Millennium ancestor," a 6-million-year-old bipedal hominid from Kenya. *Comptes rendus de l'Académie des Sciences de Paris, série 2a* 332:134–44.

Pickford, M., B. Senut, D. Gommery, and J. Treil, J. 2002. Bipedalism in *Orrorin tugenensis* revealed by its femora. *Comptes Rendus Palévol* 1:1–132.

Pilbeam, D. 1996. Genetic and morphological records of the Hominoidea and hominid origins: A synthesis. *Molecular Biology and Evolution* 17:1081–90.

———. 2004. The anthropoid postcranial axial skeleton: Comments on development, variation, and evolution. *Journal of Experimental Zoology, Part B: Molecular and Developmental Evolution* 302B:241–67.

Pilbeam, D. R., and S. J. Gould. 1974. Size and Scaling in Human Evolution. *Science* 186:892–901.

Pilbeam, D. R., and N. M. Young. 2004. Hominoid evolution: Synthesizing disparate data. *Comptes Rendus Palévol* 3:303–19.

Plavcan, J. M. 2000. Inferring social behavior from sexual dimorphism in the fossil record. *Journal of Human Evolution* 39:327–44.

Polanski, J. M., and R. G. Franciscus. 2006. Patterns of craniofacial integration in extant *Homo, Pan,* and *Gorilla. American Journal of Physical Anthropology* 75: 195–96.

Ponce de León, M. S., and C. P. E. Zollikofer. 2001. Neanderthal cranial ontogeny and its implications for late hominid diversity. *Nature* 412:534–38.

Ponce de León, M. S., L. Golovanova, V. Doronichev, G. Romanova, T. Akazawa, O. Kondo, H. Ishida, and C. P. E. Zollikofer. 2008. Neanderthal brain size at birth provides insights into human life history evolution. *Proceedings of the National Academy of Sciences USA* 105:13764–68.

Pontzer, H., and J. M. Kamilar 2009. Great ranging associated with greater reproductive investment in mammals. *Proceedings of the National Academy of Sciences USA* 106:192–96.

Pontzer, H., and R. W. Wrangham. 2006. The ontogeny of ranging in wild chimpanzees. *International Journal of Primatology* 27:295–309.

Pontzer, H., D. E. Lieberman, E. Momin, M. J. Devlin, J. D. Polk, B. Hallgrímsson, and D. M. L. Cooper. 2006. Trabecular bone in the bird knee responds with high sensitivity to changes in load orientation. *Journal of Experimental Biology* 209: 57–65.

Pontzer, H., J. H. Holloway, D. A. Raichlen, and D. E. Lieberman. 2009. Control and function of arm swing in human walking and running. *Journal of Experimental Biology* 212:523–34.

Porter, J., B. Craven, R. M. Khan, S. J. Chang, I. Kang, B. Judkewicz, J. Volpe, G. Settles, and N. Sobel. 2007. Mechanisms of scent tracking in humans. *Nature Neuroscience* 10:27–29.

Potts, R. 1996. Humanity's Descent: The Consequences of Ecological Instability. New York: William Morrow and Co.

———. 1998. Environmental hypotheses of hominin evolution. *Yearbook of Physical Anthropology* 41:93–136.

Potts, R., A. K. Behrensmeyer, A. Deino, P. Ditchfield, and J. Clark. 2004. Small mid-Pleistocene hominin associated with East African Acheulean technology. *Science* 305:53–4.

Poulsen, P. H., Smith, D. F., L. Ostergaard, E. H. Danielsen, A. Gee, S. B. Hansen, J. Astrup, and A. Gjedde. 1997. *In vivo* estimation of cerebral blood flow, oxygen consumption and glucose metabolism in the pig by [15O]water injection, [15O]oxygen inhalation and dual injections of [18F]fluorodeoxyglucose. *Journal of Neuroscience Methods* 77:199–209.

Pozzo, T., A. Berthoz, and L. Lefort. 1990. Head stabilization during various locomotor tasks in humans, I: Normal subjects. *Experimental Brain Research* 82: 97–106.

Pozzo, T., Y. Levik, and A. Berthoz. 1995. Head and trunk movements in the frontal plane during complex dynamic equilibrium tasks in humans. *Experimental Brain Research* 106:327–38.

Prescott, J., N. Ripandelli, and I. Wakeling. 2001. Binary taste mixture interactions in PROP non-tasters, medium-tasters and super-tasters. *Chemical Senses* 26: 993–1003.

Preuschoft, H. B. Demes, M. Meier, and H. F. Bär. 1986. The biomechanical principles realized in the upper jaw of long-snouted primates. In *Primate Evolution*, vol. 1, ed. J. G. Else and P. C. Lee, 229–37. Cambridge: Cambridge University Press.

Preuss, T. M., H. Qi, and J. H. Kaas. 1999. Distinctive compartmental organization of human primary visual cortex. *Proceedings of the National Academy of Sciences USA* 96:11601–6.

Prinz, J. F., and P. W. Lucas. 1997. An optimization model for mastication and swallowing in mammals. *Proceedings of the Royal Society of London, Series B* 264: 1715–21.

Proctor, D. F. 1977. The upper airways, I: Nasal physiology and defense of the lungs. *American Review of Respiratory Disease* 115:97–129.

———. 1982. The upper airway. In *The Nose,* ed. D. F. Proctor, and I. Andersen, 23–43. Amsterdam: Elsevier Biomedical.

Prothero, J. 1997. Scaling of cortical neuron density and white matter volume in mammals. *Journal of Brain Research* 38:513–24.

Proust, M. 1913–27. *Remembrance of Things Past,* vol. 1, *Swann's Way* and *Within a Budding Grove.* Trans. C. K. Scott Moncrieff and Terence Kilmartin. New York: Vintage Press.

Prugnolle, F., A. Manica, and F. Balloux. 2005. Geography predicts neutral genetic diversity of human populations. *Current Biology* 15: R159–R160.

Pulaski, P. D., D. S. Zee, and D. A. Robinson. 1981. The behavior of the vestibulo-ocular reflex at high velocities of head rotation. *Brain Research* 222:159–65.

Purslow, P. P. 1985. The physical basis of meat texture: Observations on the fracture behaviour of cooked bovine *M. semitendinosis. Meat Science* 12:39–60.

Pusey, A. E, A. E. Oehlert, G. W. Williams, and J. Goodall. 2005. Influence of ecological and social factors on body mass of wild chimpanzees. *International Journal of Primatology* 26:3–31.

Quam, R. M., and F. H. Smith. 1998. A reassessment of the Tabun C2 mandible. In *Neandertals and Modern Humans in in Western Asia,* ed. T. Akazawa, K. Aoki, and O. Bar-Yosef, 405–21. New York: Plenum.

Raadsheer, M. C., T. M. van Eijden, F. C. van Ginkel, and B. Prahl-Andersen. 1999. Contribution of jaw muscle size and craniofacial morphology to human bite force magnitude. *Journal of Dental Research* 78:31–42.

Rae, T. C., and T. Koppe. 2004. Holes in the head: Evolutionary interpretations of the paranasal sinuses in catarrhines. *Evolutionary Anthropology* 13:211–23.

Raff, R. 1996. The Shape of Life: Genes, Development, and the Evolution of Animal Form. Chicago: University of Chicago Press.

Rafferty, K. L., S. W. Herring, and C. D. Marshall. 2003. Biomechanics of the rostrum and the role of facial sutures. *Journal of Morphology* 257:33–44.

Raichle, M. E., and D. A. Gusnard. 2002. Appraising the brain's energy budget. *Proceedings of the National Academy of Sciences USA* 99:10237–39.

Raisz, L. G. 1999. Osteoporosis: Current approaches and future prospects in diagnosis, pathogenesis, and management. *Journal of Bone and Mineral Metabolism* 17:79–89.

Rak, Y. 1983. *The Australopithecine Face.* New York: Academic Press.

———. 1986. The Neanderthal: A new look at an old face. *Journal of Human Evolution* 15:151–64.

———. 1988. On variation in the masticatory system of *Australopithecus boisei.* In *Evolutionary History of the "Robust" Australopithecines,* ed. F. Grine, 193–98. New York: Aldine de Gruyter.

Rak, Y., and R. J. Clarke. 1979. Ear ossicle of *Australopithecus robustus. Nature* 279: 62–63.

Rak, Y., A. Ginzburg, and E. Geffen. 2002. Does *Homo neanderthalensis* play a role in modern human ancestry? The mandibular evidence. *American Journal of Physical Anthropology* 119:199–204.

———. 2007. Gorilla-like anatomy on *Australopithecus afarensis* mandibles suggests *Au. afarensis* link to robust australopiths. *Proceedings of the National Academy of Sciences USA* 104:6568–72.

Rak, Y., W. H. Kimbel, and E. Hovers. 1994. A Neanderthal infant from Amud Cave, Israel. *Journal of Human Evolution* 26:313–24.

Rakic, P., and D. R. Kornack. 2001. Neocortical expansion and elaboration during primate evolution: A view from neuroembryology. In *Evolutionary Anatomy of the Primate Cerebral Cortex,* ed. D. Falk and K. Gibson, 30–56. Cambridge: Cambridge University Press.

Ramprashad, F., J. P. Landolt, K. E. Money, and J. Laufer. 1984. Dimensional analysis and dynamic response characterization of mammalian peripheral vestibular structures. *American Journal of Anatomy* 169:295–313.

Ramirez Rozzi, F. V., and J. M. Bermúdez de Castro. 2004. Surprisingly rapid growth in Neanderthals. *Nature* 428:936–39.

Ranly, D. M. 1984. *A Synopsis of Craniofacial Growth.* 2nd ed. Norwalk, CT: Appleton and Lange.

Ratiu, P., I. F. Talos, S. Haker, D. E. Lieberman, and P. Everett. 2004. The tale of Phineas Gage, digitally remastered. *Journal of Neurotrauma* 21:637–43.

Raven, H. C. 1950. *The Anatomy of the Gorilla.* New York: Columbia University Press.

Ravosa, M. J. 1988. Browridge development in Cercopithecidae: A test of two models. *American Journal of Physical Anthropology* 76:535–55.

———. 1999. Anthropoid orgins and the modern symphysis. *Folia primatologica* 70:65–78.

———. 2000. Size and scaling in the mandible of living and extinct apes. *Folia Primatologica* (Basel) 71:305–22.

Ravosa, M. J., and W. L. Hylander. 1994. Function and fusion of the mandibular symphysis in primates: Stiffness or strength? In *Anthropoid Origins,* ed. J. G. Fleagle and R. F. Kay, 447–68. New York: Plenum.

Ravosa, M. J., K. R. Johnson, and W. L. Hylander. 2000a. Strain in the galago facial skull. *Journal of Morphology* 245:51–66.

Ravosa, M. J., C. J. Vinyard, M. Gagnon, M., and S. A. Islam. 2000b. Evolution of anthropoid jaw loading and kinematicpatterns. *American Journal of Physical Anthropology* 112:493–516.

Rayfield, E. J. 2007. Finite element analysis and understanding the biomechanics and evolution of living and fossil organisms. *Annual Review of Earth and Planetary Sciences* 35:541–76.

Reby, D., K. McComb, B. Cargnelutti, C. Darwin, W. T. Fitch, and T. Clutton-Brock. 2005. Red deer stags use formants as assessment cues during intrasexual agonistic interactions. *Proceedings of the Royal Society of London, Series B: Biological Sciences* 272:941–7.

Reed, K. E. 1997. Early hominid evolution and ecological change through the African Plio-Pleistocene. *Journal of Human Evolution* 32:289–322.

Reed, K. E., and S. M. Russak 2009. Tracking ecological change in relation to the emergence of *Homo* near the Plio-Pleistocene boundary. In *The First Humans: Origin and Early Evolution of the Genus* Homo, ed. F. E. Grine, J. G. Fleagle, and R. E. F. Leakey, 159–71. New York: Springer.

Rees, J. S. 2001. An investigation into the importance of the periodontal ligament and alveolar bone as supporting structures in finite element studies. *Journal of Oral Rehabilitation* 28:425–32.

Reid, D. J., and M. C. Dean. 2006. Variation in modern human enamel formation times. *Journal of Human Evolution* 50:329–46.

Reidenberg, J. S., and J. T. Laitman. 1991. Effect of basicranial flexion on larynx and hyoid position in rats: An experimental study of skull and soft tissue interactions. *Anatomical Record* 230:557–69.

Relethford, J. H. 1994. Craniometric variation among modern human populations. *American Journal of Physical Anthropology* 95:53–62.

———. 2002. Apportionment of global human genetic diversity based on craniometrics and skin color. *American Journal of Physical Anthropology* 118:393–98.

Rice, S. H. 1997. The analysis of ontogenetic trajectories: When a change in size or shape is not heterochrony. *Proceedings of the National Academy of Sciences USA* 94:907–12.

———. 2002. The role of heterochrony in brain evolution. In *Human Evolution through Developmental Change,* ed. N. Minugh-Purvis and K. J. McNamara, 154–70. Baltimore, MD: Johns Hopkins University Press.

Richardson, M. K. 1999. Vertebrate evolution: The developmental origins of adults variations. *BioEssays* 21:604–13.

Richmond, B. G., and W. L. Jungers. 2008. *Orrorin tugenensis* femoral morphology and the evolution of hominin bipedalism. *Science* 319:1662–65.

Richmond, B. G., B. W. Wright, I. Grosse, P. C. Dechow, C. F. Ross, M. A. Spencer, and D. S. Strait. 2005. Finite element analysis in functional morphology. *Anatomical Record* 283A:259–74.

Richtsmeier, J. T. 2002. Cranial vault morphology and growth in craniosynostoses. In *Understanding Craniofacial Anomalies: The Etiopathogenesis of Craniosynostoses and Facial Clefting,* ed. M. P. Mooney and M. I. Siegel, 321–41. New York: Wiley-Liss.

Richtsmeier, J. T., K. Aldridge, V. B. DeLeon, J. Panchal, A. A. Kane, J. L. Marsh, P. Yan, and T. M. Cole III. 2006. Phenotypic integration of neurocranium and brain. *Journal of Experimental Zoology* 306B:360–78.

Rightmire, G. P. 1988. *Homo erectus* and later Middle Pleistocene humans. *Annual Review of Anthropology* 17:239–59.

———. 1990. *The Evolution of* Homo erectus: *Comparative Anatomical Studies of an Extinct Species* . Cambridge: Cambridge University Press.

———. 1998. Human evolution in the Middle Pleistocene: The role of *Homo heidelbergensis. Evolutionary Anthropology* 6:218–27.

———. 2004. Brain size and encephalization in early to mid-Pleistocene *Homo. American Journal of Physical Anthropology.* 124:109–23.

———. 2008. *Homo* in the Middle Pleistocene: Hypodigms, variation, and species recognition. *Evolutionary Anthropology* 17:8–21.

Rightmire, G. P., and H. J. Deacon. 1991. Comparative studies of late Pleistocene human remains from Klasies River Mouth, South Africa. *Journal of Human Evolution* 20:131–56.

Rightmire, G. P., D. Lordkipanidze, and A. Vekua. 2006. Anatomical descriptions, comparative studies and evolutionary significance of the hominin skulls from Dmanisi, Republic of Georgia. *Journal of Human Evolution* 50:115–41.

Rilling, J. K. 2006. Human and non-human primate brains: Are they allometrically scaled versions of the same design? *Evolutionary Anthropology* 15:65–77.

Rilling, J. K., and R. A. Seligman. 2002. A quantitative morphometric comparative analysis of the primate temporal lobe. *Journal of Human Evolution* 42:505–33.

Rivera-Pérez, J. A., M. Mallo, M. Gendron-Maguire, T. Gridley, and R. R. Behringer. 1995. Goosecoid is not an essential component of the mouse gastrula organizer but is required for craniofacial and rib development. *Development* 121:3005–12.

Robertshaw, D. 2006. Mechanisms for the control of respiratory evaporative heat loss in panting animals. *Journal of Applied Physiology* 101:664–68.

Robinson, G. W., and K. A. Mahon, K. A. 1994. Differential and overlapping expression domains of Dlx-2 and Dlx-3 suggest distinct roles for Distal-less homeobox. *Evolution and Development* 1:172–79.

Robling, A. G., F. M. Hinant, D. B. Burr, and C. H. Turner. 2002. Improved bone structure and strength after long-term mechanical loading is greatest if loading is separated into short bouts. *Journal of Bone and Mineral Research* 7:1545–54.

Robling, A. G., A. B. Castillo, and C. H. Turner. 2006. Biomechanical and molecular regulation of bone remodeling. *Annual Review of Biomedical Engineering* 8:455–98.

Robson, K. L., and B. A. Wood. 2008. Hominin life history: Reconstruction and evolution. *Journal of Anatomy* 212:394–425.

Roche, A. F. 1953. Age increases in cranial thickness during growth. *Human Biology* 25:81–92.

Roche, H., A. Delagnes, J.-P. Brugal, C. Feibel, M. Kibunjia, P.-J. Texier, and V. Mourre. 1999. Evidence for early hominids lithic production and technical skill at 2.3 Mya, West Turkana, Kenya. *Nature* 399:57–60.

Rodriguez, I., and P. Mombaerts. 2002. Novel human vomeronasal receptor-like genes reveal species-specific families. *Current Biology* 12:R409–11.

Rodriguez, A. A., B. R. Myers, and C. N. Ford. 1990. Laryngeal electromyography in the diagnosis of laryngeal nerve injuries. *Archives of Physical Medicine and Rehabilitation* 71:587–90.

Roff, D. A. 1995. The estimation of genetic correlations from phenotypic correlations: A test of Cheverud's conjecture. *Heredity* 74:481–90.

Rohen, J. W., and A. Castenholz. 1967. Über die Zentralisation der Retina bei Primaten. *Folia Primatologica* 5:92–147.

Rohlf, F. J., and D. E. Slice. 1990. Extensions of the Procrustes methods for the optimal superimposition of landmarks. *Systematic Zoology* 39:40–59.

Rohrer, B., and W. K. Stell. 1994. Basic fibroblast growth factor (bFGF) and transforming growth factor beta (TGF-beta) act as stop and go signals to modulate postnatal ocular growth in the chick. *Experimental Eye Research* 58:553–61.

Rönning, O. 1995. Basicranial synchondroses and the mandibular condyle in craniofacial growth. *Acta Odontologica Scandinavica* 53:162–66.

Roorda, A., and D. R. Williams. 1999). The arrangement of the three cone classes in the living human eye. *Nature* 397:520–22.

Rosas, A., and J. M. Bermúdez de Castro. 1998. On the taxonomic affinities of the Dmanisi mandible (Georgia). *American Journal of Physical Anthropology* 107: 145–62.

Rose, M. R., and G. V. Lauder, ed. 1996. *Adaptation.* San Diego, CA: Academic Press.

Roseman, C. C. 2004. Detecting interregionally diversifying natural selection on modern human cranial form by using matched molecular and morphometric data. *Proceedings of the National Academy of Sciences USA* 101:12824–29.

Rosenberg, K. R., and W. Trevathan. 1996. Bipedalism and human birth: The obstetrical dilemma revisited. *Evolutionary Anthropology* 4:161–68.

Roseman, C. C., and T. D. Weaver. 2004. Multivariate apportionment of global human craniometric diversity. *American Journal of Physical Anthropology* 125: 257–63.

———. 2007. Molecules vs. morphology? Not for the human cranium. *BioEssays* 29:1185–88.

Rosenberger, A. L. 1986. Platyrrhines, catarrhines and the anthropoid transition. In *Major Topics in Primate and Human Evolution,* ed. B. A. Wood, L. Martin, and P. Andrews, 66–88. Cambridge: Cambridge University Press.

Rosowski, J. J. 1994. Outer and middle ears. In *Comparative Hearing: Mammals,* ed. R. R. Fay and A. N. Popper, 172–247. Berlin: Springer-Verlag.

Ross, C. F. 1996. Adaptive explanations for the origin of the Anthropoidea (Primates). *American Journal of Primatology* 40:205–30.

———. 2000. Into the light: The origin of Anthropoidea. *Annual Review of Anthropology* 29:147–94.

———. 2001. *In vivo* function of the craniofacial haft: The interorbital "pillar." *American Journal of Physical Anthropology* 116:108–39.

———. 2005. Finite element analysis in vertebrate biomechanics. *Anatomical Record,* 283A:253–58.

Ross, C. F., and M. Henneberg. 1995. Basicranial flexion, relative brain size, and facial kyphosis in *Homo sapiens* and some fossil hominids. *American Journal of Physical Anthropology* 98:575–93.

Ross, C. F., and W. L. Hylander. 1996. *In vivo* and *in vitro* bone strain in the owl monkey circumorbital region and the function of the postorbital septum. *American Journal of Physical Anthropology* 101:183–215.

———. 2000. Electromyography of the anterior temporalis and masseter muscles of owl monkeys (*Aotus trivirgatus*) and the function of the postorbital septum. *American Journal of Physical Anthropology* 112:455–68.

Ross, C. F., and R. D. Martin. 2006. The role of vision in the origin and evolution of primates. In *Evolution of Nervous Systems,* vol. 4, *Primates,* ed. J. H. Kaas, 59–78. Oxford: Academic Press.

Ross, C. F., and M. J. Ravosa. 1993. Basicranial flexion, relative brain size, and facial kyphosis in nonhuman primates. *American Journal of Physical Anthropology* 91: 305–24.

Ross, C. F., M. Henneberg, M. J. Ravosa, and S. Richard. 2004. Curvilinear, geometric and phylogenetic modeling of basicranial flexion: Is it adaptive, is it constrained? *Journal of Human Evolution* 46:185–213.

Rouquier, S., A. Blancher, and D. Giorgi. 2000. The olfactory receptor gene repertoire in humans and mouse: Evidence reduction of the functional fraction in primates. *Proceedings of the National Academy of Sciences USA* 97:2870–74.

Roux, W. 1881. *Der Kampf der Theile im Organismus.* Leipzig: Engelmann.

Rowlerson, A., F. Mascarello, A. Veggetti, and E. Carpene. 1983. The fibre-type composition of the first branchial arch muscles in Carnivora and Primates. *Journal of Muscle Research and Cell Motility* 4:443–72.

Rowlett, R. M. 2000. Fire control by *Homo erectus* in Africa and East Asia. *Acta Anthropoligica Sinica* 19:198–208.

Rubenson, J., D. B. Heliams, S. K. Maloney, P. C. Withers, D. G. Lloyd, and P. A. Fournier. 2007. Reappraisal of the comparative cost of human locomotion using gait-specific allometric analyses. *Journal of Experimental Biology* 210:3513–24.

Rubin, C. T. 1984. Skeletal strain and the functional significance of bone architecture. *Calcified Tissue International* 36 (Supplement 1): 11–18.

Rubin, C. T., and L. E. Lanyon. 1984. Regulation of bone formation by applied dynamic loads. *Journal of Bone and Joint Surgery* 66:397–402.

———. 1985. Regulation of bone mass by mechanical strain magnitude. *Calcified Tissue International* 37:411—417.

Ruff, C. B., and A. Walker. 1993. Body size and body shape. In *The Nariokotome* Homo erectus *Skeleton,* ed. A. Walker and R. E. F. Leakey, 221–65. Cambridge, MA: Harvard University Press.

Ruff, C. B., A. Walker, and E. Trinkaus. 1994. Postcranial robusticity in *Homo,* III: Ontogeny. *American Journal of Physical Anthropology* 93:35–54.

Ruff, C. B., E. Trinkaus, and T. W. Holliday. 1997. Body mass and encephalization in Pleistocene *Homo. Nature* 387:173–76.

Russell, A. P., and J. J. Thomason. 1992. Mechanical analysis of the mammalian head skeleton. In *The Skull,* vol. 3, ed. J. Hanken and B. K. Hall, 345–83. Chicago: University of Chicago Press.

Ruvolo, M. 1997. Molecular phylogeny of the hominoids: Inference from multiple independent DNA sequence data sets. *Molecular Biology and Evolution* 14: 248–65.

Ryan, A. S., and D. C. Johanson. 1989. Anterior dental microwear in *Australopithecus afarensis. Journal of Human Evolution* 18:235–68.

Saketkhoo, K., I. Kaplan, and M. A. Sackner. 1979. Effect of exercise on nasal mucous velocity and nasal airflow resistance in normal subjects. *Journal of Applied Physiology* 46:369–71.

Sakka, M. 1984. Cranial morphology and masticatory adaptations. In *Food Acquisition and Processing in Primates,* ed. D. J. Chivers, B. A. Wood, and A. Bilsborough, 415–27. New York: Plenum Press.

Salazar-Ciudad, I., and J. Jernvall. 2005. Graduality and innovation in the evolution of complex phenotypes: Insights from development. *Journal of Experimental Zoology, Part B: Molecular and Developmental Evolution* 304B:619–31.

Santa Luca, A. P. 1978. A re-examination of presumed Neandertal-like fossils. *Journal of Human Evolution* 7:619–36.

Santonja, M., and P. Villa. 2006. The Acheulean in southwestern Europe. In *Axe Age: Acheulean Toolmaking,* ed. N. Goren-Inbar and G. Sharon, 429–77. London: Equinox.

Sarnat, B. G. 1982. Eye and orbital size in the young and adult: Some postnatal experimental and clinical relationships. *Ophthalmologica* 185:74–78.

Sarnat, B. G., and M. Wexler. 1967. Rabbit snout growth after resection of central linear segments of nasal septal cartilage. *Acta Oto-laryngologica* 63:467–78.

Sasaki, C. T., P. A. Levine, J. T. Laitman, and E. S. Crelin Jr. 1977. Postnatal descent of the epiglottis in man: A preliminary report. *Archives of Otolaryngology* 103: 169–71.

Sasaki, K., A. G. Hannam, and W. W. Wood. 1989. Relationships between the size, position, and angulation of human jaw muscles and unilateral first molar bite force. *Journal of Dental Research* 68:499–503.

Satokata, I., and R. Maas. 1994. Msx1 deficient mice exhibit cleft palate and abnormalities of craniofacial and tooth development. *Nature Genetics* 6:348–56.

Saunders, J. B. D. M., V. T. Inman, and B. Bresler. 1953. The major determinants in normal and pathological gait. *Journal of Bone and Joint Surgery* 35A:543–55.

Sausse, F. 1975. La mandibule atlanthropienne de la Carrière Thomas I (Casablanca). *L'anthropologie* 79:81–112.

Schaefer, M. S. 1999. Foramen magnum–carotid foramina relationship: Is it useful for species designation? *American Journal of Physical Anthropolgy* 110:467–71.

Schenker, N. M., A. M. Desgouttes, and K. Semendeferi. 2005. Neural connectivity and cortical substrates of cognition in hominoids. *Journal of Human Evolution* 49:547–69.

Scherer, P. W., I. I. Hahn, and M. N. Mozell. 1989. The biophysics of nasal airflow. *Otolaryngologic Clinics of North America* 22:265–78.

Scheuer, J. L., and S. M. Black. 2000. *Developmental Juvenile Osteology.* London: Academic Press.

Schick, K. D., and N. Toth. 1993. *Making Silent Stones Speak: Human Evolution and the Dawn of Technology.* New York: Simon and Schuster.

Schilling, T. F., and P. V. Thorogood. 2000. Development and evolution of the vertebrate skull. In *Development, Growth and Evolution: Implications for the Study of the Hominid Skeleton,* ed. P. O'Higgins and M. Cohn M, 57–83. London: Academic Press.

Schmid, C. 1995. Insulin-like growth factors. *Cellular Biology International* 19: 445–57.

Schmidt-Nielsen, K. 1984. *Scaling: Why is animal size so important?* Cambridge: Cambridge University Press.

———. 1990. *Animal Physiology: Adaptation and Environment.* 4th ed. Cambridge: Cambridge University Press.

Schmidt-Nielsen, K., W. L. Bretz, and C. R. Taylor. 1970. Panting in dogs: Unidirectional air flow over evaporative surfaces. *Science* 169:1102–4.

Schmittbuhl, M., J. Rieger, J. M. Le Minor, A. Schaaf, and F. Guy. 2007. Variations of the mandibular shape in extant hominoids: Generic, specific, and subspecific quantification using elliptical Fourier analysis in lateral view. *American Journal of Physical Anthropology* 132:119–31.

Schneider, R. A., D. Hu, J. L. Rubenstein, M. Maden, and J. A. Helms. 2001. Local retinoid signaling coordinates forebrain and facial morphogenesis by maintaining FGF8 and SHH. *Development* 28:2755–67.

Schneiderman, E. D. 1992. *Facial Growth in the Rhesus Monkey.* Princeton, NJ: Princeton University Press.

Schoenemann, P. T., M. J. Sheehan, and L. D. Glotzer. 2005. Prefrontal white matter volume is disproportionately larger in humans than in other primates. *Nature Neuroscience* 8:242–52.

Schowing, J. 1968. Demonstration of the inductive role of the brain in osteogenesis of the embryonic skull of the chicken. *Journal of Embryology and Experimental Morphology* 19:83–94.

Schultz, A. H. 1940. The size of the orbit and the eye in primates. *American Journal of Physical Anthropology* 26:389–408.

———. 1942. Conditions for balancing the head in primates *American Journal of Physical Anthropology* 29:483–97.

———. 1960. Age changes in primates and their modification in man. In *Human Growth,* ed. J. M. Tanner, 1–20. New York: Pergamon Press.

————. 1961. Vertebral column and thorax. *Primatologica* 4:1–66.

————. 1962. Metric age changes and sex differences in primate skulls. *Yearbook of Physical Anthropology* 10:129–54.

————. 1969. The skeleton of the chimpanzee. In *The Chimpanzee,* vol. 1, ed. G. H. Bourne, 50–103. Basel: Karger.

Schutte, S., S. P. van den Bedem, F. van Keulen, F. C. van der Helm, and H. J. Simonsz. 2006. A finite-element analysis model of orbital biomechanics. *Vision Research* 46:1724–31.

Schwartz, G. T. 2000. Taxonomic and functional aspects of the patterning of enamel thickness distribution in extant large-bodied hominoids. *American Journal of Physical Anthropology* 111:221–44.

Schwartz, G. T., and M. C. Dean. 2000. Interpreting the hominid dentition: Ontogenetic and. phylogenetic aspects. In *Development, Growth and Evolution,* ed. P. O'Higgins and M. Cohen, 85–122. London: Academic Press.

Schwartz, G. T., J. F. Thackeray, C. Reid, and J. F. van Reenan. 1998. Enamel thickness and the topography of the enamel-dentine junction in South African Plio-Pleistocene hominids with special reference to the Carabelli trait. *Journal of Human Evolution* 35:523–42.

Schwartz, G. T., D. J. Reid, M. C. Dean, and M. S. Chandrasekera. 2000. Aspects of tooth crown development in common chimpanzees (*Pan troglodytes*) with a note on the possible role of sexual dimorphism in canine growth. *Proceedings of the 11th International Symposium on Dental Morphology* 323–37.

Schwartz, G. T., D. J. Reid, and M. C. Dean. 2001. Developmental aspects of sexual dimorphism in hominoid canines. *International Journal of Primatology* 22: 837–60.

Schwartz, J. H. 2000. Taxonomy of the Dmanisi crania. *Science* 289:55–56.

Schwartz, J. H., and I. Tattersall. 1996. Significance of some previously unrecognized apomorphies in the nasal region of *Homo neanderthalensis. Proceedings of the National Academy of Sciences USA* 93:10852–54.

————. 2000. The human chin revisited: What is it and who has it? *Journal of Human Evolution* 38:367–409.

————. 2003. *The Human Fossil Record,* vol. 2, *Craniodental Morphology of Genus* Homo *(Africa and Asia).* New York: Wiley-Liss.

————. 2005. The Human Fossil Record, vol. 4, *Craniodental Morphology of Early Hominids (Genera* Australopithecus, Paranthropus, Orrorin*) and Overview.* New York: Wiley-Liss.

Schlosser, G., and G. P. Wagner. 2005. *Modularity in Development and Evolution.* Chicago: University of Chicago Press.

Scott, J. E., and C. A. Lockwood. 2004. Patterns of tooth crown size and shape variation in great apes and humans and species recognition in the hominid fossil record. *American Journal of Physical Anthropology* 125:303–19.

Scott, J. H. 1958. The cranial base. *American Journal of Physical Anthropology* 16: 319–48.

Scott, J. H. 1967. *Dento-facial Development and Growth.* Oxford: Pergamon Press.

Scott, R. S., P. S. Ungar, T. S. Bergstrom, C. A. Brown, F. E. Grine, M. F. Teaford, and A. Walker. 2005. Dental microwear texture analysis shows within-species diet variability in fossil hominins. *Nature* 436:693–95.

Segal, M. B. 2000. The choroid plexuses and the barriers between the blood and the cerebrospinal fluid. *Cellular and Molecular Neurobiology* 20:183–96.

Seidler, H., D. Falk, C. B. Stringer, H. Wilfing, G. Muller, D. zur Nedden, G. Weber, W. Recheis, and J. L. Arsuaga. 1997. A comparative study of stereolithographically modelled skulls of Petralona and Broken Hill: Implications for future studies of middle Pleistocene hominid evolution. *Journal of Human Evolution* 33: 691–703.

Semaw, S. 2000. The world's oldest stone artifacts from Gona, Ethiopia: Their implications for understanding stone technology and patterns of human evolution between 2.6–2.5 million years ago. *Journal of Archaeological Science* 27:1197–214.

Semaw, S., P. Renne, J. W. K. Harris, C. Feibel, R. Bernor, N. Fesseha, and K. Mowbray. 1997. 2.5 million-year-old stone tools from Gona, Ethiopia. *Nature* 385: 333–38.

Semaw, S., M. J. Rogers, J. Quade, P. R. Renne, R. F. Butler, D. Stout, M. Dominguez-Rodrigo, W. Hart, T. Pickering, and S. W. Simpson. 2003. 2.6-million-year-old stone tools and associated bones from OGS-6 and OGS-7, Gona, Afar, Ethiopia. *Journal of Human Evolution* 45:169–77.

Semaw, S., S. W. Simpson, J. Quade, P. R. Renne, R. F. Butler, W. C. McIntosh, N. Levin, M. Dominguez-Rodrigo, and M. J. Rogers. 2005. Early Pliocene hominids from Gona, Ethiopia. *Nature* 433:301–5.

Semendeferi, K. 2001. Advances in the study of hominoid brain evolution: Magenetic resonance imaging (MRI) and 3-D imaging. In *Evolutionary Anatomy of the Primate Cerebral Cortex,* ed. D. Falk and K. Gibson, 257–89. Cambridge: Cambridge University Press.

Semendeferi, K., E. Armstrong, A. Schleicher, K. Zilles, and G. W. Van Hoesen. 2001. Prefrontal cortex in humans and apes: A comparative study of area 10. *American Journal of Phyisical Anthropology* 114:224–41.

Semendeferi, K., A. Lu, N. Schenker, and H. Damasio. 2002. Humans and great apes share a large frontal cortex. *Nature Neuroscience* 5:272–76.

Senut, B., M. Pickford, D. Gommery, P. Mein, C. Cheboi, and Y. Coppens Y. 2001. First hominid from the Miocene (Lukeino Formation, Kenya). *Comptes rendus de l'Académie des sciences de Paris, série 2a* 332:137–44.

Shang, H., H. Tong, S. Zhang, F. Chen, and E. Trinkaus. 2007. An early modern human from Tianyuan Cave, Zhoukoudian, China. *Proceedings of the National Academy of Sciences USA* 104:6573–78.

Shapiro, D., and J. T. Richtsmeier. 1997. Brief communication: A sample of pediatric skulls available for study. *American Journal of Physical Anthropology* 103:415–16.

Shapiro, R. 1960. *The Normal Skull: A Roentgen Study.* New York: Hoeber.

Shapiro, R., and F. Robinson. 1980. *The Embryogenesis of the Human Skull.* Cambridge, MA: Harvard University Press.

Sharpe, P. T. 1995. Homeobox genes and orofacial development. *Connective Tissue Research* 32:17–25.

Shaw, R. M., and G. S. Molyneux. 1994. The effects of mandibular hypofunction on the development of the mandibular disc in the rabbit. *Archives of Oral Biology* 39:747–52.

Shea, B. T. 1977. Eskimo craniofacial morphology, cold stress and the maxillary sinus. *American Journal of Physical Anthropology* 47:289–300.

———. 1983. Paedomorphosis and neoteny in the pygmy chimpanzee. *Science* 222:521–22.

———. 1985. Ontogenetic allometry and scaling: A discussion based on the growth and form of the skull in African apes. In *Size and Scaling in Primate Morphology,* ed. W. L. Jungers, 175–205. New York: Plenum Press.

———. 1986. Ontogenetic approaches to sexual dimorphism in anthropoids. *Human Evolution* 1:97–110.

———. 1989. Heterochrony in human evolution: The case for neoteny reconsidered. *Yearbook of Physical Anthropology* 32:69–101.

———. 1992. Neoteny. In *The Cambridge Encyclopedia of Human Evolution,* ed. S. Jones, R. Martin, and D. Pilbeam, 401–4. Cambridge: Cambridge University Press.

Shea, J. J. 2006. The origins of lithic projectile point technology: Evidence from Africa, the Levant, and Europe. *Journal of Archaeological Science* 33:823–46.

Shellis, P. 1981. Human tissues: Dentine. In *Dental Anatomy and Embryology,* ed. J. W. Osborn, 166–74. Boston: Blackwell.

Shellis, P. 1984. Variations in growth of the enamel crown in human teeth and a possible relationship between growth and enamel structure. *Archives of Oral Biology* 29:697–705.

Shellis, R. P., and K. M. Hiiemae. 1986. Distribution of enamel on the incisors of Old World monkeys. *American Journal of Physical Anthropology* 71:103–13.

Shepherd, G. M. 2004. The human sense of smell: Are we better than we think? *PLoS Biology* 2:E146.

———. 2006. Smells, brains and hormones. *Nature* 439:149–51.

Shepherd-Barr, K., and G. Shepherd. 1998. Madeleines and neuromodernism: Reassessing mechanisms of autobiographical memory in Proust. *Auto/Biography Studies* 13:39–60.

Sherwood, C. C., P. W. Lee, C. B. Rivara, R. L. Holloway, E. P. Gilissen, R. M. Simmons, A. Hakeem, J. M. Allman, J. M. Erwin, and P. R. Hof. 2003. Evolution of specialized pyramidal neurons in primate visual and motor cortex. *Brain, Behavior and Evolution* 61:28–44.

Sherwood, C. C., C. D. Stimpson, M. A. Raghanti, D. E. Wildman, M. Uddin, L. I. Grossman, M. Goodman, et al. 2006. Evolution of increased glia-neuron ratios in the human frontal cortex. *Proceedings of the National Academy of Sciences USA* 103:13606–11.

Sherwood, R. J., S. C. Ward, and A. Hill. 2002. The taxonomic status of the Chemeron temporal (KNM-BC 1). *Journal of Human Evolution* 42:153–84.

Shi, P., M. A. Bakewell, and J. Zhang. 2006. Did brain-specific genes evolve faster in humans than in chimpanzees? *Trends in Genetics* 22:608–13.

Shipman, P. 2001. *The Man Who Found the Missing Link: Eugène Dubois and his Lifelong Quest to Prove Darwin Right.* New York: Simon and Schuster.

Shosani, J., C. P. Groves, E. L. Simons, and G. F. Gunnell. 1996. Primate phylogeny: Morphological vs. molecular results. *Molecular Phylogenetics and Evolution* 5: 101–53.

Simpson. S. W., J. Quade, N. E. Levin, R. Butler, G. Dupont-Nivet, M. Everett, and S. Semaw. 2008. A female *Homo erectus* pelvis from Gona, Ethiopia. *Science* 322: 1089–92.

Singer, R., and J. Wymer. 1982. *The Middle Stone Age at Klasies River Mouth in South Africa.* Chicago: University of Chicago Press.

Sirianni, J. E., and L. Newell-Morris. 1980. Craniofacial growth of fetal *Macaca nemestrina:* A cephalometric roentgenographic study. *American Journal of Physical Anthropology* 53:407–21.

Sirianni, J. E., and D. R. Swindler. 1979. A review of postnatal craniofacial growth in Old World monkeys and apes. *Yearbook of Physical Anthropology* 22:80–104.

———. 1985. *Growth and Development of the Pigtailed Macaque.* Boca Raton, FL: CRC Press.

Sirianni, J. E., and A. L. Van Ness. 1978. Postnatal growth of the cranial base in *Macaca nemestrina. American Journal of Physical Anthropology* 49:329–39.

Skelton, R. R., and H. M. McHenry. 1998. Trait list bias and a reappraisal of early hominid phylogeny. *Journal of Human Evolution* 34:109–13.

Sleigh, M. A., J. R. Blake, and N. Liron. 1988. The propulsion of mucus by cilia. *American Review of Respiratory Disease* 137:726–41.

Small, D. M., and J. Prescott. 2005. Odor/taste integration and the perception of flavor. *Experimental Brain Research* 166:345–57.

Smeathers, J. E. 1989. Transient vibrations caused by heel strike. *Proceedings of the Institute of Mechanical Engineers* [H] 203:181–86.

Smith, B. H. 1989. Dental development as a measure of life history in primates. *Evolution* 43:683–88.

———. 1994. Patterns of dental development in *Homo, Australopithecus, Pan,* and *Gorilla. American Journal of Physical Anthropology* 94:307–25.

Smith, E. P., J. Boyd, G. R. Frank, H. Takahashi, R. M. Cohen, B. Specker, T. C. Williams, D. B. Lubahn, and K. S. Korach. 1994. Estrogen resistance caused by a mutation in the estrogen-receptor gene in a man. *New England Journal of Medicine* 331:1056–61.

Smith, F. H., and S. P. Paquette. 1989. The adaptive basis of Neandertal facial form, with some thoughts on the nature of modern human origins. In *The Emergence of Modern Humans: Biocultural Adaptations in the Later Pleistocene,* ed. E. Trinkaus, 181–210. Cambridge: Cambridge University Press.

Smith, F. H., and G. C. Raynard. 1980. Evolution of the supraorbital region in Upper Pleistocene fossil hominids from south-central Europe. *American Journal of Physical Anthropology* 53:589–609.

Smith, H. F., and F. E. Grine. 2008. Cladistic analysis of early *Homo* crania from Swartkrans and Sterkfontein, South Africa. *Journal of Human Evolution* 54:684–704.

Smith, M. M., and Z. Johanson. 2003. Separate evolutionary origins of teeth from evidence in fossil jawed vertebrates. *Science* 299:1235–6.

Smith, R. J. 1978. Mandibular biomechanics and temporomandibular joint function in primates. *American Journal of Physical Anthropology* 49:341–9.

———. 1984. The plan of the human face: A test of three general concepts. *American Journal of Orthodontics* 85:103–8.

———. 1985. Functions of condylar translation in human mandibular movement. *American Journal of Orthodontics* 88:191–202.

———. 2005. Species recognition in paleoanthropology: Implications of small sample sizes. In *Interpreting the Past: Essays on Human, Primate, and Mammal Evolution in Honor of David Pilbeam,* ed. D. E. Lieberman, R. J. Smith, and J. Kelley, 207–19. Boston: Brill Academic Publishers.

Smith, R. J., and W. L. Jungers. 1997. Body mass in comparative primatology. *Journal of Human Evolution* 32:523–59.

Smith, T. D., and K. P. Bhatnagar. 2000. The human vomeronasal organ, II: Prenatal development. *Journal of Anatomy* 197:421–36.

Smith, T. D., T. A. Buttery, K. P. Bhatnagar, A. M. Burrows, M. P. Mooney, and M. I. Siegel. 2001. Anatomical position of the vomeronasal organ in postnatal humans. *Annals of Anatomy* 183:475–79.

Smith, T. M., D. J. Reid, and J. E. Sirianni. 2006. The accuracy of histological assessments of dental development and age at death. *Journal of Anatomy* 208:125–38.

Smith, T. M., D. J. Reid, M. C. Dean, A. J. Olejniczak, and L. B. Martin. 2007a. Molar development in common chimpanzees (*Pan troglodytes*). *Journal of Human Evolution* 52:201–16.

Smith, T. M., M. Toussaint, D. J. Reid, A. J. Olejniczak, and J.-J. Hublin, J.-J. 2007b. Rapid dental development in a Middle Paleolithic Belgian Neanderthal. *Proceedings of the National Academy of Sciences USA* 104:20220–25.

Smith, T. M., P. T. Tafforeau, D. J. Reid, R. Grün, S, Eggins, M. Boutakiout, and J.-J. Hublin. 2007c. Earliest evidence of modern human life history in North African early *Homo sapiens*. *Proceedings of the National Academy of Sciences USA* 104: 6128–33.

Sockol, M. D., D. Raichlen, and H. D. Pontzer. 2007. Chimpanzee locomotor energetics and the origin of human bipedalism. *Proceedings of the National Academy of Sciences USA* 104:12265–69.

Sohmer, H., and S. Freeman S. 2000. Basic and clinical physiology of the inner ear receptors and their neural pathways in the brain. *Journal of Basic and Clinical Physiology and Pharmacology* 11:367–74.

Sokal, R. R., and J. F. Rohlf. 1981. *Biometry: The principles and practice of statistics in biological research*. 2nd ed. San Francisco: W. H. Freeman and Company.

Spector, J. A., J. A. Greenwald, S. M. Warren, P. J. Bouletreau, F. E. Crisera, B. J. Mehrara, and M. T. Longaker. 2002. Co-culture of osteoblasts with immature dural cells causes an increased rate and degree of osteoblast differentiation. *Plastic and Reconstructive Surgery* 109:631–42.

Spencer, F. 1990. *Piltdown: A Scientific Forgery*. Oxford: Oxford University Press.

Spencer, M. A. 1998. Force production in the primate masticatory system: Electromyographic tests of biomechanical hypotheses. *Journal of Human Evolution* 34:25–54.

———. 1999. Constraints on masticatory system evolution in anthropoid primates. *American Journal of Physical Anthropology* 108:483–506.

Spencer, M. A., and B. Demes. 1993. Biomechanical analysis of masticatory systems configuration in Neandertals and Inuits. *American Journal of Physical Anthropology* 91:1–20.

Spencer, M. A., and P. S. Ungar. 2000. Craniofacial morphology, diet and incisor use in three Amerind populations. *International Journal of Osteoarchaeology* 10: 229–41.

Sperber, G. 2001. *Craniofacial Development*. Hamilton, Ontario: B. C. Decker.

Spoor, C. F. 1993. The comparative morphology and phylogeny of the human bony labyrinth. PhD diss., University of Utrecht.

———. 1997. Basicranial architecture and relative brain size of Sts 5 (*Australopithecus africanus*) and other Plio-Pleistocene hominids. *South African Journal of Science* 93:182–86.

Spoor, F., and J. G. M. Thewissen. 2008. Comparative and Functional Anatomy of Balance in Aquatic Animals. In *Sensory Evolution on the Threshold: Adaptations in secondarily aquatic vertebrates*, ed. J. G. M. Thewissen and S. Nummela, 257–84. Berkeley: University of California Press.

Spoor, F., and F. Zonneveld. 1995. Morphometry of the primate bony labyrinth: A new method based on high-resolution computed tomography. *Journal of Anatomy* 186:271–86.

————. 1998. Comparative review of the human bony labyrinth. *Yearbook of Physical Anthropology* 27:211–51.

Spoor, F., B. Wood, and F. Zonneveld. 1994. Implications of early hominid labyrinthine morphology for evolution of human bipedal locomotion. *Nature* 369: 645–48.

Spoor, F., S. Bajpai, S. T. Hussain, K. Kumar, and J. G. Thewissen. 2002a. Vestibular evidence for the evolution of aquatic behaviour in early cetaceans. *Nature* 417: 163–66.

Spoor, F., F. Esteves, F. Tecelão Silva, and R. Pacheco Dias. 2002b. The bony labyrinth of Lagar Velho 1. In *Portrait of the Artist as a Child: The Gravettian Human Skeleton from the Abrigo do Lagar Velho and Its Archeological Context,* ed. J. Zilhão and E. Trinkaus, 287–92. Lisbon: Instituto Português de Arqueologia.

Spoor, F., J.-J. Hublin, M. Braun, and F. Zonneveld. 2003. The bony labyrinth of Neanderthals. *Journal of Human Evolution* 44:141–65.

Spoor, F., M. G. Leakey, P. N. Gathogo, F. H. Brown, S. C. Antón, I. McDougall, C. Kiarie, F. K. Manthi, and L. N. Leakey. 2007a. Implications of new early *Homo* fossils from Ileret, east of Lake Turkana, Kenya. *Nature* 448:688–91.

Spoor, F., S. T. Garland, G. Krovitz, T. M. Ryan, M. T. Silcox, and A. Walker. 2007b. The primate semicircular canal system and locomotion. *Proceedings of the National Academy of Sciences USA* 104:10808–12.

Srivastava, H. C. 1992. Ossification of the membranous portion of the squamous part of the occipital bone in man. *Journal of Anatomy* 180:219–24.

Standring, S., J. Healy, D. Johnson, and A. Williams, eds. 2004 *Gray's Anatomy: The Anatomical Basis of Clinical Practice.* 39th ed. Edinburgh: Churchill–Livingstone.

Stanford, C. M., F. Welsch, N. Kastner, G. Thomas, R. Zaharias, K. Holtman, and R. A Brand. 2000. Primary human bone cultures from older patients do not respond at continuum levels of *in vivo* strain magnitudes. *Journal of Biomechanics* 33:63–71.

Starck, D., and R. Schneider. 1960. Respirationsorgane. In *Primatologia,* vol. 3, part 2, ed. H. Hofer, A. H. Schultz, and D. Starck, 423–587. Basel: Karger.

Stedman, H. H., B. W. Kozyak, A. Nelson, D. M. Thesier, L. T. Su, D. W. Low, C. R. Bridges, J. B. Shrager, N. Minugh-Purvis, and M. A. Mitchell. 2004. Myosin gene mutation correlates with anatomical changes in the human lineage. *Nature* 428:415–18.

Steiper, M. E., and N. M. Young. 2006. Primate genomic divergence dates and their paleontological context. *Molecular Phylogenetics and Evolution* 41:384–94.

Stephan, H., H. Frahm, and G. Baron. 1981. New and revised data on volumes of brain structures in insectivores and primates. *Folia Primatologica* 35:1–29.

Stephan, H., G. Baron, and H. D. Frahm. 1988. Comparative size of brains and brain structures. In *Comparative Primate Biology,* ed. H. Steklis and J. Erwin, 1–38. New York: Alan R. Liss.

Stevens, K. N. 1972. The quantal nature of speech: Evidence from articulatory-acoustic data. In *Human Communication: A Unified View,* ed. E. E. David and P. B. Denes, 51–66. New York: McGraw-Hill.

———. 1989. On the quantal nature of speech. *Journal of Phonetics* 17:3–45.

Stevens, K. N, and A. House. 1955. Development of a quantitative description of vowel articulation. *Journal of the Acoustical Society of America* 27:484–93.

Stewart, T. D. 1962. Neanderthal cervical vertebrae with special. attention to the Shanidar Neanderthals from Iraq. *Bibliotheca Primatologica* 1:130–54.

Stiner, M. C., N. D. Munro, T. A. Surovell, E. Tchernov, and O. Bar-Yosef. 1999. Paleolithic population growth pulses evidenced by small animal exploitation. *Science* 283:190–94.

Stiner, M. C., N. D. Munro, and T. A. Surovell. 2000. The tortoise and the hare: Small game use, the broad spectrum revolution, and Paleolithic demography. *Current Anthropology* 41:39–73.

Stockard, C. R. 1941. The Genetic and Endocrinic Basis for Differences in Form and Behavior, as Elucidated by Studies of Contrasted Pure-Line Dog Breeds and Their Hybrids. Philadelphia: Wistar Institute of Anatomy and Biology.

Strait, D. S. 1999. The scaling of basicranial flexion and length. *Journal of Human Evolution* 37:701–19.

———. 2001. Integration, phylogeny, and the hominid cranial base. *American Journal of Physical Anthropology* 114:273–97.

Strait, D. S., and F. E. Grine. 2004. Inferring hominoid and early hominid phylogeny using craniodental data: The role of fossil taxa. *Journal of Human Evolution* 47: 399–452.

Strait, D. S., and C. F. Ross. 1999. Kinematic data on primate head and neck posture: Implications for the evolution of basicranial flexion and an evaluation of registration planes. *American Journal of Physical Anthropology* 108:205–22.

Strait, D. S., F. E. Grine, and M. A. Moniz. 1997. A reappraisal of early hominid phylogeny. *Journal of Human Evolution* 32:17–82.

Strait, D. S., Q. Wang, P. C. Dechow, C. F. Ross, B. G. Richmond, M. A. Spencer, and B. A. Patel. 2005. Modeling elastic properties in finite-element analysis: How much precision is needed to produce an accurate model? *Anatomical Record* 283A:275–87.

Strait, D. S., B. G. Richmond, M. A. Spencer, C. F. Ross, P. C. Dechow, and B. A. Wood. 2007. Masticatory biomechanics and its relevance to early hominid phylogeny: An examination of palatal thickness using finite-element analysis. *Journal of Human Evolution* 52:585–99.

Strait, D. S., G. W. Weber, S. Neubauer, J. Chalk, B. G. Richmond, P. W. Lucas, M. A. Spencer, et al. 2009. *Proceedings of the National Academy of Sciences USA* 106:16010–15.

Strand-Vidarsdóttir, U., O'Higgins, P., and C. Stringer. 2002. The development of regionally distinct facial morphologies: A geometric morphometric study of population-specific differences in the growth of the modern human facial skeleton. *Journal of Anatomy* 201:211–29.

Stringer, C. B. 1986. The credibility of *Homo habilis*. In *Major Topics in Primate and Human Evolution,* ed. B. A. Wood, L. Martin, and P. Andrews, 266–94. Cambridge: Cambridge University Press.

Stringer, C. B., and C. Gamble. 1993. *In Search of the Neanderthals.* London: Thames and Hudson.

Stringer, C. B., J.-J. Hublin, and B. Vandermeersch. 1984. The origin of anatomically modern humans in Western Europe. In *The Origins of Modern Humans: A World Survey of the Fossil Evidence,* ed. F. H. Smith and F. Spencer, 51–135. New York: Alan R. Liss.

Stringer, C. B., M. C. Dean, and R. D. Martin. 1990. A comparative study of cranial and dental development within a recent British sample and among Neandertals. *Primate Life History and Evolution:* 14:115–52.

Su, D. F., S. H. Ambrose, D. DeGusta, and Y. Haile-Selassie. 2009. Paleoenvironment. In Ardipithceus kadabba: *Late Miocene Evidence from the Middle Awash, Ethiopia,* ed. Y. Haile-Selassie and G. WoldeGabriel, 521–47. Berkeley: University of California Press.

Sugimura, T., J. Inada, S. Sawa, and Y. Kakudo. 1984. Dynamic responses of the skull caused by loss of occlusal force. *Journal of Osaka Dental University* 18:29–42.

Sun, Z., E. Lee, and S. W. Herring. 2004. Cranial sutures and bones: Growth and fusion in relation to masticatory strain. *Anatomical Record* 276A:150–61.

Suwa, G., and R. T. Kono. 2005. A micro-CT based study of linear enamel thickness in the mesial cusp section of human molars: Reevaluation of methodology and assessment of within-tooth, serial, and individual variation. *Anthropological Science* 113:273–89.

Suwa, G., B. A. Wood, and T. D. White. 1994. Further analysis of mandibular molar crown and cusp areas in Pliocene and Early Pleistocene hominids. *American Journal of Physical Anthropology* 93:407–26.

Suwa, G., T. White, and F. C. Howell. 1996. The mandibular postcanine dentition from the Shungura Formation, Ethiopia: Crown morphology, taxonomic allocations and Plio-Pleistocene hominid evolution. *American Journal of Physical Anthropology* 101:247–82.

Suwa, G., R. T. Kono, S. Katoh, B. Asfaw, and Y. Beyene. 2007. A new species of great ape from the late Miocene epoch in Ethiopia. *Nature* 448:921–24.

Suwa, G., B. Asfaw, R. T. Kono, D. Kubo, C. O. Lovejoy, and T. D. White. 2009a. The *Ardipithecus ramidus* skull and its implications for hominid origins. *Science* 326:68e1–7.

Suwa, G., R. T. Kono, S. W. Simpson, B. Asfaw, C. O. Lovejoy, and T. D. White. 2009b. Paleobiological implications of the *Ardipithecus ramidus* dentition. *Science* 326:94–9.

Swindler, D. R., and C. D. Wood. 1973. *An Atlas of Primate Gross Anatomy.* Seattle: University of Washington Press.

Swisher, C. C., W. J. Rink, H. P. Schwarcz, and S. C. Antón. 1997. Dating the Ngandong Humans. *Science* 276:1575–76.

Swisher, C. C., III, W. J. Rink, S. C. Antón, H. P. Schwarcz, G. H. Curtis, A. Suprijo, and Widiasmoro. 1996. Latest *Homo erectus* of Java: Potential contemporaneity with *Homo sapiens* in southeast Asia. *Science* 274:1870–74.

Swisher, C. C., G. H. Curtis, T. Jacob, A. G. Getty, A. Suprijo, and Widiasmoro. 1994. Age of earliest known hominids in Java, Indonesia. *Science* 263:1118–21.

Takacs, Z., J. C. Morales, T. Geissmann, and D. J. Melnick. 2005. A complete species-level phylogeny of the Hylobatidae based on mitochondrial ND3-ND4 gene sequences. *Molecular Phylogenetics and Evolution* 36:456–67.

Takahashi, I., I. Mizoguchi, M. Nakamura, M. Kagayama, and H. Mitani. 1995. Effects of lateral pterygoid muscle hyperactivity on differentiation of mandibular condyles in rats. *Anatomical Record* 24:328–36.

Takahata, N., and Y. Satta. 1997. Evolution of the primate lineage leading to modern humans: Phylogenetic and demographic inferences from DNA sequences. *Proceedings of the National Academy of Sciences USA* 94:4811–15.

Tam, P. P., and P. A. Trainor. 1994. Specification and segmentation of the paraxial mesoderm. *Anatomy and Embryology* 189:275–305.

Tanne, K., J. Miyasaka, Y. Yamagata, R. Sachdeva, S. Tsutsumi, and M. Sakuda. 1988. Three-dimensional model of the human craniofacial skeleton: Method and preliminary results using finite element analysis. *Journal of Biomedical Engineering* 10:246–52.

Tattersall, I. 1992. Species concepts and species identification in human evolution. *Journal of Human Evolution* 22:341–49.

Tattersall, I., and G. J. Sawyer. 1996. The skull of *Sinanthropus* from Zhoukoudian, China: A new reconstruction. *Journal of Human Evolution* 22:79–108.

Tattersall, I., and J. Schwartz. 2000. *Extinct Humans.* Boulder, CO: Westview Press.

———. 2008. The morphological distinctiveness of *Homo sapiens* and its recognition in the fossil record: Clarifying the problem. *Evolutionary Anthropology* 17: 49–54.

Taylor, A. B. 2006. Size and shape dimorphism in great ape mandibles and implications for fossil species recognition. *American Journal of Physical Antropology* 129: 82–98.

Taylor, C. R., and V. J. Rowntree. 1973. Temperature regulation and heat balance in running cheetahs: A strategy for sprinters? *American Journal of Physiology* 224: 848–51.

Taylor, C. R., K. Schmidt-Nielsen, R. Dmi'el, and M. Fedak. 1971. Effect of hyperthermia on heat balance during running in the African hunting dog. *American Journal of Physiology* 220:823–7.

Teaford, M. F., and P. S. Ungar. 2000. Diet and the evolution of the earliest human ancestors. *Proceedings of the National Academy of Sciences USA* 97:13506–11.

Terhune, C. E., W. H. Kimbel, and C. A. Lockwood. 2007. Variation and diversity in *Homo erectus:* A 3D geometric morphometric analysis of the temporal bone. *Journal of Human Evolution* 53:41–60.

Thexton, A. J. 1992. Mastication and swallowing: An overview. *British Dental Journal* 173:197–206.

Thexton, A. J., and A. W. Crompton. 1998. The scientific basis of eating. In *Frontiers of Oral Biolology,* ed. R. W. A. Linden, 9:168–222. Basel: Karger.

Thomas, S. N., T. Schroeder, N. H. Secher, J. H. Mitchell. 1989. Cerebral blood flow during submaximal and maximal dynamic exercise in humans. *Journal of Applied Physiology* 67:744–8.

Thomason, J. J., and A. P. Russell. 1986. Mechanical factors in the evolution of the mammalian secondary palate: A theoretical analysis. *Journal of Morphology* 189:199–213.

Thompson, D'Arcy W. [1917]. *On Growth and Form.* Cambridge: Cambridge University Press.

Thomson, A., and D. Buxton. 1923. Man's nasal index in relation to certain climatic conditions. *The Journal of the Royal Anthropological Institute* 53:92–122.

Thorne, A. G., and M. H. Wolpoff. 1981. Regional continuity in Australasian Pleistocene hominid evolution. *American Journal of Physical Anthropology* 55:337–49.

Thorogood, P. 1988. The developmental specification of the vertebrate skull. *Development* 103: S141–53.

Thorpe, S. K. S., R. L. Holder, and R. H. Crompton. 2007. Origin of human bipedalism as an adaptation for locomotion on flexible branches. *Science* 316:1328–31.

Thorstensson, A., J. Nilsson, H. Carlson, and M. R. Zomlefer. 1984. Trunk movements in human locomotion. *Acta Physiologica Scandinavica* 121:9–22.

Tillier, A. M. 1977. La pneumatisation du massif cranio-facial chez les hommes actuels et fossiles. *Bulletins et mémoires de la Société d'Anthropologie de Paris* 4:177–89.

———. 1999. Les enfants moustériens de Qafzeh: Interprétation phylogénétique et paléoauxologique. Paris: CNRS Éditions.

Tilton, F. E., J. J. C. Degioanni, and V. S. Schneider. 1980. Long term follow-up of Skylab bone demineralization. *Aviation Space and Environmental Medicine:* 51:1209–13.

Tobias, P. V. 1966. Fossil hominid remains from Ubeidiya, Israel. *Nature* 211:130–33.

———. 1967. Olduvai Gorge, vol. 2, *The Cranium and Maxillary Dentition of Australopithecus (Zinjanthropus) boisei.* Cambridge: Cambridge University Press.

————. 1971. *The Brain in Hominid Evolution.* New York: Columbia University Press.

————. 1991. *Olduvai Gorge,* vol. 4, *The Skulls, Endocasts and Teeth of* Homo habilis. Cambridge: Cambridge University Press.

Tocheri, M. W., C. M. Orr, S. G. Larson, T. Sutikna, Jatmiko, E. W. Saptomo, R. A. Due, T. Djubiantono, M. J. Morwood, and W. L. Jungers. 2007. The primitive wrist of *Homo floresiensis* and its implications for hominin evolution. *Science* 317:1743–45.

Tompkins, R. L. 1996. Relative dental development of Upper Pleistocene hominids compared to human population variability. *American Journal of Physical Anthropology* 99:103–18.

Torii, M. 1995. Maximal sweating rate in humans. *Journal of Human Ergology* 24: 137–53.

Toth, N., K. D. Schick, E. S. Savage-Rumbaugh, R. A. Sevcik, and D. M. Rumbaugh. 1993. *Pan* the tool-maker: Investigations into the stone tool-making and tool-using capabilities of a bonobo (*Pan paniscus*). *Journal of Archaeological Science* 20:81–91.

Trainor, P. A., and R. Krumlauf. 2000. Patterning the cranial neural crest: Hindbrain segmentation and Hox gene plasticity. *Nature Reviews: Neuroscience* 1:116–24.

————. 2001. Hox genes, neural crest cells and branchial arch patterning. *Current Opinion in Cell Biology* 13:698–705.

Trainor, P. A., and P. P. L. Tam. 1995. Cranial paraxial mesoderm and neural crest cells of the mouse embryo: Co-distribution in the craniofacial mesenchyme but distinct segregation in the branchial arches. *Development* 121:2569–81.

Trauth, M. H., M. A. Maslin, A. Deino, and M. R. Strecker. 2005. Late Cenozoic moisture history of East Africa. *Science* 309:2051–53.

Trenouth, M. J. 1989. Craniofacial shape in the anencephalic human fetus. *Journal of Anatomy* 165:215–24.

Trinkaus, E. 1981. Neandertal limb proportions and cold adaptation. In *Aspects of Human Evolution,* ed. C. B. Stringer, 187–224. London: Taylor and Francis.

————. 1983. *The Shanidar Neandertals.* New York: Academic Press.

————. 1987. The Neandertal face: Evolutionary and functional perspectives on a recent hominid face. *Journal of Human Evolution* 16:429–43.

————. 2002. The mandibular morphology. In *Portrait of the Artist as a Child: The Gravettian Human Skeleton from the Abrigo do Lagar Velho and Its Archeological Context,* ed. J. Zilhão and E. Trinkaus, 312–25. Lisbon: Instituto Português de Arqueologia.

————. 2003. Neandertal faces were not long; modern human faces are short. *Proceedings of the National Academy of Sciences USA* 100:8142–45.

————. 2005. Early modern humans. *Annual Review of Anthropology* 34:207–30.

———. 2007. European early modern humans and the fate of the Neandertals. *Proceedings of the National Academy of Sciences USA* 104:7367–72.

Trinkaus, E., and M. LeMay. 1982. Occipital bunning among later Pleistocene hominids. *American Journal of Physical Anthropology* 57:27–35.

Trinkaus, E., and P. Shipman. 1993. *The Neandertals: Changing the Image of Mankind.* New York: Alfred A. Knopf.

Trinkaus, E., and J. Zilhão. 2002. Phylogenetic implications. In *Portrait of the Artist as a Child: The Gravettian Human Skeleton from the Abrigo do Lagar Velho and Its Archeological Context,* ed. J. Zilhão and E. Trinkaus, 497–518. Lisbon: Instituto Português de Arqueologia.

Trinkaus, E., S. E. Churchill, and C. B. Ruff. 1994. Postcranial robusticity in *Homo,* II: Bilateral asymmetry and bone plasticity. *American Journal of Physical Anthropology* 93:1–34.

Trinkaus, E., O. Moldovan, Ş. Milota, A. Bîlgăr, L. Sarcina, S. Athreya, S. E. Bailey, et al. 2003a. An early modern human from Peştera cu Oase, Romania, *Proceedings of the National Academy of Sciences USA* 100:11231–36.

Trinkaus, E., Ş. Milota, R. Rodrigo, M. Gherase, and O. Moldovan. 2003b. Early modern human cranial remains from Peştera cu Oase, Romania, *Journal of Human Evolution,* 45:245–53.

Trulsson, M., and R. S. Johansson. 2002. Orofacial mechanoreceptors in humans: Encoding characteristics and responses during natural orofacial behaviors. *Behavioural Brain Research* 135:27–33.

Tucker, A., and P. T. Sharpe. 2004. The cutting-edge of mammalian development: How the embryo makes teeth. *Nature Reviews: Genetics* 5:499–508.

Tuominen, M., T. Kantomaa, and P. Pirttiniemi. 1994. Effect of altered loading on condylar growth in the rat. *Acta Odontologica Scandinavica* 52:129–34.

Turner, C. H., Y. Takano, and I. Owan. 1995. Aging changes mechanical loading thresholds for bone formation in rats. *Journal of Bone and Mineral Research* 10: 1544–49.

Uhthoff, H. K., and Z. F. G. Jaworski. 1978. Bone loss in response to long-term immobilization. *Journal of Bone and Joint Surgery* 60:420–29.

Uhthoff, H. K., G. Zekaly, and Z. F. G. Jaworski. 1985. Effects of long-term nontraumatic immobilization on metaphyseal spongiosa in young adult and old beagle dogs. *Clinical Orthopedics and Related Research* 192:278–83.

Unda, F. J., A. Martin, C. Hernandez, G. Perez-Nanclares, E. Hilario, and J. Arechaga. 2001. FGFs-1 and -2, and TGF beta 1 as inductive signals modulating in vitro odontoblast differentiation. *Advances in Dental Research* 15:34–37.

Ungar, P. S. 1994. Patterns of ingestive behavior and anterior tooth use differences in sympatric anthropoid primates. *American Journal of Physical Anthropology* 95: 197–219.

———. 1996. The relationship of incisor size to diet and anterior tooth use in sympatric Sumatran anthropoids. *American Journal of Primatology* 38:145–56.

———. 2004. Dental topography and diets of *Australopithecus afarensis* and early *Homo. Journal of Human Evolution* 46:605–22.

Ungar, P., and M'Kirera, F. 2003. A solution to the worn tooth conundrum in primate functional anatomy. *Proceedings of the National Academy of Sciences USA* 100:3874–77.

Ungar, P. S., and R. S. Scott. 2009. Dental evidence for the diets of early *Homo.* In *The First Humans: Origin and Early Evolution of the Genus* Homo, ed. F. E. Grine, J. G. Fleagle, and R. E. F. Leakey, 121–34. New York: Springer.

Ungar, P. S., K. J. Fennell, K. Gordon, and E. Trinkaus. 1997. Neandertal incisor beveling. *Journal of Human Evolution* 32:407–21.

Ungar, P. S., F. E. Grine, M. F. Teaford, and S. El Zaatari. 2006. Dental microwear and diets of African early *Homo. Journal of Human Evolution* 50:78–95.

Ungar, P. S., F. E. Grine, and M. F. Teaford. 2008. Dental microwear and diet of the Plio-Pleistocene hominin *Paranthropus boisei. PLoS ONE* 3:E2044.

Vallois, H. V. 1926. La sustentation de la tête et le ligament cervical postérieur chez l'homme et les anthropoides. *L'anthropologie* 36:191–207.

Vandaele, D. J., A. L. Perlman, and M. D. Cassell. 1995. Intrinsic fibre architecture and attachments of the human epiglottis and their contributions to the mechanism of deglutition. *Journal of Anatomy* 186:1–15.

Van den Berg, J. 1958. Myoelastic-aerodynamic theory of voice production. *Journal of Speech and Hearing Research* 1:227–44.

van den Eynde, B., I. Kjaer, B. Solow, N. Graem, T. Kjaer, and M. Mathiesen. 1992. Cranial base angulation and prognathism related to cranial and general skeletal maturation in human fetuses. *American Journal of Orthodontics* 88:433–38.

van der Klaauw, C. 1948–52. Size and position of the functional components of the skull. *Archives Neerland Zoologie* 9:1–559.

Van der Linden, E. J., A. R. Burdi, and H. J. de Jongh. 1987. Critical periods in the prenatal morphogenesis of the human lateral pterygoid muscle, the mandibular condyle, the articular disk, and medial articular capsule. *American Journal of Orthodontics and Dentofacial Orthopaedics* 91:22–8.

Vandermeersch, B. 1981. *Les hommes fossiles de Qafzeh (Israel).* Paris: CNRS Éditions.

Van Essen, D. C. 1997. A tension-based theory of morphogenesis and compact wiring in the central nervous system. *Nature* 385:313–18.

Van Valen, L. 1973. A new evolutionary law. *Evolutionary Theory* 1:1–30.

Van Valkenburgh, B. 2001. The dog-eat-dog world of carnivores: A review of past and present carnivore community dynamics. In *Meat-Eating and Human Evolution,* ed. C. B. Stanford, and H. T. Bunn, 101–21. Oxford: Oxford University Press.

Varene, P., L. Ferrus, G. Manier, and J. Gire. 1986. Heat and water respiratory exchanges: Comparison between mouth and nose breathing in humans. *Clinical Physiology* 6:405–14.

Varrela, J. 1992. Dimensional variation of craniofacial structures in relation to changing masticatory-functional demands. *European Journal of Orthodontics* 14:31–36.

Vasavada, A. N., B. W. Peterson, and S. L. Delp. 2002. Three-dimensional spatial tuning of neck muscle activations in humans. *Experimental Brain Research* 147: 437–48.

Vekua, A., D. Lordkipanidze, G. P. Rightmire, J. Agusti, R. Ferring, G. Maisuradze, et al. 2002. A new skull of early *Homo* from Dmanisi, Georgia. *Science* 297:85–89.

Verborgt, O., G. J. Gibson, and M. B. Schaffler. 2000. Loss of osteocyte integrity in association with microdamage and bone remodeling after fatigue in vivo. *Journal of Bone and Mineral Research* 15:60–67.

Verrelli, B. C., and S. A. Tishkoff. 2004. Signatures of selection and gene conversion associated with human color vision variation. *American Journal of Human Genetics* 75:363–75.

Verwoerd, C. D., N. A. Urbanus, and D. C. Nijdam. 1979. The effects of septal surgery on the growth of nose and maxilla. *Rhinology* 17:53–63.

Verwoerd, C. D., N. A. Urbanus, and G. J. Mastenbroek. 1980. The influence of partial resections of the nasal septal cartilage on the growth of the upper jaw and the nose: An experimental study in rabbits. *Clinical Otolaryngology* 5:291–302.

Vidal, P. P., W. Graf, and A. Berthoz. 1986. The orientation of the cervical vertebral column in unrestrained awake animals, I: Resting position. *Experimental Brain Research* 1:549–59.

Vieille-Grosjean, I., P. Hunt, M. Gulisano, E. Boncinelli, and P. Thorogood. 1997. Branchial HOX gene expression and human craniofacial development. *Developmental Biology* 183:49–60.

Vignaud, P. Duringer, H. T. Mackaye, A. Likius, C. Blondel, J. R. Boisserie, L. de Bonis, et al. 2002. Geology and palaeontology of the Upper Miocene Toros Menalla hominid locality, Chad. *Nature* 418:152–55.

Viire, E. S., and J. L. Demer. 1996. The human vertical vestibulo-ocular reflex during combined linear and angular acceleration with near-target fixation. *Experimental Brain Research* 112:313–24.

Vinicius, L. 2005. Human encephalization and developmental timing. *Journal of Human Evolution* 49:762–76.

Vinkka-Puhakka, H., and I. Thesleff. 1993. Initiation of secondary cartilage in the mandible of the Syrian hamster in the absence of muscle function. *Archives of Oral Biology* 38:49–54.

Vogel, S. 2003. *Comparative Biomechanics: Life's Physical World.* Princeton, NJ: Princeton University Press.

Vollmer, D., U. Meyer, U. Joos, A. Vegh, and J. Piffko. 2000. Experimental and finite element study of a human mandible. *Journal of Cranio-maxillofacial Surgery* 28:91–96.

Von Baer, K. E. 1828. Entwicklungsgeschichte der Tiere: Beobachtung und Reflexion. Königsberg: Bornträger.

Von Meyer, G. H. 1867. Die Architektur der Spongiosa. *Archiv für Anatomie, Physiologie und wissenschaftliche Medizin* 34:615–28.

Vorperian, H. K., R. D. Kent, M. J. Lindstrom, C. M. Kalina, L. R. Gentry, and B. S. Yandell. 2005. Development of vocal tract length during early childhood: A magnetic resonance imaging study. *Journal of the Acoustical Society of America* 117:338–50.

Vrba, E. S. 1998. Multiphasic growth models and the evolution of prolonged growth exemplified by human brain evolution. *Journal of Theoretical Biology* 190: 227–39.

Wagner, G. P. 1996. Homologues, natural kinds and the evolution of modularity. *American Zoologist* 36:36–43.

Wagner, G., and L. Altenberg. 1996. Complex adaptations and the evolution of evolvability. *Evolution* 50:967–76.

Wakeley, J. 2008. Complex speciation of humans and chimpanzees. *Nature* 452: E3–4.

Walker, A. 1981. Diet and teeth: Dietary hypotheses and human evolution. *Philosophical Transactions of the Royal Society of London, Series B* 292:57–64.

Walker, A., and R. E. F. Leakey. 1993. *The Nariokotome Skeleton.* Cambridge, MA: Harvard University Press.

Walker, A., Leakey, R. E., J. M. Harris, and F. H. Brown. 1986. 2.5-Myr *Australopithecus boisei* from west of Lake Turkana, Kenya. *Nature* 322:517–22.

Walker, R., K. Hill, O. Burger, and A. M. Hurtado. 2006a. Life in the slow lane revisited: Ontogenetic separation between chimpanzees and humans. *American Journal of Physical Anthropology* 129:577–83.

Walker, R., O. Burger, J. Wagner, and C. Von Rueden. 2006b. Evolution of brain size and juvenile periods in primates. *Journal of Human Evolution* 51:480–89.

Wall, C. E., C. J. Vinyard, K. R. Johnson, S. H. Williams, and W. L. Hylander. 2006. Phase II jaw movements and masseter muscle activity during chewing in *Papio anubis. American Journal of Physical Anthropology.* 129:215–24.

Wall, J. D., and S. K. Kim. 2007. Inconsistencies in Neanderthal genomic DNA sequences. *PLoS Genetics* 3:1862–66.

Walsberg, G. E. 1988. Consequences of skin color and fur properties for solar heat gain and ultraviolet irradiance in two mammals. *Journal of Comparative Physiology, B* 158:213–21.

Wandsnider, L. 1997. The roasted and the boiled: Food composition and heat treatment with special emphasis on pit-hearth cooking. *Journal of Anthropological Archaeology* 16:1–48.

Ward, C., M. G. Leakey, and A. Walker. 2001. Morphology of *Australopithecus anamensis* from Kanapoi and Allia Bay, Kenya. *Journal of Human Evolution* 41: 255–368.

Warren, D. W., V. A. Hinton, H. C. Pillsbury, and W. M. Harfield. 1987. Effects of size of the nasal airway on nasal airflow rate. *Archives of Otolaryngology* 113: 405–408.

Weaver, A. H. 2005. Reciprocal evolution of the cerebellum and neocortex in fossil humans. *Proceedings of the National Academy of Sciences USA* 102:3576–80.

Weaver, T. D., and K. Steudel-Numbers. 2005. Does climate or mobility explain the differences in body proportions between Neandertals and their Upper Paleolithic successors? *Evolutionary Anthropology* 14:218–23.

Weaver, T. D., C. C. Roseman, and C. B. Stringer. 2007. Were Neandertal and modern human cranial differences produced by natural selection or genetic drift? *Journal of Human Evolution* 53:135–45.

Weaver, T. D., C. C. Roseman, and C. B. Stringer. 2008. Close correspondence between quantitative- and molecular-genetic divergence times for Neandertals and modern humans. *Proceedings of the National Academy of Sciences USA* 105: 4645–49.

Wegner, R. N. 1955. Studien über Nebenhöhlen des Schädels. *Wissenschaftliche Zeitschrift der Universität Greifswald.* 5:1–111.

Weidenreich, F. 1937. The dentition of *Sinanthropus pekinensis:* A comparative odontography of the hominid. *Palaeontologica Sinica, New Series D* 1:1–180.

———. 1941. The brain and its rôle in the phylogenetic transformation of the human skull. *Transactions of the American Philosophical Society* 31:328–442.

Weijs, W. A. 1994. Evolutionary approach of masticatory motor patterns. In *Biomechanics of Feeding in Vertebrates,* ed. M. C. Chardon, V. L. Bels, and P. Vandewalle, 281–320. Berlin: Springer-Verlag.

Weijs, W. A., and B. Hillen. 1984. Relationship between the physiological cross-section of the human jaw muscles and their cross-sectional area in computer tomograms. *Acta Anatomica* (Basel) 118:129–38.

———. 1986. Correlations between the cross-sectional area of the jaw muscles and craniofacial size and shape. *American Journal of Physical Anthropology* 70:423–31.

Weinbaum, S., Cowin, S. C., and Y. Zheng. 1994. A model for the excitation of osteocytes by mechanical loading-induced bone fluid shear stresses. *Journal of Biomechanics* 27:339–60.

Weiner, J. S. 1954. Nose shape and climate. *American Journal of Physical Anthropology* 12:615–18.

———. 1955. *The Piltdown Forgery.* Oxford: Oxford University Press.

Weiner, S., and W. Traub. 1992. Bone structure: From ångstroms to microns. *Journal of the Federation of American Societies for Experimental Biology* 6:879–85.

Weinmann, J. P., and H. Sicher. 1941 *Bone and Bone.* St. Louis, MO: Mosby.

Weiss, P., and R. Amprino. 1940. The effects of mechanical stress on the differentiation of scleral cartilage *in vitro* and in the embryo. *Growth* 4:245–58.

Weller, R. O. 2005. Microscopic morphology and histology of the human meninges. *Morphologie* 89:22–34.

Wells, H. G. 1898. *The War of the Worlds*. London: William Heinemann.

Wells, S. 2002. *The Journey of Man: A Genetic Odyssey*. Princeton, NJ: Princeton University Press.

West-Eberhard, M. J. 2003. *Developmental Plasticity and Evolution*. Oxford: Oxford University Press.

Weston, E. M., and A. M. Lister. 2009. Insular dwarfism in hippos and a model for brain size reduction in *Homo floresiensis*. *Nature* 459:85–88.

Wheatley, J. R., T. C. Amis, and L. A. Engel. 1991. Oronasal partitioning of ventilation during exercise in humans. *Journal of Applied Physiology* 71:546–51.

Wheeler, P. E. 1984. The evolution of bipedality and loss of functional body hair in hominids. *Journal of Human Evolution* 13:91–98.

———. 1985. The loss of functional body hair in man: The influence of thermal environment, body form and bipedality. *Journal of Human Evolution* 14:23–28.

———. 1991. The thermoregulatory advantages of hominid bipedalism in open equatorial environments: The contribution of increased convective heat loss and cutaneous evaporative cooling. *Journal of Human Evolution* 21:107–15.

White, A. A., and M. M. Panjabi. 1978. *Clinical Biomechanics of the Spine*. Philadelphia: J. B. Lippincott.

White, T. D. 2003. Early hominins: Diversity or distortion? *Science* 299:1994–97.

White, T. D., D. C. Johanson, and W. H. Kimbel. 1981. *Australopithecus africanus:* Its phylogenetic position reconsidered. *South African Journal of Science* 77: 445–70.

White, T. D., G. Suwa, and B. Asfaw. 1994. *Australopithecus ramidus,* a new species of early hominid from Aramis, Ethiopia. *Nature* 371:306–12.

White, T. D., G. Suwa, S. Simpson, and B. Asfaw. 2000. Jaws and teeth of *Australopithecus afarensis* from Maka, Middle Awash, Ethiopia. *American Journal of Physical Anthropology* 111:45–68.

White, T. D., B. Asfaw, D. DeGusta, W. Gilbert, G. Richards, G. Suwa, and F. C. Howell. 2003. Pleistocene *Homo sapiens* from Middle Awash, Ethiopia. *Nature* 423:742–47.

White, T. D., G. WoldeGabriel, B. Asfaw, S. Ambrose, Y. Beyene, R. L. Bernor, J.-R. Boisserie, et al. 2006. Asa Issie, Aramis and the origin of *Australopithecus*. *Nature* 440:883–89.

White, T. D., B. Asfaw, Y. Beyene, Y. Haile-Selassie, C. O. Lovejoy, G. Suwa, and G. WoldeGabriel. 2009. *Ardipithecus ramidus* and the paleobiology of early hominids. *Science* 326:75–86.

Whitfield, J. F. 2003. The primary cilium: Is it an osteocyte's strain-sensing flow-meter? *Journal of Cellular Biochemistry* 89:233–37.

Whitfield, J. F., P. Morley, and G. E. Willick. 2002. The control of bone growth by parathyroid hormone, leptin, and statins. *Critical Reviews in Eukaryotic Gene Expression* 12:23–51.

Whitford, W. G. 1976. Sweating responses in the chimpanzee (*Pan troglodytes*). *Comparative Biochemistry and Physiology, Part A* 53:333–36.

Wilkie, A. O., and G. M. Morriss-Kay. 2001. Genetics of craniofacial development and malformation. *Nature Reviews: Genetics* 2:458–68.

Wilkins, A. S. 2002. *The Evolution of Developmental Pathways.* Sunderland, MA: Sinauer Associates.

Williams, F. L., L. R. Godfrey, and M. R. Sutherland. 2003. Diagnosing heterochronic perturbations in the craniofacial evolution of *Homo* (Neanderthals and modern humans) and *Pan* (*P. troglodytes* and *P. paniscus*). In *Patterns of Growth and Development in the Genus* Homo, ed. J. L. Thompson, G. Krovitz, and A. J. Nelson, 405–41. Cambridge: Cambridge University Press.

Willmore, K., L. Leamy, and B. Hallgrímsson. 2006 The effects of developmental and functional interactions on mouse cranial variability through late ontogeny. *Evolution and Development* 8:550–67.

Wilson, S. L., B. T. Thach, R. T. Brouillette, and Y. K. Abu-Osba. 1981. Coordination of breathing and swallowing in human infants. *Journal of Applied Physiology* 50:851–58.

Winter, D. A., K. G. Ruder, and C. D. MacKinnon. 1990. Control of balance of upper body in gait. In *Multiple Muscle Systems: Biomechanics and Movement Organization,* ed. J. M. Winters and S. L. Y. Woo, 534–41. New York: Springer.

WoldeGabriel, G., S. H. Ambrose, D. Barboni, R. Bonnefille, L. Bremond, B. Currie, D. DeGusta, et al. 2009. The geological, isotopic, botanical, invertebrate, and lower vertebrate surroundings of *Ardipithecus ramidus. Science* 326: 65e1–5.

Wolf, K. 2002. Visual ecology: Coloured fruit is what the eye sees best. *Current Biology* 12:R253–55.

Wolff, J. 1986 [1892]. *The Law of Bone Remodeling.* Berlin: Springer-Verlag.

Wolpert, L. 1991. *The Triumph of the Embryo.* Oxford: Oxford University Press.

Wolpoff, M. H. 1968. Climatic influence on the nasal aperture. *American Journal of Physical Anthropology* 29:405–23.

———. 1975. Dental reduction and the probable mutation effect. *American Journal of Physical Anthropology* 43:307–8.

———. 1986. Describing anatomically modern *Homo sapiens:* A distinction without a definable dfference. *Anthropos* (Brno) 23:41–53.

———. 1999. *Paleoanthropology.* 2nd ed. Boston: McGraw-Hill.

Wolpoff, M. H., X. Wu, and A. G. Thorne. 1984. Modern *Homo sapiens* origins: A general theory of hominid evolution involving the fossil evidence from East Asia. In *The Origins of Modern Humans: A World Survey of the Fossil Evidence,* ed. F. H. Smith and F. Spencer, 411–83. New York: Alan R. Liss.

Wolpoff, M. H., J. Hawks, D. W. Frayer, and K. Hunley. 2001. Modern human ancestry at the peripheries: A test of the replacement theory. *Science* 291:293–97.

Won, Y. J., and J. Hey. 2005. Divergence population genetics of chimpanzees. *Molecular Biology and Evolution* 22:297–307.

Wood, B. A. 1984. The origin of *Homo erectus. Courier Forschungsinstitut Senckenberg* 69:99–111.

———. 1985. Early *Homo* in Kenya and its systematic relationships. In *Ancestors: The Hard Evidence,* ed. E. Delson, 206–14. New York: Alan Liss.

———. 1991. *Koobi Fora Research Project,* vol. 4, *Hominid Cranial Remains.* Oxford: Oxford University Press.

———. 1992. Origin and evolution of the genus *Homo. Nature* 355:783–90.

———. 1993. Early *Homo:* How many species? In *Species, Species Concepts, and Primate Evolution,* ed. W. H. Kimbel and L. B. Martin, 485–522. New York: Plenum Press.

———. 1994. Taxonomy and evolutionary relationships of *Homo erectus. Courier Forschungsinstitut Senckenberg* 171:159–65.

Wood, B. A., and L. C. Aiello. 1998. Taxonomic and functional implications of mandibular scaling in early hominins. *American Journal of Physical Anthropology* 105:523–38.

Wood, B. A., and M. Collard. 1999a. The human genus. *Science* 284:65–71.

———. 1999b. The changing face of genus *Homo. Evolutionary Anthropology* 8: 195–207.

Wood, B. A., and P. Constantino. 2007. *Paranthropus boisei:* Fifty years of evidence and analysis. *Yearbook of Physical Anthropology* 50:106–32.

Wood, B. A., and B. G. Richmond. 2000. Human evolution: Taxonomy and paleobiology. *Journal of Anatomy* 197:19–60.

Wood, B. A., S. A. Abbott, and H. Uytterschaut. 1988. Analysis of the dental morphology of Plio-Pleistocene hominids, IV: Mandibular postcanine root morphology. *Journal of Anatomy* 156:107–39.

Wood, J. N., and J. Grafman. 2003. Human prefrontal cortex: Processing and representational perspectives. *Nature Reviews: Neuroscience* 4:139–47.

Wooding, S., B. Bufe, C. Grassi, M. T. Howard, A. C. Stone, M. Vazquez, D. M. Dunn, W. Meyerhof, R. B. Weiss, and M. J. Bamshad. 2006. Independent evolution of bitter-taste sensitivity in humans and chimpanzees. *Nature* 440:930–34.

Wood Jones, F. 1940. The nature of the soft palate. *Journal of Anatomy* 74:147–70.

Woodside, D. G., S. Linder-Aronson, A. Lundstron, and J. McWilliam. 1991. Mandibular and maxillary growth after changed mode of breathing. *American Journal of Orthodontics and Dentofacial Orthopaedics* 100:1–18.

Wrangham, R. W. 1977. Feeding behaviour of chimpanzees in Gombe National Park, Tanzania. In *Primate Ecology*, ed. T. H. Clutton Brock, 503–38. London: Academic Press.

———. 2006. The cooking enigma. In: *Early Hominin Diets: The Known, the Unknown, and the Unknowable*, ed. P. Ungar, 308–23. New York: Oxford University Press.

———. 2009. *Catching Fire: How Cooking Made us Human*. New York: Basic Books.

Wrangham, R. W., and N. L. Conklin-Brittain. 2003. The biological significance of cooking in human evolution. *Comparative Biochemistry and Physiology, Part A: Molecular and Integrative Physiology* 136:35–46.

Wrangham, R. W., J. H. Jones, G. Laden, D. Pilbeam, and N. L. Conklin-Brittain. 1999. The raw and the stolen: Cooking and the ecology of human origins. *Current Anthropology* 40:567–94.

Wu, X. 1992. The origin and dispersal of anatomically modern humans in East and Southeast Asia. In *The Evolution and Dispersal of Modern Humans in Asia*, ed. T. Akazawa, K. Aoki, and T. Kimura, 373–78. Tokyo: Hokusen-sha.

Wu, X., and F. E. Poirier. 1995. *Human Evolution in China*. Oxford: Oxford University Press.

Wysocki, C. J., and G. Preti. 2004. Facts, fallacies, fears, and frustrations with human pheromones. *Anatomical Record* 281A:1201–11.

Yamada, K., and D. B. Kimmel. 1991. The effect of dietary consistency on bone mass and turnover in the growing rat mandible. *Archives of Oral Biology* 36:129–38.

Yamei, H., R. Potts, B. Yuan, Z. Guo, A. Deino, W. Wang, J. Clark, G. Xie, and W. Huang. 2000. Mid-Pleistocene Acheulean-like Stone Technology of the Bose Basin, South China. *Science* 287:1622–26.

Yang, A., and J. Hullar. 2007. Relationship of semicircular canal size to vestibular-nerve afferent sensitivity in mammals. *Journal of Neurophysiology* 98:3197–205.

Yang, X., and G. Karsenty. 2002. Transcription factors in bone: Developmental and pathological aspects. *Trends in Molecular Medicine* 8:340–45.

Yilmaz, S., H. N. Newman, and D. F. G. Poole. 1977. Diurnal periodicity of von Ebner growth lines in pig dentine. *Archives of Oral Biology* 22:511–13.

Yokley, T. R. 2006. The functional and adaptive significance of anatomical variation in recent and fossil human nasal passages. PhD diss., Duke University.

———. 2009. Ecogeographic variation in human nasal passages. *American Journal of Physical Anthropology* 138:11–22.

Yokoyama, Y., C. Falguères, R. Grün, T. Jacob, and F. Sémah. 2008. Gamma-ray spectrometric dating of late *Homo erectus* skulls from Ngandong and Sambungmacan, Central Java, Indonesia. *Journal of Human Evolution* 55:274–77.

Yoshida, T., P. Vivatbutsiri, G. Morriss-Kay, Y. Sagac, and S. Isekia. 2008. Cell lineage in mammalian craniofacial mesenchyme. *Mechanisms of Development* 125: 797–808.

Young, R. W. 1959. The influence of cranial contents on postnatal growth of the skull in the rat. *American Journal of Anatomy* 105:383–90.

Yu, J. C., J. S. McClintock, F. Gannon, X. X. Gao, J. P. Mobasser, and M. Sharawy. 1997. Regional differences of dura osteoinduction: Squamous dura induces osteogenesis, sutural dura induces chondrogenesis and osteogenesis. *Plastic and Reconstructive Surgery* 100:23–31.

Zajac, F. E. 1989. Muscle and tendon: Properties, models, scaling, and application to biomechanics and motor control. *Critical Reviews in Biomedical Engineering* 17: 359–411.

Zelditch, M. L. 1988. Ontogenetic variation in patterns of phenotypic integration in the laboratory rat. *Evolution* 42:28–41.

Zelditch, M. L., and A. C. Carmichael. 1989. Ontogenetic variation in patterns of developmental and functional integration in skulls of *Sigmodon fulviventer*. *Journal of Mammalology* 70:477–84.

Zelditch, M. L., and W. L. Fink. 1995. Allometry and developmental integration of body growth in a Piranha, *Pygocentrus nattereri* (*Teleostei: Ostariophysi*). *Journal of Morphology* 223:341–55.

———. 1996. Heterochrony and heterotopy: Stability and innovation in the evolution of form. *Paleobiology* 22:241–54.

Zelditch, M. L., D. O. Straney, D. L. Swiderski, and A. C. Carmichael. 1990. Variation in developmental constraints in Sigmodon. *Evolution* 44:1738–47.

Zelditch, M. L., W. L. Fink, and D. L. Swiderski. 1995. Morphometrics, homology, and phylogenetics: Quantified character as synapomorphies. *Systematic Biology* 44:179–89.

Zelditch, M. L., D. Swiderski, D. Sheets, and W. L. Fink. 2004. *Geometric Morphometrics for Biologists: A Primer*. London: Elsevier Academic Press.

Zenker, W., and S. Kubik. 1996. Brain cooling in humans: Anatomical considerations. *Anatomy and Embryology* 193:1–13.

Zhang, S. C., B. Ge, and I. D. Duncan. 1999. Adult brain retains the potential to generate oligodendroglial progenitors with extensive myelination capacity. *Proceedings of the National Academy of Sciences USA* 96:4089–94.

Zhang, X., and S. Firestein. 2002. The olfactory receptor gene superfamily of the mouse. *Nature Neuroscience* 5:124–33.

Zhao, H., L. Ivic, J. M. Otaki, M. Hashimoto, K. Mikoshiba, and S. Firestein. 1998. Functional expression of a mammalian odorant receptor. *Science* 279:237–42.

Zhao, K., P. W. Scherer, S. A. Hajiloo, P. Dalton. 2004. Effect of anatomy on human nasal air flow and odorant transport patterns: Implications for olfaction. *Chemical Senses* 29:365–79.

Zhao, K., P. Dalton, G. C. Yang, and P. W. Scherer. 2006. Numerical modeling of turbulent and laminar airflow and odorant transport during sniffing in the human and rat nose. *Chemical Senses* 31:107–18.

Ziablov, V. 1982. Structure and physicomechnical properties of the human dura mater from the age aspect. *Arkhiv Anatomii, Gistologii, i Émbriologii* 82:29–36.

Zihlman, A. 1984. Body build and tissue composition in *Pan paniscus* and *Pan troglodytes* with comparison to other hominoids. In *The Pygmy Chimpanzee,* ed. R. Sussman, 179–200. New York: Plenum Press.

Zihlman, A., C. Boesch, and D. Bolter. 2004. Wild chimpanzee dentition and its implications for assessing life history in immature hominin fossils. *Proceedings of the National Academy of Sciences USA* 101:10541–43.

Zilhão, J. 2006. Neandertals and Moderns Mixed, and It Matters. *Evolutionary Anthropology* 15:183–95.

Zilhão, J., F. d'Errico, J. G. Bordes, A. Lenoble, J. P. Texier, and J. P. Rigaud. 2006. Analysis of Aurignacian interstratification at the Chatelperronian-type site and implications for the behavioral modernity of Neandertals. *Proceedings of the National Academy of Sciences USA* 103:12643–48.

Zollikofer, C. P., and M. S. Ponce de León. 2004. Kinematics of cranial ontogeny: Heterotopy, heterochrony, and geometric morphometric analysis of growth models. *Journal of Experimental Zoology, Part B: Molecular and Developmental Evolution* 302:322–40.

Zollikofer, C. P., M. S. Ponce de León, F. Esteves, T. F. Silva, and R. P. Dias. 2002. In Portrait of the Artist as a Child: The Gravettian Human Skeleton from the Abrigo do Lagar Velho and Its Archeological Context, ed. J. Zilhão and E. Trinkaus, 326–41. Lisbon: Instituto Português de Arqueologia.

Zollikofer, C. P., M. S. Ponce de León, D. E. Lieberman, F. Guy, D. Pilbeam, A. Likius, H. T. Mackaye, P. Vignaud, and M. Brunet. 2005. Virtual cranial reconstruction of *Sahelanthropus tchadensis. Nature* 434:755–59.

Zucconi, M., L. Ferini-Strambi, S. Palazzi, C. Curci, E. Cucchi, and S. Smirne. 1993. Craniofacial and cephalometric evaluation in habitual snorers with and without obstructive sleep apnea. *Otolaryngology: Head and Neck Surgery* 109:1007–13.

Zuckerman, S. 1928. Age changes in the chimpanzee, with special reference to the growth of the brain. *Proceedings of the Zoological Society of London* 1–42.

Zwislocki, J. J. 1986. Analysis of cochlear mechanics. *Hearing Research* 22:155–69.

Acknowledgments

I am fortunate to have benefited enormously from the suggestions of many colleagues who have read and critiqued various portions of this book: Ofer Bar-Yosef, Dennis Bramble, Fuzz Crompton, Iain Davidson, Chris Dean, Maureen Devlin, Virginia Diewert, Nate Dominy, Rebecca German, Brian Hall, Benedikt Hallgrímsson, Ralph Holloway, Jean-Jacques Hublin, Bill Kimbel, Richard Klein, Philip Lieberman, Peter Lucas, Charles Marshall, Robert Martin, Brandeis McBratney-Owen, John McGrath, Gillian Morriss-Kay, Paul O'Higgins, David Pilbeam, Philip Rightmire, Gary Schwartz, Gordon Shepherd, Tanya Smith, Geoffrey Sperber, Fred Spoor, Dave Strait, Peter Ungar, Chris Vinyard, Adam Wilkins, and Todd Yokely. I thank them all, but I am especially grateful to my father, Philip Lieberman, and to my former professors, now colleagues: Ofer Bar-Yosef, Fuzz Crompton, and David Pilbeam. Their decades of support have helped to guide my intellectual development. I also owe very special thanks to Chris Dean, Alan Walker, Adam Wilkins, and Bernard Wood, for reading and critiquing the entire manuscript. Finally, Michael Fisher has been the very model of a patient, helpful, and trusting editor.

I also thank the many colleagues and students who have helped me in various ways by sharing data, by providing access to CT scans, radiographs, fossils, and casts, and by engaging in many stimulating discussions. These individuals include all of the above, as well as Brian Addison, Susan Antón, Juan Luis Arsuaga, Felicitas Bidlack, Andy Biewener, José Braga, Alison Brooks, Michel Brunet, Julian Carlos, Judy Chupasko, Bob Corruccini, Adam Daoud, Terry Deacon, Viktor Deak, Peter Ellison, Carolyn Eng, Craig Feibel, Bob Franciscus, Giselle Garcia, Fred Grine, Franck Guy, David Haig, Maureen Devlin Hamalainen, Bill Hylander, Farish A. Jenkins Jr., William Jungers, Jason Kaufman, George Koufos, George Lauder, Meave Leakey, Steve Leigh,

Karel Liem, David Lordkipanidze, Peter Lucas, Lu Zune, Masrour Makaremi, Giorgio Manzi, Rob McCarthy, Michèle Morgan, Ken Mowbray, Bob Mucci, John Polk, Marcia Ponce de León, Herman Pontzer, Rick Potts, Dave Raichlen, Yoel Rak, Brian Richmond, Neil Roach, Campbell Rolian, Chris Ruff, Maryellen Ruvolo, John Shea, Chris Stringer, Ian Tattersall, Francis Thackeray, Noreen Tuross, Gary Sawyer, Gen Suwa, Alan Walker, Gerhardt Weber, William Werbel, Katherine Whitcome, Tim White, Milford Wolpoff, Richard Wrangham, Sara Wright, Franklin Yates, John Yellen, Nathan Young, Katie Duncan Zink, and Christoph Zollikofer. I apologize to anyone I have inadvertently left out.

Index

Page numbers for entries in figures are followed by an *f*, those for entries in notes by an *n*, and those for entries in tables by a *t*.